Chemicals, Environment, Health

A Global Management Perspective

Chemicals, Environment, Health

A Global Management Perspective

Edited by
Philip Wexler • Jan van der Kolk
Asish Mohapatra • Ravi Agarwal

CRC Press is an imprint of the
Taylor & Francis Group, an **informa** business

Philip Wexler, as part of the editorial team, contributed to this book in his capacity as a private citizen, not as a government employee. The views expressed are strictly his own. No official support or endorsement by the U.S. National Library of Medicine or any other agency of the U.S. Federal Government was provided or should be inferred.

CRC Press
Taylor & Francis Group
6000 Broken Sound Parkway NW, Suite 300
Boca Raton, FL 33487-2742

First issued in paperback 2019

ISBN-13: 978-1-4200-8469-6 (hbk)
ISBN-13: 978-0-367-38252-0 (pbk)

Visit the Taylor & Francis Web site at
http://www.taylorandfrancis.com

and the CRC Press Web site at
http://www.crcpress.com

For my parents (Yetty and Will), my son (Jacob), and dear Nancy

Philip Wexler

For my wife, Jeanette, who accompanied me in many ways during a 40 year journey promoting chemical safety

Jan van der Kolk

For my loving wife, Sarah, my Mom (Kanak) and Dad (Mahendra)

Asish Mohapatra

To my father

Ravi Agarwal

Contents

SECTION III Global/Multilateral Instruments

SECTION IV SAICM

SECTION V Organizations

Foreword

The past 40 years have seen a phenomenal growth in globally oriented public and private initiatives related to chemical and other environmental issues. The groundbreaking 1972 United Nations Conference on the Human Environment held in Stockholm, in which I was honored to play a role, ushered in a veritable sea change in international environmental policies. It gave rise to the first World Environment Day and the creation of the United Nations Environment Programme. It put the environment on the international agenda as a global concern, which must be and could be reconciled with economic development as two sides of the same coin. This led the way to the acknowledgment that sustainable development is the most logical and viable pathway to the human future.

Over the years, one conference, or rather milestone, led to another—Stockholm to Rio to Johannesburg—with many intervening activities. Stakeholders who played an influential though limited role in 1972 from outside of the Conference were brought into the fold to offer their unique perspectives. The developing world, which suffered most from environmental degradation, yet did not have the resources to ameliorate it, asserted its insistence that more developed countries provide the new and additional resources they would need to address their environmental problems. Although steps have been taken, the ever widening gap between South and North has still to be successfully addressed. By many measures, we are better off than we were in 1972. With more emerging issues, such as new technologies, and much greater knowledge, formidable challenges remain. What is most important is that the dialogue has been established, is continuing and mechanisms have been created that contribute to solutions of many problems. Governance has now also become a subject in its own right, as critical to many successful approaches. So are linkages between chemicals management and other health and environmental problems. Indeed, on an even broader level, so is the concept of environmental mainstreaming, in which considerations about the environment (including chemicals) are integrated fully into decision making in the economic, social, and physical realms of development.

The very capable editors of this book have assembled a distinguished roster of contributors to create a valuable guidebook to global chemicals management cooperation as it stands today and is projected to evolve in the future. An opening background chapter setting the historical and contextual framework is followed by chapters covering the major conferences, international treaties and conventions, and organizations. Select regional and national activities round out the scope of the book. Invited essays in such diverse areas as emergencies, information resources, global financial instruments, and governance further supplement the core text. Naturally, today's major policy framework, the Strategic Approach to International Chemicals Management (SAICM) and its associated International Conference on Chemicals Management (ICCM), are also highlighted. Finally, a concluding chapter analytically presents a look at the future of global chemicals policy.

This is not the first book to consider chemicals, environment, and health from a global perspective, but it is the first to do it so thoroughly and concisely, without getting bogged down in a litany of legalistic detail. The chapters consistently offer precision, perspective, and reflection, and will be appreciated not only by the professional policy community, but to anyone wanting a clear look through the complex maze. I applaud this contribution to global environmental knowledge and understanding, and appreciate the opportunity to introduce it to both experienced hands at the subject and the new generation of researchers and practitioners. It is a book to read straight through or savor a section at a time, and belongs on the bookshelf of anyone interested in making a difference in the way we and future generations will live our lives in a world in which our health, and the environments, will no longer be endangered by potentially hazardous chemicals at any stage of their life cycle.

Maurice F. Strong
Founder and Chairman of Cosmos International Inc.,
Honorary Professor of Peking University (Beijing), and
Honorary President, Oriental Environment Research Institute (China)

Maurice F. Strong has had a remarkable career in both business and public service, primarily in the fields of international development, the environment, energy and finance. He has played a unique and pioneering role in globalization of the environment movement as Secretary-General of both the 1972 United Nations Conference on the Human Environment, which first put the environment on the international agenda, and the 1992 Rio Earth Summit. He was the first Executive Director of the United Nations Environment Programme. Strong continues to be active in environment and related fields, particularly in China (a country he has had a long relationship with and where he now spends much of his time). To learn more about Maurice Strong's illustrious career, visit http://www.mauricestrong.net.

Preface

INTRODUCTION

Chemicals are ubiquitous. Man has been aware of naturally derived chemicals for thousands of years, while synthetic chemicals have been with us for perhaps 200 years, with both performing a variety of important functions in our lives.

The toxicity of certain chemicals to man and animals has also been known since antiquity.

Chemical industry as producer of many new molecules for different purposes is however, relatively young, a little more than a century. In that period of time, the number of new molecules and the volumes of their production have increased tremendously. With more than 56 million organic and inorganic chemicals registered with the Chemical Abstracts Services of the American Chemical Society as of late 2010, and nearly 100,000 in commerce, they are having an increasingly huge impact on our lives.

AWARENESS

With this rapid growth in chemical synthesis, distribution, use, and subsequent exposure, society eventually realized the need to manage chemicals in a sound way, albeit often with delays. The need to protect workers, globally, from potentially harmful effects began as early as 1921 with the ILO (International Labour Organization) Convention to ban white lead in most paints. This convention subsequently proved to be beneficial for the larger population groups (e.g., countries that had implemented this Convention had hardly any problems with children affected by white lead from paint indoors).

As a class of chemicals, pharmaceuticals were subject to early regulatory controls, long after it became clear that the same substances that could cure illnesses could also be toxic for the patient.

Pesticides were another group to receive wider attention, both to assure that their beneficial effects were optimized and to control unwanted side effects. For example, in the United States, the Pure Food and Drugs Act and the Federal Insecticide Act were passed in 1906 and 1910 respectively. In fact, many early regulations, including those for foodstuffs, pharmaceuticals, and pesticides, centered first on quality and later on safety. Pesticide regulations were introduced in many countries in the first half of the twentieth century. The awareness of the potentially harmful effects of chemicals, particularly pesticides on the environment reached, in a sense, its watershed moment soon after the publication of Rachel Carson's *Silent Spring* in 1962.

Further efforts to control worker exposure to hazardous industrial chemicals started in the mid-1960s. These resulted in schemes for classification and labeling, as well as regulations for transport, to prevent accidents.

The 1960s witnessed a number of environmental problems traceable to chemical waste. One of the most well known is the Love Canal site in the United States. Later, many thousands of heavily contaminated sites were discovered. In the Netherlands, entire urban areas were destroyed in the 1970s after it had become clear that houses had been built on a previous chemical waste dump. In the same decade, numerous instances were uncovered of contaminated sites in developing countries. These were due to dumping of chemical and pesticide waste and/or the uncontrolled import of chemical waste.

MANAGEMENT, INSTRUMENTS, AND CHALLENGES

International attempts to address chemical safety could be said to truly begin with the 1972 Stockholm Conference, which resulted in the creation of UNEP (United Nations Environment Programme) and, shortly thereafter, the predecessor of what is today UNEP Chemicals.

Since this Conference, and in particular since the UNCED (United Nations Conference on the Environment and Development) in 1992 in Rio de Janeiro, many initiatives have been undertaken to address the potentially harmful effects of chemicals. Most recently, in 2002, in Johannesburg, South Africa, the World Summit on Sustainable Development, building upon the Rio conference, articulated the often cited goal of ensuring that chemicals are produced and used in ways that minimize significant adverse impacts on the environment and human health.

This book seeks to give a full overview of these developments and their impacts at international, and select regional and national levels. Further, it offers an outlook for the next 5–10 years—specifically, of the current challenges that need to be addressed to meet the goals that the international community set in Johannesburg in 2002.

The United Nations will be holding another conference on sustainable development in Rio in May 2012, 20 years after the historic 1992 conference, and thus already being informally referred to as Rio + 20. The conference will seek three objectives: securing renewed political commitment to sustainable development, assessing the progress and implementation gap in meeting already agreed commitments, and addressing new and emerging challenges. The members have agreed to the following two themes: green economy within the context of sustainable development, and poverty eradication, and institutional framework for sustainable development.

INTEGRATION OF CHEMICALS IN WIDER SUSTAINABLE DEVELOPMENT

Chemicals are but one, albeit important, topic on the international environment and health agenda. However, they touch upon a host of other issues of critical importance to health and the environment: climate change (several chemicals are important contributors to global warming or the depletion of the ozone layer), biological diversity (certain chemicals are known to affect ecosystems in various regions of the world), transmissible diseases (consider the continuing debate of the role of certain pesticides in malaria control), poisoning incidents (most cases of poisoning worldwide are attributed to abuses of pesticides). In this sense, almost all the Millennium Development Goals, established by the United Nations, with a target date of 2015, have a direct or indirect link with chemicals management. To cover all these, though, would take several more volumes.

Several key lessons have gradually been learned over the years.

First, many instruments have developed separately and without much coordination. This has significantly complicated work both at international and national levels. Only recently, attempts to streamline have resulted in more concrete and integrated actions (i.e., enhancing synergies between the Basel, Rotterdam, and Stockholm Conventions). Much work still remains to be done in this area.

Second, chemicals management and its instruments have mostly been looked at as a technical and specialized niche area, somehow removed from other societal concerns. This, notwithstanding the key importance of chemicals and chemistry for all areas of the economy and development, and the scientifically indisputable negative effects of several dozen chemicals on health or the environment worldwide. Very few attempts have been made to link chemicals management to the wider sustainable development agenda, or to broader mechanisms, including financial instruments, or development planning in general.

This book addresses not only the individual instruments and their implementation in several regions and countries, but also underpins the need for such further integration.

OVERVIEW

Since the 1972 Stockholm Conference, a number of books, papers, and monographs have discussed problems and situations related to chemicals management. These publications have largely dealt with key issues as seen from a specific perspective.

This book is the first to bring together, in a cohesive manner, history, legal and other instruments, roles of international organizations, capacity strengthening initiatives, and accomplishments at all governmental levels.

Philip Wexler
Jan van der Kolk
Asish Mohapatra
Ravi Agarwal

Editors

Philip Wexler is a Technical Information Specialist at the National Library of Medicine's (NLM's) Toxicology and Environmental Health Information Program. He is the federal liaison for the Toxicology Education Foundation and the World Library of Toxicology. He coordinates and manages NLM's risk-assessment information databases and online tools on the TOXNET system, and is project manager of the LactMed file on drugs and lactation. He is team leader for the development of the ToxLearn online tutorials, a joint activity with the U.S. Society of Toxicology (SOT). Served as chair, for two years, of SOT World Wide Web Advisory Team, he is President of the Society's Ethical, Legal, and Social Issues Specialty Section in 2009–2010. He has coorganized the Toxicology History Room for a variety of professional meetings. He was a member of the Education and Communications Work Group of the CDC/ATSDR's National Conversation on Public Health and Chemical Exposure project. Wexler has published numerous papers on toxicology information and has lectured and taught widely on the subject in the United States and abroad. He is editor-in-chief of the *Encyclopedia of Toxicology*, 2nd edition, 2005, with a third edition in progress, and *Information Resources in Toxicology*, 4th edition, 2009, both published by Elsevier Science. He is currently working on a major review article on toxicology informatics for *Critical Reviews in Toxicology*. He is the recipient of the SOT's 2010 Public Communications Award.

Jan van der Kolk has a background in chemistry and microbiology. He served as deputy director of Environmental Health in the Ministry of Environment of the Netherlands until 2005. Since, he has been working as an independent expert, under the company named Eco Conseil, mainly in the field of implementing International Environmental Agreements, mostly in countries of Africa, Asia, and the Caribbean. He has worked extensively with the European Union and most international organizations that have programs for the sound management of chemicals (WHO, FAO, UNEP, UNITAR, OECD) and pesticides. He was one of the founding fathers of UNEP Chemicals. He has been chair of the Codex Committee on Pesticides Residues, under the Codex Alimentarius Commission and of the Working Group on Pesticides of the OECD. He was an initiator of the review of the European Chemicals Management rules, which ultimately resulted in the REACH regulation.

Asish Mohapatra is a health risk assessment and toxicology specialist for Health Canada (Alberta/Northern Region) Environmental Health Program (contaminated sites). He has 15 years of experience in the public and private sectors in the areas of life sciences, environmental public health sciences, chemical and computational toxicology, health risk-assessment and management, and environmental management. He has postgraduate and predoctoral degrees in life sciences (toxicology) and environmental sciences (industrial toxicology and hemato-toxicology), respectively. He has extensively reviewed and analyzed projects on chemical risk assessment and management, and numerous human health risk-assessments and management projects. He has also reviewed several environmental impact-assessment projects related to air, soil and groundwater, biotic effects and community health-assessment issues around residential, commercial, and industrial contaminated sites. Additionally, he has conducted critical reviews of air, water and soil toxicology, indoor and outdoor air quality health effects assessment and dynamics, and exposure analysis and health risks from everyday exposure to emerging physical, chemical, biological, and psychosocial stressors. He has conducted uncertainty analysis, quantitative risk-assessment modeling, and toxicological evaluations of petroleum, chlorinated, and polyaromatic hydrocarbons. He has been evaluating existing and emerging tools and computational technologies (e.g., semantic Web informatics, data fusion tools) to effectively use them to analyze, interpret, disseminate, and share toxicological and health risk-assessment data from disparate sources under public health toxicology and risk-assessment frameworks.

Ravi Agarwal is founder director of Toxics Link, a key environmental NGO located in New Delhi and working on issues of chemical safety and waste for more than 15 years. A Communications Engineer by training, he pioneered public advocacy based work in this area, after more than 15 years of professional experience as an entrepreneur and engineer. He has been part of several policy and legislative processes in India as member of Standards Expert Groups on Biomedical Waste, Hazardous Waste technologies, Plastics Waste management, amongst others. He has lectured extensively on chemical safety issues besides helping in on the ground initiatives as well as the formulation of new policy. He has written widely on these issues, both in journals as well as in the popular media. Internationally he has worked closely with agencies like WHO and UNEP for initiatives on hazardous waste trade, mercury, technological options for biomedical waste treatment, and lead in paints. He has participated as an NGO representative in the formulation of several International multilateral treaty processes, including the Stockholm Convention on POPS, the Basel Convention, the SAICM process, as well as the ongoing intergovernmental negotiations for a Mercury Treaty. He is an Executive Board member of the International POPS Elimination Network (IPEN), a global network with over 600 members mostly from the global south as its Treasurer, besides being a Steering Committee member, and has been a member of the Zero Mercury Working Group, and the Basel Action Network since their inception. He was the first India chair of the Global Greengrants Foundation, and initiated the Environmental Equity and Justice Partnership fund in India to support grassroots work on chemical safety. He was awarded the IFCS–WHO Special Recognition Award for Chemical Safety in 2008 and the Ashoka Fellowship in 1998.

Contributors

Ravi Agarwal
Toxics Link
New Delhi, India

Mohamed Tawfic Ahmed
Suez Canal University
Ismailia, Egypt

Melanie Ashton
Independent Chemicals Consultant
London, United Kingdom

Pavan Baichoo
International Labour Office
Geneva, Switzerland

Åke Bergman
Stockholm University
Stockholm, Sweden

Viveka Bohn
Formerly Swedish Environment Ambassador
Stockholm, Sweden

Craig Boljkovac
The United Nations Institute for Training and Research (UNITAR)
Geneva, Switzerland

Arlindo Carvalho
Department of Environment
São Tomé and Príncipe

Cheryl Chang
David and Lucile Packard Foundation
San Francisco, California, the United States

and

Formerly of The United Nations Institute for Training and Research (UNITAR)
Geneva, Switzerland

Marta Ciraj
Ministry of Health of the Republic of Slovenia
Republic of Slovenia

Mark Davis
AGPM
The Food and Agriculture Organization of the United Nations (FAO)
Rome, Italy

Chris Dijkens
Inspectorate, Ministry of Infrastructure and Environment
The Hague, the Netherlands

David Downie
Fairfield University
Fairfield, Connecticut, the United States

John Duffus
The Edinburgh Centre for Toxicology
Edinburgh, United Kingdom

Lars-Göran Engfeldt
Formerly Swedish Environment Ambassador and Liaison Officer in the 1972 Stockholm Conference
Stockholm, Sweden

Heidelore Fiedler
United Nations Environment Programme (UNEP) Chemicals
Châtelaine (GE), Switzerland

John A. Haines
Retired from The World Health Organization (WHO)
Divonne-les-Bains, France

Achim Halpaap
The United Nations Institute for Training and Research (UNITAR)
Geneva, Switzerland

Georg Karlaganis
Federal Office for the Environment
Bern, Switzerland

and

University of Bern
Bern, Switzerland

Shelley Kath
Helios Centre
Montreal, Quebec, Canada

Boitumelo V. Kgarebe
Organisation for the Prohibition of Chemical Weapons (OPCW)
The Hague, the Netherlands

Ebeh Adayade Kodjo
Alliance Nationale des Consommateurs et de l'Environnement
Lomé, Togo

Pia M. Kohler
Department of International Relations in the Political Science
University of Alaska Fairbanks
Fairbanks, Alaska, the United States

Jan van der Kolk
Eco Conseil
Voorburg, the Netherlands

and

Retired from the Ministry of the Environment
The Hague, the Netherlands

Heinz Leuenberger
United Nations Industrial Development Organisation (UNIDO)
Vienna, Austria

Mariann Lloyd-Smith
International POPS Elimination Network (IPEN)

and

National Toxics Network Inc.
New South Wales, Australia

Naglaa M. Loutfi
Suez Canal University
Ismailia, Egypt

Gamini Manuweera
Secretariat Stockholm Convention
Geneva, Switzerland

and

Formerly Registrar of Pesticides
Department of Agriculture
Colombo, Sri Lanka

Ernest Mashimba*
Government Chemist Laboratory Agency
Dar-es-Salaam, Tanzania

Bert Metz
European Climate Foundation
The Hague, the Netherlands

and

Formerly cochair of the IPCC Working Group III

Asish Mohapatra
Contaminated Sites, Environmental Health Program
Health Canada (Alberta Region/Northern Region)
Calgary, Alberta, Canada

Sergio Peña Neira
School of International Commerce
Universidad del Mar
Viña del Mar, Chile

* Deceased.

and

School of Law
Universidad de Valparaíso
Valparaíso, Chile

DaeYoung Park
Young & Global Partners SPRL
Brussels, Belgium

Franz Perrez
International Affairs Division
Federal Office for the Environment (FOEN)
Bern, Switzerland

and

University of Bern School of Law
Bern, Switzerland

Linn Persson
Stockholm Environment Institute
Bangkok, Thailand

Pierre Portas
Formerly of Secretariat Basel Convention
Ste Cécile-Les-Vignes, France

John A. Pwamang
Environmental Protection Agency
Accra, Ghana

Lakshmi Raghupathy
SWITCH-ASIA Project
Deutsche Gesellschaft für International Zusammenarbeit (GIZ) GmbH
Gulmohar Park, New Delhi, India

Jody A. Roberts
Center for Contemporary History and Policy
Chemical Heritage Foundation
Philadelphia, Pennsylvania, the United States

Cristina B. Rodrigues
Organisation for the Prohibition of Chemical Weapons (OPCW)
The Hague, the Netherlands

Martin Scheringer
ETH Zürich
Zürich, Switzerland

Hamoudi Shubber
Secetariat Stockholm Convention
Geneva, Switzerland

and

Formerly of the SAICM Secretariat
Geneva, Switzerland

Richard Sigman
The Organisation for Economic Co-operation and Development (OECD)
Paris, France

Johan Sliggers
Ministry of Infrastructure and Environment
The Hague, the Netherlands

Ibrahima Sow
Global Environment Facility (GEF)
Washington, DC, the United States

Michael Stanley-Jones
Secretariat Stockholm Convention
Geneva, Switzerland

and

Formerly of the United Nations Economic Commission for Europe (UNECE)
Geneva, Switzerland

Elisa Tonda
United Nations Environment Programme (UNEP)

and

Formerly of United Nations Industrial Development Organisation (UNIDO)
Vienna, Austria

Michael Walls
Regulatory and Technical Affairs
American Chemistry Council
Washington, DC, the United States

Jack Weinberg
International POPS Elimination Network (IPEN)

Peter Westerbeek
Inspectorate, Ministry of Infrastructure and the Environment
The Hague, the Netherlands

Philip Wexler
National Library of Medicine
Bethesda, Maryland, the United States

Arnold van der Wielen
Retired from the Ministry of Infrastructure and the Environment
The Hague, the Netherlands

Section I

The Context

1 Creating and Controlling Chemical Hazards
A Brief History

Jody A. Roberts

CONTENTS

OVERVIEW

We live in a thoroughly chemical world. Chemicals, quite literally, comprise everything. This has always been the case, but perhaps now we are more in touch with this fact than at any previous time in human history. This is due, in part, to our increasing ability to exert influence over the chemicals available in the world, and even more powerfully, to put new chemicals of our own making into that world. Thus the knowledge of and presumed control over chemicals has created a moment in which we feel as though we have finally mastered our environment. But if the uniqueness of this moment results in part from this new-found creative power, it is equally due to our developing comprehension of how well we understand the consequences of those actions. Molecules produced decades ago, whose production has long since ceased, continue to pervade our environments and our bodies. Compounds created in one hemisphere travel the ecological currents to arrive unannounced in distant places. Entirely new vocabularies have been invented in recent years just to begin accounting for all of the new things we now know and to mark the places of the things we still do not. These new words—biopersistence, bioaccumulation, endocrine disruption, chemical mutagenesis, toxicogenomics, and nanotoxicology—stand as historical markers of our time. A century ago, a scientist would have had no understanding of, let alone familiarity with, a word like bioaccumulation not simply because of a lack of knowledge but because of an entirely different conceptual framework for thinking through the risks posed by chemicals to organisms and their environments. Just as our understanding of chemistry has evolved over the centuries (perhaps millennia, if we consider activities that existed long before the word chemistry ever appeared), so too have our understandings of the interrelations between chemicals, our environment, and our health.

Despite an increased appreciation of the hazards (and our lack of knowledge about them) posed by some chemicals, production grows. Each year, we produce more chemicals than in the previous year. We continue to invent new chemicals. And we create altogether new methods for creating these new chemicals. Our creative pursuits generally far outpace our efforts to fully understand these new substances or to control them adequately. The evolution of chemistry is also then an evolution of the means by which we seek to control, or "manage," the risks and harms associated with the development, use, and disposal of these molecules.

This brief introductory chapter, which explores the development of new tools and efforts to understand the intimate link between our chemical pursuits and the risks that emerge to humans and our environment, will be positioned in relatively long historical context. Indeed, it is an impossibly long history given the brevity of the narrative here. But given the context of this volume, it might be useful to appreciate the ways in which humans have dealt with the consequences of chemical adventures and resulting exposures.

ANCIENT ROOTS/ROUTES

While we are accustomed to thinking about our physical experience of the world, and navigating the dangers it poses to us, we are less familiar with thinking about the ways in which our chemical bodies come into intimate daily chemical contact with the world. With each breath, each gulp of drink, and each mouthful of food, molecules from "outside" come "inside" where simple, fundamental, but potentially risky reactions take place. While our understanding of these interactions has become more sophisticated through the development of chemistry, toxicology, pharmacology, and the like, human interest in navigating these risky interactions is as old as the species itself (if only because every organism, in order to survive, must find a way to safely interact with the world).

Foraging, farming, and herding—all early forms of procuring food—are sophisticated ways of sorting things out; we organize into groups, things that are safe to eat and keep them separate from things that are not. These safeguards become more elaborate when we include systems of food preparation and consumption. The anthropologist Mary Douglas demonstrates the simple, common, beauty in these systems, which cross cultures (Douglas, 1966; see also Douglas and Wildavsky, 1980). It is important to keep the clean separate from the unclean; the pure away from the impure. It would be anachronistic and reductionist to read these stories of sorting, classifying, and organizing, as simply tales of navigating risks of our chemical environment. But it would also be naïve to ignore these ancient roots to our species' need to develop simple systems for protecting ourselves from our living environment.

FROM HARVESTING TO PERFECTING POISONS

Knowing which plants or parts of animals are healthful or harmful allows one to enjoy those elements of the world that will help one to thrive while avoiding those that will cause harm. It also means that one can more purposefully harness those ingredients which can harm others (or oneself should one so chose). As this knowledge became more sophisticated and more specialized, fewer people could be entrusted with it. It is from this situation that we see the emergence of perhaps the first in the lineage that will eventually become our modern chemists, toxicologists, or pharmacologists: the herbalist.

In the figure of the herbalist, we have someone who represents specialized knowledge of the world for treatment of maladies, boosting health, and when necessary providing forms of nature's poisons. Interestingly, the herbalist is a figure that spans cultures even if the person goes by different names in those traditions. Despite geographic and cultural distance, their prominent traits are amazingly similar. And, indeed, so is the knowledge held by this person, which is perhaps one reason this person has endured to become a contemporary of the modern day scientist. This person continues to be an important one for helping us to navigate through the world safely. Although the practice cannot be said to be rooted in chemistry explicitly (especially for those traditions that exist within a very different picture of the human body), herbal treatments of many kinds for everyday problems easily move into the domain of the molecular sciences when the door is opened for them (as is evidenced in the adoption of many "alternative" therapies).

Chemistry is as much practice as it is theory. Here again we have links between past and present. The herbalist embodies knowledge not just of the world, but of how to prepare the world for proper human consumption. Crucial elements are extracted and distilled from their natural reservoirs. Treatments are prepared with mortar and pestle into powders, pastes, and pills. If specific knowledge of the world does not provide a link between these traditions, than surely the practices and material culture do.

THE PROTO CHEMISTS

It is at this point in our history that our *protochemists* emerge: the alchemist, the iatrochemist, and the metallurgist (Brock, 1992). The beauty of these professions is that they span the globe, demonstrating the multiple ways in which the chemical sciences developed in different corners of the world, and that they are all thoroughly hands-on activities, which emphasizes the ways in which manipulation of matter and substance have been at the root of this long tradition. Yet, despite their similarities, these three practices possess their own unique attributes.

The alchemist has emerged from history as the storied predecessor to todays chemist. Shrouded in mystery (and often depicted in paintings working in darkness), we have embraced the idea of the solitary scholar, probing the depths of the universe for ways to unlock the secrets of matter. What we often forget (or ignore) are the ways in which these individuals developed sophisticated means for notation, cataloging substances and reactions, and tools of the trade that long outlived the figure of the alchemist. The iatrochemist is to the apothecary what the alchemist is to the chemist, but with less intrigue. These figures developed skills that would give rise to fields more closely aligned with pharmacy and toxicology. Indeed, it was the iatrochemist Theophrastus Bombastus von Hoenheim, otherwise known as Paracelsus, who famously offered the dictum that the dose makes the poison, which has lived as the mantra (for better or for worse) of modern day toxicology.* The insight captured in this lesson, offered in defense of his seemingly unorthodox practices, highlight a key moment in understanding relationships between health, disease, and yet to be articulated chemical interventions. The metallurgist/smelter completes this triumvirate. Many of these practices have a lineage longer than that of recorded history, but they found new expression and new appreciation around the same time as the alchemists and iatrochemists entered the scene. Smelting and the working of ores distinguished cultures around the globe, especially those in the modern day Middle East. Indeed, legendary Damascus Steel, characterized by unsurpassed strength, is considered by many to be perhaps the earliest material to be embedded with a nano-sized microstructure. With the publication of Agricola's *De Re Metallica*, the mining and smelting trades became more tightly interwoven with the other emerging molecular sciences.

These practices, however, had other consequences as well. As the scale of mining and metal working increased in scope and scale, so too did the hazards of the job. Mine tailings and contact with heavy metals extracted during ore processing increased. Unique diseases associated with mining surfaced along with these precious raw materials. Signs of asbestosis and silicosis were already recognizable in the eighteenth and nineteenth centuries (Markowitz and Rosner, 2002; Michaels, 2008; Rosner and Markowitz, 1994). The ailments suffered by coal miners were well recognized, if not entirely understood. The relationship between mining and disease generated a pattern that would later be recognized as the basis of vastly different disorders, but all linked to the common problem of occupational exposure.

Increased availability of these substances also meant that they were finding new (or expanded) uses. Drawing on knowledge from previous traditions, the use of heavy metals for medical purposes, such as the application of mercury to treat venereal diseases, continued to grow. Mercury was also famously used by hatters to stiffen the felt being used. Lead found wide application in pigments and pipes. This, despite the long history of lead poisoning dating back at least to the Roman Empire and stretching across the centuries when lead was used for almost everything, from building water transport infrastructures and preserving wine (Warren, 2001). Perhaps amazingly, many of these practices continued unabated into the twentieth century (Markowitz and Rosner, 2002).

* The full quote reads: "All things are poison, and nothing is without poison: the *Dosis* alone makes a thing not poison," taken from *The Reply to Certain Culminations of His Enemies (Seven Defensiones)* (Paracelsus 1996 (1941), p. 22).

A "CHEMICAL" REVOLUTION

The "Chemical Revolution" of the late eighteenth century is typically categorized as a theoretical debate, one pitting Georg Stahl's theory of phlogiston versus Antoine Lavoisier's new theory of oxygen (Brock, 1992). Stahl's ideas about why materials could burn, due to the phlogiston contained within, were seen to be out of step with new Lavoisier's experiments. The story, of course, is more complicated than that and it has within it the seeds of other revolutions. Lavoisier's theory of oxygen was given greater countenance and strength because of two additional key components: the broad new system within which Lavoisier placed oxygen and tools by which he demonstrated, defended, and propagated his new theories (Kim, 2003; Levere, 2001).

Lavoisier's chemistry introduced a new way of understanding chemical reactions, which allowed for a more complex set of reactions to be possible and to account for those reactants and products. Equally important were Lavoisier's contributions to the practice of chemistry and the development of tools for quantifying chemical reactions. This early work, quickly adopted in Germany and more slowly in Britain, laid a foundation for thinking differently about how chemical species interact with one another, and how they might be analyzed. Methods for analyzing complex organic liquids and gases which developed over the succeeding generation, created new methods that would find application in agricultural as well as pharmaceutical settings. Thus, the chemical revolution was at least as important for its analytical breakthroughs as it was for its theoretical reframing of the chemical sciences.

BIRTH OF AN INDUSTRY

The nineteenth century's industrial revolution was not isolated to the development of motion and mechanisms. It was also a time when chemistry became the basis for a brand new industry. Chemists were already busy at work linking their laboratories with farm fields (Brock, 1992). But the creation of a new color dye, mauve, from coal tar extracts in the laboratory of Charles Perkins, marks a change in the ways in which chemistry and industry coexisted—and it marks an important moment for occupational and environmental toxicology (Garfield, 2001; Travis, 1993).

Perkins' creation of the first synthetic dye, and its quick application to commercial industries, precipitated a race among chemical powerhouses across Europe in search of other dyes that may be hidden in coal waste (Travis, 1993). Understanding why these particular compounds acted as dyes was secondary to finding more of them. This model of innovation, search first and understand later, arguably guided the chemical industry through the twentieth century. Of course, what made Perkin's molecule display such brilliant colors was the abundance of conjugated bonds available in the aromatic compounds that would come to characterize the azo dyes. While Perkin isolated the first of these compounds and put it to commercial use, it was the Germans who created an industry around these organic dyes. Work done with these organic compounds helped to make Germany a leader in industrial chemistry. It also made it a site for the emerging fields of occupational health and exposure.

Before the century's end, solid links had been established between worker exposure in the dye industry and the development of rare cancers. Unions representing workers in these new industries worked with their own medical and health professionals to document these cases. But knowledge of the practice of organic chemistry and knowledge of the effects these practices might have in occupational settings did not travel at the same rate. While Europe gathered experience and knowledge in both of these domains, less-developed chemical industries, like those in the United States, gathered unequal bits of information from across the ocean (Michaels, 2008).

THE WAR YEARS

The onset of war in the twentieth century altered this landscape more rapidly and more dramatically than at any previous time in this history. The geopolitics of the chemical industry shifted as Europe fell into ruin and the United States collected and exploited the spoils of war (in the form of patents and other trade secrets). Chemicals became the basis for new industries, new weapons, and new materials. And knowledge of and around the toxicity of these materials expanded as these toxic properties were sought out purposefully to combat foes, both domestic and foreign.

With world war erupting in the second decade of the twentieth century, radical changes to industrial infrastructure were required to keep the war machine running. Global supply chains were being severed. Coupled with increased strain on raw materials pressure increased on chemists to find suitable synthetic alternatives for crucial materials and processes. Perhaps the most important of these was the development of the Haber–Bosch process for the synthesis of ammonia (Smil, 2004). If the Haber–Bosch process was the most important breakthrough in those early years of war, then Fritz Haber's transformation of chlorine gas into a weapon of war was the most infamous (Russell, 2001). Haber's contributions made him a national hero in Germany. For the countries that witnessed the events at Ypres where French soldiers were gassed in their trenches, the future was being written. Countries such as the United States quickly mobilized academic chemists through their Chemical Warfare Service to begin research and development of new potential chemical weapons (Russell, 2001). The first of the World Wars ended before these tools could be put to use, but the knowledge gleaned from the process proved invaluable. With the end of the war came the spreading of its spoils; in this case, the patents that had made the chemical industry in Germany the dominant figure globally. The budding chemical industry in the United States now had practical experience of their own garnered during the war and the patents of its German rivals. All that was needed was sustained support from the U.S. government to maintain the research initiatives begun during the war.

The industry did succeed in keeping money flowing for research by, in part, coupling research for chemical weapons into research on new pesticides (Russell, 2001). As the historian Ed Russell uncovers, the similarities in projects were more than coincidental. More often than not, a bad chemical weapon made for a good pesticide, and vice versa. Additionally, the support technologies would be remarkably similar—one could simultaneously prepare for dusting fields and trenches. In the case of pesticides, the model for innovation was a bit different than what emerged with synthetic dyes. Preparing poisons meant also wanting to understand what would be poisonous and why. If efficiency and potency were not motives, then the understanding that we would likely need to protect our own soldiers proved enough to begin more serious toxicity testing. And so with the rise of more potent poisons between the wars also came a parallel effort that would provide the basis for our understanding of the toxicity of these materials.

As was the case in the World War I, these chemical concoctions devised during the interwar period never made it to the battle field when war broke out again. And so as with many wartime innovations, producers had surpluses of product in search of new markets. The parallel developments of poisons for humans and pests made this transition smoother than might have been otherwise. But pesticides were not the only chemical innovations looking for markets. Demand for materials during World War II had necessitated the creation of a host of new chemical products that needed new users/consumers (Ndiaye, 2007).

Behind this push of wartime products into the civilian market were the skills developed in those war years by chemists and chemical engineers in the field of mass production. Stories of innovation of new materials, like radioactive material, typically dominate our fables. Missing are the engineers who figured out how to produce these materials on scales previously unfathomed (Ndiaye, 2007). Indeed, scale becomes one of the defining characteristics of the modern chemical enterprise with rippling effects on economies and environments. And it set the stage for that most pervasive of modern chemical wonders: plastic. The combination of the creation of new synthetic materials with the engineering capability of mass production made these new artifacts possible. It is difficult to imagine a day in modern life where plastic is not present. Their plasticity in form and function has made them ubiquitous in our lives. Their durability and their sheer endless quantities have made them ubiquitous in our environment. Despite the various and intimate ways in which these materials shape our lives, little thought was given to what might result from those constant contacts.

RACHEL CARSON AND THE DECADES OF DISASTERS

For most folks, 1962 is the turning point for our contemporary concerns and preoccupations with synthetic chemicals. In that year, Rachel Carson published *Silent Spring*, altering the political landscape with her warnings about the ways in which synthetic chemicals produced during and in the wake of World War II were changing the chemical composition of our environment and our bodies (Carson, 1962). While Carson's legacy continues to be debated, the fact remains that she succeeded in drawing attention to the ways in which our understandings of the natural world were being radically remade in this post-war era. While her fame and notoriety more commonly flow from her discussions of dichloro-diphenyl-trichloroethane (DDT), in truth Carson identified many of the emerging problems that became hallmarks of the decades to follow (and which, by and large, remain our key points of concern). She worried not simply about DDT, per se, but its ability to persist in local ecologies, to move across the ecological landscape, and to accumulate as it moved through the environment and the food chain. Today we call compounds of this type as PBTs—persistent, bioaccumulative or toxic chemicals. Carson could not point out that compounds like DDT are PBTs; she was in a sense defining them as a class of chemicals as she wrote. Her legacy is perhaps better remembered not through the ban on DDT, but through the establishment of PBTs as a class of chemicals of concern; through the Stockholm Convention in 2001; through her work to link the human body with its environment; and perhaps most importantly through the attention she gave to observing nature not simply as a place separate from humans, but as a place where we see directly the ways in which our actions join with the environment, and eventually come back to us.

Carson's writings echoed all the more forcefully as the 1970s and 1980s produced a litany of names of places to mark one chemical disaster after another, which helped to mark a new era of vigilance, activism, and regulation. Images of a Cuyahoga River in flames and cities blanketed in smog helped to instigate a national (and eventually global) conversation about the state of nature. But if water and air were the visible poster-children for environmental regulation and industry reform, the more silent spills, leaks, and contaminated sites were nonetheless receiving increased scrutiny. Events in Love Canal, Times Beach, and Seveso introduced words like dioxin into the global vernacular. But it was the events of December 3, 1984 that rewrote the relationships between governments, industries, citizens, and activists. The escape of methyl isocyanate from a Union Carbide plant in Bhopal, India killed thousands in a single night and left scores of thousands more ill and debilitated for decades. Beyond the enormity of the tragedy that unfolds from there, the incident marked an important moment in thinking about chemical hazards.

The tragedy at Bhopal reconfigured the relationships between citizens, corporations, and the state in dramatic ways. Corporations were made to confront the meaning of being both local and global citizens. In succeeding years, the chemical industry banded together realizing that the weakness of one could be end of them all. They worked to become global partners in establishing new standards for operation at their facilities, largely through the establishment of the Responsible Care program. Beyond operational procedures, the program also encouraged more engagement at the local level with neighboring communities. Communities, too, sought new avenues for cooperation and new partnerships. Direct interaction with plant mangers and operators helped to fill the void left by many disengaged state apparatuses. Nongovernmental organizations (NGOs), too, seized the moment and began advocating with and for communities as they sought safer living conditions. With the growth in online activism in the closing years of the twentieth century, local communities linked with one another in ways that created bridges across geographies, while also making geography less important. For all the new modes of advocacy and interventions created in response, Bhopal remained a tragedy seeking justice. As we passed the quarter century mark since the first insult, the waters around Bhopal remained polluted, communities continued to suffer elevated incidents of a plethora of diseases, and the factory itself slowly introduced new problems as its decaying remnants leached into the land. Perhaps, then, Bhopal served not so much as a symbol of an industry that was, but as a symbol of what a new globalized industry would become (Fortun, 2001).

AT THE CLOSE OF THE TWENTIETH CENTURY AND BEYOND

As the twentieth century closed, humans found themselves sitting at the nexus of this long and diverse history. Our practices, products, and pollution represented an uneasy mix of recent innovations and old hazards. While the blunt fact of these hazards may not be new in human history, stark differences do separate what was from what is becoming. New tools—conceptual, political, and material—are emerging to confront the hazards that our industrial heritage has born out. Three topics in particular warrant further discussion.

First, our understanding of risk and the links between human and environmental health have become more sophisticated, more nuanced, and more powerful. That is, the picture has become more complex. As the previous stories outline, our understandings between our contact with the world, our manipulations of that world and our health have been evolving for millennia. In more recent times, our scientific enterprise has become fractured and specialized. Experts exploring human health rarely come into contact with those exploring similar issues and questions in the nonhuman worlds. But recent decades have brought about a convergence, some stemming from those thoughts penned by Carson, others as a result of keen observations, and still more thanks to those minds that see connections in the world where others see only differences. Seemingly "old" sciences and end points like developmental biology, chemical mutagenesis, and carcinogenesis have found new meaning within new sciences such as endocrine disruption, epigenetics, toxicogenomics, and other offspring of the -omics revolutions. These mergers have provided opportunities to reexamine the connections between humans and their environment. And through increased analytical capabilities, which have led to the development of a plethora of new studies in human biomonitoring, our intimate connection with our environment is once again being made tangible. These insights have changed the ways in which we perceive time and space in terms of chemical contamination.

The flow of chemicals follows ecological boundaries, not political boundaries. Likewise, then, systems for controlling chemicals must respect and privilege the ecological over the political. Managing the hazards presented by chemical exposures is not new, of course. When Duke Eberhard Ludwig of Würtemberg issued an edict in 1696 promising "loss of life, honour, and fortune" to all those who dealt in wine adulterated with lead oxide, he and his advisors were instituting a system of chemical management to protect human health (quoted in Eisinger, 1982). But as the business of the chemical sciences grew, with its processes more intensive, its volumes expanding, and its products traveling the globe, such localized measures have been replaced with more serious forms of global governance.

Unions served as one of the early conduits for the internationalization of chemical and industrial hazards in the late nineteenth and early twentieth centuries (Markowitz and Rosner, 2002; Michaels, 2008). Their networks helped to unite workers in chemical plants whose exposures and diseases were not confined by political spaces. Since many chemical companies, too, existed beyond these political boundaries, the work of the unions helped to spur a new era of investigation into the hazards of the workplace. But while some companies may have crossed these boundaries, operators in each nation still played by local rules. And so while these channels proved crucial for moving knowledge across the Atlantic, their actions were not immediately successful in uniting governments.

INTERNATIONAL COOPERATION ON MANAGEMENT

The harmonization of national policies finally became a topic when, in 1972, representatives from the global community met in Stockholm to discuss the intersections of humans with their environments. This conference was a milestone in thinking about the ways in which humans from all countries are required to come together as a global community to take collective ownership of the emerging environmental problems. That is, it perhaps marks the beginning of a global perspective on the need to protect shared, common, resources and to create healthy environments for all. The decades following the first meeting in Stockholm have witnessed a continued proliferation of international efforts: conferences in Rio (1992) and Johannesburg (2002); legal instruments arising from meetings in Basel, Rotterdam, and Stockholm; the creation of the United Nations Environment Programme and the Organization for Cooperation and Economic Development; and perhaps more importantly organized institutions and communities that have kept up the work during the long pauses in between. Since such highlights in the global management of chemicals are the subject matter for this volume, I will leave the details to the experts that follow. But it is important to note the diverse nature of these organizations: they represent nations, industries, scientific communities, policy experts, environmental professionals and more. They represent the four corners of the world and a spectrum of expertise. But the question remains: what will it take to create a truly global system of chemical management that can adequately protect *all* peoples *and* our living environment? It is here that we see the one component of recent decades that has played such a key role, but one that is still largely missing from our public dialogues about chemical management and governance: the role of social movements.

ENVIRONMENTAL JUSTICE AND GOVERNANCE

The environmental justice movement in the United States of America, for example, has historical roots intertwined with that of the civil rights movement, which emerged about the same time as our modern environmental movement (Pellow and Brulle, 2005). Despite these commonalities, we traditionally treat the development of these three events as largely separate and distinct. The more familiar origin stories for the environmental justice movement place it in more recent decades, arising out of an increasing awareness about the proximity of neighborhoods comprised of racial minorities and low income to industries that presented environmental health hazards. In the decades since, the environmental justice movement has brought renewed attention to the local, place-based hazards that confront many communities located near chemical facilities. These experiences, often articulated through direct action or coordinated outreach, must inform any chemical management system—local or global. Examples abound in recent decades: leaded gasoline, asbestos, lead in paint—taboos of the United States and many European nations become surplus stock, which become cheap goods that continue to find significant markets in other countries. From time to time, these goods circulate back to us, perhaps in the form of children's toys, but by and large many of us presume these artifacts of an earlier industrial age have gone the way of the dodo. But our materials and their constituent chemicals travel the globe on commercial and ecological winds requiring vigilance locally and globally (Ottinger, 2010).

The results of these efforts are not limited to social movements. The work undertaken and continuing by a variety of activist groups tied to environmental justice, environmental health, and globalization have created new tools for both local and global chemical management. The Bucket Brigades that arose in the refinery towns of Louisiana's chemical corridor, South Philadelphia, and the Bay Area have become globalized networks of environmental justice (EJ) activists sharing tools, tactics, and information (Casper, 2003; Washington et al., 2006). The buckets, themselves—an "ordinary" five gallon bucket turned into a cheap, portable, air monitoring canister—have changed the way instruments for community monitoring have developed. And their users, "citizen scientists," have challenged ideas of authority and expertise in creating their own information about chemical health and risks. Consider, too, the continued efforts to bring justice to Bhopal through organizations like the International Campaign for Justice in Bhopal, which brings together a coalition of NGOs and individuals seeking support for the survivors of that lingering event. But while such examples offer a glimpse at what has been and could be done, there remains very little coordination between the official experts and community activists—despite the treasure trove of data which the latter have collected.

All of which becomes dramatically more important when we consider the geography of chemical production that will unfold in the coming century. Refining and production facilities have already begun to relocate to the global South. The trend will accelerate in coming years as the cost of doing business in the North comes up and companies seek to locate their facilities closer to their new consumer base in countries like China, India, and Brazil. As we debate the effects of Toxic Substances Control Act reform in the United States and the effects of Registration, Evaluation, Authorisation and Restriction of Chemicals in the European Union, we must also be thinking more globally and seeking answers about the ways in which these changes in national dialogues will change what happens in the new centers of production. What will a chemical management program for the twenty first century look like? What will it *need* to look like to bring harmony to our systems of oversight that protects *all* citizens from harm, no matter how far downwind or downstream they may be? Finding ways to merge the national and global sentinels of health and regulation with the views from citizens in the street might lead to a system of management that can actually succeed.

REFERENCES

Brock, W.H. 1992. *The Chemical Tree*. New York, NY: W. W. Norton.

Carson, R. 2002 (1962). *Silent Spring*. New York, NY: Houghton Mifflin.

Casper, M.J., ed. 2003. *Synthetic Planet: Chemical Pollutants and the Hazards of Modern Life*. New York, NY: Routledge.

Douglas, M. and A. Wildavsky. 1980. *Risk and Culture*. Berkeley, CA: University of California Press.

Douglas, M. 2002 (1966). *Purity and Danger*. New York, NY: Routledge.

Eisinger, J. 1982. Lead and wine: Eberhard Gockel and the Colica Pictonum. *Medical History*, 26: 279–302.

Fortun, K. 2001. Advocacy *after Bhopal: Environmentalism, Disaster, and New Global Orders.* Chicago, IL: University of Chicago Press.

Garfield, S. 2001. *Mauve: How One Man Invented a Color that Changed the World.* New York, NY: W. W. Norton.

Kim, M. G. 2003. *Affinity, that Elusive Dream: A Genealogy of the Chemical Revolution.* Cambridge, MA: MIT Press.

Levere, T. H. 2001. *Transforming Matter: A History of Chemistry from Alchemy to the Buckyball.* Baltimore, MD: The Johns Hopkins Press.

Markowitz, G. and D. Rosner 2002. *Deceit and Denial: The Deadly Politics of Industrial Pollution.* Berkeley, CA: University of California Press.

Michaels, D. 2008. *Doubt is Their Product.* Oxford: Oxford University Press.

Ndiaye, P. A. 2007. *Nylon and Bombs: DuPont and the March of Modern America.* Baltimore, MD: The Johns Hopkins University Press.

Ottinger, G. 2010. Buckets of resistance: Standards and the effectiveness of citizen science. *Science, Technology, and Human Values*, 35(2): 244–270.

Paracelsus. 1996 (1941). *Four Treatises.* H. E. Sigerist (ed.). Baltimore, MD: The Johns Hopkins University Press.

Pellow, D. N. and R. J. Brulle, eds. 2005. *Power, Justice, and the Environment: A Critical Appraisal of the Environmental Justice Movement.* Cambridge, MA: MIT Press.

Rosner, D. and G. Markowitz 1994. *Deadly Dust: Silicosis and the Politics of Occupational Disease in Twentieth-Century America.* Princeton, NJ: Princeton University Press.

Russell, E. 2001. *War and Nature: Fighting Humans and Insects with Chemicals from World War I to Silent Spring*. Cambridge: Cambridge University Press.

Smil, V. 2004. *Enriching the Earth: Fritz Haber, Carl Bosch, and the Transformation of World Food Production.* Cambridge, MA: MIT Press.

Travis, A. S. 1993. *The Rainbow Makers: The Origins of the Synthetic Dyestuffs Industry in Western Europe.* Bethlehem, PA: Lehigh University Press.

Warren, C. 2001. *Brush with Death: A Social History of Lead Poisoning.* Baltimore, MD: The Johns Hopkins University Press.

Washington, S. H., P. C. Rosier, and H. Goodall, eds. 2006. *Echoes from the Poisoned Well: Global Memories of Environmental Injustice.* Lanham, MD: Lexington.

Section II

Conferences

2 Stockholm 1972
Conference on the Human Environment

Lars-Göran Engfeldt

The unanimous adoption on December 3, 1968 of United Nations (UN) General Assembly Resolution 2398 (XXIII) was a seminal event. In accepting the proposal made by Sweden for the convening of a United Nations Conference on the Human Environment in 1972, the UN took on a vast new cross-sectorial area, the human environment, for international cooperation at the highest global political level. Earlier, such cooperation had been fragmentary, scientifically oriented, and mainly based on nature conservation. Environment diplomacy was born as a new and distinct type of diplomacy. A unique multilateral process followed, marked by strong continuity and agenda strength, which has been carried on up to the present day. Its high points were the pioneering Stockholm Conference in 1972, the 1992 United Nations Conference on Environment and Development in Rio de Janeiro and the 2002 World Summit on Sustainable Development in Johannesburg.

The initiative of Sweden was a reaction to the visibly increased ecosystem disturbances that occurred when human activities started to impact the entire planet around the middle of the twentieth century. Several factors converged in the 1960s that laid the basis for national and international political responses to the environmental problems. They included strong public reaction in industrialized countries, influential publications such as Rachel Carson's *Silent Spring* in 1962 (Carson, 1962), and spectacular environment-related accidents.

In several industrialized countries, the period up to 1972 saw the enactment of environmental legislation and the setting up of government machinery for environmental protection. This process accelerated particularly fast in the United States, which held a preeminent international policy leadership role since the end of World War II.

Two main factors dominated the landscape of the UN in the 1960s: the Cold War and a shift toward emphasis on economic and social development issues.

The Cold War imposed severe limitations on the work of the organization to maintain international peace and security. The Security Council was often paralyzed as the two opposing world powers, the United States and the Soviet Union and their allies, locked in fierce competition for geopolitical influence which played out in the organization itself and in the rapidly decolonizing parts of the world.

The decolonization also led to the UN's increased involvement in economic and social development. With decolonization, the number of member states rose from 51 at the UN's outset in 1946 to 123 in 1967.* The situation in many of these new states was often desperate, prompting the start of international development assistance. Initial hopes that these would be only temporary efforts to support self-help soon began to fade.

Two seminal initiatives were taken to increase the role and leverage of developing countries. The first was the founding of the nonaligned movement in 1961, which had a general political role. The second was the establishment by 77 developing countries of a joint negotiation mechanism, the Group of 77 (G77), at the first United Nations Conference on Trade and Development (UNCTAD) in 1964 with the aim of safeguarding the economic interests of the members in UN negotiations on international economic issues. The creation of the G77 was an important political development for the Stockholm process, even if the model of group to group negotiations was formalized only at the time of the preparations for Rio.

The rapid developments in science and technology saw the emergence of new types of agenda items in areas where the United States and the Soviet Union did not have colliding interests, such as the peaceful uses of outer space and the seabed outside areas of national jurisdiction. The possibility of using scientific discourse to promote détente was being explored both in the United States and the Soviet Union by the time of the Swedish initiative. This included the environment which was seen then as a largely scientific and technological issue.

Favorable UN dynamics combined with the geopolitical position of Sweden and the key role of the chief architect of the initiative, the Permanent Representative of Sweden to the UN, Sverker Åström, were the main assets supporting Sweden's role as initiator and facilitator during the initial years.

In light of later experiences it can be recalled that the prevailing serious institutional shortcomings and limitations figured prominently already in the considerations preceding the initiative. The post-World War II international system, strongly anchored in the overriding principle of national sovereignty, had not been equipped to respond to the demands of the rapidly changing and ever more interdependent world. It was deemed unlikely that this would change anytime soon as no major country questioned this system and its underlying principle of sectorial organization and decision making. The only realistic possibility to deal with the new cross-sectorial environmental issue was to promote increased coordination between the parts of the system. Within the fragmented UN structure, the specialized agencies could not be expected to accomplish such a transformation. This would require strong initiatives from governments, which in turn was unlikely as they themselves had a long way to go with regard to coordination within their own administrations. The effect was that the specialized agencies operated as independent entities in the international system, supported by their national interest groups, and jealously guarding their roles and prerogatives. The conclusion drawn in 1968 was that if anything were to be accomplished in the international field, it had to be done within the existing context with all its limitations.

* The present number is 192.

The concept that was launched by Sweden was visionary for its time and, at the same time, politically realistic. It had the following main building blocks.

- *Global scope:* The global character of the environmental problems as well as the need for increasing public interest in the issue made it natural to consider convening a UN Conference.
- *Broad involvement:* The strong need to increase awareness about the full economic, social, and political effects of these problems required bringing together actors from different sectors and disciplines. Through greater insight, it would be easier to gain acceptance for the necessary measures at national and international levels.
- *Action orientation:* The Conference needed to focus on certain concrete problem areas in order to gain an overview of those problems that could only, or best, be solved through international cooperation. It would also be useful to define an international division of work for taking the appropriate regulatory measures. An action-oriented perspective was thus clearly present from the beginning.
- *Interagency coordination:* The need for interagency coordination was crucial. The only way to mobilize enough political support and strength for this to happen was to try to establish a comprehensive framework by ensuring that a broad discussion could take place at the central UN level and through this vehicle achieve a common outlook and direction of the efforts of the UN system. A negative approach toward the specialized agencies would have been counterproductive. Instead, it needed to be made clear that the activities of the agencies would continue as before, and the best possible cooperation would be sought with them.
- *Cause–effect focus:* The Conference needed to focus both on the deleterious impact of human activities on nature and on the effects on humans. In the first category, pollution of various kinds and chemical contamination were to be highlighted while the second looked at issues such as negative consequences of rapid urbanization.
- *Current institutions:* No new international institutions were to be proposed. At the time, this was the internal consensus view.

The preparatory process for the Stockholm Conference took place in two phases. The first was a political anchoring stage under Swedish leadership and lasted to the summer of 1970. The second, substantive, stage started in the autumn of 1970 when the importance moved to the newly appointed Secretary-General of the Conference, Maurice Strong of Canada, and his independent Conference secretariat.

The initiative had its core supporters among a group of western industrialized countries, notably the Nordic countries, Canada, the Netherlands, and the United States. Their focus was on international cooperation dealing with global environmental problems deemed to be of common interest to all countries, especially various forms of pollution. They argued that developing countries might be able to avoid the kinds of costly mistakes made earlier by industrialized countries in their own process of economic and social development. This meant that the focus was on the self-interest of developing countries, with the underlying assumption that no major changes in the international economic system were called for.

In the first phase, a series of political obstacles was overcome through diplomatic efforts. This included the active opposition of UK, France, and the specialized agencies. Ideological constraints by the Soviet Union were also overcome. The almost nonexistent capacity of the UN Secretariat in the field of environment was another serious threat which was averted due to the link which the Secretariat established with the Swedish UN mission for advice and assistance. The Secretary-General of the UN was entrusted with the overall responsibility and an advisory Preparatory Committee was established. This secured a wide mandate for Maurice Strong.

Several important issues for the entire Stockholm–Rio–Johannesburg process were identified during these initial years. These included the complex environment–development linkages, the special needs of developing countries, the need for prevention and precaution, and the uncertainties of climate change. The process manifested a clear potential for normative development and measures to increase overall understanding of the environmental threat. The first proposals were made for legally binding agreements dealing with various sources of pollution.

Different views emerged about the future role of the UN in the new area of the environment. The key proponents wanted an action-oriented and broad consideration of the issue with the UN at the center in a clear coordinating role. UK and France echoed views of the specialized agencies that the central authority should remain with the agencies and kept a restrictive position. The Soviet Union also had strong reservations about central coordination.

There remained a lack of clarity as to the real political commitment and involvement of developing countries which comprised the vast majority of member states. This problem took an ominous turn when Brazil attacked the initiative in the spring of 1970, characterizing it as a "rich man's show" to divert attention from the development problems of developing countries (Engfeldt, 2009, p. 41). When it became clear that the Conference project was so well anchored in the UN that it could not be dismantled, Brazil refocused its political energy on safeguarding what it felt to be the real interests of developing countries in the preparatory process. One reflection was the introduction, later in 1970, of the demand that rich countries had a duty to put additional resources needed for environmental protection measures at the disposal of developing countries—the additionality concept. This triggered a gradual development of the entire process with its culmination in the preparations for the Rio Conference when the environment issue was incorporated in the overall framework of North–South relations.

As the second stage got under way, Maurice Strong quickly managed to gain the confidence of delegations. He managed to exercise a remarkable level of personal influence, while at the same time maintaining the confidence of delegations. The design of the preparatory process, the methods of accelerating political agreement and the innovative role of the Conference secretariat were key ingredients of his leadership.

He conceived a three-level preparatory process that would avoid the seeming contradiction between the desire by governments for both comprehensiveness and action.

Level 1: *Intellectual and conceptual framework*: designed to provide a comprehensive review of the existing state of knowledge and opinion on the relationship between man and his environment. The main contribution was an unofficial report prepared by Barbara Ward and René Dubos with the title of the motto of the Stockholm Conference, "Only One Earth" (Ward and Dubos, 1972).

This innovative publication constituted the world's first state of the environment report when it was published in 1972 and had a major impact on public opinion and elites in industrialized countries and also, to some extent, in the developing world.

Level 2: *Action plan for future work*: producing an action plan and work program for the years ahead was the centerpiece of the substance considered at the Stockholm Conference. The plan would contain those items that had sufficient consensus to enable agreement: (i) on concrete recommendations for further action, and (ii) on institutional arrangements for taking such action.

Level 3: *Issues for immediate action*: consisted of specific issues that required immediate initiation of international action that could be completed, at least through an initial stage, by the time of the Conference.

Maurice Strong's "process is the policy" concept was an innovative tool to increase the quality and level of consensus gradually by promoting constant interaction between the substantive and political aspects of an issue. This was undertaken through a complex series of consultations and negotiations, parallel or additional to the official proceedings. The concept was assisted by the deadline presented by the Conference itself, which served as a powerful stimulus to achieve results.

Strong also broke new ground that would have major repercussions in the future by opening up the process and inviting the active involvement of civil society, the scientific community, and the corporate sector.

The Conference secretariat combined innovation with a high level of ambition and thoroughness never before seen at a UN Conference. The amount of documentation was drastically reduced, and the resources released were used to set up an impressive network of influential consultants from all geographic regions. The secretariat played a significant role with lasting effects in the preparations of national reports and their analysis. This process was greatly assisted by the visits by Strong, or his representatives, to many countries. It was also important for the outcome that the conference papers were the exclusive responsibility of the secretariat and not of the specialized agencies. This ensured that a unified and coherent perspective was presented, in line with the original aim to provide a common outlook and direction for the international environmental efforts.

The task of positively engaging developing countries emerged as a main challenge when a thinly veiled threat of a developing country boycott of the Conference was articulated in the spring of 1971. There was deep dissatisfaction that the preparations in their view were too oriented toward the interests of industrialized countries. In a strategic breakthrough, Strong managed in June 1971 to persuade the Prime Minister of India, Indira Gandhi, to come to Stockholm where she was the only foreign Head of State at this ministerial level Conference. Her address to the Conference that mass poverty is the greater polluter of all had a profound and lasting political effect. Substantively, the special seminar held in Founex, Switzerland, in June 1971 was a defining moment and paved the way for the attendance and active involvement of developing countries in the Conference. Its report was also a major intellectual contribution to the further international discourse on environment and development.

The key message from Founex was contained in the following sentence: "If the concern for human environment reinforces the commitment to development, it must also reinforce the commitment to international aid" (United Nations, Development and Environment, 1972).

The three main substantive decisions taken by the Conference consisted of the Stockholm *Declaration*, *Action Plan*, and the *Resolution on Institutional and Financial Arrangements*. Thanks largely to the organization of the preparatory process they had been widely agreed before the Conference.

As the role of the Preparatory Committee was an advisory one, the intergovernmental process was to a large degree of a consultative character. The preparation of the *Declaration* was an exception with governments fully in charge. This became a milestone document, which provided the first agreed global set of basic normative principles for future work in the field of the human environment, and made a considerable contribution to the development of international environmental law. Its concepts were further elaborated by the 1992 Rio Declaration (see Chapter 3 of this book). *Principle 21* has been widely referred to as the most important legal point in the Declaration. This Principle included an affirmation of the responsibility of States to ensure that activities within their jurisdiction do not cause damage to the environment of other States, or of areas beyond the limits of national jurisdiction.

The *Action Plan* was conceptually based on the knowledge theme. It focused on strengthening ongoing activities within the UN system, particularly research and studies of various kinds, supplemented by calls for normative developments in some areas. Beside the Declaration, one other Level 3 process achieved a concluded international action by the time of the Conference. That was in the field of marine pollution, the London Dumping Convention formally adopted later in 1972, followed by the MARPOL Convention of 1973.

Most of the 150 recommendations contained in 109 points were adopted by consensus. Among the most controversial ones that had to be adopted by vote, with reservations or watered down, were the issues of family planning and a proposal for a 10-year moratorium on commercial whaling.

The Action Plan was an impressive achievement for its time and became a major stimulus to ongoing activities, and to several new ones, including a more systematic monitoring of the state of the environment.

The new activities included the convening of the Habitat I Conference in Vancouver in 1976, the establishment of a warning system relating to natural disasters, an expansion of international cooperation in marine pollution, an international program for the protection of the world's genetic resources, the development of standards for measuring and limiting noise emissions, and activities related to the control and recycling of wastes in agriculture. The Plan also was instrumental in establishing an international program in environmental education.

The Plan further recommended that plans should be developed for an international registry of data on chemicals in the environment (International Registry of Potentially Toxic Chemicals (IRPTC)), for monitoring and epidemiological research programs aiming at early warning and prevention of deleterious effects of environmental agents and for the establishment of an international referral system for sources of environmental information (INFOTERRA). Also, a worldwide network of monitoring stations was set up.

An important contribution of the Plan was also that it contained language that was a precursor to the precautionary principle embodied in the Rio Declaration on Environment and Development adopted in 1972. This included a warning to be mindful of activities in which there was an appreciable risk of effects on climate.

The new activities included the convening of the Habitat I Conference in Vancouver in 1976, the establishment of a warning system relating to natural disasters, an expansion of international cooperation in marine pollution, an international program for the protection of the world's genetic resources, the development of standards for measuring and limiting noise emissions, and activities related to the control and recycling of wastes in agriculture. The Plan also was instrumental in establishing an international program in environmental education.

As the substantive preparations of the Conference got under way, there was increasing agreement that an organized, *institutional follow up* would be necessary. The concept chosen was heavily influenced by Maurice Strong's thinking. He foresaw Stockholm providing the impetus that could link all components of cooperation from different organizations within a network or system. Instead of a new specialized agency or executive organ, a policy evaluation and review mechanism could become the institutional center or brain of the international environment network. Its operational influence should be exercised through financing. An early consensus emerged among member states for various reasons that no new specialized agency should be established.

The end result was a recommendation to establish the United Nations Environment Programme (UNEP) with the following four entities.

An intergovernmental committee (Governing Council) is set up under the General Assembly with the task of providing general policy guidance for direction and coordination of environmental programs within the UN system and to keep the world environmental situation under review.

A *small environment secretariat* to serve as the focal point for environmental action and coordination within the UN system led by an Executive Director (ED), elected by the General Assembly on the nomination of the Secretary-General. Normally, officials with the same rank (Under-Secretary-General) were appointed, not elected, by the General Assembly. This placed the ED on the same level as the heads of the specialized agencies. The ED was further mandated to advice UN intergovernmental bodies, a unique role for an international civil servant. Following the Stockholm Conference, the General Assembly decided in the autumn of 1972 to locate the secretariat in Nairobi after a debate that at times became acrimonious with industrialized countries arguing for a location of such a catalytic organization in the center of the UN system, and not in the periphery.

A voluntary *environment fund* created to provide additional financing for environmental programs in order to finance, either wholly or partly, the costs of the new environmental initiatives undertaken within the UN system. The underlying rationale was that the specialized agencies would also increase their own financial resources in the field of the environment, which they failed to do for many years. There was a general realization that the fund would not play any significant role in overall development assistance. The United States had pledged to offer US$ 100 million during a five-year period on a 40% matching basis.

An Environment Coordinating Board (ECB) is set up under the chairmanship of the ED in order to have maximum efficiency in coordinating UN environmental programs. Governments were called upon to ensure their own coordination of environmental action, both national and international. The ECB was abolished in 1977.

The Conference faced considerable managerial and political challenges when it convened on June 5, 1972, a day that has since been commemorated as World Environment Day.

The Conference attracted many more direct and indirect participants than had been foreseen. There were 1350 delegates from 113 countries, 850 observers from the specialized agencies, and 250 nongovernmental organizations (NGOs), as well as an unprecedented attendance of 1600 representatives of media. There were also many nongovernmental representatives present who did not engage directly in the Conference. The total influx from abroad was between 4000 and 5000 persons.

Also, delegates had only 10 working days at their disposal to consider the massive agenda. This was the case even after deciding to limit the consideration only to recommendations for international action, and forward recommendations that had been developed for national action to governments for their consideration. This reflected the political sensitivities at the time to what could been seen as interference in internal affairs of states.

The organizational arrangements stood up excellently to an extraordinary challenge of which there had been no previous experience.

Several important political factors contributed to the favorable dynamics of the Conference.

The issue of nonparticipants was particularly sensitive because of tensions between NGO's and governments. In the end the feared, violent incidents did not materialize. An important contribution was the establishment of the Environment Forum, and a more informal NGO facility outside Stockholm. The NGO factor played a considerable role in how the Conference proceedings evolved, particularly in the issue of commercial whaling.

The decision-making procedures were marked by the novelty of the situation, and by the presence of so many delegates who were not familiar with UN procedures. If the substantive issues had not been cleared to such a major extent in advance of the Conference, it would have been very difficult to ensure success.

The decision at the last moment by the Soviet Union and its allies not to participate because of a long controversy related to the Cold War, the noninvitation of the German Democratic Republic, was a setback, but did not negatively affect the proceedings. Understandings had been reached before the Conference not to let the invitation problem affect considerations of substantive question.

A most important result was that a feared environment-development conflict was prevented, which was largely due to the Founex initiative and its effects. However, the bitterness of specialized agencies over having been rebuffed at an earlier stage in their demands to introduce the substantive areas took an ominous turn during the Conference and was only barely contained.

Overall, there was a strong desire by several key delegations to demonstrate their own constructive roles to their home constituencies who were able to watch the Conference proceedings closely through the massive media reporting. China was a particular case. This was the first important international event in which it participated as a new UN member. China quickly established itself as an undisputed leader among developing countries and took a high profile. Its involvement in the substantive negotiations remained limited to the draft Declaration, which was reopened at its insistence.

The Conference ended on a high note after overcoming the hurdles surrounding the Declaration, particularly the issue of weapons of mass destruction. There was a feeling of important accomplishment, and even that history had been made.

The results by far surpassed the original objectives. Through the Stockholm Conference, the environment was legitimized as an area of both national and international concern and cooperation and the first steps were taken to give practical effect to this new recognition. A visible point of reference and authority for all future international work in the environment field had been established. Stockholm also initiated the drive to widen the environment agenda beyond concerns about conservation and pollution to include issues such as development assistance, trade and development.

Further, the Conference was instrumental in building national and international institutions in the environmental field, and in establishing a framework for treaty making. It was the first event in which civil society participation was directly supported and had a concrete impact in an intergovernmental negotiation process, and it became a model for many global UN Conferences.

There were also issues and areas where political and structural constraints had a restrictive effect, or when developments proceeded in unforeseen directions.

Restrictive effects included the controversy over the population question and the difficulty of adapting economic policies to the environmental challenge in both industrialized and developing countries. This would have determining effects for the further process in both cases.

The Action Plan focused on the symptoms rather than the causes of environmental problems. This was in line with prevailing thinking and administrative methods but, in practical terms, it established the environment as an add-on issue, and did not support the integrative message that had been developed so successively during the preparatory processes, such as the Founex seminar. As a consequence, environmental measures still continue to be considered additional rather than integrated parts of economic policy.

Also, the Action Plan did not address the issues of costs or relative priorities. There was a prevailing perception among delegations that the Conference and its follow up would not affect national interests and priorities in any fundamental way. Significantly, there was relatively little specific focus on environmental policies of industrialized countries. This was a result partly of the activities of a secret group of some industrialized countries (the Brussels group) and partly of the financial dependence of developing countries on the North.

There were also serious limitations imposed on UNEP from the beginning. It had not been possible to secure a strong and binding general coordinating mandate for the new environmental body. Significantly, the specialized agencies, supported by

France, the Soviet Union, and the United Kingdom, managed to specifically weaken UNEP from the outset in a key role—coordination of assessments. This issue remains closely linked to the dilemmas of sectorial decision-making structures at the national level, and the continuing lack of will and capability to significantly reform the obsolete structure of the UN system.

The limits of political space had been clearly outlined even before the final negotiation phase of the Stockholm Conference began when an attempt to set up a powerful UN Charter body with the working name "Biosphere Council" received practically no support.

The problem of issue fragmentation that followed after Stockholm is the prime example of unforeseen effects. Initial progress in reducing complex ecological problems to manageable levels by negotiating a multitude of agreements (some 300 new multilateral environment agreements after Stockholm) led over time to serious loss of coherence and policy control, compounded by an increasing implementation deficit and a general lack of enforcement.

Despite these serious challenges, discussed further on in this book, many of which remain with us today, the Stockholm Conference has proved to be the real start of the international management of chemicals, resulting in a set of instruments that shape today's and tomorrow's work in this field.

REFERENCES

Carson, R. 1962. *Silent Spring*, Fawcett World library, New York (reprinted in September 1967).

Engfeldt, L.-G. 2009. *From Stockholm to Johannesburg and Beyond*, The Government Offices of Sweden, p. 41.

United Nations, Development and Environment. 1972. Report and Working Papers of a Panel of Experts Convened by the Secretary-General of the United Nations Conference on the Human Environment (Founex, Switzerland, June 4–12, 1971), Mouton, The Hague, Paris.

Ward, B. and R. Dubos. 1972. *Only One Earth—The Care and Maintenance of a Small Planet*, George Mcleod Ltd, Toronto.

3 Rio 1992
The UN Conference on Environment and Development (The "Earth Summit")

Shelley Kath[*]

CONTENTS

[*] The opinions expressed in this chapter do not necessarily reflect the views of the Helios Centre.

INTRODUCTION

Many of the core concerns underlying policy instruments and international regulatory regimes in the realm of chemical and hazardous waste management have their roots in the 1992 United Nations Conference on Environment and Development (UNCED) held in Rio de Janeiro, Brazil and known informally as the Rio Earth Summit. As a result of the agreements and consensus-building achieved during the conference, the underlying notion that people have rights to a healthy environment was ensconced in both the language and foundations of international agreements and a certain collective consciousness. This was nothing less than a paradigm shift, signaling not only an increase in expectations, but intergovernmental support for the legitimacy of those expectations.

Consequently, the tenets we now consider crucial to environmental regulation and advocacy flowed directly from the Summit. Three specific concepts, the application of the precautionary principle, the polluter-pays principle, and the notion of free access to environmental information (Huismans and Halpaap, 1998; Peterson, 2004), emerged directly from the *Rio Declaration on Environment and Development*,[*] in *Principles* 15, 16, and 10, respectively. Another well-known product of Rio, the massive and comprehensive implementation plan known as *Agenda 21*,[†] contains in Chapters 19 and 20 the key values, guidelines, and policy recommendations that serve as fundamental elements of important regimes such as the Rotterdam and

[*] Declaration on Environment and Development, UN Doc. A/ CONF.151/5, reprinted in 31 I.L.M. 874, 877 (1992).

[†] http://www.un.org/esa/dsd/agenda21/

Stockholm Conventions and the Globally Harmonized System (GHS) of Classification and Labeling of Chemicals.*

Thus, to assist in understanding the history, development, and future direction of chemical and hazardous waste management at the international level, we review the products, organizations, and legacies forged at and as a result of the Rio Earth Summit. In addition to examining the written agreements produced and organisms launched at Rio, we will also provide a brief analysis of the effectiveness of these products in furthering the goals of sustainable development and in advancing the state of chemical and hazardous waste management.

HISTORICAL OVERVIEW

Twenty years after the first global Conference on the Human Environment, held in Stockholm, Sweden in 1972, over 170 nations convened in Rio de Janeiro, Brazil, from June 3 to 14, 1992. The intent of the conference was to continue the environmental work begun at Stockholm, and begin a close global examination of the concept of sustainable development. This concept grapples with and seeks to reconcile the intertwined needs and reciprocal impacts of environmental quality and economic development. The UN Conference that was to affect all future such meetings was this one, the Rio Earth Summit.

The original purpose of UNCED flowed from recommendations made in the famous report by the World Commission on Environment and Development (commonly known as the "Brundtland Commission," named for Gro Harlem Brundtland, who served as its chair). Entitled *Our Common Future* (World Commission on Environment and Development, 1987),† the report made recommendations designed to rectify what the Commission had identified as the root cause of many of the world's woes, namely, the fact that environmental and economic matters were so tightly intertwined that environmental polices that ignored economic concerns, and economic policies that ignored environmental issues, were destined for difficulty or defeat.

In an effort to make concrete the budding concept of sustainable development,‡ the report set out a slate of measures for improving planning, decision-making, transparency, and other fundamental aspects of governance in ways that would balance economic and environmental concerns. Published in 1987, it also called for a high-level international conference to be held by 1992 to evaluate progress on implementation of the Brundtland Commission's recommendations. Maurice Strong, a Canadian businessman and one of the world's most influential political activists, was a Commission member and the leader of the effort to organize the 1992 conference. Having served as Secretary-General of the Stockholm Summit, he was well-suited to the task, and in 1990 was appointed Secretary-General of UNCED.

---

* UN *Globally Harmonized System of Classification and Labelling of Chemicals (GHS)*, 3rd ed., 1992, online at the UN Economic Commission for Europe, at: http://213.174.196.126/trans/danger/publi/ghs/ghs_rev03/03files_e.html

† This publication is often referred to as "the Brundtland Report."

‡ The oft-quoted definition of "sustainable development" reads: "Sustainable development is development that meets the needs of the present without compromising the ability of future generations to meet their own needs." Ibid, p. 43.

As a venue, Brazil provided a stark contrast to the 1972 Summit in Stockholm. It was during their stay in South America that many environmental groups of the North witnessed for the first time the harsh realities of life in countries of the Southern hemisphere and began to understand that the priorities of people in such countries were not pollution abatement or protection of scenic wonders, but food, shelter, protection from hazardous waste among other fundamental necessities.

The UNCED conference at Rio was, at the time, an event like no other in the history of environmental assemblies. Indeed, the formal agreements made at Rio were but one part of the story, and perhaps more significant was the mobilization of tens of thousands of participants from across the globe—grass-roots activists, governments and heads of state, intergovernmental organizations such as the World Bank, nongovernmental organizations (NGOs), and business interests—all focused on plotting a course for planetary health (French, 2002). In addition to the main conference for delegates and other officials, a parallel forum for civil society was held also in Rio but some 20 miles away from the UNCED conference center (Brooke, 1992). The "Global Forum" (Esty, 1993), as it was named, drew some 17,000 people,* most of them connected with NGOs from around the world, who used the forum to share concerns, strategies, and frustrations surrounding the alleviation of the planet's problems.

As this chapter will document, opinions vary widely on whether the Earth Summit can be characterized as a success, a failure, or some combination of the two. Since one reasonable measure of a global summit's success is the extent to which it fulfilled its stated goals and objectives, we will begin our analysis with a brief look at why the Rio Summit was convened and what it was intended to achieve.

RIO EARTH SUMMIT GOALS AND OBJECTIVES

The UN Resolution adopted in December 1989 establishing UNCED outlines the official mandate for the Summit: to "elaborate strategies and measures to halt as follows and reverse the effects of environmental degradation in the context of increased national and international efforts to promote sustainable and environmentally sound development in all countries."† Prior to this, the World Commission on Environment and Development recommended in its report to the United Nations General Assembly (UNGA) in 1987 that the issues of environment and development were inextricably linked, and that no real progress on one of these issues could be made without progress on the other. Among the Commission's other recommendations were that the UN prepare a universal declaration and convention on environmental protection and sustainable development. As has been observed, the over-arching tenet of the commission, later echoed by both UNCED's secretariat and chairperson, was that "environment and development issues be fully integrated" (Haas et al., 1992, p. 7).

* United Nations, Department of Public Information, *UN Conference on Environment and Development* (1992), May 23, 1997, "Summary," p. 2 of 4, at: http://www.un.org/geninfo/bp/enviro.html

† UN Resolution 44/228, part 1.3, New York, NY, December 22, 1989.

The specific objectives of the Rio Earth Summit were many: UNGA Resolution 44/288 establishing UNCED lists a total of 23 separate aims, literally too many to mention. Among these, however, the following were particularly important:

- "To examine the state of the environment and the changes that have occurred" since the 1972 Stockholm Conference on the Human Environment.*
- "To recommend measures to be taken at the national and international levels to protect and enhance the environment ... through the development and implementation of policies for sustainable and environmentally sound development" that pay special attention to environmental issues within economic and social development processes, and which focus upon "preventive action at the sources of environmental degradation, clearly identifying the sources of such degradation and appropriate remedial measures, in all countries," all while "taking into account the specific needs of developing countries."†
- "To promote the further development of international environmental law," and to examine "the feasibility of elaborating general rights and obligations of States, as appropriate, in the field of the environment, and taking into account relevant existing international legal instruments," while allowing for the "special needs and concerns of the developing countries."‡
- "To examine ways and means further to improve cooperation in the field of protection and enhancement of the environment between neighbouring countries, with a view to eliminating adverse environmental effects."§
- "To examine the relationship between environmental degradation and the international economic environment, with a view to ensuring a more integrated approach to problems of environment and development in relevant international forums without introducing new forms of conditionality."¶
- "To examine strategies for national and international action with a view to arriving at specific agreements and commitments by Governments and by intergovernmental organizations for defined activities to promote a supportive international economic climate conducive to sustained and environmentally sound development in all countries, with a view to combating poverty and improving the quality of life, and bearing in mind that the incorporation of environmental concerns and considerations in development planning and policies should not be used to introduce new forms of conditionality in aid or in development financing and should not serve as a pretext for creating unjustified barriers to trade."**

While most of the listed objectives were of a general level, bracketing substantive issues, one made specific mention of toxics and hazardous wastes:

* UN Resolution 44/228, part 1.3, New York, NY, December 22, 1989, para. 15 (a).
† Ibid., para. 15 (c).
‡ Ibid., para. 15 (d).
§ Ibid., para. 15 (e).
¶ Ibid., para. 15 (h).
** Ibid., para. 15 (i).

> To examine strategies for national and international action with a view to arriving at specific agreements and commitments by Governments for defined activities to deal with major environmental issues in order to restore the global ecological balance and to prevent further deterioration of the environment, taking into account the fact that the largest part of the current emission of pollutants into the environment, *including toxic and hazardous wastes,* originates in developed countries, and therefore recognizing that those countries have the main responsibility for combating such pollution.* (emphasis added).

As these objectives show, the Earth Summit was extremely ambitious: it was to do no less than review the world's environmental problems, examine the relationship between environment and development, promote the theoretical and practical integration of these two subjects through the concept of "sustainable development," launch new international agreements and policies, create new processes and structures for international cooperation on environmental and socio-economic issues, and do all these things while taking into account the special needs and concerns of developing countries. In retrospect, it seems clear that the attainment of the Summit's numerous specific objectives was hobbled from the outset by the sheer magnitude of the task it had set for itself in attempting to resolve such a wide-ranging set of issues. Still, it is agreed by critics and supporters alike that the Rio Earth Summit was a ground-breaking event.

RIO EARTH SUMMIT OUTCOMES

OVERVIEW

The Earth Summit's importance lay in the fact that it generated a new level of governmental consensus and public awareness as to the fundamental needs required for planetary recovery. Though the conference galvanized on multiple fronts a multitude of accomplishments, impacts, and initiatives, focus here is on the formal outcomes, namely, the written, legal, and policy instruments, and the creation and modification of specific organizations flowing from negotiations and agreements.

KEY RIO DOCUMENTS: PRINCIPLES, CONVENTIONS, AND A PLAN

What propelled the Rio Summit to its legendary status in the history of environmental policy is that Rio produced no fewer than five international agreements, which, as treaties, continue to play important roles for public policy. The five "Rio documents," also referred to as the "Rio agreements," include two sets of nonbinding principles (the *Rio Declaration on Environment and Development*† and the *Statement of Principles for the Sustainable Management of Forests*, or "Forest Principles"‡), two

* Ibid., para. 15 (f).

† Declaration on Environment and Development, UN Doc. A/CONF.151/5, reprinted in 31 I.L.M. 874, 877 (1992).

‡ Nonlegally Binding Authoritative Statement of Principles for a Global Consensus of the Management, Conservation and Sustainable Development of all Types of Forests, UN Doc. A/CONF.151/6, reprinted in 31 I.L.M. 881 (1992). Hereinafter "Forest Principles."

treaties (the *UN Convention on Biological Diversity, 1992*,[*] and the *UN Framework Convention on Climate Change, 1993*[†]), and one comprehensive action plan (*Agenda 21*[‡]). Additionally, it should be noted that the *UN Convention to Combat Desertification*[§] can be held to have been born in Rio, as it was during the Earth Summit that the process was put in place for negotiating this agreement. The five Rio agreements are summarized below, followed by a brief mention of Rio's contribution to the *Convention to Combat Desertification*.

Rio Declaration on Environment and Development

The Rio Declaration on Environment and Development ("Rio Declaration") lists the 27 major principles for sustainable management of the planet and in doing so sums up the philosophy of "sustainable development" (Weiss, 1992). As the United States and Israel refrained from reopening the debate during preparatory sessions regarding "people under occupation," a sensitive phrase and issue for both countries, the Declaration was adopted more easily than might have been anticipated and was the first document adopted at Rio (UNCED, 1992). As the International Institute for Sustainable Development stated in a summary of the proceedings:

> The approved text represents to a large extent, an attempt to balance the key concerns of both Northern and Southern countries. Far from a perfect text, each side achieved success in enshrining those specific principles that are of particular importance to their respective political agendas.[¶]

The Rio Declaration distinguished itself from its 1972 Stockholm predecessor in three ways. First, it acknowledged the need for global awareness and government consensus in achieving sustainable development. Many of these principles had not, until that time, been universally accepted.[**] Second, it highlighted the notion of common but differentiated responsibility for developing nations in contrast to industrialized states. Finally, it endorsed on an international level a precautionary approach to environmental protection (Weiss, 1992).

There are several principles in the Declaration that warrant specific review here, as their significance for international environmental law in general, and chemical and hazardous waste management in particular, cannot be understated. Four of the Declaration principles now represent key tenets of international environmental law: the human-centric tenet of sustainable development (*Principle* 1), the sovereign right of each state to exploit its own resources (*Principle* 2), the notion of "common but

---

[*] June 5, 1992, reprinted in 31 I.L.M. 818 (1992). Hereinafter, "Convention on Biological Diversity."

[†] UN Doc. A/AC.237/18, reprinted in 31 I.L.M. 849 (1992). Hereinafter "UNFCCC."

[‡] UN Doc. A/CONF.15 1/4.

[§] Full text available on the Web site of the Secretariat for the Convention to Combat Desertification, at: http://www.unccd.int/convention/text/convention.php

[¶] International Institute for Sustainable Development, *Earth Negotiations Bulletin*, Vol. 2, No. 13. "A Summary of the Proceedings of the United Nations Conference on Environment and Development 3–14 June 1992," full text available at: http://www.iisd.ca/vol02/0213001e.html [hereinafter "IISD Summary of Proceedings"]. The quoted passage appears at "Part III: Rio Declaration," full text available at: http://www.iisd.ca/vol02/0213032e.html

[**] Ibid, at:http://www.iisd.ca/vol02/0213032e.html

differentiated responsibilities" (CBDR) (*Principle* 7) and the precautionary principle (*Principle* 15). As mentioned in the Introduction, the precautionary principle plays a fundamental role in chemical and hazardous waste issues, along with three others: the "polluter pays" principle (*Principle* 16), the principle of intergenerational equity (*Principle* 3), and the notion of public access to environmental information and justice (*Principle* 10). We will look briefly at each of these principles.

Principle 1 states a fundamental tenet of sustainability: "Human beings are at the centre of concerns for sustainable development. They are entitled to a healthy and productive life in harmony with nature." The Principle reflects the view of many developing nations that the *raison d'être* of "sustainable development," and environmental protection in general, must be to improve the quality of life for human beings in the sense that the meeting of basic human needs must be paramount in environmental protection activities.

Principle 2 enshrines the state's "sovereign right to exploit their own resources pursuant to their own environmental policies,"* so long as the right is carried out in line with the state's own developmental policies. As in the Stockholm Declaration, Principle 2 also observes that the right to resource exploitation is accompanied by the "responsibility to ensure that activities within their jurisdiction or control do not cause damage to the environment of other States or of areas beyond the limits of national jurisdiction."

Principle 7 encapsulates the principle of CBDR, a concept frequently thought to have been born at Rio, and one which continues to play an important role in framing the obligations of developed versus developing countries under a number of international instruments. But, like the precautionary principle, the CBDR principle predates Rio, having been an important concept underlying the 1987 *Montreal Protocol on Substances that Deplete the Ozone Layer,*† and recognized in other important international agreements (Sands, 1994, pp. 295–296). Specifically, *Principle* 7 states:

> States shall cooperate in a spirit of global partnership to conserve, protect and restore the health and integrity of the Earth's ecosystem. In view of the different contributions to global environmental degradation, States have common but differentiated responsibilities. The developed countries acknowledge the responsibility that they bear in the international pursuit to sustainable development in view of the pressures their societies place on the global environment and of the technologies and financial resources they command.

Principle 15 articulates the "Precautionary Principle"—the notion that when an activity raises threats of harm to human health or the environment, precautionary measures should be taken even if there may still be some doubt about cause and effect relationships. Thus, *Principle* 15 states:

> In order to protect the environment, the precautionary approach shall be widely applied by States according to their capabilities. Where there are threats of serious OR

* Rio's *Principle* 2 replicates very closely *Principle* 21 of the Stockholm Declaration.

† Montreal Protocol on Substances That Deplete the Ozone Layer, Sept. 16, 1987, 26 I.L.M. 1541 (entered into force Jan. 1, 1989).

> irreversible damage, lack of full scientific certainty shall not be used as a reason for postponing cost-effective measures to prevent environmental degradation.

Contrary to popular opinion, the Precautionary Principle did not make its debut at the Earth Summit but rather first appeared in the World Charter for Nature that was adopted by the UN General Assembly in 1982.* Furthermore, some writers distinguish between a precautionary "principle" and a precautionary "approach," with the latter being viewed as somewhat softer and not as likely to receive serious treatment in international law (Garcia, 1995; FAO, 1996; Recuerda, 2008).

Principle 16 has played an important role in creating a basis in international law for holding accountable companies and organizations whose activities result in pollution, for the environmental damage they cause. This principle states:

> National authorities should endeavour to promote the internalization of environmental costs and the use of economic instruments, taking into account the approach that the polluter should, in principle, bear the cost of pollution, with due regard to the public interest and without distorting international trade and investment.

Principle 3, expressing the notion of intergenerational equity, is particularly important in matters of chemical management in light of the fact that certain chemicals, such as dichloro-diphenyl-trichloroethane (DDT), are known to have health impacts that may extend across generations. This principle is articulated as follows:

> The right to development must be fulfilled so as to equitably meet developmental and environmental needs of present and future generations.

Finally, *Principle* 10 on access to environmental information, contains the seeds of nearly all systems for reporting and making public various kinds of data and policy information that are required by international legal instruments. This principle states:

> Environmental issues are best handled with participation of all concerned citizens, at the relevant level. At the national level, each individual shall have appropriate access to information concerning the environment that is held by public authorities, including information on hazardous materials and activities in their communities, and the opportunity to participate in decision-making processes. States shall facilitate and encourage public awareness and participation by making information widely available. Effective access to judicial and administrative proceedings, including redress and remedy, shall be provided.

Together, these and others among the Rio Declaration's presentation of 27 principles have influenced and informed countless instruments and policies of international environmental law.

---

* United Nations General Assembly Resolution A/Res/37/7, October 28, 1982, Article 11. Available online at: http://www.un-documents.net/a37r7.htm

Statement of Principles for the Sustainable Management of Forests ("Forest Principles")

The Rio "Forest Principles," as this agreement is commonly called, consist of a non-legally binding "statement of principles" articulating the need to preserve forests without setting timetables or standards for doing so (Raloff, 1992). Essentially, the Principles lay out the basic policy requirements for realizing the goal of sustainable forest management.

Fifteen years after their introduction at Rio, the "Non-Legally Binding Instrument (NLBI) on All Types of Forests" emerged to strengthen the message of the Forest Principles. Together, these two instruments serve as the launching point for important new regime proposals, such as REDD (reducing emissions from deforestation and forest degradation) (Ikkala, 2009).

It may be said that the Forest Principles put forward at Rio were, as Haas, Levy and Parson noted, "salvaged from the wreckage of a failed earlier attempt to negotiate a treaty on forests" (Haas et al., 1992, pp. 6–11), an effort which had been ongoing for many years. Nonetheless, the general guidelines contained in the Forest Principles still play a valuable role as "soft law" in this area. According to Ken Wan, these principles "formed the foundation for the contextually specialized 'Sustainable Forest Management' concept, which recognizes that 'forest resources and forest lands should be sustainably managed to meet the social, economic, ecological, cultural, and spiritual needs of the present and the future generations'" (Wan, 2009; citing UN Department of Economic and Social Affairs, 2005).

UN Framework Convention on Climate Change

While generally understood as one of the Rio documents, the UN Framework Convention on Climate Change (UNFCCC) was actually negotiated prior to Rio and in a separate process from UNCED. The Convention was in fact adopted at UN Headquarters in New York on May 9, 1992, roughly one month before the opening of the Earth Summit. That said, the Convention was opened for signature at Rio, and on June 12, 1992 did gather signatures from 154 countries. As a result, the UNFCCC was, and is today, seen as an important outcome of the Earth Summit. The Convention entered into force on March 21, 1994 and there are now 194 parties to the Convention (193 states plus the European Community).

The UNFCCC, the first international effort of its kind, represented a monumental step, in addressing the multitude of serious problems caused by global warming and the concomitant changes in the Earth's climate. The Convention sets out an overall framework within which national and international actions are to be taken to tackle the environmental and social challenges posed by these impacts.

While the long-term goal for stabilizing greenhouse gas (GHG) emissions contained no specific targets, the UNFCCC's near-term, nonbinding goal included a specific objective for developed countries: they were to reduce GHG emissions to 1990 levels by 2000.* In a concrete demonstration of Rio Declaration *Principle* 7 on

* By 2006, 16 of the 38 Annex I countries had met their targets while 19 had not. "World Ahead of Kyoto Emissions Targets," New Scientist, November 19, 2008, at: http://www.newscientist.com/article/mg20026833.400-world-ahead-of-kyoto-emissions-targets.html

"common but differentiated responsibility," the Convention stipulated different obligations for different countries, depending primarily on their level of development and industrialization. The specific obligations for each country are laid out in Annexes to the Convention.

See Chapter 9 of this book on the UN Framework Convention on Climate Change.

Convention on Biological Diversity

The UN Convention on Biological Diversity (CBD)* was a watershed event in the conservation of biological diversity, the sustainable use of biological resources, and the equitable sharing of benefits from the use of those resources. The objectives of the CBD are "the conservation of biological diversity, the sustainable use of its components, and the fair and equitable sharing of the benefits arising out of the utilization of genetic resources, including, by appropriate access to, genetic resources and by appropriate transfer of relevant technologies, taking into account all rights over those resources and to technologies, and by appropriate funding."† Prior to Rio, interest had been mounting in finding a way to protect the Earth's biological resources and stem problems associated with species extinction and harm to supporting ecosystems. As a result, the United Nations Environment Programme (UNEP) in November 1988 convened the Ad Hoc Working Group of Experts on Biological Diversity to explore the need for an international CBD.‡ At Rio, on June 5, 1992, the Convention was opened for signature and by June 4, 1993 had received 168 signatures. It entered into force on December 29, 1993 and was ratified on December 30, 1993, only 90 days after the 30th ratification, making it one of the most rapidly implemented environmental treaties ever.

The CDB adheres to three main objectives; the conservation of biological diversity, the sustainable use of the components of biological diversity, and the fair and equitable sharing of the benefits arising out of the utilization of genetic resources.§ It offers broad guidelines for national-level protection of biological diversity and requires the formulation of national biodiversity strategies. As well, the Convention acknowledges the importance of national sovereignty over biological resources and the need for prior informed consent (PIC) prior to the transfer of resources out of a country. The CBD also stipulates that biodiversity use must be sustainable and that benefits from such use must be equitably shared between the source and receiving countries (French, 2002).

---

* Full text of the *UN Convention on Biological Diversity* is available from the CBD Secretariat. Web site at: http://www.cbd.int/doc/legal/cbd-un-en.pdf. As set out in Art. 2, "Biological diversity" refers to "the variability among living organisms from all sources including, *inter alia*, terrestrial, marine and other aquatic ecosystems and the ecological complexes of which they are part: this includes diversity within species, between species and of ecosystems," whereas "biological resources" refers to any "genetic resources, organisms or parts thereof, populations, or any other biotic component of ecosystems with actual or potential use or value for humanity."

† *UN Convention on Biological Diversity, supra*, Article 1.

‡ Secretariat of the UN Convention on Biological Diversity, at: http://www.cbd.int/history/

§ From the Convention on Biological Diversity, at: http://www.cbd.int/convention/about.shtml

Since the coming into force of the Convention, one protocol to the CBD has been adopted. The *Cartagena Protocol on Biosafety** was adopted in 2000, with the objective of governing the movements of living modified organisms (LMOs) resulting from modern biotechnology, from one country to another. It entered into force on 11 September 2003 and essentially allows nations the choice whether to allow imports of products containing genetically modified organisms.

Agenda 21

Due to the unwieldy size of *Agenda 21* (nearly 300 pages) and the need for reporters at the Earth Summit to communicate news of the plan quickly, the UN issued a 45-page "*Agenda 21*—Press Summary" that provides a very useful review. Thus, the brief review of the key chapters relating to chemicals and hazardous wastes—namely, Chapters 19, 20, 6, and 8—makes extensive reference to that document. *Agenda 21*, a title derived from "Agenda for the 21st Century" (Sandbrook, 1997), is a wide-ranging action plan for implementing the principles in the Rio Declaration, the objectives articulated in the other Rio agreements, and a comprehensive plan for attaining the objectives of environmental protection and sustainable development.† *Agenda 21* covers an enormous number of global concerns, spelling out in great detail paths for improving the state of affairs for each one. Philippe Sands observes that taken together, the 40 chapters of *Agenda 21* "constitute the framework for international law in the field of sustainable development" (Sands, 2003, p. 11). UNCED Secretary-General Maurice Strong characterized *Agenda 21* as "... the broadest consensus ever achieved on a text and is a *political commitment* prior to a legal commitment" (UNCED, 1992).

Agenda 21 addresses four program areas: (1) Social and Economic Dimensions, (2) Conservation and Management of Resources for Development, (3) Strengthening the Role of Major Groups, and (4) Means of Implementation. The text of each program area describes the rationale for the actions prescribed, the objectives sought, and specific activities and means of implementation. The plan "sets out the objective for achieving sustainable development by the 20th century, sector by sector: how to act to protect the atmosphere, slow down deforestation, stop the erosion of arable land, protect the ocean and marine resources, protect fresh water and achieve better management to prevent disease, take account of waste management (nuclear, toxic, chemical or dangerous)" (UNCED, 1992). It covers a wide range of issues including conservation and resource management (e.g., atmosphere, forests, water, waste, chemical products), socioeconomic concerns (e.g., human habitats, health, demography, consumption and production patterns, etc.), the strengthening of NGOs other social groups such as unions, women, youth, and funding mechanisms and other means of implementation.

As mentioned earlier, several chapters in *Agenda 21* served and continue to serve as the foundation for a number of international legal instruments, policies, and regimes in the area of chemical and hazardous waste management. Chapter 19, titled

---

* Full text on the CBD, at: http://www.cbd.int/doc/legal/cartagena-protocol-en.pdf

† The full text of *Agenda 21* is available on the Commission for Sustainable Development, at: http://www.un.org/esa/dsd/agenda21/res_agenda21_00.shtml

the “Environmentally Sound Management of Toxic Chemicals, Including Prevention of Illegal International Traffic in Toxic & Dangerous Products,” is critical in this regard, but Chapter 20 (“Environmentally Sound Management of Hazardous Wastes, Including Prevention of Illegal International Traffic in Hazardous Wastes”), Chapter 6 (“Protecting and Promoting Human Health”), and Chapter 8 (“Integrating Environment and Development in Decision-making”) also play important roles. We will look very briefly at the key features of these chapters.*

Chapter 19, titled “Environmentally Sound Management of Toxic Chemicals, Including Prevention of Illegal International Traffic in Toxic & Dangerous Products” and presented in Section II of *Agenda 21*, recognizes that chemical contamination can be a source of “grave damage to human health, genetic structures and reproductive outcomes, and the environment.”† This chapter specifically addresses the special challenges and needs of developing countries in managing toxic chemicals. Additionally, the treatment acknowledges that many countries lack national systems to cope with chemical risks, and/or the scientific means of collecting evidence of misuse and of judging the impact of toxic chemicals on the environment.‡

Chapter 19 sets out the following six program areas: expanding and accelerating international assessment of chemical risks; harmonization of classification and labeling of chemicals; information exchange on toxic chemicals and chemical risks; establishment of risk reduction programs; strengthening of national capabilities and capacities for management of chemicals; and prevention of illegal international traffic in toxic and dangerous products.§ These six program areas involve, to varying degrees, “hazard assessment (based on the intrinsic properties of chemicals), risk assessment (including assessment of exposure), risk acceptability and risk management.”¶

As mentioned earlier, Chapter 19 has played a critical role in the design of several important international chemical and toxics conventions and instruments, including the Strategic Approach to International Chemical Management (SAICM; see Section IV of this book). Similarly, Chapter 21 provided the impetus for the establishment in 1994 of the Intergovernmental Forum on Chemical Safety (IFCS, see Section V, Chapter 21 of this book), a forum that proved to be very useful to many country officials tasked with managing toxics in their countries. Chapter 21 also influenced the UN’s GHS initiative facilitating a standardized, global system for classifying and labeling chemicals. The GHS was first published in 2003 (see Section III, Chapter 11 of this book).

* The United Nations conveniently issued a 45-page Press Summary of *Agenda 21*, and since it provides a more accessible presentation of *Agenda 21* (the full document comprises nearly 300 pages and a laborious format), our examination of the key chapters relating to chemicals and hazardous wastes (namely, Chapters 6, 8, 19, and 20) makes use of and rests on the summary. The Press Summary is available online at: http://www.un.org/esa/sustdev/documents/agenda21/english/A21_press_summary.pdf

† Chapter 19, para. 2.

‡ UN, *Agenda 21* Press Summary, at: http://www.un.org/esa/sustdev/documents/agenda21/english/A21_press_summary.pdf

§ Chapter 19, para. 4.

¶ Ibid., para. 5.

Chapter 20, titled "Environmentally Sound Management of Hazardous Wastes, Including Prevention of Illegal International Traffic in Hazardous Wastes" and presented in Section II of *Agenda 21*, recognizes that effective controls over the generation, storage, treatment, recycling and reuse, transport, recovery, and disposal of hazardous wastes is critical for the protection of human health, the environment, effective natural resource management and sustainable development in general.[*] The chapter promotes integrated life-cycle management and states that "the overall objective is to prevent to the extent possible, and minimize, the generation of hazardous wastes, as well as to manage those wastes in such a way that they do not cause harm to health and the environment,"[†] through an integrated approach to hazardous waste management.

Among other things, the chapter highlights the importance of and need for international cooperation in a variety of areas, including but not limited to the dissemination of information on risks and new technologies for reducing production of hazardous wastes, improvement of methods for handling and disposal of hazardous wastes, design and development of individual nations' hazardous waste programs and centers, and the importance of cooperation in controlling transboundary shipping.[‡] Indeed, the impetus for this chapter came in no small part from concern about illegal international movement of hazardous wastes, in contravention of existing national legislation and international legal instruments, and thus it includes specific recommendations for reducing illegal traffic in toxic and dangerous wastes.[§] In response to these concerns, Chapter 6, described below, proposes a ban on the export of wastes to nations that cannot demonstrate the capacity to deal with them in an environmentally sound fashion (see Section III, Chapter 8 of this book).

Chapter 6 of *Agenda 21, Section I*, titled "Protecting and Promoting Human Health," looks at the task of protecting and promoting human health from two angles: while development activities often affect the environment in ways that cause or exacerbate health problems, it is equally true that a lack of development can and frequently does affect human health in negative ways. Thus, proposals in this chapter focus on issues such as "meeting primary health care needs, controlling communicable diseases, coping with urban health problems, reducing health risks from environmental pollution, and protecting vulnerable groups such as infants, women, indigenous peoples and the very poor."[¶]

Acknowledging the link between waste generation and its risk to human health, Chapter 6 also advocates increased use of health risk assessments, particularly in large cities, and emphasizes preventative as opposed to simply damage control measures in order to reduce man-made disasters, such as those involving toxic

---

[*] Chapter 20, para 1.

[†] Chapter 20, para. 6.

[‡] UN, *Agenda 21* Press Summary, at: http://www.un.org/esa/sustdev/documents/agenda21/english/A21_press_summary.pdf

[§] Chapter 20, para 5.

[¶] UN, *Agenda 21* Press Summary, p. 6, at: http://www.un.org/esa/sustdev/documents/agenda21/english/A21_press_summary.pdf

wastes or other industrial by-products.* Chapter 6 also recommends various actions for minimizing the pollution hazards in workplaces and individual dwellings, including but not limited to development of pollution control technologies for air and water pollution (including indoor air), limiting the use of pesticides, and promoting the introduction of environmentally sound technologies in the industry and energy sectors.†

Chapter 8, titled "Integrating Environment and Development in Decision-making" and presented in Section 1 of *Agenda 21*, focuses on effective environmental decision-making, and calls upon governments to explore how, through cooperation between government and business and industry, effective use can be made of economic instruments and market mechanisms in connection with a variety of issues, including waste management.‡ The chapter proposes the "full integration of environmental and developmental issues for government decision-making" on a variety of policies affecting environment and development, and encourages governments to seek a broader range of public participation.§ It underscores the need for comprehensive information-gathering activities and improved assessment of environmental risks and benefits.¶ Of particular relevance for the area of chemical and hazardous wastes, the chapter also underscores that environmental costs should be incorporated in decisions made by producers and consumers alike, and that prices should, among other things, contribute toward preventing environmental degradation.**

The massive scope of *Agenda 21* as a plan was matched only by the enormous challenge of how to fund its programs and initiatives. Following Rio, it was stated that "in order to achieve the objectives set out in *Agenda 21*, the United Nations estimates that 600 billion Dollars a year would need to be invested until the year 2000, that is US$ 125 billion from international aid coffers. Until the opening of the Conference, roughly 55 billion Dollars was being provided each year by way of Official Development Aid (ODA)" (UNCED, 1992). In the end, however, the industrialized nations "failed to agree on the much-touted objective of devoting 0.7% of their GNP to ODA by the year 2000" (UNCED, 1992).

In the years following Rio, the massive action plan represented by *Agenda 21* received a great deal of attention and thought. In the context of the five-year review of progress made after Rio, mandated by the UNGA and held by the UN in 1997 in New York City, a Resolution was adopted that outlined the progress—or lack

---

* The Global Development Research Center, "Waste Management in Agenda 21," at: http://www.gdrc.org/uem/waste/waste_in_agenda21.html

† UN, *Agenda 21* Press Summary, at: http://www.un.org/esa/sustdev/documents/agenda21/english/A21_press_summary.pdf

‡ The Global Development Research Center, "Waste Management in *Agenda 21*," at: http://www.gdrc.org/uem/waste/waste_in_agenda21.html

§ UN, *Agenda 21* Press Summary, p. 9, at: http://www.un.org/esa/sustdev/documents/agenda21/english/A21_press_summary.pdf

¶ UN, *Agenda 21* Press Summary, p. 9, at: http://www.un.org/esa/sustdev/documents/agenda21/english/A21_press_summary.pdf

** UN, *Agenda 21* Press Summary, at: http://www.un.org/esa/sustdev/documents/agenda21/english/A21_press_summary.pdf

thereof—on Rio objectives.* It was replete with somber statements highlighting the need for renewed efforts, such as the following:

> Five years after the United Nations Conference on Environment and Development, the state of the global environment has continued to deteriorate, as noted in the Global Environment Outlook of the United Nations Environment Programme, and significant environmental problems remain deeply embedded in the socio-economic fabric of countries in all regions.

Having admitted the serious nature of the environmental, social, and economic problems that remained after Rio, the UN General Assembly concluded that the primary problem in realizing Rio's goals and objectives lay with implementation. Thus, in a supplemental planning document, the "Programme for Further Implementation of *Agenda 21*," it tried to address how to implement a plan of such great proportions and ambitions.† That plan promised renewed and vigorous efforts but failed to make serious inroads on improving the human or environmental situation.

Agenda 21 has continued to be the focus of much discussion and reflection at subsequent environment and development summits, such as the World Summit on Sustainable Development (WSSD) held in Johannesburg, South Africa from August 26 to September 2002. (An examination of the WSSD is found in Section II, Chapter 4 of this book.) More importantly, however, progress on *Agenda 21* continues to be the focus of the inter-governmental organization created to track and facilitate its implementation: the UN Commission on Sustainable Development (CSD). The work of the CSD is summarized later in this chapter.

Convention to Combat Desertification

As mentioned earlier, the Convention to Combat Desertification was not actually adopted at Rio, but could certainly be said to have taken root there since it was at Rio that the initial steps to create the Convention were made. This agreement targeted four regions: Africa, Asia, Latin America and Caribbean, and the Northern Mediterranean, encouraging each region to design and implement a plan for halting desertification that would recognize and work within specific regional needs. National and subregional action plans were also encouraged. The Convention was adopted in June 1994 and entered into force in December 1996, three months after receiving the 50th ratification.‡

At UNCED, the question of how to approach the serious problem of desertification was a major concern, and, largely at the insistence of the African countries, "the Conference supported a new, integrated approach to the problem, emphasizing action to promote sustainable development at the community level" and called upon the UNGA to establish an Intergovernmental Negotiating Committee Desertification

* UN General Assembly Resolution A/RES/S-19/2, September 19, 1999, at: http://www.un.org/documents/ga/res/spec/aress19-2.htm

† UN General Assembly Resolution A/RES/S-19/2, September 19, 1999, paragraph 9.

‡ Full text available on the Web site of the Secretariat for the Convention to Combat Desertification, at: http://www.unccd.int/convention/text/convention.php

(INCD), with the goal of preparing, by June 1994, a Convention to Combat Desertification, particularly in Africa. The General Assembly agreed, and in December 1992 adopted Resolution 47/188, which created the INCD.* Following a series of five negotiating sessions over the next two years, the Convention was adopted in Paris on June 17, 1994. It entered into force on December 26, 1996.

Organizations and Institutions Established or Restructured

Rio can be said to have led, directly or indirectly, to the creation of numerous intergovernmental and nongovernmental agencies focused on environment, energy, and sustainable development. That said, only a few were specifically created to function as follow-up mechanisms to the Earth Summit: the Commission on Sustainable Development; Inter-agency Committee on Sustainable Development; High-level Advisory Board on Sustainable Development; World Business Council on Sustainable Development (WBCSD).†

Commission on Sustainable Development

The creation of the CSD was envisioned in *Agenda 21* and executed by the UN General Assembly following Rio through the adoption on Jan 29, 1993 of Resolution A/RES/47/191 establishing the CSD.‡ It was created in December 1992 to ensure effective follow-up of UNCED. The Commission was created in December 1992 to ensure effective follow-up of UNCED, to monitor and report on implementation of the agreements at the local, national, regional, and international levels. It was agreed that a five-year review of Earth Summit progress would be made in 1997 by the United Nations General Assembly meeting in a special session.§

The CSD operates as a functional commission of the UN Economic and Social Council (ECOSOC). Its role includes: (1) the review of progress at the international, regional, and national levels in the implementation of recommendations and commitments contained in *Agenda 21*; (2) the elaboration of policy guidance and options for future activities to achieve sustainable development; and (3) the promotion of dialogue and partnership building for sustainable development with governments, the international community, and major groups identified in *Agenda 21*.

A "High-level Advisory Board on Sustainable Development" was formed at Rio to oversee the work of the CSD and provide guidance on issues related to the implementation of *Agenda 21*. The Board was also tasked with the responsibility of providing expert advice to the UN Secretary General, the CSD, the Economic and Social Council, and the General Assembly.¶

---

* UN General Assembly Resolution A/RES/47/188, December 22, 1992, full text available at: http://www.unccd.int/convention/history/GAres47_188.php

† http://www.un.org/geninfo/bp/enviro.html

‡ http://www.un.org/documents/ga/res/47/ares47-191.htm

§ UN CSD, at: http://www.un.org/esa/dsd/csd/csd_aboucsd.shtml

¶ UNGA Resolution A/RES/47/191, January 29, 1993, article 30, at: http://www.un.org/documents/ga/res/47/ares47-191.htm

Interagency Committee on Sustainable Development

Another organization whose launch was made official at the Rio Conference was the Interagency Committee on Sustainable Development (IACSD), an intergovernmental organization designed to facilitate coordination between various agencies and on a number of initiatives and programs.

According to a description of IACSD by UNEP, the agency structure involved a system of task managers set up for thematic areas, and it was UNEP that served as task manager for toxic chemicals and hazardous wastes (along with atmosphere, desertification and drought, and biodiversity).[*] The IACSD was decommissioned in October 2001, when the UN decided to replace it with other coordinating bodies within the UN.[†]

See also Chapter 23 of this book on the IOMC.

World Business Council on Sustainable Development

NGOs of all types were created in anticipation of the Earth Summit and many more emerged in the years following the event, but among the most powerful of these groups is the business and industry organization known as the World Business Council on Sustainable Development (WBCSD). The WBCSD is a CEO-led, international association of now more than 200 business concerns, having as its mission "to provide business leadership as a catalyst for change toward sustainable development, and to support the business license to operate, innovate and grow in a world increasingly shaped by sustainable development issues."[‡] In the run-up to Rio, Secretary General Maurice Strong invited Swiss industrialist Stephan Schmidheiny to coordinate the business participation in the Summit, specifically to "involve business in sustainability issues and give it a voice in the forum."[§] Following the Summit, Mr Schmidheiny and number of his colleagues launched the WBCSD, and after merging with the World Industry Council on the Environment in 1995, opened a secretariat in Geneva, Switzerland.[¶] Today, the WBCSD wields strong influence in the area of national and international sustainability policy.

See also Chapter 20 of this book on ICCA.

Restructuring of the Global Environment Facility

Originally established in 1990 as a pilot program of the World Bank (Sjoberg, 1999), the Global Environment Facility (GEF) was substantially reformed and restructured following Rio. The GEF serves as a funding source and funding mechanism serving primarily to assist developing countries in finding the means they require to implement programs that help protect the global environment. In its own words, the GEF is currently the "largest funder of projects to improve the global environment."[**]

At Rio, during last-minute negotiations on financial agreements relating to *Agenda 21*, a strong consensus developed on the need to restructure the GEF, addressing to

---

[*] UNEP, New York Office, at: http://www.nyo.unep.org/emg2.html

[†] UNEP, New York Office, at: http://www.nyo.unep.org/emg2.htm

[‡] World Business Council on Sustainable Development, at: http://www.wbcsd.org/

[§] Ibid.

[¶] Ibid.

[**] GEF, at: http://www.thegef.org/gef/whatisgef

calls from developing countries for increased financial resources and concerns about how aid was being limited by donor countries who argued that sustainable development aid should not be considered separately from standard development aid. Various actors, including the European Community, lobbied to replenish the GEF with fresh and additional resources for specific *Agenda 21* programs, and ultimately, the decision was made to move the GEF out of the World Bank and let it operate as an independent and permanent organization, but it was not executed until some two years later.*

As the GEF notes, at Rio, "the decision to make the GEF an independent organization enhanced the involvement of developing countries in the decision-making process and in implementation of the projects."† But in 1994, when the GEF was officially made a permanent entity, the new arrangement made the World Bank the Trustee of the GEF Trust Fund—a move which did not represent the kind of independence that those calling for reform at Rio had had in mind. This led a number of observers to question the integrity and effectiveness of the GEF, arguing that the World Bank, through myriad activities, has assisted a number of environmentally destructive projects over the years (see, e.g., Chatterjee and Finger, 1994). On the other hand, it was through the restructuring at Rio that the GEF came to serve as the financial mechanism for various multilateral environmental agreements.

See also Chapter 49 of this book on Global Financial Instruments.

DIRECT IMPACTS OF THE RIO EARTH SUMMIT: SUCCESSES AND SHORTCOMINGS

Since 1992, the Rio Earth Summit has been the subject of extensive review, analysis, and critique. Indeed, entire books have been devoted to the subject (see, e.g., Haas, 1992, pp. 6–11; Sandbrook, 1992; Grubb et al., 1993; Chatterjee and Finger, 1994; Middleton et al., 1994; Freestone, 1994; Pallemaerts, 1994; Porras, 1994). In his 1997 essay reviewing progress on Rio objectives after five years, Richard Sandbrook states:

> Anyone setting out to give a global assessment of progress since the events in Rio in 1992 could well be considered as either a fool, or arrogant or both. There is so much one could report on and in so many places that the task seems absurd. So all that can be done is to attempt a comparison of expectations then with realities now. Hardly a scientific exercise based on empirical research to be sure (Sandbrook, 1997).

This caution is even truer today, nearly 20 years after the Earth Summit. Thus, it is not the intention of the current chapter to provide a comprehensive synthesis of all such examinations and critiques to date. Nonetheless, what is offered here is a brief identification and description of some of the common themes and threads running throughout earlier assessments of Rio.

---

* The GEF, at: http://www.gefweb.org

† The GEF, at: http://www.thegef.org/gef/whatisgef

Prior to engaging in that exercise, however, one critical albeit obvious point must be made about evaluations of the success or failure of Rio: nearly all evaluations of the Rio Earth Summit are strongly influenced by, if not a product of, the theoretical and/or political paradigms of the author engaging in the analysis. Thus, for example, commentators who follow the thinking that growth and development lead inexorably to environmental disturbance and degradation, report that the Earth Summit was, by and large, a failure for the environment and a great success for business and industry, which artfully managed to turn public and political attention away from problems caused by industrialization (pollution, depletion of natural resources) and toward the many human benefits to be found in development (e.g., Chatterjee and Finger, 1994). On the other hand, commentators who believe that "sustainable development" is a truly laudable concept, containing the seeds for the dual existence of prosperity and environmental integrity, view the Earth Summit as a success—even if the primary contribution of the Summit was simply dissemination and popularization of the idea of "sustainable development" (e.g., Dernbach, 2002).

But these disparate and wide-ranging views on successes and failures are, themselves, an indication that, despite being heralded as an event that brought many people from many areas of the world and many walks of life together to work on common problems, the Rio Earth Summit was not truly a unifying event. On the contrary, some have argued that, "[i]n effect, existing positions were polarized at Rio by the experience of meeting together under the media spotlight, not reconciled" (Seyfang and Jordan, 2002). Similarly, those whose interests were most served by the Summit—some would say business and/or northern hemisphere environmental organizations (see, e.g., Chatterjee and Finger, 1995)—tended to look back at the Summit with fonder memories than those who felt they left with little or nothing, namely, many of the poorer countries of the Southern hemisphere, or more accurately, the civil society groups and organizations within them.

Given the context just described, the present description of Rio's successes and failures takes as its starting point the simple question whether the Summit succeeded or failed in realizing the specific objectives articulated by the General Assembly when UNCED was provided with its original mandate (these objectives are summarized in the section "Rio Earth Summit Goals and Objectives"). Thus, the analysis below reviews briefly how the Earth Summit fared as measured against these objectives. It would be disingenuous, however, to end the analysis there, since many commentators in the years following Rio have offered enlightening and useful views on the Summit's impact which go beyond the question of whether it met its stated objectives. Thus, the analysis below also outlines in summary fashion some of the most frequently cited "successes" and "failures" of the Rio Summit identified by commentators over the years.

Successes ... with Caveats

Rio's Success in Light of Its Official Objectives

Measured against the fundamental aims in its primary stated objectives as outlined in UNGA Resolution 44/288 (see the section "Rio Earth Summit Goals and Objectives"),

Rio was, on the whole, a solid success. Simply put, this is because the stated objectives, not surprisingly, tend to be objectives mandating examination, study, policy development, the identification of problems and solutions, and the promulgation of recommendations and plans. Thus, the objective requiring the Earth Summit "to examine the state of the environment" and the changes that occurred after the Stockholm Summit, was clearly met: many scientific reports covering the gamut of environmental issues were made available and studied at the Summit. Similarly, the objective of recommending measures to "protect and enhance the environment through the development and implementation of policies for sustainable and environmentally sound development" was largely met by *Agenda 21*, a hugely detailed action plan. Another stated objective, promoting the "further development of international environmental law," was also fulfilled via the negotiation and completion of the two conventions (UNFCCC and CBD), the Statement of Forest Principles, and the Rio Declaration. While the later two were not binding legal agreements, they still exert legitimate influence in international environmental law as "soft law." Rio also succeeded to some extent in examining ways and means of furthering cooperation between neighboring countries, if only by providing a context for neighboring countries to begin dialogues with each other about common problems and cross-border issues. As well, the Earth Summit certainly was successful in examining "the relationship between environmental degradation and the international economic environment" though it certainly cannot be said that such examination led to anything resembling a consensus. Finally, it can be argued that, through its production of *Agenda 21*, and through negotiations to restructure the GEF,* the Earth Summit succeeded at least in part in examining "strategies for national and international action" aimed at promoting a "supportive international economic climate conducive to sustained and environmentally sound development."

We now turn to some of the "successes" and "shortcomings" commonly ascribed to the Earth Summit, restating the fact that most evaluations of Rio did not seek to measure its results against its official objectives.

Awareness Raised on Environment, Development and the Connection between the Two

As the delegates at Rio carried out their examination of the state of the environment and the changes since Stockholm, a task named among the Summit's stated objectives, media reports echoed the findings to the world, resulting in one of the most massive awareness-raising events to date. The articulation of the various environmental and human problems plaguing the planet to an audience that was as broad in its socioeconomic status as it was in its geographic distribution, counts as a milestone in the history of environmentalism. As Haas, Levy and Parsons point out, one of the most important contributions of the Earth Summit was the process used for consciousness raising at the summit and its subsequent popularization in society (Haas et al., 1992, pp. 28–29). Thus in Rio, those concerned with planetary problems used the Summit as a vehicle to "spread the word" about various issues to a global audience for the first time—an audience ranging from politicians and scientists to

* This topic is treated in detail in the section "Restructuring of the Global Environment Facility."

poor, indigenous laborers, and third-world mothers. In short, one of the key tools used by environmental advocates, activists, and their opponents, to wit, education of the public, was utilized and popularized at Rio as never before.

While environment as an issue was certainly well-rooted prior to the Summit, it is undeniable that, as an event, Rio went farther in popularizing the environment as a *cause célèbre* than any event before or after. As Daniel Esty notes: "... perhaps most important, the 1992 Earth Summit will be remembered for its remarkable role in worldwide environmental education" (Esty, 1993). While the 1972 Stockholm Summit "served to educate governmental elites around the world about environmental issues," the Earth Summit at Rio "was an event of the masses" (Esty, 1993).

Even within months of the end of the Rio Summit, some experts stated that Rio would be "much less remembered for the agreements produced and much more remembered for the symbolic emergence of the environment as a global issue of first-order importance," and that the Summit would "be seen as establishing irrevocably the connection between environmental protection and economic growth" (Esty, 1993).

Rio also made clear and popularized the notion that specific environmental issues such as forestry, endangered species, oceans, and atmosphere were highly interconnected, and that inevitably in the course of coping with problems faced in one area, problems in other areas would arise.

Formalization and Popularization of Concept of Sustainable Development

If the Earth Summit did nothing else, it propelled the concept of "sustainable development" into the spotlight of public attention, and ultimately injected the concept into common parlance and thought (see, e.g., Koh, 1997; Dernbach, 2002). Since Rio, the number of agencies, organizations, NGOs, and associations, as well as corporate and government policies organized around or supporting "sustainable development" is truly staggering. But today, even as at Rio, it is clear that not everyone thinks about the same thing when speaking about the concept. Some see sustainability as the essential idea in that concept, viewing development as something that must be carefully tempered, even constrained, by the need to ensure that the environment is not harmed in a permanent way. Hence, development practices having negative impacts on the environment are not "sustainable" practices, in that they are not reconcilable with the realities of the planet's finite resources and human demands on "the commons." Others, however, view the controlling term in the concept of "sustainable development" as development: taking the view that the environment is a resource to be used in the pursuit of prosperity (or simply climbing out of poverty), but a resource to be used carefully and efficiently, so that it may remain available for development for future generations as well as the present.

Indeed, much has been written about the various often conflicting meanings ascribed to the term "sustainable development" (see, e.g., Stone, 1993, 1994), and it is certainly not within the scope of the present discussion to develop this point. Rather, the point to be made here is that the lack of consensus on the meaning of sustainable development today, or what could also be described as a case of severe "mission drift" from the notion as it was originally expressed in the Brundtland report,* is something that can be traced back to the Earth Summit.† Thus, despite whatever notions people had about the concept on the way to Rio, by the end of the Summit most observers of the event, on and off site, understood that while the term "sustainable development" was not something that necessarily put the environment—or people, for that matter—first, the exact definition and how to operationalize the concept remained elusive. As such, while Rio made "sustainable development" a household word, it also left a legacy of confusion over the concept, or at least a situation in which the term is used by different constituencies to serve competing objectives.

No matter what view or attitude one takes about the concept of sustainable development, it remains an undeniable fact that the Earth Summit at Rio was responsible for rooting this concept firmly and deeply in the mind of conference-goers and the public alike. As testimony to the enduring nature of the concept, one of the aims announced for the "Rio Plus 20" Summit in 2010 was to renew political commitment toward, and the public's interest in sustainable development (see, e.g., UN Deptartment of Public Information, Background Release, 2010). The themes of that Summit will be: building a green economy and an institutional sustainable development framework.

Launching and Popularization of the Notion of Sustainable Consumption

One recent and comprehensive review emphasized that, the notion of sustainable consumption was largely born and made popular at the Rio Earth Summit (Jackson and Michaelis, 2003). As Tim Jackson and Laurie Michaelis observe:

> The term sustainable consumption itself can be dated more or less to Agenda 21—the main policy document to emerge from the Rio Earth Summit in 1992. Chapter 4 of Agenda 21 was entitled "Changing Consumption Patterns." It called for "new concepts of wealth and prosperity which allow higher standards of living through changed lifestyles are less dependent on the Earth's finite resources." In so doing, it provided a potentially far-reaching mandate for examining, questioning, and revising consumption patterns—and, by implication, consumer behaviours, choices, expectations and lifestyles. (Jackson and Michaelis, 2003, p. 2)

Unfortunately, the enthusiasm and sincerity with which the concept of sustainable consumption was originally discussed waned significantly in later years. Taylor and Michaelis observe that this decline stemmed in no small part from disagreement over whether the concept implied "consuming more efficiently, consuming more responsibly, or quite simply consuming less," and that because of this variability, "... by the time of the second Earth Summit in 2002, many of the organizations who had grasped the dialogue on sustainable consumption so enthusiastically began to distance themselves from its more radical implications. Some of them dropped it completely" (Jackson and Michaelis, 2003, p. 3).

---

* [S]ustainable development is development that meets the needs of the present without compromising the ability of future generations to meet their own needs (World Commission on Environment and Development, 1987).

† For a thorough discussion of the confusion over the term "sustainable development" in the early 1990s, on the heels of the Earth Summit, see Christopher D. Stone (1993–1994).

Today, the concept of sustainable consumption as treated by international agencies such as UNEP, is understood to mean consuming differently and efficiently, not necessarily consuming less (Jackson and Michaelis, 2003, p. 3). Nonetheless, the noting of consuming less has retained its significance in certain circles with namely, with environmentalists, as witnessed by the huge popularity of *The Story of Stuff* on the internet—an animated documentary about the origins and ultimate impacts of consumerism.*

Multiple International Agreements and High Level of Political Commitment

As highlighted earlier, perhaps one of the features of the Rio Summit that has led many to color the event as a success is the fact that with three major international agreements (the UNFCCC, CBD, and Forest Principles), as well as a set of guiding principles (the Rio Declaration) and a detailed action plan (*Agenda 21*), Rio can easily be described as the most prolific global environmental summit to date. Certainly, any number of problems may be identified in the structure, scope, wording and/or implementation of these documents, and one may criticize the fact that the Forest Principles and Rio Declaration were nothing more than political commitments to sets of principles, but given the long and arduous process involved in the negotiation and signing of treaties and other agreements (see, e.g., Firestone),† it is still remarkable that Rio was able to produce final products from five separate sets of negotiations. Furthermore, the power of achieving unified political commitment on issues must not be overlooked: political commitment is always a necessary precursor to the conclusion of full-fledged, legally binding treaties. On the level of political commitment, Rio was successful. In his 1997 critique of Rio, Richard Sandbrook, states, "On balance, the Earth Summit in 1992 can still be seen as a high point of political commitment to solving global environment and development problems" (Sandbrook, 1997).

Mobilization and Coalition Building within Civil Society

A number of commentators and observers of the Rio event have pointed to the role of the Summit in mobilizing ever greater numbers and types of people to the causes of environmental protection and/or sustainable development (see, e.g., Haas, 1992, p. 28). In contrast to the earlier conference at Stockholm, the Earth Summit provided the context for an unprecedented number of civil society groups to meet face-to-face and network on an enormous variety of topics, from environment and natural resources, to energy, poverty, population, and education, to name just a few. The event provided civil society attendees with a valuable opportunity to learn about issues other than their own, to see the connections among issues, and to lobby each other (often as much as they lobbied Summit delegates) in an effort to broaden their circles of influence. As such, Rio launched an era of coalition-building among and within the various interests represented by civil society groups that continues to this day.

---

* Short film by Annie Leonard, launched online December 2007, at: http://www.storyofstuff.com/international/. *The Story of Stuff* Web site claims that since the film's launch, it has been viewed by millions of people in more than 224 countries and territories.

† Firestone mentions that negotiation and conclusion of treaties is typically a multiyear exercise: Witness the Law of the Sea Convention, which took nine years to conclude. Ibid.

The Summit also significantly raised the profile and status of civil society organizations in the eyes of governments and international agencies. Despite the fact that the vast majority of NGOs were corralled toward the "alternative venue" a number of miles away from the main summit proceedings, the Earth Summit gave credibility to these groups in a way that ultimately served to strengthen the position of NGOs in relation to formal UN processes, both to the media and to the public as well. In 1993, Esty wrote: "Never have representatives of so many nongovernment [sic] organizations (NGOs) attended a major international event, presenting such a broad array of views and perspectives" (Esty, 1993).

Creation of the Commission on Sustainable Development and National Reporting

As mentioned earlier, the CSD grew out of recommendations in *Agenda 21* (Esty, 1993, Chapter 38) that there should be an international agency charged with monitoring and reporting on progress on *Agenda 21* objectives, implementation of the agreements made at Rio and implementation of sustainable development goals generally. The creation of the CSD may be considered a successful outcome of Rio because the CSD still functions today as a permanent UN body dedicated to the pursuit of sustainable development objectives and the implementation of *Agenda 21*. More importantly, it collects information on national plans and progress on implementing sustainable development, and periodically issues summary reports about worldwide progress within specific thematic areas, based on the reports from participating countries. The information CSD collects comes to it largely by way of the CSD's National Reporting system, under which countries submit periodic reports to the CSD on their progress within the thematic areas. While the program is strictly voluntary, many countries opt to report on the status of sustainable development within their borders.

The "birds-eye view" that the CSD enjoys by virtue of examining the national reports allows it to make some very important observations. In the spring of 2010, for example, the CSD was able to conclude the following after reviewing the latest crop of national reports on the topics of sustainable consumption and production, chemicals, mining, transport, and waste management:

> Only a few countries have managed to weaken the link between economic activity and resource extraction, pollution and waste generation. The projected growth in population, income and wealth over the next 40 years is expected to put increasing pressure on resources. If rising middle classes of emerging economies were to emulate the consumption patterns of rich countries, two planets would be needed by 2040. (UN Commission on Sustainable Development, Press Release, 2010)

Such reports and pronouncements obviously have an important role to play in keeping critical environmental and development issues in the public eye.

Shortcomings

In general, criticisms of Rio far outnumber instances of praise. As Paul Harris stated, in a 1996 examination of equity and international environmental institutions:

> Many assessments of the Earth Summit have been written and almost as many are negative. There is almost an endless supply of skepticism regarding the ability of the institutions emanating from UNCED to produce significant environmental benefits. (Harris, 1996, p. 294)

Further, Najam Adil, states in "The View from the South: Developing Countries in Global Environmental Politics": "Rio's legacy probably owes as much to the many disappointments since that conference as it does to its actual achievements" (Najam, 2005, p. 234). Indeed, the criticisms and complaints made about Rio since 1992 are too numerous to afford a full accounting—or even a brief mention of each one. That said, it is possible to identify certain themes that occur with some frequency among evaluations and critiques of the Earth Summit's process, products, and general impact. These themes are addressed briefly below.

North–South Tensions

Many authors have highlighted the conflicts at the Earth Summit between Northern and Southern hemisphere constituencies, as well as the Summit's failure to resolve them. Elizabeth R. DeSombre provides a succinct description of the fundamental issues in North–South tensions as they played out at the Rio Summit:

> Developing countries were concerned that international environmental regulations would impact their ability to develop. They wanted acknowledgement that most of the damage to the global environment had been done by the developed countries, and assurance that they would not be preventing from developing and using the same technology their predecessors had. Developed states, on the other hand, wanted acknowledgement of the role of population growth in environmental degradation, and an equal allocation of responsibility for addressing environmental problems. (DeSombre, 2006, pp. 25–26)

The challenge for the Earth Summit was to try to find some way to bring the North and South to the table in a way that provided some movement toward a resolution of these differences. But it was not to be. As Gill Seyfang and Andrew Jordan state in their 2002 examination of environmental mega-conferences, "[a]s was the case with Stockholm, Rio conspicuously failed to reconcile the conflicting demands of industrialized and industrializing countries" (Seyfang and Jordan, 2002). In a 1997 essay, Richard Sandbrook even identified the failure to resolve the North–South dilemma as the key shortcoming of Rio:

> But on balance, the Earth Summit in 1992 can still be seen as a high point of political commitment to solving global environment and development problems. What failed was the "bargain" that some sought at Rio: This was broadly that, in return for addressing the big environmental issues of climate change, biodiversity loss and deforestation, the wealthy world would help to finance, and support with technology, accelerated but "sustainable" development for the "South" (Sandbrook, 1997; Najam, 2002, pp. 46–50; Conca, 2005, pp. 127–128).*

* A number of authors have spoken about this "bargain": see, for example, Adil Najam (2002) and Ken Conca (2005).

The lasting impact of the North–South difficulties at Rio is explained in cogent fashion by Adil Najam:

> Much of the attention since UNCED has focused on the failure of the North to deliver the "goodies" that had been promised or implied at Rio—such as additional resources, technology transfers, and capacity building. Indeed, the inability of the North to fulfill these commitments has been a major contributor to the growing sense of malaise. However, the erosion of the conceptual building blocks of the Rio Bargain is an even more telling indictment of the fast deteriorating state of North–South relations. As the concept of sustainable development loses its policy edge, and as the key principles of additionality, common but differentiated responsibility, and polluter-pays are steadily eroded with each new [multilateral environmental agreement], the developing countries have a diminishing interest in staying engaged in these processes. These issues defined the *raison d'être* for the South's engagement in global environmental negotiations. (Najam, 2002)

Whether or not the failure to reconcile North–South demands and interests was the primary problem with Rio, one thing is certain: this issue was linked to a number of other major shortcomings, as we shall see below.

Lack of Funding

New money for sustainable development programs—that was the rallying cry of developing countries at Rio, but it was a cry largely unheeded. As Andrew Jordan explains, from very early on, the developing countries made clear the need for the industrialized countries to offer substantial amounts of additional finances—that is, financial aid that was not simply to be redirected from previous programs or aid transfers from the North to the South (Jordan, 1994). This was known as the concept of "additionality," and Jordan labels the finance issue in general and additionality in particular as "one of the most contentious issues raised" at Rio (Jordan, 1994).

The absence of funding necessary to carry out the enormously ambitious action plan in *Agenda 21* was clearly one of the most serious shortcomings of the Summit (see, e.g., Haas et al. 1992, pp. 26–27). Indeed, as Seyfang and Jordan mention, "little of the 'new and additional' money mentioned in *Agenda 21* for sustainable development in developing countries ever materialized" (Seyfang and Jordan, 2002/03 citing UNEP, 1999; United Nations, 2001). The funding problem was exacerbated by a global recession that was posing great challenges for the world economy at the time, and by already weak Official Development Assistance (ODA) (see, e.g., Haas et al., 1992, p. 26). Moreover, complaints were about waste, inefficiency and lack of interdepartmental coordination at the UN were on the rise. As Richard Sandbrook states: "... any evaluation of how well the UN and its family of agencies have done in following up on the Earth Summit are complicated by the general anti-UN environment that has developed in the OECD and elsewhere" (Sandbrook, 1997). However, part of the problem stemmed from the drafting of Chapter 33 of *Agenda 21*, the chapter dealing with measures for implementing the programs therein. According to Andrew Jordan, "Chapter 33 is an adroitly crafted diplomatic compromise that fails to bind anyone to anything" as all funding measures were essentially voluntary

(Jordan, 1994). Jordan also raises the noteworthy point that "the myriad actions prescribed in *Agenda 21* are neither prioritized nor properly costed" (Jordan, 1994). Given the larger dynamics at play in that era, however, it is easy to doubt whether drafting alone could have solved the problem.

Funding for environmental programs and sustainable development "was still a divisive issue at Rio Plus Five in 1997" (Long, 2000), and that trend has continued to the present day with no end in sight.

Weak or Unenforceable Agreements

Even shortly after the Earth Summit, the climate change treaty (the UNFCCC) was widely regarded as an agreement with no teeth, the Biodiversity agreement (the CBD) was seen by many as extremely weak, and the Agreement on Forest Principles was not even a treaty. Typical of the majority of comments made about the Rio agreements at the time was the following:

> A watered down Convention on Climate Change, a Convention on Biodiversity weakened by the absence of the United States' signature, and declarations of intent: this is the upshot of the two-week Rio Conference on the Environment and Development, which brought together 178 national delegations and ended on June 14 with speeches from 117 Heads of State and Government. (UNCED: Mixed Bag of Results at Rio Conference, 1992).

Furthermore, the North–South issues discussed above played an important part in explaining why these agreements were as weak as they were. As Elizabeth DeSombre notes, "Disagreement between developed and developing states led to weak language in the Declaration and *Agenda 21* that satisfied neither bloc" (DeSombre, 2006).

Accompanying the problems with content are problems of implementation. By the time of Rio Plus Five in 1997, participants in that conference were complaining that little meaningful work had been done in either implementing the international conventions signed at Rio in 1992 (or in putting *Agenda 21* into action) (Long, 2000). However, it is the responsibility of states to sign, ratify, and implement treaties within their own borders; thus in that context, it is unfair to blame the Earth Summit at Rio for the slow pace of implementing the Rio agreements. Rather, foot-dragging by certain countries, especially the United States, in relation to signing and ratifying treaties, has been a serious obstacle to progress (DeSombre, 2005, p. 187). For example, at Rio, the United States refused to sign the CBD, although it was signed later in 1993 under the Clinton Administration. And while the United States signed and ratified the UNFCCC, it never ratified the Kyoto Protocol negotiated in later years.

Still, it must be recognized that despite these problems with the content and process of the two Rio Conventions (CBD and UNFCCC), the contribution of the Rio Declaration and *Agenda 21*, particularly in relation to chemical management issues, cannot be easily dismissed. The UN Institute for Training and Research (UNITAR) states: "The legal framework of chemical safety was promoted by the Rio Declaration. Three chapters of *Agenda 21* outline action plans for the environmentally sound management of toxic chemicals and hazardous wastes as well as the associated international legal instruments and mechanisms."* Moreover, as UNITAR observes, before the end of the 1990s, two of the key chemical and hazardous waste management agreements, the Basel and Rotterdam Conventions, entered into force and negotiations were strongly underway on the Stockholm Convention (POPS).† Thus, it seems not unlikely that Rio played a part in fostering the enthusiasm and political work to bring those agreements to life.

Corporate Influence, Cooptation, and Minimization of Corporate Responsibility

Much has been made, from many angles, about the prominent role that business had at Rio. Many question why and how business interests were there in the first place, given the fact that apart from delegates representing their states, and intergovernmental organizations, the UN generally recognizes as official observers of the process only those groups within civil society that are recognized by the UN as NGOs. Granted, business associations formed as NGOs existed then as they do today, but the corporate lobby at Rio found a way to the table that did not always comport with observer status as traditionally known. In a critical retrospective on Rio, Neil Middleton, Phil O'Keefe, and Sam Moyo give a vivid example of how the corporate interests were treated relative to their NGO counterparts:

> There was a moment in the preparatory commissions when a major Northern trade association was to be recognized while a number of the grassroots African organizations were to be excluded. That particular insanity was stopped, but it is clear that any programme in which it is necessary to fight that kind of battle is deeply flawed. (Middleton et al., 1995)

Indeed, while some authors have casted the participation of the business sector at Rio in a positive light (because business seemed ready to engage in dialogue on environmental issues) (e.g., Koh, 1997), most of the commentary has been critical and much of it has been harsh. In *The Earth Brokers: Power, Politics and World Development*, Pratap Chatterjee and Matthias Finger explain that at Rio, multinational interests lobbied UNCED negotiators hard, became respectable participants in the UNCED process, helped fund the event and ultimately conveyed the view quite successfully on many levels, that they were much more part of the solution than they were part of the problem (Chatterjee and Finger, 1995).

Similarly, Elizabeth DeSombre states:

> Those who believe that true environmental protection requires a fundamental change from business-as-usual decry the extent to which the Rio Conference fully institutionalized the shift from seeing industry and wealth as the cause of environmental degradation to viewing them as the solution to environmental problems. Whatever one believes about this relationship, the conference at Rio helped to cement an international policymaking process in which industry and economic growth is central to the way environmental protection is negotiated. (DeSombre, 2006)

---

* UN Institute for Training and Research, at: http://www.unitar.org/ilp/waste-management

† Ibid.

Finally, Chatterjee and Finger, go so far as to say that "UNCED has promoted business and industry" and that, when all was said and done, it was the corporate interests that prevailed at Rio (Chatterjee and Finger, 1995). When we look at how well business interests and the corporate world have fared since 1992, in comparison with how well the environment and those living in poverty have fared, it is rather easy to admit that such conclusions are plausible, even probable.

RIO EARTH SUMMIT: LEGACY, LONG-TERM IMPACTS, AND INFLUENCE ON CURRENT AND EMERGING ISSUES

In a 2002 survey of 252 scholars and practitioners from 71 countries, almost 70 percent viewed the Earth Summit at Rio as having been "very significant" or "monumental." Only about 6%, however, held the view that significant progress had been made on implementing Rio goals (Najam et al., 2002). The study's primary author notes that: "The survey suggests that Rio's greatest impact came from its indirect outputs: its success in giving a higher global profile to issues of environment and development; spurring the growth of national and international institutions, policies, projects, and multilateral agreements for environment and development; and giving more prominence to the views of developing countries on global environmental policy" (Najam, 2005, p. 229).

Certainly, the many objectives and ideals of Rio's *Agenda 21* still inspire and influence international agreements and collaborative efforts on environmental issues, including chemical management. The *Dubai Declaration on International Chemicals Management* (2006),* for example, specifically references Chapter 19 of *Agenda 21*, which deals with the environmentally sound management of toxic chemicals.

Nonetheless, it is undeniable that Rio's impact and its potential for continuing influence was greatly altered and limited by the establishment of the World Trade Organization (WTO) following the Marakkesh Agreement of 1994 and the ensuing interest in globalization. Martin Khor, Executive Director of the South Centre and former Director of the Third World Network, summarizes well how globalization has largely overshadowed its rival paradigm, sustainable development:

> Globalisation found a new institutional house with its many rooms in the WTO's several agreements. Moreover the WTO's dispute settlement system based on retaliation and sanctions gave it a strong enforcement capability. The WTO agreements rivalled the chapters of Agenda 21 and the Rio Declaration. The UNCED did not have a compliance system or a strong agency for following up its agreements. As the 1990s drew on, and the WTO agreements became more and more operational, the globalisation paradigm far outstripped the sustainable development paradigm. Marakkesh 1994 overrode and undermined Rio 1992.†

---

* As presented in Strategic Approach to International Chemicals Management, Comprising the Dubai Declaration on International Chemicals Management, the Overarching Policy Strategy and the Global Plan of Action, issued by the Secretariat for the Strategic Approach to International Chemicals Management, June 6, 2006, pending formal publication, at: http://www.saicm.org/documents/saicm%20texts/standalone_txt.pdf

† "Globalization and the Crisis of Sustainable Development," undated, at: www.unu.edu/interlink/papers/WG1/Khor.doc

Regardless of this dynamic, however, it is clear that the Rio Earth Summit of 1992 ushered in an unprecedented era of global summits on environmental, energy, and sustainable development issues. Skeptics of the processes, viewing the paucity of concrete or lasting results following Rio, might say this was Rio's only real contribution. But as we have seen, Rio came by its reputation as the cradle of concern about sustainable development and as some would say, environmental activism, honestly: the event truly represented the first time in human history that the subjects of environment and development took the world stage in such a visible and connected form.

Nonetheless, at the risk of encouraging ever more time- and resource-consuming global summits that tend to instigate change at a glacial pace surprising in context of consensus-driven, negotiated international processes,* Rio must not be romanticized. Thankfully, early signs from the UN preparatory process for the "Rio plus 20" summit in 2012 give some hope in this regard. The first session of the Preparatory Committee for the Summit, held May 17–19, 2010, reviewed a report on progress to date in implementing the outcomes of the major summits since 1992 that did not paint a rosy picture. For example, the report states:

> While progress has been made on the economic front and in the amelioration of poverty in some regions, the dividends have been unequally shared between and within countries, many countries are not on track for achieving key Millennium Development Goals, and most of the environmental indicators have continued to deteriorate. (UN General Assembly, 2010)

Further, the UN General Assembly Background Release unveiling of the report, observes that: "The world is facing unprecedented multifaceted crises: climate change; [problems with] food water and energy supplies; financial and economic uncertainty; unemployment; unsustainable consumption patterns; disappearing species; over fishing; sanitation [problems], and many more" (UN General Assembly Department of Public Information, 2010). Reacting to the report, UN Under-Secretary-General for Economic and Social Affairs, Sha Zukangin, commented on the failure of traditionally conceived development to improve the lives of millions, stated that: "The sad truth is that, despite two centuries of spectacular growth on our planet, we have failed to eradicate the scourge of poverty," and warning that: "If we continue on our current path we will bequeath material and environmental poverty, not prosperity, to our children and grandchildren." (UN News Service, 2010)

Interestingly, however, Under-Secretary-General Sha Zukang puts his faith in the successful implementation of sustainable development as the solution to this dire situation. He states:

> Our stopgap solutions in response to these crises, with short-term timeframes and sector-based approaches, can no longer suffice in tackling the multiple crises. Only sustainable development, with its inherent emphasis on interlinkages to address social, economic and environmental challenges in a balanced and integrated manner, can provide long-term and durable solutions to the crises. (UN General Assembly Department of Public Information, 2010, p. 2)

* Seyfang and Jordan (2002) provide an interesting and thorough study of global summits, from Stockholm to Johannesburg.

This view of sustainable development as the lifeboat from which we can navigate the world's highly challenging and often intractable problems seems somewhat surprising in light of the weak results produced by efforts made in the name of this concept over the last two decades. However, the strength with which the notion of sustainable development continues to drive institutional and individual thinking about how to solve global environment and human problems has not waned since Rio, and it is this dominant paradigm—however imperfect—that is perhaps the Earth Summit's greatest legacy.

CONCLUSION

It may be that the ultimate contribution of the Rio Earth Summit, given the painfully slow progress on improving the environment and human lives since 1992, is that it underscored the limitations of efforts to solve the Earth's and humanity's problems by way of global, internationally-negotiated solutions to produce concrete results in useful timeframes. This point is well-summarized by then Executive Director of the International Institute for Environment and Development, Richard Sandbrook, in the following excerpt from his 1997 essay, "Rio Plus Five—What Has Happened and What Next" (Sandbrook, 1997):

> After all of this, one must conclude that Rio did very little to change the world at a global level towards a concerted style of development based around the environment: sustainable development if you will. But I would maintain that the Earth Summit was a watershed in terms of how we tackle such issues. Many governments came at last to appreciate that there are no global solutions to sort out various worldwide problems. There are some truly global issues such as climate change deserving of international negotiation. But so many others, such as deforestation, are worldwide problems deserving of national resolve first. The onus of proof was shifted away from the supranational to the national and even the local level. (Sandbrook, 1997, pp. 31–32)

It is an undeniable fact that the Earth Summit at Rio represented a watershed event in terms of international environmental law and policy. Even today, memories of Rio and what was accomplished there are raised as standards to which more recent and future environmental conferences should be held, but as "Rio plus 20" approaches, many people, from experts to citizens, will be questioning the true utility of global summits and global solutions, asking whether we are simply living Albert Einstein's definition of insanity: "doing the same thing over and over again and expecting different results."

Indeed, before simply repeating the past, and relying primarily on international instruments requiring global consensus and current views of "sustainable development," perhaps it is time to consider new avenues of collaboration and cooperation at the global level.

ACKNOWLEDGMENT

The author acknowledges Erica Brown, with gratitude, for her invaluable assistance in the editing, correction, and other aspects of the preparation of this chapter.

REFERENCES

Brooke, J. 1992. The Earth Summit; four of the varied faces in the global crowd at the Rio gathering. *New York Times Online*, June 11, 1992. Available at: http://www.nytimes.com/1992/06/11/world/the-earth-summit-4-of-the-varied-faces-in-the-global-crowd-at-the-rio-gathering.html

Brown, R. M. 1983. *Sudden Death.* New York, NY: Bantam Books, p. 68.

Chatterjee, P. and M. Finger. 1994. *The Earth Brokers.* London: Routledge, pp. 155–156. Chatterjee, P. and M. Finger. 1995. *The Earth Brokers: Power, Politics and World Development.* London: Routledge, p. 3.

Conca, K. 2005. Environmental governance after Johannesburg: From stalled legalization to environmental human rights? *Journal of International Law and International Relations*, 1(1–2): 121–138 at pp. 127–128.

Dernbach, J. C. 2002. Sustainable development: Now more than ever. In *Stumbling toward sustainability.* J. C. Dernbach, Ed. Washington, DC: Environmental Law Institute, pp. 45–61.

DeSombre, E. R. 2005. Understanding United States Unilateralism: Domestic Sources of U.S. International Policy. In *The Global Environment: Institutions, Law, and Policy*, 2nd Edition. R. S. Axelrod, D. L. Downie and N. J. Vig, Eds. Washington, DC: CQ Press, pp. 181–199 at p. 187.

DeSombre, E. R. 2006. *Global Environmental Institutions.* New York, NY: Routledge.

Esty, D. C. 1993. Beyond Rio: Trade and the environment. *Environmental Law*, 23: 387–396 at p. 389.

FAO. 1996. *Fisheries Technical Paper* No. 350, Part 2, 210pp. Rome: FAO.

Freestone, D. 1994. The road from Rio: International law after the Earth summit. *Journal of Environmental Law*, 6(2): 201.

French, H. 2002. Reshaping global governance. In *State of the World 2002—Progress towards a Sustainable Society (Worldwatch Institute Report).* Linda Starke, Ed. London: Worldwatch Institute and W.W. Norton & Co., pp. 174–198 at p. 178.

Garcia, S. M. 1995. The precautionary approach to fisheries and its implications for fishery research, technology and management: An updated review. In *FAO, Precautionary Approach to Fisheries. Part 2: Scientific Papers.* Prepared for the Technical Consultation on the Precautionary Approach to Capture Fisheries (including species introductions). Lysekil, Sweden, June 6–13. (A scientific meeting organized by the Government of Sweden in cooperation with FAO.)

Grubb, M. et al. 1993. *The Earth Summit Agreements: A Guide and Assessment.* London: Royal Institute of International Affairs and Earthscan.

Haas, P. M., M. A. Levy and E. A. Parson. 1992. Appraising the Earth Summit: How should we judge UNCED's success? *Environment*, 34(8): 6–11, 26–33.

Harris, P. G. 1996. Considerations of equity and international environmental institutions. *Environmental Politics*, 5(2): 274–301 at p. 294.

Huismans, W. and A. Halpaap. 1998. *International Environmental Law: Hazardous Materials and Waste.* Geneva, Switzerland: UNITAR.

Ikkala, A.-L. 2009. *The Upcoming REDD Mechanism in the Light of Existing Forest Instruments.* M.Sc. Thesis, Lund University, International Master's Programme in Environmental Studies and Sustainability Science (LUMES), Lund University Center for Sustainability Studies, May. Available at: http://www.lumes.lu.se/database/alumni/07.09/thesis/Ikkala_Anna-Lena.pdf.2

Jackson, T. and L. Michaelis. 2003. *Policies for Sustainable Consumption: A Report to the Sustainable Development Commission*, September 17.

Jordan, A. 1994. Financing the UNCED agenda: The controversy over additionality. *Environment*, 36: 16.

Koh, T. 1997. Five after Rio and fifteen years after Montego Bay: Some personal reflections. *Environmental Policy and Law*, 27(4): 242.

Long, B. L. 2000. *OECD—International Environmental Issues and the OECD, 1950–2000: An historical perspective*. OECD Publishing, p. 21.

Middleton, N., P. O'Keefe and S. Moyo. 1995. *The Tears of the Crocodile: From Rio to Reality in the Developing World.* London: Pluto Press at p. 27.

Najam, A. 2002. Unraveling of the Rio bargain. *Politics and the Life Sciences*, 21 at p. 49.

Najam, A. 2005. The view from the South: Developing countries in global environmental politics. In *The Global Environment: Institutions, Law, and Policy*, 2nd Edition. R. S. Axelrod, D. L. Downie and N. J. Vig, Eds. Washington, DC: CQ Press, pp. 225–243.

Najam, A., J. M. Poling, N. Yamagishi, D. G. Straub, J. Sarno, S. M. DeRitter and E. M. Kim. 2002. From Rio to Johannesburg: Progress and prospects. *Environment*, 4(7): 26–38.

Pallemaerts, M. 1994. International environmental law from Stockholm to Rio: Back to the future? In *Greening International Law*. P. Sands, Ed. New York, NY: The New Press, pp. 1–19.

Peterson, P. 2004. *International Environmental Law: Hazardous Materials and Waste*, 2nd Revised Edition. Geneva, Switzerland: UNITAR.

Porras, I. M. 1994. The Rio Declaration: A new basis for international cooperation. In *Greening International Law*. P. Sands, Ed. New York, NY: The New Press, pp. 20–33.

Raloff, J. 1992. Rio Summit launches two "Earth" treaties. *Science News,* 141(June 20): 407.

Recuerda, M. A. 2008. Dangerous interpretations of the precautionary principle and the foundational values of the European Union Food Law: Risk versus risk. *Journal of Food Law and Policy*, 4:1.

Sands, P. 1994. The "greening"of international law: Emerging principles and rules. *Indiana Journal of Global Legal Studies*, 1(2): 293–324 at 295–296.

Sands, P. 2003. *Principles of International Environmental Law*, 2nd Edition. New York, NY: Cambridge University Press, at p. 11.

Sandbrook, R. 1997. Rio plus five—What has happened and what next? An essay published in *Winners of the Blue Planet Prize 1992*. International Institute for Environment and Development (IIED), May 1997. Available at: http://www.af-info.or.jp/blueplanet/doc/essay/1992essay-iied.pdf

Seyfang, G. and A. Jordan. 2002. The Johannesburg Summit and sustainable development: How effective are environmental mega-conferences? In *Yearbook of International Co-operation on Environment and Development, 2002/2003*. London: Earthscan, pp. 19–26.

Sjoberg, H.1999. Restructuring the Global Environment Facility: Working Paper 13, September. The Global Environment Facility.

Stone, C. D. 1993–1994. Deciphering "Sustainable Development. *69 Chicago-Kent Law Review*, 69: 977–985.

United Nations. 2001. Implementing Agenda 21: Report of the Secretary-General, ECOSOC, E/CN.17/2002/PC.2/7. New York, NY: United Nations.

UN Commission on Sustainable Development, Press Release. 2010. *Report Finds Few Countries Able to Break Link between Drive toward Prosperity and Environmental Stress, May 3.* Available at: http://www.un.org/esa/dsd/newsmedi/nm_pdfs/csd-18/pr_trends_report.pdf

UN Department of Economic and Social Affairs. 2005. *Implementation of Proposals for Action Agreed by Intergovernmental Panel on Forests and by Intergovernmental Forum on Forests (IPF/IFF): Action for Sustainable Forest Management.* Available at: http://www.un.org/esa/forests/pdf/publications/proposals-for-action.pdf, 16 August 2006.

UN Dept. of Public Information, Background Release. 2010. *2010 Rio Conference Not Earth Summit Commemorative Event; Rather Aims to Renew Political Commitment to Sustainable Development, Preparatory Committee Told, ENV/DEV/1139, May 17.* Available at: http://www.un.org/News/Press/docs/2010/envdev1139.doc.htm

UNEP. 1999. *Global Environmental Outlook.* London: Earthscan Publications.

UNCED. 1992. Mixed bag of results at Rio Conference. *Europe Environment*, June 19.

UN General Assembly Dept. of Public Information. 2010. *Countries Start Preparations for Rio + 20: Holistic Response Could Still Achieve Vision Set at Historic Earth Summit*, ENV/DEV/1138, May 17.

UN General Assembly. 2010. *Progress to date and remaining gaps in the implementation of the outcomes of the major summits in the area of sustainable development, as well as an analysis of the themes of the Conference: Report of the Secretary General*, A/CONF. 216/PC/2, April 1, at pp. 5–6.

UN News Service. 2010. *Preparations Begin for 20-Year Review of Landmark UN Environmental Conference*, May 17. Available at: http://unclef.com/apps/news/story.asp?NewsID=34731&Cr=sustainable+development&Cr1=

Wan, K. 2009. Global sustainable development "from above" and local injustice "below": The governance challenge facing the Congo Basin Rainforest. *Law, Social Justice and Global Development Journal*, 1: 1–28, http://www.go.warwick.ac.uk/elj/lgd/2009_1/wan

Weiss, E. B. 1992. *United Nations Conference on Environment and Development*, I.L.M. 814.

World Commission on Environment and Development. 1987. *Our Common Future.* Oxford, UK: Oxford University Press.

4 Johannesburg 2002
The World Summit on Sustainable Development

*Shelley Kath**

CONTENTS

* The opinions expressed in this chapter do not necessarily reflect the views of the Helios Centre.

INTRODUCTION AND HISTORICAL OVERVIEW

> I think we have to be careful not to expect conferences like this to produce miracles. But we do expect conferences like this to generate political commitment, momentum and energy for the attainment of the goals.
>
> **UN Secretary-General Kofi Annan**
> *World Summit on Sustainable Development, Johannesburg*[*]

From August 26 to September 4, 2002, a decade after the United Nations Conference on Environment and Development (UNCED, or the "Rio Earth Summit"),[†] world leaders and over 8000 individuals representing over 900 organizations met at the World Summit on Sustainable Development (WSSD) in Johannesburg, South Africa. The aims of the WSSD, known informally as "Rio Plus 10," or "Earth Summit II," were to evaluate and build on Rio's achievements, commitments, and targets, and to increase implementation of the goals raised in Rio's *Agenda 21*.[‡] Participants at the WSSD again tackled the enormous issues centered on the environment raised at Rio and in the Millennium Development Goals (MDGs),[§] this time addressing environmental protection as connected to and influenced by the two other key facets of sustainable development: social development and economic growth.

The primary motivation for holding the WSSD was to address the widespread concern that the environmental protections promised at Rio were not materializing sufficiently quickly or effectively to match the pace of problems raised by unchecked development, pollution, and resource depletion. Also fuelling the call for the WSSD was a growing realization among world leaders that many of the root causes of environmental degradation could be traced to poverty and lack of access to basic necessities such as clean water, sanitation, and clean energy. Gro Harlem Brundtland, the then Director General of the UN World Health Organization, emphasized these links with a vivid analogy in her address to the Summit. She observed that if a news report emerged that every 45 minutes a plane full of children crashed somewhere in the world, a massive outcry for action would follow. Yet no call for preventative measures occurs in the case of children dying from illnesses related to environmental degradation, despite the fact that the same number of children perish from such causes.[¶] Clearly then, the difference between immediate and incremental catastrophes influences public awareness: the question for the WSSD was then how to work

---

[*] United Nations News Centre, "UN summit 'major leap forward' for partnerships in global fight against poverty–Annan," 4 September 2002, full text available at: http://www.un.org/apps/news/story.asp?NewsID=4615&Cr=johannesburg&Cr1=summit

[†] See Chapter 3 of this book, "Rio 1992: The UN Conference on Environment and Development (The "Earth Summit")."

[‡] *Agenda 21* was the program of action adopted during the Rio Summit, the result of three years development of local and global goals for sustainable development for the twenty-first century.

[§] The Millennium Development Goals were the set of eight international goals and their subsets agreed to by world leaders in 2000, which articulated the interrelated range of social, economic, and environmental challenges facing the world, to be achieved by 2015. The WSSD therefore constituted an important barometer, in summit form, for the targets agreed upon two years earlier.

[¶] "Is the Future Sustainable? The Johannesburg Summit" AWC Zurich Newsletter (November 2002) online at: http://www.fawco.org/index.php?option=com_content&view=article&id=171:johannesburg-summit-2002&catid=44&Itemid=100484

this reality to effect, through the UN, nongovernmental organizations (NGOs), and partnership agreements, the necessary changes to improve environmental health worldwide (Krugman, 2010).

A key factor driving the UN to convene the WSSD was the continuing connection between poverty and environmental degradation. A 1997 review of progress in the five years following the Rio Summit* uncovered a disturbing lack of progress on many fronts, chiefly in the areas of poverty alleviation and social equity. In relation to the management of hazardous chemicals the lack of momentum since Rio in implementing the Basel, Rotterdam, and Stockholm Conventions, spurred great interest in renewing the commitments made in those agreements on risk reduction and promotion of improved health and safety.† Thus, in late 1999, the United Nations General Assembly (UNGA) called for a 2002 WSSD that would serve as a 10-year review of progress since Rio. Unlike the 1997 review, the 10-year review occurred in the form of a series of meetings of global preparatory committees ("Prep Comms") that preceded the Summit itself (Gardiner, 2002).

The WSSD was held in South Africa, which created an opportunity to point up a number of severe social, environmental, and economic problems on that continent, including but not limited to food security, access to safe water and sanitation, desertification, and HIV/AIDS. Speeches by world leaders during the high-level segment reflected many of these concerns and highlighted the quality-of-life inequities experienced by people in the developing world. Certain dignitaries emphasized the need to look beyond discussions about meeting basic human needs and address directly issues of social equity such as "global apartheid" (as it was termed by WSSD Secretary-General Nitin Desai) in North/South relations, the principle of common but differentiated responsibilities, inequities associated with globalization, unsustainable patterns of consumption and production, and the correlation between poverty and environmental degradation.‡

The final tally on Summit participation revealed attendance by 104 heads of State and government, over 9000 delegates, over 8000 NGO representatives, and 4000 members of the press.§ A total of 191 countries participated in the event, which at the time was the largest UN conference ever held (Rutsch, 2002). It is noteworthy that the WSSD in Johannesburg was the first major conference of its kind to try to be "carbon neutral," and to this end it proposed that participating nations fund renewable energy

* 19th Special Session of the United Nations General to Review and Appraise the Implementation of *Agenda 21*, New York City, June 23–27, 1997 (Earth Summit +5), summarized by the International Institute for Sustainable Development (IISD) at: http://www.iisd.ca/csd/ungass.html

† *Basel Convention on the Control of Transboundary Movements of Hazardous Wastes and their Disposal* (1989), text available on Basel Convention Web site at: http://www.basel.int/text/17Jun2010-conv-e.pdf; *Rotterdam Convention on the Prior Informed Consent Procedure for Certain Hazardous Chemicals and Pesticides in International Trade* (1998), text available on Rotterdam Convention Web site at: http://www.pic.int/en/ConventionText/RC%20text_2008_E.pdf; *Stockholm Convention on Persistent Organic Pollutants (POPS)* (2001), text available on Stockholm Convention Web site at: http://chm.pops.int/Convention/tabid/54/language/en-US/Default.aspx#convtext

‡ UN World Summit on Sustainable Development, Department of Public Information, "World Summit Declares 'Fault Line' between rich and poor threatens prosperity, adopts broad measures to alleviate poverty, protect environment," ENV/DEV/J/35 (September 4, 2002).

§ UN, "With a sense of urgency, Johannesburg Summit sets an action agenda," UN Press Release (September 3, 2002).

projects in order to compensate for delegates' use of fossil fuels in connection with the Summit. Ultimately, however, only a few nations participated in this effort and the project fell far short of its $5 million target (Doyle, 2007). As a result, efforts to reduce the Summit's carbon footprint consisted primarily in minor actions such as the donation of paper documents left over from the Summit to South African libraries.*

WSSD SUMMIT GOALS AND OBJECTIVES

The official goals and objectives of the WSSD are rooted in a UNGA resolution adopted in December 2000, mandating a 10-year review of progress since UNCED.† Specifically, the resolution called for a comprehensive follow-up Summit that would, through a series of meetings prior to and during the conference, examine progress on Rio's *Agenda 21*, identify areas in need of further efforts for implementation, assess progress made in ratifying various international conventions agreed to during and since Rio, review obstacles to progress on these fronts, and develop "action-oriented decisions"‡ to encourage new initiatives and vehicles for spurring action on sustainable development. Furthermore, the WSSD was to identify areas for future action in light of new developments, challenges, and opportunities. The ultimate goal was "to reinvigorate the global commitment to sustainable development,"§ with UNGA stipulating that the WSSD "result in renewed political commitment and support for sustainable development, consistent, *inter alia*, with the principle of common but differentiated responsibilities."¶

In sum, politically, the WSSD was to be a venue for strengthened governmental commitment for the overarching goal of sustainable development, and a means to develop further partnerships among the triad of governments, NGOs, and businesses in order to implement the policies and agreements in support of that goal.

In terms of substantive focus, the organizing concept for the 10-year review was sustainable development, and to this end UNGA directed the WSSD to "ensure a balance between economic development, social development, and environmental protection, as these are interdependent and mutually reinforcing components of sustainable development."** The end goal remained that of sustainable development, with concern focused on how states could be encouraged and supported in moving beyond policy to the practical matters of implementing plans for achieving targets.

WSSD SUMMIT OUTCOMES

Overview

The primary outcomes of the WSSD can be divided into two general categories: (1) key documents produced, resolutions adopted and agreements passed, and (2) new

* UN "Summit print materials to be donated to South African Library System," UN Press Release (September 4, 2002).
† UNGA A/RES/55/199 (February 5, 2001).
‡ Ibid.
§ Ibid.
¶ Ibid.
** Ibid.

partnerships and organisms established. We will look briefly at each in turn, focusing on results that occurred during or immediately after the WSSD. Before examining these direct outcomes, some clarifying comments on terminology are in order.

Prior to the Summit, negotiations and discussions in the Prep Comms and other meetings led to suggestions by the UN Secretariat that the WSSD aim to produce two types of outcomes:* "*Type 1* outcomes"—negotiated and proposed for adoption by all member states at the Summit, and "*Type 2* outcomes"—nonnegotiated, voluntary initiatives created for the purpose of implementing the sustainable development objectives of *Agenda 21*. In the discussion which follows, the *Type 1* outcomes—namely, the Johannesburg Plan of Implementation (JPOI) and the Johannesburg Declaration, are covered under the section "Key Documents Produced," and the *Type 2* outcomes, consisting almost entirely of what came to be known as "*Type 2 partnerships*," are covered under the section "New Partnerships and Organisms Established."

Key Documents Produced

The negotiations prior to and during the WSSD produced two documents, generally viewed as two of the key outcomes of the WSSD: the JPOI and the Johannesburg Declaration on Sustainable Development.

Johannesburg Plan of Implementation

The JPOI† was intended to serve as a plan of action for implementing the commitments originally agreed in Rio at UNCED. Though drafted during the final preparatory meeting before the Summit, the JPOI was the focus of negotiations at WSSD itself, because "round-the-clock negotiations by ministers during the last three days of the [PREPCOM IV] session failed to produce consensus on key aspects of the plan, particularly on energy, trade, finance and globalization" (IISD, 2002, p. 2). The JPOI remains an active plan, serving as one of the bases for the activities of the UN's Commission on Sustainable Development, which is charged with its implementation. Furthermore, the JPOI is frequently referenced in the modification or creation of various international instruments. Recently, for example, it was referenced in connection with UN discussions on the preparation of a globally legally binding instrument on mercury.‡

* Stakeholder Forum for Our Common Future, "Comments on the proposed framework of outcomes documents for Earth Summit 2002," Stakeholder Forum, 2002, p. 2.

† UN (A/CONF. 199/20) *Report of the World Summit on Sustainable Development, Johannesburg, South Africa, August 26 to September 4, 2002* (United Nations publication, Sales No. E.03.II.A.1 and corrigendum), chap. I, resolution 2, annex ("Johannesburg Plan of Implementation of the World Summit on Sustainable Development" hereinafter referred to as JPOI), Text available online from the Commission for Sustainable Development at: http://www.un.org/esa/sustdev/documents/WSSD_POI_PD/English/WSSD_PlanImpl.pdf

‡ UNEP, "Synergies and institutional cooperation and coordination with related multilateral environmental agreements and policies" [Note by the secretariat], Intergovernmental negotiating committee to prepare a global legally binding instrument on mercury, First session, Stockholm, June 7–11, 2010, Item 4 of the provisional agenda UNEP(DTIE)/Hg/INC.1/17.

The final version of the plan laid out ten substantive chapters covering: (1) poverty eradication; (2) changing unsustainable patterns of consumption and production; (3) protecting and managing the natural resource base; (4) sustainable development in a globalizing world; (5) health; (6) Small Island Developing States (SIDS); (7) Africa; (8) other regional initiatives; (9) means of implementation (e.g., education, finance); and (10) the institutional framework for sustainable development. Highlights of the commitments made in these chapters are summarized below.

Some of the most frequently discussed commitments made at the WSSD concerned quality-of-life issues, and included targets and timelines. In the JPOI chapter on poverty eradication, concrete commitments included consensus on agreeing, by 2015 to halve the number of people who have no access to safe drinking water or basic sanitation, those living on less than $1 a day, and those suffering from hunger. As well, there was a commitment to achieve by 2020 a significant improvement in the lives of at least 100 million slum dwellers. Several of these objectives were reaffirmations of MDGs.* In the JPOI's health chapter, two of the more widely-touted commitments were (1) to reduce by 2015 the mortality rate for infants and children below 5 years of age by two-thirds, and (2) reduce by 2015 maternal mortality rates by three-quarters, compared to 2000 levels. There were also commitments to reduce by 2010 HIV prevalence among young men and women by 25%, and to fight malaria, tuberculosis, and other diseases.

The JPOI chapter on changing patterns of sustainable production and consumption pointed to a key WSSD commitment, encouraging the development of a set of 10-year programs designed to speed a shift toward sustainable consumption and production by delinking economic growth and environmental degradation. This commitment applied to all nations, with developed nations taking the lead role.

This JPOI chapter also articulates a number of important commitments made at the WSSD concerning chemicals and their impacts on human health (see also the section "Reinforcement of Sustainable Development Ideals through the JPOI and Declaration"). At a general level, member states agreed to renew the *Agenda 21* commitment to sound management of hazardous wastes chemicals throughout their life cycle as well as *Agenda 21*'s commitment to protection of human health and the environment:

> aiming to achieve, by 2020, that chemicals are used and produced in ways that lead to the minimization of significant adverse effects on human health and the environment, using transparent science-based risk-assessment procedures and science-based risk-management procedures, taking into account the precautionary approach, as set out in principle 15 of the Rio Declaration on Environment and Development, and support developing countries in strengthening their capacity for the sound management of chemicals and hazardous wastes by providing technical and financial assistance.†

---

* Criticism of the WSSD included accusations by some that targets and agreements articulated there were not ground-breaking. While true, this may conversely be viewed positively, praising collective consensus on crucial and fundamental needs. The JPOI therefore provided in 2002 important reaffirmations and a barometer for continued achievement.

† Supra note 20, JPOI, para. 23.

In order to achieve these goals within the 2020 timeframe, the following specific commitments were made in paragraph 23 of the JPOI:

a. Promote the ratification and implementation of relevant international instruments on chemicals and hazardous waste, including the Rotterdam Convention on Prior Informed Consent Procedures for Certain Hazardous Chemicals and Pesticides in International Trade so that it can be implemented by 2003 and the Stockholm Convention on Persistent Organic Pollutants so that it can be implemented by 2004, and encourage and improve coordination as well as supporting developing countries in their implementation.
b. Further, develop a strategic approach to international chemicals management (SAICM) based on the Bahia Declaration and Priorities for Action beyond 2000 of the Intergovernmental Forum on Chemical Safety by 2005, and urge that the United Nations Environment Programme, the Intergovernmental Forum, other international organizations dealing with chemical management and other relevant international organizations and actors closely cooperate in this regard, as appropriate.
c. Encourage countries to implement the new globally harmonized system for the classification and labeling of chemicals as soon as possible with a view to having the system fully operational by 2008.
d. Encourage partnerships to promote activities aimed at enhancing environmentally sound management of chemicals and hazardous wastes, implementing multilateral environmental agreements, raising awareness of issues relating to chemicals and hazardous waste, and encouraging the collection and use of additional scientific data.
e. Promote efforts to prevent international illegal trafficking of hazardous chemicals and hazardous wastes and to prevent damage resulting from the transboundary movement and disposal of hazardous wastes in a manner consistent with obligations under relevant international instruments, such as the Basel Convention on the Control of Transboundary Movements of Hazardous Wastes and Their Disposal.
f. Encourage development of coherent and integrated information on chemicals, such as through national pollutant release and transfer registers.
g. Promote reduction of the risks posed by heavy metals that are harmful to human health and the environment, including through a review of relevant studies, such as the United Nations Environment Programme of global assessment of mercury and its compounds.*

While this is not the place to provide a full evaluation of progress on these objectives since the WSSD, it is useful at this point to provide some observations about the movement on some of these objectives made recently by the UN Commission on Sustainable Development (CSD). At the 18th session of the CSD, held in New York City from May 3–14, 2010, one of the key themes discussed as part of the CSD's

* Ibid., footnotes in original omitted.

regular "thematic review" process, was chemical management and hazardous waste. The issues discussed included the synergies process between the Basel, Stockholm, and Rotterdam Conventions, promotion of the SAICM, and its role as a potential model for discussions on the 10-year framework of programs on Sustainable Consumption and Production (IISD, 2010). Indeed, this CSD session provided an opportunity to consider whether the world would succeed or fail in meeting the WSSD's 2020 goal of ensuring that chemicals are produced and used in ways that minimize significant adverse impacts on human health and the environment. No predictions of consequence were made on this subject, however (IISD, 2010, p. 13). It is also worth noting that on general issues of sustainable development, a number of delegates at the session commented on the need to make more visible progress more quickly, in decoupling economic development from resource extraction/pollution/waste generation (IISD, 2010).

In relation to protection of natural resources, many JPOI commitments were in the areas of water, fisheries, and oceans. For water resource management in general, member states committed to develop integrated management and efficiency plans by 2005. On ocean health, commitments were made to encourage the application of the ecosystem approach in order to (1) further the sustainable development of the oceans by 2010, and (2) create international marine sanctuaries by 2012. The JPOI also announced a commitment to establish a regular process for global reporting and assessment of the marine environment by 2004. With respect to ocean fisheries, commitments were made to maintain or restore the health of depleted fish stocks to sustainable levels by, if possible, 2015, and to implement international plans of action for the management of fishing capacity by 2005.

Commitments on biodiversity were also addressed in the JPOI chapter on natural resources. Member states committed to significant reduction by 2010 in the rate of biodiversity loss. Furthermore, the JPOI "underlined the need to strengthen collaboration within and between the [UN] system and other relevant international organizations, to build better synergies among the various biodiversity-related conventions, for better recognition of the linkages between trade and biodiversity, to establish cooperation to achieve synergies and mutual supportiveness with the framework of the World Trade Organization, and for increased scientific and technical cooperation between relevant international organizations."*

The JPOI chapter on natural resources also contained commitments to protect the atmosphere. Concerning ozone layer protection, states committed to improve access by developing countries to alternatives to ozone-depleting substances by 2010, and to assist in compliance with the Protocol's phase-out schedule. With respect to climate change, it was agreed that those states which had not yet ratified the Kyoto Protocol would "ratify it in a timely manner." Additionally, states committed to "meet all the commitments and obligations under the United Nations Framework Convention on Climate Change" (UNFCCC) and "work cooperatively toward achieving the objectives of the Convention."

* Cooperation with Other Organizations, Initiatives and Conventions, Note by The Executive Secretary, UNEP/CBD/COP/7/19, December 10, 2003 (Conference of the Parties to the Convention On Biological Diversity, Seventh Meeting, Kuala Lumpur, February 9–20, 2004), pp. 1–2.

The topic of energy was hotly contested during the WSSD, but negotiations did result in a few notable commitments in the JPOI. Under "poverty eradication," the Plan contained a commitment to improve access to reliable, affordable, economically viable, socially acceptable, and environmentally sound energy services and resources. In the JPOI chapter on "patterns of consumption and production," the Plan committed states to diversify energy supply and "substantially increase the global share of renewable energy sources in order to increase its contribution to total energy supply."* It also committed states to remove distortions in and improve the functioning and transparency of energy markets, establish domestic energy efficiency programs, and accelerate research, development, and dissemination of energy efficiency and conservation technologies.

In the JPOI chapter on sustainable development in a globalizing world, it was agreed that "urgent action at all levels" was needed to "promote open, equitable, rules-based, predictable, and nondiscriminatory multilateral trading and financial systems that benefit all countries in the pursuit of sustainable development."† As part of these commitments, it was agreed that states work toward the successful completion of the Doha Ministerial Declaration's work program, the program for carrying out negotiations on a wide range of subjects dealing with trade, development and the environment, and implement measures in the Monterrey Consensus. While the slow pace of the Doha Round of negotiations over the past few years has frustrated many, progress on these commitments may be accelarated now that the Doha talks have restarted, a situation facilitated perhaps by the world economic and financial crisis.‡ Another focal commitment concerned corporate responsibility: States agreed to "[a] ctively promote corporate responsibility and accountability, based on the Rio principles, including through the full development and effective implementation of intergovernmental agreements and measures, international initiatives and public-private partnerships and appropriate national regulations, and support continuous improvement in corporate practices in all countries."§

Several chapters of the JPOI dealt with specific geographic regions. The chapter on sustainable development of SIDS included commitments to speed up the implementation of several action plans by 2004, a program that included a commitment to support the availability of adequate, affordable, and environmentally sound energy services in small-island developing States. In the chapter on Africa, States committed to improve sustainable agricultural productivity and food security in accordance with the MDGs, and to help Africa implement one extremely modest objective of the New Partnership for Africa's Development namely, to secure access to energy for at least 35% of the African population, particularly in rural areas, within 20 years.

Of the various commitments contained in the "means of implementation" chapter of the JPOI, the most widely proclaimed dealt with education as a cornerstone of sustainable development. In one example, states agreed to ensure that by 2015, all children would be able to complete a full course of primary schooling. States

* Supra note 20, JPOI, para. 20(e).

† Supra note 20, JPOI, para. 47(a).

‡ See, for example, World Trade Organization, "Doha Development Agenda: Negotiations, implementation and development," at: http://www.wto.org/english/tratop_e/dda_e/dda_e.htm

§ Supra note 20, JPOI, para. 49.

committed as well to achieve equal access to all levels of education for both genders, and agreed to eliminate gender disparity in primary and secondary education by 2005.

In the JPOI chapter on the institutional framework for sustainable development, an important commitment was to enhance the role of the UNCSD in the areas of monitoring progress on implementation of *Agenda 21* and foster coherence of implementation, initiatives and partnerships. Commitments were also made to facilitate the integration of the environmental, social, and economic facets of sustainable development into UN led regional commissions, and to take immediate steps toward formulating national strategies for sustainable development, with intention to begin implementation by 2005.

Johannesburg Declaration on Sustainable Development*

The Declaration was intended to be the vehicle for expressing political consensus at the WSSD, and through it, member states reaffirmed their commitment to sustainable development as embodied in *Agenda 21* and the Rio Declaration on environment and development. In the Declaration, states vowed to "advance and strengthen the interdependent and mutually reinforcing pillars of sustainable development."† In emphasizing political commitment to implementation, the Declaration explicitly commits states to expediting the achievement of its socioeconomic and environmental targets in the JPOI, and to the task of monitoring progress at regular intervals.

The Declaration contains strong language acknowledging the North–South tensions that so frequently interfered with progress not only at the WSSD but at other major international summits and conferences. Thus, the Declaration states: "The deep fault line that divides human society between the rich and the poor and the ever-increasing gap between the developed and developing worlds pose a major threat to global prosperity, security and stability."‡ Furthermore, the Declaration warns that member states "risk the entrenchment of these global disparities" and advises that "unless we act in a manner that fundamentally changes their lives, the poor of the world may lose confidence in their representatives and the democratic systems to which we remain committed…" With these statements, the Declaration reflected the frustrations of many WSSD participants concerning effectiveness and intransigence as a result of the rich-poor nation dynamic, making a brave admission about the need for increased cooperation and collaboration.

It is worth noting that, in addition to reaffirming member states' commitment to the outcomes contained in the Rio Summit, the Declaration reaffirms their commitment to the principles and purposes of the Charter of the UN specifically, and to international law generally.

---

* UN, A/CONF.199/L.6/Rev.2 (Corr. 1) (September 4, 2002). *Report of the World Summit on Sustainable Development, Johannesburg, South Africa, August 26 to September 4, 2002* (United Nations publication, Sales No. E.03.II.A.1 and corrigendum), chap. I, resolution 1, annex ("Johannesburg Declaration").

† Ibid., "Johannesburg Declaration," para. 5.

‡ Ibid., "Johannesburg Declaration," para. 11.

New Partnerships and Organisms Established

New Partnerships

The concept of partnerships among governments and businesses played an enormous role at the WSSD, having been conceptualized in preparatory meetings as the prime product for "Type 2 outcomes." In fact, Type 2 outcomes were essentially synonymous with "Type 2 partnerships"—those between governments and "major groups,"* including business and industry organizations and NGOs from other sectors. These voluntary, multistakeholder partnerships were viewed by proponents as a new vehicle to help fill the implementation gap resulting from the paucity of traditional "Type 1" partnerships, which involved relationships among government actors only. The distinction in partnership types and the fostering of Type 2 partnerships was rooted in UNGA Resolution 56/226, adopted prior to the WSSD, which encouraged "global commitment and partnerships, especially between Governments of the North and the South, on the one hand, and between Governments and major groups on the other."† Today, these arrangements are generally referred to as "partnerships for sustainable development," the term used by the CSD, which facilitates and tracks these partnerships.‡ To date, little information exists about the effectiveness of these partnerships in advancing the objectives of sustainable development, but it is clear that the WSSD served to popularize and encourage them (see also the section discussing partnerships as a legacy of the WSSD).

By Summit's end, over 220 Type 2 partnerships between governments, business, and civil society (Type 2) had been announced and/or launched, representing some $235 million in resources,§ though not all of it representing fresh funding streams, as will be discussed later. One example was the $41 million "West Africa Water" campaign to provide potable water and sanitation to rural villages, a campaign whose partners included, among others, the U.S. Agency for International Development, World Vision, and the World Chlorine Council (Schmitt, 2002). Another example was the Congo Basin Forest Partnership, aimed at promoting economic development, poverty alleviation, improved local governance, and improved resource management in central Africa. The U.S. government committed to spending up to US$ 53 million on the partnership, while NGO partners, including Conservation International, Wildlife Conservation Society, and World Wildlife Fund, agreed to raise over US$ 37 million (Scherr et al., 2006).

New Organisms and New Roles

Network of Regional Governments for Sustainable Development

The Network of Regional Governments for Sustainable Development (Nrg4SD) "... is the only global network devoted to promoting sustainable development. It has the

---

* The term "major groups" flows from *Agenda 21*, which recognized nine "major groups" of civil society, and stressed the need for fresh forms of participation at all levels so that nations' economic and social sectors could feasibly engage in sustainable development.

† UNGA Resolution 56/226 (December 24, 2001).

‡ See CSD at: http://www.un.org/esa/dsd/dsd_aofw_par/par_csdregipart.shtml

§ UN, "Key Outcomes of the Summit," UN/DESA Press Release (September 2002).

unique feature of being governed jointly between the North and the South and is currently chaired by Wales and the State of Sao Paulo."* In fact, the Nrg4SD was created in an effort to acknowledge the role of subnational governments in sustainable development.

Today, the Nrg4SD is busy and particularly active in the area of climate change. It played a central role in the second Governors' Global Climate Summit held in Los Angeles in September 2009, highlighting the critical role of subnational governments in reducing greenhouse gas emissions and implementing adaptation measures.† The Nrg4SD participated in various discussions leading up to the Copenhagen Climate Conference (COP 15/CMP 5) in order to raise awareness among European regions of the issues on the table at the COP,‡ and is currently preparing for participation in COP 16.

UN-Energy

A new UN body, UN-Energy, was established in connection with the WSSD in order to "... help ensure coherence in the UN system's multi-disciplinary response to the WSSD and to ensure the effective engagement of non-UN stakeholders in implementing WSSD energy-related decisions."§ The body, UN-Energy, strives to "promote system-wide collaboration in the area of energy with a coherent and consistent approach since there is no single entity in the UN system that has primary responsibility for energy."¶ UN-Energy remains engaged in international energy discussions, particularly in the area of energy efficiency, although the then Director General of the International Atomic Energy Agency, Dr. Mohammed El Baradei, observed that with "no budget or authority," UN-Energy "has had minimal impact" and thus should be reformulated to be, or replaced by, a truly global energy authority akin to the World Health Organization.**

UN CSD: New Role in Monitoring and Facilitating Partnerships for Sustainable Development

One outcome of the WSSD was to modify the operation and role of the CSD, particularly in connection with voluntary, multistakeholder initiatives. The JPOI designated the commission as the focal point for discussion on partnerships which promote sustainable development, and emphasized that such arrangements actually facilitate the implementation of intergovernmental commitments. Further, the JPOI, supported by the Summit Secretary-General Nitin Desai, emphasized that Type 2 partnerships were not intended to replace government commitments or allow governments to circumnavigate commitments, but to supplement and strengthen them and help implement the targets in *Agenda 21*.

---

* Nrg4SD, "Nrg4sd in Los Angeles for the Second Governors' Global Climate Summit" Nrg4SD Press Release, September 29, 2009.

† Ibid.

‡ http://www.nrg4sd.org/default.aspx

§ UN-Energy at: http://esa.un.org/un-energy/

¶ Ibid.

** "Remarks on Energy and Development at IAEA Scientific Forum," made at the 12th Scientific Forum of the 53rd Session of the IAEA General Conference (September 15, 2009).

DIRECT IMPACTS OF THE WSSD: SUCCESSES AND SHORTCOMINGS

In this section, we examine some of the successes and shortcomings of the Summit, taking into account the immediate outcomes discussed above.

Successes

Reinforcement of Sustainable Development Ideals through the JPOI and Declaration

The major part of discussion, commitments, and action at the WSSD involved the reaffirmation of the principles of sustainability in *Agenda 21* and the Rio Declaration. If nothing else, both the JPOI and the Johannesburg Declaration clearly solidified consensus on this concept as the central tool for dealing with the traditionally competing interests of environmental protection, economic growth, and human well being.

Some have characterized as a success the fact that *Agenda 21* and the Rio Declaration were not renegotiated. There are some indications prior to the Summit that such retrenchment, at least on certain issues, was a potential danger. Therefore, in light of the fact that the governments, NGOs, and the private sector reaffirmed their commitment to Rio's program, and that the basic objectives set out for the WSSD by UNGA were reached at least at a general level, the prevention of policy slippage may be seen as a legitimate, though modest success of the Summit.

In relation to chemicals and their impacts on human health and safety, the WSSD agreed to aim to achieve by 2020 the goal that chemicals should be "used and produced in ways that lead to the minimization of significant adverse effects on human health and the environment (. . .)."* As previously discussed in the section "Johannesburg Plan of Implementation," the WSSD called upon nations to advance this objective by stepping up efforts to implement the Basel, Rotterdam, and Stockholm Conventions and stimulating the SAICM process (Downie et al., p. 128).

Commitment to Education as a Tool of Sustainable Development

The central role of education and government commitment to education, as a vehicle for positive change in sustainable development may be identified as a positive outcome of the WSSD. This was a forward-looking outcome, designed to address future environmental and sustainability concerns as well as those of the present.

Specific Commitments from Summit Nations: An Additional Boost to Implementation

Specific commitments of funding or other assistance were announced by various governments attending the Summit and generally, these commitments involved key areas of concern, such as access to water or poverty reduction. As with the Type 2 partnerships, these commitments were voluntary in nature, a fact that drew criticism from a variety of civil society participants. Nonetheless, it is important to recognize that the Summit provided the venue for substantial commitments by individual

* Supra note 20, JPOI, para. 23. See also the section "Johannesburg Plan of Implementation" of this chapter.

states, which helped the WSSD realize its goal of transforming words into action. The $970 million "Water for the Poor" initiative announced by the U.S. to improve drinking water and sanitation in many countries, as well as the European Union's $700 million partnership initiative on energy are but two examples. As is so often the case, however, some of the monies pledged were already in the pipeline, having been pledged in part at previous conferences.*

Shortcomings and Problem Areas

The WSSD saw significant tensions among its participants, both in terms of the general approaches advocated by various constituencies and groups, and dissatisfaction over what some saw as insufficient targets being set, largely on energy and poverty reduction, lack of timetables on certain issues, the lack of cooperation between North and South, and mistrust and disagreements over problem-solving methods among business and many environmental NGOs. In addition to these structural tensions, there were several substantive areas on which very little progress was made during the Summit, in particular, renewable energy, climate change, corporate accountability, and regulation of multinational corporations.

Renewable Energy Sources

As previously mentioned, the question of access to clean, renewable sources of energy was one of the most hotly contested issues at the WSSD. While delegates from Europe, Brazil, and the Philippines, among others, argued for firm targets for the development of renewable energy sources such as wind and solar, specifically, 15% by 2015, the United States, Japan, and the oil-producing countries strongly opposed this move. Despite significant momentum on this issue, the European Union ultimately backed down to avoid imperiling the entire Summit, and the final text involved a minimal commitment on the subject. States agreed to "substantially increase the global share of renewable energy sources with the objective of increasing its contribution to total energy supply" and to do so "with a sense of urgency"† but no deadline. The weak stance on renewables was further diminished by the fact that on the topic of diversity of energy supply, the JPOI text featured fossil-fuel technologies more prominently than it featured renewable energy technologies.‡ Consequently, "[e]nvironmental and development groups were furious that what seemed an imminent deal to set firm targets and a timetable to encourage the spread of wind, solar, and other renewable energies in developing countries suddenly was watered down in favour of fossil fuel energies" (Fox, 2002).

Global Warming, Climate Change, and Kyoto

The twin issues of global warming and climate change saw rather minimal treatment on the Summit's official agenda, much to the dismay of environmentalists alarmed

* See, for example, UN "Countries commit to Major Partnership Initiatives on Water, Energy," August 29, 2002, UN Dept. of Economic and Social Affairs, Division for Sustainable Development, at: http://www.un.org/jsummit/html/whats_new/feature_story29.htm

† Supra note 20, JPOI, para. 20(e).

‡ Ibid., paras. 9, 20, and 62.

by the rapid pace of deterioration and weak progress in this area. As ecologist George M. Woodwell of the Woods Hole Research Center in Massachusetts observed, "Climate change was and is the big issue, [but] it was ignored" (Pickrell, 2002, p. 164). Despite undeniable foot-dragging on this issue, the WSSD did offer some positive movement. Both Russia and Canada announced intentions to ratify the Kyoto Protocol, a move that created hope that the Protocol would come into effect despite the dampening efforts of the United States, among others. During the final hours of the Summit, British Prime Minister Tony Blair, French President Jacques Chirac, and German Chancellor Gerhard Schroeder all called attention to the pressing issue of climate change, and urged final ratification of the Kyoto Protocol (Fox, 2002).

Corporate Accountability and the Regulation of Multinational Corporations

The WSSD was, by all accounts, more inclusive of business than were previous summits, and this brought to the foreground tensions between the corporate world and civil society on whether the focus should be on corporate responsibility (the business position) or corporate accountability (the NGO position) (Hamann et al., 2003). A number of NGOs and their allies pushed hard to create an International Convention on Corporate Accountability, but this effort was unsuccessful (Hamann et al., 2003). It is not difficult to see why concerns grew among NGOs about the sincerity and credibility of the business world in pursuing sustainable development. As one author stated: "Post-Enron, it's difficult to believe that companies can be trusted even to keep their own books, let alone save the world" (Klein, 2002). The results of a poll of business actors conducted shortly after the WSSD substantiate some of these concerns: 88% of those surveyed indicated that the Summit decisions in the JPOI would not affect their corporate operations at all, or have only a slight effect (Russel, 2002). Statistics such as these cast doubt about the sincerity of corporate engagement in partnerships, as well as about the effectiveness of the partnerships themselves.

Many NGOs viewed the resistance to targets and timetables on issues such as renewable energy, as rooted in the pro-corporate stance of various powerful governments. Not surprisingly, much of the focus of these concerns fell upon the United States, whose delegation at the Summit "belligerently" blocked proposals that involved directly regulating multinational corporations (Klein, 2002). Summing up the sentiments of many outside the United States, one participant noted: "The U.S. position on a range of global issues had a common denominator: The United States would accept no timetables or targets. It would offer no pledges. Whether the issue was water, sanitation, energy or pollution, the Bush team held firm, often finding allies in oil-producing nations or among poor nations that said they could not afford renewable fuels" (Fox, 2002).

LONG-TERM LEGACIES OF THE WSSD AND IMPACTS ON CURRENT AND EMERGING ISSUES

The fundamental question for any critical evaluation of the Summit's success is simply: did the WSSD do what it was supposed to do? To answer this question effectively, it is as important to look at the Summit's legacies as it is to review its immediate outcomes. A brief analysis shows that both positive and negative effects of the WSSD are still being felt today.

On the positive side, the Summit's strong view of environmental protection as inextricably linked with concerns about social development and economic growth has evolved into an institutional paradigm and normative approach that today still consistently informs and affects policies in both the environmental and developmental arenas. As testimony to this fact, the CSD, as reconstituted following recommendations made at the WSSD, continues its efforts to actualize the concept of sustainable development through a variety of activities and mechanisms. Thus, attention to sustainable development and what it means in practice finds its way frequently into government programs, into university courses, to and into corporate responsibility statements and initiatives, to name a few. Furthermore, the two new organisms created at WSSD, UN-Energy and the Nrg4SD, both continue to function and carry out WSSD objectives. Additionally, the WSSD and its primary outcome, the JPOI, are frequently referenced in relation to current work in environmental and health areas, including water and sanitation issues, biological diversity, ocean quality, and the survival of ocean fisheries.

Another arguably positive legacy is the Type 2 partnership model launched at the WSSD. Given that one of the primary objectives of the Summit was to foster partnerships in the service of sustainable development implementation, it must be recognized that the WSSD succeeded in this task (Scherr and Gregg, 2006). To a large extent, these partnerships represented the Summit's response to UNGA's demand for progress in real-world, concrete terms for implementation of sustainable development. The Type 2 partnership promoted heavily at the WSSD has continued to attract participants, and the CSD's efforts subsequent to the Summit have been instrumental in encouraging such activity. As a result, many new partnerships have been created since the WSSD.

Since February 2004, the CSD has maintained a registry of these "Partnerships for Sustainable Development," and following the intent of the WSSD and JPOI to make voluntary, multistakeholder partnerships a legitimate and reliable vehicle for implementing sustainable development, the CSD requires all registered partnerships to follow specific guidelines designed to encourage transparent and credible reporting. Nonetheless, CSD Partnerships are "self-governing bodies with their own accountability mechanisms" and, as such, are not required to adhere to externally imposed targets or submit to mandatory monitoring procedures.[*] Today, the registry information is presented in the "CSD Partnerships Database," which contains information on 347 partnerships.[†] Despite their apparent popularity, the successfulness of these partnerships in promoting the original goals of sustainability has yet to be assessed.[‡] It is still too early to provide an accurate evaluation of their effectiveness.

[*] Commission for Sustainable Development, "Frequently Asked Questions," online at: http://www.un.org/esa/dsd/dsd/dsd_faqs_partnerships.shtml#PQ12

[†] Commission for Sustainable Development, online at: http://webapps01.un.org/dsd/partnerships/public/welcome.do

[‡] UN, "Secretary-General's Report on Partnerships for Sustainable Development," (E/CN.17/2010/13), February 2010, paras 69–70. This report mentions that the UN is considering undertaking a systematic evaluation of such partnerships within the framework of the preparations for the 20-year review of *Agenda 21* to occur in 2012.

The WSSD also bequeathed the ecosystem approach to solving environmental problems. In the seventh meeting of the Conference of the Parties to the Convention on Biological Diversity, held in February 2004 in Kuala Lumpur, the WSSD's contribution to the strength of this approach was noted and participants at a meeting on synergies between the Rio Conventions "... underscored the importance of the 'ecosystem approach' as an instrument to achieve synergy."[*]

Furthermore, the WSSD stimulated the further development of Sound Chemicals Management, in particular with the 2020 goal (see the section "Johannesburg Plan of Implementation") and its support to the SAICM development.

On the negative side, the failure of the WSSD to confront and make real progress on reconciling the views and interests of rich and poor countries is a fundamental problem that continues to challenge international negotiations. The opportunity lost at Johannesburg to make headway on healing the North–South rift is problematic in that it allowed the growth of misunderstanding and mistrust that, left unresolved, will hinder global collaboration.

Finally, it must be said that the state of the planet and its inhabitants since the world summit does not reflect large-scale improvements of the type envisioned at Johannesburg. Despite the good intentions at the WSSD and the comprehensive approach of the JPOI, the disturbing conditions affecting people and the environment that originally gave rise to the concept of and need for sustainable development, have not abated since the WSSD. On the contrary, many situations continue to deteriorate. Safe drinking water and proper sanitation continue to evade the lives of billions.[†] Due to the lack of sustainable energy sources, over 1.5 million women and girls die every year from inhaling poisonous fumes as they cook or heat their homes with traditional stoves that burn wood, leaves, or dung.[‡] As well, "[s]ome 80 percent of the world's fish stocks have been fished to their limits."[§] The list of planetary and human maladies continues. With such news, it is difficult to claim that the WSSD, having set its sights on reducing these and other problems, was successful in a meaningful way.

CONCLUSION

There are nearly as many different views on the success or failure of the Johannesburg Summit as there are participants and observers. Perhaps the most useful way to sum up the impact of the WSSD is to repeat the International Institute for Sustainable Development's (IISD's) comment in its final report on the Summit that, "... if measured against the UNGA's stated objectives, the WSSD produced both advances and

---

[*] Cooperation with Other Organizations, Initiatives and Conventions, Note by the Executive Secretary, UNEP/CBD/COP/7/19, December 10, 2003 (Conference of the Parties to the Convention on Biological Diversity, Seventh Meeting, Kuala Lumpur, February 9–20, 2004), p.13.

[†] "Experts Warn of Impending Global Crisis as Commission on Sustainable Development Continues Review of Decisions Relating to Water, Sanitation." *M2 Presswire* (2008); "India below Bangladesh and Pakistan in sanitation facilities." *The Hindustan Times* (2006).

[‡] "Environment: Indoor Air Pollution—Silent Killer of Women." *Inter Press Service* (2007).

[§] "Trade: Subsidies depleting fish stock and hurting small fishers." *Inter Press Service English News Wire* (2009).

setbacks" (IISD, 2002, p. 18).* Interestingly, both of these implicate global collaboration in important ways.

Perhaps one of the most solid advances was conceptual; it cannot be denied that the WSSD went farther than ever before in raising awareness about the close connections among the three pillars of sustainable development and in so doing, fostered the realization of the need for collaboration among people, agencies and governments working in the different spheres of environmental protection, social development, and economic growth. Another major advance was the popularization of voluntary, multistakeholder partnerships (Scherr and Gregg, 2006). Regardless of the skepticism with which some viewed these arrangements, their proliferation during and since the WSSD underlined the point that the magnitude of the implementation challenge requires creative, collaborative solutions beyond intra- and intergovernmental action.

With respect to setbacks, it can be argued that the WSSD offered an unprecedented yet lost opportunity to confront the ever growing divide between North and South. The lack of progress on that front represented a major stumbling block to effective global collaboration that, as mentioned above, continues to hobble the identification of global solutions to environmental problems. The other setback of the WSSD, which provides a challenge still to be overcome, is the implementation gap between plans and results, words and actions. This gap must be rectified in order to heal credibility concerns among various stakeholders and between the North and South, since such concerns also keep global collaboration from becoming truly effective.

In conclusion, we can do no better than echo the Summit's Secretary-General Desai, who said, "The question is, will Johannesburg make a genuine difference? That has to be the test for an implementation conference."†

ACKNOWLEDGMENT

The author acknowledges Erica Brown, with gratitude, for her invaluable assistance in the editing, correction, and other aspects of the preparation of this chapter.

REFERENCES

Downie, D. L., J. Kruger and H. Selin. 2005. Global policy for hazardous chemicals. In *Global Environmental Policy: Institutions, Law and Policy*, 2nd Edition. R. Axelrod, D. L. Downie and N. Vig, Eds. Washington, DC: CQ Press, pp. 125–145.

Doyle, A. 2007. Norway, UK try to tackle planes' Greenhouse Gases. Reuters, Environmental News Network, January 3, 2007. Available at: http://www.enn.com/top_stories/article/5759/print

* IISD, "A Brief Analysis of the WSSD," *Earth Negotiations Bulletin*, 22(51) (September 6, 2002), p. 18.

† United Nations, Department of Economic and Social Affairs, Division for Sustainable Development "The Johannesburg Summit Test: What Will Change," September 25, 2002, available online at: http://www.un.org/jsummit/html/whats_new/feature_story.html

Fox, T. C. 2002. Environmentalists take harsh blow: U.N. Summit on Sustainable Development accedes to U.S. demands: No timetables, no targets on weaning from fossil fuels. *National Catholic Reporter*, September 13. Available at: http://www.thenation.com/print/article/booby-traps-rio-10

Gardiner, R. 2002. Earth Summit 2002: Briefing Paper, a paper written as part of the Stakeholder Forum's Toward Earth Summit 2002 Project, p. 3, January.

Hamann, R., N. Acutt, and P. Kapelus. 2003. Responsibility versus accountability? Interpreting the World Summit on Sustainable Development for a synthesis model of corporate citizenship. (Turning Point). *The Journal of Corporate Citizenship,* 9:20–36.

IISD. 2002. A brief history of the WSSD. *Earth Negotiations Bulletin*, September 6, 22(51), 2.

Klein, N. 2002. Booby traps at Rio + 10. (Comments). (World Summit on Sustainable Development). *The Nation*. Available at: http://www.thenation.com/print/article/booby-traps-rio-10

Krugman, P. 2010. Drilling, disaster, denial, May 2. *New York Times*, A25. Available at: http://www.nytimes.com/2010/05/03/opinion/03krugman.html

Pickrell, J. 2002. Small steps: World Summit delegates wrangle over eco-friendly future. *Science News*, 162(11):164–165.

Russel, T. 2002. Business and the environment (World Summit on Sustainable Development). *Greener Management International,* 39:3–5.

Rutsch, H. 2002. Undoing the damage we have caused: Ensuring environmental sustainability: World Summit on Sustainable Development. *UN Chronicle*, December 1. Available at: http://www.un.org/wcm/content/site/chronicle/home/archive

Scherr, S. J. and R. J. Gregg. 2006. Johannesburg and beyond: The 2002 World Summit on Sustainable Development and the Rise of Partnerships, *Georgetown International Environmental Law Review,* 18(3):425–463.

Schmitt, B. 2002. Industry details plans for global aid partnerships, *Chemical Week*, 164(35):65.

5 International Conference on Chemicals Management 1, ICCM-1

Linn Persson and Viveka Bohn

CONTENTS

INTRODUCTION

The first International Conference on Chemicals Management (ICCM-1) was held in Dubai on February 4–6, 2006. This chapter reviews the expectations and visions for the Conference, as well as its outcome and contribution to the development of a coherent global chemicals management.

The ICCM-1 was held in response to the decision of the United Nations Environment Programme (UNEP) Governing Council in 2002, which called for the development of a Strategic Approach to International Chemicals Management (SAICM). This decision was later endorsed by the World Summit on Sustainable Development (WSSD) in September 2002 as well as by the High-level Plenary Meeting of the United Nations General Assembly in September 2005 (SAICM, 2006).

The WSSD also set the goal that, by 2020, chemicals should be "used and produced in ways that lead to the minimization of significant adverse effects on human health and the environment" (UN, 2002). The challenges contained in this "2020 goal" were, and still are, significant. The piecemeal approach to managing chemicals, the fragmented and differing responses to problems encountered, the lack of capacity in developing countries, the uneven playing field for the industry as well as a lack of financial and technical resources, are some of the factors leading to the current use and production of chemicals with harmful effects on human health and the environment. The reason for developing SAICM was to help countries and other stakeholders meet those challenges in an effective and coordinated manner.

The ICCM-1 was preceded by extensive work by a Preparatory Committee, which held three sessions between 2003 and 2005. Furthermore, numerous regional and other meetings were convened (SAICM, 2006).

EXPECTATIONS AND VISION FOR THE ICCM-1

Before the ICCM-1, representatives of more than 140 countries, as well as numerous international organizations, private sector, and civil society representatives, had participated in the Preparatory Committee sessions as well as in other meetings and had reached agreements on the building blocks of SAICM. The three components agreed upon were a high-level political declaration, later known as the Dubai Declaration, an Overarching Policy Strategy, and a Global Plan of Action (SAICM, 2005). The main expectation of the ICCM-1 was the adoption of SAICM. This in itself was not to be taken for granted after difficulties at the third and final Preparatory Committee meeting (ENB, 2006, p. 8). There was disagreement on what the text should say on financial matters, for instance whether or not to mention international financial institutions in the text, since this was coupled to the issue of new and additional financial resources for the implementation of SAICM (SAICM, 2005). Related to this was also the need for decisions to be taken on other implementation issues, such as the proposed Quick Start Programme and its Trust Fund.

Already during the preparatory phase it had been agreed that SAICM would be a nonlegally binding policy framework. However, at the start of the ICCM-1, countries disagreed on the use of the term "voluntary" in the introductory text to SAICM as well as in its Global Plan of Action (SAICM, 2005). This discussion also included concerns regarding the relation to WTO, which led to disagreement on the so-called "savings clause." The concept of precaution had been an issue for discussion during the preparatory phase and countries still disagreed on how to deal with precaution as a principle.

There were thus several important hurdles to get over before a decision could be taken to adopt SAICM.

Main Outstanding Issues before the ICCM-1

- Final adoption of SAICM with three components, the Dubai Declaration, the Overarching Policy Strategy, and the Global Plan of Action
- The "savings clause"
- Financial matters
- The scope of SAICM
- The voluntary nature of the agreement
- The precautionary principle
- The implementation arrangements including the proposed Quick Start Programme

---

THE OUTCOME OF THE ICCM-1

The main outstanding issues after the work of the Preparatory Committee were those that dominated the deliberations at the ICCM-1. On the last day of the Conference, the assembled ministers, heads of delegation, and representatives of civil society and the private sector, could finally adopt SAICM, consisting of the Dubai Declaration, The Overarching Policy Strategy, and the Global Plan of Action. The adoption of SAICM by all present participants confirms the multistakeholder commitment to SAICM. The overall aim of SAICM was set to "achieve the sound management of chemicals throughout their life cycle so that, by 2020, chemicals are used and produced in ways that lead to the minimization of significant adverse effects on human health and the environment" (SAICM, 2006). The details of the complete SAICM agreement are covered in Chapter 17 of this book. This chapter looks primarily at the solution at ICCM-1 of the outstanding issues from the preparatory phase.

In the *Dubai Declaration,* countries acknowledged that sound management of chemicals is essential for achieving sustainable development and agreed that fundamental changes are needed in the way chemicals are managed (SAICM, 2006). They committed to achieving chemical safety, and by this, also to assist in combating poverty and disease and to contribute to the improvement of human health and the environment. This is based on the understanding that achieving chemical safety is indeed a necessary component in the overall goal of eradicating poverty and disease and safeguarding human health and the environment. A community which does not handle chemicals in an overarching, preventive, and precautionary manner, cannot achieve sustainable development and thus not offer roads out of poverty for those with the most pressing needs. With this the 2020 goal from the WSSD was reaffirmed.

There were several changes made to the text of the Dubai Declaration during the negotiations at the ICCM-1. Multilateral Development Banks were included together with UN organizations and others as being important players in the implementation of SAICM. However, after pressure from one delegation, all references to the World Bank were taken out in spite of several delegations arguing that the World Bank should be mentioned, having substantial activities in the field of chemicals management. Furthermore, there was a lengthy discussion on the "savings clause." For the US delegation it was seen as important to include a text stating that as a nonbinding agreement, SAICM "does not change rights and obligations under legally-binding international agreements." The reason for this was mainly trade-related concerns. Under WTO, also nonbinding but internationally negotiated agreements may be considered as justified cause for action. The European Union (EU) delegation and others were for the same reason strongly opposing the inclusion of such a text and the final paragraph (number 28 of the Dubai Declaration) states that "we acknowledge that as a new voluntary initiative in the field of chemicals management, the Strategic Approach is not a legally-binding instrument."

The negotiations regarding the *Overarching Policy Strategy* were most difficult in the section on *financial considerations.* The developing countries were asking for commitments on new and additional resources for the implementation of SAICM, whereas the developed countries in general were more reluctant to such language. There was also a gap in position between the developed countries with certain delegations opposing all text on financial assistance for SAICM implementation. The final text reads, "The Strategic Approach should call upon existing and new sources of financial support to provide additional resources." There is also an invitation to the UN specialized agencies, funds, and programmes to include the Strategic

Approach objectives in their activities, and to the Global Environment Facility (GEF) and the Multilateral Fund under the Montreal Protocol to consider how they, within their respective mandates, can support SAICM implementation (SAICM, 2006). The G77 and China argued for including the option of opening a new GEF focal area for chemicals management, this was opposed by the EU and the United States and was eventually not included in the final text (ENB, 2006, p. 5; SAICM, 2006). Another difficulty in the negotiations was the *scope of SAICM* as lined out in the Overarching Policy Strategy. Eventually, negotiations led to the scope of the agreement being limited to agricultural and industrial chemicals (Paragraph 3b of the Overarching Policy Strategy).

Furthermore, the Overarching Policy Strategy pinpointed the need for action to be taken at national and subnational levels to help finance SAICM. Funds for the activities will come from industry partnerships, but also through integration of the SAICM objectives into development cooperation activities. The ICCM-1 also agreed on establishing a Quick Start Programme with a time-limited voluntary trust fund from which countries may apply for funding for SAICM implementation. The strategic priorities of the fund were discussed, for instance, the balance between funding analytical work, and enabling more concrete capacity building efforts (ENB, 2006, p. 5). Several countries made pledges to support the fund at the Conference.

Apart from financial considerations, the Overarching Policy Strategy also made a statement of needs and spelled out the objectives of SAICM in five categories; risk reduction, knowledge and information, governance, capacity building and technical cooperation, and illegal international trade.

The negotiations of the Third Preparatory Committee and the ICCM-1 had stumbled over the use of the word *voluntary* in the introduction to the *Global Plan of Action* as well as in the Overarching Policy Strategy. The agreement reached was to refer to "activities that may be undertaken voluntarily" in the Global Plan of Action, but to remove text on the voluntary nature of SAICM in the Overarching Policy Strategy (ENB, 2006, p. 4; SAICM, 2006). It can be noted that since the whole SAICM agreement is nonbinding the discussion of the wording was actually of less importance for the actual implementation outcome. The Global Plan of Action finally agreed upon contained 273 activities divided into 36 work areas and was structured in accordance with the five categories of SAICM objectives. The understanding was that the Global Plan of Action will continue to evolve, opening for possibilities to add new actions to it.

The disagreement over the *precautionary principle* also caused lengthy deliberations. Some countries made the case that the text should be explicit about the connection between precaution and human health, whereas other delegations argued that SAICM should not redefine already agreed instruments, such as the Rio Declaration and its *Principle* 15 on precaution in an environmental context. One reason for some delegations to oppose a more comprehensive understanding of the precautionary principle was that they feared it may be used as an unjustified limitation in international trade. The compromise reached gave a final text stating that one of the objectives in relation to risk reduction is to appropriately apply the "precautionary approach, as set out in *Principle* 15 of the Rio declaration" while aiming to minimize adverse effects on human health and the environment (ENB, 2006, p. 8).

Apart from SAICM itself, the ICCM-1 also adopted several resolutions. One resolution dealt with the implementation arrangements (UNEP/GCSS.IX/6/Add.2, Annex III, Resolution 1). It calls on all stakeholders to contribute to achieving the aims of SAICM and invites UNEP and WHO to provide staff and other resources to assist in the implementation. Another resolution (UNEP/GCSS.IX/6/Add.2, Annex III, Resolution 4) gives the priorities and the institutional arrangements for the Quick Start Programme as well as the terms of reference for the Trust Fund (ENB, 2006, p. 7).

DISCUSSION AND CONCLUSIONS

With the ICCM-1, the decisive step to adopt SAICM was taken. This meant the establishment of a framework for global actions and coordinated measures aiming at significantly reducing the risks associated with the whole life cycle of chemicals by 2020. The ICCM-1 managed to solve the outstanding issues from the preparatory phase and reach agreement among participants.

After the closure of the Conference, views differed on the success of the deliberations. Some were of the opinion that substantial changes would have to be made to SAICM in order for it to actually lead to the fulfilment of the 2020 goal. Others argued that considering the difficult negotiations on financial matters it had to be seen as a success that SAICM was at all adopted and that this in itself should be seen as a good start for the further global collaborations (ENB, 2006, p. 9).

A general idea behind the start of the SAICM negotiations was that SAICM would fill a gap in international chemicals policy by setting a comprehensive framework that would include all chemicals and all aspects of chemicals management. The agreement finally adopted by the ICCM-1 offered this to a certain extent, but the scope was limited to industrial and agricultural chemicals. Some were also disappointed that no new ground was broken when it comes to the principles and approaches guiding SAICM, the final text not referring to the precautionary principle as it has evolved in other fora, such as the negotiations on biosafety and persistent organic pollutants (ENB, 2006, p. 8). Concern may also be raised that in opposition to the integration efforts behind the adoption of SAICM, the single issue approach still persists, for example, with the recent decision to start negotiations of a specific convention on mercury.

However, the SAICM framework had been put in place, and with the exceptions mentioned above it is still the most encompassing agreement on chemicals management that the global community has at its disposal for improving chemical safety. Now remains the implementation of the agreement, which often turns out to be the most challenging part of a process. From other Multilateral Environmental Agreements (MEAs) we know that also seemingly easy, straightforward, and well-defined implementation tasks may be very difficult indeed.

SAICM aims at a comprehensive approach, enabling countries to take a holistic and preventive grip on chemicals management. Furthermore, it argues that efforts from a large range of stakeholders are essential for its full implementation. This is a fundamental strength of SAICM but it also places the agreement in a category of MEAs that are more difficult to implement than others (Underdal, 2002). Countries are already struggling to mainstream and integrate a long row of important issues in their planning and in development activities. Sustainable development as an overarching goal still needs to be implemented. The gap between developed and developing countries is growing in many aspects and the international community still needs a sharper focus on the implementation of the large set of existing agreements (e.g., Engfeldt, 2009).

Achieving integration of chemicals management as prescribed by SAICM is a daunting task since it includes considering chemical safety concerns when they arise in all political fields. Even if the political will is present, it may be difficult to accomplish because we may not always know how to do it. Integrating chemicals management at all levels from the political sphere to the on-the-ground implementation means treading new ground for most countries and political fora.

A positive development regarding the need for a holistic and preventive approach to chemicals management is that in parallel with the development of SAICM, some countries are making national commitments in the same direction. The REACH (Registration, Evaluation, Authorisation and Restriction of Chemicals) in the EU is one such example. Recently, there have been indications that the United States is working along similar lines (US EPA, 2009), and Asian countries have also been reported to adopt chemical legislation to be more in line with the REACH (Park, 2009). The question that remains to be answered is whether SAICM can contribute to spreading this development also to other regions and countries, or whether the legislative divide will persist and grow.

The outcome on the ground in terms of chemical safety will depend not only on the continued efforts of the ICCM forum but also on actual implementation in all countries. This in turn will depend on the resources that are made available for developing countries in order to allow them to narrow the gap in chemicals management compared to developed countries. The current trend of chemical production being moved from developed to developing countries is adding urgency to the need to decrease the legislative divide between countries. Narrowing the gap will also be of outmost importance in order to combat illegal production and trade of chemicals.

With SAICM in place, we now have the tool to shift global chemicals management from regulating specific problems to addressing generic issues. However, the nonbinding character of SAICM, together with the daunting implementation challenge, gives at hand that SAICM still needs to prove itself as a useful tool in the strive toward the 2020 goal and the ICCM will have to give clear directions in the implementation of SAICM for chemicals safety to become a reality.

REFERENCES

ENB, 2006. Summary of the International Conference on Chemicals Management and the Ninth Special Session of the UNEP Governing Council/Global Ministerial Environment Forum. *Earth Negotiations Bulletin*, February 4–9, 16(54). Available at: www.iisd.ca/unepgc/unepss9

Engfeldt, L.-G. 2009. *From Stockholm to Johannesburg and beyond—The Evolution of the International System for Sustainable Development Governance and Its Implications.* Västerås: Edita Västra Aros.

Park, D. 2009. REACHing Asia Continued (September 16). Available at SSRN: http://ssrn.com/abstract=1474504

SAICM. 2005. SAICM/ICCM.1/INF/1, November 23. Scenario note for the International Conference on Chemicals Management. Available at: http://www.chem.unep.ch/ICCM/meeting_docs/default.htm

SAICM. 2006. SAICM/ICCM.1/7, March 8. Report of the International Conference on Chemicals Management on the work of its first session. Available at: http://www.chem.unep.ch/ICCM/meeting_docs/default.htm

UN. 2002. A/CONF.199/20*, Report on the World Summit on Sustainable Development, Johannesburg, South Africa, August 26 to September 4. Available at: http://www.un.org/jsummit/html/documents/summit_docs.html

Underdal, A. 2002. Conclusions: Patterns of regime effectiveness. *Environmental Regime Effectiveness: Confronting Theory with Evidence*. Miles, E., Underdal, A., Andresen, S., Wettestad, J., Skjaerseth, J., and Carlin, E., Eds. Cambridge, MA: MIT Press.

US EPA. 2009. Essential Principles for Reform of Chemicals Management. Available at: http://www.epa.gov/oppt/existingchemicals/pubs/principles.html

6 International Conference on Chemicals Management 2, ICCM-2

Cementing Process and Making Progress toward the 2020 Goal

Melanie Ashton and Pia M. Kohler

CONTENTS

INTRODUCTION

As they reached an agreement on the Strategic Approach to International Chemicals Management (SAICM) in 2006 in Dubai, stakeholders of the International Conference on Chemicals Management (ICCM) recommitted to the goal, agreed at the 2002 World Summit on Sustainable Development (WSSD), to minimize the adverse impacts of chemicals on human health and the environment globally by 2020 (SAICM, 2006a, "Objectives"). It is worth recalling that at the Rio Conference in 1992, delegates had agreed on the goal of achieving the sound management of chemicals by 2000.

Prior to the second meeting of ICCM-2 in Geneva, Switzerland in May 2009, participants in this high-level International Multistakeholder Forum for Chemicals Management issues held several regional and subregional coordination meetings as well as convening a meeting of the Open-ended Legal and Technical Working Group (OELTWG) in October 2008 in Rome, Italy. In particular, the OELTWG discussed the rules of procedure for the ICCM, using the rules of procedure for the preparatory committee for SAICM as a guide (Ashton et al., 2008).

Delegates at ICCM-2 expected to finalize and agree on the rules of procedure for ICCM and to finalize other institutional arrangements, including agreed modalities for reporting, and intersessional arrangements to ensure that processes were in place to support substantive work toward the WSSD goal. ICCM-2 was also required to address a request from the Intergovernmental Forum on Chemical Safety (IFCS) to be integrated into the ICCM as a subsidiary body (SAICM, 2008). While many developing country governments and nongovernmental organization (NGO) representatives hoped that this integration would move forward, most developed countries felt strongly to the contrary. In addition, ICCM-2 provided the opportunity for participants to take stock of the first phase of SAICM's implementation (Ashton et al., 2009), and to address substantive issues including the process for addressing emerging issues under the ICCM, the need for ongoing finance for SAICM implementation, and opportunities for cooperation with other institutions.

The following sections discuss the finalization of institutional arrangements, the progress made on substantive issues, and the issue of financing and potential financing, as stakeholders work toward meeting their 2020 goal.

SETTING SAICM IN MOTION: FINALIZING THE ICCM PROCESS

As ICCM-2 marked the first substantive meeting of ICCM since it had adopted SAICM in 2006, stakeholders had before them several agenda items to finalize and had to put in place the procedures that would not only impact future ICCM meetings but also the preparatory process, including leading up to ICCM-3 (2011).

Agreed Rules of Procedure

Discussion on rules of procedure began in Rome, Italy, in October 2008 at the meeting of the OELTWG for ICCM-2. It was intended that the rules of procedure for ICCM would be agreed at this meeting, allowing more time for substantive discussion at ICCM-2. Although progress was made, the work was not completed at OELTWG (Ashton et al., 2008). Deliberations continued at ICCM-2 and in the resolution, participants agreed to the rules including procedures for representation, credentials, and accreditation and the conduct of business at ICCM, as applying to any session of the ICCM. Agreement was not reached, however, on rule 33 on the adoption of decisions, with participants divided over whether decisions should be made by consensus or by a two-third majority vote. Participants will consider this issue again at ICCM-3. In the interim, the default procedure is that decisions will be taken by a consensus of government representatives (SAICM, 2009).

Institutional Arrangements

ICCM-2 considered institutional arrangements, including modalities for reporting by stakeholders on progress in implementing SAICM. ICCM-2 adopted a proposal on modalities for reporting containing three chapters on overall guidance, on indicators for reporting by stakeholders on progress in the implementation of the Strategic Approach, and on the preparation of reports (Ashton et al., 2009; SAICM, 2009).

The chapter on overall guidance elaborates the need to use a simple electronic data collection tool, as well as a single set of indicators for all stakeholders; structure indicators to take advantage of existing reporting mechanisms and avoid duplication with reporting to Multilateral Environmental Agreements (MEAs); and publish all reports on the SAICM Web site (Ashton et al., 2009).

The proposal identifies 20 indicators, which are organized into five groupings on risk reduction, knowledge and information, governance, capacity building and technical cooperation, and illegal international traffic. Data for these indicators are to be collected nationally and monitored at both the regional and global levels (SAICM, 2009).

The SAICM Secretariat was requested to finalize the overall guidance and prepare a baseline estimates report for comment by the meeting of the open-ended working group (OEWG) for intersessional work scheduled to convene in 2011 (see below). The conference is likely to evaluate the use of these indicators at ICCM-3, make adjustments where necessary and consider their adoption.

Open-Ended Working Group for Intersessional Work

ICCM-2 agreed to establish an OEWG as a subsidiary body for intersessional work. There was broad agreement that comprehensive preparation was key to getting the most benefit of ICCM meetings and that, given the triennial schedule for ICCM meetings, a preparatory meeting was necessary. The OEWG will consider implementation, development, and enhancement of SAICM, including reviewing and prioritising emerging issues for discussion at ICCM-3 (Chynoweth and Ashton, 2009). Several governments, including the United States, were hesitant to establish a permanent body, due to its long-term financial implications. As such, it was agreed that the status of the OEWG would be confirmed at ICCM-3. The OEWG will meet once prior to ICCM-3 in 2011, and the Bureau for ICCM-3 will also serve as the Bureau for OEWG.

Procedure for Adding New Activities to the Global Plan of Action

The Global Plan of Action (GPA) is intended to be a guidance document that includes activities that may be undertaken voluntarily by stakeholders to achieve the 2020 goal. The plan is to be reviewed by stakeholders during the implementation of SAICM (SAICM, 2006a). Currently, the GPA includes over 247 activities, agreed at ICCM-1. At ICCM-2, some governments preferred a focus on prioritizing current GPA activities, as opposed to developing a procedure allowing for the addition of new activities (Ashton et al., 2009). Despite this reluctance, a procedure for including new activities, which applies as of the close of ICCM-2, was agreed and included as an annex to the report of ICCM-2 (SAICM, 2009a). The procedure sets out that proposed additional activities are to be presented by a stakeholder, or group of stakeholders. The method for the discussion and endorsement of proposals comprises the preparation and circulation of a justification document by those proposing the activity, a list of priority proposals for inclusion in the agenda by the regional consultation, and the posting of the proposals on the SAICM Web site, and compilation of any comments from stakeholders (Ashton et al., 2009). Proposals for additions to the GPA will be presented, justified, and considered by the OEWG meeting, and then selected activities will be forwarded to the Conference for further consideration. According to the agreement, the proposal must be relevant to SAICM's Overarching Policy. The extent to which the issue identified has adverse effects on human health and the environment must be demonstrated, as must the activity's consistency with and complementarity to the existing international policies or agreements (SAICM, 2009).

MEETING THE 2020 GOAL: GROUNDWORK AND PROGRESS

Throughout ICCM-2, several stakeholders stressed that by ICCM-3 the 2020 goal would only be 10 years away and called for mobilizing the means of meeting the goal, not only through financing but also through enhanced cooperation with intergovernmental organizations (IGOs). Extensive discussions also focused on how to address emerging issues put forward by stakeholders in addition to priorities already identified in the GPA.

FINANCING

Discussions on financing at ICCM-2 focused on the Quick Start Programme (QSP) (under which disbursement of funds is limited to 2013) and on the need for putting in place a longer-term financial mechanism for implementing SAICM and achieving the 2020 goal.

The QSP was established in 2006 by ICCM-1 as a voluntary, time-limited fund to "support initial enabling capacity building and implementation activities in developing countries, least developed countries, small island developing states and countries with economies in transition" (SAICM, 2006b, Decision I/4). The fund has since received over US$ 23 million from 23 donors (SAICM, 2009b). ICCM-2 participants acknowledged successes of the QSP but many delegates also underscored the need for its review. The resolution on financial and technical resources for implementation requests the QSP Executive Board to evaluate the QSP, its effectiveness and the efficiency of its implementation, and to make recommendations to ICCM-3. Furthermore, stakeholders are invited to report on the steps taken to implement SAICM financial arrangements and ICCM-3 is called upon to review the adequacy of existing SAICM financial and technical arrangements.

Donations to the QSP from developing countries, notably South Africa and Madagascar, were held up as examples of the QSP attracting funds from nontraditional countries, and several calls were made for further broadening the donor base to the QSP. In particular, industries were encouraged to step up their contributions, even as their in-kind contributions and partnership activites were acknowledged. The resolution urges potential donors, including governments in a position to do so, IGOs, the private sector (including industry, foundations, and NGOs) and other stakeholders to contribute to the QSP, and current donors are called upon to continue and strengthen their support.

In addition to the provision of reviewing the QSP and other financial arrangements, participants kept an eye toward putting in place a financial mechanism after the QSP's expiration (the deadline for disbursements is currently set for the end of 2013, while ICCM-2 did resolve to extend the deadline for contributions until ICCM-3). Several visions were put forward for a long-term SAICM financial mechanism. The Multilateral Fund (MLF) under the Ozone regime was held up as an example, and some supported creating opportunities to fund SAICM implementation activities under the MLF, while others preferred modeling the financial mechanism on the MLF. Many discussions centered on the ongoing negotiations for the fifth replenishment of the Global Environmental Facility (GEF), and in particular on proposals being made to expand its support for chemicals management beyond its responsibilities as financial mechanism for the Stockholm Convention on Persistent Organic Pollutants. The ICCM-2 resolution on the issue urges the GEF to consider expanding its activities related to the sound management of chemicals to facilitate SAICM implementation while respecting its responsibilities as the financial mechanism to the Stockholm Convention (SAICM, 2009).

A brief analysis of activities under the QSP indicates that the fund has provided finance for 92 activities across 84 countries, to a total value of US$ 18,993,655.* Considering the QSP is tasked to support "enabling capacity building and implementation activities," the specific activities are funded under the QSP warrant consideration. More than US$ 8 million of QSP funds have been used for countries wishing to develop, or update National Chemical Profiles. Preparation of National Profiles began in the mid-1990s under the IFCS. These profiles allow countries to assess all aspects of the chemical life cycle including production, import, export, storage, transport, distribution, use and disposal of chemicals. They are intended to contribute to a better understanding of which problems or potential problems related to chemicals exist in a country and what mechanisms are available to address these problems. Although National Profile development was initiated in the mid-1990s, 61 countries accessed QSP funds for the development or updating of profiles. The need to update profiles indicates that the work completed in the 1990s was not sustainable, and that prior to national-level implementation of sound chemicals management, significant reassessment and planning were necessary.

In addition, more than US$ 2 million of QSP funds were used to assist countries in integrating chemicals management into national planning processes. In total, US$ 10 million of the U$ 19 million of QSP funds have been allocated to update the work discontinued in the 1990s and to prepare national planning processes to include the issue of chemicals management in the future. While both aspects of this preparatory and planning work were no doubt necessary to build an enabling environment for sound chemicals management, both activities are preliminary and will not lead, without significant further implementation work, to meeting the 2020 goal.

As of February 2010, the negotiations for the Fifth Global Environment Facility have been completed in June 2010 (see Chapter 49 of this book). The revised

* Information on QSP projects is based on the table of projects funded by the QSP of SAICM, as updated on October 2009 and available on the SAICM Web site at: http://www.saicm.org/documents/_menu_items/QSP%20trust%20fund%20approved%20projects%20Dec%2009.pdf

programming document (GEF, 2009) acknowledged that the international chemicals agenda has expanded considerably in quantity and scope, requiring an enhanced response from the GEF. Specific reference is made to SAICM, as well as to the goal of GEF's chemicals program "to promote the sound management of chemicals throughout their life-cycle in ways that lead to the minimization of significant adverse effects on human health and the global environment" (GEF, 2009). It also notes that some funding for the objectives and activities of the SAICM that contribute to global environmental benefits, beyond persistent organic pollutants (POPS), would therefore ensure that the GEF can fully maximize the delivery of global environmental benefits from sound chemicals management activities.

The programming document also acknowledges that the five main objectives in the SAICM overarching, policy strategy, risk reduction, knowledge and information, governance, capacity building, and illegal traffic, include elements that allow for the generation of global environmental benefits. As such the three following objectives are proposed for Chemicals under GEF-5: phase out POPS and reduce POPS releases; phase out ozone-depleting substances (ODSs) and reduce ODS releases; and pilot sound chemicals management and mercury reduction (GEF, 2009). It is also noted that GEF would support some of the SAICM priority "work areas" and activities that generate global environmental benefits, and references the SAICM GPA global priorities, including risk reduction from mercury and other chemicals of global concern; hazardous waste reduction; illegal traffic; and contaminated sites.

Under the GEF-5 replenishment negotiation, three replenishment scenarios (total replenishments of US$ 4.5, 5.5, and 6.5 billions, respectively) are being considered (GEF, 2009). Under these scenarios the Chemicals focal area is allocated US$ 450, 550, and 650 millions, respectively. GEF-4 saw a Chemicals allocation of US$ 300 million, hence all the scenarios result in a signifant increase for Chemicals funding. Within the Chemicals allocation, support to mercury and sound chemicals management is allocated with US$ 20, 40, and 100 millions, under the respective scenarios, with approximately half of this allocation predicted for activities related to sound chemicals management. With the potential to add an additional US$ 50 million to current financing for SAICM, the GEF may become a major player in the implementation of SAICM. However, this potential will only be realised if GEF cofinance requirements can be met by project proponents. Under GEF-4 some major POPS activities with GEF-funding were unable to be implemented, due to lack of sufficient cofinance.

In light of the above, and dependent on the outcomes of the GEF-5, discussions at ICCM-3 are therefore likely to continue to focus on the need for sustained finance for chemicals management and the role of GEF in assisting with this, to allow developing countries, and countries with economies in transition, to arrive at a point of full support for implementation work, and to avoid (re)planning and (re)assessment work, which is necessary when international financial support is unstable.

Emerging Issues

Four emerging issues were also tabled for in-depth consideration at ICCM-2: lead in paint, nanotechnology and nanomaterials, chemicals in products, and electronic waste (e-waste). Resolutions were adopted on all of these. A resolution was adopted also on perfluorinated compounds (PFCs) as an emerging issue.

On lead in paint, it was underscored that this concern was still an emerging challenge in many developing countries with significant impacts on children's health in particular (Ashton et al., 2009). ICCM-2 endorsed a global partnership to promote phasing out the use of lead in paints. This global partnership, whose Secretariat would be provided by United Nations Environment Programme (UNEP) and World Health Organisation (WHO), was tasked with preparing, as a first step, a business plan articulating milestones for progress in several areas, including awareness raising on the toxicitiy of lead and on alternatives, assistance to industry, prevention programmes to reduce exposure, and promotion of national regulatory efforts (SAICM, 2009).

On nanotechnology and manufactured nanomaterials, there was some debate as to whether this issue was only emerging for a small group of countries (Ashton et al., 2009). Nevertheless, ICCM-2 recognized their potential benefits and their potential risks to human health and the environment invited further research on realizing benefits and understanding risks, and encouraged capacity building for these materials' responsible use and management and requested governments and industry to promote appropriate action to safeguard human health and the environment. The resolution on nanotechnology and nanomaterials recognizes the role of regulatory, voluntary, and partnership approaches in promoting the responsible management of nanotechnologies and nanomaterials throughout their life cycles. While recognizing the need to protect confidential business information, ICCM-2 encouraged the wide dissemination of human health and environmental safety information in relation to these products (SAICM, 2009).

On chemicals in products, stakeholders stressed the paucity of available information. ICCM-2 agreed to further consider the need to improve the availability of and access to information on chemicals in products in the supply chain and throughout their life cycle. The resolution invites UNEP to lead and facilitate a project on information sharing for chemicals in products (including collecting and reviewing existing information on information systems pertaining to chemicals in products, including regulations, standards, and industry practices) and to develop specific recommendations to promote SAICM implementation (SAICM, 2009).

On e-waste, African countries in particular highlighted the constraints and challenges they face in coping with the onslaught of "near end-of-life" and "end-of-life" electrical and electronic products being exported to their countries (Ashton et al., 2009). In the resolution, ICCM-2 recognizes the lack of capacity to handle e-waste in an environmentally sound manner, the pressing need for the continued development of clean technology and the environmentally friendly design and recycling of electronic and electrical products, and the importance of considering product stewardship and extended producer responsibility aspects in the life-cycle management of electronic and electrical products. ICCM-2 also invites the Inter-Organisation Programme for the Sound Management of Chemicals (IOMC) and the Basel and Stockholm Convention Secretariats to convene a workshop on e-waste, based on a life-cycle approach, and addressing green design, green chemistry, and recycling and disposal. Governments, IGOs, the industry sector, and NGOs are requested to provide expertise and financial and in-kind resources on a voluntary basis to support the workshop's organization (SAICM, 2009).

On managing PFCs and the transition to safer alternatives, ICCM-2 recognized the decision by the Stockholm Convention to list perfluorooctane sulfonate (PFOS), perfluorooctane sulfonyl fluoride (PFOS-F) and its salts, and that further scientific research may be needed to show whether certain other PFCs are persistent and possibly cause adverse effects in humans and the environment. The resolution invites the IOMC and the OECD, together with governments and stakeholders, to consider developing a stewardship program to gather information on PFCs in product and PFC releases, alternatives currently in use, potentially safer alternative substances or technologies, criteria for alternatives on the necessity and possibility of technology transfer, progress in and examples of regulatory actions and voluntary programs, monitoring, emissions, exposure, environmental fate, and transport, and on the potential effects of PFCs and their alternatives on human health and the environment (SAICM, 2009).

Cooperation and Integration with Other Intergovernmental Organizations

ICCM-2 considered three areas of cooperation, including the health sector; with the Commission on Sustainable Development (CSD); and whether to integrate the IFCS into the ICCM as a subsidiary body.

On the health aspects of SAICM, the resolution emphasizes the need to fully engage the health sector in national, regional, and international SAICM fora, the essential cross-sectoral responsibilities of national focal points, and the importance of regional health and environmental interministerial processes as a springboard for effective intersectoral actions. It also calls on the health sector to actively participate in actions to implement the decisions of the ICCM-2, and invited the WHO to intensify its activities in the sound management of chemicals in support of SAICM (2009a).

On cooperation with the CSD, the resolution welcomed CSD's thematic focus on chemicals in 2010–2011 and emphasized the need to mainstream chemicals management in development strategies. The resolution also invites the CSD to explore the role of the private sector in supporting the global sound management of chemicals (SAICM, 2009a).

The IFCS had proposed that it would be renamed the IFCS and, in an advisory role to ICCM, provide an independent, objective source of information about chemicals management issues, including potential health, environmental, and socioeconomic impacts and possible response actions (Chynoweth and Ashton, 2009). Regarding the IFCS, ICCM-2 decided, in light of the establishment of an OEWG as an ICCM subsidiary body, not to integrate IFCS into ICCM at this time. The resolution commends IFCS for its historic contributions to the environmentally sound management of toxic chemicals, and says that IFCS should itself determine whether and how it may continue to serve its functions.

This emphasis on coordination with IGOs is just one illustration of the broad range of participants in the multistakeholder forum that is ICCM. In addition to governments and IGOs, ICCM-2 brought together industry groups, organizations representing the science community, and a broad array of nongovernmental organizations around the goal of sound chemicals management. While disagreement still remains on the extent to which these stakeholder groups should influence ICCM decisions, ICCM continues to provide a unique forum for bringing together those actors who will need to cooperate to achieve the 2020 goal.

CONCLUSION

The substantive work carried out at ICCM-2 focused on putting in place many of the building blocks and procedures necessary for productive implementation of SAICM during the intersessional period. Yet, with less than a decade left to meet the 2020 goal of achieving the sound management of chemicals throughout their life cycle to minimize significant adverse effects on human health and the environment, ICCM-3, scheduled for 2012, will represent the real test of SAICM as a regime. At ICCM-3, SAICM's diverse stakeholders will be looking for measurable progress toward meeting the 2020 goal and for a clear path forward, including the commitment to provide the financial and technical resources necessary to follow that path.

Resolution on the issue of financing will also be a key signal of the political will to meet this goal under a voluntary regime. On the need for additional finance, several governments and NGO representatives at ICCM-2 urged the industry to contribute financial resources for the implementation of SAICM, but although industry representatives stated their commitment to provide in-kind resources to assist in the implementation of SAICM, no commitments of additional funding were made at ICCM-2. Many underscored the need for permanent and stable finance from multilateral sources, but with the pending fifth replenishment of the GEF, and the potential for funds to be dedicated to SAICM implementation with it, uncertainty remains as to the type and scale of resources which will be available to meet the 2020 goal.

REFERENCES

Ashton, M., P. Barrios, T. Kantai, and P. M. Kohler. 2009. Summary of the second session of the *International Conference on Chemicals Management*. May 11–15, 2009. *Earth Negotiations Bulletin*, 15(175):1–19.

Ashton, M., W. Mwangi, and O. Pasini. 2008. Summary of the first meeting of the open-ended legal and technical working group of the ICCM and informal discussions on preparations for ICCM-2. October 21–24, 2008. *Earth Negotiations Bulletin*, 15(62):1–11.

Chynoweth, E. and M. Ashton. 2009. ICCM2 lays groundwork but clock ticking for 2020 goal. *Chemical Watch*. Available at: http://chemicalwatch.com

GEF. 2009. Revised GEF-5 Programming Document. Available at: http://www.gefweb.org/uploadedfiles/R.5.22%20-%20Revised%20GEF-5%20Programming%20Document.pdf

SAICM. 2006a. Strategic Approach to International Chemicals Management. Available at: http://www.saicm.org/documents/saicm%20texts/SAICM_publication_ENG.pdf

SAICM. 2006b. SAICM/ICCM.1/7: Report of the International Conference on Chemicals Management on the work of its first session. March 8. Available at: http://www.chem.unep.ch/ICCM/meeting_docs/iccm1_7/7%20Report%20E.pdf

SAICM. 2008. SAICM/ICCM.2/INF/21: Documents submitted by the Intergovernmental Forum on Chemical Safety concerning the future relationship of the Forum to the Strategic Approach. December 10, 2008. Available at: http://www.saicm.org/documents/iccm/ICCM2/meeting%20documents/ICCM2%20INF21%20IFCS%20future.pdf

SAICM. 2009a. SAICM/ICCM.2/15: Report of the International Conference on Chemicals Management on the work of its second session. May 27, 2009. Available at: http://www.saicm.org/documents/iccm/ICCM2/meeting%20documents/ICCM2%2015%20meeting%20report%20E.pdf

SAICM. 2009b. Summary table of QSP Trust Fund Contributions. December. Available at: http://www.saicm.org/documents/_menu_items/summary%20QSP%20TF%20contributions%2016.12.09.pdf

Section III

Global/Multilateral Instruments

7 The Aarhus Convention
Impact on Sound Chemicals Management with Special Emphasis on Africa

Ebeh Adayade Kodjo[*]

CONTENTS

INTRODUCTION

Over the past several decades, the international community has established a range of voluntary instruments and legally binding agreements designed to tackle the world's most serious environmental challenges. Each treaty focuses on a specific problem such as hazardous waste, trade in endangered species, chemicals management, climate change, or some other pressing concerns at the time the agreement was developed.

[*] The opinions expressed in this chapter are those of the author and do not necessarily represent the views of any organization.

Despite their varying themes, these agreements also have a great deal in common: they all promote sustainable development and they all use similar principles, processes, and policy tools to achieve their goals. Legal systems across the globe are responding to environmental concerns in surprising new ways. Environmental problems increasingly are viewed as transcending national borders. Some, including global warming, hazardous chemicals, and climate change, pose significant risks to the very health of the planet. Throughout the world, nations are upgrading their environmental standards by transplanting law and regulatory policy innovations derived from the experience of other countries, including nations with very different legal and cultural traditions. Law has become a critical part of efforts to combat global environmental problems and to improve living conditions in developing countries and, in particular, in Africa. The Convention on Access to Information, Public Participation in Decision Making and Access to Justice in Environmental Matters (Aarhus Convention), adopted under the auspices of the United Nations, Economic Commission for Europe during the Fourth Ministerial Conference "Environment for Europe" in Aarhus, Denmark, on June 25, 1998,* demonstrates the real engagement of parties to promote environmental law in the convention geographical area.

The Aarhus Convention is a new kind of environmental agreement that links environmental rights and human rights. The Convention establishes that sustainable development can be achieved only through the involvement of all stakeholders. The Convention links government accountability and environmental protection and focuses on interactions between the public and public authorities in a democratic context. It is forging a new process for public participation in the negotiation and implementation of international agreements. The Convention is not only an environmental agreement, but also a Convention about government accountability, transparency, and responsiveness (The United Nations, Economic Commission for Europe, 2000). The Convention goes further than any other convention in imposing clear obligations on Parties and public authorities toward the public as far as access to information, public participation, and access to justice are concerned (The United Nations, Economic Commission for Europe, 2000, p. 1).

Although regional in scope, the significance of the Aarhus Convention is global (Annan, 2000, p. 2). It is by far the most impressive elaboration of *Principle* 10 of the Rio Declaration, which stresses the need for citizens' participation in environmental issues and for access to information on the environment held by public authorities.

This chapter deals with the impact of the Aarhus Convention on chemicals management issues. Specifically, does this Convention have any potential or significant impact on the development of the sound chemicals management legislation in Africa? What are the concrete achievements of public authorities on the promotion of access to information, public participation in decision making, and access to justice in environmental matters in the region? What are the challenges and what recommendations can be made to improve access to information, public participation in decision making, and access to justice in environmental issues in the region? To what extent have the Aarhus principles been incorporated into international conventions and national law?

---

* http://www.unece.org/env/pp/documents/cep43e.pdf

OVERVIEW OF AFRICAN CHEMICALS LEGISLATIONS

Before independence, African countries were colonized by some occidental countries and therefore the politics of colonies were related to the politics of the colonial powers. After independence, several dictatorial regimes were established, many in 1960. This situation has seriously affected the relationship between public authorities and public interest groups. The role of public interest groups has been recognized at the international level and progressively integrated at the national level.

The point of the departure of the affirmation of the role of the public on environmental issues was the United Nations Conference on the Human Environment of 1972* (see Chapter 2 of this book). *Principle* 1 of the Stockholm Declaration states that:

> Man has the fundamental right to freedom, equality and adequate conditions of life, in an environment of a quality that permits a life of dignity and well-being, and he bears a solemn responsibility to protect and improve the environment for present and future generations.†

The first sentence of *Principle* 1 links environmental protection to human rights norms and raises environmental rights to the level of other human rights.

Paragraph 23 of the World Charter for Nature,‡ fully supported by African countries, further discusses public participation, while also stressing the importance of access to justice mechanisms:

> All persons, in accordance with their national legislation, shall have the opportunity to participate, individually or with others, in the formulation of decisions of direct concern to their environment, and shall have access to means of redress when their environment has suffered damage or degradation.

Also, paragraph 24 of the Charter states:

> Each person has a duty to act in accordance with the provisions of the present Charter; acting individually, in association with others or through participation in the political process, each person shall strive to ensure that the objectives and requirements of the present Charter are met.

This paragraph clearly states the individual obligation to protect the environment, which is concomitant to the enjoyment of a healthy environment.

Resolution 45/94 of the General Assembly§ recognized that:

> all individuals were entitled to live in an environment adequate for their health and well-being and called upon Member States and intergovernmental and non-governmental organizations dealing with environmental questions to enhance their efforts towards ensuring a better and healthier environment.

---

* The resolution 45/94 of the General Assembly, Stockholm, June 5–16, 1972, A/CONF.48/14, http://www.un.org/documents/ga/.../a45r094.htm

† http://www.unep.org/Documents.Multilingual/Default.asp?documentid=97&article

‡ UN General Assembly, Resolution 37/7 of October 28, 1982, http://treaties.un.org/doc/Treaties/1998/06/19980625%2008-35%20AM/Ch_XXVII_13p.pdf

§ General Assembly Resolution 45/94 of December 14, 1990, http://www.un.org/documents/ga/.../a45r094.htm

This resolution has a real impact on African environmental legislation. Several African countries have integrated the right to a healthier environment in their Constitution. The right to a healthier environment has been proclaimed by the African Charter on human rights (Article 24), as well as other African countries' Constitutions, such as Togo (Article 41 of the Fundamental law of October 14, 1992, which states "any citizen has the right to a healthier environment," Benin (Article 35 of the fundamental law), and Senegal (Article 27). Other countries have similar provisions.

The United Nations Conference on Environment and Development* (see Chapter 3 of this book) was another serious step on the road of sustainable environment management and, in particular, chemicals management issues in Africa. *Principle* 10 of the Rio Declaration on Environment and Development states clearly:

> Environmental issues are best handled with the participation of all concerned citizens, at the relevant level. At the national level, each individual shall have appropriate access to information concerning the environment that is held by public authorities, including information on hazardous materials and activities in their communities, and the opportunity to participate in decision-making processes. States shall facilitate and encourage public awareness and participation by making information widely available. Effective access to judicial and administrative proceedings, including redress and remedy, shall be provided.†

Agenda 21, the global action plan for sustainable development issued from the Rio Conference and fully supported by African countries, has clearly demonstrated from Chapters 19 to 22 the importance of access to information and public participation in hazardous chemicals and waste management. Chapter 20, paragraph 40(c) "encourage(s) institutionalization of communities' participation in planning and implementation procedures for solid waste management." Chapter 21, paragraph 25(d) also

> encourage(s) non-governmental organizations, community-based organisations and women's, youth and public interest group programmes, in collaboration with local municipal authorities, to mobilize community support for waste reuse and recycling through focused community-level campaigns.

Chapter 21, paragraph 46(c) invites governments to:

> launch campaigns to encourage active community participation involving women's and youth groups in the management of waste, particularly household waste.‡

Chapter 27, paragraph 9 invites governments and international agencies to institute flexible mechanisms to allow the participation of Civil Society Organisations (CSOs). All these important dispositions have been integrated to several international treaties ratified by African countries.

---

* Rio, June 3–14, 1992 (A/CONF.151/26), http://www.un.org/documents/ga/conf151/aconf15126-1annex1.htm

† http://www.unep.org/Documents.Multilingual/Default.asp?DocumentID=78&Article

‡ http://www.un.org/esa/dsd/agenda21/res_agenda21_21.shtml

The Convention on the Prior Informed Consent Procedure for Certain Hazardous Chemicals and Pesticides in International Trade (Rotterdam Convention)* (see Chapter 14 of this book) contains such obligations. Article 14, paragraph 1 states that each party shall facilitate the "provision of publicly available information on domestic regulatory actions relevant to the objectives of this convention." The same obligation is contained in Article 15, paragraph 2 of the Convention, which states that:

> each party shall ensure, to the extent practicable that the public has appropriate access to information on chemical handling and accident management and on alternatives that are safer for human health or the environment than the chemicals listed in Annex III.

The Convention on Persistent Organic Pollutant (Stockholm Convention)† (see Chapter 15 of this book) has also contributed to improve African countries' chemicals legislation. Article 10, paragraph b of the Convention states that "each party shall, within its capacities, promote and facilitate provision to the public of all available information on persistent organic pollutants." Article 13, paragraph 3 of the Convention on the Ban of the Import into Africa and the Control of Transboundary Movement and Management of Hazardous Wastes Within Africa (Bamako Convention)‡ also obliges African governments to set up information collection and dissemination mechanisms on hazardous wastes. Even if the process is very slow, the African countries are trying to implement and translate these international agreements into their national laws.

KEY ACHIEVEMENTS OF PUBLIC AUTHORITIES

Article 1 of the Aarhus Convention requires Parties to guarantee the rights of access to information, public participation in decision making, and access to justice in environmental matters in order to contribute to the protection of the right of every person of "present and future generations" to live in an environment adequate to his or her health and well-being. The Convention stands on three "pillars," (1) access to information, (2) public participation, and (3) access to justice.

The Access to Information Pillar in Africa

The access to information stands as the first of the pillars. It is the first in time, since effective public participation in decision making depends on full, accurate, up-to-date information. It can also stand alone, in the sense that the public may seek access to information for any purpose, not just to participate. The access to information

* Convention on the Prior Informed Consent Procedure (PIC) for Certain Hazardous Chemicals and Pesticides in International Trade, adopted on September 11, 1998 and entered into force on February 24, 2004, http://www.pic.int/en/ConventionText/ONU-GB.pdf

† The Convention on Persistent Organic Pollutant (Stockholm convention) adopted on May 22, 2001 in Stockholm and entered into force on May 17, 2004, http://chm.pops.int/Portals/0/Repository/convention_text/UNEP-POPS-COP-CONVTEXT-FULL.English.pdf

‡ Convention on the Ban of the Import into Africa and the Control of Transboundary Movement and Management of Hazardous Wastes within Africa adopted in Bamako on January 30, 1991, entered into force on March 10, 1999, http://www.africa-union.org/root/AU/Documents/Treaties/Text/hazardous wastes.pdf

is split into two parts. The first part is the right of the public to seek information from public authorities and the obligation of public authorities to provide information in response to a request. This type of access to information is called "passive," and is covered by Article 4. The second part of the information pillar concerns the right of the public to receive information and the obligation of authorities to collect and disseminate information of public interest without the need for a specific request (Article 5).

The Right to Access to Information

In Africa, more than 40% of the population is illiterate and cannot read or write French, English, or Portuguese. Illiteracy is a serious obstacle for the promotion of the right to access to information and in particular the access to information on chemicals. Very few people in Africa can seek information from public authorities on environmental issues and in particular on chemicals. No effort has been made to promote national languages, and if some are promoted, there are no environmental data available in national languages.

A survey conducted in Togo revealed that very few people were aware about persistent organic pollutant chemicals. Only 15% of one thousand farmers surveyed using or handling hazardous pesticides were aware of the hazards. The survey also revealed only 2% of the workers in contact with polychlorinated biphenyl (PCB) sources of release in Togo were informed about their existence and even less about their harmful effects on health and the environment.* This situation is a real obstacle to the exercise of the right to access to information, as people must be aware of the existence of a particular environmental or chemicals issue before seeking information from public authorities.

Furthermore, even if the relationship between civil society and public authorities has improved over the last 20 years as the continent has democratized, the right to access information tends not to be respected. Retrieval of public documents, for example, often is limited to those who have a personal relationship with the holder of such documents.

The Right to Receive Information

The right of the public to receive information and the obligation of authorities to collect and disseminate information of public interest is the second component of the "access to information."

This right is also proclaimed by several African environmental laws. The Togolese Framework-Law on Environment Management of 2008 states in its Article 30 that "the State shall take the provision to ensure the access to information of the citizens." Article 31, paragraph 1 states that "public authorities have the obligation to sensitize, educate and inform citizens on environmental problems." According to paragraph 2 of the same article, they shall do this to "allow the participation of the public to the environment management." However, this law seems to be in conflict with other

* ANCE-Togo, survey on POPS in Togo (2005) http://www.ancetogo.org/ance_anglais/pop_inventaire.html, National Implementation Plan of the Stockholm Convention, Version 03, February 2006; Survey on the Health Impact of POPS in Togo, Ministry of Environment, 2005.

public laws. Indeed, according to the public law in several African countries, the silence of a public authority to a special request in any matter should be considered as a refusal.* In other words, if a public authority can keep silent to a request of a citizen seeking information, it means that the public authority is not obliged to provide the information needed. This situation is a serious obstacle to the promotion of the right to receive information proclaimed by the Aarhus Convention.

Regarding the obligation of the authorities to collect and disseminate information of public interest, it is important to note that several African countries are trying to achieve this obligation in the framework of the implementation of several international chemicals agreements. Article 14, paragraph 3 of the Convention on the Prior Informed Consent Procedure for Certain Hazardous Chemicals and Pesticides in International Trade (Rotterdam Convention) declares most information as "public" and thus should not be considered as confidential.†

The information that cannot be considered as confidential concerns those referred to in Annexes I and IV, submitted pursuant to Articles 5 and 6, respectively, of the Convention. Annex I regulates the information requirements for the notification made pursuant to Article 5, which includes the properties, identification, and use of pesticides, such as common name, chemicals name according to an internationally recognized nomenclature, trade names and names of preparations, code numbers, information on hazard classification, use or uses of the chemical and physicochemical, toxicological, and ecotoxicological properties. Annex IV concerns the information and criteria for listing severely hazardous pesticide formulations in Annex III. Also, the information contained in the safety data sheet referred to in paragraph 4 of Article 13 are the expiration date of the chemical, information on precautionary measures, including hazard classification, the nature of risk, and the relevant safety advice, and the summary results of the toxicological and ecotoxicological tests. Finally, the production date of the chemical should not be considered as confidential according to the Convention.

Article 15, paragraph 1 obliges parties to establish national registers and databases including safety information for chemicals. Article 10, paragraph (b) of the Convention on Persistent Organic Pollutant (Stockholm Convention)‡ states that each party shall, within its capacities, promote and facilitate "provision to the public of all available information on persistent organic pollutants." Paragraph (c) states that each party shall promote the

> development and implementation, especially for women, children and the least educated, of educational and public awareness programmes on POPS, as well as on their health and environment effects and on their alternatives.

However, the real impact of these conventions at the national level is still problematic. Surveys conducted in several African countries demonstrated that there is a lack of information on chemicals in the region.

---

* Professor Vigon, Droit public fondamental en Afrique Francophone, Revue Scientifique de l'Université de Lomé, Mars 2007.

† Article 14, paragraph 3 of the PIC Convention, http://www.pic.int/en/ConventionText/ONU-GB.pdf

‡ Article 10, paragraph (b) of the Stockholm Convention, http://chm.pops.int/Portals/0/Repository/convention_text/UNEP-POPS-COP-CONVTEXT-FULL.English.pdf

Regarding hazardous pesticides, an independent survey conducted in Togo revealed that 56% of the farmers could not distinguish insecticides (to combat insects) from fungicides (to combat diseases caused by microscopic funguses); 72% knew the difference between insecticides and herbicides; and 95% were unaware of persistent organic pollutants (POPS) pesticides.[*] In Benin, 71% of the farmers were unaware of harmful effects of Endosulfan.[†] In Mali, another survey conducted in 2003 revealed 61% of intoxications related to the use of Endosulfan among women and 39% among men. Among the victims, 39% were between the ages of 5 and 18, 22% between 22 and 35, 22% between 45 and 56, and 17% between 63 and 65.[‡] They were also unaware that pesticides are used for other purposes. Indeed, pesticides such as Endosulfan, massively used in cotton production in some countries, have been found among market gardeners. In addition, many pesticides used could not be identified because they were put in unlabeled containers.[§]

The United Nations Directorial Principles for Consumers Protection[¶] recognize to each consumer the right to access information. Paragraph 12 of these Directorial Principles proclaims that each country shall take necessary measures to provide to consumers information on precautionary measures, including hazard classification, the nature of risk, and the relevant safety advice. Such information shall be provided to consumers through international symbols. The adoption of the Globally Harmonized System (GHS) by the United Nations Economic and Social Council (ECOSOC) in 2003 (see Chapter 11 of this book) was also guided by the need to provide to consumers suitable information on chemicals hazards. The Johannesburg Action Plan on Sustainable Development[**] invites countries to implement the GHS. The decision to establish a network for information exchange on chemicals during the Fourth Session of the International Forum on Chemicals Safety[††] aims to promote information exchange on chemicals at the international, regional, and national levels. In the framework of this initiative, a national network on information exchange is established in several African countries but very few of these networks are operational.

Regarding PCBs, some surveys conducted in the region have also clearly indicated that several users of PCBs equipment are unaware of their existence and workers handle fluids potentially contaminated by PCBs without any precaution or

---

* ANCE-Togo, Survey on the use of hazardous pesticides in Togo (2007), http://www.ancetogo.org/ance_anglais/etude_impact.html

† OBEPAB, Survey on the use of Endosulfan in Benin (2003), http://www.ejfoundation.org/page246.html

‡ PAN Africa, Annual report, 2003, http://www.pan-afrique.org

§ ANCE-Togo, Survey on Endosulfan (2008), http://www.ancetogo.org/ance_anglais/endosulfan.html

¶ The United Nations Guidelines for Consumers Protection adopted in June 1999, http://www.un.org/esa/sustdev/publications/consumption_en.pdf

** The World Summit on Sustainable Development (WSSD), Domain of Activity III, Paragraph 23 (c), adopted in Johannesburg (2002), http://www.un.org/esa/sustdev/documents/WSSD_POI_PD/English/POIChapter3.htm

†† International Forum on Chemicals Safety, Bangkok, November 1–7, 2003, Document IFCS/FORUM IV/ 3 INF, http://www.who.int/ifcs/documents/forums/forum4/en/11inf_en.pdf

safety.* According to the initial inventory of PCBs in Togo,† all the transformers containing PCBs do not include any security device aimed at avoiding contamination and therefore there is a serious risk for release to soil and exposure of human beings. There is also a dissemination of PCBs or PCB-containing devices in the environment. Some local companies still sell such equipment and this leads to inappropriate use in all sectors of activity: iron welding, transportation, arts and crafts, household work, and so on. PCB-contaminated oils are also used for the manufacturing of kitchen utensils or skin-care products without any notification. Furthermore, PCBs are used for other outdoor applications such us painting, fireproof protection, adhesive products, and coatings.

These surveys have clearly demonstrated that people are still unaware of chemicals issues in Africa and the public authorities do not play a part in dissemination of information, education, and awareness-raising. Some public authorities approached in Togo said they need financial support to allow the implementation of international agreements related to chemicals. Even if it is true that this financial assistance is necessary, we believe that African governments could implement some of their obligations related to access to information by using their national communications facilities such as television, newsletters, radio, and other public communication services. The lack of information on chemicals seriously compromises public participation in decision making in the region.

The Public Participation in Decision Making

The second pillar of the Aarhus Convention is the public participation in decision making. It relies upon the other two pillars for its effectiveness: the information pillar to ensure that the public can participate in an informed fashion, and the access to justice pillar to ensure that participation happens in reality and not just on paper. The public participation pillar is divided into three parts. The first part concerns participation by the public that may be affected by or is otherwise interested in decision making on a specific activity, and is covered by Article 6. The second part concerns the participation of the public in the development of plans, programs, and policies relating to the environment, and is covered by Article 7. Finally, Article 8 covers participation of the public in the preparation of laws, rules, and legally binding norms.

Several voluntary and legally binding international instruments approved by African countries recognized the important role of public participation in the elaboration, implementation, and monitoring of environmental regulations. The Rio Declaration (*Principle* 10) insists on the necessity of promoting the participation of the public at the national, regional, and international levels. The World Charter of Nature (*Principle* 23) recognizes the right to CSOs to be understood and to influence political decisions. This public participation must include the participation of women (*Principle* 20), youth (*Principle* 21), and local communities (*Principle* 22).

* ANCE-Togo, Survey on the impact of PCB on health and environment (2008), http://www.ancetogo.org/ance_anglais/launch_pcb_project.html

† Inventaire National des PCBs au Togo, Ministry of Environment (2005).

The Strategic Approach for International Chemicals Management (SAICM), also approved by African governments, sets forth strategies, policies, and plans of action to implement a visionary chemicals management goal that was initially adopted by heads of state at the 2002 WSSD in Johannesburg.* This goal is restated in the SAICM Overarching Policy Strategy document as follows:

> The overall objective of the Strategic Approach is to achieve the sound management of chemicals throughout their life-cycle so that, by 2020, chemicals are used and produced in ways that lead to the minimization of significant adverse effects on human health and the environment.† (See Chapter 17 of this book.)

This goal could not be achieved without the full participation of all stakeholders. That is why Mr. Koffi A. Annan, Secretary-General of the United Nations, has proclaimed:

> A key element in that quest is the strengthening of citizens' environmental rights so that members of the public and their representative organizations can play a full and active role in bringing about the changes in consumption and production patterns which are so urgently needed. The active engagement of civil society, both in the formulation on policies and in their implementation, is a prerequisite for meaningful progress towards sustainability. (Annan, 2000, p. V)

The Participation by the Public in Decision Making on a Specific Activity

The first part of public participation concerns the participation by the public that may be affected by or is otherwise interested in decision making on a specific activity. This kind of participation is covered by Article 6 of the Aarhus Convention.

The main domain of application of this kind of public participation is regulated by the Convention on Environmental Impact Assessment (EIA) in a Transboundary Context (Espoo Convention).‡ According to Article 2, paragraph 6 of the Espoo Convention,

> the party of origin shall provide an opportunity to the public in the areas likely to be affected to participate in relevant environmental impact assessment procedures regarding proposed activities and shall ensure that the opportunity provided to the public of the affected Party is equivalent to that provided to the public of the Party of the origin.

* See the WSSD, Johannesburg Plan of Implementation, paragraph 23 http://www.un.org/esa/sustdev/documents/WSSD_POI_PD/English/POIChapter3.htm

† Paragraph 13, SAICM Overarching Policy Strategy in http://www.chem.saicmunep.ch//standalone_txt.pdf

‡ Convention on EIA in a Transboundary Context, adopted on February 25, 1991 in Espoo and entered into force on September 10, 1997, http:/ec.europa.eu/environment/international_issues/pdf/agreements_en.pdf, also see the link:http://www.unep.org/dpdl/Law/Law_instruments/multilateral_instruments.asp

This obligation has also been formulated by several international conferences* and agreements.† Article 9, paragraph 1‡ of the Convention on the Transboundary Effects of the Industrial Accidents states that:

> parties shall ensure that adequate information is given to the public in the areas capable of being affected by an industrial accident arising out of a hazardous activity.

Article 2, paragraph 2 states:

> the party of origin shall, in accordance with the provision of this Convention and whenever possible, and appropriate, give the public in the areas capable of being affected an opportunity to participate in relevant procedures with the aim of making known its views and concerns on prevention and preparedness measures, and shall ensure that the opportunity given to the public of the affected Party is equivalent to that given to the public of the Party of origin.

EIA has been integrated into several national environmental laws in Africa. Section 2 of the Framework-Law on Environment Management of Togo and, in particular, Article 38 have proclaimed the obligation for EIA. Paragraph 2 of this article states that:

> this authorisation is provided on the base of an environmental impact assessment appreciating the negative or positive consequences on the environment which could generate the envisaged activities, projects, programmes or plans.

The ministerial decree N° 9469 MJEHP-DEEC of 28 November 2001 of the Ministry of Environment of Senegal regulating the organization and functioning of the Environment Technical Committee§ recommends the participation of the public in EIA. According to this decree, the Environment Technical Committee shall ensure that the public has been consulted and involved in any EIA activity.

---

* Conference on security and cooperation in Europe, International conference of Sofia on environment protection (October–November, 1989) (CSCE/SEM 36, November 2, 1989), http://www.unece.org/env/pp/implementation%20guide/.../part3.pdf, Ministerial Declaration on healthier and safe environment and the sustainable development in Asia and Pacific, Bangkok, October 16, 1990 (A/CONF.151/PC/38), http://www.unhchr.ch/environment/bp1.html. Paragraph 27 of this declaration confirms "the right of individuals and non-governmental organisations to be informed about environmental problems to which they are concerned, to access to necessary information, and to participate to the formulation and implementation of decisions susceptible to affect their environment." The Arab Declaration on environment and development and future perspectives, adopted by the Arab Ministerial Conference on environment and development (Cairo, September 1991), A/46/632, referred in the United Nations Document E/CN.4/Sub.2/1992/7, 20), http://www.unhchr.ch/environment/bp1.html

† *Principle* 23 de la Charte Mondiale de la nature, Article 4, Paragraph 1, de la convention des nations unies sur les changements climatiques, http://unfccc.int/resource/docs/convkp/conveng.pdf et la Section III de l'Agenda 21, http://www.un.org/esa/sustdev/documents/agenda21/english/agenda21toc.htm

‡ Convention on the Transboundary Effects of Industrial Accidents, adopted in Helsinki on March 17, 1992 and entered into force on April 19, 2000, http://www.ecolex.org/server2.php/libcat/docs/multilateral/en/TRE001143.txt

§ *Official Journal* N° 6025, pp. 795 and 796 of the Republic of Senegal of January 12, 2002.

Public participation in EIA is also proclaimed by several other national laws in Africa.* However, public participation in EIA is still very limited by the lack of information and expertise among African CSOs.

The Public Participation in the Development of Plans and Programs

The right to public participation in the development of action plans and programs is recognized by Article 7 of the Aarhus Convention. The implementation of this right is essential to achieving sustainable development defined as "development that meets the needs of the present without compromising the ability of future generations to meet their own needs."† The Rio Declaration's *Principle* 3 adds "the right to development must be fulfilled so as to equitably meet developmental and environmental needs of present and future generations" (see Chapter 3 of this book).

The right to public participation has been introduced into the Stockholm Convention. Article 10, paragraph (d) encourages parties to promote public participation in addressing POPS and their health and environmental effects and developing adequate responses, including opportunities for providing input at the national level regarding implementation of this Convention. Article 7, paragraph 2 of this Convention adds:

> the parties shall, where appropriate, cooperate directly or through global, regional and subregional organizations, and consult their national stakeholders, including women's groups and groups involved in the health of children, in order to facilitate the development, implementation and updating of their implementation plan.

IFCS IV (see Chapter 21 of this book) had also, with a special resolution,‡ invited countries to establish appropriate measures to allow the full participation of the public to the development of actions plans and programs.

In several African countries, this right has been introduced into national laws. Article 61 of the Togolese Forestry Code§ recognizes the need to associate public interest groups to achieve sustainable management of water, soils, and sites. The Ministerial decree of the ministry of environment of Togo regulating the role and the composition of the National Committee for Chemicals Safety¶ states that the Committee shall involve all stakeholders in preparation, elaboration, adoption, and implementation of chemicals programs and policies.

The Labour Code of Togo** has also established a technical committee called Safety and Health in Workplaces to be in charge of the homologation, importation,

---

* Decree n° 94-086/PRES promulgating the Law n° 002/94/ADP of January 10, 1994, the Law n° 002/94/ADP of January 19, 1994 promulgating the Environment Code, part 1 (2.32MB) of Burkina Faso, the Law n° 83-05 of January 28, 1983 relating to Environment Code l' Part 1 (2.33MB) and Part 1 (1.21 MB) of Senegal, http://www.unep.org/Padelia/publications/Comp1.htm

† The World Commission on Environment and Development in the Brundtland Report, Our Common Future, p. 6, http://www.wikilivres.info/wiki/index.php/Rapport_Brundtland_-_3

‡ Resolution IFCS/FORUM IV/ 3 INF, Fourth Forum on Chemicals Safety, http://www.who.int/ifcs/documents/forums/forum4/.../11inf_en.pdf

§ Togolese forestry code adopted on June 13, 2008, http://www.legitogo.tg

¶ The Ministerial decree N° 07/MERF of August 2001, http://www.legitogo.tg

** Labour Code N° 2006-010 of Togo of December 13, 2006, http://www.legitogo.tg

and exportation of hazardous substances. Article 43 of the Code invites this committee to involve the public interests groups in all programs of action. The Law regulating the genetic plant resources in Togo* in its Article 13 obliges the technical committee, in charge of the elaboration of programs of actions, to associate public interest groups. The same formulation is proclaimed in the national Law regulating medicines and pharmacy in Togo.† It is clear that this right is integrated into the legislations of African countries. However, the effective participation of the public in the preparation of legally binding instruments seems to be problematic.

Public Participation in the Preparation of Laws, Rules, and Legally Binding Norms

The right to public participation in the preparation of laws, rules, and legally binding norms is proclaimed by Article 8 of the Aarhus Convention. In several African laws, this right is not as clearly stated as it is in Article 8. This right can be considered as contained in the general "right to public participation."

In several African countries, the initiative of a law belongs both to parliamentarians and Government. Article 83 of the Togolese fundamental Law clearly states that *the initiatives of laws belong both to parliamentarians and Government.* Usually, when a law is initiated by the Government, its preparation is done through ministries and then there is a possibility to associate public interest groups. Another possibility for the public to be involved in the preparation of laws, rules, and legally binding norms is when it is a ministerial decree, which regulates a specific aspect. Usually the preparation of such regulations comes from a technical committee of which public interest groups are members. However, this participation is not automatic and there are many regulations to which the public is not associated.

The situation is more complicated when a law is initiated by the Parliament. The general procedure at this case is that the draft regulation is sent to the government for comments and inputs; this work is done through a very closed ministerial or interministerial committee to which generally the public is not associated.

To allow the full participation of the public in the preparation of legally binding instruments on sound chemicals management in Africa, it is urgent to state clearly this right in national laws.

The Right to Access to Justice

The third pillar of the Aarhus Convention is the access to justice pillar. It enforces both the information and the participation pillars in domestic legal systems, and strengthens enforcement of domestic environmental law. It is covered by Article 9. Specific provisions in Article 9 enforce the provisions of the Convention that convey rights onto members of the public. The justice pillar also provides a mechanism for the public to enforce environmental law directly. The right to access to justice has also been

* Law of 3 July 1996, No. 96-007/PR regulating the Genetic Plant Resources in Togo, http://www.legitogo.tg

† Law N° 2001-002 of January 23, 2001 regulating medicines and pharmacy in Togo, http://www.legitogo.tg

recognized by paragraph 37 (d) of the United Nations Principles for consumers' protection, which states that each government shall ensure that national laws recognizing the right of each abused consumer to be compensated are promulgated and enforced. Article 19, paragraph 5 of the Basel Convention also states that "each party shall introduce appropriate national/domestic legislation to prevent and punish illegal traffic."

Access to justice under the Convention means that the public has the ability to go to court or another independent and impartial review body to ask for review of potential violations of the Convention. The access to justice provisions provide the right to individuals and public interest groups to go to court or another review body.

In several African countries, there are no specific provisions for access to justice in environmental matters. However, it is still possible for an individual or organization to go to court by using general laws such as the right to a healthier environment. Concerning the question of the nature of the right to a healthy environment, the Supreme Court of the Philippines has said:

> Although the rights to a decent environment and to health were formulated as State policies, that is, imposing upon the State a solemn obligation to preserve the environment, such policies manifest individual rights not less important than the civil and political rights enumerated under the Bill of Rights of the Constitution.* (La Vina, 1994, p. 246)

In some European countries, other cases considering the existence of a right to a healthy environment can be found in Belgium† and Slovenia. The Constitutional Court of Slovenia‡ recognized a legal interest of individuals on the basis of the constitutional right to a healthy living environment "to prevent actions damaging the environment." Similar cases have been brought under Article 8 of the Convention for the Protection of Human Rights and Fundamental Freedoms.§

The penal code also provides some possibilities of action to the court on environmental matters and, in particular, on chemical matters in Africa. The illicit trade of pesticides, for example, can be reprimanded by penal code provisions. However, access to justice in environmental matters has not really improved in the region. For example, only one case related to water pollution by hazardous chemicals by a company called the International Fertiliser Group (IFG)¶ has been introduced to the court of Lomé by an environmental protection group in 2006.

The lack of information of public interest groups and the lack of expertise and financial resources are serious obstacles for improving the right to access to justice in environmental matters in Africa.

---

* Minors OPOSA v. Sec'y of the Department of Environment and Natural Resources, 33 ILM 168 (1994). This case is described in A. G. M. La Vina (1994).

† Pres. Trib. First Inst. Antwerp, Decision of April 20, 1999.

‡ Constitutional Court of Slovenia, Dec. No. U-I-30/95-26, 1/15-1996, also see Milada Mirkovic, Legal and Institutional Practices of Public Participation: Slovenia, in Doors to Democracy: Current Trends and Practices in Public Participation in Environmental Decision-making in Central and Eastern Europe (Szentendre, REC, June 1998), http://www.unece.org/env/pp/implementation%20guide/.../part3.pdf

§ November 4, 1950, 213 U.N.T.S. 222, as amended. Article 8, titled Right to Respect for Private and Family Life, http://www.law-lib.utoronto.ca/.../Final_Paper_Women's%20International%20Human%20Rights_Howard

¶ Tribunal of Lomé, case ATA c IKG, Rev. Palais 2008, p. 33.

CHALLENGES

The following challenges remain to promote the effective implementation of the right to access to information, public participation, and access to justice in environmental matters in Africa and in many other regions.

CREATE AND MAINTAIN AN INFORMATION/KNOWLEDGE BASE

Certain chemicals-related information/data are vital for implementing and enforcing many aspects of chemicals legislation. The Aarhus Convention has proposed the development of systems of pollution inventories or registers. These mechanisms for information are covered by Article 5, paragraph 9. The Convention takes an affirmative approach to the development of such systems of inventories or registers. National legislation should provide the necessary legal basis for obtaining such data. Legal provisions requiring the automatic reporting to government of new information regarding substantial or significant risks, incidents of poisonings, pollutant releases and transfers, and so on should also be considered. Opportunities to enforce and control compliance with chemicals legislation can be expected to increase once record-keeping and registering practices are established or strengthened. Also, effective flow of hazard and chemical safety information on chemicals and chemical products from suppliers to end users is important for reducing risks and should be embedded in national legislation. Some countries, with the technical assistance of United Nations Institute for Training and Research (UNITAR), are establishing national pollutants and transfer registers. This kind of information system, if it is done in a transparent and participatory process, with the involvement of all stakeholders, will really improve the knowledge and information systems in the region.

PROMOTE NATIONAL AND REGIONAL CHEMICALS POLICY REFORM

African chemicals laws need real improvement to achieve sustainable development. The European Union (EU) has finalized the adoption of new chemicals legislation called REACH* (Registration, Evaluation, Authorisation and Restriction of Chemicals) (see Chapter 44 of this book) whose preamble states: the European Union is aiming to achieve that, by 2020, chemicals are produced and used in ways that lead to the minimization of significant adverse effects on human health and the environment.

This makes REACH one of the first governmental efforts to reform national law, policy, and practice toward the aim of achieving the chemical safety goals agreed upon in 2002 at WSSD and elaborated in SAICM.† Under REACH, chemical producers and importers who wish to market a chemical for use in the EU will be

* REACH will enter force later in 2007 and its operational requirements will start to be applied in 2008; REACH press release from the European Union dated December 13, 2005. http://europa.eu/rapid/pressReleasesAction.do?reference=IP/05/1583&format=HTML&aged=0&language=EN&guiLanguage=en; the text of the REACH legislation can be found at: http://register.consilium.europa.eu/pdf/en/06/st07/st07524.en06.pdf#search=%22%22no%20data%20no%20market%22%20EU%20REACH%22

† The Strategic Approach to International Chemicals Management, how IPEN and civil society can contribute to its implementation, an IPEN discussion paper (2006), http://www.ipen.og

required first to generate and make available data on the chemical's properties, including its hazardous characteristics. They will also be required to make available information on the chemical's uses and safe ways of handling. The REACH legislation has proclaimed several chemicals principles such "no data, no market," the right to know, and other fundamental principles. In this context, information on chemicals relating to the health and safety of humans and the environment should not be regarded as confidential. Additionally, as called for in the Kiev Protocol of the Aarhus Convention on Access to Information (see also Chapter 13 of this book), information on pollutant releases and transfers should be freely available to the public. The adoption of a regional chemicals law similar to REACH in Africa is highly recommended for a sound chemicals management in the region.

To Promote Other Chemicals Information Systems

The obligation to educate and inform the public could be done by African governments by establishing a national Web site on chemicals that could be updated daily. The government of Togo, with the financial support of the United Nations Development Programme (UNDP), has established a Web site containing a database of all national laws published from 1960 onward including chemicals legislation.* This database has also been produced in a CD-ROM and in a printed version. The Aarhus Convention also gives special attention to new forms of information, including electronic information. This is referred to in the preamble and in Article 3 on the general provisions and in Articles 4 and 5 on access to information. The Convention takes into account the changing information technology, which is moving toward electronic forms of information, and the ability to transfer information over the Internet and other systems (The United Nations, Economic Commission for Europe, 2000, p. 7). This kind of information system will contribute to improve access to information, public participation, and access to justice in environmental matters and, in particular, in chemicals issues.

CONCLUSION

The right to access to information, public participation, and justice in environmental matters and, in particular, in chemicals issues needs improvement in the African region to achieve the SAICM overall objective, which is

> to achieve the sound management of chemicals throughout their life cycle so that, by 2020, chemicals are used and produced in ways that lead to the minimization of significant adverse effects on human health and the environment.†

Although regional in scope, the significance of the Aarhus Convention has reached the African continent. It is by far the most impressive elaboration of *Principle* 10 of the Rio Declaration, which stresses the need for citizens' participation in

* See http://www.legitogo.org

† Paragraph 13, SAICM Overarching Policy Strategy in http://www.chem.saicmunep.ch//standalone_txt.pdf

environmental issues and for access to information on the environment held by public authorities. African countries are trying to establish appropriate measures to promote education and awareness for the public on chemicals issues, but they need technical and financial resources to reform their chemicals legislation to meet the Aarhus Convention obligations and guidelines.

REFERENCES

Annan, K. A. 2000. Secretary-General of the United Nations, Foreword, p. V, *The Aarhus Convention: An Implementation Guide*. The United Nations. Economic Commission for Europe, New York and Geneva: United Nations, ECE/CEP/72, http://www.unece.org/env/pp/implementation%20guide/.../part1.pdf

The United Nations. 2000. Economic Commission for Europe, *The Aarhus Convention: An Implementation Guide.* New York and Geneva: United Nations. ECE/CEP/72, p. 1.

Vina, A. G. M. L. 1994. The right to a sound environment in the Philippines: The significance of the Minors Oposa case. *RECIEL*, 3(4): 246, http://www.unece.org/env/pp/implementation%20guide/.../part3.pdf

8 The Basel Convention—A Promising Future

Pierre Portas

CONTENTS

The Basel Convention and its sister chemical conventions and protocols represent the foundation upon which to build and consolidate an adequate global response for protecting human health and the environment from the adverse effects of chemicals and waste and for planting the seeds to manage future emerging problems soundly and safely on a sustainable basis.

PREAMBLE

Global environmental issues are becoming more and more complex and are evolving rapidly. They are interweaving profoundly and permanently with social situations and economic processes. As a consequence, it is no more possible nor plausible to dissociate the human world from the natural world.

While critical global environmental issues are being treated separately for operational reasons and necessity, it is crucial to elaborate cross-sectoral policies to ensure that no one global issue is being left behind with the consequence of potentially undermining efforts devoted to the "top" issues.

The international community has both the capacity and the means to give equitable treatment to the most important global environmental issues which are intermingled, namely issues concerning climate change, biodiversity loss, unsustainable use and management of natural resources (land erosion, deforestation, over-fishing, etc.), chemical poisoning and waste. By engaging international efforts in dealing with all these issues in a more balanced way, without questioning the necessity to further increase efforts devoted to climate change issues, it would help bridge gaps between nations, between different levels of economic development and would trigger more focus on social vulnerability to a wide range of anthropogenic pollution and contamination.

We need to go into a higher level of prospective to confront our current way of acquiring knowledge compared to tomorrow's needs. Political and industry leaders have the opportunity to launch ambitious projects that would provide breathing space for hope and enthusiasm. It would require a new deal in international cooperation and a profound transformation of the relationship between politics, economy and society.

A new way of protecting the environment would be desirable that fosters a sharing of responsibilities, know-how and experience capable of initiating, accelerating, accompanying, facilitating or supporting changes in national and global policies. The prevention of pollution should become a societal objective; how the society in her wholeness can decide to prevent or limit the cause(s) of such pollution and mitigate its effects. This would imply new institutional arrangements to enable the diverse public and private actors to address the issue jointly and decide on specific measures that would enhance the society's ability to release the intellectual, financial and institutional resources required to solve the problems and prevent further occurrence of avoidable harms.

INTRODUCTION

The Basel Convention on the Control of Transboundary Movements of Hazardous Wastes and Their Disposal (1989) entered into force in 1992. The Convention has 175 (2011) Parties. Its main purpose is to minimize both the quantity and hazard potential of waste, to reduce its transboundary movements to a minimum and to treat and dispose of such waste as close as possible to its source of generation. The underlying principle of the Convention is the environmentally sound management of the waste. Its operational arm is a global control system of export, transit and import. The Conference of the Parties has established 14 regional and coordinating centers in Africa, Asia, Europe, Latin America and the Caribbean and in Eastern Europe to facilitate the implementation of the Convention.*

The Basel Convention is at the core of the international architecture designed to aim at a world with less hazardous waste. The Convention is the backbone of a global initiative taken 20 years ago to reduce and then eliminate the adverse effects of the generation and management of hazardous waste on human health and the environment. Through its implementation it has routed the concept of environmentally sound management (ESM) of waste into the reality of trade, economic development, social equity, environmental protection, and international cooperation.

* http://www.basel.int

The fast changing political, economic, and social environment coupled with an increase in energy price, natural resources, a redefinition of the place of agriculture in development, and the consequences of major in-depth disturbances of climate systems, as well as financial constraints and the diversity of strategies used by market players, pose new challenges for the governance of the Basel Convention as a pillar of international environmental governance.

Parties to the Convention have tools in their hands that they could use to respond to these challenges, address emerging difficulties, and build workable and sustainable solutions. The Convention provides a comprehensive policy framework. Through the Conference of Parties' work, the required policy tools have been elaborated. It concerns the principles of harmonization of law and regulations, the promotion of best practices, the exchange of information, the facilitation of the transfer of environmentally sound technologies, the reduction of gaps or deficiencies in the international rules and procedures regarding transport of waste, the design of a strategic framework, the development of partnership initiatives or the promotion of level-playing fields in hazardous waste management, and recycling at the regional and global level.

The Parties have recognized that, when dealing with any kind of waste, the precautionary principle, the principle that the polluter pays and for the generator to engage his or her extended individual producer responsibility are important principles to ensure the sound and safe management of hazardous waste.

A NORMATIVE INSTRUMENT

The control system and the ESM principles are the principal pillars of the Basel Convention. Since 1999, the Parties have revigorated the ESM pillar to respond to the need for effective implementation. As a normative instrument, the Convention generates universal norms for the sound and safe management of hazardous waste and other wastes. Its implementation is guided by the concept of the waste hierarchy where minimization is the preferred option and final disposal the last choice. In-between, reuse, recycling, and recovery are recommended as preferable to landfilling or incineration.

Four main policy directions have been articulated by the Parties as a roadmap toward ensuring effective and efficient implementation. It concerns waste minimization, life-cycle approach to materials, integrated waste management (to separate hazardous waste from other solid nonhazardous waste), and regional approach. These policy directions guide the implementation of the Strategic Plan until 2010. A new strategic framework is under preparation to cover the next 10 years.

SEARCH FOR SOUND GOVERNANCE

Many issues remain to be clarified to aim at achieving ESM, in particular:

- What are the capacities in the world to manage waste in an environmentally sound and efficient way today and tomorrow?
- Can the costs of waste management be internalized into prices of consumer goods and of waste management services to promote waste minimization, prevention, and avoidance, and to move toward life-cycle management of

harmful chemicals and products and extended or individual producer responsibility?

- What are the needs for international cooperation to meet the demand for ESM?
- Do we know how much waste (municipal, industrial, hazardous, agricultural, demolition and construction, and radioactive) is being produced, its characteristics, and by whom, and what is happening to this waste?

As it stands today, the development of a global strategy for the prevention of environmental risks and damages arising from waste and hazardous waste suffers from a lack of or insufficient data, immature policy comprehensiveness, insufficient monitoring of waste flows and related management practices, lack of understanding of the effects of waste management on human health and the environment, imprecise knowledge on emerging trends, lack of level-playing field, and inadequate international cooperation.

Within recurrent uncertainties regarding waste issues worldwide, the Basel Convention provides certainty and stability. Although attempts to drift away from its robust control by some economic actors, because it is perceived as an obstacle to free trade, are occurring on a regular basis, the Convention has demonstrated its resilience, solidity, and relevance.

Such resistance results from the fact that the world cannot do without it. There is no other global instrument that controls transboundary movements of hazardous and other problematic waste. ESM norms used worldwide have been generated through the Basel Convention. The Convention has an in-build capacity to adapt to new scientific and technological changes through the revision of its technical Annexes.

In a world of growing complexities and uncertainties, there is a need for more transparency in waste flows and practices as a mean to protect human health and the environment. Economic globalization results in a globalized trade of waste. This is the reason why it is important to strengthen the Basel Convention as the international legal and normative instrument providing all countries in the world with the required stability, comprehensiveness, and transparency and predictability necessary to deal with waste, especially hazardous waste, in a way to protect human health and the environment, and to seize opportunities to transform waste into resources.

The Waste-Chemical Nexus

The Basel Convention is an important component of the international rules and procedures put in place to address the entire life cycle of harmful chemicals. It constitutes with the Stockholm Convention on Persistent Organic Pollutants (POPS) (see Chapter 15 of this book), the Rotterdam Convention on Prior Informed Consent procedure (PIC) (see Chapter 14 of this book), SAICM (see Chapter 17 of this book), the Montreal Protocol on ozone-depleting substances (see Chapter 16 of this book), the IMO/MARPOL Convention* controlling discharges of ship's

* http://www.imo.org/About/Conventions/ListOfConventions/Pages/International-Convention-for-the-Prevention-of-Pollution-from-Ships-(MARPOL).aspx

operational residues, the London Convention 1972* on the prevention of dumping of waste at sea and the future Mercury convention a set of comprehensive normative and guiding instruments that are unique. Their cooperation is essential to strengthen the implementation of each treaty on the ground, to fill gaps, to ensure efficiency and effectiveness and improve the protection of both the marine and terrestrial environment. Such architecture is further deepened through the implementation of regional environmental agreements such as the Bamako (Africa) or Wangani (Pacific) Conventions.

In particular, Parties to the Basel, Rotterdam, and Stockholm Conventions have approved measures to enhance cooperation and collaboration among the three instruments. This process represents one of the few concrete attempts to put into action a long overdue change in the way the intergovernmental institutions address environmental issues.

The simultaneous extraordinary meetings of the Conference of Parties (ExCOPs) to the Basel, Rotterdam and Stockholm Conventions were held 22–24 February 2010 in Nusa Dua, Bali, Indonesia. At the ExCOPs, delegates adopted an omnibus synergies decision on joint services, joint activities, synchronization of the budget cycles, joint audits and joint managerial functions. This joint meeting heralded a new era of multilateralism with constructive implications for the ongoing international environmental governance debate within the United Nations.†

Sound global environmental governance requires that efforts and resources are invested in building regional level-playing field as a base for aiming at a global level-playing field. What does this mean? It implies that governments, with other stakeholders such as industry, business, municipalities, or civil society, work together to design workable and sound systems supported by adequate norms and procedures. For instance, the waste dimension should be taken into account when deciding on land-use management, proposing energy shifts, developing infrastructures, and selecting technologies. Failing to do that may lead to displacing environmental problems to the weaker part of the chain.

In this context, it is worth mentioning the existence of regional waste conventions or protocols such as the Bamako Convention for Africa, the Waigani Convention for the Pacific, or the Izmir Protocol for the Mediterranean Sea. Over the years, interregional cooperation has remained at a low level. Their implementation is often problematic. The Bamako Convention, for example, lacks political support while the Izmir Protocol is not endorsed by key countries in the Mediterranean. Overall, there would be a need to review the situation of regional conventions and protocols, especially those experiencing difficulties, to see how best to support them and make them operational. Clearly, enhanced cooperation with the global instruments would seem both necessary and logical.

It also makes sense not to dissociate environmental issues when drafting strategies or plans. Issues on waste are transverse. They concern all sectors of society and every individual. From that point of view, waste issues are both global and universal. Improved cooperation among the waste and chemical conventions will

* http://www.londonprotocol.imo.org

† http://www.basel.int/meetings/meetings.html

help countries to organize themselves better in putting in place the appropriate institutional architecture to implement the conventions in a more efficient and effective manner.

The combination of domestic, regional, and international efforts is central to improving global environmental governance but is not sufficient. Indeed, different types of partnerships should be considered (e.g., public–private, region to region, and land–sea interface). These partnerships are critical to address the entire life cycle of harmful chemicals from design, production, commercialization, reuse, recycling, and recovery to final disposal.

WHERE ARE SOME OF THE STUMBLING BLOCKS?

There is a significant increase in the recycling, recovery, and reuse of waste worldwide with a corresponding accelerated development of a global and intraregional trade in recyclables and of used or end-of-life equipment. Environmentally sound recycling can make a positive contribution to sustainable development in terms of reducing pressure on virgin materials, safeguarding landscapes from expanding mining activities or by reducing environmental problems and risks and economic costs associated with the disposal of waste. Sound recycling across national boundaries can bring environmental benefits when, for instance, there are economies of scale so that a number of companies in different countries can share a facility, avoiding use of low-standard technologies, final disposal, or long-distance shipments.

There is, however, a negative face to the economic boom for recyclables. Countries are at different levels of economic development, and recycling facilities operate at different standards depending on the country. There is no level-playing field at the regional or global level leaving room for unscrupulous operators to do their business.

Electronic waste, for instance, is one of the fastest growing waste stream in the world: it represents both a high asset and a big problem. The quick economic gain from exporting or importing electronic wastes too often overshadows its potential harm. The high quantities of electronic waste exported to Asia or Africa are overwhelming importing countries' capacity to deal properly with this waste. A sizable part of local recycling is done in the informal sector where recycling of electronic waste takes place with dramatic consequences on human health and the environment.

Many governments, because of uncertainties in the characterization of electronic waste, are reluctant to impose the Basel Convention's strict control procedure on trade of hazardous electronic waste, in particular where such trade brings in revenues and generates jobs.

Globally, more and more governments give priority to recycling strategies. As a consequence, several governments are working toward reducing barriers to trade in recyclables or recycled materials and encouraging reuse and recycling of materials. This current trend toward establishing a loop for electronic waste and other recyclables has the consequence of increasing the international flow of used or end-of-life products, part of which is illegal or carried out on the fringe of law.

The growing production of chemicals is one of the main contributors to the increase in hazardous waste. More and more of hazardous chemicals find their way into products, and these products, in turn, become hazardous waste at the time of disposal. The proliferation of chemicals means that among the used or end-of-life electronic equipment exported to Asia or Africa, for instance, the chances of finding electronic hazardous waste or components such as cathode ray tubes with lead-containing glass, printed circuit boards with heavy metals, fluorescent tubes (from crystal displays) with mercury, nickel–cadmium batteries, or plastic components with brominated flame retardants are highly likely.

Rapidly developing international and regional recycling schemes should be combined with mechanisms capable of providing information about the potential hazardous implications, understanding and monitoring such schemes to ensure their accountability, and soundness from an environmental, health, and economic perspective.

The globalization of trade can have the consequence, in the absence of comprehensive and coherent rules and regulations or due to lack of implementation of these rules and regulations, to make the waste follow the path of least resistance. A large number of obsolete ships on their last voyage will be recycled in yards located in Asia, many of which use techniques such as beaching that are unsound and unsafe. The rolling steel recovered from the recycling of old ships is essential for the development of the countries infrastructures and represent a major economic sector generating revenues and jobs. However, the conditions under which such recycling is done are often dramatic for the workers and for the environment exposing them to a mix of very hazardous materials, including asbestos, heavy metals, and complex combustion products.

Another example of the adverse effects of globalization is some of the indirect consequences of the trade in used tires for recycling. When used tires are transported, they may contain water that cannot be emptied from the tires. In tropical zones, such tires become breeding ground for mosquitoes responsible for infectious diseases such as malaria or Chikunguniya. International transport of used tires may contribute to the spreading of infectious diseases worldwide, a situation that may be aggravated with changes in climate conditions.

In all these situations, the Basel Convention has a significant role and should be applied. If no serious attempts are made to better regulate the trade in hazardous electronic waste, it is the control system of the Basel Convention that may become irrelevant. The Convention is very clear in legal terms but Parties do argue on definition and classification. It is therefore quite urgent for Parties to agree on what sort of electronic waste or parts are to be controlled under the Convention and what should be left out of its regulatory requirements.

Similarly, although it may not be economically sound to stop the trade in used tires, the precautionary principle, that is part of the ESM of waste, should be exercised in a more systematic way to protect people from infectious diseases. When it comes to ship recycling, there are two factors to consider. One relates to the importance to implement the IMO Convention for the Safe and Environmentally Sound Recycling of Ships which was adopted in May 2009 and when necessary, to accompany such implementation with the ESM principles adopted by the Basel Convention. Failing to do that, efforts to reach a level-playing field between those operating at sea and on land may be undermined. The other point concerns the necessity to properly manage at shore the waste resulting from both the dismantling of the ship and the recycling of its parts. The issue of ship recycling is exemplary of the essential dimension of sustainable development where the environmental, social, and economic dimension should be enhanced together.

A BETTER FUTURE BUILT ON ESM NORMS

Waste generation and management have the capability to impact environment and health adversely everywhere and at any time. This is the reason why ESM embraces the "all waste concept," requires constant improvement of practices, and serves as a catalyst for international cooperation. Everyone is exposed to waste but not in an equal manner. Children are very vulnerable as well as people surviving in poverty. ESM therefore carries a social equity component and a drive for modernization of an international architecture conducive of making waste an asset (keeping in mind that a priority is to minimize the production of waste) instead of a liability.

The concept of ESM has been elaborated in the 1980s. The ESM definition can be summarized as taking all practical steps to protect human health and the environment from the adverse effects of waste. Because of the cross-border nature of the trade in waste, national standards will not be enough. In order to bring consistency to environmental standards and best practices among countries, a global, or at least regional, level-playing field should be achieved. ESM provides a solid basis for reaching such a level-playing field.

The ESM of waste represents a significant contribution to sustainable development while the improper handling of waste may act against developmental goals. Broadening the resource base for the ESM of waste at the international level means providing a solid and lasting support to the efforts of the international community to conserve natural resources and energy, to protect biological diversity, to reduce the impacts of climate change, to reduce pollution and contamination of waters and soils, to improve sanitation and food security, to provide economic opportunities, to transfer cleaner technologies, to promote technological innovation, and to reduce risks from chemicals.

Minimization, prevention, and the sound recycling and management of waste are the means to protect human health and the environment while creating social values and economic opportunities.

ESM strategies can help people acquire working skills and decent jobs, especially for those living in poor areas. It could be achieved, for instance, by promoting safe activities geared toward improving collection, segregation of municipal waste, or dismantling of end-of-life equipment or postconsumer goods, in particular obsolete electrical and electronic equipment. Social values and economic opportunities created through ESM strategies will provide incentives for children living in poor conditions to go to school and be taken away from indecent activities.

It will also provide stimuli to stop uncontrolled disposal of household waste and other garbage or residues, such as backyard burning or littering that is a source of disease and contamination of the environment. Separate collection of certain waste, such as organic waste from households or wood waste from gardens or parks could be promoted with a view to using them as a source of energy, under proper control; it could expand the use of renewable energy for heat or power supply.

Sound waste management such as reducing the volume of municipal waste and diverting biodegradable waste away from landfills is now recognized as contributing significantly to reducing emissions of greenhouse gases. However, the increase in waste quantities being treated in facilities as well as an expanding transport for collecting waste will increase energy consumption and greenhouse gas emissions.

Overall, a better management of municipal waste and using waste as a resource, in the context of the application of the ESM principles, will help in decoupling environmental pressure from economic growth. It will, however, not be sufficient to stop the continuing growth in the quantities of waste generated nationally and globally. Other approaches, such as cradle-to-cradle, are needed to reduce quantities of waste and increase recyclability.

A promising development is intimately linked to a sustained development of and access to investments in building ESM facilities, in developing new waste management capacity to keep up with the increased production of waste, in innovations to reduce waste volumes, avoiding harmful chemicals in products (waste avoidance) and hazard potential and in facilitating the access to and transfer of sound, safe, and proven technologies.

ARTICULATING A COMPREHENSIVE GLOBAL ESM SCHEME OR MOVING FROM WASTE TO RESOURCES

There is a pressing need to reinforce or strengthen international instruments of relevance to the protection of the environment and human health and to ensure sufficient coherence among these instruments, especially the waste and chemicals conventions. But this is not enough. A more proactive determination is necessary in addressing global environmental challenges; achieving ESM objectives through strategic sound policy, adequate rules, and procedures and public–private partnerships.

It has become evident that not a single country can deal properly with the waste generated domestically or imported. No one country is self-sufficient in waste management. Trade in waste and hazardous waste is motivated by three factors. One is that a country needs to export waste because it does not have facilities to treat or dispose of certain types of materials. Another factor is related to the economic value of certain waste materials and recyclables in the macroeconomic context.

There is a third factor. It concerns illegal traffic and unscrupulous trade in waste. Waste follows the path of least resistance in the absence of safeguards and traffic lights. Organized crime is heavily involved in such trades that have become global. It represents a high danger for people, the environment, and fair trade.

Implementation of rules and procedures as well as market transactions will benefit from more certainty. This is the reason why new approaches that would consolidate the implementation of the Basel Convention could be designed to achieve an improved:

- Transparency in the export and import of waste,
- Traceability of the waste materials, including their final disposal,

- Predictability in transactions to ensure waste is properly handled,
- Certainty in the legal coverage, if any, of the waste traded.

Sound policies, adequate legal and regulatory framework, fair competition, and international cooperation are essential to enable recycling and other waste activities to contribute to environmental protection or to reduce the introduction of harmful chemicals in products.

Efficient recycling is dependent, to a large degree, on the possibility to trade recyclables internationally because no one country possesses the skills, capacity, or infrastructure to reuse, recycle, or recover the immense variety of recyclable materials. As a consequence, the market is driving recyclables across borders faster than the development of policies, safeguards, and legislation. In turn, such dichotomy is at the source of many difficulties encountered today, which are exacerbated by the increase in the globalization of trade in waste.

Economic actors have set the scene regarding the shape of the international trade in recyclables. Policy and legislation are behind and, internationally, a well organized coherence of action is still to be put in place. There is currently no level-playing field at the global level. And tools to ESM principles worldwide to reduce the potential negative effects of trade of recyclables, including the presence of harmful chemicals, on human health and the environment, are not sufficiently well developed.

In some countries, export of waste for reuse, recovery, or recycling represents one of the major foreign trade sectors. Such economic drivers have the consequence of a more pressing demand for fair trade in recyclables and recycled materials. This in turn will impact on the macroeconomy of trade for these materials.

Currently, there is no satisfactory situation because not all the safeguards are in place to avoid the economic recycling activities that may pollute the environment, impact negatively on human health, or generate unscrupulous dealings and illicit traffics on a worldwide scale. Moving from waste to resources will not, by itself, be sufficient to avoid the undesirable effects of improper recycling nor will it eliminate the hazardous properties of certain waste streams and the need for taking all precautionary measures.

It is therefore important and pressing to design an effective international policy architecture supported by adequate norms within which market forces will operate in a fair way. Such global framework should be conducive to improving environmental protection, enhancing cooperation between the Basel Convention and the chemicals conventions, programs or initiatives, driving innovations, and reducing business risks.

To progress in the design of coherent ESM-derived tools, it is important to respond to several issues, in particular

- How to eliminate unsound practices that cause health hazards and environmental damages?
- How to introduce environmentally friendly processes or technologies that minimize the quantity and hazard potential of waste?
- How to improve energy efficiency, resource productivity, natural resources conservation, and material recycling or recovery?

- How to ensure a sustainable trade in recyclables that is environmentally friendly?
- How to accompany the transition to a society which is less resource demanding?

Urgency

Globally, there is urgency in consolidating and reinforcing the international public environmental legal architecture controlling transboundary movements of waste, especially hazardous waste, and providing the necessary support to implement adequate norms for their ESM.

Several tools could be considered as useful. The following tools are complementary and should not be seen in isolation. They represent a set of proactive measures that both public and private stakeholders could develop together to ensure that their design corresponds to real needs and that their implementation is feasible and workable. In no circumstances these tools should be undermining the control system of the Basel Convention, be used to circumvent some of its legal provisions or promote transboundary movements of hazardous waste for final disposal.

International ESM Standard

The development of an international ESM standard could serve two distinct purposes, namely

1. It could provide a management process for companies wanting to establish waste or recycling facilities so that building contractors would have to fulfill the requirements of the international standard.
2. It could or should provide an operational standard, so that users, customers, and regulators could be satisfied that the facilities are operated in an environmentally sound way in accordance with internationally agreed criteria such as those embodied in the Basel Convention.

In order to support the development and implementation of an international standard there would be a need for crafting technical guidelines for the ESM of the different types of waste materials subject to recycling or other waste management operations. The preparation of such guidelines would benefit from the work done so far for the ESM of wastes internationally (i.e., Basel Convention and Organisation for Economic Cooperation and Development (OECD)).

An international standard would provide a specification for operational management systems for safe and environmentally sound facilities. It would encourage best practices and facilitate the selection of, for instance, recycling facilities by users or customers. An international standard would provide requirements for those bodies auditing and certifying adequacy of recycling or other waste management facilities.

ESM Certification

The idea would be to develop an internationally recognized ESM certification system, based on an international standard, that would demonstrate that recycling companies or facilities and other similar waste management companies, in different parts of the world, are operating in accordance with ESM principles, Such an international certification scheme would entail a process of audit, verification, and certification recognized by both governments and industry. It would trigger a series of check points from design of a product to its reuse, recycling, and final disposal when it has reached the end of its useful life.

Actually, some governments, industry, and nongovernmental organizations (NGOs) have or are developing national or industry-based recycling certification schemes that are important to consider in the exploration for the development of an international ESM standard. The development of such certification schemes will by itself provide information on where problems lie and where the gaps are found.

Traceability Systems

A certification system is industry driven or facility specific. It would need to be complemented by a traceability system to achieve an improved transparency in waste flow. There is a growing demand by manufacturing companies and others for more responsible management of products, goods, or components at the end of their operational or useful life. The design and implementation of traceability solutions could start from the place of manufacture to final disposal covering the assembly line, the finished product, the used product, transport, processing, and final disposal. Such traceability systems would provide enhanced certainty with regard to the proper, safe, and sound management of used or end-of-life products and components, their destination, and the way they are handled when reaching the end of their useful life. It would also assist in ensuring the sound management of hazardous waste, especially used or end-of-life hazardous products, equipment, or components in accordance with domestic and international norms, rules, and procedures. Finally, it could help in reviewing where harmful chemicals could be avoided in products in order to reduce risks at a time of disposal.

A Global ESM Normative Framework

The tools to assist in building a level-playing field for the sound and safe management and flow of waste and recyclables worldwide need to be sustained by norms and law. The ESM principles, applicable to every waste, could be incorporated into normative requirements. This would provide regulators with the necessary legal framework to implement ESM principles in a coordinated and comprehensive way.

The design and development of a practical international ESM normative framework could be envisaged to respond to the demand for more certainty, predictability, and transparency in the flow of waste and recyclables. Such instruments would capitalize on the existing use of the ESM principles throughout the world and the work done so far by the Parties to the Basel Convention and OECD Members in the adoption of tools and methodology to assist national authorities to implement these principles. It could also benefit from the practical approach developed by the Bureau of International Recycling to assist its member companies to improve environmental performance through the use of environmental management tools.

The development of a comprehensive normative framework could contribute to improving environmental policy coherence at the international level. The building of a global level-playing field would guarantee a fair trade respectful of people and nature. It could also act as an incentive to facilitate and promote the transfer of environmentally sound technologies and stimulate international cooperation. An international ESM framework would provide a vehicle to reduce the environmental footprint of the trade in waste and recyclables through the improvement of operational standards and safeguards and by making the intercontinental and intraregional flow of recyclables more transparent.

This global framework would provide an environment conducive to promoting the importance of the Basel Convention as its center of gravity. It is essential that the Basel Convention appears on the radar screen of politicians not only when there is an accident or an incident but because of its true and significant contribution to protecting the global environment and improving the livelihood of people. The framework would complement and reinforce the implementation of the Basel Convention and its sister regional treaties and enhance the implementation of the global chemicals conventions. In turn, such global frameworks would be nourished by and enriched through the implementation of the Basel Convention and related chemical conventions and protocols.

FUTURE OF THE BASEL CONVENTION

The Basel Convention is at the crossroad. The challenges ahead are immense. In a world shaken by uncertainties and political, social, economic, financial, and environmental disturbances, the Convention remains a stronghold of stability and solidity, a lighthouse to guide further development of a global ESM architecture.

The fast-evolving trade in chemicals, waste, and recyclables, the discrepancy among countries in terms of their capacity to manage waste or chemicals in a safe and sound way, the trend toward moving from waste to resources, and the continued increase in the generation of waste call for sound governance based on the designing of a comprehensive and coherent set of ESM-derived tools that would capture the global reality of waste management in the context of the life-cycle approach to the management of harmful chemicals

Governments have taken the measure of the challenge. But they are incapacitated by the energy crisis, the financial and economic turmoils associated with social unrest, and the food disaster. Wastes, especially hazardous wastes, do not appear on their radar screen as a priority. This is the reason why initiatives taken outside government circles should be envisaged to facilitate a transition to a more environmentally friendly and socially acceptable economic development. Global environmental governance represents one response by governments but much needs to be done.

The development of ESM-derived tools, such as a certification scheme supported by an international standard, improved traceability of waste flows together with the design of an international ESM framework—one global normative system—represent the core of a forward-looking policy action that would enhance coherence and comprehensiveness, extend ESM principles to "all waste," support the objectives of the Basel, Rotterdam, and Stockholm Conventions, and provide for a higher degree of certainty and transparency in the global trade in chemicals, waste, and recyclables necessary to ensure that trade objectives and environmental protection support each other overtime and everywhere.

A paradigm shift is necessary to accomplish the goals of the international environmental agreements regulating waste and chemicals. Instead of looking at them as barriers or obstacles, it would be essential to accept that their environmental objectives build bridges between people and nature by addressing ecological interconnectivity that supports life on Earth.

9 The Climate Change Convention and the Kyoto Protocol*

Bert Metz

CONTENTS

INTRODUCTION

Climate change is one of the biggest challenges to the international community. It has many common elements with management of chemicals and environmental health, such as certain chemicals involved, development of new and clean technologies, investigation of significant efficiency gains, and sustainable increased food production.

* This is a shortened version of the section on the Climate Change Convention and the Kyoto Protocol, in Metz (2010); see http://www.controllingclimatechange.info

This chapter provides general insight into the UN Climate Change Convention and its relation to international management of many problems, several having close links with chemicals and waste management. Many of its mechanisms are also highly relevant for advanced systems of chemicals management, such as technology transfer, sharing the burden, and collaboration between developing and industrialized countries.

There is now an established set of international agreements to deal with the problem of climate change. In the first place there is the United Nations Framework Convention on Climate Change and its Kyoto Protocol. Related to these, but completely independent, are many other international agreements between states and/or private entities: agreements on research & development (R&D) in the framework of the International Energy Agency (IEA), financial arrangements of multilateral development banks to invest in emission reduction projects, programs to promote energy efficiency and renewable energy, CO_2 capture and storage and other mitigation technologies, as well as joint regional expert centers.

CLIMATE CHANGE CONVENTION

The Climate Change Convention, officially called the United Nations Framework Convention on Climate Change (UNFCCC), was agreed in 1992 at the World Summit of Environment and Development in Rio de Janeiro. It had been negotiated in a period of about two years after the concerns of scientists about the changing climate and the global impacts of it had convinced political leaders that is was time to act. The first assessment report of the Intergovernmental Panel on Climate Change (IPCC), published in 1990, galvanized those concerns. This led to the UNFCCC that was agreed in 1992 and entered into force in 1994 after 55 countries (representing 55% of industrialized countries emissions) had ratified it (i.e., officially approved through their national parliaments or other mechanisms).

The UNFCCC is, as the title says, a framework agreement. It has only limited specific obligations to reduce emissions of greenhouse gases (GHGs), but formulates principles, general goals, and general actions that countries are supposed to take. It also establishes institutions and a reporting mechanism, as well as a system for review of the need for further action. Over time it has received almost universal subscription* (see Box 9.1).

Country obligations differ, for that reason a distinction is made between Annex-I, Annex-II, and non-Annex-I countries. The Annex-I and Annex-II countries are listed in Figure 9.1, together with other relevant groupings. Former Eastern European and former Soviet Union countries have a special status under the Convention as the so-called "countries with economies in transition" (EIT).

KYOTO PROTOCOL

At the first COP to the Convention in 1995, a decision was taken that further action was needed to address climate change. It was agreed to start negotiations toward a protocol (an annex to the Convention) that would commit industrialized countries to

* http://unfccc.int/resource/docs/convkp/conveng.pdf

BOX 9.1: THE UNITED NATIONS FRAMEWORK CONVENTION ON CLIMATE CHANGE (UNFCCC): KEY ELEMENTS

Principles:

- Common but differentiated responsibility
- Special consideration for vulnerable developing countries
- Precautionary principle
- Polluter pays
- Promote sustainable development

Goals: The ultimate goal (Article 2) is to "stabilize greenhouse gas concentrations in the atmosphere at a level that would prevent dangerous anthropogenic interference with the climate system. Such a level should be achieved within a time frame sufficient to allow ecosystems to adapt naturally to climate change, to ensure that food production is not threatened and to enable economic development to proceed in a sustainable manner."

Regulated Substances: CO_2, CH_4, N_2O, HFCs, PFCs, SF_6

Participation: Almost universal (191 countries and the European Union (EU), September 1, 2008)

Actions Required:

- Minimize emissions and protect and enhance biological carbon reservoirs, the so-called "sinks" (all countries); take action with the aim to stop growth of emissions before 2000 (industrialized, the so-called Annex-I countries)
- Promote development, application, and transfer of low-carbon technologies; Annex-I countries to assist developing countries
- Cooperate in preparing for adaptation
- Promote and cooperate in R&D
- Report on emissions and other actions (the so-called "national communications," annually for Annex-I countries and less frequent for others)
- Assist developing countries financially in their actions (rich industrialized countries, the so-called Annex-II countries)

Compliance: Review of reports by the secretariat and by visiting expert review teams

Institutions:

- Conference of the Parties (COP), the supreme decision making body; rules of procedure for decisions never agreed
- Bureau (officials, elected by the COP, responsible for overall management of the process)

- Two Subsidiary Bodies (for Implementation and for Scientific and Technological Advice) to prepare decisions by the COP
- Financial mechanism, operated by the Global Environment Facility of Worldbank, UNDP, and United Nations Environment Programme (UNEP), filled by Annex-II countries on voluntary basis; two special funds: a Least Developed Country Fund and Special Climate Change Fund, mainly to finance adaptation plans and capacity building, but also technology transfer and economic diversification
- Expert groups on Technology Transfer, Developing Country National Communications, Least Developed Country National Adaptation Plans
- Secretariat (located in Bonn, Germany)

Other Elements:

- Requirement to regularly review the need for further action

Source: Adapted from Metz, B. 2010. *Controlling Climate Change*, Cambridge University Press, Cambridge and New York.

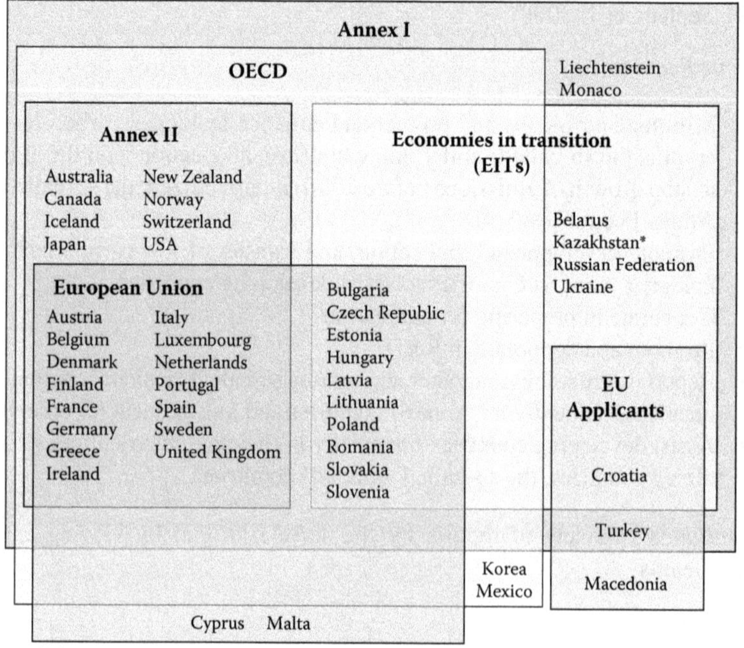

FIGURE 9.1 Country groupings under the UNFCCC, Organisation for Economic Cooperation and Development (OECD), and EU. (From IPCC. 2010. Brochure of the Intergovernmental Panel on Climate Change on Understanding Climate Change, 22 years of IPCC Assessment. With permission.)

further reduce their GHG emissions. The industrialized countries had not yet done much in terms of emission reductions at the time. Therefore, developing countries were deliberately exempted from further action in light of the "common and differentiated responsibility" principle of the Convention. The United States explicitly agreed with that as one of the Parties to the Convention.

These negotiations led in 1997 to the agreement of the Conference of Parties on the so-called Kyoto Protocol. It reaffirms in fact the basic agreement of the Convention and adds a number of elements: quantified emission caps for Annex-I countries, the so-called flexible mechanisms to allow for cost-effective implementation by using cheap emission reductions elsewhere (emission trading between Annex-I countries, a clean development mechanism (CDM) on projects done in developing countries, and joint implementation on projects in Annex-I countries), a compliance mechanism, and a new Adaptation Fund that gets its funding from a levy on CDM projects (see Box 9.2).

Why the United States Pulled Out of Kyoto

The United States agreed with the agreement reached in Kyoto in December 1997 after Vice President Al Gore came personally to Kyoto to instruct the U.S. negotiators to be more accommodating on the reduction targets for GHG emissions. As a result the United States agreed to reduce its emissions with an average of 7% below 1990 by 2008–2012. The EU accepted −8% and Japan and Canada, −6%. The United States got a lot of what it had asked for: a so-called basket of gases, allowing countries the flexibility to decide what kind of reductions they would prefer to meet their target; inclusion of afforestation and reforestation as a "sink" for CO_2; and the so-called flexibility mechanisms that allow countries to trade emission allowances between them and to use investments in projects in developing countries to compensate for reductions they would not realize at home (through the CDM).

But this happened against the background of a strong anti-Kyoto sentiment in the U.S. Congress. Nevertheless, the Clinton Administration agreed with the outcome in Kyoto.

When the Bush Administration took office in January 2001, international negotiations on the details of the Kyoto Protocol were still going on and the newly appointed officials of the United States were participating in them. In March 2001, President George Bush, materialising his campaign stance about climate change, announced that the United States would not ratify the Protocol. The reasons given were that it would seriously harm the U.S. economy and developing countries were exempt from emission reductions. The economic argument was surprising, because the IPCC's Third Assessment Report that was about to be published clearly showed the economic costs of implementing the Kyoto Protocol to be very modest.* Special interests, that is, the coal and oil industry, apparently had a lot of influence. The other argument, that developing countries were exempt from emissions reductions, was a direct

* The IPCC Working Group III report of 2001 estimated the economic cost of implementing the Kyoto Protocol for the United States at a GDP reduction of less than 0.5% by 2010 in a system of global emissions trading, compared to what it otherwise would have been; in other words, the economy would not grow with say 25% over the period 2000–2010, but with something like 24.5%.

consequence of the negotiating mandate of 1995. Australia followed suit in not ratifying the Kyoto Protocol, although it in fact was implementing climate policy to meet its agreed target (of +8% compared to 1990 by 2008–2012).

BOX 9.2: THE KYOTO PROTOCOL

Principles: Same as the Convention
Goals: Same as the Convention
Regulated Substances: CO_2, CH_4, N_2O, HFCs, PFCs, SF_6
Participation: 180 countries and the EU (United States is not a Party)

Actions:

- Annex-I countries together reduce emissions to 5% below 1990 level, on average over the period 2008–2012; specific emission caps for individual countries[a]
- Option to use flexible mechanisms, that is, international trading of emission allowances between Annex-I countries, or using the emissions reductions from projects in developing countries (through the CDM) or other Annex-I countries (Joint Implementation)
- Option to develop coordinated policies and measures
- Strengthened monitoring and reporting requirements for countries with reduction obligations

Compliance: Shortage in emission reduction to be compensated in period after 2012, with 30% penalty

Institutions:

- COP of the Convention, acting as the Meeting of the Parties of the Protocol (CMP) as decision-making body
- Use all other Convention institutions
- Compliance Committee, with consultative and enforcement branch
- Executive Board for the CDM
- Joint Implementation Supervisory Committee
- Adaptation Fund, managed by the Adaptation Fund Board and administration by GEF and Worldbank; fund gets its money from a 2% levy on CDM projects

Other Elements:

- Requirement to review the need for strengthening the actions

[a] See http://unfccc.int/kyoto_protocol/items/3145.php

Source: Adapted from Metz, B. 2010. *Controlling Climate Change*, Cambridge University Press, Cambridge and New York.

These withdrawals gave a shock to the international community and to the ongoing negotiations. Maybe due to this shock-effect, an agreement was reached in June 2001 between all other countries on the outstanding details of the Kyoto Protocol implementation. Speeches at that meeting frequently mentioned the victory of multilateral approaches to solving global problems. The United States answered that it would follow its own policies to tackle the problem, but everybody knew there was no credible U.S. federal policy to reduce GHG emissions. This situation has not fundamentally changed since, although the Obama Administration likes very much to cooperate internationally.

ARE COUNTRIES MEETING THEIR EMISSION REDUCTION OBLIGATIONS?

The countries that ratified Kyoto are collectively on track to meet the agreed emissions reduction of 5% below 1990 by 2008–2012. In 2005 their emissions were 15% below the 1990 level* (see Figure 9.2). There are large differences however, EITs were about 35% below and the non-EIT countries 3% above. Individual countries show even much larger differences: Latvia was 59% below its 1990 level (with a 2008–2012 target of −8%), while Spain was 53% above (with a target of +15%†). Including land-use changes does not change this picture radically, except for Latvia that had negative overall emissions, that is, fixation of CO_2 in forests was bigger than the emissions of GHGs to the atmosphere from all other sources. Striking is that Canada's emissions were 54% above 1990 in 2005 (with a target of −6%), while the United States, which is not a Party to the Kyoto Protocol, only saw a 16% increase above

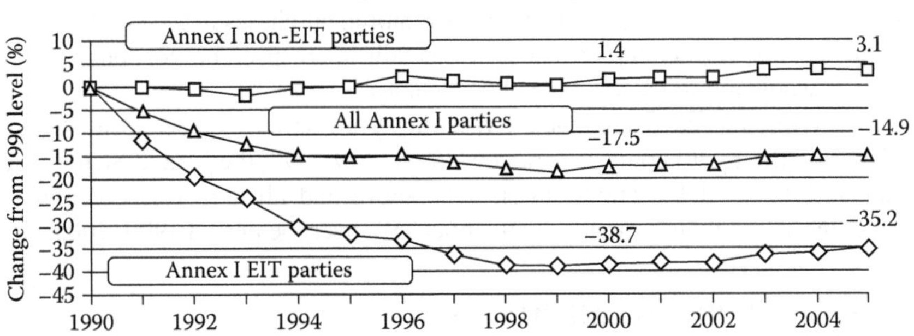

FIGURE 9.2 GHG emissions from Annex-I Kyoto Parties 1990–2005; excluding sinks and sources from land use and land-use change. (From UNFCCC greenhouse gas emission trends 1990–2005. With permission.)

* Excluding land-use change emissions (relatively small); see http://unfccc.meta-fusion.com/kongresse/071120_pressconference07/downl/201107_pressconf_sergey_konokov.pdf

† The 15 EU Member States redistributed their collective −8% target among individual Member States to allow for specific national circumstances; see http://reports.eea.europa.eu/eea_report_2007_5/en/Greenhouse_gas_emission_trends_and_projections_in_Europe_2007.pdf

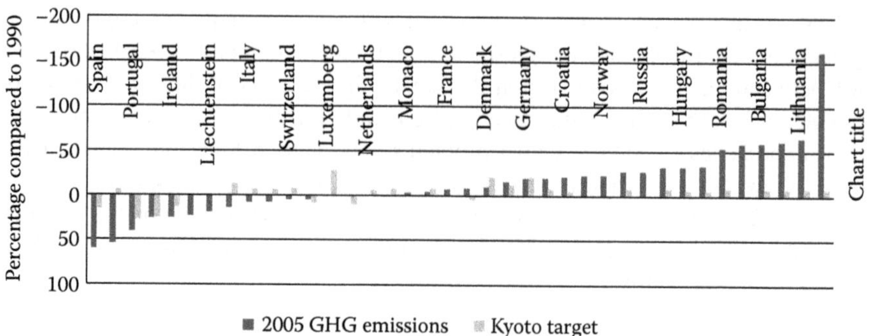

FIGURE 9.3 Performance of the individual Kyoto Annex-I Parties till 2005 and their Kyoto Protocol obligations; total net GHG emissions, including sources from land use and land-use change. (From UNFCCC greenhouse gas emission trends 1990–2005. With permission.)

1990 in 2005.* It confirms the complete lack of implementation of the Kyoto Protocol obligations in Canada. For comparison, China and India roughly doubled their emissions over that same period.† Projections for the period 2008–2012 show that the overall picture will roughly remain the same.‡ Figure 9.3 shows the performance of individual countries.

These numbers do not necessarily mean that countries will not meet their obligations. The Kyoto Protocol has provisions to trade emission allowances, or in other words, to buy emission allowances on the carbon market in case of a domestic shortfall. That can be done through country-to-country deals (such as Russia selling part of its surplus to Japan or Canada), or through CDM projects in developing countries and Joint Implementation projects in other Annex-I countries. So, in theory, any country could still meet its obligations by making the required purchases in time for the 2008–2012 targets. For most countries that seems a realistic prospect, but for some, such as Canada, it would require a very big political change.

Global emissions have continued to rise: they grew about 25% between 1990 and 2005. While the Kyoto countries are now 15% below 1990, the United States (good for about 25% of global emissions) is 16% above and all non-Annex-I countries together increased their emissions with about 75% over the period 1990–2005.

CLEAN DEVELOPMENT MECHANISM

One of the successes of the Kyoto Protocol is the CDM. It creates the possibility for Annex-I countries to meet part of their commitments through emissions reductions from projects in developing countries. These projects would at the same time contribute to sustainable development in developing countries (the so-called "host

* http://unfccc.int/ghg_data/ghg_data_unfccc/time_series_annex_i/items/3814.php
† http://cait.wri.org/
‡ http://unfccc.meta-fusion.com/kongresse/071120_pressconference07/downl/201107_pressconf_katia_simeonova_part1.pdf

countries"). The principle is simple: any emission reduction project in a developing country that otherwise would not have happened is lowering global emissions and could therefore replace a comparable action in an industrialized country. It is a market mechanism. If it is cheaper to realize reductions in developing countries, it lowers the costs for industrialized countries. The clause "that otherwise would not have happened" is of course crucial. If projects would have happened anyway, trading the resulting emission reductions no longer is a net global reduction. So the effectiveness of the CDM depends strongly on this so-called "additionality" issue.

How has the CDM Developed?

As of January 1, 2009 there were 4474 CDM projects in the pipeline (i.e., either submitted to or registered by the CDM Executive Board). Out of these, 1370 were registered and for 465 certified emission reduction units (CERs) were issued.* Together they are good for a reduction of about 0.3 $GtCO_2eq$ (Gigatons CO_2 equivalent/year) in the period 2008–2012 and about 0.7 $GtCO_2eq$/year from 2013–2020. Given their relatively low price, the CDM CERs are very likely to be bought by Annex-I countries to meet their obligations. To put things in perspective, the 0.3 $GtCO_2eq$/year is about 50% of the total reduction that Kyoto Annex-I countries are supposed to realize.† In other words, domestic emissions reductions in these countries will be only half of what they would have been without the CDM, if indeed all

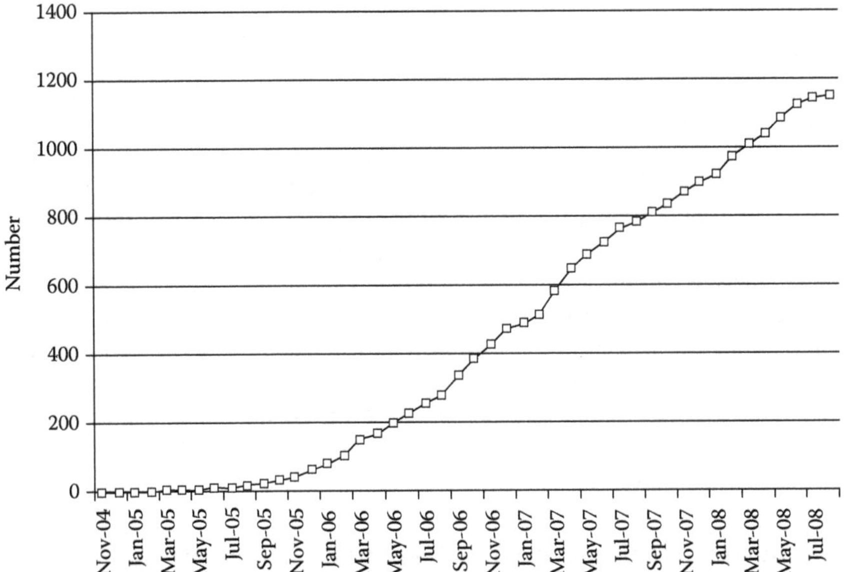

FIGURE 9.4 Number of registered CDM projects over time. (Reproduced from UNEP Risoe CDM/JI Pipeline Overview, http://www.cdmpipeline.org/overview.htm#4. With permission.)

* http://www.cdmpipeline.org/overview.htm#4
† http://www.feem.it/NR/rdonlyres/2C130D3B-124F-427E-9FAE-E958F0E83263/838/0800.pdf

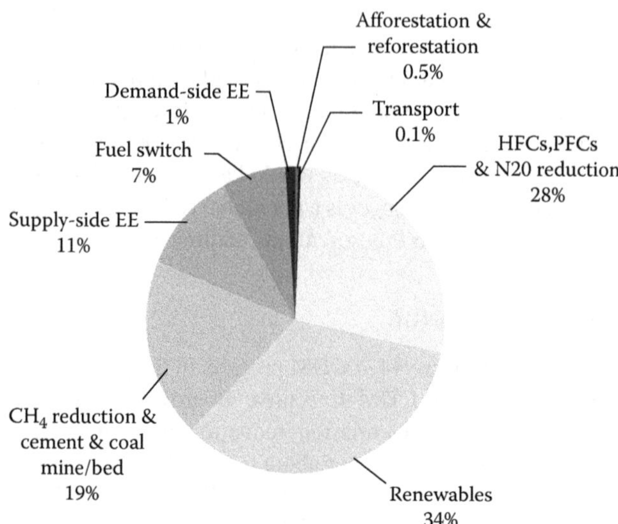

FIGURE 9.5 CERs expected until 2012 from CDM projects in each sector. (Reproduced from UNEP Risoe CDM/JI Pipeline Overview, http://www.cdmpipeline.org/overview.htm#4. With permission.)

available CERs are bought. The rest of the required reductions is offset by CDM credits.

CDM projects are covering a wide range of mitigation activities. The number of projects on renewable energy is the highest, with much smaller numbers for landfill gas (methane) recovery and destruction of hydrofluorocarbon-23 (HFC-23) at hydrochlorofluorocarbon (HCFC) plants and nitrous oxide (N_2O) at chemical plants. In terms of tons of CO_2eq reduction expected before the end of 2012, renewable energy projects represent 34%, HFC-23, perfluorocarbon (PFC) and N_2O projects 28%, reflecting the high Global Warming Potential of HFC-23. Figures 9.4 and 9.5 give an impression of the strong growth of the CDM (number of projects registered) and the relative contributions of various types of projects.

Projects are concentrated in a limited number of countries. Figure 9.6 shows that China, India, Brazil, and Mexico together host about 70% of all CDM projects. These countries have organized their CDM activities well, making it easier for foreign buyers to get substantial tonnage without excessive administrative efforts. This also means that many countries hardly benefit from the CDM. All African countries together, for instance, only host 2% of the projects.

How Much of the Projected CDM Emission Reductions are Additional to What Otherwise would have Occurred?

According to the CDM rules, emissions reductions from CDM should be 100% additional. There is even a specific requirement to demonstrate that additionality in applying for an approval of a CDM project. But how is the real situation? This depends on what is considered to be the business as usual (BAU) (or baseline)

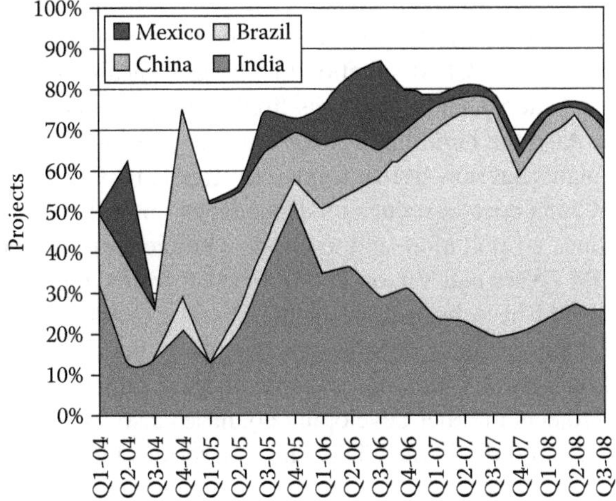

FIGURE 9.6 CDM projects in the pipeline in Brazil + Mexico + India + China as a fraction of all projects. (Reproduced from UNEP Risoe CDM/JI Pipeline Overview, http://www.cdmpipeline.org/overview.htm#4. With permission.)

development. A number of hydropower projects have been approved under the CDM, many of which were already under development before the CDM came into being. Hydropower has been commercially attractive in many places for a long time. So why would certain hydropower projects be considered additional? A possible justification would be that the economic profitability (in terms of the time it takes to recoup the investment) might be less than what investors find acceptable. The CDM revenues can then make the difference between an unattractive and an attractive investment. But this is unlikely the case for most of the hydropower projects registered under CDM, given prior approval and comparable projects that were realized without CDM money.*

Another interesting case is the destruction of HFC-23 from HCFC-22 production facilities. It is technically feasible to destroy HFC-23 in off-gas by using incinerators. The cost of this destruction, including investment and operating costs, is less than US$ 0.20/tonne of CO_2eq destroyed.† A number of HCFC-22 plants in the world have installed these devices. It is thus very hard to argue that this is something that cannot be seen as "state of the art." Nevertheless, HFC-23 destruction at 10 existing plants in China, India, and Korea was approved as a CDM project. Worse is that the CER's from these projects were sold at market prices of up to US$ 15–20/tonne of CO_2eq avoided, meaning that a substantial profit was made. And even worse is that attempts are being made to get CDM approval also for HFC destruction at newly built HCFC-22 plants. The counterargument from proponents of these CDM projects is that new

* Haya, B., Failed mechanism: how the CDM is subsidizing hydro developers and harming the Kyoto Protocol, International Rivers, 2007; see http://internationalrivers.org/en/climate-change/carbon-trading-cdm/failed-mechanism-hundreds-hydros-expose-serious-flaws-cdm; see also http://www.indiatogether.org/2008/jul/env-cdm.htm

† IPCC Special Report on Safeguarding the Ozone Layer and the Global Climate System, 2000.

HCFC-22 plants in developing countries are simply not being equipped with HFC destructors, because there is no economic or regulatory reason to do so.

Really worrisome is the CDM situation in China. Basically all new investments in hydropower, wind energy, and natural gas-fired power plants are cofunded through the sale of CERs. Also the building of more efficient (the so-called "supercritical") coal-fired power plants has now been accepted as eligible for CDM.* This means that almost anything China does to reduce its dependency on coal (which it is now also importing), to reduce air pollution, and to improve efficiency of power plants is now done through CDM (Wara and Victor, 2008). In other words, the assumption is that nothing of this would have been done in the absence of the CDM. That is hard to believe, since many of these installations have been built before without CDM funding and self-interest of China makes most of these projects completely viable. Given that such projects in other developing countries also will be eligible and the huge role of China and India in the CDM, this is a serious blow to the additionality of the CDM.

This issue is therefore on the table at the ongoing negotiations for a new international agreement for the period after 2012 (see below). There are strong voices calling for a serious reform of the CDM to repair these weaknesses.

INSTITUTIONAL INFRASTRUCTURE

Another key achievement of the Climate Change Convention and the Kyoto Protocol is that an elaborate institutional infrastructure has been built to deal with climate change. Apart from a carbon market with a wide range of players and institutions, there is a whole machinery of reporting on emissions, vulnerabilities to climate change, and planning and implementation of adaptation and mitigation activities (mandated by the Convention and the Protocol). Countries have implemented registries of GHG emissions and policies to control emissions. Because of the CDM, many developing countries have done that as well.

NEW AGREEMENTS BEYOND 2012

It is obvious that further steps are needed to curb global emissions after the Kyoto Protocol's commitments for 2008–2012 expire. Negotiations for such an agreement formally started in December 2007 with the adoption of the so-called Bali Action Plan by the Conference of Parties to the UNFCCC, after the United States had given up its resistance to start negotiating.

What Gases will be Controlled?

There are very few proposals to expand the list of controlled substances. The more serious are to add a few fluorinated gases: NF_3, hydrofluoroethers and fluorinated ethers, and perfluoropolyethers. Informally, some people float the idea of adding

* The acceptance of the methodology for supercritical coal fired power plants was limited to a maximum of 15% of a country's power supply; see http://cdm.unfccc.int/index.html

black carbon, an aerosol from incomplete combustion that has a warming effect, but the emissions are quite uncertain and measures to reduce them are not straightforward. An interesting proposal is to move the control of HFCs to the Montreal Protocol, since the main driver for increased emissions is the replacement of CFCs and HCFCs.

How Much Should Emissions be Reduced?

The long-term goal of the Climate Convention is to stabilize GHG concentrations in the atmosphere at "safe" levels. There is growing support to translate that into a reduction of global emissions with 50% by 2050. But there is still a difference of opinion about the base year. Is it compared to 1990, which would be just about consistent with a stabilization at an average global temperature increase in the long term of 2°C (Meinshausen, 2009)? Or is it compared to 2005, which would make it more like a 3°C scenario?

The other important point is the interim target for 2020. To keep the possibility open of staying on a 2 degree course, global CO_2 emissions should start declining not later than about 2015. That is a serious additional constraint on the longer term emission reduction goal. And it is only for global emissions. What would it mean for emission reductions of developed and developing countries?

Who Does What?

Table 9.1 shows the summary from IPCC* of the various studies that looked into that question, with different assumptions about what is an equitable distribution of the effort. For the 450 ppm CO_2eq scenario, the resulting numbers for allowable emissions fall in a fairly narrow band: 25–40% below 1990 level for developed (Annex-I) countries by 2020 and 80–95% by 2050. For developing countries in Latin America, Middle-East, and East Asia, a deviation from the baseline emissions is needed. Since the deviation for all developing countries together is about 15–30% (Den Elzen and Hoehne, 2008), for the more advanced regions mentioned it will be about one-third higher: 20–40%. This is a deviation from the baseline; so growth of emissions would still be possible. China, for instance, could under this regime still increase its emissions between 1990 and 2020 with 2–3 times instead of 3–4 times if it would not take any action.

Under no circumstance can developed countries alone reduce emissions sufficiently to achieve stabilization at any level, since emissions eventually have to go down to almost zero.

This immediately raises concerns about the ability of China and other developing countries to do this without harming their social and economic development. Is it fair to ask such an effort from China and other developing countries? The numbers mentioned above do come from studies where equity was an explicit requirement; so the answer in principle should be "yes." However, this ignores practical problems of access to the latest technology, capacity in the country to organize drastic change, and finan-

* IPCC AR4, WG III, Chapter 13.

TABLE 9.1
Range of Allowed Emissions[a] Compared to 1990 for Stabilization at Different Levels for Annex-I and Non-Annex-I Countries, as Reported by Studies with Different Assumptions on Fair Sharing Efforts

Scenario Category	Region	2020	2050
A-450 ppm CO_2-eq[b]	Annex I	−25% to −40%	−80% to −95%
	Non-Annex I	Substantial deviation from baseline in Latin America, Middle East, East Asian and Centrally-Planned Asia	Substantial deviation from baseline in all regions
B-550 ppm CO_2-eq	Annex I	−10% to 30%	−40% to 90%
	Non-Annex I	Deviation from baseline in Latin America and Middle East, East Asia	Deviation from baseline in most regions, especially in Latin America and Middle East
C-650 ppm CO_2-eq	Annex I	0% to −25%	−30% to −80%
	Non-Annex I	Baseline	Deviation from baseline in Latin America and Middle East, East Asia

Source: From IPCC. 2010. Brochure of the Intergovernmental Panel on Climate Change on Understanding Climate Change, 22 years of IPCC Assessment. With permission.

[a] The aggregate range is based on multiple approaches to apportion emissions between regions (contraction and convergence, multistage, Triptych and intensity targets, among others). Each approach makes different assumptions about the pathway, specific national efforts and other variables. Additional extreme cases—in which Annex I undertakes all reductions, or non-Annex I undertakes all reductions—are not included. The ranges presented here do not imply political feasibility, nor do the results reflect cost variances.

[b] Only the studies aiming at stabilization at 450 ppm CO_2-eq assume a (temporary) overshoot of about 50 ppm (see Den Elzen and Melnahausen, 2006).

cial resources to do the necessary investments. International assistance from developed countries would therefore be needed.

What Kind of Actions can Countries Commit to?

Emission ceilings are the simplest form of commitment to action. That is the way things were done under the Kyoto Protocol (see the section "Kyoto Protocol"). Absolute ceilings can, however, be problematic for countries with strongly varying economic growth rates and the costs of meeting them cannot be predicted accurately. Alternatives exist in the form of emissions per unit of production, for instance, per unit of GDP for a country as a whole or per tonne of steel or per kWh electricity produced for a sector. Moving to such targets of course means that the resulting emissions are no longer certain. With strong economic growth or growth of production, emissions come out higher. Nevertheless, such relative or dynamic targets are being considered for developing countries, where uncertainties of growth are high and increasing costs can be problematic. Emission trading systems can handle such dynamic targets, although the system becomes more complex.

Another way is to commit to the use of the cleanest technology. For instance, by requiring minimum fuel efficiency standards for cars, and energy efficiency standards for washing machines, refrigerators, televisions, computers, and other products. This can also be applied to steel-, glass-, and cement-making processes and other manufacturing processes. These so-called "best available technology" approaches have been used widely in controlling other environmental problems.* Technology commitments could also be in the form of information sharing or joint demonstration programs, although the impact of those on emission reductions would be hard to measure.

Yet another approach is to commit to implementing policies and measures. When doing that the result in terms of GHG emissions would not be exactly known, but could be estimated roughly. In fact, the Kyoto Protocol has an almost forgotten Article 2 that creates the possibility for coordinated action on policies and measures. This would make sense to make standards for traded products more effective.

Finance

To get a meaningful agreement for the period after 2012, substantial financing will be needed to assist developing countries with mitigation and adaptation measures and to promote technology development and diffusion.

For mitigation in a scenario where global emissions are back at 2005 levels by 2030 (more or less equivalent to stabilization at 450–500 ppm CO_2eq) additional investments of about US$ 200 billion/year in 2030 will be needed, according to estimates of the UNFCCC secretariat. For adaptation, the amount needed will be about US$ 50–180 billion/year in this scenario in which the most serious damages from climate change are avoided. Together this would represent about US$ 250–400 billion/year by 2030. More recent numbers from the IEA suggest that this amount might be twice as high.† This looks like a huge number. Compared to the total annual investments in the world it is, however, only 1–2% (2–4% if IEA is right). As percentage of world GDP, it is even lower: less than 1%, even for the higher numbers.

Investment is not the same as costs. A considerable part of the investments in emission reductions has benefits that make these investments profitable (through saved energy, improved air quality and lower health care costs, and reduced oil imports or otherwise). It means that the net incremental costs (see Chapter 49 of this book on Global Financial instruments) of these investments are much lower or even zero. For 2020, an estimate is US$ 65–100 billion/year in incremental costs for emission reductions, adaptation and technology research, and development and demonstration together.‡

* See Chapter 11 of Metz (2010).

† IEA, WEO 2008.

‡ Climate Works Foundation, project Catalyst interim results, http://www.project-catalyst.info

Current flows from Climate Change Funds under the Climate Change Convention and the Kyoto Protocol are very modest: Adaptation Fund: US$ 80–300 million/year, and Least Developed Country Fund and Special Climate Change Fund together something like US$ 15 million/year. Huge increases in these and other funds will be needed.

TECHNOLOGY

Modern low-carbon technology is essential for controlling climate change. We know that a large part of the opportunities for emissions reduction can be found in developing countries. We also know that there are many barriers to the use of these modern low-carbon technologies in these countries. Removing these barriers is therefore critical. This is commonly called the problem of "technology transfer."

What can international agreements do to remove these barriers? It is helpful to make a distinction between diffusion of existing technologies and the development of new ones. Existing technologies, such as energy efficient cars, appliances. and industrial equipment, are readily available in developed countries (although they may not be universally applied there). But these technologies are much rarer in developing countries. Exceptions are recently built large-scale manufacturing plants for steel, cement, or fertilizer, where often the most modern and efficient technology is being used. Lack of knowledge of investors, high initial investments, insufficient maintenance expertise, banks that shy away from investments they are not familiar with, and absence of government regulations are some of the most important reasons for this.*

International agreements can do something about creating the need for investments (in the form of countries committing to action), making it easier to access international financing (see the section "Finance") and sharing the experience of countries by creating databases and best practice examples.† Much of the international action to assist countries in implementing modern low-carbon technology is happening outside the Climate Change Convention. IEA operates a series of the so-called Implementing Agreements that allow IEA Member and nonmember countries and other organizations to engage in sharing information about implementing specific low-carbon technologies.‡ There are currently 42 of these cooperative arrangements. There are also many public–private partnerships active in this field, such as the Renewable Energy and Energy Efficiency Partnership (REEEP), funded by national governments, businesses, development banks and NGOs§ and the Renewable Energy Network for the 21st century (REN21), connecting governments, international institutions, NGOs and industry associations.¶

Development of low-carbon technology is different from diffusion. Development means scaling up of promising results from research and demonstrating it at semi-

* IPCC, Special Report on Methodological and Technological Aspects of Technology Transfer, 2000.
† http://unfccc.int/ttclear/jsp/index.jsp
‡ http://www.iea.org/textbase/techno/index.asp
§ http://www.reeep.org/31/home.htm
¶ http://www.ren21.net/

commercial scale. It also means technology improvements based on R&D that can significantly reduce costs.

Traditionally, a new technology was developed in industrialized countries and then diffused to developing countries. Although this is still happening, it is no longer the only mechanism. Technology is now also developed in more advanced developing countries. Japan moved from a country good in copying and cheaply producing electronic products in the 1960s to the place where much of the innovation in these products is taking place today.* China is following that pattern and has already become the producer of the best and lowest-cost supercritical coal-fired power plants, the main manufacturer of electric bikes, solar water heaters, and solar panels.† It has taken the number 1 position in wind turbine manufacturing in 2009.‡ Innovation capacity is rising fast, reflected by the tripling of R&D expenditures from 0.5% to 1.5% of GDP since 1990.§ In India, Suzlon, one of the world's biggest wind turbine manufacturers, acquired a German firm, strengthening its market power and its innovative capacity.

What does this mean for the role of international agreements in promoting the development of new low-carbon technologies? This role is probably limited. Arrangements such as the IEA Implementing Agreements can help share information. In the precompetitive research stage, higher government R&D budgets can help. Doubling or tripling global energy-related R&D budgets (aiming at low-carbon technologies and energy efficiency) could be made part of the financial arrangements of a new agreement. Providing support to developing countries to build up their innovation capacity should be part of that effort. The model of the Consultative Group on International Agricultural Research (CGIAR) might be useful. It is a network of 16 international research centers, spread over all regions of the world, aiming at providing food security to all people. It is funded by bilateral and multilateral donors and private foundations.¶

When technologies enter the stage of development, scaling up, and market introduction, commercial interest will dominate and the role of governments changes. International cooperation (not necessarily within the UNFCCC) could speed up the market introduction of new low-carbon technologies by setting up larger demonstration programs, with supporting government funding (this could also be one of the purposes of a new funding system).

Measuring, Reporting, and Verifying

The accountability for commitments in a new agreement has received a lot of attention during the negotiations on the Bali Action Plan. It is captured in the so-called "MRV clause": actions committed to should be *measurable, reportable, and verifiable*. Since this applies to both developed and developing countries, this is certainly

* Patent applications from Canon multiplied 10-fold between the 1960s and the 1980s/1990s; see Suzuki and Kodama (2004).
† See Chapters 5 and 6 of Metz (2010).
‡ http://www.climatechangecorp.com/content.asp?contentid=5344
§ http://www.oecd.org/dataoecd/54/20/39177453.pdf
¶ http://www.cgiar.org/

a step up from the current arrangements in the Convention and the Kyoto Protocol. There we do have a system of the so-called national communications and review, but the requirements for developing countries are not very stringent. What is also important is that the MRV requirement for a new agreement applies to financial and technical support by developed countries as well. It is likely that a new system of reporting of those actions will be set up. This was seen as a major step forward by developing countries.

How these MRV clauses are going to be implemented is yet unclear. It is likely reporting systems will build on the existing system of national communications, by making them more frequent and provide more direct guidance on what they should contain. An obvious improvement would be to have frequent inventories of GHG emissions from developing countries (currently not required and most developing countries have only submitted one inventory that is completely outdated by now). Frequent reporting on actions already taken in developing countries* would ensure that such actions can be taken into account when discussing appropriate actions by developing countries.

The "measurable" clause will have an impact on the form of agreed actions, because they should indeed be measurable, ruling out vague, noncommittal formulations.

The "verification" part gets us into the discussion on review and compliance. The current review requirements for developed countries require an administrative and a so-called "in-depth review," involving a team of experts visiting the country. This is still a relatively soft review process. It is also increasingly difficult to find qualified experts for such country visits. For a developing country's national communications, there is only a limited administrative review. Upgrading the system of review, by a more rigorous administrative review, as well as by a professionalized country visit program, would make a lot of sense.

Copenhagen and Beyond

The negotiations on a new agreement as described above were supposed to come to a fruitful end in December 2009, at the Copenhagen Conference of Parties. That did not happen however. Negotiations got stuck on a whole range of fundamental differences of opinion, mostly between developed countries on the one hand and developing countries on the other. Developing countries refused to enter into legally binding action to curb emissions, arguing developed countries had not done enough and were putting far too little money on the table to assist them. What did come out of Copenhagen was a political declaration, the so-called Copenhagen Accord, which was not formally adopted by all Parties to the UNFCCC, but almost 140 countries have associated themselves with the declaration as of end of July 2010.

The Copenhagen Accord contains a number of significant elements. First, it has two appendices where countries could enter their national pledges, one for developed and one for developing countries. In all, 42 developed countries and 40 developing countries have registered their pledges for national action to address emissions of

* See Chapter 4 of Metz (2010).

GHGs.* The impact of these pledges of course varies, with some countries pledging not much more than what would happen anyway under a BAU scenario.† However, other countries pledged significantly lower emissions by 2020 than BAU: Brazil, for example, pledged emissions by 2020 to be 38% below BAU. Indonesia pledged a reduction of 26% below BAU by 2020 and 41% if additional financing would be provided by developed countries. The caveat is that the high end or all of many of the pledges are conditional upon certain requirements to be fulfilled. For most developed countries that condition is related to all major emitters making an appropriate contribution. For many developing countries, the high end or all of their pledges are conditional upon financing to be provided to capture the incremental costs. Pledges are quite diverse in the way they express the pledge: most developed countries use absolute emission levels compared to some base year, and for developing countries it ranges from general policy pledges to numbers for reducing their carbon intensity (emissions per unit of GDP). China, for instance, pledged a 40–45% reduction of carbon intensity by 2020 compared to 2005.‡

Together the pledges mean a greatly enhanced level of activity across the world in curbing emissions of GHGs. However, they fall far short of the goal formulated also in the Copenhagen Accord that global temperatures should not go beyond 2°C above the preindustrial level. Current pledges are putting us on a track to 2.5–5°C by the end of the century (UNEP, 2010).

Another significant element of the Copenhagen Accord is the commitment from developed countries to put US$ 30 billion of additional climate finance on the table for the period 2010–2012, to be used by developing countries for action to address climate change. Although these funds are often relocated from other budgetary allocations on development assistance, they constitute a significant increase of available climate finance. For the longer term the Accord contains a somewhat vague statement of the need to mobilize at least US$ 100 billion/year of additional climate finance from public or private sources by 2020. A UN High-Level Advisory Group has looked into the possible sources of these large financial flows.§ At the Cancun COP16 all elements of the Copenhagen Accord were formally adopted as part of the COP decisions taken.

Outlook

It looks like a comprehensive new climate change agreement will not be possible for some years to come. Partial agreements inside or outside the UNFCCC will probably cement national action by countries. A fully legally binding treaty looks very unlikely. Voluntary action will be the dominant driver. Climate finance will probably come in piecemeal. The one thing that can radically change this in the coming years

* See http://unfccc.int/home/items/5262.php

† See, for instance, http://www.climateactiontracker.org/

‡ This 40–45% improvement in carbon intensity is not much better than BAU, according to many analyses, see, for example, http://www.pbl.nl/nl/publicaties/2010/Evaluation-of-the-Copenhagen-Accord-Chances-and-risks-for-the-2degreeC-climate-goal.html

§ http://www.un.org/wcm/webdav/site/climatechange/shared/Documents/AGF_reports/AGF_Final_Report.pdf

is the spreading of the conviction that a transformation to a low-carbon and climate resilient economy is in the best interest of countries. When that belief reaches a critical mass, international negotiations will fundamentally change. Countries will seek cooperation to benefit as much as possible from this transition, rather than resist climate action because they think it harms their economic development.

REFERENCES

Den Elzen, M. and N. Hoehne. 2008. Reductions of greenhouse gas emissions in Annex I and non Annex I countries for meeting concentration stabilisation targets. *Climatic Change*, 91: 249–274.

Meinshausen, M. et al. 2009. Greenhouse gas emission targets for limiting warming to 2°C. *Nature*, 458: 1158–1162.

Metz, B. 2010. *Controlling Climate Change.* Cambridge University Press, Cambridge and New York.

Suzuki, J. and F. Kodama. 2004. Technological diversity of persistent innovators in Japan: Two case studies of large Japanese firms. *Research Policy* 33: 531–549.

UNEP. 2010. The Emissions Gap Report—Are the Copenhagen pledges sufficient to limit global warming to 2 or 1.5°C? A preliminary assessment. Nairobi.

Wara, M. and D. Victor. 2008. A realistic policy on international carbon offsets, Stanford University Programme on Energy and Sustainable Development, Working paper no. 74, April 2008, http://iis-db.stanford.edu/pubs/22157/WP74_final_final.pdf

10 Convention on Long-Range Transboundary Air Pollution

Johan Sliggers

CONTENTS

INTRODUCTION

The 1979 Convention on Long-Range Transboundary Air Pollution (CLRTAP) celebrated its 30th anniversary in 2009. The CLRTAP, a very successful Convention, is the oldest multilateral environmental agreement (MEA) addressing air pollution. Under this Convention there are eight protocols, all of which have entered into force. These protocols have helped bring about drastic cuts in air pollution in the United Nations Economic Commission for Europe (UNECE) region (Sliggers and Kakebeeke, 2004).

All Parties to the CLRTAP are committed to reducing air pollution in general, and in particular, long-range transboundary air pollution. Article 2 of the Convention sets out the objective of the Convention:

> The Contracting Parties, taking due account of the facts and problems involved, are determined to protect man and his environment against air pollution and shall endeavour to limit and, as far as possible, gradually reduce and prevent air pollution including long-range transboundary air pollution.*

The Convention entered into force in 1983. There are currently 51 Parties to the Convention, including almost all UNECE member countries (56). The UNECE region consists of all European countries, extending to the east as far as Kazakhstan and to the west to the United States of America and Canada. The European Community is also Party to the Convention and all of its protocols.

The first section of this chapter on the CLRTAP outlines the eight protocols under the Convention, and is followed by a description of the organization and operation of the Convention. The third section deals with the science under the Convention and how policy makes use of this science. The last section looks into the future and discusses the challenges ahead.

PROTOCOLS

The EMEP Protocol (Protocol on Long-Term Financing of the Cooperative Programme for Monitoring and Evaluation of the Long-Range Transmission of Air Pollution in Europe) is the first of the Convention's protocols. This Protocol, to which almost all Parties to the Convention are Party, secures the funding of the international coordination work of the five program centers under the EMEP Steering Body. The budget for the coordination costs is shared between the Parties, which pay mandatory contributions according to the UN scale of assessment based on each country's GDP.

The other seven protocols deal with the reduction of emissions of air pollutants. They gradually step up to more advanced protocols with higher ambitions. The first Sulphur Protocol and the NO_x Protocol called for an initial cut in SO_2 and NO_x emissions to abate acidification, and were followed by the VOC (volatile organic compounds) Protocol, which addressed ozone. The Second Sulphur Protocol increased the reduction in SO_2 emissions by setting national ceilings. The Gothenburg Protocol targeted acidification, eutrophication, and ground-level ozone. This protocol included reductions for ammonia and further reduced emissions of the substances covered by the first three reduction protocols by setting national emission ceilings (NECs). The Heavy Metals (HM) Protocol and Persistent Organic Pollutants (POPS) Protocol focused on reducing the effects of these

* http://www.unece.org/env/lrtap/welcome.html has a link to the text of the Convention and the text of the protocols. It also provides information on the state of ratifications of the Convention and its protocols and has separate sections on the Executive Body and all the Working Groups and Task Forces under it, including their session documents. There is also a link to the scientific centres under the Convention.

substances by banning, restricting, and abating them. After the signing of these two protocols the Executive Body wrote to United Nations Environment Programme (UNEP) to encourage it to develop global legally binding instruments for POPS and cadmium (Cd), lead (Pb), and mercury (Hg). (See Table 10.1 for more information on the reduction protocols.)

ORGANIZATION AND OPERATION

Executive Body and Its Main Subsidiary Bodies

Figure 10.1 shows the organization of the CLRTAP. At the top is the Convention's decision-making body, the Executive Body. The Executive Body, which is the Conference of Parties to the Convention, normally meets once a year in December. The Parties to the protocols in force also use the Executive Body meetings to take decisions on matters such as evaluations of the protocols or amendments to the protocols.

The Implementation Committee assists the Executive Body. The nine members of the Implementation Committee are appointed by the Executive Body and act on a personal title. The Committee monitors compliance with the obligations in the protocols by the Parties and reports to the Executive Body. As with all international environmental conventions, the tools for enforcing obligations are limited. Nonetheless, the Convention has a good record with respect to the implementation of obligations, and the number of cases of Parties failing to comply is falling steadily. The Committee operates on the principle that a cooperative and facilitative approach to those in breach of their protocol commitments is more likely to produce positive results for the Convention and for the environment than a confrontational approach. For the European Union (EU) countries, compliance is somewhat different. Since the European Community is Party to the Convention and all of its protocols, the European Commission implements all of its obligations in EU legislation. EU member states are thus also bound to the protocol obligations through EU legislation. It is important to note that EU countries do not ratify protocols as a whole. Each EU country is responsible for its own implementation and ratification.

The three main subsidiary bodies under the Executive Body are the Working Group on Effects (WGEs), the EMEP Steering Body, and the Working Group on Strategies and Review (WGSR). The first two bodies focus on providing scientific grounds for the decisions of the strategy group, which is the policy-making body and therefore negotiates the development and review of protocols. The Convention is well known for its policy making in close cooperation with science and is often held up as an example (Raes and Swart, 2007). The three bodies and the task forces and coordinating centers under them are discussed in the section "Science and Policy."

Reporting

The Parties to the Convention and its protocols report their emissions every year. Every 2 years they also report their implementation of the obligations in the protocols. The Implementation Committee is an important user of these reports.

TABLE 10.1
Protocols under the CLRTAP

Protocol	Key Obligations	Remarks
EMEP Protocol, Geneva, 1984	Financing of EMEP program by mandatory contributions from the Parties to the Protocol. The EMEP budget covers the coordination costs of the monitoring network, emission data collection, modeling concentrations and depositions, and integrated assessment modeling	The scientific cooperation started many years before the Convention was signed in 1979 and could be seen as the forerunner of the Convention. At that time this cooperation was unique in the Cold War era
Sulphur Protocol, Helsinki, 1985	Reduction of 30% SO_2 by 1993 from the base year 1980	The first "flat rate" reduction protocol. Protocols of this kind with uniform reduction percentages could be called "first-generation protocols"
NO_x Protocol, Sofia, 1988	Stabilization of 1987 NO_x emissions by 1994. Mandatory ELVs for new combustion plants and application of BATs for new and existing large combustion plants and new mobile sources	Twelve countries declared at the signing of the Protocol that they would reduce their NO_x emissions by 30% by 1998
VOC Protocol, Geneva, 1991	Emission reduction of 30% of VOC by 1999 compared to 1988 for all countries except Bulgaria, Greece, and Ukraine (stabilization). Application of ELVs based on BATs for new large combustion plants and new mobile sources. Existing large combustion installations are required to apply BATs. Measures to prevent vapor losses in fuel distribution for motor vehicles and for products containing solvents	VOC is defined as all volatile organic compounds, except methane. Methane does contribute to ozone formation, however, and both gases are greenhouse gases. Methane falls under the Kyoto Protocol, ozone does not. Of the substances contributing to the greenhouse effect, ozone is third in line after CO_2 and methane
Second Sulphur Protocol, Oslo, 1994	Emission ceilings for all member states in 2000/2005/2010. ELVs for large stationary sources and restrictions on fuels	The first protocol with differentiated emission obligations based on the "critical loads approach," a "second-generation protocol"

Heavy Metals (HMs) Protocol, Aarhus, 1998	National emission levels of Cd, Pb, and Hg are to be reduced below 1990 levels (or an alternative base year between 1985 and 1995). Mandatory ELVs and BATs for major stationary sources and unleaded petrol and restrictions on mercury in batteries	ELVs are directed toward dust/particles and therefore also reduce other HMs. Apart from the chlor-alkali industry the Protocol does not address mercury from stationary sources After the signing of the HMs Protocol UNEP explored global measures for Hg, Cd, and Pb. In February 2009 the Governing Council agreed to initiate a process aimed at producing a legally binding agreement for mercury in 2013
Persistent Organic Pollutants (POPS) Protocol, Aarhus, 1998	Elimination of production and use restrictions for 11 pesticides and two industrial chemicals. Mandatory BATs for major sources of three unintentionally released substances (dioxins/furans, polycyclic aromatic hydrocarbons (PAHs), and hexachlorobenzene) and reduction of national emission levels of these substances to below 1990 levels (or an alternative base year between 1985 and 1995). ELVs for waste incineration The POPS Protocol has been revised considerably. The amended Protocol has been adopted by the EB in December 2009 (see section "Challenges and the future")	The POPS Protocol can be seen as the father of the UNEP Stockholm Convention on POPS. The Stockholm Convention is clearly inspired by the Protocol. Its structure resembles that of the Protocol. Yet there are clear differences such as the number of substances (now 23 for the Protocol and 19[a] for the Stockholm Convention, with 5 and 3 in the pipeline, respectively), the submission of new substances (review of a dossier or of a substance), the extent of the measures, and a financial mechanism to dispose of stocks
Protocol to Abate Acidification, Eutrophication, and Ground-level Ozone, Gothenburg, 1999	NEC for SO_2, NO_x, NH_3, and VOC in 2010 for all member states (except Canada and the USA which have no ceilings for NH_3). Mandatory ELVs for new and existing stationary sources and new mobile sources. Restrictions on fuels. Mandatory application of BATs for stationary and new mobile sources. Measures to reduce NH_3 from intensive cattle, pig, and poultry breeding	Like the Second Sulphur Protocol, the emission ceilings are based on effects (acidification, eutrophication, and ozone) and cost-effectiveness, the so-called "critical loads approach." The EU NEC Directive of 2001 implements the emission ceilings of the Protocol. The NEC Directive does not cover the ELVs and BATs of the Gothenburg Protocol

[a] All 19 substances of the Stockholm Convention are all covered by the POPS Protocol. The three isomers of hexachlorocyclohexane (HCH) are counted as one substance (see Table 10.3).

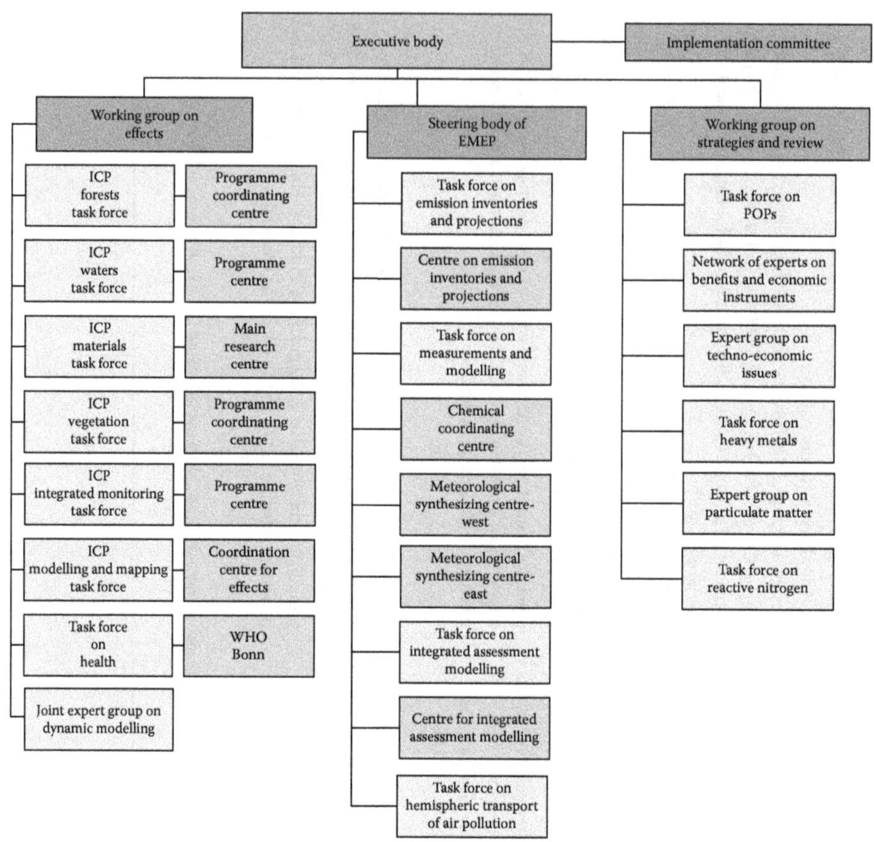

FIGURE 10.1 Organizational setup of the CLRTAP (as of September 2010).

To assist the Parties with emission reporting, the Convention regularly updates the Emission Inventory Guidebook in which Parties can find simple and more advanced methodologies and emission factors for calculating emissions and emission projections for many substances. The Guidebook provides information on how to calculate emissions. The Guidelines on Emission Inventories specify what data should be delivered and when. From 2009 onward, the Parties are obliged to produce an Informative Inventory Report to underpin their emission figures. The EU uses the same emission data for its NEC Directive. The Parties also report emission data to the Climate Change Convention. Because there is a significant overlap with climate change emission reporting (some substances are reported to both the CLRTAP and UNFCCC, energy use, activities, sources etc.), the methodology, the emission data, and the inventory report required under the Convention are fully harmonized with UNFCCC requirements.

A "Strategies and Policies" questionnaire is used for reporting on compliance. Part I of the questionnaire contains questions on the implementation of the obligations of the protocols and has to be answered every 2 years. Part II of the questionnaire is more general and only has to be answered every 4 years. Information related to more general policies, such as types of fuels used, energy conservation, economic instruments, and so on is collected in this part. In both types of reporting, electronic aids such as templates, the Internet, and so on are used.

Consensus Decisions

From the start, decisions have been taken by consensus, although decision making in the Convention is officially by majority voting unless the Convention or its protocols specify that consensus decision is warranted, for example, for the adoption of amendments to protocols. Despite its tendency to dilute and slow down action in tackling the major environmental problems, consensus decision making is "efficient" because the Parties are more likely to respect an Executive Body decision if they subscribe to its terms than if they are driven reluctantly into observance by means of a majority vote. Consensus decision making has proven itself but it can only work if Parties use it to find solutions. It often means that one has to find innovative and flexible solutions and the Convention is well known for that. However, sometimes Parties use it as a veto to block progress.

Implementation and Ratification

The strength of the Convention and its protocols can be deduced from its implementation at the national level (see Table 10.2). Prior to ratification, countries take all legal and other appropriate measures to implement the obligations in order to ensure compliance with them at the time of entry into force. Ratification is therefore a simple indicator for demonstrating implementation of the legal obligations. The contents of Table 10.2 are discussed in more depth in the section "Challenges and the future."

Methodology

The Convention is serviced by a small UNECE secretariat, which administers the sessions and their documentation and helps the bodies carry out their activities under the Convention. What makes the Convention tick is the motivation of the Parties and the people taking initiatives and taking the lead in the work to be done. The work plans of all the working groups, task forces, expert groups, and centers are discussed annually throughout the Convention and adopted by the Executive Body. The work in the groups and task forces is usually led by one or two Parties who often contribute significantly to its work, both in terms of manpower and financially. In addition, Parties hosting centers usually contribute additional manpower and financial resources. One can say that the Convention really works bottom-up. Parties' initiatives are usually taken up in scientific programs and policymaking (protocols, guidelines etc.). Parties that are active are rewarded by their influence on the course the Convention is taking.

Financing

The Convention has two important financial mechanisms to support the scientific work being carried out under it: the EMEP Protocol with its compulsory contributions,

TABLE 10.2
Ratification of the Convention and Protocols in Different UNECE Subregions (as of September 2, 2009)

Instrument—Year of Adoption	EU[a] (28)	EECCA[k] (12)	SEE[l] (7)	Other Europe (7)	North America[b] (2)	Total (56)
LRTAP—1979	28	9	7	5	2	51
EMEP[c]—1983	28	3	5	4	2	42
First sulfur[d] —1985	16	3	1	3	1	24
NO_x[e]—1988	23	3	2	3	2	33
VOC[f]—1991	18	0	1	4	0	23
Second sulfur[g]—1994	22	0	1	4	1	28
HMs[h]—1998	21	1	1	4	2	29
POPS[i]—1998	22	1	1	4	1	29
Gothenburg[j]—1999	21	0	1	2	1	25

Source: Data from ECE. 2007. Implementation of UNECE multilateral environmental agreements, sixth ministerial conference "Environment for Europe," Belgrade, October 10–12, p. 169, 2007, ECE/Belgrade.conf/2007/12.

Note: Numbers indicate the number of states that have ratified each instrument. Numbers in parentheses show the total number of countries in each subregion.

[a] These figures include the European Community.
[b] The United States has existing national instruments with similar provisions.
[c] Protocol on Long-Term Financing of the Cooperative Programme for Monitoring and Evaluation of the Long-range Transmission of Air Pollution in Europe (EMEP).
[d] Protocol on the Reduction of Sulfur Emissions or Their Transboundary Fluxes by at least 30%.
[e] Protocol Concerning the Control of Emissions of Nitrogen Oxides or Their Transboundary Fluxes.
[f] Protocol Concerning the Control of Emissions of Volatile Organic Compounds or Their Transboundary Fluxes.
[g] Protocol on Further Reduction of Sulfur Emissions.
[h] Protocol on Heavy Metals.
[i] Protocol on Persistent Organic Pollutants.
[j] Protocol to Abate Acidification, Eutrophication and Ground-level Ozone.
[k] Eastern Europe, Caucasus and Central Asia (EECCA).
[l] South East Europe (SEE).

and the voluntary Trust Fund for the core activities not covered by the EMEP Protocol. The EMEP Protocol sets the financial requirements of EMEP and divides the budget for the coordination work of the five centers under the EMEP program into Party contributions. Currently, the budget for EMEP is almost $2.4 million/year.

The voluntary Trust Fund sets the budget for the seven centers for effects-oriented work and integrated assessment modeling activities. This Fund also divides its budget into contributions by the Parties, but not all Parties to the Convention pay their attributed contributions. The hosting countries usually pay directly to the center they host. This accounts for 50% of the total budget of the centers. Roughly 25% of the budget is donated by other Parties directly to the fund. This 25% entering the voluntary

Trust Fund has up to now been distributed evenly between the centers. The resulting situation is that some centers receive most of their budget while others are not so well off. The budget for this fund is just over $2.1 million/year.

A third Trust Fund provides support for countries with economies in transition (Eastern Europe, Caucasus and Central Asia (EECCA) and South-Eastern Europe (SEE) countries) to facilitate their participation in the activities and their implementation and ratification of the protocols. This Fund pays the travel expenses of participants from countries with economies in transition to enable them to take part in the Executive Body and the Working Groups, as well as funding workshops, translations of documents, and so on. The Fund is made up of voluntary donations. However, only a few Parties donate to this Fund. The Trust Fund is also used for some bilateral projects between several member states of the Convention.

SCIENCE AND POLICY

EMEP Steering Body

Science has always underpinned the Convention. As long ago as 1974, well before the CLRTAP was signed in 1979, there were meetings of a task force charged with developing a program for monitoring and evaluating the long-range transmission of air pollutants. Since then the work of the EMEP Steering Body has evolved to what it is today: an important scientific network on emissions, dispersion modeling, coordination of air monitoring networks, and integrated assessment modeling, covering five centers (see Figure 10.1). The Centre on Emission Inventories and Projections (Austria) collects the Parties' emission data and coordinates the review of the data. The two Meteorological Synthesizing Centres (MSC-West, Norway, and MSC-East, Moscow) use the emission data to model concentrations and depositions of air pollutants, the West center focusing on traditional air pollutants and the East center focusing on HMs and POPS. The models are verified and calibrated using monitoring data collected by the Chemical Coordination Centre in Norway. This center coordinates the pan-European monitoring network. The budget of the EMEP Protocol is for the coordination of the work. The Parties participate at their own expense, that is, they pay their own travel expenses for attending Task Force meetings. The Parties also pay for the monitoring sites themselves.

Integrated Assessment Modeling

The Centre for Integrated Assessment Modelling is located at the International Institute for Applied Systems Analysis (IIASA) in Austria. IIASA operates the well-known RAINS/GAINS model for calculating cost-effective scenarios, which are used for policy making (IIASA, 2009). Both the Convention and the EU use this model, which brings together all the knowledge assembled in the Convention. The model consists of the EMEP dispersion model for calculating concentrations and depositions. The model is fed with activities such as energy use, types of industrial plant, cars, lorries, livestock farming and so on, together with their emissions and abatement possibilities, including costs and emission reductions. The model also contains data on critical loads and levels (sustainable depositions and sustainable concentration levels) for the whole of Europe. The model can calculate the most cost-efficient abatement solution for all kinds of environmental targets. The RAINS model has been used for the Second Sulphur Protocol, the Gothenburg Protocol, and the EU's NEC Directive. In recent years the RAINS model, which covers the major traditional air pollutants, has been expanded to include greenhouse gases—the GAINS model—with the result that it is now possible to calculate integral solutions for both problems or map out the consequences of air pollution policy measures on climate change and, conversely, of climate change measures on air pollution.

Working Group on Effects

Soon after the signing of the first SO_2 Protocol in 1985, the Convention started work on the effects of air pollutants. The WGE is in charge of the effects work under the Convention and is assisted by six International Cooperative Programmes and a Task Force on Health. The Task Force on Health is led by the WHO in Bonn (Germany). Five of the six programs coordinate an effects monitoring network: Forests, Waters, Vegetation, Materials, and Integrated Monitoring. The coordinating centers for these monitoring networks are hosted by Germany (Forests), Norway (Waters), United Kingdom (Vegetation), Sweden (Materials), and Finland (Integrated Monitoring). The ICP on Mapping and Modelling is responsible for "critical loads" maps that are used for policy making. The Coordinating Centre for Effects hosted by the Netherlands gathers the information from its National Focal Centres and partly from other ICPs, fills in the gaps, and produces pan-European maps that are used in integrated assessment modeling. Again, the budget of the voluntary Trust Fund only covers the coordinating activities. The Parties participate at their own expense, that is, they pay their own travel expenses for attending Task Force meetings. And again, the Parties pay for the monitoring sites themselves. The effects networks play an important role in the policy process. They produce the critical loads and levels that are the basis for further reductions. Also, they monitor the effects related to concentrations and depositions and thus show trends and assess whether policy is successful. Other activities include dynamic modeling to incorporate the time needed for ecosystems to recover, dose–response functions, stock at risk, and valuation of stock. These data can be used to calculate damage in monetary terms and thus also the benefits to be obtained when concentrations and depositions decrease.

The EMEP Steering Body and the WGEs hold regular joint bureaux meetings to harmonize their scientific work. Both bodies have developed a new long-term strategy for 2010–2019. For this purpose they consult the Working Group on Strategy and Review and the Executive Body who set the priorities from a policy point of view. These two scientific legs under the Convention therefore deliver the scientific basis that is used to design and assess policies to abate transboundary air pollution.

Working Group on Strategies and Review

All policy matters are negotiated under the WGSR. The main task of the WGSR is to review and revise existing protocols and develop new ones. The WGSR also deals with other matters involving policy choices, such as the Guidelines on Emission Inventories and the Guidelines on Effects. The WGSR is supported by the Task Forces on POPS and HMs with respect to matters related to these two protocols. The other four groups play a role in the more traditional air pollutants covered by the Gothenburg Protocol.

CHALLENGES AND THE FUTURE

After thirty years one can say that the Convention is mature and has proved to be an important instrument in the abatement of air pollution in the UNECE region. Many initiatives developed under the Convention have found their way into other conventions and other regions of the world. This is a satisfying observation, but it should not be an excuse to rest on one's laurels. Air pollution is not solved yet. Among the immediate challenges lying ahead for the Convention is the failure of countries with economies in transition to implement and ratify protocols and the revision of the three last protocols. Looking further ahead, when these protocols are revised the Convention will need to update its strategy to cope with the future, the interaction with climate change being the most pressing issue.

Improving the Participation of EECCA and SEE Countries

An important challenge for the Convention is to encourage Eastern and South-eastern European countries to participate. These countries have emerged from the collapse of the Soviet system, and their current geopolitical context is completely different from what it was in the first 10 years of the Convention. These new countries have had to contend with serious economic problems and political instability. As a consequence, they lag behind in implementing and ratifying the protocols of the Convention (See EECCA and SEE in Table 10.2). The Convention is taking up this challenge in many ways, with projects, bilateral cooperation, capacity building, meetings in these countries, and above all with the EECCA action plan. A project to stimulate five SEE countries to ratify the three most recent protocols was developed in 2007. Another important initiative to help EECCA and SEE countries sign up to the protocols is to introduce greater flexibility (e.g., more time to implement Emission Limit Values (ELVs) for existing installations) in the three most recent protocols, which are currently being revised. A special questionnaire for EECCA and SEE countries published at the beginning of 2009 has initiated momentum in this respect.

The Convention and the EU

Since the start of the Convention the EU has grown from 6 to 27 countries all of which are Party to the Convention and most of its protocols. Also, the European Community is Party to the Convention and its protocols. The European Commission sees to it that all EU Member States implement all the provisions of the protocols. This growing EU clearly influences the operation of the Convention. With half of the total Parties the EU dominates what happens in the Convention. Most initiatives in the Convention are taken by individual EU countries that first have to persuade the other EU countries and the European Commission before the EU as a whole can submit, for instance, an amendment to a protocol. EU rules prescribe this just as "speaking with one tongue" in negotiations. This development over the last few years resulted in much prework and coordination between EU Member States and the European Commission. Yet, the Convention largely depends on the EU. Reasons for this are that

- The countries East and Southeast of the EU have faced severe problems in their economic development and have difficulty in setting up their environmental laws and regulations.
- The United States and Canada often find it hard to agree to international regulations.
- Norway and Switzerland, the largest of the rest of the countries within the ECE, already follow the directives and regulation of the EU related to environment.

The European Commission increasingly questions the added value of the Convention. As a first priority the Commission feels that especially the EECCA and SEE countries should implement and ratify the existing protocols. Many EU countries share the vision that more ratifications are essential but they also feel that there is still an important role to play for the Convention both in the further development of policy and in the development of the science under the Convention.

Revision of the Protocols

The Convention is in the process of updating the three most recent protocols, the POPS Protocol, the HMs Protocol, and the Gothenburg Protocol. The objective is to further reduce the effects of air pollutants by taking more measures, updating Best Available Techniques (BATs) and ELVs, modernizing and streamlining obligations, building in more flexibility for countries with economies in transition, and so on.

The negotiations to amend the *POPS Protocol* have been finalized in 2009. The Executive Body adopted an amended Protocol in December 2009. The major revisions are the incorporation of seven substances (the Protocol now covers 23 substances), the updating of BATs and ELVs including ELVs for a few new sources, the introduction of flexibility for EECCA and SEE countries, and an expedited procedure to amend annexes in the future. This expedited procedure will enable Parties to opt out of the amendment instead of ratifying it (opting in). Five new substances that were submitted by the EU and Norway in 2008 will not form part of the current amendment of the POPS Protocol. The review of these five substances has been finalized in 2010, after which their inclusion in the annexes of the Protocol is currently being negotiated. The expedited procedure to amend annexes will be used for these five substances. The POPS Protocol is the father of the Stockholm Convention (2004), which is modeled after the POPS Protocol. The POPS Protocol continues to drive the Stockholm Convention as can be seen by the substances overview in Table 10.3. In this respect it is peculiar to see that the European Commission and some Member States of the EU question further progress of the Protocol. One could judge

TABLE 10.3
Substances in the POPS Protocol and in the Stockholm Convention

Substance	Type of Substance[a]	POPS Protocol[b]	Stockholm Convention[b]
Aldrin	P	X	X
Chlordane	P	X	X
Chlordecone	P	X	X
DDT	P	X	X
Dieldrin	P	X	X
Endrin	P	X	X
Heptachlor	P	X	X
Hexabromobiphenyl	C	X	X
Hexachlorobenzene	C, U	X	X
Mirex	P	X	X
Polychlorinated biphenyls (PCBs)	C, U	X	X
Toxaphene	P	X	X
Lindane (γ-HCH) and α- and β-HCH	P	X	X
Polycyclic aromatic hydrocarbons (PAHs)	C, U	X	
Dioxins/Furans	U	X	X
Hexachlorobutadiene (HCBD)	P, C	X	
Polychlorinated naphthalenes (PCNs)	P, C, U	X	
Pentachlorobenzene (PCB)	P, C, U	X	X
Pentabromodiphenylether (PeBDE)[c]	C	X	X
Perfluorooctanesulfonate (PFOS)[d]	C	X	X
Octabromodiphenylether (OctaBDE)[e]	C	X	X
Short-chain chlorinated paraffins (SCCP)	C	X	Under review
Dicofol	P	A	
Endosulfan	P	A	Under review
Hexabromocyclododecane (HBCD)	C	A	Under review
Pentachlorophenol	P	Under review	
Trifluralin	P	Under review	

[a] The substances can be categorized as P: pesticide; C: chemical use, for example, flame retardant; and/or U: unintentional release, for example, burning of wastes.

[b] X: incorporated in one of the annexes of the POP Protocol or Stockholm Convention. A: accepted as POP but not yet taken up in the POP Protocol.

[c] Tetrabromodiphenyl ethers and pentabromodiphenyl ethers present in commercial-PentaBDE.

[d] PFOS and related substances.

[e] Hexabromodiphenyl ethers and heptadiphenyl ethers present in commercial-OctaBDE.

this as a typical example of parricide. Yet, if the Stockholm Convention develops into a well functioning institution and encompasses the substances of the POPS Protocol, then there is no more need to update the Protocol with more substances in the future after the incorporation of the five substances in the pipeline.

The review of the *HMs Protocol* was completed in 2006. Work started on the costs and benefits for a revised HMs Protocol in 2007. In 2008 further work was done to update BATs and ELVs and to address mercury emissions from stationary sources, which are currently not included in the Protocol. In 2008 the EU submitted a proposal to add a number of mercury-containing products to the Protocol. Furthermore, in 2009 the EU and Switzerland submitted proposals to update the HMs Protocol. In December 2009 the Executive Body mandated the WGSR to start negotiations to revise and amend the HMs Protocol in 2011. A revision of the Protocol could be finalized rather quickly, since all the preparatory work have already been done. Furthermore, negotiations for the revision could benefit from the work done for the amended POPS Protocol. The revision could potentially lead to the same kind of improvements as for the POPS Protocol: the updating of (BATs and ELVs including ELVs for mercury from stationary sources, the addition of certain mercury-containing products, the introduction of flexibility for EECCA and SEE countries and an expedited procedure to amend annexes in the future. A revised HMs Protocol with mercury ELVs for stationary sources and more mercury product measures would lead to substantial reductions of mercury emissions in the UNECE region. In addition, the revised HMs Protocol could again have a major impact on the UNEP process to address mercury emissions. UNEP started work on HMs after the 1998 HMs Protocol, and in February 2009 the Governing Council of UNEP opted for a legally binding instrument to be finalized in 2013.

In 2007 the review of the *Gothenburg Protocol* was finalized and work started on its revision. The Parties stated that they wished to incorporate particulate matter (PM) in the Protocol. It is intended that the revised Protocol will have NEC for 2020 for five substances: SO_2, NO_x, VOC, NH_3, and $PM_{2,5}$. The Parties also called for the synergies and trade-offs with climate change to be taken into account and the nitrogen cycle included in the scenario calculations. In terms of the five substances, this would mean a second (NH_3), third (NO_x, VOC), or even fourth (SO_2) round of reductions for land-based sources. This is in contrast to emission from ships, an area in which little has so far been done to abate emissions. Further reductions for land-based sources would have been difficult had this situation persisted. Fortunately, IMO decided in 2008 to reduce ship emissions at sea. Two new developments are to attempt to set nonbinding aspirational targets for 2050 in line with targets set for climate change and to explore a climate goal for air pollutants because some air pollutants effect our climate as "short-lives climate substances." The GAINS model will be used to prepare for the negotiations on the NEC for 2020. And, of course, the Protocol will be revised with the updated BATs and Emission Limit Values, flexibility for EECCA and SEE countries, and an expedited procedure to amend annexes in the future. The revised Gothenburg Protocol should be ready for adoption by the Executive Body by the end of 2011.

Long-Term Strategy

In 1999, after the finalization of the three most recent protocols, the Convention produced a long-term strategy. Following the revision of the POPS, HMs, and Gothenburg Protocols in 2009–2011, this strategy needs to be modified. The first few years would obviously be devoted to the implementation and ratification of the revised protocols with special attention to the EECCA and SEE countries. Now that acidification to a large extent is solved, the Convention will need to focus more on health (ozone, particles), nitrogen (biodiversity), and climate change. This calls for a new setup of the effects related work under the Convention also to contribute more to integrated assessment modeling and cost–benefit analyses. The Convention also has to evaluate how to continue with POPS and HMs. An interesting idea is to follow up on the three protocols around 2020 with just one "Multieffect/Multipollutants (M&M) Protocol" with BATs and ELVs per sector but without provisions on the production and use of POPS of the current POPS Protocol. Such an "M&M Protocol" would also have national ceilings for the major pollutants for 2030 based on integrated assessment modeling and cost–benefit analyses. A very important element in the new strategy would be the relationship between and linkage of air pollution policy with climate change policy. An interesting question in this respect is whether there should still be separate policies for air pollution and climate change in 2030.

ACKNOWLEDGMENT

I would like to thank the Secretariat of the CLRTAP for their input and review of this chapter.

REFERENCES

ECE, 2007, Implementation of UNECE multilateral environmental agreements, sixth ministerial conference "Environment for Europe," Belgrade, October 10–12, 2007, ECE/Belgrade.conf/2007/12.

IIASA, 2009, http://www.iiasa.ac.at/rains/index.html, IIASA Web site on the RAINS and GAINS model.

Raes, F. and R. Swart, 2007, Climate assessment, What's next? *Science*, 138: 1386.

Sliggers, J. and W. Kakebeeke, eds., 2004, *Clearing the Air: 25 Years of the Convention on Long-Range Transboundary Air Pollution*. UN Publication Sales no. E.04.11.E.20, ISBN 92-1-116910-0.

11 The Globally Harmonized System of Classification and Labelling of Chemicals

The System, Its History and Context, and the Future of Implementation

Cheryl Chang[*]

CONTENTS

[*] The author has contributed this work in her individual capacity. Any views or opinions presented in this chapter are solely those of the author and do not necessarily represent those of the organization(s).

INTRODUCTION

Chemicals play an integral part of our daily lives and are a major component of our economy. The role of chemicals in international trade continues to grow, as globalization expands its reach and complexity. While chemicals can greatly enhance our quality of life, in the absence of effective management measures, they also have the potential to harm us and our environment. To protect human health and the environment, at every point in the chemical life cycle—in production, storage, transport, and consumption, we must make a concerted effort to manage chemicals in a sound and responsible way. This includes communicating information on chemical hazards, and their safe handling and use through consistent and comprehensible means, so that people can take the necessary actions to protect themselves, and to prevent adverse release of chemicals into the environment.

In support of the increasing need for sound chemicals management, the Globally Harmonized System for Classification and Labelling (GHS) was developed. The GHS is a system for classifying and labeling chemical hazards, or, in other words, for identifying the intrinsic hazards of a chemical substance or mixture, and for communicating these hazards to target audiences. The GHS is an international chemicals management tool developed through international discussions and consensus. It is a consistent and comprehensive system that supports international, regional, and national goals for protecting human health and the environment and facilitating trade.

The overall objectives of the GHS are to

- Enhance the protection of human health and the environment by providing an internationally comprehensible system for hazard communication.
- Provide a recognized framework for countries without existing hazard classification and communication systems to adopt an internationally recognized and consistent system.
- Reduce the need for testing and evaluation of chemicals.
- Facilitate international trade in chemicals whose hazards have been properly assessed and identified on an international basis.

The GHS can be adopted by countries without an existing hazard classification and/or communication system, and can also be a tool for countries with existing systems to streamline or regularize several competing systems within their own countries.

WHY WAS THE GHS DEVELOPED AND WHY IS IT IMPORTANT?

To illustrate the importance of the GHS, consider some hypothetical examples which reflect the reality of chemical use around the world:

Imagine you are a furniture manufacturer in China. One of your workers accidentally spills a drum of wood coating. Unfortunately, the safety data sheet (SDS) accompanying the chemical provides no information on treatment in the case of accidental release, and you are not sure of the best way to clean up the spill while minimizing harm to your workers and the environment.

Imagine you are a farmer in Ghana. You purchase some pesticides at the local market to protect your crops against pests. However, you have no idea how to apply them, as the bottle comes with no accompanying information on how to use or handle the chemical. Therefore, you simply dilute the solution and pour a little over each crop with a cup you dip in a bucket of the solution. Later on your hand feels swollen and itchy.

Imagine you are a mother of two in the suburbs of Japan. You go to the store to buy some household cleaner to clean the kitchen. On one bottle there is a label with a red X, which you guess indicates danger. Another cleaner is labeled with a skull and cross bones, which you also guess indicates something toxic or deadly. However, you are not sure. Which chemical is dangerous? Which chemical is more dangerous?

These examples highlight the current lack of consistency and availability of information on chemicals and their safe use and disposal, both in developed and developing countries.

Consider also the current national and international situation of chemical hazard classification and communications around the world. In some cases, countries do not have existing systems, and this can expose their citizens and the environment to potential harm from inadvertent misuse of chemicals. These countries may receive only the chemical hazards information provided by exporters, which can be insufficient, incorrect, or incomprehensible. Other countries have developed regulatory systems for dealing with hazardous chemicals classification and communication. However, in some countries these efforts may be limited or insufficient, leaving citizens vulnerable to chemical hazards. In countries that do have comprehensive chemical management systems, there can be different regulations and standards for the various stages of chemicals' life cycle and use. For example, many countries have separate regulations for managing hazardous chemicals in industrial workplaces, transport, or agriculture. In many cases, this leads to duplication of efforts, confusion about jurisdiction, or inconsistent coverage within a country.

This effect is multiplied when chemicals trade between countries is taken into account. And even in cases where national chemicals hazard classification and labeling systems between countries are similar, slight differences in criteria can lead to divergent information on the degree of hazard for the same product. For example, what might be considered as flammable in one country could be considered combustible in another. This has resulted in a "patchwork" of diverse and, at times, conflicting national and international systems for hazard classification and communication, which can result in miscommunication on chemicals use and handling, decreased safety for human health and the environment, as well as create unnecessary barriers to international trade. As the above examples and description make clear, a system, such as the GHS, is vital for supporting the responsible and consistent management of chemicals.

HISTORY OF THE GHS

The GHS was developed as a result of an international mandate agreed at the 1992 United Nations Conference on Environment and Development (UNCED) to develop and implement "a globally harmonized hazard classification and compatible labeling system, including material SDSs and easily understandable symbols" (see Chapter 3 of this book for full reference to the Rio Conference). In support of this, the International Labor Organisation (ILO), in the context of its Convention and Recommendation on Safety in the Use of Chemicals at Work, reviewed the requirements necessary to ensure successful achievement of this mandate. Their research found that there were four major existing systems that could be regularized: the UN Transport Recommendations; the US Requirements for Workplace, Consumer and Pesticides; the European Union (EU) Dangerous Substance and Preparations Directives; and the Canadian Requirements for Workplace, Consumers and Pesticides.

As a next step, the Inter-Organisation Programme for the Sound Management of Chemicals (IOMC) created the Coordinating Group for the Harmonization of Chemical Classification Systems (CG/HCCS) to coordinate development of the system. The CG/HCCS was comprised of representatives from national governments, industry, and labor. They worked on a consensus basis by a set of principals which applied to all related groups working to develop different aspects of the system. These principals were that: protection would not be reduced, the system would be based in intrinsic properties (hazards) of chemicals, all types of chemicals would be covered, all systems would have to be changed, involvement of all stakeholders should be ensured, and comprehensibility must be addressed.

Three technical focal points were designated for developing the system. The Organisation for Economic Cooperation and Development (OECD) was assigned the work of health and environmental hazards and mixtures, the UN Committee of Experts on the Transport of Dangerous Goods (UNCETDG) in cooperation with the ILO, were selected to take the lead on developing physical hazards, and the ILO was chosen to lead the work on hazard communication. These groups continue to play an active role in the further development of the system.

Around the efforts to create the GHS, an international infrastructure developed to support the system as it evolved. In 1999, the United Nations Economic and Social Council (ECOSOC) enlarged the mandate of the UNCETDG, and the group was renamed the UN Committee of Experts on the Transport of Dangerous Goods and on the Globally Harmonized System of Classification and Labelling of Chemicals (UNCETDGGHS). With this enlarged role, the Committee formed two sub-groups: the UN Sub-Committee of Experts on the Globally Harmonized System of Classification and Labelling of Chemicals (UNSCEGHS) and the UN Sub-Committee of Experts on the Transport of Dangerous Goods (UNSCETDG). The United Nations

Institute for Training and Research (UNITAR) and the ILO were also nominated as the focal points for capacity building to support GHS implementation.

In 2002 the IOMC CG/HCCS completed the development of the GHS, and presented it to the UNSCEGHS where it was adopted in December 2002. Subsequently, the UNCETDGGHS further endorsed the system, and by July 2003 the system had also received the endorsement of ECOSOC.

The UNSCEGHS plays an important role in maintaining the GHS. It is comprised of representatives from participating countries, with a chair and two vice chairs. The subcommittee meets twice a year to discuss relevant issues related to the revision, updating, and promotion of the GHS. It is supported by a Secretariat based in the UN Economic Commission for Europe. The responsibilities of the UNSCEGHS are to

- Act as a custodian of the GHS, through management and direction of the harmonization process.
- Maintain the system by keeping it up-to-date, and introducing changes and updates as necessary.
- Promote understanding and use of the system worldwide.
- Encourage feedback on the system.
- Make guidance available on the system, including application, interpretation, and use of technical criteria.
- Prepare programs of work and submit recommendations to the UNCETDGGHS.

INTERNATIONAL CONTEXT ON THE GHS

Implementation of the GHS supports efforts related to international and national sustainable development. Of the United Nations Millennium Development Goals,* Number 7† is to "ensure environmental sustainability." As part of this, it was recommended that efforts should be made to reduce "exposure to toxic chemicals in vulnerable groups" and to "improve frameworks for chemical management." An important goal agreed at the World Summit on Sustainable Development in Johannesburg, South Africa in 1992 was to "achieve by 2020 that chemicals are used and produced in ways that lead to the minimization of significant adverse effects on human health and the environment" (see Chapter 4 of this book). Finally, Chapter 19 of *Agenda 21* recognizes the need to protect vulnerable groups from toxic chemicals.

The GHS also supports a number of international agreements and efforts related to sound chemicals management, including several of which are written in detail in this book. For example, the Strategic Approach to International Chemicals Management (SAICM), a framework for international action on chemicals management, notes the importance of the GHS in its Overarching Policy Strategy (OPS). GHS is also included as a work area in the SAICM Global Plan of Action, with eight distinct activities directly related to the GHS. The Rotterdam Convention has close links to hazard identification and communication issues and the GHS through its efforts to have countries monitor and control trade in certain hazardous chemicals. This includes requirements

* http://www.un.org/millenniumgoals/

† http://www.un.org/millenniumgoals/environ.shtml

for countries to ensure that chemicals used for occupational purposes have SDSs that follow an internationally recognized format, which can be considered a reference to the GHS. In order to promote further coordination, the UNSCEGHS and the Basel Convention on the transboundary movement of hazardous waste have established a correspondence working group. The Stockholm Convention also refers to the use of SDS and other means of hazard communication. ILO Convention 170 refers to the evaluation of chemical hazards and provision of hazard information in the workplace, and the objective of ILO Recommendation 177 is to protect workers against the risks associated with the use of chemicals in the workplace through requirements for classification and labeling. Both the Food and Agriculture Organization (FAO) and World Health Organization (WHO) have stated that they plan to update their relevant pesticide and chemicals management tools to be in line with the GHS, including the International Code of Conduct on the Distribution and Use of Pesticides, the WHO Recommended Classification of Pesticides by Hazard, and the IPCS (International Programme on Chemical Safety) Chemical Safety Cards.

THE GHS PURPLE BOOK

The GHS is described in a book, informally known, as the "Purple Book." The cover of the GHS Purple Book is shown in Figure 11.1. This book contains information on all the classification classes and categories, and outlines all the requirements for hazard communication. In a series of annexes, it also provides guidance on applying the GHS, including guidance on the preparation of SDSs, consumer product labeling based on the likelihood of injury, and comprehensibility testing methodology. This document is the primary information source on the GHS, but other technical assistance tools have been and will be developed to assist and promote implementation. The Purple Book is revised on a regular basis to incorporate agreed changes and updates to the GHS. Currently, it is in its third revised edition.

For the transport sector, GHS is integrated into the UN Recommendations on the Transport of Dangerous Goods, Model Regulations (UNRTDG). Therefore, the implementation of the GHS in the transport sector is through the UNRTDG. These model regulations are updated regularly and incorporate the latest version of the GHS Purple Book. The most recent version of the regulations is the 16th revised edition, which incorporates the third revised edition of the GHS Purple Book.*

THE SCOPE OF THE GHS

The GHS standardizes and harmonizes classification and labeling of all types of chemicals by defining

- The physical, health, and environmental hazards of chemicals.
- Classification processes for using existing data on chemicals to compare with defined hazard criteria.
- The format and information to be provided on labels and SDSs.

* http://www.unece.org/trans/danger/publi/unrec/rev16/16files_e.html

FIGURE 11.1 Cover of the GHS Purple Book.

The GHS covers all hazardous chemicals, and has the goal of identifying the intrinsic hazards of chemical substances and mixtures, and to communicate hazard information about these chemicals.

It should be noted that the GHS does not require or define testing. It is only required that tests to determine hazardous properties be conducted according to internationally recognized scientific principles. Further, the GHS criteria are test-method neutral, allowing different approaches as long as they are scientifically sound and validated according to international procedures and criteria already accepted in existing systems for a particular hazard class. Since the GHS is based on the use of existing data, compliance with these criteria will not require retesting of chemicals for which accepted test data already exists.

GHS CLASSIFICATION

The first step of the GHS is to identify the hazards of a chemical. The GHS evaluates the physical, health, and environmental hazards of chemicals in the following hazard classes:

Physical Hazards

- Explosives
- Flammable gases
- Flammable aerosols
- Oxidizing gases
- Gases under pressure
- Flammable liquids
- Flammable solids
- Self-reactive substances and mixtures
- Pyrophoric liquids
- Pyrophoric solids
- Self-heating substances and mixtures
- Substances and mixtures which, in contact with water, emit flammable gases
- Oxidizing liquids
- Oxidizing solids
- Organic peroxides
- Corrosive to metals

Health Hazards

- Acute toxicity
- Skin corrosion/irritation
- Serious eye damage/eye irritation
- Respiratory or skin sensitization
- Germ cell mutagenicity
- Carcinogenicity
- Reproductive toxicity
- Specific target organ toxicity—single exposure
- Specific target organ toxicity—repeated exposure
- Aspiration hazard

Environmental Hazards

- Hazardous to the aquatic environment
- Hazardous to the ozone layer

Within each of these classes, the GHS requires the identification of the hazard category, that is, the severity of each hazard. The number of categories and how they are divided depends on the type of hazard class. Therefore, any application of the GHS would require careful review of the system.

GHS and Classification of Mixtures

Classification of mixtures is given special discussion within the GHS because it can be difficult to identify the hazards of the separate substances in the context of a solution. The GHS takes a tiered approach to classification of mixtures based on the following steps:

1. Where test data is available for the mixture, classification should be based on that data. (There are exceptions for carcinogens, mutagens, and reproductive toxins.)
2. Where test data is not available for a particular mixture, then bridging principles relevant to the specific hazard class should be used.
3. If neither applicable test data is available, nor the bridging principles can be applied, then calculation or cut off values from a particular hazard class can be used.

For further details and description of bridging principles, calculation, and cutoff values, see the Purple Book, Section 1.3.

GHS HAZARD COMMUNICATION

The second important part of the GHS is the communication of chemical hazards. Based on the hazard classes and categories identified, the GHS requires that these hazards be communicated through chemicals' labels and SDSs.

GHS LABELS

For a GHS label, the information required includes

- Signal word: A signal word denotes the relative level of severity of a hazard and alerts the reader to a potential hazard on the label. The signal words used in the GHS are "Danger" and "Warning," where "Danger" indicates a more severe hazard category and "Warning" indicates a less severe hazard category.
- Hazard statement: A hazard statement is a phrase assigned to a hazard class and category that describes the nature of the hazards of a chemical, including, where appropriate, the degree of hazard. Examples of hazard statements include, "extremely flammable gas," "fatal if swallowed," or "toxic to aquatic life."
- Precautionary statements and pictograms: A precautionary statement is a phrase and/or pictogram that describes recommended measures that should be taken to minimize or prevent adverse effects resulting from exposure to a chemical, or improper storage or handling of a chemical. A pictogram is a graphical composition that may include a symbol plus other graphic elements, such as a border, background pattern, or color that is intended to convey specific information. The GHS provides information on what pictograms to use for given hazard classes and categories in the sectors related to supply and use. The UNRTDG defines the pictograms for the transport sector. Examples of GHS pictograms are shown in Figure 11.2.

FIGURE 11.2 Examples of GHS pictograms.

- Product and supplier identification: A GHS label should include a product identifier and it should match the product identifier used on the SDS. The name, address, and telephone number of the manufacturer or supplier of the substance or mixture should also be provided on the label.

The way the label is formatted depends on the intended user of the label. The GHS defines minimum standards for the actual format or layout of the label. Authorities may chose to specify how the information should appear, or leave it to the discretion of the supplier.

GHS Safety Data Sheets

An SDS is a comprehensive document provided for workplace chemical management. The GHS provides guidelines on when SDSs should be used. For GHS SDSs, the system requires a 16 heading format in the specific order listed below:

1. Identification
2. Hazard(s) identification
3. Composition/information on ingredients
4. First-aid measures
5. Fire-fighting measures
6. Accidental release measures
7. Handling and storage
8. Exposure controls/personal protection
9. Physical and chemical properties
10. Stability and reactivity
11. Toxicological information
12. Ecological information
13. Disposal considerations
14. Transport information
15. Regulatory information
16. Other information

Product Life Cycle and Application of the GHS

The target audience and stage in a product's life cycle are important considerations when applying the GHS hazard communication elements. While all hazardous chemicals should be classified, how these hazards are communicated will depend on the stage of a product's life cycle including development, manufacturing, storage, transport, sale, use, and disposal.

Target Audiences

While the GHS has the benefit of improving overall chemicals management, the system specifically targets transporters, workers, consumers, and emergency responders. For transport, the implementation of the GHS is based on the application of the UNRTDG. Containers of hazardous chemicals will be marked with pictograms that address acute toxicity, physical hazards, and environmental hazards. For workers and the workplace, it is recommended that all GHS elements will be adopted, including labels and SDSs. For all workers, including transport and emergency responders, training should be provided to help ensure effective communication and comprehension of hazard communication elements. For the consumer sector and certain industrial workplaces, in particular smaller enterprises, labels are the primary focus of GHS application. These labels will include the core elements of the GHS, subject to some sector-specific considerations in certain systems.

IMPLEMENTING THE GHS

The GHS is a tool for countries without existing systems to adopt an internationally recognized system for chemicals classification and labeling. Even within a country, it can be a basis for streamlining and providing further consistency to often disconnected systems within the same country among different agencies. Between countries, it can help to improve the flow of safe chemicals-handling information and improve efficiencies in international trade.

Building Block Approach

While the overall goal is for the GHS to be implemented consistently worldwide, developers of the system wanted to make a chemical management tool that could be adapted to the particular needs and circumstances of different countries. Therefore, the GHS was designed as a series of modules or "blocks" which could be built upon to form a regulatory approach that is adapted to various audiences and sectors. With this "Building Block Approach" (BBA), while the whole system is available, countries can chose not to adopt the full range. For example, while physical hazards are important in the workplace and for transport, certain physical hazards may not be relevant for consumers for the way they use a product. In general, the GHS notes that

a. Hazard classes are building blocks: Within their jurisdiction and keeping in mind the goal of full harmonization as well as international conventions, competent authorities may decide which hazard classes they may not apply.
b. Within a hazard class, each hazard category can be seen as a building block: For a given hazard class, competent authorities have the possibility not to apply all categories.

The GHS describes the BBA, but there continues to be discussion at the international level as to the exact definition of the "BBA." This could mean that countries only adopt certain parts of the GHS, or that they take a stepped approach to implementing the whole system, integrating parts of the system over time. Others assert that the GHS could not truly be a harmonized system unless it is full and consistently applied.

Timeframes and Transition Periods

In general, there is no mandated schedule for implementing the GHS worldwide. The agreed international target date for implementation of the GHS was 2008. Many countries and the private sector used this as a general goal to direct and motivate their efforts. Although this target date has passed, there is now sufficient momentum among countries to advance to the goal of global implementation. Countries may choose to implement the GHS based on their own particular circumstances. However, the timeframes and transition periods of regional and international trade partners, as well as the varying needs of the different sectors that will implement the GHS within a given country, also are important considerations when planning for GHS implementation.

Defining GHS Implementation

As countries around the world progress toward implementing the GHS, the actual definition of "implementation" has also been discussed. According to some, GHS implementation is simply defined as completed once the legislation based on the system is in place within a country. Others argue that implementation can only be defined as achieved once legislation is in place and industry consistently applies the system. Still others take it further and make the case that the GHS cannot be fully implemented until legal instruments are established, industry fully uses the system, and stakeholders, including chemicals users, are fully able to comprehend the GHS labels and SDSs.

In all cases, enforcement is the key to an effective implementation.

GHS IMPLEMENTATION AROUND THE WORLD

Countries around the world are now in the process of working toward GHS implementation. Because each country is at a different stage of implementation, it is worth it to look at some of the key players in chemicals production and consumption. Many countries are looking especially to align their own implementation efforts with these leaders because of their role in international trade.*

Brazil

Since 2001, Brazil has been active in GHS capacity building, and has conducted numerous GHS awareness raising activities to inform industry and the public of the

* The latest information on worldwide implementation and a full list of available information on implementation in various countries can be found at: http://www.unece.org/trans/danger/publi/ghs/implementation_e.html

system. In 2007, the President of Brazil signed a decree formalizing a GHS Working Group, which has the responsibility to elaborate and propose strategies, guidelines, programs, and plans of action for GHS implementation. The GHS Working Group meets regularly to consider implementation issues including revision of legislation to be in line with GHS, transitional periods, application of the BBA, and so on. The Brazilian Association of Technical Standards (ABNT) issued standards for terminology, classification, labeling, and SDS based on the GHS. These standards are currently out for public comment and the approved standards will be published in the coming year. Related to the workplace, in 2009, the Brazilian Health and Safety Tripartite Commission approved the revision of the Ordinance 26 (1978), which deals with Hazard Communication of Chemicals to be in line with GHS. Based on this, Brazil will begin to elaborate a technical text and raise further awareness. The text has gone to public consultation in 2010 and it is expected that decisions will be taken in 2011. Currently, the Ministry of Health is in the process of translating the Purple Book into Portuguese and consults with other stakeholders regarding the translation of expressions and certain words. It is expected that the translation will be finished in the first quarter of 2009.

China

For the implementation of GHS in China, 26 Standards on classification, labeling, and precautionary statements of chemicals have been developed since 2006. They came into force on January 1, 2008 in the production sector and in the distribution sector on December 31, 2008. The Standards for Classification and Hazard Communication of Chemicals (CNH 475) and the Standards on Preparation of Precautionary Label for Chemicals (CNH 477) are being updated to be in line with the technical requirements provided in the 26 standards. CNH 475 would apply to the signs in chemicals production sites and the signs of consumer goods. CNH 477 stipulates the definitions, contents, requirements, and methods for use of the label for chemicals at workplace.

European Union

On December 16, 2008 the new regulation on classification, labeling, and packaging of substances and mixtures (CLP Regulation) was adopted by the European Parliament and the Council, and on January 20, 2009 it was entered into force. This CLP Regulation aligns existing EU legislation to the GHS (Regulation (EC) 1272/2008, OJ L 353) and complements the regulation on the Registration, Evaluation, Authorisation and Restriction of Chemicals (REACH). Based on the new rules, the deadline for substance classification is December 1, 2010 and for mixtures June 1, 2015. Along with the CLP Regulation, which replaces previous rules on classification, labeling and packaging of substances (Directive 67/548/EEC) and mixtures (Directive 1999/45/EC), two related acts were adapted to the new rules: *Directive 2008/112/EC, OJ L 345* and *Regulation (EC) 1336/2008, OJ L 354*. The EU is currently leading international efforts on implementation of the GHS.

Japan

To implement GHS-based labeling and SDS requirements in Japan, the Industrial Safety and Health Law was amended and put into force in 2006. In parallel, the Japanese Industrial Standards, JIS Z7250:2005 for SDSs and JIS Z7251:2006 for labeling were amended and published in accordance with the GHS. Japan is in the process of revising their classification manual and technical guidance to facilitate the classification of the 1500 chemicals regulated under the law, and to eliminate any discrepancies in classification among experts. This will be done according to the second revised edition of GHS. Furthermore, Japan will complete a new manual and guidance for mixtures. These will be available also in English. The classification results of the 1500 substances are now available in English on the Web site at: <http://www.safe.nite.go.jp/english/ghs_index.html>. Japan is now working to classify additional substances outside of the 1500. Software to classify mixtures was developed in Japanese and this was expected to be available in English before the end of 2009.

South Africa

To support GHS implementation in South Africa, an interdepartmental committee was established by the Department of Labour to ensure coordination and avoid duplication of activities including the development of GHS legislation. The committee has developed a legal implementation strategy which includes compliance and enforcement requirements, appropriate budget allocations, support to industry for transition, and establishment of a permanent approach to ongoing input into international discussions and alignment of the effective dates of all legislative amendments. In 2007, Standards South Africa, a division of the South African Bureau of Standards developed a draft national standard: SANS 10234:2007—"Globally Harmonized System of classification and labelling of chemicals." (This standard is also being used as the basis for the development of a harmonized regional standard for the Southern African Development Community (SADC). This work is being done by the region's standardization body, SADC Cooperation in Standardization (SADCSTAN)). For South Africa, this draft GHS implementing legislation includes a five-year transition period and provisions for facilitating a national transition while still accommodating international trade. The draft was released for public comment, and the results are currently being consolidated. Following, stakeholder workshops and awareness raising activities were completed throughout 2008 to inform the public about the new system. The finalized draft GHS legislation has been submitted to the Minister for confirmation and further dissemination in 2010. In support of implementation, the GHS will be included in a number of industry and labor training programs.

United States of America

The Department of Labor's Occupational Safety and Health Administration (OSHA) has completed a draft proposal to change its existing Hazard Communication Standard to adopt the provisions of the GHS. This Notice of Proposed Rulemaking

(NPRM) for the GHS is under review with the Office of Management and Budget. After a review period of up to 90 days, the proposal will be published in the Federal Register as law. The proposal, including an accompanying preliminary economic analysis, is based on comments received following publication of an Advance Notice of Proposed Rulemaking (September 2006). OSHA continues work on several guidance products and compliance assistance tools that will facilitate the transition from the current HCS to the GHS. Other agencies, including the Environmental Protection Agency and the Department of Transport, are closely following discussions on the GHS.

THE FUTURE OF GHS

Increasingly, countries and stakeholders are acknowledging the importance of the GHS as a tool for sound chemicals management. As a consequence, there is increasing support for countries to implement the GHS as a foundation for supporting safe chemical use. However, GHS implementation worldwide is very heterogeneous and the stages of implementation for countries vary greatly. A survey by UNITAR and the OECD in 2008 revealed that the majority of respondent countries have started implementing the GHS, but that they have encountered numerous obstacles that have challenged the implementation process. This includes lack of resources, capacity, awareness, and expertise on the GHS.

Successful worldwide implementation will require collaboration between governments, civil society groups, and the private sector. Further, the goal of a "globally harmonized system" will require significant international cooperation and support. While global discussions on the GHS as a system occur within the context of the UNSCEGHS, a number of global efforts have been launched to support GHS capacity building, including the UNITAR/ILO Global GHS Capacity Building Programme and bilateral efforts from aid agencies, such as the Swedish Chemicals Agency (KemI) and the Ministry of Trade and Industry (METI) in Japan. However, there is still much work before the ideal of GHS implementation becomes a reality. In particular, leading chemical producing and consuming countries must lead the way in adopting the system, which will then lead to other countries to follow their example. As the impacts of chemical use on our health and environment become increasingly clear, the sound management of chemicals through tools such as the GHS will become an ever greater priority for countries and the global community.

Useful Web Sites

GHS official text and corrigenda: GHS (Rev.3) (2009)
http://www.unece.org/trans/danger/publi/ghs/ghs_welcome_e.html
UN Subcommittee of Experts on the GHS
http://www.unece.org/trans/main/dgdb/dgsubc4/c4age.html
UNECE's site on GHS: Status of implementation
http://www.unece.org/trans/danger/publi/ghs/implementation_e.html
UNITAR/ILO GHS Global GHS Capacity Building Programme
http://www.unitar.org/cwm/ghs

WSSD Global Partnership for Capacity Building to Implement the GHS http://www.unitar.org/cwm/ghs_partnership/index.htm

REFERENCES AND FURTHER READING

Developing a National GHS Implementation Strategy: A Guidance Document to Support Implementation of the GHS. 2005. Geneva, United Nations Institute for Training and Research.

GHS Implementation. (n.d). Retrieved June 1, 2009, http://live.unece.org/trans/danger/publi/ghs/implementation_e.html, United Nations Economic Commission for Europe (UNECE).

Globally Harmonized System of Classification and Labelling of Chemicals (GHS). 2007. New York and Geneva, United Nations.

Historical Background on the GHS. (undated) Retrieved March 2, 2008, from http://www.unece.org/trans/danger/publi/ghs/histback_e.html

Implementation of the GHS: Reports from Governments or organizations, Transmitted by the expert from South Africa. (July 2008) Sub-Committee of Experts on the Globally Harmonized System of Classification and Labelling of Chemicals. UN/SCEGHS/15/INF.23.

Information about the status of the implementation of GHS in Japan. 2008. Retrieved March 2, 2009, from http://www.safe.nite.go.jp/english/ghs_index.html

Report on the Preparation for GHS Implementation in Non-OECD Countries. 2007. Geneva, United Nations Institute for Training and Research/International Labour Organisation/Organisation for Economic Cooperation and Development.

Understanding the Globally Harmonized System of Classification and Labelling of Chemicals (GHS): A Companion Guide to the Purple Book. 2008. Geneva, United Nations Institute for Training and Research.

WSSD Global Partnership for Capacity Building to Implement the GHS: Annual Report 2008. WSSD GHS Capacity Building Annual Reports. May 2009. Geneva, United Nations Institute for Training and Research.

12 The International Code of Conduct on Distribution and Use of Pesticides

*Gamini Manuweera**

CONTENTS

* The opinions expressed in this chapter do not necessarily reflect the views of the Government of Sri Lanka.

INTRODUCTION

Efficient utilization of all available options for increased food production and reduced risk in public health has become more important than ever in finding solutions for the global issues of population increase, shrinking resources, and heterogeneous distribution of knowledge. Among the other aspects, control of pests is a major challenge in both food production and public health. Integrated pest and vector management, public sanitation, and biotechnology are some of the approaches available for sustainable solutions to the problem. However, use of pesticides has become one of the most popular alternatives employed at field level in pest control. Study of recent trends indicates a marked shift in the use of highly hazardous pesticides from developed countries to developing countries. A properly established regulatory control system, which is essential to ensure human and environmental safety of pesticide use, is often not found in developing countries. Established standards in distribution and use of pesticides are therefore necessary to complement the situation to serve as set of norms for the observance by stakeholders, especially the pesticide industry.*

The International Code of Conduct on Distribution and Use of Pesticides jointly implemented by the Food and Agriculture Organization (FAO) of the United Nations (UN) and the World Health Organization (WHO) is a dynamic instrument designated to represent globally accepted, up-to-date standards for pesticide management with modern approaches in risk reduction, protection of human and environmental health. It facilitates sustainable development in agriculture through judicious and effective use of pesticides while supporting Integrated Pest Management (IPM) strategies and effective interventions in public health risk management. An important feature of the Code is that it embraces the life-cycle concept of pesticide management.

SCOPE

The Code functions as the international framework and point of reference for the judicious use of pesticides with respect to all aspects relating to their use. It is of special significance for countries yet to establish or in the process of establishing an adequate and effective regulatory infrastructure for the sound management of pesticides. The Code addresses major stakeholders of all public and private entities engaged in, or associated with the distribution and use of pesticides in agriculture, public health, and environmental protection, including the users of pesticides, commerce and trade, food storage, and pesticide application equipments.

HISTORICAL PERSPECTIVE

Pesticides came to the limelight in agriculture in the nineteenth century with developing industries, and the more modern synthetic pesticides, in particular, after World War II, and an increased use especially in the developing world, with the advent of the green revolution in the early 1960s forming an essential component in the promotion of agricultural production. Proliferation of indiscriminate use

* Pesticides industry is understood to include the whole chain from producers to retail.

combined with lack of precautions in protecting human health and the environment in developing countries led to serious problems compelling the need for intervention by international agencies.

The Code was first adopted in 1985 by FAO of the UN at its 23rd conference, as a voluntary instrument in support of increased food security, while protecting human health and the environment. At the 25th session of the FAO conference in 1989, the Code was amended to include provisions on the Prior Informed Consent Procedure (PIC). Consequent to the adoption of the Rotterdam Convention on the PIC for Certain Hazardous Chemicals and Pesticides in International Trade in 1998 and the changes in the international policy framework along with the emergence of new challenges in pesticide management, the FAO commenced revising the Code again and, the present version was adopted in November 2002 by the FAO council at its 123rd session.

Although the first version of the Code was established intending to address agricultural pesticides many provisions in the text were equally applicable to public health pesticides and subsequent revisions have been able to capture most of the related missing elements. In 2007, WHO signed an MOU with FAO to establish a joint program on pesticide management in implementing the Code which gave the Code more recognition and improved the much needed strength at country level. So, from initially agricultural pesticides oriented Code it became an international Code that includes all pesticides.

OBJECTIVES

Considering the magnitude of diversity in issues and scopes related to sound management of pesticides between individual countries, the most pragmatic role of the Code is to efficiently facilitate assuring the establishment of an effective system to regulate aspects associated with the life cycle of pesticides specific to the particular nation and its sustainable implementation. To this end, the Code has identified number of specific objectives in its Article 1:

i. Established as voluntary standards of conduct for all public and private entities engaged in or associated with the distribution and use of pesticides, particularly where there is lack of adequate regulatory control.
ii. Designed for use as a basis within the national legislation for any concerned party to judge whether any proposed action constitutes acceptable practices.
iii. Shared responsibility by many sectors of society in working together to achieve the benefits to be derived by the necessary and acceptable use of pesticides.
iv. Cooperation between the governments of exporting and importing countries to promote practices that reduce health and environmental risks associated with pesticides, while ensuring their effective use.
v. Addresses the entities of international organizations, governments of exporting and importing countries, pesticide industry, the application equipment industry, traders, food industry, users, and public sector organizations.

vi. Recognition of training at all appropriate levels as an essential requirement in implementing and observing the provisions in Code by all stakeholders.

MAJOR PROVISIONS OF THE CODE

Sound management of pesticides is addressed by the Code under specific themes of the life cycle of pesticides in 12 separate articles including articles describing its *objectives* and clarifying the *terms and definitions* relating to the text of the Code.

Pesticide Management

The management of pesticides in the Code mainly focuses on those countries lacking appropriate legislation or with limited success in regulation or enforcement. In addressing the aspects related to sound management of pesticides in a country, the Code recognizes responsibilities under four regimes; the government, governments of pesticide exporting countries, the pesticide industry including traders and national and international organizations.

The national government has the overall responsibility of regulating the availability, distribution, and use within the country. In countries where appropriate legislation and advisory services are yet to establish, the pesticide industry is required to adhere to the provisions of the Code as a standard for practice while the exporting countries are required to provide technical assistance and ensure good trading practices. The Code further identifies that in practicing good trading, the pesticide industry should pay special attention to the areas of product quality, packaging and labeling, local procurement, the choice of formulations, information, and instructions that should carry with the product, provision of technical support and retaining of active interest in following their products to the end-user.

Apart from the management aspects related to distribution and availability of pesticides, the Code identifies research on development and promotion of IPM, alternatives, resistance management, personal protective equipments and low risk, more efficient, environmental friendly, and cost-effective pesticide application methods especially in relation to small-scale users under tropical climates, as important elements of a sound management system.

In implementing the provisions on sound management of pesticides, the Code has so far relied heavily on promoting establishment of a product registration system, which is the logical option that needs to be ensured in place, at first. Since the last revision of the Code nearly a decade ago, most countries have now successfully established at least some sort of control system on pesticides or are in the process of developing one. Thus some of the basis on which the Code was last revised needs to be revisited to accommodate surfacing issues.

Of the elements of sound management, effective implementation of use of personal protective equipments and proper pesticides application are extremely important aspects but very poorly put in place in developing countries though promoted by the Code under sound management. Developing countries often pay very little attention to these two aspects in chemical pest control leading to many health, environmental, economic, and social problems. One of the reasons for this scenario is that the potential consequences are not apparent and difficult to assess under the prevailing local conditions. Data collection on practices in the field, including cases of abuse or poisoning, is weak or even nonexistent.

Testing of Pesticides

Pesticide testing, according to the Code, involves all major aspects of assessment of the chemical including its potency, usage, chemistry and adverse impact on the environment and on humans. The objective is to provide required scientific data and information to adequately evaluate associated risk under the proposed conditions of use. The reports required for this purpose under the Code could be classified into three major segments in product development life cycle; that is, molecular properties of the active ingredient, technical grade product, and final marketable formulation. The responsibility of development of satisfactory test reports using sound scientific procedures and the principles of good laboratory practices is vested on the industry.

In sound management efforts, validation of bioefficacy under local conditions and verification of basic physico-chemical parameters of pesticide formulations are those that most developing countries strive to establish. Yet, in many cases the required expertise and laboratory facilities fall short of meeting the international or reasonably acceptable standards in operation for a satisfactory outcome. Though the Code has adequately emphasized the need, by identifying how stakeholders should contribute toward achieving the goals, satisfactory progress is often not found in reality.

The other aspect addressed under this article that requires more attention is postregistration surveillance or monitoring health and environmental effects under field conditions. Assessment of field situations, vital for sound management interventions, is severely handicapped in many developing countries due to lack of a proper monitoring and feedback system. In support of postregistration surveillance the Code developed (2009) a new guideline on incident reporting, as a tool to facilitate the initiatives by national governments.

Health and Environmental Risks in Developing Nations

For reducing the risks, the Code heavily depends on implementing the related regulatory provisions including registration and other control systems, supported by health surveillance, guidance on pesticide poisoning treatments, statistics on incidents, and effective communication with end users. The major stakeholder for this purpose is the respective government. It also promotes aspects related to residues in food and the environment, alternative pest control options, and all other possible risk reduction options, covering the entire life cycle of pesticides.

Unlike the aspects discussed in the preceding sections, reducing health and environmental risks directly link to the field conditions prevailing in the country. A majority of developing countries share a similar scenario, detrimental to any favorable risk-reduction efforts. The presence of a significantly wider spectrum of pests inherent in tropical environments, a farming community representing one of the most vulnerable sectors of society with very weak economic and educational status, traditional and strongly inherited social behaviors coupled with the extremely weak or nonexistence of administrative infrastructure down to the end-user retard effective intervention. These are constraints limiting the development and implementation of efficient programs for risk reduction. There are very few risk reduction options that are pragmatic under such conditions. Out of them, the control of availability of pesticides is the most effective risk-reduction tool available for the governments. Some of the common risk-reduction options, such as proper labeling, education and training, and regulated distribution often do not deliver the anticipated outcomes. In some instances they may cause adverse consequences; the color bands on the labels depicting the acute mammalian toxic hazard of the formulation, provided to discourage choosing highly toxic pesticides, have in fact increased farmers using more of hazardous pesticides. For example, in Sri Lanka farmers anticipated that highly toxic pesticides are more potent on the pest compared to less toxic alternatives. Therefore, choosing the right blend of risk-reduction options in the developing countries is a challenge and requires in-depth understanding about prevailing local field situations. One of the pragmatic approaches is to prioritize the issues in relation to effective interventions available. Apart from the occupational health risks, use of pesticides for self-harm (suicide) is a very significant problem in a number of developing countries for which due attention has not been paid as yet.

Regulatory and Technical Requirements

Introduction of necessary legislation and provisions for their effective enforcement is identified in the Code as the initial step of regulatory requirements. The national governments are required to establish a pesticide registration scheme as the controlling instrument of the regulatory program while the industry is required to facilitate by providing necessary technical information developed conforming to international standards.

With a varying degree, most of the developing countries now at least have legislative control provisions in place, if not a comprehensive product registration scheme. What is most challenging for developing countries is the satisfactory evaluation of risk under local conditions to facilitate sound management decisions in pesticide registration, the aspect that has been identified by the Code as the next important step in regulatory control. The two key drawbacks in addressing these important aspects are, finding suitable technical expertise on risk evaluation under developing country situations and the availability of vital scientific information generated representing the local conditions. Except for mammalian toxicity and product chemistry, most of the data generated by the industry require extrapolation to the local conditions, as in many cases they are generated in developed countries under temperate climatic conditions. Often the accuracy of such attempts are challenged, based on acceptable scientific norms. The Code addresses the key elements of a regulatory control instrument including alternative implementing options in great details. It includes equivalence, harmonized registration requirements, procedures and evaluation criteria, re-registration system, collecting and recording statistical data, standardization of application and protective equipments, illegal trading, good agricultural practices, and so on. The Code does not explicitly address the issue of the need for sound data under tropical climate conditions and their proper evaluation, including the possibility that use may not be safe under prevailing conditions.

Equivalence

With more and more pesticide active ingredients becoming generic in the market, the number of production facilities has significantly increased in last few decades. In most cases the purity levels of the active ingredient and composition of the formulants of generic pesticides from different sources vary from those of the basic manufacturers. Therefore, toxicological, biological, and environmental profiles of generic pesticides need not essentially be comparable.

Severe economic and resource limitations prevailing in developing countries prevent the respective governments to respond to the situation in effectively assessing the risk. In order to assess whether supposedly similar pesticides originating from different manufacturers present similar levels of risk, the similarity of the impurities and toxicological profile, as well as the physical and chemical properties of the product are used as the means of determining the equivalence. The principles and their application procedure in determining equivalence of pesticides are described in the *Manual on Development and Use of FAO and WHO Specifications for Pesticides* (2002). Two technical grade pesticides from different manufacturers or manufacturing processes are considered equivalent if the materials meet the requirements of the existing FAO/WHO specifications including assessments of the manufacturing process used and the impurity profile together with assessments of the equivalence of toxicological and ecotoxicological profiles.

Apart from the approach of the equivalence concept, the Code promotes cooperation with other governments for harmonization of pesticide registration requirements, procedure, and evaluation criteria. Such initiatives are very pragmatic and effective, especially in the context of developing countries, in achieving the objectives of sound management of pesticides. The complexity in integrating the priorities and policies of individual countries with respect to national pesticide management is one of the challenges that retard the efforts in regional or bilateral harmonization of registration requirements. Careful selection of elements for harmonization is therefore very important for sustainability.

Availability and Use

The major thrust of the Code in availability and use of pesticides is on the promotion of WHO hazard classification as the basis for regulation. Though this article is the shortest in the Code it contains some recommendations that are focused, pragmatic, and proactive. Perhaps the most proactive recommendation of the Code is prohibition of WHO class Ia and Ib pesticides, appearing in this article. The provision was later raised into the level of program with the FAO council decision in 2006* to promote progressive banning of WHO class Ia and Ib pesticides. Consequently the FAO/WHO panel of experts of pesticides in 2008 recommended a strategic plan for FAO and WHO for implementation. Among the main barriers of banning highly hazardous pesticides are the relatively low cost of the products and the fear of retardation of agricultural production with removal of commonly used highly hazardous pesticides from the market. The belief of potential yield losses associated with banning has been proven unrealistic, if the farmers have affordable alternatives (Manuweera et al., 2008). With the GHS (see Chapter 11 of this book) gradually taking over the WHO classification, this provision will need to be amended to reflect similar hazard classes under GHS.

For those countries where field implementation of comprehensive management programs is not viable, this article provides a basis for practical risk reduction options for the authorities as it promotes restricting the availability to certain groups of users as a clear and simple option of dealing with hazardous pesticides to minimize the potential risks.

Distribution and Trade

Implementing a licensing procedure for pesticide dealers is promoted by the Code to facilitate provision of sound risk reduction and efficient use. Though the licensing of dealers is useful in bringing certain control over the distribution channel with respect to accessibility to pesticides and spurious products, their role as facilitators of risk reduction is not realistic under the conditions prevailing in developing countries. Repacking or decanting of pesticides to food or beverage containers, although illegal in many countries, is a common practice in many developing countries mainly due to lack of enforcement, of social responsibility and ignorance of potential risk. Governments enforcing regulatory measures to prevent this practice as promoted by the Code are not very effective as the implementing instrument is generally weak. Effective awareness campaigns, on the other hand are more efficient in rural communities than taking measure through regulatory control to arrest the situation, as well as proper training of field inspectors and extension service workers.

Obsolete Stocks

The issue of accumulation of obsolete stocks of pesticides is a common and serious problem in many developing countries. The Code identifies market driven supply process as a better option as against the centralized purchasing, especially by government agencies, which has been a common practice in developing countries in the past. However, the market driven processes also could produce obsolete stocks, unless the industry practices certain ethics in reaching their marketing targets and responsible stock management strategies in the distribution channel. The Code is not specific about the responsibilities of the industry on dealing with obsolete stocks but

* Recommendations of 131st session of the FAO Council held in 2006, http://www.fao.org/unfao/bodies/council/cl131/index_en.htm

is mainly focused on quality assurance, risk reduction, and so on which would indirectly link to the management of obsolete pesticides.

Another aspect that the Code has not given much prominence to is the accumulation of unused stocks in household environments leading to the risks not only related to the environment but also to accidental and deliberate self-poisonings.

Counterfeits and Illegal Trafficking

In some manufacturing processes, the end-product has different grades of purity, even if it is from the same source. If different chemical reaction pathways were employed it is also common to find marked variation of the quality standards of different sources. In some instances, some branded products are seen counterfeited by cheap low-grade material. Importing countries with no effective quality control system are the main destinations of those substandard products. There are a number of recommendations in the Code assigning responsibility mainly on the pesticide industry to observe corrective measures.

Countries with land boarders often encounter incidents of illegal trafficking often with labels in foreign languages leading to many problems. This is an issue that could be effectively addressed by intergovernmental collaboration where the Code is not adequately strong. Field level adulteration is one of the other common problems found in developing countries. Effective enforcement of regulatory control mechanisms and compliance monitoring schemes by respective governments are necessary to address such problems but often lacking due to resource constraints.

Dispensing pesticides into other containers at retail level, high-risk storage conditions at retail outlets as well as in the household environment are serious problems at field level which regulators find challenging to address. A more supportive role in responsible marketing practices by the industry is key to make any progress. The observance of the Code by the industry should focus more on these aspects as they are vital in risk reduction. It is a common sight in agricultural communities in developing countries of left-over chemicals and used containers posing serious accidental and environmental hazards, where awareness is one of the most efficient options.

Information Exchange

Information exchange in the Code is dealing with interagency and intergovernmental sharing of information and provision of legal mandates for public access to information on risk and regulatory issues. The Code explicitly discusses establishing a system to facilitate information sharing related to control actions on pesticides between national regulatory authorities similar to the information exchange mechanism of the Rotterdam Convention (see Chapter 14 of this book). This is a very useful tool in improving the management status in developing countries. Except for the Rotterdam Convention initiative, the success in establishing effective information exchange between regulators is very limited among the developing countries. One of the major impediments is the lack of visible opportunities in place to initiate coordination between the countries in establishing an information exchange mechanism. The existing regional and sub regional bodies are the best option for this purpose, if its importance is emphasized and priority is duly identified by the respective international instruments such as the Strategic Approach to International Chemicals Management (SAICM).

Labeling, Packaging, Storage, and Disposal

This section deals with important field aspects related to risk reduction. It provides a comprehensive coverage relating to good labeling practices that should be adopted by the industry and governments dealing with packaging, storage, disposal, decanting into other containers, and accumulation of obsolete stock.

However, the expectations are yet to be realized to a satisfactory level in many developing countries. Lack of adequate legal provisions or their enforcement for proper control over labeling, resources for their effective implementation, multilingual social settings requiring larger space to accommodate mandatory label information and not having the label in the local dialects are found in rural settings.

Advertising

This is an area where the industry has a challenge in making a delicate balance between achieving marketing goals while conforming to the standards recommended by the Code. With high competition between the companies in promoting products driven by sales targets, the observance of the Code is often not prevalent, especially among the local traders who are not affiliated with multinational players in the industry or regional/international associations of the industry. Unethical promotions such as lotteries offering very valuable household items lead to increased potential of risks and/or indiscriminate use. Most Governments from developing countries that adopt open market policies have failed to address this aspect effectively.

IMPLEMENTATION OF THE CODE

Publicity and collaborative actions in observing the provisions are the two key implementing tools the Code describes as the approach of its implementation. The responsibility of implementing these recommendations is vested on the stakeholders ranging from the UN system, governments, NGOs, and regional groupings to the pesticide industry.

A joint panel of FAO and WHO experts recommends the implementing strategies, guidance documents, and priorities in action based on field status of pesticides and observance of the Code. The panel includes subject matter specialists and pesticide regulators representing both developed and developing countries. The expert panel meetings are also attended by other related international agencies, the pesticide industry and NGOs. It monitors and reviews the role of the Code in the international context for sound management of pesticides and sets the standards of practice for reference by the stakeholders. The pesticide industry in the panel comprises CropLife International representing the so-called research-based pesticide industry along with generic pesticide producers, the European Crop Protection Association

(ECPA) in Europe, and the Asociación Latinoamericana de la Industria Nacional de Agroquímicos in Latin America.

FAO and WHO periodically obtain the status of management of pesticides and observance of the Code at national level through a structured survey addressing all key issues addressed by the Code.

Strategic Program

In support of efficient implementation of the Code, the panel has developed a strategic plan for the years 2006–2011.* The scope of the strategic program has been based on the Inter-Organisation Programme for the Sound Management of Chemicals (IOMC; see Chapter 23 of this book) which was established to strengthen cooperation and increase international coordination in the field of chemical safety. Some of the general principles on which the Strategic Programme has been developed include pesticide use as part of integrated pest or vector management, multistakeholder collaboration, coordination of interventions at the country level, evidence, and need-based priority setting. The capacity building in implementation of the Code at country level has been identified under technical guidance, awareness building, monitoring and evaluation, and resource mobilization.

STRENGTHS OF THE CODE

There are number of unique strengths associated with the Code:

- Ability to play a proactive role in sound management of pesticides; for example, initiatives on progressive banning of WHO Hazard class I pesticides; revision of criteria for inclusion of newly emerging human health and environmental concerns.
- FAO and WHO, as implementing agencies, have the global mandate to set policy directions on agriculture and public health, respectively, complementing the objectives of the Code.
- FAO supports international legally binding instruments that addresses issues related to pesticides and chemical pest control (e.g., Rotterdam Convention and International Plant Protection Convention) and programs related to agriculture which are directly related to the aspects of sound management of pesticides and extremely effective in facilitating the achievement of the objectives of the Code. Some of FAO's initiatives and programs of importance include IPM, standardization of pesticide application instruments, EMPRES (Emergency Prevention System for Transboundary Animal and Plant Pests and Diseases) activities of locust control, and Africa Pesticide Stockpiles.

* Strategic Programme for the implementation by FAO of the revised version of the International Code of Conduct on the Distribution and Use of Pesticides 2006–2011; Food and Agriculture Organization of the United Nations September 2006.

- The activities of WHO that facilitate pesticides management include hazard classification of pesticides, public health initiatives at national and regional level specially focusing on the developing countries such as Global malaria programme, International Programme on Chemical Safety (IPCS), Inter-Governmental Forum on Chemical Safety (IFCS), Global Collaboration for Development of Pesticides for Public Health (GCDPP), Integrated vector management, and Poison Control initiatives and support.
- UN interagency chemicals and pesticide related programs such as JMPR and JMPS (see Chapter 18 of this book).
- Extensive field network of the WHO and FAO, especially among the developing countries.
- Dynamicity of the implementing guidelines of the Code to meet the changing Global trends.
- Coverage of wider range of stakeholders; farmers and farmer associations, IPM researchers, extension agents, crop consultants, food industry, manufacturers of biological and chemical pesticides and application equipment, environmentalists and representatives of consumer groups, governments, Inter-governmental Organizations (IGOs), and related UN agencies.

BENEFITS

- The Code promotes the settings and actions at field level that facilitates other global and local initiative related to chemicals management; for example, RC, IPM, SAICM.
- Allows harmonized management of pesticides in the international context for effective interventions and efficient resource mobilization.
- Sets international reference standards for sound management of pesticides and practices associated with distribution, marketing and use.
- At country level:
  - Use of the provisions of the Code and its guidelines as the standards for setting local regulatory control option.
  - Obtaining cooperation of the local counterparts of key international stakeholders collaborated in implementing the Code for national regulatory initiatives.
  - Efficient interagency collaboration and harmonization.

CHALLENGES

Observance on the Code

The two main implementing strategies, publicity and collaborative action, promoted by the Code are too passive under the circumstances of the voluntary nature of the Code. For effective use the Code and its up-to-date implementing guidelines by the participating governments and other stakeholders require development of a pragmatic approach.

Incorporation into an effective environmental program at field level has not been satisfactory so far; this demands strengthening of collaboration with other agencies mandated for environmental protection.

An effective mechanism for satisfactory collaboration of the key players at international level (industry, exporting countries and international chemical related instruments) for efficient collaboration as recommended by the Code is yet to be seen. The expert panel meeting provides a forum for addressing the issues but requires more integrated efforts in dealing with the many varying situations.

Social expectations, policies, and field situations are highly variable between countries. Therefore, development of standards for sound management, including risk evaluation criteria to suit all the scenarios, is a challenge. Efforts of addressing such a wider scope might, however, have negative consequences for reaching the goals of the Code and its standards. If one has to take into account the concerns of different stakeholders, compromises are unavoidable. On the other hand in developing countries, especially in least developed countries, the demands in sound managements are often too ambitious. Priorities are not properly understood due to confusing signals from the developed world on sound management options leading to unrealistic expectations.

Challenges in implementing and observance of the Code in developing countries:

- Poor political interest, often due to ignorance on the nature of many of the adverse impacts associated with pesticides.
- Conflicting political interests in regulation of pesticides.
- Significant contrast in resource strength between the government regulatory program when compared to the local pesticide industry.
- Weak linkage between the country level international agencies and the national implementing agency on pesticide management in promoting the Code.
- Less emphasis on promotion of the Code by other international chemicals management instruments and programs active in the field that could play an important role in facilitating the objectives of the Code.
- Absence of pragmatic options for seeking or providing assistance in meeting the objectives of the Code from exporting countries.

Management of Pesticides

Observance of the Code should ensure sound management of pesticides. But in many developing countries full implementation of the provisions of the Code and their adequate management is not feasible because of several types of constraints. Many countries inherently lack some vital elements in the basic management of pesticides. Such countries are unable to fully rely on their own resources for proper risk assessment in pesticide registration systems. One of the options promoted by the Code is a regionally harmonized system for registration requirements and registration. The capacity and required political commitment by the countries for the development of such a harmonized system is often lacking. Availability of pragmatic risk reduction options under the conditions prevailing in developing countries is extremely limited.

The main drawback is fragmented infrastructure. Weaknesses in interagency coordination and cooperation at national and field level aggravate the situation.

Pesticides are a set of chemicals intentionally toxic to the environment and living organisms but meant to be used in the environment by farmers, majority of whom represents rural, less privileged communities of developing countries. Those farmers associate strongly with poverty, low level of education, and lack of minimum technological requirements in handling pesticides. Achieving the objectives of sound management under these circumstances requires much more concerted and committed actions by the key players, that is, governments, exporting countries government, pesticide industry, and regional and international instruments of the UN.

FUTURE

Though the Code has served the developing countries for many years in establishing sound management systems on pesticides, its usefulness could be enhanced in many areas for furtherance of its benefits.

The Code requires revision, to be more focused within articles addressing specific aspects of sound management and to be more contemporary. The present version is 10 years old and mainly focused on substances used against pests in agriculture and veterinary fields. Joining of WHO with FAO in 2008 as implementing partners of the Code requires revision of some sections to be more comprehensive in their scope addressing both agriculture and public health applications of pesticides. Although the Code is revised regularly to take into account any shortcomings, these still needs to be addressed more clearly.

The science of chemical pest control, knowledge base on adverse effects of pesticides, marketing strategies of pesticide products and use practices by the farming communities are changing at a rapid rate. The review process of implementing guidelines of the Code requires a more strategic approach to serve the international community in a timely manner. The present system adopted by the panel of experts on development and review of guidelines needs fundamental changes.

When reviewing the focus and thrust of the Code, the changing status of management of pesticides in developing countries should be taken into consideration. At present there are only a few countries lacking any control over pesticides. Some of the major challenges in achieving sound management of pesticides in developing countries include:

- Quality assurance of pesticides products
- Proper labeling
- Judicious marketing and advertising
- Surveillance and reporting of incidents
- Monitoring of levels of pesticide residues in the environment and treated crops
- Dealing with smuggling and border control of illegal pesticides
- Unethical international and local trade practices leading to problems in product quality, identity, accumulation of obsolete stocks
- Developing skills on risk evaluation and management

Several of these elements are currently partially or fully covered by the Code, but may need to be further discussed and strengthened in view of more effective implementation of the Code in many countries.

The Code should develop means to obtain full collaboration of all chemicals related to international programs, especially UNEP and make the Code the point of reference when it comes to the issues addressing pesticides in respective disciplines. SAICM could play as the main instrument for this purpose. The Code is an ideal basis for SAICM to develop integrated approach in sound management of chemicals at country level. Because it is the most comprehensive and established instrument available at international level dealing with sound management of chemicals, especially addressing the needs of developing countries. Further, chemicals management in many developing countries mainly involves pesticide management, when considering the magnitude of use and distribution of different groups of chemicals. A possible approach to obtain the full status of the Code among other international initiatives and programs is to promote the Code at national level to create a demand for international collaboration. This could be achieved through integration of the Code and its Guidelines as the basis, where applicable, into pest and pesticide management initiatives of FAO and WHO including IPM, management of obsolete stocks, used containers, public health diseases.

Donor agencies and special project-type assistance programs should consider incorporation of the standards prescribed by the Code as a prerequisite where pesticides are involved as an input. The national regulatory controls schemes related to occupational and environment exposure require integration of the relevant provisions of the Code in their decision-making criteria so that life cycle management of pesticides is effectively facilitated. This demands strengthened coordination among national agencies dealing with agriculture, health, environment, and labor, which is inherently poor among developing countries.

REFERENCES

Manual on Development and Use of FAO and WHO Specifications for Pesticides. 2006. Revision of the 1st Edition in 2002. Rome: FAO.

Manuweera, G., M. Eddleston, S. Egodage, and N.A. Buckley. 2008. Do targeted bans of insecticides to prevent deaths from self-poisoning result in reduced agricultural output? *Environmental Health Perspectives*, 116(4): 492–495, 18414632 (P,S,E,B,D).

13 The Kiev Protocol on Pollutant Release and Transfer Registers

*Michael Stanley-Jones**

CONTENTS

INTRODUCTION

The newest international chemical treaty, the Kiev Protocol on Pollutant Release and Transfer Registers (PRTRs), like its parent, the Aarhus Convention,† traces its origin to the historic 1992 Rio Earth Summit.

* The opinions expressed in this chapter do not necessarily reflect the views of the United Nations Economic Commission for Europe (UNECE), the United Nations Environment Programme (UNEP), or their member countries; or the Parties to the Kiev Protocol on Pollutant Release and Transfer Registers to the UNECE Convention on Access to Information, Public Participation in Decision-making or Access to Justice in Environmental Matters (Aarhus Convention).

† United Nations Economic Commission for Europe (UNECE) *Convention on Access to Information, Public Participation in Decision-making and Access to Justice in Environmental Matters* (Aarhus Convention), adopted in Aarhus, Denmark, at the Fourth Ministerial Conference "Environment for Europe" (June 1998) and *Protocol on Pollutant Release and Transfer Registers (PRTRs) to the Convention on Access to Information, Public Participation in Decision-making and Access to Justice in Environmental Matters*. The Aarhus Convention has 44 Parties (as of August 30, 2010). Included among these is the European Community as a regional economic integration organization. The Protocol on Pollutant Release and Transfer Registers was adopted at the Fifth Ministerial Conference "Environment for Europe" in May 2003. Thirty-six member states and the European Community signed the Protocol upon its adoption. Following the declaration of independence of the Montenegro and its succession to the treaties to which the State Union of Serbia and Montenegro was a signatory, the number of Signatories to the Protocol rose to 38.

The 1992 United Nations Conference on Environment and Development (UNCED) in Rio de Janeiro (Brazil) recognized the importance of public access to information on environmental pollution, including emissions inventories. *Principle* 10 of the Rio Declaration on Environment and Development, adopted by UNCED, states that "each individual shall have appropriate access to information concerning the environment that is held by public authorities" (see Chapter 3 of this book) as well as "the opportunity to participate in decision-making processes," and that countries shall "encourage public awareness and participation by making information widely available."*

Chapter 19 of *Agenda 21*† recommends that governments should collect sufficient data about various environmental media while providing public access to the information. Governments, with the cooperation of industry and the public, were to implement and improve databases about chemicals, including inventories of emissions. Chapter 19 of *Agenda 21* further states that the broadest possible awareness of chemical risks is a prerequisite for chemical safety.

To implement these *Agenda 21* commitments, under the auspices of the United Nations Economic Commission for Europe (UNECE) negotiation of the Convention on Access to Information, Public Participation in Decision-making and Access to Justice in Environmental Matters were launched in 1996. The Convention was adopted in June 1998 in the Danish city of Aarhus at the Fourth Ministerial Conference in the "Environment for Europe" process. It is widely known today as the "Aarhus Convention" (see also Chapter 7 of this book).

Negotiation of the Protocol on PRTRs to the Aarhus Convention began in 2001 and concluded with adoption of the Protocol in May 2003 at the fifth Ministerial Conference "Environment for Europe," held in Kiev.

Another major milestone in the development of pollutant registers which grew out of the UNCED commitments was the ratification of the Protocol on PRTRs by the European Community (EC) in February 2006. That same month, the EC adopted a regulation implementing the Protocol across the 27 countries of the European Union (EU). A side agreement between the EU and European Free Trade Association (EFTA) extends the obligation to report annually on PRTRs to an additional four countries.

The Kiev Protocol on PRTRs is the first legally binding international instrument on PRTRs. As of January 2011, 37 states and the EC were signatories and 25 states and the EC had ratified the instrument.

The Protocol entered into force on October 8, 2009, 90 days after France had deposited its instrument of ratification, the 16th member state of the UN to do so, thereby ensuring the instrument's entry into international law.

THE AARHUS CONVENTION AND THE KIEV PROTOCOL

PRTRs are a singular tool for public access to environmental information to inventories of pollution from industrial sites and other sources. The Aarhus Convention requires its member states to track the release and transfer of pollutants. The Convention includes broad, flexible provisions calling on parties to establish nationwide, publicly

* http://www.unece.org/env/pp/documents/cep43e.pdf

† See http://www.iisd.org/rio + 5/agenda/agenda21.htm

accessible "pollution inventories or registers" covering inputs, releases and transfers of substances and products. States must progressively establish PRTRs.

The 2003 Protocol on PRTRs regulates the establishment of the registers envisaged by the Aarhus Convention in more detail. Although strongly linked by their unique approach to public involvement in environmental decision making, the Protocol on PRTRs and the Aarhus Convention are separate and independent legal instruments.

The Protocol is open to all member states of the UN, including those that are not members of UNECE or parties to the Aarhus Convention. It is also open to all regional economic integration organizations, which may join and exercise their privileges under the Protocol collectively on behalf of their constituent members if their members so choose.

The Protocol's first Meeting of the parties, held from April 20 to 22, 2010 in Geneva, adopted a set of decisions establishing the main institutions and procedures to ensure the effective implementation of the Protocol, including rules of procedure, a compliance mechanism, a system of reporting on implementation, a scheme of financial arrangements, and a four-year work program. The compliance mechanism follows the approach used for the compliance mechanism of the Aarhus Convention, whereby any member of the public may trigger a review of compliance by a compliance review body by submitting a complaint.

The meeting also established an intergovernmental working group and elected a Bureau to oversee the functioning of the work program between the sessions of the meeting of the parties. The Bureau is made up of representatives of Belgium (Chair), Norway (Vice-Chair), United Kingdom (Vice-Chair), Czech Republic, Spain, Sweden and the European Commission (representing the EU).

A party to the Protocol which is not a party to the Aarhus Convention will not face any limitation on the exercise of its privileges under the Protocol. The activities undertaken on behalf of the Protocol will be financed separately from the activities undertaken through the Convention's program of work. The Protocol's rules of procedure and other governance measures are fully independent of the Convention's own rules and procedures.

HOW THE PROTOCOL WORKS

The Protocol on PRTRs is intended to guarantee public access to information on releases and transfers of certain pollutants through national registers that can be searched through the Internet. Facilities will be required to report annually on their releases (into the environment) and transfers of certain pollutants. This information is placed on a public register, known as a PRTR.

A PRTR is an inventory of pollutant releases and transfers from industrial sites and other sources. PRTRs contain information on pollutants from specific industrial sites and information on releases of certain pollutants to air, land, and water considered to pose the most significant threats to environment or health. PRTRs also contain information on waste which is transferred from one facility to a waste disposal site or recovery facility.

PRTRs may also contain information on pollution from other sources, such as pollution from traffic to air, from agriculture to water, and from small polluting enterprises to land, water, or air. The public can search through the Internet for specific information related to companies, geographical locations, and so on. Such registers facilitate public participation in decision making and contribute to the prevention and reduction of environmental pollution.

PRTRs have proven to be effective environmental management tools and to provide benefits to national stakeholders. Addressing chemicals wastes, awareness raising, prioritization of sources of chemicals and facilitating national inventories are natural uses of PRTRs. They provide multiple functions, for example,

- Regularly gather data for major pollutants, including Hg
- Reduce costs to government and industry from a coordinated reporting system
- Hold and manipulate data to allow updating/tracking of releases
- Provide a portal for information to civil society
- Provide comprehensive data for reporting on releases and tracking of hazardous chemicals and for identifying priority chemicals management areas in national action plans
- Trigger cleaner production initiatives in industries

Although the Protocol regulates information on pollution, rather than the pollution itself, it is expected to reduce pollution because companies will not want to be identified as major polluters.

POLLUTANTS COVERED UNDER THE PROTOCOL

PRTRs under the Protocol cover information on at least 86 pollutants, including major greenhouse gases, acid rain pollutants, heavy metals, dangerous pesticides, and cancer-causing chemicals. The Protocol on PRTRs Annex II list of pollutants covers all but one of the persistent organic pollutants (POPS) which are inventoried under the Stockholm Convention on POPS.

The national registers also provide a framework for reporting on pollution from diffuse sources such as traffic, agriculture, aquaculture, and small and medium-sized enterprises.

The emphasis of the Protocol is on the amount of pollution. The Protocol tries to find a balance between the reporting burden and the relevance of the information provided. Instead of covering broad number of pollutants, the Protocol concentrates on releases of a limited number of specific pollutants and pollutant categories in order to present an overall picture of the amount of pollution.

The Protocol identifies a number of important groups of substances, such as total organic carbon (TOC), halogenated organic compounds, phenols, particulate matter (PM_{10}), dioxins, polycyclic aromatic hydrocarbons (PAHs), cyanides, fluorides, non-methane volatile organic compounds (NMVOCs), perfluorocarbons (PFCs), and hydrochlorofluorocarbons (HCFCs), as well as key individual pollutants. These groups cover potentially thousands of single substances.

The Protocol sets minimum requirements. Parties developing PRTRs may go further, in light of their national priorities and concerns. For example, if a local industrial facility emits significant amounts of a substance not covered under the Protocol, it may be important to include that substance in the national reporting requirements.

Many of the substances included in the Protocol are severely restricted, banned, or being phased out under international agreements. They are included in the Protocol for the sake of completeness and to help countries track legacy pollutants.

TRANSFERS OF POLLUTANTS

The concept of "releases" is generally understood to cover situations where pollutants are emitted or introduced into the environment from a facility or other sources. The concept of "transfers" applies instead to movement of pollutants within or between facilities. The Protocol covers only "off-site" transfers.

Movements of pollutants/waste between two installations of the same facility on the same site or adjoining sites will be an on-site transfer, and therefore not subject to reporting.

The Protocol gives parties an option on how to report off-site transfers of waste. Under the Protocol, each party has to choose between the pollutant-specific approach and the waste-specific approach for reporting off-site transfers of waste.

If the party chooses the pollutant-specific reporting approach, it must require facilities to report when individual thresholds for the chemical substances in the waste are exceeded during the reporting year.

If the party chooses a waste-approach, it must require facilities to reporting when thresholds for the total volume of waste are exceeded during the reporting year.

Under the latter approach, the threshold is set at 2000 tons for nonhazardous waste and two tons for hazardous waste.

If the destination of the hazardous waste is a transboundary one, that is, the waste is headed to export to another country, the Protocol requires the parties to collect from the operator information on the actual destination of the waste and the manner of its disposal or reuse. The information collected under the Protocol on the transboundary movement of hazardous waste supports the objectives of the Basel Convention* (see Chapter 8 of this book).

WHO REPORTS?

Mandatory annual reporting is required of facility operators for a wide range of activities. These include refineries, thermal power stations, the chemical and mining industries, waste incinerators, wood and paper production and processing, waste water treatment facilities, animal and vegetable products, shipbuilding, and intensive agriculture, and aquaculture among others, where these activities are carried out on a significant scale.

In all, 64 activities grouped by sectors are covered.

* The Basel Convention on the Control of Transboundary Movements of Hazardous Wastes and their Disposal. The Basel Convention entered into force in 1992.

The national legal framework defines the obligations of the administrative authorities who will be collecting, validating, and managing the register, as well as dealing with accessibility of the data and confidentiality issues.

In countries that already have systems of pollutant reporting, the two most common structures in use for collecting the data needed to establish national emissions registers are

a. Information requirements set in environmental permits
b. Compulsory self-monitoring and reporting

Many countries, especially in Western Europe, already have well-developed systems for the permitting of large industrial installations, including mandatory self-monitoring and reporting of polluting emissions. To avoid duplication of effort, they have linked the collection of data required for their national PRTRs to requirements already in place in their permitting system.

Parties may take "appropriate enforcement measures" to implement the Protocol's against operators as well as officials responsible for the registers where they are found to have acted in bad faith, fraudulently, or negligently, and such actions have hampered the implementation of the Protocol.

PUBLIC ACCESS TO INFORMATION

As the first goal of the Protocol is to enhance public information, PRTR information should be available via direct electronic access, such as an open Web site. Parties must provide "other effective means" for members of the public who do not have electronic access.

PRTRs should provide information on individual facilities, on diffuse pollution and on aggregate pollution levels. The Protocol allows limited provision for polluters to request that their data remain confidential.

The PRTRs must be publicly accessible through the Internet free of charge, and must be searchable, user-friendly and timely. They must have limited confidentiality provisions and allow for public participation in their development and modification.

The Protocol's objective is to enhance public access to information and to facilitate public participation as well as to encourage pollution reduction. PRTRs are intended first to serve the general public. PRTRs can also assist governments in tracking pollution trends, setting priorities, and monitoring compliance with international commitments, and they can benefit industry through improved environmental management.

There are many potential users of PRTRs. These include, first of all, the general public and citizens' organizations interested in obtaining information on local, regional, or national pollution. Health professionals can use the information in public health decisions. PRTRs can be a valuable tool for environmental education. Environmental authorities can use PRTRs to review both the permit compliance of local facilities as well as national progress toward international commitments. For polluting facilities, both the exercise of monitoring or estimating pollution levels as well as their publication can encourage efforts to improve efficiency and reduce pollution levels.

Public awareness of the register is actively being promoted and parties to the Protocol provide assistance to the public in accessing the register and in understanding and using its information.

PUBLIC PARTICIPATION

The Protocol calls for public participation in the development and modification of PRTRs. The negotiations or the Protocol itself provide an example, as they involved technical experts from governments, environmental nongovernmental organizations (NGOs), international organizations, and industry. Participation of all interested parties was considered crucial to guarantee the transparency and acceptance of the Protocol.

The Protocol refers to two instances when public participation is relevant: (a) during the establishment of the PRTR, and (b) in the modification of the PRTR. In either instance, opportunities for public participation should be provided at an early stage when it can influence the decision-making process.

Although the minimum requirements of the Protocol must always be met, input from the public may influence how they are met and whether the national PRTR goes further.

A participatory process for establishing or developing PRTR is essential for the future success of the system. Involvement of all stakeholders, that is, reporting facilities, NGOs and civic organizations, workers in the facilities, health officials, pollution control officials, local authorities, and academia, is important. Those countries having to develop their PRTRs from the beginning will especially benefit from the experiences of other countries.

To ensure that the public is given sufficient opportunity to participate, some parties may set in place detailed rules on how to inform the public, how the opportunity for consultation should be publicized, for example, in mass media or regional media and official journals; information panels in city halls or other relevant buildings; or by post, Internet postings, or pod casts.

Public participation can take different forms, including through public meetings, working groups or other standing committee, or through community forums. The rules for public participation may ensure that comments made by the public can be sent by both electronic and nonelectronic means.

INTERNATIONAL COORDINATION AND RELEVANCE FOR OTHER INTERNATIONAL AGREEMENTS

Broad international cooperation is also an important element for the Protocol's implementation, in areas such as sharing information and providing technical assistance to parties that are developing countries or countries with economies in transition.

The Protocol is designed as a dynamic instrument that can be revised to better serve users' needs and in keeping with new technical developments. Parties should strive to achieve convergence between their national registers so that information on the registers can be compared worldwide.

The application of information collected under the Protocol to other international agreements and processes has been promoted by the International PRTR Coordinating Group, the UN interagency body which seeks to improve coordination between international organizations, Governments, and other interested parties in their ongoing and planned efforts to the further development and implementation of PRTR systems.*

The role of PRTRs in achieving the Strategic Approach to International Chemicals Management (SAICM) was recognized in the SAICM Global Plan of Action, adopted at the first International Conference on Chemicals Management (ICCM-1) in Dubai UAE in February 2006. The Plan contained no less than seven different activities related to PRTR development.†

Countries with national PRTRs are encouraged to make the appropriate linkages with global POPS monitoring projects being developed by United Nations Environment Programme (UNEP). The UNEP Global Environment Facility (GEF) funded POPS monitoring projects will deal with the presence and reduction of POPS in the environment and humans, and look at the reduction of POPS emissions at the sources, stockpiles, POPS in use, and contaminated sites.

The global project, "POPS monitoring, reporting and information dissemination using PRTRs," was initiated under the GEF in 2008, with the participation of 13 countries (see Table 13.1).

PRTRs under the Protocol also include the major greenhouse gas emissions. The Protocol requires governments to collect annual reports on major greenhouse gas emissions (among other pollutants) by industry on a facility-by-facility basis and to share this information with the public. All of the substances identified in the UNFCCC and Kyoto Protocol are also contained in the PRTR Protocol's Annex II list of threshold pollutants.‡

In 2009, the ECE at its biennial session debated the relevance of the Aarhus Convention and Protocol on PRTRs to achieving the objectives of a post-Kyoto

---

* See http://www.unece.org/env/pp/prtr/Intl%20CG%20images/about.html

† SAICM Global Plan of Action, possible work areas and associated activities No. 124–126 and 177–180. Details of the work areas and their activities were not negotiated. Thus, the target contained in the Global Plan of Action of establishing PRTRs in every country by 2015 cannot be said to have been endorsed by the ICCM. Such a target is, in any case, highly unrealistic.

‡ UNFCCC emissions data are generally considered more accurate than emissions data from other sources because the data are reported directly by governments using national statistics, using Intergovernmental Panel on Climate Change (IPCC) methodologies, and according to agreed reporting guidelines. In addition, national inventories of Annex I Parties are subject to expert review, which also contributes to Parties' collective confidence in the quality of these emission estimates. Because of the detailed inventory reporting requirements for Annex I Parties, the UNFCCC data cover more GHGs, and at a higher level of sectoral disaggregation, than do external sources. The UNFCCC emissions data are, however, limited in geographic and temporal coverage. Data for Annex I Parties only include the period covered by the Convention, and do not yet include a full time series for all Parties. For non-Annex I Parties, the data set is even more limited: only a single year estimate is available for most reporting Parties, and for several countries no data are available yet.

TABLE 13.1
List by Region of Countries with Existing PRTRs (in Bold) and Countries which are Developing PRTRs

North America, Europe, and the Caucasus	Asia and Pacific	Latin America and the Caribbean	Africa
Austria	**Australia**	**Chile**	Togo[c]
Belgium	Cambodia[a]	Costa Rica[b]	
Bulgaria	**Japan**	Dominican Republic[b]	
Canada	Kazakhstan[a,d]	Ecuador[a]	
Croatia	**Korea**	El Salvador[b]	
Czech Republic	Thailand[a]	Guatemala[b]	
Denmark		Honduras[b]	
Finland		**Mexico**	
France		Nicaragua[b]	
Georgia		Panama	
Germany		Peru[a]	
Greece			
Hungary			
Iceland			
Ireland			
Italy			
Netherlands			
Norway			
Poland			
Portugal			
Slovak Republic			
Slovenia			
Spain			
Sweden			
Switzerland			
Ukraine[a]			
United Kingdom			
United States			

[a] Through the global project, "POPS monitoring, reporting and information dissemination using PRTRs," initiated under the GEF in 2008.
[b] Exploring development of a regional PRTR and cooperating with the GEF global project.
[c] Developing a national PRTR adapted to specific needs.
[d] Also UNECE member state.

agreement on climate change.* It found that "GHG data reported to national Protocol on Pollutant Release and Transfer Registers under the ECE Protocol on Pollutant Release and Transfer Registers to the Aarhus Convention could contribute to the development of relevant climate change statistics."†

PRTRs offer an established reporting network on emissions and transfers in many countries, a developing network in others and in general a proven and widely established mechanism for collecting emissions and transfer data. The international PRTR Coordinating Group reported to the *ad hoc* open-ended working group to prepare for the intergovernmental negotiating committee on mercury‡ that "[w]ith respect to reporting of releases of mercury within PRTRs, all established PRTRs and all those under development include mercury and its compounds as substances to be reported."§ They concluded that PRTRs could be foreseen as an integral part of the eventual mercury-emissions reporting and tracking which would be an anticipated element of a global instrument on mercury.¶

CONCLUSION

At the end of 2009, some 50 national PRTR systems were in operation or under development around the globe. With the entry into force of the Kiev Protocol on PRTRs and the coordinated efforts of UN agencies and its allied regional organizations to promote PRTRs in other regions, the future role of PRTRs in international chemicals and waste management appears to be secure.

---

* "... to ensure that transparency, accountability and public engagement were guaranteed in climate decision-making. ... [t]he incorporation of provisions on access to information, public participation and access to justice in relation to climate change was urged in any post-Kyoto agreement. An agreement should build on the broad consensus reflected in Article 6 or the UNFCCC and *Principle* 10 of the Rio Declaration on Environment and Development. Such provisions on access to information and public participation in decision-making could draw upon elements of the Aarhus Convention and the experience of its implementation in the region." Report of the 63rd session of the Commission. available at: http://www.unece.org/commission/2009/ANNUAL_REPORT_2009_E.pdf 285k 02/Sep/2009

† Report of 2009 ECE session, available at: http://www.unece.org/env/documents/2009/ECE/CEP/ece.cep.2009.5.e.pdf 380k 09/Mar/2011

‡ Meeting in Bangkok, October 19–23, 2009.

§ The Protocol on PRTR thresholds by facility for reporting releases of mercury and mercury compounds to air is 10 kg/year; for releases to water and land, 1 kg/year; for off-site transfers as waste, 5 kg/year; and for manufacture, process, or use (MPU), 5 kg/year. The US Environmental Protection Agency, in recognition of the highly persistent, highly bioaccumulative toxic nature of mercury metal and mercury compounds, lowered the reporting threshold for these to 10 lb/year (~5 kg/year) in 1999.

¶ UNEP(DTIE)/Hg/WG.Prep.1/INF/2. See http://www.chem.unep.ch/mercury/WGprep.1/Documents/k10_2)/English/WG_Prep_1_INF2_PRTRs.doc

14 The Rotterdam Convention

*Ernest Mashimba**

CONTENTS

* Late Chief Government Chemist, Dar es Salaam, Tanzania, http://www.gcla.go.tz, and former member of the Chemicals Review Committee of the Rotterdam Convention, died suddenly September 18, 2010.

INTRODUCTION

The growth of world trade in chemicals leads to several risks associated with improper use and handling of those chemicals and pesticides especially in developing countries. There are about 1500 new chemicals introduced to the world market every year, which influence the governments on the process of supervision of potentially dangerous chemical substances crossing the borders every day. In Tanzania there are about eight borders that we share with neighboring countries; hence there is a remarkable movement of different goods, and chemicals are among those goods. As a result the Rotterdam Convention came as a means to overcome the risks associated with chemicals import and utilization.*

The Rotterdam Convention is a multilateral environmental agreement that promotes shared responsibilities and cooperative efforts among parties in relation to importation of hazardous chemicals. The Convention presents a valuable tool to parties in managing chemicals that they want to import or export through monitoring and controlling its trade by using proper labeling, include directions on safe handling, and inform purchasers of any known restrictions or bans.† The Convention applies to pesticides and chemicals banned or severely restricted to protect human health and the environment from potential harm and contribute to the sound use of their environmental.

The essence of the Rotterdam Convention is information exchange which includes two main components: a legally binding prior informed consent (PIC) procedure and an information exchange procedure. The PIC procedure is a mechanism of formally obtaining and disseminating the decisions of importing countries as to whether they wish to receive future shipments of specified chemicals and for ensuring compliance with these decisions by the exporting countries.‡ In the information exchange procedure under the Convention, Parties are obliged to inform other Parties about each ban or severe restriction on a chemical it implements nationally.

Further, a Party planning to export a chemical that is banned or severely restricted for use within its territory, must inform the importing Party whether such export can take place. This component of the Convention involves only governments and it does not involve any action from exporters.§ However, exporters have to comply with decisions taken by the importing country.

The Convention is one of a series of agreements focusing on chemicals safety and environmental protection and is sometimes referred to as a "first line of defense"

* http://www.pic.int

† http://www.ec.gc.ca/international/multilat/rotterdam_e.htm

‡ Green Customs initiative, 2005.

§ http://www.ec.gc.ca/international/multilat/rotterdam_e.htm

because it seeks to prevent problems from happening at the first place. Other Conventions related to chemical activities includes the Stockholm Convention which is aimed at halting the use of hazardous persistent organic pollutants (POPS) and coping with their widespread presence in the environment while the Basel Convention targets international trade and the effective management and clean up of hazardous wastes. However, the Bamako Convention is concerned with the ban of the import into Africa and the control of transboundary movement and management of hazardous wastes within Africa. It is a treaty of African nations prohibiting the import of any hazardous (including radioactive) waste. Most of African countries including Tanzania have acceded to and ratified the main international Conventions associated with chemicals management.

BACKGROUND INFORMATION

The dramatic growth in chemicals production and trade led to increasing concern about the potential risks related to the inappropriate use and handling of hazardous chemicals and pesticides, especially in developing countries like Tanzania. This is due to the fact that most of these developing countries lack expertise, adequate infrastructure, and capacity to monitor the import and safe use of these chemicals as a result they are mostly at risk.*

In the mid-1980s governments started to address chemicals and pesticide issues by developing and promoting voluntary information exchange programs on dangerous chemicals and pesticides. The Food and Agriculture Organization (FAO) launched its "International Code of Conduct on the Distribution and Use of Pesticides" in 1985 and the United Nations Environment Programme (UNEP) set up the "London Guidelines for the Exchange of Information on Chemicals in International Trade" in 1987. In 1989, the two organizations jointly introduced the voluntary PIC procedure into these two instruments (whereby FAO operates the Secretariat for pesticides while UNEP operates the chemicals Secretariat). From the above the governments got the necessary information enabling them to assess the risks of hazardous chemicals and to take informed decisions on their future import.†

The officials attending the Rio Earth Summit in 1992 adopted Chapter 19 of *Agenda 21*, and called for the adoption of a legally binding instrument on the PIC procedure by 2000. Hence, the Convention came into force on February 24, 2004 after being adopted in 1998. On May 30, 2007 there were 115 Parties and two more states (Bosnia-Herzegovina and Vietnam) that had ratified‡ (Party means a state or regional economic integration organization that has consented to be bound by this Convention and for which the Convention is in force). As of April 2011, there were 140 Parties to the Convention. As a party to the Convention the country participates in the Convention by ensuring that chemicals listed in the Convention are not exported to countries that do not wish to receive them. The initial list of chemicals covered by the PIC procedure when the Rotterdam Convention was adopted included 22 pesticides and 5 industrial chemicals. More chemicals have been added and

* http://ec.europa.eu/environment/chemicals/pic/index.htm
† http://www.fao.org/newsroom/en/news/2004/37667/index.html
‡ http://www.ec.gc.ca/international/multilat/rotterdam_e.htm

currently there are a total of 40 chemicals in Annex III, the list is expected to expand in the years to come as more chemicals will be added because of their potential risks discovered.

THE OBJECTIVES OF THE ROTTERDAM CONVENTION

The Rotterdam Convention enables the world to monitor and control the trade in certain hazardous chemicals.

The responsibility of the importing countries among these is to ensure that

- Stakeholders (e.g., importers, users, workers, etc.) are informed on notifications received by the designated national authority (DNA).
- Import decisions are clearly communicated to all exporting countries and domestic manufactures of the chemical concerned.

The export countries are required to ensure that

- No export of a chemical takes place without consent of the importing country unless the chemical is approved in the countries in question.
- Import decisions are communicated to exporters, industry, and other relevant authorities by the DNA.
- Shared responsibility and cooperative efforts among Parties is promoted in the international trade of hazardous chemicals in order to protect human health and the environment from potential harm.
- They contribute to the environmentally sound use of those hazardous chemicals, by facilitating information exchange about their characteristics, by providing national decision-making information regarding their use, import, and export and by disseminating these decisions to Parties.

Meeting the Objectives

The objectives of the Rotterdam Convention are met through

- Provision of early warnings on potentially hazardous chemicals
- Provisions of the basis for decisions regarding future imports of chemicals
- Enforcement of import decisions
- Exchange of information on decisions regarding ban and severe restrictions in countries

The PIC Procedure*

The Convention focuses on the export of banned or severely restricted chemicals (BSR) that are included in Annex III to the Convention, which can only take place

* http://www.pic.int/proceedings/Kiev1.PDF+challenges+of+rotterdam+Convention

with the PIC of the importing party. Before applying this system some procedural steps and conditions should be followed and met.

Provided that the obligations are fulfilled, a chemical that has been banned or severely restricted by at least two Parties belonging to different geographic regions may be added to Annex III (Parties are obliged to notify all such domestic bans/severe restrictions).

A procedure is established for formally obtaining and making known the decisions of importing countries as to whether they wish to receive future shipments of listed chemicals and for ensuring compliance with these decisions by exporting countries.

For every chemical listed in Annex III of the Convention a decision guidance document (DGD) is prepared and sent to all Parties with a request that they take a decision as to whether they will allow future import of the chemical. The resulting decisions on future import of these chemicals (import responses) are published by the Secretariat and made available to all Parties every six months through the PIC Circular.

The PIC procedure provides all Parties with an opportunity to make informed decisions as to whether they will consent to future imports of the chemicals listed in Annex III of the Convention. All Parties are required to ensure that their exports do not take place contrary to an importing Party's import decision.

Scope of the Rotterdam Convention

The Convention covers pesticides and industrial chemicals that have been banned or severely restricted considering health or environmental reasons by Parties and which have been notified by Parties for inclusion in the PIC procedure. One notification from each of two specified regions triggers consideration of addition of a chemical to Annex III of the Convention. Severely hazardous pesticide formulations (SPHFs) that present a hazard under conditions of use in developing countries or countries with economies in transition may also be nominated for inclusion into Annex III.*

The Convention does not Cover

This Convention does not apply to

- Narcotic drugs and psychotropic substances
- Radioactive materials
- Wastes
- Chemical weapons
- Pharmaceuticals, including human and veterinary drugs
- Chemicals used as food additives
- Food
- Chemicals in quantities not likely to affect human health or the environment provided they are imported
  i. For the purpose of research or analysis
  ii. By an individual for his or her own personal use in quantities reasonable for such use

* http://www.ec.gc.ca/international/multilat/rotterdam_e.htm

Current Situation

There are currently 39 chemicals listed in Annex III of the Convention and subject to the PIC procedure, including 24 pesticides, 4 SPHFs, and 11 industrial chemicals. More chemicals continue to be added. The Conference of the Parties (COP) decides on the inclusion of new chemicals (see footnote 7).

Once a chemical is included in Annex III, a DGD containing information concerning the chemical and the regulatory decisions to ban or severely restrict the chemical for health or environmental reasons is circulated to all Parties.

The Parties are given a period of nine months to respond concerning the future import of that particular chemical. The response can be in a form of a final decision (to allow import of the chemical, not to allow import, or to allow import subject to specified conditions) or an interim response. Decisions by an importing country must be trade neutral (i.e., apply equally to domestic production for domestic use as well as to imports from any other source).

The import decisions are circulated and exporting country Parties are obligated under the Convention to take appropriate measure to ensure that exporters within their jurisdiction comply with the decisions.

Information Exchange

The Convention also promotes the exchange of information on a very broad range of chemicals.

It does so through

- The requirement for a Party to inform other Parties of each national ban or severe restriction of a chemical
- The possibility for a Party which is a developing country or a country in transition to inform other Parties that it is experiencing problems caused by an SPHF under conditions of use in its territory
- The requirement for a Party that plans to export a chemical that is banned or severely restricted for use within its territory, to inform the importing Party that such export will take place, before the first shipment and annually thereafter
- The requirement for an exporting Party, when exporting chemicals that are to be used for occupational purposes, to ensure that an up-to-date safety data sheet is sent to the importer
- Labeling requirements for exports of chemicals included in the PIC procedure, as well as for other chemicals that are banned or severely restricted in the exporting country

PIC COORDINATION

Conference of the Parties

The COP is the highest authority of the Convention that oversees the implementation of the Convention through its meetings that are scheduled every two years.

It decides on the inclusion of a chemical into Annex III of the Convention. The COP also establishes subsidiary bodies like the Chemical Review Committee (CRC).

Secretariat

The Secretariat is provided jointly by UNEP and FAO. This assists Parties in the implementation of the Convention, coordinates with regional and international partners and organizes meetings of the COP and CRC.

Designated National Authority

In order for PIC to operate a respective government must appoint a DNA (Ndiyo, 2008) to perform the administrative functions required by the Convention, and acting on its behalf. DNAs are government departments responsible for policy decisions and regulation of pesticides or industrial chemicals. In some countries there is one DNA for both the industries, chemical and pesticide, whereas in some countries there are two DNAs, one for industrial chemicals and separate one for the pesticides. As an example, Tanzania has two DNAs, one for industrial chemicals and a separate one for pesticides (see Chapter 42 of this book). These two DNAs cooperate well and both are members of the national PIC Committee.

Chemical Review Committee: Establishment, Organization, and Functions

Establishment

The CRC was established in September 2004 by the COP, as a subsidiary body of the COP; it was established by the COP according to Article 18 Paragraph 6, for the purpose of performing functions assigned by the Convention.

Functions and Mandate of CRC

The main functions and mandate of CRC are as follows:

- To make recommendations on inclusion of chemicals in Annex III by reviewing notifications of BSR.
- To make recommendation on inclusion of SPHF by reviewing proposals for SHPF according to Annex IV Part 3 of the Convention.
- To draft DGDs for chemicals/SHPF that have been recommended and decided by the committee for listing in Annex III.
- The committee makes recommendations on removal of chemicals from Annex III after reviewing the information that was not available at the time of decision to list the chemical in Annex III which indicates that its listing may no longer be justified. A new revised draft DGD is then prepared.

Organization and Operations

The meeting of CRC members is held annually in either Geneva or Rome. However, there is no formal reason why the meeting should not meet in other places.

The rules of procedure for COP also apply to the CRC, that is, to strive for consensus with the possibility of voting (2/3) if the consensus cannot be reached; however, there is no voting in the COP.

Members

The CRC is composed of 31 government designated experts in chemicals management from different geographical locations that balance across five UN regions (namely Africa, Asia-Pacific, Central and Eastern Europe, Latin America and the Caribbean, Western Europe, and others). Half the members of CRC are changed every two years resulting in average membership period of four years. According to decisions RC-1/7 members are required to submit a declaration of conflict of interest. The members also serve in their individual capacity and not as national representatives. Hence, before appointment, each prospective member has to declare his/her conflict of interest. Experts serve on the CRC for a period of two or four years.

Bureau

The Bureau is usually appointed at the first meeting of a new CRC meeting. It is composed of a president appointed by all CRC members and vice presidents from each of the PIC regions. The office of Bureau lasts until a new CRC has been formed. A new CRC is formed after the two-year term of service of members expires. In addition to these tasks the president of the Bureau chairs the CRC meetings.

Overall the Bureau has the task of ensuring that the CRC meeting runs smoothly. The Bureau may hold a meeting prior to the meeting of CRC or when a specific need arises.

The CRC Task Groups

The CRC has developed working procedures and policy guidance covering a broad range of issues related to the work of CRC. These working procedures are intended to facilitate the operation of the committee and to help ensure consistency and transparency. In that light task groups are established to enhance the work of CRC through carrying out initial reviews prior to the CRC. All CRC members participate in the work of one or more task groups.

Each intercessional task group is composed of one or two coordinators and a representative group of members of the committee, proposed by the Secretariat, in consultation with the Bureau.

Once the task group is established, all CRC members are informed of its composition and designation of the coordinators that normally take the lead in reviewing the documentation and drafting the report. Members are free to participate in as many task groups as they wish, informing the coordinators and the Secretariat of their interest.

The task group undertakes an initial detailed review of the notifications and supporting documentation for a chemical in light of the information requirements and criteria set out in Annexes I and II of the Convention, normally relevant working papers and policy guidance are available to direct the work of the committee. The initial review facilitates the work of the committee by ensuring that there is a clear understanding of the scope of the regulatory action, and whether or not the criteria in Annex III of the Convention have been met. After completion, the draft report is circulated to all members of the task group for comments, prior to the meeting where the task group has to present to the committee their assessment of whether the notifications and supporting documentation meet the requirements for inclusion in the Convention.

SCHEMES OF INCLUDING A CHEMICAL INTO THE ANNEXES OF THE CONVENTION

There are two schemes that are followed to include a chemical into the Annexes of the Convention. The schemes are, one for BSR and the other for SHPFs.

Banned or Severely Restricted Chemicals

- At least two parties from at least two different countries propose the inclusion of a chemical in Annex III (see Figure 14.1).
- They should notify the Secretariat when they take a final regulatory action to ban or severely restrict a chemical.
- For national regulatory actions prior to the ratification of the Convention the DNA has to notify the Secretariat after the Convention has entered into force in that country. This excludes those submitted under the previous voluntary PIC procedure.
- For new regulatory actions the DNA has to notify the Secretariat within 90 days after the decision has been taken.
- The Secretariat verifies whether the notification is complete as per Annex I. Upon receipt of a second complete notification, the notifications and supporting documents are forwarded to the CRC.

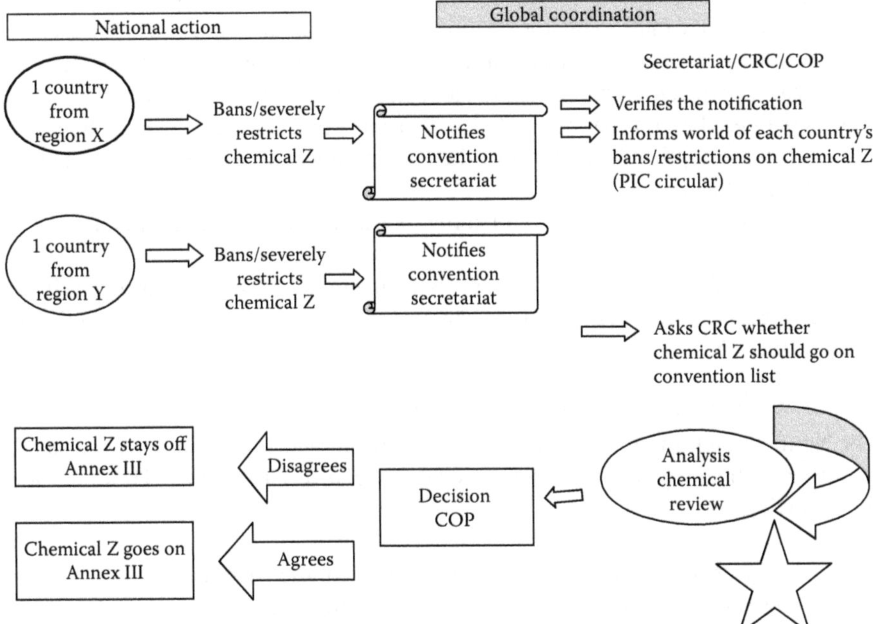

FIGURE 14.1 Adding a chemical to Annex III.

- CRC reviews the information provided in such notifications in accordance with the criteria set out in Annex II.
- CRC recommends to the COP whether the chemical should be listed in Annex III and drafts a DGD.
- The COP decides whether the chemical should be made subject to the PIC procedure by its inclusion into Annex III.

Review of Notification

In reviewing the notifications the CRC has to confirm that the regulatory action has been taken in order to protect human health or the environment.

The regulatory action has been taken as a consequence of a risk assessment based on review of scientific data in the context of the conditions prevailing in the Party in question. The data should be generated according to scientifically recognized methods and data reviews should have been performed and documented according to recognized scientific principles and procedures.

The CRC should consider whether the regulatory action provides a sufficiently broad basis to merit inclusion of the chemical in Annex III through consideration of the following:

- Whether the final regulation was based on risk evaluation in the country in question.
- Whether the final regulatory action led, or would be expected to lead to a significant decrease in the quantity of the chemical used or the number of its uses.
- Whether the final regulatory action led to an actual reduction of risk or would be expected to result in a significant reduction of risk for human health or the environment of the Party that submitted the notification.
- Whether the considerations that led to the final regulatory action being taken are applicable only in a limited geographical area or in other limited circumstances.
- Whether there is evidence of ongoing international trade in the chemical.

Also CRC should take into account that intentional misuse is not in itself an adequate reason to list a chemical in Annex III. There is a notification form that is filled by the DNA.

Common Challenges for DNAs to Complete Forms*

The Notification Form is a template for standardized reporting of national decisions. However, difficulties are encountered by DNAs in filling the forms. These include the following:

- Clearly understanding the nature and identity of the chemical
- Understanding what a severely restricted chemical is

* In Report of Meeting of Regional Experts, Geneva, Switzerland, December 2005.

- The need for a risk or hazard evaluation
- Lack of sufficient background information on regulatory action (little information found; elapse of time since decision was taken or poorly documented, etc.)
- Information available is voluminous and therefore making it difficult to prepare summaries.
- Action not considered because of deliberate misuse.

To alleviate these problems, policy guidelines have been developed by the Secretariat. These include the use of bridging information (see the section "Bridging Information") where a risk evaluation cannot be carried out or focused summaries (see the section "Challenges of Using Bridging Information") when the information is voluminous. Training of relevant staff is also necessary.

Severely Hazardous Pesticide Formulations

- A developing country or country with economy in transition may propose a pesticide formulation that is causing health or environmental problems under conditions of use in its territory.
- The proposal should contain the information set out in Part 1 of Annex IV (DNA–Party).
- The Secretariat verifies whether notification is complete and collects additional information as set out in Part 2 of Annex IV.
- The proposal and supporting information are forwarded to the CRC for review.
- CRC reviews the information provided in accordance with the criteria set out in Part 3 of Annex IV.
- CRC makes recommendations as to whether the formulation should be listed in Annex III and prepares a DGD.
- The recommendations and DGD are forwarded to the COP.
- The COP decides whether the chemical should be made subject to the PIC procedure.

Reviewing Proposals for SHPFs by CRC

In reviewing proposals for severely hazardous pesticide formulations (SHPFs) the CRC considers

- The reliability of the evidence indicating that use of the pesticide formulation, in accordance with common or recognized practices within the proposing Party, resulted in the reported incidents.
- The relevance of such incidents to other states with similar climate, conditions, and patterns of use of the formulation.
- The existence of handling or applicator restrictions involving technology or techniques that may not be reasonably or widely applied in states lacking the necessary infrastructures.

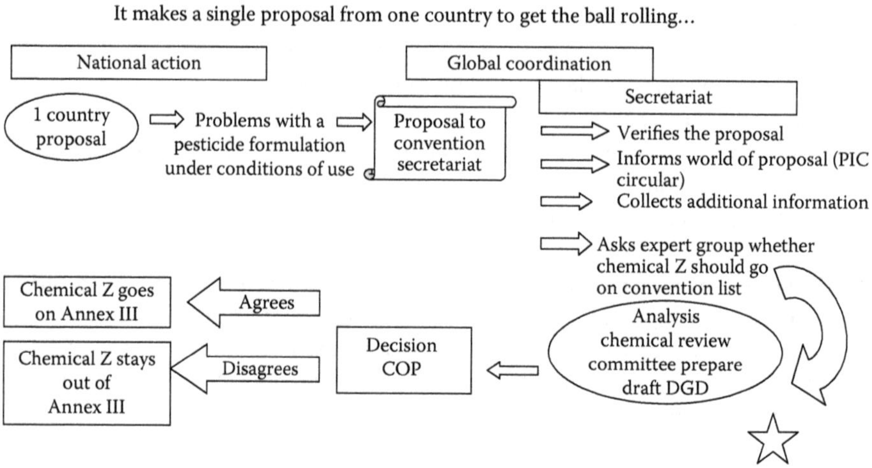

It makes a single proposal from one country to get the ball rolling...

National action

Global coordination

Secretariat

1 country proposal

Problems with a pesticide formulation under conditions of use

Proposal to convention secretariat

Verifies the proposal

Informs world of proposal (PIC circular)

Collects additional information

Asks expert group whether chemical Z should go on convention list

Chemical Z goes on Annex III

Agrees

Chemical Z stays out of Annex III

Disagrees

Decision COP

Analysis chemical review committee prepare draft DGD

FIGURE 14.2 Adding a "SHPF" to Annex III.

- The significance of reported effects in relation to the quantity of the formulation used.
- Intentional misuse is not in itself an adequate reason to list a formulation in Annex III.

A summary of the notification process is shown in Figure 14.2.

Why Not Many Proposals for SHPF are Submitted to the Secretariat*

In many developing countries the following problems among others are encountered by parties:

- Poor reporting and collection of pesticide poisoning incidents
- Where incidences/poisonings happen they are not communicated to the DNA or information is linking between exposure and effects
- Poor understanding by farmers/end-users of the risks of pesticides use to health and the environment resulting in not reporting symptoms to medical staff
- Often only anecdotal information on poisoning incidences exists

To alleviate these problems, a good reporting system both for acute and for chronic incidents has to be established in such countries. This should include an efficient and effective collection of information at all levels. In addition, medical personnel and the population (through community health monitoring) should be trained on how to act

* Mashimba, ENM and Katagira F: National Experience in collecting Information of Severely Hazardous Pesticide Formulations (SHPF); A paper presented during the joint Workshop of the Rotterdam and Stockholm Convention for effective participation in the Chemical Review Committee's Work—CRC and POPRC, Cairo, Egypt 17–19 November, 2009.

on pesticide incidents and incidents with other hazardous chemicals. The community should thus be empowered.

Issues and Challenges Encountered by CRC

- Many proposals for SHPF submitted to the Secretariat by developing countries are not based on a systematic mechanism for collecting and reporting information of poison incidents with SHPFs from the field to the DNA, especially in rural areas where most of the incidents occur. This is exacerbated by the absence of poison centers or presence of only weak poison centers.
- Generally most of the countries reporting incidents fail due to poor documentation of their notifications as well as limited risk evaluation reports.
- Completed pesticide incident forms may lack some important information that is required, such as the link between exposure and effect.

To resolve these problems, financial assistance and capacity building is important to developing countries in order to be successful and meet the requirements of the Convention. In addition, the decision for the inclusion of the chemical into Annex III should be based on health or environmental effects. In certain cases, COP decisions seem to be driven by economic or political interest.

BRIDGING INFORMATION

Meaning and Use

Conducting a risk evaluation may be a difficult task for most developing countries and countries with economies in transition; as part of a proposal to qualify, criteria of Annex II should be met. As such there is an opportunity for using bridging information for a notifying Party as a basis for its national decision. Bridging is when the notifying Party uses a risk-evaluation report and/or exposure assessments completed by another country or international body capable of performing risk evaluations as a basis of its national decision.

The CRC determines whether the conditions in the country which completed the original risk assessment and exposure assessments carried out under other international agreements or Conventions are similar to and compatible with those prevailing in the notifying country.

The CRC considers such bridging information on a case by case basis. In reviewing the information the committee takes into account:

- Exposure or potential exposure as a key element
- The information is science based and includes the best available knowledge
- The information is sufficiently detailed to enable the CRC to make an assessment

Thus countries that lack capacity and resources for conducting risk evaluations can still use this for inclusion of hazardous chemicals and pesticides in Annex III of the Convention. The elements provided in the bridging policy paper address both human health and environmental exposure.

Examples of countries which have successfully used bridging information include:

- Sahelan Countries (Burkina Faso, Cape Verde, Gambia, Mali, Mauritania, Niger, and Senegal) for Endosulfan.[*,†] These countries compared the product applications in Australia and in the United States and the decisions taken in Europe and France for their risk evaluation.
- Jamaica for notification of Aldicarb, used comparable information regarding the comparability of the conditions in Jamaica and the United States in relation to risk evaluation.[‡]

Challenges of Using Bridging Information

The conditions of the notifying Party should be the similar to and compatible with the country which completed the original risk assessment and exposure assessments carried out under other international agreements or Conventions, otherwise the notification would be rejected.

Focused Summaries

Focused summaries summarize briefly the main elements of a risk evaluation (see also the section "Common Challenges for DNAs to Complete Forms"). They need to

- Demonstrate how the notification fulfils the criteria in Annex II, summarize key decisions and key findings, with references to the associated documents
- Be translated into English where necessary
- Be accompanied by supporting documentation

The challenge that is obvious is the translation of such information from a local language in a notifying country into English. In recognition of this fact the secretariat has trained a member of staff in this context.

SUMMARY OF THE MAIN STEPS IN THE PIC PROCEDURE

- Governments should notify that they have banned or severely restricted a chemical or are experiencing problems with a SHPF.
- A DGD is prepared and circulated.
- Governments take up to nine months to transmit their decisions to the Secretariat, but the lack of response does not mean approval.

---

* UNEP/FAO/RC/CRC.5/16.
† UNEP/FAO/RC/CRC.6/16.
‡ www.pic.int/INCS/CRC5/pInf6)/English/K0842549%20CRC-5-INF-6.doc

- The Secretariat disseminates importing countries' decisions every six months by means of the PIC circular.
- Exporting countries have to ascertain that their exporters comply with importing countries' decisions.

CONCLUSION AND OUTLOOK FOR THE FUTURE OF THE CONVENTION

The Rotterdam Convention, one of the major instruments for sound chemicals management, has been adopted more than 10 years ago and came into force in 2004. In the decade of its implementation it has shown to be a very useful instrument to promote information exchange on hazardous chemicals and to raise awareness of the import and use of certain hazardous chemicals.

Chemicals in the PIC list are known to be hazardous to health and/or the environment, for which reason the chemicals have been banned or severely restricted in some countries. However, adding other chemicals to the initial list has shown to be a complicated process, in which sometimes economic or political considerations also play a major role. Also, in the absence of voting by the COP, adding other chemicals has been limited to those few substances for which consensus was reached at the COP level. Another issue that merits attention is that requirements to add SHPFs are often beyond the capabilities of those countries for which this part of the Convention is specifically intended.

Although the Convention does not ban any substance, but has the objective to create awareness about hazards and risks, some countries see inclusion of a chemical as a *de facto* ban. Also, import decision making and provision of import responses from individual countries has been of utmost importance because it gives opportunity for countries to accept or reject PIC chemicals.

Parties to the Convention should continually make efforts to meet obligations stated under Article 10 of the Convention. However, making use of assistance from the Secretariat and other collaborative partners, increased assistance for capacity building, and sustainable funding seem major prerequisites for a successful continuation of the implementation of the Convention.

Also, further integration with other multilateral environmental agreements will assist in broadening the scope of sound chemicals management, both at international and at national levels.

REFERENCE

Ndiyo, D. 2008. Notification of Final Regulatory Action for Implementation of the Rotterdam Convention, Paper Presented at Stakeholders National Seminar on Review of National Action Plan for Implementation of The Rotterdam Convention.

15 Implementing the Stockholm Convention

*An Increasingly Expensive Challenge**

Pia M. Kohler and Melanie Ashton

CONTENTS

INTRODUCTION

Persistent organic pollutants (POPS) as a class of pollutants have been the focus of international negotiations since the early 1990s. A POP is a chemical exhibiting several characteristics, including persistence and bioaccumulation once released in the environment, a propensity for long-range environmental transport, and adverse effects on human health and/or the environment. The Stockholm Convention on POPS, which entered into force in 2004, set out control measures for 12 POPS known as the "dirty

* An earlier version of this chapter has been included in the following online publication: Anne Petitpierre-Sauvain, editor. 2010. The Basel, Rotterdam, and Stockholm Conventions on Chemicals and Wastes—Regulation, Sound Management and Governance. *EcoLomic Policy and Law* Special Edition 2008–2010, 233pp. http://www.ecolomics-international.org/headg_ecolomic_policy_and_law.htm

dozen." These include pesticides, such as dichloro-diphenyl-trichloro-ethane (DDT), industrial chemicals, such as polychlorinated biphenyls (PCBs), and unintentional byproducts, such as dioxins and furans.

In this chapter we first provide an overview of the origins, and a brief history of the Stockholm Convention, and related negotiations. We then examine progress made in implementing the Convention's provisions, and the financial implications of recent developments under the Convention following parties' decision, in May 2009, to add nine chemicals to the treaty's scope. Finally, we look forward to the likely finance scenarios under the upcoming fifth replenishment of the Global Environment Facility (GEF), and related activities on expanding the available financial resources for implementing the Convention.

THE STOCKHOLM CONVENTION

Origins of the Convention

Concerns with the potential health and environmental effects of chemicals being released in the environment were brought to the fore with the publication of Rachel Carson's *Silent Spring* in 1962 which warned against the far reaching impacts of DDT use. Since then, the ever growing number of chemicals in use and being released into the environment has outpaced regulations on many chemicals' production and use. Nevertheless, the use of chemicals has long been the focus of international regulatory responses. In 1972, the UN Conference on the Human Environment resulted in the creation of United Nations Environment Programme (UNEP), and it adopted the Stockholm Action Plan, which addressed hazardous chemicals, in particular calling on states to minimize their environmental releases, and, 20 years later, as countries prepared for the 1992 UN Conference on Environment and Development, chemicals management was the focus of Chapters 19 and 20 of *Agenda 21** (Selin, 2010, see also Chapter 3 of this book). This emphasis on chemicals management continued into the 1990s as several institutions were established providing fora for a broad range of stakeholders to address issues related to chemicals management, including, for example through the Intergovernmental Forum on Chemical Safety (IFCS, see Chapter 21 of this book) in 1994.

A regional treaty, the 1979 Convention on Long-Range Transboundary Air Pollution (CLRTAP, see Chapter 10 of this book), played a significant role in bringing the issue of POPS to the fore of international attention. This Convention, which entered into force in 1983, was negotiated under the auspices of the UN Economic Commission for Europe (UNECE), and brings together parties from Europe and North America. The CLRTAP includes several protocols, including a Protocol on POPS adopted in 1998. In particular the CLRTAP played an essential role in providing a forum for framing the problem of POPS, and in 1990 a task force on POPS began assessing the issue (Selin, 2010). This agenda item was at first largely driven by Canada and Sweden as researchers in those countries had begun detecting unexpectedly high concentrations of organic substances in their Arctic areas (Fenge, 2003;

---

* Available at http://www.un.org/esa/dsd/agenda21/res_agenda21_20.shtml

Downie and Fenge, 2003; Selin, 2003). In 1991, Arctic countries also came together to establish the Arctic Monitoring and Assessment Programme (AMAP) to "monitor the levels of pollutants and to assess their effects in the Arctic environment" (Reiersen et al., 2003; Selin and Selin, 2008). By late 1994, the CLRTAP executive body decided to form a preparatory working group to discuss drafting a protocol, and negotiations on the CLRTAP Protocol on POPS were concluded in June 1998 (Selin, 2010).

At the global level, POPS too were rising in salience. In 1995, the UNEP Governing Council began assessing a list of 12 POPS and negotiations on a global POPS Convention began in 1998. Arctic indigenous people's research demonstrating them to be particularly vulnerable to POPS contamination (Selin, 2010), also played an active role in negotiating a global POPS treaty. While Arctic indigenous groups had not participated in the early CLRTAP POPS negotiations, they had played an active role in the AMAP process and in March 1997 formed a coalition (the Northern Aboriginal Peoples' Coordinating Committee on POPS) to participate in the later CLRTAP negotiations (Fenge, 2003). Several coalitions of Arctic indigenous groups played a significant role in the global POPS treaty negotiations, including the Russian Association of Indigenous Peoples of the North (RAIPON) and the Inuit Circumpolar Council (ICC). In particular Sheila Watt-Cloutier, then vice-president of the ICC, representing the Inuit of Greenland, Alaska, Russia, and Canada, is credited with emphasizing the public health threat from POPS, through interventions but also by presenting the Executive Director of UNEP with an Inuit carving of a mother and child. This carving was present on the dais at all subsequent negotiations (Watt-Cloutier, 2003; Fenge, 2003; Selin, 2010).

In 1998, over 400 advocacy groups came together to form the International POPS Elimination Network (IPEN, see Chapter 25 of this book). IPEN was established with the aim of supporting the elaboration of global POPS controls and also played a key role in bringing together arctic indigenous groups and indigenous peoples of Africa (Watt-Cloutier, 2003). These connections proved significant in bridging two key concerns surrounding the POPS negotiation: the adverse health impacts of POPS in Arctic indigenous populations, and the adverse health impacts of malaria in countries still relying on DDT for malaria vector-control.

In Stockholm in May 2001, 92 States and the European Community signed the final text of the Convention on Persistent Organic Pollutants. The preamble to the Convention addresses several key elements shaping the Convention, including acknowledgments that "Arctic ecosystems and indigenous communities are particularly at risk because of the biomagnification of (POPS) and that contamination of their traditional foods is a public health issue," and that "precaution underlies the concerns of all the Parties*." Furthermore, it emphasizes an awareness of "the health concerns, especially in developing countries, resulting from local exposure to [POPS], in particular impacts upon women and, through them, upon future generations." Another key element of the final Convention was that, while nine of the substances in the "dirty dozen" were listed for elimination (under Annex A), and while Annex C identifies those POPS subject to control from unintentional production, Annex B provides for restrictions on the production and use of DDT. In particular, the Annex identifies

* The full text of the 2001 Stockholm Convention on Persistent Organic Pollutants is available at http://www.pops.int

disease vector control in accordance with WHO recommendations and guidelines as an acceptable purpose for both the production and use of DDT.

A Precautionary and Dynamic Convention

The Convention lists the "dirty dozen" in three different annexes according to Parties' responsibilities for control measures. Annex A lists nine substances slated for elimination.* Annex B lists DDT as a substance for restriction, and Annex C lists three substances that are produced unintentionally† and outlines guidance for preventing their production. The Convention was structured so as to allow for additions to each of these Annexes.

The preamble to the Stockholm Convention‡ acknowledges "that precaution underlies the concerns of all the Parties and is embedded within this Convention." There are several references to precaution in the Convention, especially relating to the listing of new chemicals, and Article 8 (Listing of chemicals in Annexes A, B, and C) concludes by stating that "The Conference of the Parties, taking due account of the recommendations of the Committee, including any scientific uncertainty, shall decide, in a precautionary manner, whether to list the chemical, and specify its related control measures, in Annexes A, B and/or C."

The Convention provided for Conference of Parties (COP) 1 to establish a POPS Review Committee (POPRC) to undertake the review of any chemicals nominated for listing under the Convention. The Intergovernmental Negotiating Committee extensively negotiated the terms of reference of this expert body, and at COP1 parties agreed to establish a 31-member Committee (Kohler, 2006). Members are government-designated experts: eight from African States, eight from Asian and Pacific States, three from Central and Eastern European States, five from Latin American and Caribbean States, and seven from Western European and other States. The UN's five regional groups are entrusted with identifying the countries eligible to designate experts to the Committee, and provisions were made to ensure half the membership of the Committee would rotate every two years.

The POPRC terms of reference agreed at COP1 left much of the organization of work to the discretion of the Committee itself. The salient points of the COP guidance include: that meetings shall be open to parties and other observers, the establishment of open *ad hoc* working groups, a conflict of interest procedure, annual meetings and timelines for making documents available, and interpretation at meetings (Decision SC-1/7).

The POPRC follows a three-stage review process detailed in Annexes D, E, and F of the Convention, beginning with an assessment of whether a chemical nominated for listing meets initial screening criteria set out in the Convention. The Committee decides whether a global ban is warranted prior to taking into account socio-economic

* It is important to note here that the Stockholm Convention commitments do not provide differentiated timelines nor differentiated targets for developed and developing countries, in contrast to other well-known treaties such as the Montreal Protocol on ozone depleting substances and the Kyoto Protocol on climate change.

† PCB (polychlorinated biphenyl) is listed both under Annexes A and C.

‡ The full text of the 2001 Stockholm Convention on Persistent Organic Pollutants is available at http://www.pops.int

considerations, therefore, in practice potential health and environmental adverse effects essentially trump adverse socio-economic consequences of a global ban. From 2005 to 2008, the POPRC met four times and completed the review process for nine nominated chemicals, recommending their listing under the Convention to COP4 in May 2009.

Financial Mechanism

Financing the implementation of the Convention was a key dimension of the negotiations leading to the finalization of the Convention. Prior to the Convention's entry into force, a POPS Club, which attracted funding from governments and from non-governmental actors, provided funding for activities under the Convention (Earth Negotiations Bulletin, 2002). Throughout the negotiation process, developing countries and countries with economies in transition raised concerns regarding their access to the financial and technical assistance necessary for them to be able to implement the Convention's requirements. In particular, under Article 13.4 of the Convention, signatories agree that

> the extent to which the developing country Parties will effectively implement their commitments under this Convention will depend on the effective implementation by developed country Parties of their commitments under this Convention relating to financial resources, technical assistance, and technology transfer.

This provision would later have implications for compliance with the Convention, as at COP4 countries argued they could not agree to establish a compliance mechanism without being satisfied that adequate resources are available to comply.

During the negotiations, developing countries were vocal in calling for a stand-alone financial mechanism, akin to the Multilateral Fund under the Montreal Protocol on Ozone Depleting Substances. In contrast, developed countries much preferred working with existing international institutions, including the GEF. From the perspective of developing countries, a stand-alone financial mechanism would ensure that the COP itself would assess the needs for replenishing the fund and would set priorities for activities and countries eligible for financial assistance (Selin, 2010). This concern was addressed in part by the requirement, under Article 13.6 of the Convention, that "Contributions to the mechanism shall be additional to other financial transfers" to developing countries and countries with economies in transition. The Convention also sets out, in Article 14, provisions for the GEF to serve as the Convention's interim financial mechanism until the COP should decide upon another structure to serve as the financial mechanism.

Following the Convention's adoption in 2001, the GEF established a POPS focal area in 2003 and the GEF reported to COP1 on progress achieved in putting in place procedures for serving as the financial mechanism to the Convention. At COP1 the question of whether GEF would remain the financial mechanism as parties began their implementation process remained controversial. In the end, with some countries noting that GEF was the "only game in town," the GEF remained as the Convention's interim financial mechanism, a role discussed in greater detail below. The COP routinely provides guidance to the GEF, yet developing countries have repeatedly raised concerns relating to the financial mechanism, including relating to the lengthiness of the funding process (the project pipeline) and the difficulty of meeting cofinancing requirements (Earth Negotiations Bulletin, 2005; Earth Negotiations Bulletin, 2009).

Implementation and Compliance

Article 7 of the Convention states that parties must develop and transmit a national implementation plan (NIP) setting out the activities planned to meet their obligations under the Convention, to the COP within two years of the date on which the Convention enters into force for that party. NIPs are intended to assess the presence of scheduled POPS in each country, as well as the legislative measures in place to regulate POPS use, and to develop a list of prioritized actions to meet the requirements of the Convention. NIPs generally include an assessment of POPS import, export, and production data, inventories of POPS stockpiles, and calculations of POPS produced unintentionally through incomplete combustion processes. Guidance on the development of NIPs was provided by the Stockholm Convention Secretariat, and funding for these "enabling" activities was provided by the GEF.

Parties' obligations under the Convention require them to institute measures to reduce or eliminate unintentional production of POPS; manage POPS contaminated sites, and POPS stockpiles and wastes; eliminate or reduce intentional production of POPS chemicals; and increase information exchange and public awareness.

As nine of the "dirty dozen" chemicals (aldrin, endrin, dieldrin, chlordane, heptachlor, hexachlorobenzene, mirex, toxaphene, and PCBs) were considered to be "dead," that is, they are chemicals with very little remaining production and use, elimination was considered to be feasible, and therefore required for these chemicals which are listed in Annex A of the Convention. It is important to note that the Convention provides for identifying acceptable purposes for chemicals. The 2001 Convention text identifies the production and use of DDT for disease vector control as the acceptable purpose for the chemical, this is the only chemical of the "dirty dozen" listed under Annex B. A register of specific exemptions is also maintained by the Stockholm Convention Secretariat and specific exemptions expire after five years, but can be renewed. Parties must justify continuing need for the registration of exemptions.

Regarding those chemicals listed under Annex C arising from unintentional production, COP1 established an expert group on Best Available Techniques (BAT) and Best Environmental Practices (BEP) to continue work on draft guidelines on BAT and provisional guidance on BEP prepared by an expert group established by INC6 in 2002. These guidelines on BAT and provisional guidance on BEP were adopted by COP3 (Decisions SC3/5) in 2007 to assist parties in implementing their NIPs.

COP4: From "Dirty Dozen" to "Toxic 21"

The fourth meeting of the COP4 in Geneva in May 2009* represented the first occasion on which parties were required to consider adding chemicals to the scope of the

* http://chm.pops.int/Convention/COP/hrMeetings/COP4/tabid/404/mctl/ViewDetails/EventModID/870/EventID/23/xmid/1673/language/en-US/Default.aspx

Stockholm Convention. The POPRC concluded that nine chemicals met the criteria to be considered as POPS, and recommended these chemicals for scheduling in the Stockholm Convention.

As well as the addition of nine new chemicals to the Convention, parties at COP4 also considered: financial matters, based on the outcomes of a needs assessment, and the opportunity to provide guidance for the Fifth GEF replenishment (addressed in greater detail below), as well as the development of a compliance mechanism. These issues were negotiated in parallel working groups, but it was widely acknowledged by delegates that the interlinkages between them were significant. Addressing the links was fundamental to enabling a decision on each of the issues, as adding chemicals to the Convention meant additional finance was necessary for Convention implementation, and finance is also directly linked to the ability to comply (the Convention's Article 13 (Financial resources and mechanisms), which makes developing countries' abilities to comply contingent on financial and technical assistance). While Parties eventually agreed to add chemicals to the Convention, no progress was made on the development of a compliance mechanism (Earth Negotiations Bulletin, 2009).

Taking Stock of Implementation

As noted above, the Convention requires parties to prepare and submit NIPs within their first two years as parties. Special provisions were put in place (at COP1) to provide technical and financial assistance to developing countries and countries with economies in transition in meeting this commitment (Earth Negotiations Bulletin, 2005). The Stockholm Convention Secretariat developed guidance on the development of NIPs and funding for these enabling activities, that is, the development of NIPs was provided by the GEF.

As of March 2011, 131 NIPs had been submitted to the Secretariat of the Stockholm Convention. Up to October 31, 2008, the GEF financed development of 135 NIPs. The total value of NIP funding was US$ 58 million (Stockholm Convention, 2009a), 12% of GEF funds over the five-year period from 2003 to 2008. Despite the low percentage of funding, NIP activities represented the majority of GEF-funded POPS focal area activities.

Between January 2007 and October 2008 the focus of GEF funding shifted from NIP development to implementation. In this time period, GEF approved US$ 129.4 million in funds, across 22 full-sized projects (FSPs), and a further US$ 11.5 million, across 11 medium-sized projects. Just under a quarter of FSP funding was approved for four activities in China.

Broadening the Convention's Scope

Nine chemicals were recommended for addition to the Convention by their scientific assessment body, the POPRC. The chemicals included pentabromodiphenyl ether (pentaBDE), chlordecone, hexabromobiphenyl (HBB), alpha-hexachlorocyclohexane (alphaHCH), beta-hexachlorocyclohexane (betaHCH), lindane, commercial octabromodiphenyl ether (c-octaBDE), pentachlorobenzene (PeCB), and perfluorooctane sulfonate (PFOS).

Delegates agreed to list all of them except PFOS in Annex A of the Convention, for elimination, they will therefore be added to parties' implementation commitments. On the other hand, parties agreed to list PFOS in Annex B (for restriction) of the Convention. As well representing the first time that Parties considered adding new chemicals to the Convention, the chemicals nominated also provided additional challenges, as three of them remain produced for industrial use and in products, unlike the original "dirty dozen," which included chemicals that are in fact already mainly phased out of use.

Negotiations on pentaBDE and octaBDE (the BDEs) were fraught with resistance. Although production of these chemicals has essentially been phased out, they are ubiquitous in plastics and foam rubber products. Therefore, these chemicals are "live" in many commonly used products. As Article 6(d)iii of the Stockholm Convention prevents recycling of POPS, negotiators were left to grapple with the impact of listing the BDEs in light of the difficulty of separating BDE-containing plastic, from BDE-free plastic and the potential widespread fallout on the plastics recycling industry. Discussions on recycling, reuse, and trade of BDE-containing products, and the need to reduce the risks posed by new POPS in the waste stream, were extensive, and parties eventually agreed to list the BDEs and, with certain provisions, to permit recycling of products containing BDEs (Ashton, 2009). This was disappointing to many environmental NGO representatives, who stressed the danger of recycling BDEs into more products and therefore continuing to expose people and the environment to these POPS (Earth Negotiations Bulletin, 2009).

The listing of PFOS presented the challenge of listing a truly "live" chemical. PFOS is still widely produced and found extensively in products. The EU, supported by most developed countries called for immediate listing in Annex A of the Convention, but developing countries, led by China, argued that without the availability of cost-effective and environmentally-friendly alternatives they would not support the listing. Eventually Parties agreed to list PFOS in Annex B (for restriction) of the Convention. The decision on PFOS outlines several acceptable purposes including for fire-fighting foam and insect baits for leaf-cutting ants. It also outlines specific exemptions for metal plating, leather, textiles, paper, and plastics and rubber (Ashton, 2009).

Similar to the "acceptable purpose" identified for DDT for disease vector control, parties also agreed to a framework for acceptable purpose relating to PFOS, under which Parties must register uses for acceptable purposes with the Stockholm Convention Secretariat, and report every four years on the progress made to eliminate PFOS. A review of acceptable purposes will be undertaken in 2015 by the COP, and every four years thereafter in an effort to make progress on the phase out of PFOS (Stockholm Convention, 2009b).

Synergies

Another factor impacting the Stockholm Convention implementation is the ongoing international concern over the proliferation of issue-specific chemical and waste conventions, prompting calls to synergize activities and management to reduce the administrative burden, and to increase resources available for implementation. COP4 granted the final stamp of approval for increasing synergies among the Basel,

Stockholm, and Rotterdam Conventions,* and to convening an Extraordinary Meeting of the Conference of Parties (ExCOPs) of each of the Conventions to consider joint decision making and budgeting. The ExCOPs convened in February 2010 back to back with the UNEP Governing Council Special Session and Global Ministers Environment Forum.

The ExCOPs agreed to joint activities, joint managerial functions, joint services, synchronization of budgets, and joint audits of the three chemicals and wastes conventions. Perhaps most significant is the decision to undertake joint managerial functions. Parties agreed to establish a joint head position of the three conventions. The Executive Director of the UNEP was requested to immediately proceed with the appointment. It is anticipated that the joint head will lead the "synergyzation" process and act as a figure head to raise the profile of the chemicals and wastes conventions among donors, and to coordinate fund-raising efforts (Earth Negotiations Bulletin, 2010).

Noncompliance

The extent of capacity building and technical assistance available to parties, and the mechanisms through which it is made available, is closely tied to the question of noncompliance. At COP4 in May 2009, parties were considering several related issues. Article 17 of the Convention calls upon the COP to

> as soon as practicable, develop and approve procedures and institutional mechanisms for determining non-compliance with the provisions of [the] Convention and for the treatment of Parties found to be in noncompliance.

The question of noncompliance had been tackled by the COP at its third meeting and agreement remained elusive on a variety of questions, including on who would be able to trigger noncompliance procedures, what measures to apply in cases of noncompliance, and the decision-making process (Earth Negotiations Bulletin, 2007). In particular, several developing countries underscored that under the Convention in instances of noncompliance, developed countries should bear the responsibility for failing to provide adequate additional funding for developing countries' and countries with economies in transition's lack of implementation (Earth Negotiations Bulletin, 2009).

PAYING FOR POPS

Disagreement remains on the scale of funding warranted for developing countries and countries with economies in transition to be able to meet their obligations. A needs assessment had been commissioned for submission to COP4, which estimated the

* The other two chemicals-related Conventions are the 1989 Basel Convention (see Chapter 8 of this book) on the Control of Transboundary Movements of Hazardous Wastes and their Disposal and the 1998 Rotterdam Convention (see Chapter 14 of this book) on the Prior Informed Consent Procedure for Certain Hazardous Chemicals and Pesticides in International Trade.

funding needs of developing countries and countries with economies in transition to implement the provisions of the Convention from 2010 to 2014.* This needs assessment was carried out based on implementation plans submitted between June 2005 and December 2008. While as of December 2008, 137 Parties would be eligible for the Convention's financial mechanism, the needs assessment was completed based on information from only 68 parties. Further, in the needs assessment, the authors stress that the estimated demand of US$ 4.49 billion for 2010–2014 for these 68 parties is likely underestimated. Nevertheless, this total estimate was the focus of extensive discussions, with some parties questioning the methods used in calculating projected costs of activities and the uncertainties involved. Others underscored that this needs assessment was based on activities relating only to the original "dirty dozen" and not the expanded scope of 21 chemicals likely to apply at the close of COP4 (Earth Negotiations Bulletin, 2009).

Many held this estimate in stark contrast to the US$ 360 million contributed to POPS projects under GEF since 2001 (GEF, 2009), and concerns were raised about previously unmet needs likely to compound future needs. The scale of the cost burden of implementing the Convention remained as a backdrop as a contact group at COP4 discussed a range of financial issues, including a review of the financial mechanism (which parties agreed would be conducted at COP6) and elements of guidance to the GEF. In examining the GEF track record, disagreement arose on cofinancing requirements. China and other developing countries raised concerns that the cofinancing requirement was too high, flagging projects that were unable to move forward for failing to secure the necessary cofinancing. Several developed countries heralded cofinancing as a key means of meeting the needs for implementing the Convention, through the forced leveraging of additional funds, underscoring that cofinancing should be sourced from donors and development partners other than governments (Earth Negotiations Bulletin, 2009).

The fact that the Fifth Replenishment of the GEF was under negotiation as COP4 convened and also shaped discussions on guidance to the GEF. Disagreement arose over the message to convey as to the scale of funding warranted under the POPS window, several countries also raised concerns over the potential application of the resource allocation framework (RAF)† to the POPS focal area. A compromise decision on guidance to the financial mechanism was reached, calling on developed countries, to make all efforts to make adequate financial resources available (Earth Negotiations Bulletin, 2009).

As of March 2010, the negotiations for the Fifth GEF were ongoing, and have been completed in June 2010 (see Chapter 49 of this book). The revised programming document (GEF, 2009) acknowledged that the international chemicals agenda has expanded considerably in quantity and scope, requiring an enhanced response from the GEF. The document also acknowledges that the GEF's mandate as financial mechanism of the Stockholm Convention will require addressing the newly listed chemicals under the Convention. It also notes there are complex and challenging issues related to these chemicals throughout their life cycle and eligible countries

* (UNEP/POPS/COP/4/27).

† http://www.gefweb.org/operational_policies/Resource_Allocation_Framework.html

will require assistance to address these, and that this extends to environmentally sound disposal of POPS-containing waste.

Regarding the RAF, the revised programming document provides that, should the GEF Council decide to extend the resource allocation system to the POPS focal area, countries will be able to access the focal area set-aside funds (FAS) to implement enabling activities for an amount up to US$ 500,000 on an expedited basis, including for support to developing or updating NIPs and national reports. Should the resource allocation system not be extended to the POPS focal areas, enabling activities as well as regional and global projects will continue to be supported as in the past. The document envisages that under GEF-5 at least 50 countries will receive support for NIP updates.

Under the GEF-5 replenishment negotiation, three replenishment scenarios (total replenishments of US$ 4.5, 5.5, and 6.5 billions, respectively) are being considered (GEF, 2009). Under these scenarios the Chemicals focal area is allocated US$ 450, 550, and 650 millions, respectively. All of these scenarios represent a significant increase on the GEF-4 allocation of US$ 300 million to the POPS window.

Under the US$ 550 million scenario the additional resources available for POPS would also make it possible to start addressing the challenges posed by the "new" POPS recently added under the control of the Convention (GEF, 2009), with at least 10 countries implementing pilot "new" POPS reduction activities. Under the US$ 650 million scenario it is envisaged at least 12 countries would implement pilot "new" POPS reduction activities.

Under the replenishment scenarios, synergies between ozone-depleting substances (ODS) containing waste and the need to manage other hazardous wastes are also considered. Efforts to manage ODS wastes in an environmentally sound way can be supported, in parallel with managing wastes from other hazardous chemicals and efforts to mitigate climate change (GEF, 2009). Pilot destruction activities are planned for under the US$ 550 million and US$ 650 million scenarios.

CONCLUSION

Despite likely significant increases in GEF finance for Stockholm Convention implementation, available funds will be nowhere near the estimated demand of US$ 4.49 billion for 2010–2014 (for these 68 parties), a figure which is itself, likely to be significantly underestimated.

There is a clear need for innovative measures to respond to these needs that outweigh current financing resources. To be sustained in the long-term, the cost of managing chemicals requires internalization. The economic instruments to achieve this generally require institutionalization at the national level. In response to this need, UNEP Chemicals are developing a guidance document on economic instruments for chemicals and wastes management. The guidance will be tested in six national workshops in developing countries and countries with economies in transition in 2010, and pilot projects are planned thereafter (UNEP, 2010a). Capacity building activities funded under GEF will be vital in order to enable sufficient capacity in developing country governments to facilitate such internalization.

Additional efforts to increase resources for chemicals and waste management, that is, resources under the Stockholm, Rotterdam, and Basel Conventions, and the Strategic Approach to International Chemicals Management (SAICM), was initiated by the UNEP Executive Director amid growing concerns by parties at COP4, where developing countries and countries with economies in transition stressed the importance of adequate financial and technical assistance as essential requirements for the establishment of an effective compliance mechanism (UNEP, 2010b). The Consultative Process on Financing Options for Chemicals and Wastes was initiated to seek advice from governments and other stakeholders on how to respond to the growing recognition of the urgent need to secure adequate financial means and strengthened capacity building, including institutional strengthening, and technical assistance toward the implementation of the chemicals and wastes agenda, and the importance of linking obligations to financial and technical assistance (UNEP, 2010b). This informal process is occurring in parallel with the synergies process discussed above.

This informal consultative process has made several recommendations for the financing of chemicals and wastes, including the potential to leverage greater donor funding by "packaging" chemicals and wastes issues more attractively by linking the issue with human health, livelihoods, and poverty reduction. Nontraditional financing options were also considered, including the potential to: extend pilot programs on chemicals leasing, currently being trailed through National Cleaner Production Centres; developing incentives to encourage industry to reform and build on the Green Economy; and instituting economic instruments (UNEP, 2010b).

While at this stage none of the policy recommendations have been implemented, the Consultative Process is set to continue, and to report to the Third Session of the International Conference on Chemicals Management (ICCM3) in 2012. This process is likely to increase awareness and raise the profile of the financing needs of the chemicals and wastes conventions, as well as the potential to use innovative financial mechanisms to begin to bridge the mismatch between chemicals and waste management "needs" and available "resources."

While the above activities are promising, there is a need to mobilize additional financial resources for implementation expeditiously, and to prioritize the use of these resources with an eye toward the Stockholm Convention's implementation goals. Most parties have now completed their NIPs setting out how they intend to meet Convention obligations, yet some of these NIPs were completed as early as 2005. There is a risk that delayed finance will lead to the temptation to update NIPs, especially as these NIPs will have to be updated to include the newly listed POPS. Rather, a two-track process may be necessary to maximize available financial resources: one that can ensure revised planning (to address new POPS) while also prioritizing the implementation of already identified country plans.

REFERENCES

Ashton, M. 2009. Stockholm Convention moves from "Dirty Dozen" to the "Toxic 21." *Chemical Watch.* Available at: http://chemicalwatch.com

Carson, R. 1962. *Silent Spring*. Cambridge, MA: Riverside Press.

Downie, D. L. and T. Fenge. 2003. *Northern Lights against POPS: Combatting Toxic Threats in the Arctic.* Montreal, QC: McGill-Queen's University Press.

Earth Negotiations Bulletin. 2002. Summary of the *Sixth Session of the Intergovernmental Negotiating Committee for an International Legally Binding Instrument for Implementing International Action on Certain Persistent Organic Pollutants.* June 17–21. 15(69).

Earth Negotiations Bulletin. 2005. Summary of the *First Conference of the Parties to the Stockholm Convention.* May 2–6. 15(117).

Earth Negotiations Bulletin. 2007. Summary of the *Third Conference of the Parties to the Stockholm Convention.* April 30–May 4. 15(154).

Earth Negotiations Bulletin. 2009. Summary of the *Fourth Conference of the Parties to the Stockholm Convention.* May 4–8. 15(174).

Earth Negotiations Bulletin. 2010. Summary of the *Simultaneous Extraordinary COPs to the Basel, Rotterdam and Stockholm Conventions and the 11th Special Session of the UNEP Governing Council/Global Ministerial Environment Forum.* February 22–26. 16(84).

Fenge, T. 2003. POPS and Inuit: Influencing the global agenda. In *Northern Lights against POPS: Combatting Toxic Threats in the Arctic.* D. L. Downie and T. Fenge, Eds. Montreal, QC: McGill-Queen's University Press, pp. 192–213.

Global Environment Facility, 2009. Revised GEF-5 Programming Document. Available at: http://www.gefweb.org/uploadedfiles/R.5.22%20-%20Revised%20GEF-5%20 Programming%20Document.pdf

Kohler, P. M. 2006. Science, PIC and POPS: Negotiating the membership of chemical review committees under the Stockholm and Rotterdam Conventions. *Review of European Community and International Environmental Law (RECIEL),* 15(3): 293–303.

Reiersen, L.-O, S. Wilson, and V. Kimstach. 2003. Circumpolar perspectives on persistent organic pollutants: The Arctic monitoring and assessment programme. In *Northern Lights against POPS: Combatting Toxic Threats in the Arctic.* D. L. Downie and T. Fenge, Eds. Montreal, QC: McGill-Queen's University Press, pp. 60–86.

Selin, H. 2003. Regional POPS policy: The UNECE CLRTAP POPS protocol. In *Northern Lights against POPS: Combatting Toxic Threats in the Arctic.* D. L. Downie and T. Fenge, Eds. Montreal, QC: McGill-Queen's University Press, pp. 111–132.

Selin, H. 2010. *Global Governance of Hazardous Chemicals: Challenges of Multilevel Management.* Cambridge, MA: MIT Press.

Selin, H. and Selin, N. 2008. Indigenous peoples in International Environmental Cooperation: Arctic Management of Hazardous Substances. *Review of European Community and International Environmental Law,* 17(1): 72–83.

Stockholm Convention. 2009a. *Draft Report on the Second Review of the Financial Mechanism: UNEP/POPS/COP.4/INF/17* Available at: http://chm.pops.int/Portals/0/Repository/COP4/UNEP-POPS-COP.4-INF-17.English.PDF

Stockholm Convention. 2009b. *Listing of Sulfonic Acid, its Salts and Perfluorooctane Sulfonyl Fluoride: UNEP/POPS/COP.4/SC-4/17* Available at: http://chm.pops.int/Programmes/NewPOPS/DecisionsRecommendations/tabid/671/language/en-US/Default.aspx

UNEP, 2010a. *Development and Testing of Guidance on Economic Instruments.* Available at: http://www.chem.unep.ch/unepsaicm/mainstreaming/SMofChem_Economic Instruments_default.htm

UNEP, 2010b. *Policy Paper on Financing Chemicals and Wastes.* Available at: http://www.unep.org/environmentalgovernance/LinkClick.aspx?fileticket=BD1J4fyxkTk%3d&tabid=1635&language=en-US

Watt-Cloutier, S. 2003. The Inuit journey towards a POPS-free world. In *Northern Lights against POPS: Combatting Toxic Threats in the Arctic.* D. L. Downie and T. Fenge, Eds. Montreal, QC: McGill-Queen's University Press, pp. 256–267.

16 The Vienna Convention, Montreal Protocol, and Global Policy to Protect Stratospheric Ozone

David Downie

CONTENTS

The 1985 Vienna Convention for the Vienna Convention for the Protection of the Ozone Layer and its better known progeny, the 1987 Montreal Protocol on Substances that Deplete the Ozone Layer, stand at the center of international policy to protect stratospheric ozone, often called the ozone layer.* Analysts and policymakers cite the

* Ozone Secretariat (2009a), UNEP (2009a), and Ozone Secretariat (2009b) contain the current (as amended) and original text of these two treaties, as well as all amendments, adjustments, and other decisions agreed to through 2008, during all the Conference of Parties (COP) to the Vienna Convention and Meeting of the Parties (MOP) to the Montreal Protocol. You can also find these on the Ozone Secretariat's official Web site at http://www.unep.org/ozone

strong rules, innovative characteristics, global participation, and historic success of global ozone policy as breakthroughs in environmental cooperation and evidence that the international community possesses the capacity to address large, complex, and difficult global environment problems.

This chapter provides an overview of global policy to protect stratospheric ozone. The first section provides a short history. Formal policy began with the 1985 Vienna Convention, a framework treaty that mandated no action beyond scientific research, and then grew rapidly, first through the historic 1987 Montreal Protocol and then via a series of agreements that significantly expanded the Protocol. The second section outlines the current status of global ozone policy, including the control measures and institutions that exist today, and reasons that global ozone policy is widely considered a significant global success to date. The final section describes key implementation issues going forward.

A BRIEF HISTORY OF THE GLOBAL OZONE POLICY*

Ozone (O_3) is a molecule consisting of three oxygen atoms. Ninety percent of ozone exists in the upper atmosphere, or stratosphere, between 10 and 50 km (6–30 miles) above the earth. Although representing just three out of every 10 million molecules, stratospheric ozone, often called the ozone layer, protects life on earth by absorbing most of the harmful UV-B ultraviolet radiation from the sun and all of the lethal UV-C radiation.† Should the ozone shield become seriously depleted, increasing levels of UV-B radiation reaching the earth's surface would lead to more melanoma and nonmelanoma skin cancers, more eye cataracts, weakened immune systems, damage many kinds of plants, including reduced crop yields, damage ocean ecosystems and likely reduce fishing yields, negatively impact many animals, and even speed deterioration of plastics used outdoors (UNEP, 2006).

In 1974, scientists discovered that a certain set of important man-made chemicals, chlorofluorocarbons (CFCs), posed a serious threat to stratospheric ozone

* There are numerous detailed discussions of the creation and expansion of global ozone policy. I have written several short histories, so it is no surprise that this chapter draws very heavily on my most recent example in Chasek et al. (2010). For other histories and analysis that discuss different aspects of the issue's history, the chemicals that threaten the ozone layer, the scientific discoveries and, in particular, the development, content, expansion, and impact of the Vienna Convention and Montreal Protocol, see, among other sources: Downie (1993, 1995a,b, 1996, 1999); Benedick (1998); Litfin (1994); Anderson and Sarma (2002); Anderson et al. (2007). For an early specific discussion of the Vienna Convention, as a stand-alone framework convention, see Iwona Rummel-Bulska (1986). For the text of the ozone treaties, amendments, and adjustments, as well as official reports from each official meeting related to the Vienna Convention, Montreal Protocol, and other entities affiliated with the ozone regime, visit the Ozone Secretariat's Web site at www.unep.org/ozone. For daily and summary reports and analysis from many of the regime's most important meetings, see the Earth Negotiation Bulletin at http://www.iisd.ca/process/ozone_regime_intro.htm

† For detailed and authoritative scientific discussions by the Assessment Panels tasked by the country Parties to the Vienna Convention and Montreal Protocol of the ozone layer, how it operates, and how it is threatened by human activities, see WMO et al. (2007). For similar discussions of how depletion of the ozone layer would impact human health and the environment see UNEP (2006). These and other reports are available at the Web site of the Secretariat for the Vienna Convention and Montreal Protocol, at http://ozone.unep.org/

because they release chlorine atoms into the stratosphere that then act as catalysts in the destruction of ozone molecules (Molina and Rowland, 1974). CFCs at that time dominated the global markets for coolants in air-conditioning and refrigerating systems, propellants in aerosol sprays, and blowing agents for the manufacture of flexible and rigid foam, among many other profitable uses. Scientists later discovered other ozone-depleting substances (ODSs), including halons (a very effective fire suppressant), and methyl bromide (a very effective and toxic pesticide), some of which released bromine, an even more potent ozone destroyer, into the stratosphere.

The United States took aggressive action in the late 1970s, banning the use of CFCs in many uses, including aerosol spray cans—more than 40% of US CFC use at the time. Canada, Finland, Norway, Sweden, Switzerland, and a few other countries enacted similar steps. However, the countries that made up the European Community at that time refused to take meaningful Community-wide steps, expressing doubt concerning the scientific theory, noting the lack of observed ozone depletion in the atmosphere, and arguing that no substitutes existed or could easily be developed. Since the problem was clearly global in its cause and impact and because national regulation appeared stalled, advocates of protecting stratosphere ozone turned to international institutions, leading to a process that produced one of the first and most important global collaborations in managing chemical and environmental risks.

Global discussions essentially began when the Governing Council of the United Nations Environment Programme (UNEP) first took up the issue in 1976. An expert meeting then convened in 1977, after which UNEP and the World Meteorological Organization (WMO) set up the Coordinating Committee of the Ozone Layer (CCOL) to periodically assess ozone depletion. In 1983, not long after formal intergovernmental negotiations began on a possible Framework Convention, the United States, Canada, and the Nordic states formally proposed adding binding restrictions on CFC use. However, opposition by the European Community—which accounted for more than 40% of global CFC production at the time (exporting a third of that to developing countries) prevented the initial ozone agreement from including binding controls.

The 1985 Vienna Convention for the Protection of the Ozone Layer is a classic framework convention, that is, a multilateral agreement that established common principles but did not include binding commitments to specific actions. The Convention affirmed the importance of protecting the ozone layer and obligated Parties to take general measures to protect human health and the environment from human activities that might modify the ozone layer. It encouraged intergovernmental cooperation on research, systematic observation of the ozone layer, monitoring production of chemicals that might threaten it, and the exchange of information. The agreement also created the responsibility, expectation, and framework for beginning negotiations on a Protocol to control ODS production and use, should further risks be found.

Discovery of the "ozone hole" (Farman et al., 1985) provided such an impetus. Confirmation of significant and worsening ozone depletion above Antarctica galvanized proponents of CFC controls, who argued that the hole justified new negotiations on a control Protocol (despite the lack of firm evidence linking the hole to CFCs until 1989).* New negotiations began in 1986 and concluded with the landmark 1987 Montreal Protocol on Substances that Deplete the Ozone Layer, the centerpiece of global ozone policy (often referred to as the ozone regime).†

The Montreal Protocol

The Montreal Protocol established binding requirements that required industrialized countries to first freeze and then reduce their production and use of the five most widely used CFCs (CFC-11, -12, -113, -114, and -115) by 50% from 1986 levels by 2000, and to freeze halon production. The Protocol required developing countries to take the same actions but they were also given a 10-year grace period to aid their economic development, allowing them to increase their use of CFCs, which many had only recently gained access to, before taking on commitments. (The Protocol does not use the term "developing countries" but refers rather to Parties "categorized as operating under Article 5 paragraph 1 of the Montreal Protocol."‡) The Protocol also included scientific and technological assessment panels, reporting requirements, potential trade sanctions for countries that did not ratify the agreement, and a robust procedure for reviewing its effectiveness and strengthening its controls on the basis of periodic scientific and technological assessments. Importantly the Protocol could be strengthened not only by formal amendments, which was required to add chemicals for example, but also by the easier and speedier method of a Decision by an MOP, which could approve changes to the control schedules for any chemical already controlled under the Protocol.§ The Parties can also take Decisions to change other aspects of ozone policy related to the implementation of the Protocol, provided such decisions do not alter the text of the Protocol (which requires a formal amendment).

Over the years, Parties have used these mechanisms to strengthen the Protocol significantly. The process began with the 1990 London Amendment, which added controls on eight additional CFCs as well as carbon tetrachloride and methyl chloroform while a series of decisions adjusted the controls on the chemicals listed in the original protocol, mandating with complete phase-our (rather than 50% cut) by 2000.

---

* For discussion, see Downie (1996, pp. 168–181) and Litfin (1994, pp. 96–102, 125–141, and 158–188).

† Regime is the term that many international relations and some other scholars use to describe the entire set of international policies, laws, practices, and institutions that states create to manage action in a particular issue area of international relations. For a very early and influential discussion see Krasner (1983). For discussion and modification of the term's definition and how the concept can be specifically and usefully applied to global ozone policy, see Downie (1995a,b, 1996, 2005), Chasek et al. (2010, pp. 19–30).

‡ For the current official list of Parties categorized as operating under Article 5 paragraph 1 of the Montreal Protocol, see http://www.unep.ch/ozone/Ratification_status/list_of_article_5_parties.shtml

§ Although a historic agreement, the protocol did permit the continued production of ozone-depleting chemicals, neglected to specify that alternatives must not damage the ozone layer, and included no provisions for independent monitoring of production and consumption of ozone-destroying chemicals.

This represented the first binding global agreement to eliminate specific chemicals that harm the environment.

A second historic achievement in London was creation of the Multilateral Fund for the Implementation of the Montreal Protocol to assist developing countries implement the Protocol. The first major assistance fund established under a global environmental agreement (it predates and likely influenced the 1991 creation of the Global Environment Facility), the fund was necessary to attract participation by large developing countries, especially China and India but also Argentina, Indonesia, and others, which were on the verge of developing their large indigenous CFC industries and had so far refused to ratify the Montreal Protocol because without provisions for adequate financial and technical assistance giving them access to the replacement chemicals. The fund meets the incremental costs to developing countries of implementing the Protocol, in particular the extra expense of producing or using the alternatives to, for example, CFCs rather than the CFCs themselves. Thus the funds have been used to change a factory that produces or uses CFCs to one that used the replacement hydrofluorocarbons (HFCs) or and hydrochlorofluorocarbons (HCFCs).

The 1992 Copenhagen Amendment added controls for methyl bromide, Hydrobromofluorocarbons (HBFCs), and HCFCs. HCFCs are a set of transitional replacement chemicals for CFCs. They are still ozone depleting but far less damaging than CFCs. The meeting also adjusted controls on existing chemicals, accelerating the phase-out of CFCs and halons for developed countries. The 1997 Montreal Amendment and adjustment accelerated the phase-out of methyl bromide for industrialized countries to 2005, mandated that developing countries eliminate methyl bromide by 2015, earmarked specific Multilateral Fund resources for methyl bromide projects and created a new licensing system for CFC imports and exports of CFCs to help combat illegal trade. The Beijing Amendment in 1999 added bromochloromethane (BCM) to the Protocol, mandating its immediate phase-out, and also introduced production controls on HCFCs as well as controls on its trade with non-Parties. The meeting also adjusted the controls on HCFCs and increased reporting requirements on methyl bromide, in an attempt to reveal unauthorized use. In 2007, Parties significantly accelerated the controls on HCFCs, not only to protect the ozone layer more effectively but also to address climate change, as HCFCs are also strong greenhouse gases that would not exist without the Montreal Protocol's controls on CFCs.

Major Driving Forces for Ozone Policy

Four sets of large-scale causal factors shaped development of global ozone policy (Downie, 1996; Chasek et al., 2010).* First, rapidly advancing scientific knowledge in the late 1980s and early 1990s removed key arguments by control opponents and helped galvanize public opinion and policymaking. Discovery of the ozone hole

* Other discussions emphasize aspects of these factors, as well as other, but not contradictory impacts. See in particular, Litfin (1994), Downie (1995a,b, 1999), Benedick (1998), and Anderson and Sarma (2002).

helped the political conditions that led to the Montreal Protocol. Confirmation that chlorine atoms, released when CFCs break down high in the atmosphere, were the ultimate cause of the hole (although natural factors peculiar to Antarctica increase their impact) contributed to the strengthening of the ozone regime in London in 1990. Discovery of thinning ozone above the northern hemisphere, and continued worsening of the Antarctic holes, contributed to the agreements in Copenhagen in 1992 to expand and strengthen the Protocol even further. The 2007 report by the Intergovernmental Panel on Climate Change combined with evidence that a truly precautionary approach to protecting the ozone layer required further action helped create conditions for the important and surprising decision in Montreal in 2007 to accelerate the HCFC phase-out.

Second, changing patterns of economic interests impacted development of the ozone regime in several different ways. As is common in other areas, economic interests tied to CFCs and other chemicals sometimes prevented strong controls. Examples included the absence of regulations on emission in most of Europe prior to the Montreal Protocol, the exclusion of control measures in the Vienna Convention, the compromise Montreal Protocol 50% reduction target, and the very slow strengthening of the regime rules for methyl bromide and HCFCs. Yet, during several crucial periods, the development of effective substitutes, especially for CFCs, altered the economic interests of particular industries, major corporations or governments, lowering the costs associated with eliminating ODSs and allowing some actors to actually profit. This helped the regime strengthen rapidly in 1990 and 1992. The initiation of the Multilateral Fund changed the calculations of many developing countries, turning them from nonparticipants to supporters, because companies in their countries could receive support for using the newest chemical alternatives. The Fund similarly changed the economic interests and domestic political interests of some individual corporations based in developing countries as their transition to alternatives made them a supporter of stronger domestic action in their own countries as they did not want to get undercut by competitors using less expensive ozone-depleting chemicals.

Third, the need to include all countries with large levels of existing or potential ODS consumption forced policy to compromise at different stages that retarded regime strengthening. The reluctance of several key European countries to address the issue limited the cuts agreed to in the original Protocol, opposition by the United States slowed agreements to strengthen controls on methyl bromide, the need to ensure that China, India, and other large developing countries would participate in the ozone regime produced the historic and influential agreements on technical and financial assistance but increased the grace periods on some chemicals, and opposition by India and others prevented an agreement to add controls on HFCs in 2009 (see below).

Finally, the ongoing development of the regime affected, in most cases positively, subsequent opportunities to improve global management of ozone-depleting chemicals. The existence and framework of the Vienna Convention allowed governments to press forward far more quickly and effectively on a binding Protocol than if the Convention had not existed. Provisions of the Protocol allowed governments to strengthen it far more rapidly than would have been possible without such provisions.

In particular, the requirements to review the effectiveness of the control measures, the mandates for comprehensive scientific and technical assessments, and the ability for Parties to adjust existing controls through a decision by a MOP, bypassing the more time-consuming amendment procedure all created significant opportunities for the regime to strengthen quickly in response to new developments. As noted above, the creation and operation of the Multilateral Fund changed economic and political calculations within many developing countries.

CURRENT STATUS

The ozone regime enjoys very broad international support, effective implementation, and nearly universal participation. As of March 10, 2011, 196 countries were Parties to the 1985 Vienna Convention and 1987 Montreal Protocol, 195 to the 1990 London Amendment, 192 to the 1992 Copenhagen Amendment, 182 to the 1997 Montreal Amendment, and 166 to the 1990 Beijing Amendment.* As a result, the production and use of CFCs and several other ozone-depleting chemicals has declined dramatically and essentially eliminated in the United States and other industrialized countries.

Ninety-six chemicals are presently controlled by the amended Montreal Protocol, including multiple CFCs, HCFCs, and halons.† These include

- *CFCs* (chlorofluorocarbons): Discovered in the late 1920s, CFCs were revolutionary chemicals—inert, nonflammable, nontoxic, colorless, odorless, and wonderfully and profitably adaptable to a wide variety of uses, including as coolants in air-conditioning and refrigerating systems, propellants for aerosol sprays, solvents for cleaning of electronic components, and as blowing agents for the manufacture of flexible and rigid foam.
- *Halons*: Very effective fire suppressants.
- *HCFCs* (Hydrochlorofluorocarbons): The first major replacements for CFCs. While much less destructive than CFCs, HCFCs also contribute to ozone depletion and are greenhouse gases.
- *Methyl bromide* (CH_3Br): A very toxic fumigant used for high-value crops, pest control, and quarantine treatment of agricultural commodities awaiting export.
- *Carbon tetrachloride*: Widely used in fire extinguishers, as a precursor to refrigerants, and especially as a cleaning agent or solvent.
- *Methyl chloroform* (1,1,1-trichloroethane): Widely used for cleaning metal parts and circuit boards, as an aerosol propellant, and especially a solvent in the electronics industry and for paints, adhesives and other materials.

* For updates and details regarding ratification of these agreements, see details on the Ozone Secretariat's Web site at http://ozone.unep.org/Ratification_status/index.shtml or at the mirror Web site in Geneva: http://www.unep.ch/ozone/Ratification_status/index.shtml

† Information from this section is adapted from Ozone Secretariat (2008). I have used this information in Box 4.1 in Chapter 4 of Chasek et al. (2010, p. 171).

- *Hydrobromofluorocarbons* (HBFCs): Not widely used but added to the Protocol to prevent commercial introduction.
- *Bromochloromethane* (BCM): A new ODS that some companies sought to introduce into the market in 1998.

CONTROL MEASURES

Taking into account all the amendments and adjustments agreed to be Parties through 2010, industrialized country Parties to the Montreal Protocol, as amended and adjusted through the MOP in 2009, were or are currently required to*

- Phase out halons by 1994 (All phase-out schedules refer to both production and consumption. Most had interim cuts that were accelerated via adjustments. Here we list only the currently relevant controls)
- Phase out CFCs, carbon tetrachloride, methyl chloroform, and HBFCs by 1996
- Phase out methyl bromide by 2005, except for certain uses
- Reduce HCFCs by 35% by 2004, 75% by 2010, 90% by 2015, and phase out by 2020, with an allowance of 0.5% to be used only for servicing purposes during the period 2020–2030
- Phase out HBFCs by 1996
- Phase out BCM by 2002.

Exemptions exist for continuing production and consumption of small amounts of "essential uses" for some of these substances.† Such exemptions must be approved by the MOP each year. Larger exemptions exist for methyl bromide, especially for certain agricultural and quarantine and preshipment uses. All developed country Parties (with the exception of some Eastern European nations during their economic transition) have fulfilled their requirements in the past and are on pace to fulfill them in the future. As a result, implementation of the control measures in most developed countries is complete except for HCFCs and in some cases methyl bromide. Developed countries must of course continue to implement their other obligations, including those related to monitoring, reporting, and provision of technical and financial assistance to developing countries.

Developing countries have grace periods that allow them to begin and complete their phase-out schedules later than industrialized countries. The presence of different control schedules stems from recognition by the Parties that the industrialized countries had far larger ODSs emissions than developing countries when the Protocol

---

* Information from this section is adapted from Ozone Secretariat (2008) and confirmed using Ozone Secretariat (2009a, pp. 29–44), and the most recent MOP reports. I have used this information to create Box 4.1 on p. 171 and Table 4.1 on p. 177 in Chasek et al. (2010).

† For updated, official information on exemptions, see details on the Ozone Secretariat's Web site at http://ozone.unep.org/Exemption_Information/index.shtml or at the mirror Web site in Geneva: http://www.unep.ch/ozone/Exemption_Information/index.shtml

was negotiated and that developing countries needed access to some of the chemicals to aid their economic development.* Currently, developing countries are required to

- Freeze the production and consumption of CFCs, halons, and carbon tetrachloride at an average 1995–1997 levels by July 1, 1999, reduce by 50% by 2005, 85% by 2007, and phase out completely by 2010.
- Freeze methyl chloroform by 2003 at the baselines of the average 1998–2000 levels, reduce production and consumption by 30% from this baseline by 2005 and 70% by 2010, and phase out by 2015.
- Freeze methyl bromide by 2002 at average 1995–1998 levels, reduce by 20% by 2005, and phase out by 2015.
- Freeze HCFCs by 2013 at the baselines of their 2009–2010 average levels. Reduce HCFCs from this baseline by 10% by 2015, 35% by 2020, 67.5% by 2025, and phase out by 2030, allowing for an annual average of 2.5% for servicing purposes during the period 2030–2040.
- Phase out HBFCs by 1996 and phase out BCM by 2002.

Developing countries can utilize the same exemptions for continuing production and consumption of small amounts of "essential uses" for some of these substances. While implementation challenges and uncertainties remain (see discussion in the next section), most developing countries have implemented their requirements to date and are on schedule to meet future targets.

Institutions

In addition to the control measures, the Vienna Convention and Montreal Protocol have spawned a host of institutions which are critical to the operation and implementation of the global ozone policy. These include

- The COP for the Vienna Convention and the MOP for the Montreal Protocol are the decision-making authorities for their respective treaties. The Montreal MOP meets annually. The Vienna COP meets every 3 years, in coordination with that year's MOP. An Open-Ended Working Group (OEWG) holds discussions in preparation for the MOP, usually 6 months prior. All countries that ratify the treaties' can participate in the meetings with full decision-making privileges (to date, all decisions in both treaties have been reached by consensus, although the Protocol does allow supermajority voting in some situations). Nonparty governments as well

* This also reflects the principle of common but differentiated responsibilities, which many but not all governments and scholars believe has become a general principle of international environmental law. The principle states that all countries have a common responsibility to address global environmental issues but some countries have special responsibilities to act first or take special actions because of their large contribution to the problem and/or their access to greater financial and technological resources (for a representative discussion, see Chasek et al. (2010, pp. 326–329). The grace period and types of financial and technical assistance provided for developing countries in the ozone regime influenced later agreements and broader acceptance of this principle.

as representatives of international organizations, environmental nongovernmental organizations and industry groups can also attend as observers and participate in most discussions.

- The Implementation Committee under the Non-Compliance Procedure of the Montreal Protocol (Implementation Committee) addresses issues of noncompliance, including reviewing reports and other data received from Parties and holding discussions with individual Parties, and offers recommendations to the MOP. Although set up and operating in an entirely nonpunitive manner to date, the Protocol's noncompliance procedure is one of the most robust of any environmental regime.
- Three independent assessment panels—the Scientific, Environmental Effects, and Technology and Economic assessment panels—provide the Parties and the general public with periodic, comprehensive, and authoritative reviews of key issues, under instructions from the COP and MOP.
- The Ozone Secretariat provides day-to-day administration of the regime and supports the COP, MOP, OEWG, assessment panels, and Implementation Committee.
- The Multilateral Fund provides financial assistance to developing countries to aid their transition from using ozone-depleting chemicals—under rules established by the Protocol and decisions by the MOP. An "Executive Committee," composed of representatives from 14 governments—seven industrialized country donor Parties and seven developing country recipient Parties—acts as the Fund's decision-making body. The Multilateral Fund Secretariat performs provides day-to-day administration. The Fund is replenished every 3 years at levels negotiated by the MOP. The total budget for the 2009–2011 triennium is $490 million.* As of November 2010, the Fund has disbursed over $2.5 billion to support more than 6200 projects in 148 countries which in turn have phased out the consumption of more than 250,000 tons of ODSs, mostly CFCs. Designated staff within the World Bank, United Nations Development Programme, UNEP, and United Nations Industrial Development Organization act as "implementing agencies," working with host governments and managing and reporting on the implementation of the projects approved the Executive Committee. While serious debates have occurred within the MOP and Executive Committee over the years regarding specific aspects of the Fund priorities, policies, and operations, Parties consistently praise the impact of the Multilateral Fund.

* The fund has been replenished seven times: $240 million (1991–1993), $455 million (1994–1996), $466 million (1997–1999), $440 million (2000–2002), $474 million (2003–2005), $400.4 million (2006–2008), and $400 million (2009–2011). These budgets often include leftover funds from the previous triennium as well as interest accrued during the triennium. Figures in this paragraph are taken from the Fund's Web site at http://www.multilateralfund.org/ on April 24, 2011 and have also been contained in a variety of regime documents.

OZONE POLICY: A SUCCESS STORY

To date, implementation of the Vienna Convention and Montreal Protocol is widely considered a success. The treaties enjoy nearly universal ratification, possess a strong set of rules mandating the phase-out of nearly all ODSs, and have spawned an effective set of constituent institutions. The regime is often cited during negotiations on other issues as providing potential lessons in how to design, cooperate on, or implement global environmental policy.*

The most important evidence of their success is that ODS production and consumption have declined significantly. For example, in 1986 the global production of CFCs was about 1.1 million tons in 1986. In 2008 it was less than 3000 tons.† The Ozone Secretariat calculates that without the Protocol the global CFC production would have reached about 3 million tons in 2010 and 8 million tons in 2060, resulting in a 50% depletion of the ozone layer by 2035 and perhaps 130 million more cases of cataracts, 19 million additional cases of nonmelanoma skin cancer, and 1.5 million more cases of melanoma cancer (Ozone Secretariat, 2008).

Global halon production has also dropped significantly, from approximately 190,000 tons in 1986 to about 1000 tons in 2008. Methyl chloroform production dropped from more than 70,000 tons in 1989 to less than 500 tons in 2008. HCFC production has peaked. Developed country production is down nearly two-thirds from its peak, while developing country production started a downward trend in 2008. As a result of these efforts, the overall concentration of ODSs in the atmosphere has decreased significantly. The Scientific Assessment Panel has concluded that if all countries continue to implement the control regulations on schedule, then ozone depletion has likely peaked and most of the ozone layer will return to pre-1980 levels near the middle of this century and Antarctic ozone should do so in 2060–2075 (WMO et al., 2007; Ozone Secretariat, 2008).

ONGOING IMPLEMENTATION ISSUES AND FUTURE CHALLENGES

The scientific conclusion predicting a fully recovered ozone layer assumes full compliance with the Protocol's final phase-out schedules by all major Parties. While implementation in most industrialized countries is nearly complete and most trends in developing countries are positive, some challenges remain. These challenges are

* This conclusion based on personal observations during dozens of global environmental negotiations and related meetings focused on climate change, hazardous waste, trade in hazardous chemicals, persistent organic pollutants, and anthropogenic mercury emissions.

† The production and consumption figures in this section come from the official reports from Parties as tallied and aggregated by the Ozone Secretariat. These can be accessed using a database called "Data Access Centre," available at http://ozone.unep.org/Data_Reporting/Data_Access/regarding. Please note that under the rules of the protocol, "Calculated levels of Production" means the amount of controlled substances produced, minus the amount destroyed by technologies to be approved by the Parties and minus the amount entirely used as feedstock in the manufacture of other chemicals (paragraph 5 of Article 1). Thus, it represents the net increase in the amount of a controlled chemical. For methyl bromide, this does not include the amounts used by the Party for the exempted quarantine and preshipment applications (paragraph 6 of Article 2H).

well understood and regularly addressed in the MOP, Executive Committee, publications by the Secretariat, regional meetings, and other forums.* Nevertheless, it is important to outline these challenges here as some could be relevant to analogous situations in global policy on other issues,

Completing the CFC Phase-Out in Developing Countries: Of Both Legal and Black Market CFCs

Developing countries are close to eliminating production of new CFCs but the final steps face some implementation challenges. Virgin CFCs made in India and a few other countries are still cheaper than recycled CFCs or alternative chemicals. This creates incentives to keep producing CFCs legally, by applying for essential use exemptions, or illegally, by hiding production from national authorities or not reporting it to the Secretariat. Success will require a multifaceted effort to locate production while eliminating market incentives for legal and illegal production. Illegal ODSs production and trade has also been a concern for many years and while there have been a number of significant successes, including drastically reducing illegal imports to the United States and illegal exports from Russia and China, continued effort is needed to eliminate all sources, especially some in India (UNEP, 2005; IANS, 2008). One promising new implementation initiative includes cooperation with other chemical treaties. Twenty-four countries, UNEP, and the Secretariats for the Basel, Rotterdam, Stockholm, and Vienna Conventions now cooperate in the Multilateral Environmental Agreements Regional Enforcement Network (MEA-REN) to address illegal trade in ODSs, harmful chemicals, and hazardous wastes via a diverse set of initiatives including training, workshops, information exchange, and policy development.†

Essential Use Exemptions for CFCs in Metered-Dose Inhalers

Global use of CFCs in metered-dose inhalers (MDIs) is down from peak levels. However, Argentina, Bangladesh, China, Egypt, India, Iran, Pakistan, Russia, Syria, and the United States all received exemptions to continue using some pharmaceutical grade CFCs in certain MDIs in 2010. Most of these countries have signaled their intention to apply for such exemptions for 2011. Bangladesh, China, India, and Pakistan have indicated that their requests could continue for some time and even increase in size because as their economies improve, more people that need MDIs will be able to afford them—provided they have access to the proven but less expensive CFC-based MDIs. They also emphasize the difficulties in reducing the use of CFCs in MDIs as well as the broad public-health benefits of ensuring the

* These challenges have been outlined in several places in a fashion similar to how they are presented here. In particular, I draw on Ozone Secretariat (2008), Chasek et al. (2010), the official reports from the last several MOPs (each of which addressed these issues in some fashion), my personal observations of the discussions during the 2009 MOP as well as discussions during those meetings with national delegates, and the "FAQ" sections of both UNEP (2006) and WMO et al. (2007).

† For information on the MEA-REN, see the official Web site at http://www.mea-ren.org/

viability of providing low-cost, easily available options for patients and doctors (e.g., UNEP, 2009).

Some of the 2000 tons of MDI-related CFC essential use exemptions for 2010 can be met by using existing stockpiles. However, such stockpiles are declining. Thus, continuing CFC-MDIs will likely require continued CFC production. The problem of course is that when a facility remains open to produce an ODS for a certain exempted use, it might be difficult to ensure that the facility limits its output to the approved levels. Thus, truly eliminating CFC production and use, including black-market production, requires closing all production facilities as soon as possible and doing this will require a global transition away from CFC-based inhalers, something that remains problematic, particularly in certain developing countries.

HCFCs

The revised HCFCs phase-out schedules represent a challenge for countries that built their post-CFC infrastructure on HCFCs. Thus, it is possible that as the final deadlines near, some countries, particular large developing countries with major HCFC production facilities, might decide that although they have implemented significant reductions, complete elimination is not economically justified, at least for a while. In addition, timely efforts to combat climate change require that Parties replace HCFCs with energy-efficient and climate friendly chemicals, products, and processes, creating another challenge some might choose to post-pone.

ODS Banks

Every kilogram of ODS produced since the 1920s but not yet vented to the atmosphere lies trapped somewhere waiting to escape. Many millions of tons of CFCs remain in old or discarded refrigerators, air-conditioners, insulating foam, and dozens of other products and wastes, collectively known as "ODS Banks." A hundred thousand tons of ODSs enter the waste stream every year. Eventually, all the ODSs contained in these banks that we do not capture and recycle or destroy will reach the atmosphere. Parties recognize that ODS banks could delay the timely recovery of the ozone layer, and represent a source of potent greenhouse gases, and that addressing them can be a cost-effective step in pursuit of both climate and ozone protection. At the same time, many developing countries lack the necessary equipment and financial resources to destroy ODS banks in an environmentally sound manner.* This is a particular problem for many African and Caribbean countries that in the past received significant shipments of sometimes unwanted and obsolete consumer products. Thus addressing ODS banks will require difficult choices regarding resource allocation. Failing to act could lead to unnecessary emissions while focusing too many resources on ODS banks might preclude effective action in other important areas.

* For a recent discussion during the 2009 MOP on the issue of ODS Banks, see paragraphs 18–34 in UNEP (2009).

Methyl Bromide

Global methyl bromide production and use has declined significantly and production is officially phased out in developed countries. At the same time, methyl bromide is still produced and used in the United States and several other industrialized countries via the exemptions granted for certain agricultural uses as well as the blanket exemption for using methyl bromide in quarantine and preshipment applications. These exemptions represent loopholes in the control measures that retard efforts to eliminate methyl bromide completely. Most Parties argue that effective, economically viable, and environmentally friendly alternatives for all uses of methyl bromide exist and have been proven via implementation programs within their own countries as well as pilot projects undertaken in several developing countries by FAO or with the support of the Multilateral Fund. However, some maintain that such alternatives are not yet economically viable or sufficiently effective for all uses in all countries. Thus one remaining challenge is to eliminate or limit methyl bromide use to a very small amount of truly essential uses and then to trap most of all of the methyl bromide before it is emitted to the atmosphere.

Halons

Most halons have been eliminated but several essential uses continue, including in military, airplane, space, and nuclear applications. These rely largely on stockpiles to meet demand. Alternatives should be found to avoid pressure to resume large-scale production that could end up fueling less essential halon use.

Funding

"Maintaining stable and appropriate funding to assist developing countries in implementing the ozone regime will remain central to its effectiveness, its sense of constructive cooperation, and its positive impact on global environmental politics in general," (Chasek et al., 2008, p. 176). Even as implementation continues to advance, funding will be requested to help phase out the final exempted uses of CFCs and methyl bromide, to eliminate illegal production and trade, to expand the environmentally sound destruction of ODS banks, and to transition away from HCFCs. Moreover, controversial policy decisions remain with regard to what types of destruction and HCFC projects deserve funding and at what levels. While the funding levels required for these efforts might be relatively small compared to the billions promised for climate change, it is possible that economic difficulties, political differences, or honest mistakes could prevent its provision or effective application. If this occurs, the economic and political incentives that support the complete implementation of all aspects of global ozone policy could disappear.

OUTLOOK

Global ozone policy has been a significant and influential success in global environmental protection and risk management. The 25-year history of the Vienna

Convention, Montreal Protocol, and their progeny proves that global chemical control policies can work. ODS production is drastically lower on a global scale and nearly eliminated in most industrialized countries. The ozone layer is recovering and should fully recover in the next 50 years.

Some implementation challenges remain, of course, especially in large developing countries. However, the overall trends are positive and several factors argue for a positive outcome long-term, including the proven positive impacts of continued financial and technical assistance; the increasing availability of cost-effective substitutes; rapid declines in ODS production and consumption in developed countries; and changing global market conditions relevant to products and processes that use ODS versus those that do not. Thus, while it might not be possible to eliminate every kilogram of ODS, it is likely that during the next 25 years, the global community will complete the job of safeguarding human health and the environment from chemicals that threaten the ozone layer.

REFERENCES

Anderson, S. and K. Madhavea Sarma. 2002. *Protecting the Ozone Layer: The United Nations History*. London: Earthscan.

Anderson, S., K. Madhavea Sarma, and K. Taddonio 2007. *Technology Transfer for the Ozone Layer: Lessons for Climate Change*. London: Earthscan.

Benedick, R. 1998. *Ozone Diplomacy*, 2nd Edition. Cambridge, MA: Harvard University Press.

Chasek, P., D. Downie, and J. Welsh Brown. 2010. *Global Environmental Politics*, 5th Edition. Boulder, CO: Westview Press.

Downie, D. 1993. Comparative public policy of ozone layer protection. *Political Science* (NZ), 45(2): 186–197.

Downie, D. 1995a. Road map or false trail: Evaluating the precedence of the ozone regime as model and strategy for global climate change. *International Environmental Affairs*, 7(4): 321–345.

Downie, D. 1995b. UNEP and the Montreal protocol: New roles for International Organizations in regime creation and change. In *International Organizations and Environmental Policy*. R. V. Bartlett, P. A. Kurian, and M. Malik, Eds. Westport, CT: Greenwood Press.

Downie, D. 1996. Understanding International environmental regimes: The origin, creation and expansion of the ozone regime. PhD dissertation, University of North Carolina, Chapel Hill, 1996.

Downie, D. 1999. The power to destroy: Understanding stratospheric ozone politics as a common pool resource problem. In *Anarchy and the Environment: The International Relations of Common Pool Resources*. J. S. Barkin and G. Shambaugh, Eds. Albany, NY: State University of New York Press.

Downie, D. 2005. Global environmental policy: Governance through regimes. In *The Global Environment: Institutions, Law & Policy*. 2nd Edition, R. Axelrod, D. Downie, and N. Vig, Eds. Washington: CQ Press, pp. 64–82.

Farman, J. et al. 1985. Large losses of total ozone in Antarctica reveal seasonal ClO_x/NO_x interaction. *Nature*, 315: 207–210.

IANS (Indo-Asian News Service) 2008. India largest source of smuggled CFCs. *India Today*, April 27, 2008. Available at http://indiatoday.intoday.in/site/Story/7525/Master+stroke:+Sofia+Ashraf.html

Krasner, S. 1983. *International Regimes.* Ithaca, NY: Cornell University Press.

Litfin, K. 1994. *Ozone Discourses.* New York, NY: Columbia University Press.

Molina, M. and F. Sherwood Rowland. 1974. Stratospheric sink for chlorofluoromethanes: Chlorine atomic catalyzed destruction of ozone. *Nature* 249: 810–812.

Ozone Secretariat. 2008. Basic facts and data on the science and politics of ozone protection. UNEP, Press Release, September 18, 2008. Available at: http://ozone.unep.org/Events/ozone_day_2008/press_backgrounder.pdf

Ozone Secretariat. 2009a. *Handbook for the Montreal Protocol on Substances that Deplete the Ozone Layer*, 8th Edition. Nairobi: United Nations Environment Program. Available at: http://www.unep.ch/ozone/Publications/MP_Handbook/MP-Handbook-2009.pdf

Ozone Secretariat. 2009b. *Handbook for the Vienna Convention for the Protection of the Ozone Layer*, 8th Edition. Nairobi: United Nations Environment Program. Available at: http://www.unep.ch/ozone/Publications/VC_Handbook/VC-Handbook-2009.pdf

Rummel-Bulska, I. 1986. The protection of the ozone layer under the Global Framework Convention. In *Transboundary Air Pollution*. C. Flinterman, B. Kwiatkowska, and J. G. Lammers, Eds. Dordrecht, Netherlands: Martinus Nijhoff, pp. 281–296.

UNEP. 2005. Report of workshop of experts from Parties to the Montreal Protocol to develop specific areas and a conceptual framework of cooperation to address illegal trade in ozone-depleting substances. Montreal, April 3, 2005. UNEP Document UNEP/OzL.Pro/Workshop/ of April 15, 2005.

UNEP. 2006. Environmental effects of ozone depletion and its interactions with climate change: 2006 Assessment. Nairobi: United Nations Environment Program.

UNEP. 2009. Report of the twenty-first meeting of the parties to the Montreal protocol on substances that deplete the ozone layer. UNEP Document UNEP/OzL.Pro.21/8 of November 21, 2009. Available at http://ozone.unep.org/Meeting_Documents/mop/index.shtml

WMO et al. 2007. *Scientific Assessment of Ozone Depletion: 2006.* World Meteorological Organization Global Ozone Research and Monitoring Project—Report No. 50. Geneva: World Meteorological Organization, 2007.

Section IV

SAICM

17 Strategic Approach to International Chemicals Management
*Development and Opportunities**

Hamoudi Shubber†

CONTENTS

* An earlier version of this chapter has been included in the following online publication: Anne Petitpierre-Sauvain, editor. 2010. The Basel, Rotterdam, and Stockholm Conventions on Chemicals and Wastes—Regulation, Sound Management, and Governance. *EcoLomic Policy and Law* Special Edition 2008–2010, pp. 233. http://www.ecolomics-international.org/headg_ecolomic_policy_and_law.htm

† The opinions expressed in this chapter reflect only the views of the author and not of any organization.

INTRODUCTION

The Strategic Approach to International Chemicals Management (SAICM)* is a global policy framework which supports the achievement of the goal agreed in 2002 at the World Summit on Sustainable Development of ensuring that, by 2020, chemicals are produced and used in ways that minimize significant adverse impacts on the environment and human health. SAICM was adopted in February 2006 in Dubai by the International Conference on Chemicals Management (ICCM) at its first session and comprises the Dubai Declaration on International Chemicals Management, the Overarching Policy Strategy (OPS), and the Global Plan of Action (GPA).

This chapter aims to provide a perspective on the emergence of chemicals as an international concern, the development of SAICM, its features and the opportunities, and challenges that lay ahead of it. From the early stages of environmental protection and awareness to the sessions of the ICCM, chemicals management has gradually been recognized as an issue of sustainable development requiring global action. The development of SAICM was undertaken with the full involvement of all relevant sectors and stakeholders.

While the adoption of SAICM was a positive step forward, its implementation will be the indicator for measuring success against the goal set by the World Summit on Sustainable Development.

ORIGINS OF SAICM

Emergence of Chemicals Management as a Global Issue

The consumption of chemicals by all industries and our society's reliance on chemicals for virtually all manufacturing processes make chemicals production one of the major and most globalized sectors of the world economy.

Acknowledgment of the essential economic role of chemicals and their contribution to improved living standards needs to be balanced with recognition of potential harmful effects and costs. These include the chemical industry's heavy use of water and energy and the possible adverse impacts of chemicals on the environment and human health.

In June 1972, the United Nations Conference on the Human Environment held in Stockholm, Sweden (see Chapter 2 of this book) marked a turning point in the development of international environmental politics. The Conference recommended *Governments and relevant intergovernmental organizations* "to strengthen and coordinate international programs for integrated pest control and reduction of the harmful effects of agro-chemicals."† The Conference led to the creation by the United Nations General Assembly of the United Nations Environment Programme (UNEP).‡

---

* http://www.saicm.org

† Recommendation for action at the international level number 21, Chapter X: Planning and Management of Human Settlements for Environmental Quality. Available at: http://www.unep.org/Documents.multilingual/Default.asp?DocumentID=97&ArticleID=1506&l=en

‡ General Assembly resolution 2997 (XXVII), December 15, 1972, http://daccessdds.un.org/doc/RESOLUTION/GEN/NR0/270/27/IMG/NR027027.pdf

In 1983, the United Nations General Assembly established the World Commission on Environment and Development, known by the name of its Chair, Dr. Gro Harlem Brundtland, to address growing concern "about the accelerating deterioration of the human environment and natural resources and the consequences of that deterioration for economic and social development."* The report of the Commission, published in 1987 and entitled *Our Common Future* (Bruntland, 1987), was an important milestone in bringing environmental protection and sustainable development on the international political agenda. The report made numerous references to chemicals and the need for their sound management, pointing out the contribution to the improvement of living standards, as well as their risks. Sections of the report point to the possible hazardous effects of excessive use of agrochemicals, pesticides, and pest control chemicals, of the risks caused by hazardous wastes, aerosols, and refrigerating chemicals (Bruntland, 1987, Chapter 7).† The document called for the use of alternatives to chemicals, as well as the strengthening of legislation, policy, and research capacity for advancing nonchemical and less-chemical strategies (Bruntland, 1987, Chapter 5).‡

The 1992 Earth Summit

The United Nations Conference on Environment and Development, held in Rio de Janeiro, Brazil, in June 1992 was also a significant event in the creation of international environment and development frameworks and conventions. The Conference adopted *Agenda 21* and its Chapters 18 on Waste and 19 on Chemicals (see Chapter 3 of this book).

Chapter 19 of *Agenda 21* highlighted six program areas as well as relevant objectives, activities and means of implementation. The program areas identified were

1. Expanding and accelerating international assessment of chemical risks
2. Harmonization of classification and labeling of chemicals
3. Information exchange on toxic chemicals and chemical risks
4. Establishment of risk reduction programs
5. Strengthening of national capabilities and capacities for management of chemicals
6. Prevention of illegal international traffic in toxic and dangerous products

Agenda 21 stressed the need for increased coordination both within and outside the United Nations system. Intergovernmental organizations involved in chemicals safety§ established in 1995 the Inter-Organization Programme for the Sound

* General Assembly resolution A/RES/38/161, December 19, 1983. Available at: http://www.un.org/documents/ga/res/38/a38r161.htm

† *Our Common Future*, Chapter 7: Energy: Choices for Environment and Development.

‡ *Our Common Future*, Chapter 5: Food Security: Sustaining the Potential.

§ The seven participating organizations of the IOMC are: the Food and Agriculture Organization of the United Nations (FAO), the International Labour Organization (ILO), the Organisation for Economic Co-operation and Development (OECD), UNEP, the United Nations Industrial Development Organization (UNIDO), the United Nations Institute for Training and Research (UNITAR), and the World Health Organization (WHO). In addition, the United Nations Development Programme (UNDP) and the World Bank participate in the IOMC as observers.

Management of Chemicals (IOMC) with the aim of strengthening cooperation and increase coordination in the field of chemical safety among the different organizations. An Inter-Organization Coordinating Committee (IOCC) composed of representatives of the Participating Organizations coordinates relevant activities. Planning, programming, implementation, and monitoring of activities undertaken jointly or individually by the Participating Organizations are carried out by IOCC (see Chapter 23 of this book). This ensures full consultation among all those involved, with the aim to ensure effective implementation without duplication.*

The Intergovernmental Forum on Chemical Safety

Chapter 19 of *Agenda 21* called upon the governing bodies of WHO, ILO, and UNEP to convene a global forum to promote chemical safety. The organizations convened the International Conference on Chemical Safety (ICCS), which was held in Stockholm, 1994.† The Conference established the Intergovernmental Forum on Chemical Safety (IFCS) (see Chapter 21 of this book).‡

The Conference was considered to be the first session of the Forum. A key feature of the IFCS was to allow and encourage multisectoral and multistakeholder participation in an international policy process addressing chemical safety. It provided the first international open and inclusive forum concerning issues of common interest and also new and emerging issues in this area. In October 2000, the Forum met in Salvador da Bahia, Brazil, and adopted the Bahia Declaration on Chemical Safety. The Declaration reaffirmed IFCS's "commitment to Agenda 21 and recognized the importance of the provision of technical and financial assistance and technology transfer to developing countries and countries with economies in transition to accomplish Forum priorities beyond 2000."§

At this stage, voices were raised to give the comprehensive work for chemical safety an established space within the UN system. This was needed in order to give these issues more political weight and to allow the full involvement of all UN member states in a coordinated way. It was also needed in order to encourage countries to take on chemicals management in a preventive and holistic manner.

Following this, at the 1999 session, the Governing Council of UNEP adopted decision 21/7, which

> Requested the Executive Director, in consultation with Governments, the Inter-Organization Programme for the Sound Management of Chemicals, the Intergovernmental Forum on Chemical Safety and other relevant organizations and stakeholders, to examine the need for a strategic approach to international chemicals management

* http://www.who.int/iomc/en/

† The report of the Conference (document IPCS/ICCS/94.8) is available at: http://www.who.int/ifcs/documents/forums/forum1/en/FI-report_en.pdf

‡ IFCS: *Brief History & Overview*, December 2005. Available at: http://www.who.int/ifcs/documents/ifcs_overview_dec05.doc

§ The Bahia Declaration is available at: http://www.who.int/ifcs/documents/forums/forum3/en/Bahia.pdf

> and to prepare a report on this subject for detailed consideration at the seventh special session of the Governing Council/Global Ministerial Environment Forum in 2002.*

In preparation for discussion in the Governing Council and its Global Ministerial Environment Forum, UNEP used a questionnaire to solicit the views of Governments, members of the IOMC, IFCS, nongovernmental organizations, industry and environmental groups, and other stakeholders.† The Executive Director reported that

> The great majority of respondents concurred that a strategic approach was warranted, albeit with varying conceptions as to what such an approach might entail. Environmentally sound management of chemicals was seen as integral to sustainable development objectives as it is a global issue requiring a comprehensive response. A strategic approach was viewed as a means of advancing the chemical safety agenda and building on progress to date. It was envisaged that such an approach would lend greater coherence to efforts at the global, regional and national levels. One of the strongest themes to emerge was the perception that more coordinated and effective delivery of capacity-building is essential if policies and programmes relating to international chemicals management are to bear fruit. A firm belief was also expressed that any new strategic approach should not compete with or duplicate existing work, such as the valuable priority-setting exercise undertaken by IFCS and reflected in the Bahia Declaration and the Priorities for Action. Significant attention was devoted to institutional and legal coordination, issues that are under active consideration by the Global Ministerial Environmental Forum under the heading of "governance" and that will be addressed at the same February meeting as this report. Other prominent themes included the improvement of access to information on hazardous chemicals, the mobilization of greater resources to support chemicals management, and the encouragement of industry to accept increased responsibility for and play a more active role in the promotion of chemical safety.‡

In 2002, the Governing Council in its resolution SSVII/3 decided that there was a need to further develop a strategic approach and endorsed the IFCS Bahia Declaration and Priorities for Action beyond 2000 as the foundation of this approach. The Governing Council requested the Executive Director of UNEP to identify concrete projects and priorities in the context of a strategic approach, working with key partners and, together with the IFCS and the IOMC, to convene an open-ended consultative meeting involving representatives of all stakeholder groups to contribute to the further development of an SAICM.§

* The decision is reproduced in the report of the 21st session of the Governing Council http://www.unep.org/gc/gc21/Documents/K0100275-E-GC21.doc

† Views expressed are summarized in documents UNEP/GCSS.VII/INF/1, UNEP/GCSS.VII/INF/1/Add.1 and UNEP/GCSS.VII/INF/1/Add.2. Available at: http://www.unep.org/gc/GCSS-VII/

‡ Report on the implementation of the decisions adopted at the 21st session of the Governing Council/Global Ministerial Environmental Forum, report of the Executive Director (UNEP/GCSS.VII/4), presented at the seventh session of the Global Ministerial Environmental Forum, Cartagena, Colombia, February 13–15, 2002. Available at: http://www.unep.org/gc/GCSS-VII/

§ Resolution SSVII/3, SAICM can be found in the report of the seventh session of the Global Ministerial Environment Forum: http://www.unep.org/gc/GCSS-VII/Reports.htm

The 2002 World Summit on Sustainable Development

Ten years after the 1992 Earth Summit in Rio, Heads of State and Government met during the World Summit on Sustainable Development in Johannesburg to reaffirm their commitment to sustainable development, the Rio Principles and the full implementation of *Agenda 21.* Delegates adopted the Johannesburg Declaration on Sustainable Development and the Johannesburg Plan of Implementation. The Johannesburg Declaration* outlines the path taken from the 1992 Rio Earth Summit, and the Johannesburg Plan of Implementation† sets out a framework for action to implement the commitments originally agreed at Rio.

The Summit set the aim "to achieve, by 2020, the use and production of chemicals in ways that lead to the minimization of significant adverse effects on human health and the environment. [. . .]"‡ Furthermore, the WSSD endorsed the development of "a strategic approach to international chemicals management based on the Bahia Declaration and Priorities for Action beyond 2000 of the IFCS by 2005, and urge that UNEP, IFCS, other international organizations dealing with chemical management and other relevant international organizations and actors closely cooperate in this regard, as appropriate."§

Following the work of the IOMC and IFCS and the mandate of the UNEP Governing Council, the WSSD provided the objective and endorsement and time frame required for the development of an SAICM. The Johannesburg Plan of Implementation also set an ambitious and broad goal, linking the sound management of chemicals with sustainable development and acknowledging its multisectoral scope.

DEVELOPMENT OF SAICM

Sessions of the SAICM Preparatory Committee and the ICCM

In February 2003, the UNEP Governing Council agreed at its 22nd session, in decision 22/4 IV,¶ to the concept of an open-ended consultative process involving representatives of all stakeholder groups, taking the form of preparatory meetings followed by an international conference. It also proposed that the international conference be held in conjunction with the ninth special session of the Governing Council and Global Ministerial Environment Forum in early 2006 and called upon the Executive Director to strive to ensure that the process of further developing the strategic approach remained open, transparent, and inclusive, providing all stakeholders with opportunities to participate in the substantive work.

---

* http://www.un.org/esa/sustdev/documents/WSSD_POI_PD/English/POI_PD.htm

† The Johannesburg Plan of Implementation is available at: http://www.un.org/esa/sustdev/documents/WSSD_POI_PD/English/POIToc.htm

‡ See paragraph 23 of Chapter 3 of the Plan of Implementation: http://www.un.org/esa/sustdev/documents/WSSD_POI_PD/English/POIChapter3.htm

§ Ibid.

¶ The report of the meeting and decision 22/4 can be found at: http://www.unep.org/gc/gc22/REPORTS.asp

The first session of the Preparatory Committee for the Development of an SAICM (SAICM PrepCom1) was held in Bangkok, Thailand, November 2003.* The session was attended by 428 participants from 127 Governments, 19 intergovernmental organizations, and approximately 50 nongovernmental organizations drawn from a wide range of sectors including agriculture, environment, foreign affairs, health, industry, labor, and science. The Preparatory Committee considered and further developed draft SAICM elements proposed by stakeholders and compiled by the secretariat. It adopted as the overall goal of SAICM the target set down in the Plan of Implementation of the World Summit on Sustainable Development that, by 2020, chemicals be used and produced in ways that lead to the minimization of significant adverse effects on human health and the environment. Also developed at the first session were rules of procedure designed to maximize participation in the development of SAICM by all stakeholders.†

The second session of the Preparatory Committee (SAICM PrepCom2), held in Nairobi, October 2004, was again attended by approximately 400 participants, including representatives of 115 Governments, from a broad range of sectors.‡ The Committee agreed upon a tripartite structure for the SAICM documents comprising a high level declaration, an overarching policy strategy (OPS), and a Global Plan of Action (GPA). The President was mandated to prepare a draft of the declaration based on an outline agreed by the Committee and also to work with the secretariat to revise drafts of the OPS and GPA that had been developed during the session.

The third session of the Preparatory Committee (SAICM PrepCom3) was held in Vienna, September 2005. The meeting was attended by over 600 participants from 145 Governments and numerous intergovernmental and nongovernmental organizations.§ The Committee considered a draft of the high-level declaration and reached provisional agreement on most sections of the OPS and the detailed GPA, subject to final consideration by the ICCM. It was agreed that given the guidance status of the GPA, it need not be fully negotiated and would be subject to ongoing refinement in the future. The Committee provisionally agreed that the Executive Director of UNEP should be requested to perform secretariat functions to support the implementation of SAICM and that the ICCM, which was expected to adopt SAICM at its first session in February 2006, should be reconvened to undertake periodic reviews of progress in the implementation of SAICM. It also agreed provisionally on the functions of both the future SAICM secretariat and the ICCM when reconvened to exercise its proposed review role. While it was provisionally agreed that the Executive Director of UNEP should be requested to establish and assume overall responsibility for the

* PrepCom1 information and meeting documents can be found at: http://www.saicm.org/documents/prepcom1/default.htm

† The report of PrepCom1 can be found at: http://www.saicm.org/documents/meeting/prepcom1/report/en/1_7report.doc

‡ PrepCom2 information and meeting documents can be found at: http://www.saicm.org/documents/prepcom2/default.htm

§ PrepCom3 information and meeting documents can be found at: http://www.saicm.org/documents/prepcom3/default.htm

secretariat, both UNEP and WHO would take "lead roles in the secretariat in their respective areas of responsibility."*

The first session of the ICCM was held in Dubai, United Arab Emirates, February 4–6, 2006 (see Chapter 5 of this book). The Conference was held in conjunction with the 23rd session of the UNEP Governing Council and ninth session of the Global Ministerial Environment Forum. The session was the culmination of the three years process of negotiation and it had been agreed that SAICM would be embodied in a high-level declaration, an OPS and a GPA, and provisional agreement had been reached on much of the text of those documents. By the time of the first session of the ICCM, however, the final agreement had yet to be reached on these texts, and certain elements remained in square brackets to reflect a lack of consensus, in particular with regard to financial considerations and principles and approaches.†

Following intense work during the Conference and final negotiations, agreements were reached on the main documents of SAICM. The Dubai Declaration on International Chemicals Management, the OPS, and four Conference resolutions were adopted by the ICCM, while the GPA was recommended for use and further development.‡

The SAICM Framework

The three texts agreed at the first session of the ICCM, and resolutions of the Conference provide the overall outline of SAICM. The Dubai Declaration on International Chemicals Management was adopted by ministers, heads of delegation, and representatives of civil society and the private sector gathered in Dubai. The Declaration enshrines the political commitment to SAICM, as well as key principles.

The link between chemicals management and sustainable development is one of the principle features of the Declaration and SAICM.

> 1. The sound management of chemicals is essential if we are to achieve sustainable development, including the eradication of poverty and disease, the improvement of human health and the environment, and the elevation and maintenance of the standard of living in countries at all levels of development. [...]
>
> 11. We are unwavering in our commitment to promoting the sound management of chemicals and hazardous wastes throughout their life cycle, in accordance with Agenda 21 and the Johannesburg Plan of Implementation, in particular paragraph 23. We are convinced that the Strategic Approach to International Chemicals Management constitutes a significant contribution towards the internationally agreed development goals set out in the Millennium Declaration. [...]

---

* The report of PrepCom3 can be found at: http://www.saicm.org/documents/meeting/prepcom3/meeting_report/meeting_report.htm

† See the report of the first session of the ICCM, available at: http://www.saicm.org/index.php?menuid=8&pageid=7

‡ The publication of the SAICM texts and ICCM resolutions is available in Arabic, Chinese, English, French, Spanish, and Russian on the SAICM Web site: http://www.saicm.org/index.php?menuid=3&pageid=187

The Declaration also highlights the importance of the work of all stakeholders in the sound management of chemicals and in the implementation of SAICM. The special situation of developing countries and countries with economies in transition is fully recognized in the Declaration:

> We will work towards closing the gaps and addressing the discrepancies in the capacity to achieve sustainable chemicals management between developed countries on the one hand and developing countries and countries with economies in transition on the other by addressing the special needs of the latter and strengthening their capacities for the sound management of chemicals and the development of safer alternative products and processes, including non-chemical alternatives, through partnerships, technical support and financial assistance.

The Dubai Declaration also makes a number of connections between chemical safety and workers, the prevention of impacts on human health, the protection of vulnerable groups and human rights, as well as the importance of SAICM implementation and taking stock of progress.

While the overall objective of SAICM is the achievement of the 2020 goal of sound management of chemicals, the OPS defines its scope, which includes

1. Environmental, economic, social, health, and labor aspects of chemical safety.
2. Agricultural and industrial chemicals, with a view to promoting sustainable development and covering chemicals at all stages of their life cycle, including in products.*

The document also highlights needs and objectives in five work areas:

1. Risk reduction
2. Knowledge and information
3. Governance
4. Capacity-building and technical cooperation
5. Illegal international traffic

The OPS provides guidance on general principles and specific aims to be taken for each of these work areas. In addition, the GPA's 273 listed activities are also classified in relation to each work area with the assumption that their successful implementation will contribute to achieving the objectives laid out in the Strategy.

Financial considerations were a key negotiating issue during the SAICM development process. While the principle that developing countries and transition economies would need financial assistance in order to implement SAICM was generally accepted, there were varying viewpoints as to how such resources should be

---

* The Strategy also indicates that: "SAICM does not cover products to the extent that the health and environmental aspects of the safety of the chemicals and products are regulated by a domestic food or pharmaceutical authority or arrangement." Available at: http://www.saicm.org/documents/saicm%20 texts/SAICM_publication_ENG.pdf

mobilized and delivered. Ultimately, a multifaceted approach to financial considerations was agreed in paragraph 19 of the OPS, which states that

> SAICM should call upon existing and new sources of financial support to provide additional resources and should build upon, among other things, the Bali Strategic Plan for Technology Support and Capacity- building[*]. It should also include the mobilization of additional national and international financial resources, including through the Quick Start Programme and other measures set out in this paragraph, to accelerate the strengthening of capabilities and capacities for the implementation of the SAICM objectives.

The paragraph also recognizes that

> the extent to which developing countries, particularly least developed countries and small island developing States, and countries with economies in transition can make progress towards reaching the 2020 goal depends, in part, on the availability of financial resources provided by the private sector and bilateral, multilateral and global agencies or donors.

The financial arrangements for SAICM are described in a list of elements which includes, among other things

1. Actions at the national or subnational levels
2. Enhancing industry partnerships and financial and technical participation in the implementation of SAICM
3. Integration of SAICM objectives into multilateral and bilateral development assistance cooperation
4. Making more effective use of and building upon existing sources of relevant global funding, including possibly with the Global Environment Facility (GEF) (see Chapter 49 of this book) and the Montreal Protocol on Substances that Deplete the Ozone Layer and its Multilateral Fund for the Implementation of the Montreal Protocol[†]
5. Supporting initial capacity-building activities for the implementation of SAICM through the Quick Start Programme (QSP) and its voluntary, time-limited trust fund administered by UNEP
6. Inviting Governments and other stakeholders to provide resources to the SAICM secretariat

---

[*] The Bali Strategic Plan for Technology Support and Capacity-building constitutes UNEP's approach to strengthen technology support and capacity building in developing countries, as well as countries with economies in transition. The Plan was approved by the 23rd session of the UNEP Governing Council in February 2005 and is available at: http://www.unep.org/GC/GC23/documents/GC23-6-add-1.pdf

[†] The Multilateral Fund, established in 1993, is a dedicated multilateral fund for a multilateral environment agreement. It meets the agreed incremental costs of compliance activities for elimination of ozone-depleting substances (e.g., financial and technical cooperation, and technology transfer).

While the financial considerations provide a comprehensive list of different opportunities and possibilities of support, only the QSP is specific to SAICM. ICCM resolution I/4 established the QSP "to support activities to enable initial capacity building and implementation in developing countries, least developed countries, small island developing States and countries with economies in transition." In the resolution, the ICCM also called for the QSP to include a trust fund, administered by UNEP, and multilateral, bilateral, and other forms of cooperation. The trust fund is set to be open to receive contributions until 2011 and to make disbursements until 2013.

The GPA provides a list of 273 voluntary activities by stakeholders in order to pursue the commitments and objectives expressed in the Dubai Declaration and the OPS. The GPA is composed primarily of a table separated along 36 work areas consistent with the five categories of objectives defined by the OPS. For each activity, possible actors, targets and timeframes, and indicators of progress and implementation aspects are suggested. Although the GPA was not adopted, the Dubai Declaration highlights its important role:

> We recommend the use and further development of the GPA to address current and ever-changing societal needs, as a working tool and guidance document for meeting the commitments to chemicals management expressed in the Rio Declaration on Environment and Development, Agenda 21, the Bahia Declaration on Chemical Safety, the Johannesburg Plan of Implementation, the 2005 World Summit Outcome and this Strategic Approach.

For the outcome of the ICCM first session, see Chapter 5 of this book.

Characteristics of the SAICM Development Process

The adoption of SAICM marked an important step in the definition of a comprehensive and global framework for the sound management of chemicals. While its implementation and performance against the 2020 goal of the sound management of chemicals will determine its effectiveness and adequacy, the way leading to its adoption provided a number of important features. Contrary to preceding efforts to tackle chemicals-related issues, SAICM was not conceived as a legal instrument but as a voluntary mechanism. This approach allowed for greater flexibility in the definition of its objectives, engagement of stakeholders, and sectoral opportunities for implementation.

The 2020 goal of the Johannesburg Plan of Implementation allowed for SAICM to aim for an ambitious goal and a concrete framework for achieving it. Instead of relying on state-centered international law, SAICM was conceived with the different elements needed to foster international action. Political commitment was provided for by the Dubai Declaration, the OPS defined SAICM's core arrangements, and the GPA provided a suggested toolbox of particular actions. The voluntary nature of the approach allowed for a more flexible participation of all stakeholders with a focus on objectives and activities, rather than solely on rights and obligations. Building upon existing efforts, SAICM did not aim to replace or duplicate exiting programs, organizations, and treaties. Rather, SAICM aimed to provide an umbrella under which existing and future national, regional, and International Chemicals Management work could be fostered.

The multistakeholder and multisectoral engagement was one of the successes of the development of SAICM.* The shift from a legal state-centered framework to voluntary framework allowed for international nonstate actors to be involved in the development of SAICM. From its onset, SAICM was conceived as a means of linking the work of Governments, intergovernmental organizations, and civil society organizations, including industry for the sound management of chemicals. In recognition of the important role played by all stakeholders, the SAICM PrepCom rules of procedure gave equal status to all participants with decisions requiring consensus from all representatives.† This horizontal status brought a new dimension to the negotiation process, leading to trust-building and better understanding between stakeholders.

Furthermore, the SAICM development process allowed for the engagement of a maximum of sectors to be engaged in the process. This was achieved among other things through the granting of travel funding for two representatives from different Ministries of developing countries and countries with economies in transition. Different sectors were also represented by different intergovernmental organizations, as well as relevant civil society organizations, including environment and health organizations, trade unions, and industry.‡ This was further facilitated by the coconvening role of the IOMC (see Chapter 23 of this book), with Participating Organizations encouraging each of their respective sectors to actively take part in the SAICM process. Furthermore, building upon their role played during the negotiation of the Stockholm Convention, civil society organizations, and industry groups were important players in the adoption of SAICM, often collaborating together in finding solutions and compromises.

In this regard, one of the principal features of SAICM has been to link chemicals management in all sectors as an issue of sustainable development. While Chapter 19 of *Agenda 21* and the WSSD had provided a general link, SAICM offered stakeholders from all sectors concrete opportunities to tie chemicals safety with the improvement of higher living standards or achievement of Millennium Development Goals. In the context of developing countries and countries with economies in transition, SAICM, for example, aims to encourage the *mainstreaming* of chemicals management

---

* The OPS provides that the main SAICM stakeholders and sectors are understood to be "Governments, regional economic integration organizations, intergovernmental organizations, non-governmental organizations and individuals involved in the management of chemicals throughout their life-cycles from all relevant sectors, including, but not limited to, agriculture, environment, health, industry, relevant economic activity, development cooperation, labour and science. Individual stakeholders include consumers, disposers, employers, farmers, producers, regulators, researchers, suppliers, transporters and workers."

† See, for reference, the rules of procedure in document SAICM/ICCM.1/6 available at: http://www.chem.unep.ch/ICCM/meeting_docs/default.htm

‡ See, for example, an analysis of the role played by industry in the negotiations in *Business in Economic Diplomacy* by Reinhard Quick, in *The New Economic Diplomacy* (second edition), Nicholas Bayne and Stephen Woolcock (ed.), Ashgate Publishing, Ltd., 2007. See also http://books.google.ch/books?id=ELDv-26byMwC&pg=PA112&dq=NEW+ECONOMIC+DIPLOMACY+SAICM&hl=en#PPA105,M1

into national development priorities and plans. Mainstreaming activities* aim to assist countries in demonstrating the need for chemicals management using economic tools, including cost–benefits analysis.

The engagement of a large spectrum of stakeholders and sectors allowed for SAICM to receive inputs and take into account views from a variety of actors involved in chemicals management. In addition to being inclusive, SAICM's development remained transparent at all time, offering the opportunity for all participants and the external public to oversee information made available, outcomes of consultations, as well as preparatory and meeting documents.

Building on previous work and initiatives, the development of SAICM received strong high-level support. SAICM's development was endorsed by Heads of States and Government during the WSSD in Johannesburg in 2002 and during the 2005 World d Summit† as well by several Ministerial forums at the regional level.‡ During the first session of the ICCM, more than 30 ministers and senior representatives committed themselves to SAICM and the Dubai Declaration. Following its adoption, SAICM has also been formally acknowledged or endorsed by governing bodies of intergovernmental organizations and international forums.§

SAICM IMPLEMENTATION AND THE SECOND SESSION OF THE ICCM

The adoption of SAICM by ICCM-1 closed more than three years of a development process. However, this event only marked the very beginning of SAICM's implementation as its success will be measured against the 2020 goal of sound management of chemicals.

As the Dubai Declaration highlights, the implementation of SAICM will require the participation and work of all stakeholders: "We collectively share the view that implementation and taking stock of progress are critical to ensuring success . . . ."¶

While SAICM provides the policy framework and can facilitate assistance, progress depends on the initiatives of individual actors, including Governments, intergovernmental organizations, and civil society organizations.

* Activities for mainstreaming may include qualitative and quantitative analysis of links between priority chemical management issues and human health and environmental quality, research to assess the costs of inaction and benefits of action, using planning and economic terminology, of priority chemicals management issues, as well as integrating chemicals management priorities into each country's development planning processes and plans.

† In September 2005, more than 150 Heads of State and Government gathered in New York during the 2005 World Summit to follow-up to the outcomes of the Millennium Summit held in 2000. The High Level Plenary Meeting endorsed the 2005 World Summit Outcome, which endorsed the development of SAICM. See for reference: http://www.saicm.org/documents/positions/SAICM%20Para%2056k%20-%202005%20World%20Summit%20Outcome.pdf

‡ See the international and regional positions on the development of SAICM at: http://www.saicm.org/index.php?menuid=2&pageid=109&submenuheader=

§ Information on the consideration of SAICM by international forums' positions on SAICM can be found at: http://www.saicm.org/index.php?menuid=4&pageid=4

¶ Dubai Declaration, paragraph 23, p. 9, http://www.saicm.org/documents/saicm%20texts/SAICM_publication_ENG.pdf

The Enabling Phase and the QSP

SAICM can be considered as a process in which an initial enabling phase needs to be completed before full implementation can be achieved. This initial phase is aimed at addressing the needs of countries in the assessment of their capacities for the sound management of chemicals, in particular in developing countries and countries with economies in transition. While there is no definition of enabling activities, references are made in the OPS to initial activities stakeholders may undertake in preparation of their implementation of SAICM. Paragraph 22 of the Strategy provides that

> SAICM implementation could begin with an enabling phase to build necessary capacity, as appropriate, to develop, with relevant stakeholder participation, a national SAICM implementation plan, taking into consideration, as appropriate, existing elements such as legislation, national profiles, action plans, stakeholder initiatives and gaps, priorities, needs and circumstances.

The QSP was established to address some of these initial needs as its objective defined by ICCM resolution I/4 is

> to support initial enabling capacity building and implementation activities in developing countries, least developed countries, small island developing States and countries with economies in transition.

The strategic priorities of the QSP, defined in ICCM resolution I/4, provide a further indication as to the scope of enabling activities, which are to be in keeping with the work areas set out in the strategic objectives of Section IV of the OPS, namely risk reduction, knowledge and information, governance, capacity building and illegal international traffic, and relate in particular to the following strategy priorities:

- Development or updating of national chemical profiles* and the identification of capacity needs for sound chemicals management.
- Development and strengthening of national chemicals management institutions, plans, programs, and activities to implement SAICM, build upon work conducted to implement international chemicals-related agreements and initiatives.†

---

* National chemicals management profiles provide a comprehensive overview of the national chemicals management situation in a country. Their development or updating provides the opportunity to assess the existing national legal, institutional, administrative, and technical infrastructure for the sound management of chemicals. National profiles can serve as a basis for identifying national chemicals management priorities and for initiating targeted and coordinated follow-up action.

† Examples of voluntary international initiatives emanating from intergovernmental processes include the International Code of Conduct on the Distribution and Use of Pesticides developed under the auspices of the Food and Agriculture Organization (see Chapter 12 of this book) and the Globally Harmonized System of Classification and Labelling of Chemicals (see Chapter 11 of this book).

- Undertaking analysis, interagency coordination, and public participation activities directed at enabling the implementation of SAICM by integrating—that is, mainstreaming—the sound management of chemicals in national strategies, and thereby informing development assistance cooperation priorities.

Since 2006, and as of March 2010, the QSP trust fund has received pledges for an approximate total of $23,719,400 from 23 donors.* As of August 2010, 100 projects with a total value of $20,265,064 were approved. In addition, nontrust fund contributions have been provided to support bilateral and multilateral chemicals management programs, projects, and activities supporting the QSP objective and strategic priorities.†

During the second session of the ICCM (see Chapter 6 of this book), the QSP Executive Board reported that over the course of regional meetings and consultations, many stakeholders welcomed the Programme and made positive comments regarding its adequacy. Some called for more resources to be made available, for an increase in the funding available per project and per country and for the consideration of extending the duration of the Programme. Some donor Governments said that more equitable burden sharing was a precondition for maintaining their contributions to the Programme. They also observed that relying on a limited number of major donors made the Programme vulnerable to funding fluctuations and shortfalls.‡

In its decision II/3 on financial and technical resources for implementation SAICM, the ICCM requested the QSP Executive Board to evaluate the Programme, report on its effectiveness and the efficiency of its implementation, and make recommendations in the light of its findings for the consideration of the Conference at its third session. At its fifth meeting held in June 2010, the Board decided that the mid-term evaluation of the QSP would, firstly, undertake the assessment of overall outcomes and the management of the QSP and secondly evaluate individual projects.§

National and Regional Implementation

While the early successes of SAICM and of the QSP have been welcomed, the major objective of SAICM remains the achievement of the 2020 goal and full implementation by all stakeholders. At the national level, Governments are expected to take a

* Existing arrangements provide that each year two application rounds are held, during which Governments of developing countries and countries with economies in transition are eligible for projects valued between $50,000 and $250,000. Proposals may be presented by SAICM participating Governments that have given appropriate formal recognition to SAICM, at a minimum by having designated an official SAICM national focal point. On an exceptional basis, civil society networks participating in SAICM can also be eligible to present project proposals, which need to be endorsed by a SAICM national focal point.

† Additional information on the QSP can be found at: http://www.saicm.org/index.php?menuid=22&pageid=252

‡ See document SAICM/ICCM.2/5 available at: http://www.saicm.org/index.php?content=meeting&mid=42&def=2&menuid=9

§ See the report of the 5th meeting of the QSP Executive Board, document SAICM/EB.5/9, as well as other relevant meeting documents.

number of steps to ensure that SAICM's framework is translated into concrete measures. As an initial step, Governments are invited by the OPS paragraph 23 to "establish arrangements for implementing SAICM on an inter-ministerial or inter-institutional basis so that all concerned national departmental and stakeholder interests are represented and all relevant substantive areas are addressed," as well as to nominate a national focal point "to facilitate communication, nationally and internationally." Furthermore, Governments can integrate SAICM into relevant programs and plans, including those for development cooperation, as called for in OPS paragraph 19 (a).

National implementation is also aimed at other stakeholders and their engagement is important in order to cover a large scope of aspects of chemical safety. The OPS paragraph 22, for example, calls for the development, "with relevant stakeholder participation, [of] a national SAICM implementation plan, taking into consideration, as appropriate, existing elements such as legislation, national profiles, action plans, stakeholder initiatives and gaps, priorities, needs and circumstances."

At the regional level, the ICCM decided in its resolution I/1 that intersessional work should be promoted through, among other things, regional meetings. The SAICM OPS, in paragraph 26, indicates that the functions of the regional meetings will include

a. To review progress on implementation of the Strategic Approach within the regions
b. To provide guidance on implementation to all stakeholders at a regional level
c. To enable technical and strategic discussions and exchange of information to take place

Since the adoption of SAICM in February 2006, all five United Nations regions, namely the African, Asia-Pacific, Central and Eastern European and Latin American and Caribbean regions, and the Western European and Others Group, have had at least one regional meeting. The regional meetings during the first intersessional period have focused on agreeing on arrangements for regional coordination, establishing regional priorities and plans for SAICM implementation, and preparing of the second session of the ICCM. The African region adopted a regional action plan, while the Asia-Pacific and Central and Eastern European regions made first steps in this regard.*

At the second session of the ICCM, many representatives stressed the important role of the regional focal points and regional meetings. And in resolution II/2 on regional activities and coordination, the Conference recommended the establishment of regional coordination mechanisms and the development of terms of reference for regional representatives. It underlined the important role of regional meetings and coordination mechanisms in enabling stakeholders in each region to

* Further information on regional activities can be found at: http://www.saicm.org/index.php?menuid=14&pageid=294

exchange experience and identify priority needs and to develop regional positions on key issues.*

The Second Session of the ICCM

The SAICM OPS, in paragraphs 24 and 25, sets out the functions and schedule of the ICCM as follows:

The ICCM will undertake periodic reviews of SAICM. The functions of the ICCM are

1. To receive reports from all relevant stakeholders on progress in implementation of SAICM and to disseminate information as appropriate
2. To evaluate the implementation of SAICM with a view to reviewing progress against the 2020 target and taking strategic decisions, programming, prioritizing, and updating the approach as necessary
3. To provide guidance on implementation of SAICM to stakeholders
4. To report on progress in implementation of SAICM to stakeholders
5. To promote implementation of existing international instruments and programs
6. To promote coherence among chemicals management instruments at the international level
7. To promote the strengthening of national chemicals management capacities
8. To work to ensure that the necessary financial and technical resources are available for implementation
9. To evaluate the performance of the financing of SAICM
10. To focus attention and call for appropriate action on emerging policy issues as they arise and to forge consensus on priorities for cooperative action
11. To promote information exchange and scientific and technical cooperation
12. To provide a high-level international forum for multistakeholder and multisectoral discussion and exchange of experience on chemicals management issues with the participation of nongovernmental organizations in accordance with applicable rules of procedure
13. To promote the participation of all stakeholders in the implementation of SAICM

The OPS paragraph 25 also provides that the second session of the ICCM should be held in 2009 and that, "where appropriate, sessions of the ICCM should be held back-to-back with meetings of the governing bodies of relevant intergovernmental organizations in order to enhance synergies and cost effectiveness and to promote SAICM's multi-sectoral nature." ICCM2 has taken place in Geneva, May 11–15, 2009, immediately before the 62nd World Health Assembly. The second session of ICCM has been preceded by the fourth meeting of the Conference of the Parties of the Stockholm Convention† (see Chapter 6 of this book).

* See resolution II/2 in document SAICM/ICCM.2/15 available at: http://www.saicm.org/index.php?content=meeting&mid=42&def=2&menuid=9

† See http://www.pops.int

The second session of the ICCM has been an opportunity to finalize institutional arrangements, such as the adoption of its rules and Bureau (see Chapter 6 of this book).

Reporting on Progress in Implementation

Reporting on the implementation of the Strategic Approach will be a key tool in assessing progress toward the Johannesburg Plan of Implementation goal of achieving the sound management of chemicals by 2020. Paragraph 24 of the OPS provides for the Conference to carry out a number of key functions in relation to reporting, namely "to undertake periodic reviews of the Strategic Approach;" "to receive reports from all relevant stakeholders on progress in implementation of the Strategic Approach and to disseminate information as appropriate;" and "to evaluate the implementation of the Strategic Approach with a view to reviewing progress against the 2020 target and taking strategic decisions, programming, prioritizing and updating the approach as necessary."

In order to assist the development of appropriate reporting modalities, the Government of Canada has sponsored a project to develop a set of draft indicators for reporting progress on the implementation of SAICM and a baseline estimates report. The project was carried out by the consulting firm Resource Futures International, with guidance provided by an international project steering committee.

Following the adoption of reporting modalities and indicators at ICCM-2, periodic reporting will be undertaken by the Conference at its future sessions in 2012, 2015, and 2020.*

Emerging Policy Issues

One of the functions of the ICCM set out in paragraph 24 of the OPS is "to focus attention and call for appropriate action on emerging policy issues as they arise and to forge consensus on priorities for cooperative action." Paragraphs 14 (g) and 15 (g) of the OPS call, respectively, for new and emerging issues of global concern to be sufficiently addressed by means of appropriate mechanisms, and for an acceleration of the pace of scientific research on identifying and assessing the effects of chemicals on human beings and the environment, including emerging issues.

The term "emerging policy issue," may be understood to be an issue involving the production, distribution, and use of chemicals, which has not yet been generally recognized or sufficiently addressed, but which may have significant adverse effects on human beings and/or the environment.

Among the emerging policy issues are

1. Chemicals in products
2. Nanotechnology and manufactured nanomaterials
3. Electronic waste
4. Lead in paints

---

* Additional information on reporting and modalities can be found at: http://www.saicm.org/index.php?menuid=32&pageid=297

It should be noted that emerging issues on the international agenda can include those that were not adequately dealt with in the past, and which have received a new focus due to serious problems in practice that have surfaced since.

The second session of the Conference has also considered a longer-term procedure for the modalities of carrying out its functions with regard to emerging policy issues which would include revised criteria for priority setting, to be developed as necessary (see Chapter 6 of this book).

Financial Considerations

In the course of the development of SAICM, financial considerations were a crucial element of the SAICM framework. During PrepCom 3, in September 2005, a study on financial considerations for SAICM was presented. It highlighted some gaps in financing, such as the following:

- International agreements and decisions encompassed by SAICM have limited access to funding from multilateral and bilateral funding sources (e.g., the Basel Convention, the Rotterdam Convention, etc.).
- Multilateral financial mechanisms with chemicals-related mandates address only partially broader governance issues that are central to SAICM.
- Existing multilateral financial mechanisms with chemicals-related mandates are restricted to provision of support for work on a relatively limited, although important, number of chemicals.
- Integration or "mainstreaming" of the sound management of chemicals in multilateral and bilateral development assistance programming has seen slow progress with certain key exceptions.
- Despite the wealth generated by, and the growth of the chemical industry on a global basis, there are no significant mechanisms for industry financial contributions to the global agenda for the sound management of chemicals.*

Taking into consideration these elements, paragraph 19 of the OPS, which enshrines the financial arrangements for SAICM. is a comprehensive list of sources of finance and technical cooperation means. Since 2006, however, a large majority of stakeholders considered that the scope of SAICM is such that the funding necessary to achieve significant progress toward the 2020 goal far exceeded that currently available, in particular through the QSP.

Over the course of regional meetings and consultations, many stakeholders welcomed the QSP and were positive as to its adequacy for meeting its limited objective. Some called for more resources to be made available, for an increase in the funding available per project and per country, as well as the consideration of a possible extension of the duration of the QSP. Demand for QSP trust fund assistance has remained constant over the first 3 years of operation of the QSP and funds available were almost sufficient to meet the demand of all applicants. SAICM donors

* See SAICM/PREPCOM.3/INF/28: http://www.saicm.org/documents/meeting/prepcom3/en/INF28.doc

emphasized that broadening of the donor base was a crucial challenge for sustaining the Programme and its trust fund.* Some donor Governments highlighted burden-sharing as a precondition to allow present donors to maintain their contributions to the QSP and that the reliance on a limited number of important donors undermined the sustainability of the Programme. Some stakeholders noted that, thanks to the QSP, it had been possible to obtain development cooperation agency resources.

There is a shared view among a number of stakeholders that further consideration should be given to the financial framework of SAICM, in particular, as the QSP will cease to receive contributions in 2011, one year before the third session of the ICCM. Among the options considered has been the need for better use of existing resources, linking of SAICM to the GEF, the development of a standalone financial mechanisms, and better use of development assistance funding.

At ICCM-2 (see Chapter 6 of this book), a number of representatives pointed that reaching the goals of SAICM was contingent on securing sustainable financial resources and that as the QSP would end in 2013, a solution for long-term financing was needed as soon as possible (see Chapter 6 of this book). The broad scope of SAICM meant that no single solution for financing its operation existed. Some representatives argued that it was essential to seek new financing mechanisms; others considered that it would be better to make use of existing institutions such as GEF. In particular, the fifth replenishment of GEF represented a major opportunity for increasing its involvement in chemicals management (see also Chapter 49 of this book).

In Resolution II/3, the Conference recognized the need for sustainable, predictable, adequate, and accessible funding for activities in support of the sound management of chemicals and the achievement of the objectives set forth in SAICM, taking into account the priorities identified by developing countries and countries with economies in transition. It also welcomed the consideration being given to the sound management of chemicals during the fifth GEF replenishment process and urged the GEF within this process to consider expanding its activities related to the sound management of chemicals to facilitate SAICM implementation while respecting its responsibilities as the financial mechanism for the Stockholm Convention.†

CONCLUSION

The potential harmful effects on human health have gradually raised calls for their sound management. With increased use, production, and transport of chemicals, awareness of a number of related problems has gradually been on the agenda of the international community. The development and adoption of SAICM was the cumulating point of the emergence of chemicals management as a global issue. The acknowledgement of chemicals as an issue of sustainable development and the involvement of all sectors and stakeholders have also raised the profile of SAICM. SAICM recognizes the special situation of developing countries, which increase

* Information and documents on SAICM donors meetings can be found at: http://www.saicm.org/index.php?menuid=5&pageid=22

† See resolution II/3 in the report of ICCM2, document SAICM/ICCM.2/15 available at: http://www.saicm.org/index.php?content=meeting&mid=42&def=2&menuid=9

their production and consumption of chemicals and require support for their sound management.

SAICM provides an innovative mechanism for action, which has the necessary components to address the 2020 goal of the sound management of chemicals. Its comprehensiveness in scope, high-level endorsement, voluntary nature, and inclusiveness make it possible, albeit challenging. While SAICM is nor a convention, nor a forum, it may provide the example of future international multilateral initiatives. While it does not create legal obligations, it provides a framework which includes a recognized mandate, agreed texts, and a flexible plan for action.

Unlike existing chemicals conventions, SAICM benefits from its comprehensive scope, not limited to technical aspects, or specific chemicals or issues. Its flexibility in form and ambition in goal are also strengths which may in the long term make it a successful example of creative norms of the international environmental and development regimes.

However, with the international community focusing on the development and implementation of other initiatives, such as those on mercury, it is difficult, at this point in time, to assess the success or failure of SAICM. Governments' financial difficulties and the overexpansion of initiatives on International Chemicals Management seriously undermine current efforts under SAICM. Widening of focus, accompanied by reduction of resources, might diminish the global goal of SAICM and transform it into a soft initiative.

Thus, SAICM requires attention and resources to remain meaningful and retain impact. Its success requires the participation and commitment of all stakeholders. The availability of financial and technical means for implementation may be the initial indicators of a successful process. A sound, sustainable financial mechanism is a major prerequisite for a further successful implementation.

REFERENCES

Bruntland, G., ed. 1987. *Our Common Future: The World Commission on Environment and Development*, Oxford, UK: Oxford University Press.

Reinhard, Q. 2007. Business in economic diplomacy. In *The New Economic Diplomacy*, 2nd Edition, N. Bayne and S. Woolcock, eds. Aldershot, UK: Ashgate Publishing, Ltd.

Section V

Organizations

18 Food and Agriculture Organization of the United Nations

Mark Davis[*]

CONTENTS

INTRODUCTION

Food and Agriculture Organization of the United Nations (FAO) was established in 1945 to help countries produce sufficient food, feed, and fiber to sustain their own populations and provide tangible economic benefits. While also addressing fisheries and forest-based sources of food and livelihoods, nevertheless the focus of attention has consistently been on land-based production in the form of crops and livestock.

More land is devoted to agricultural production than any other human activity on earth, with 1.4×10^9 hectares of arable land and another 3.4×10^9 hectares of pasture.[†] More people work in and derive their livelihood from agriculture than any

[*] The opinions expressed in this chapter are those of the author and do not necessarily represent the views of any organization.

[†] FAOSTAT (2009).

other activity with between 50% and 70% of the global population deriving their income from rural activities.* Without agriculture, the largely urban based 50% of global population would not have food. Indeed it could be argued that without agriculture society would not exist as we know it since we would all still be hunter-gatherers.

By current estimates, the global population will increase to 9.1 billion people by 2050, at which point it is expected to plateau. The highest population growth will occur in countries where food security is most tenuous and where the impacts of climate change, encroaching desertification, water scarcity, and nutrient-depleted soils are often most acute. Feeding 48% more people will require agricultural production to increase by an estimated 70% over 2009 levels. This increase in production must be achieved in the face of shrinking land availability due to desertification, urbanization, salinity, the impacts of climate change, contamination, and other factors. In addition, as economic development progresses, more people are eating more animal products and consuming more energy, and therefore crop production resources are increasingly drawn toward animal feed and energy crops at the expense of food crops for human consumption.

Achieving the required increases in agricultural production will require an estimated investment of US$83 billion, yet in most countries and in Official Development Assistance (ODA), investment in agriculture has actually been falling since the 1970s. In 1979, 20% of ODA was allocated to agriculture while in 2007 the proportion was only 5%. Over the same period, overall ODA increased by over 200%.†

Intensification of Crop Production depends on knowledge about good agronomy and proper planning of the production cycles as well as the timely provision of high-quality inputs at the forefront of which must be good quality seeds and varieties that are adapted to maximize yields under prevalent local conditions. Other inputs would typically include fertilizer, land, water, pesticides, tools, and energy, as well as labor. However, it has long been recognized that inputs merely complement the natural processes supporting plant growth. Examples include nutrient cycles that degrade organic matter and allow plants to access key nutrients and maintain a healthy soil structure which promotes water retention and the recharge of groundwater resources, pollination services, and natural predation for pest control. Farmers who use information and knowledge about the supporting biological processes more effectively can help to maximize the efficient use of conventional inputs.

Some examples of agricultural production systems that manage ecosystem services to improve productivity and reduce environmental impact through an integrated ecosystem approach include Integrated Plant Nutrient Management (IPNM), Integrated Pest Management (IPM), Conservation Agriculture (CA), participatory plant breeding approaches, crop–livestock systems, agro-forestry systems, and integrated weed management as well as pollination management. However, these

* Sustainable agriculture in a globalized economy, Report for discussion at the Tripartite Meeting on Moving to Sustainable Agricultural Development through the Modernization of Agriculture and Employment in a Globalized Economy, Geneva, September 18–22, 2000, International Labour Office Geneva.

† Investment: Background paper for "How to Feed the World 2050—High Level Expert Forum, 12–13 October 2009," FAO, 2009.

approaches are knowledge intensive, and to assist small-holder farmers to develop the necessary knowledge and skill sets locally will require a sustained investment for capital formation (physical and human).

Agriculture, in its simplest terms, has taken useful plants and animals and gathered them together for convenience. Why wonder through a forest looking for occasional fruit trees, when several could be cultivated together, close to home? Why seek wild animals to hunt when they could be kept in an enclosure for slaughter at our convenience?

One of the problems of manipulating nature in this way is that ecosystems will always rebalance themselves. If one organism is removed, it will be replaced by another. If new organisms are introduced, they will need food and will probably be fed upon. Agriculture systematically removes natural ecosystems and replaces them with introduced species. Forests, grasslands, scrub, swamps, and sand dunes have been stripped of their natural vegetation and planted with crops worldwide. Nature responds by bringing in organisms that can thrive by consuming the new crops. These may be existing inhabitants of the environment that was replaced by agriculture or invasive species that have a particular liking for or are particularly well adapted to the crop being cultivated. The waves of fruit flies imported with agricultural produce from one region to another and adapting to local crops provide excellent examples of such invasions. Fruit flies are very selective about what they will eat, but have often found new favorites in their adopted countries. Similarly, the Western Corn Rootworm (*Diabrotica virgifera virgifera*) imported to the Balkans from the United States potentially threatens local maize as well as cucurbits and soya. In some cases organisms that are particularly well adapted to a particular crop have been introduced with that crop to countries that adopted it for purposes of broadening their supply of food and other useful crops. The maize stem borer was imported to the Balkans in aid supplies during the conflicts of the 1990s and has now proliferated in the region, and woolly white fly arrived with citrus planting material imported to Eritrea but has no natural enemy locally.

Agriculture is at the heart of sustainable development, as this depends to a large extent to the availability of adequate food resources and renewable resources of other materials, including energy.

Since the agricultural environment is now dominated by the crops being cultivated, the feeding organisms, be they fungi, insects, mammals, birds, or any other class, will thrive. This is the root of agricultural pest problems, and humans who have invested time, effort, and resources in producing the crop will want to protect it, and hence the creation of plant protection or crop protection.

The same ecological principles apply to animal production, fish farming, cultivated forests, and mushroom farms; any situation, in fact, where humans manipulate nature to produce goods for themselves. In all such situations we are in competition with other organisms, and farmers will want to protect their source of food and income.

CHEMICALS IN AGRICULTURE

For thousands of years farmers depended on ecological, cultural, or mechanical means to control pests, though there is evidence that arsenic has been used to control pests since Roman times. Most organisms consuming plants will be at the base of a food chain and will be predated or parasitized by another organism. Often this will be sufficient to keep pest populations under reasonable control. When pest populations increase, mechanical removal or destruction such as hand picking or beating could be used, when pests were known to attack at certain times of the year, planting could be adjusted to avoid them, and when pests were found to be repelled by certain conditions or other plants, the farming environment could be adapted to be surrounded by repellents.

The most basic forms of chemical control may have taken the form of smoke or plant extracts to repel insects, but chemical pest control agents as formulated pesticides began to emerge in the early part of the twentieth century. Inorganic chemical such as arsenic, mercury, copper, and sulfur were widely used.

When did FAO Start Working on Pesticides?

FAO members called for a dialogue and technical assistance in dealing with transboundary pest movements early in the organization's history. This was a direct evolution of the International Convention of Plants signed in 1929. The dialogue led to the creation of the International Plant Protection Convention (IPPC), which came into force in 1952 and whose key objectives were to prevent transboundary movement of agricultural pests and diseases, in particular through their carriage with traded agricultural produce.

The Convention was appointed in the 1989 Uruguay Round of the General Agreement of Tariffs and Trade as the standard setting body for Agreement on Sanitary Phytosanitary Measures (SPS Agreement). In 1992 the Secretariat to the Convention was established in FAO and in 1995 a process of revising and updating the Convention began which ended with the adoption of the revised text in 2005. The new IPPC Convention reflects modern phytosanitary practices and the role of the Convention in WTO Agreements. Today there are 172 parties to the IPPC which provides standards for countries to follow when exporting agricultural produce.

From the 1970s FAO has operated a program to help countries in dealing with transboundary pests by supporting networks to monitor breeding and movement of the Desert Locust (*Schistocerca gregaria*), which can affect a huge swathe of land that includes the top third of Africa, the entire Arabian Gulf, and large parts of the Indian Subcontinent. The FAO program, incorporated into the Emergency Prevention System for Transboundary Animal and Plant Pests and Diseases in 1994, also helps with national and regional efforts to control other important locust species including the Red Locust (*Nomadacris septemfasciata*) in Eastern Africa, the Brown Locust (*Locustana pardalina*) in southern Africa, the Migratory Locust (*Locusta migratoria*) throughout Africa and Asia, the Tree Locust (*Anacridium melanorhodon*) mainly in Africa, the Moroccan Locust (*Dociostaurus maroccanus*) and the Italian Locust (*Calliptamus italicus*) in North Africa, Europe, and Central Asia, and the Australian Plague Locust (*Chortoicetes terminifera*) in Australia, as well as Armyworm (*Spodoptera frugiperda*) in eastern African and *Quela Quela* birds throughout Africa.

LIFE-CYCLE MANAGEMENT

Chemical pesticides were evolving and diversifying in the twentieth century, especially after World War II. The first generation of mainstream insecticides consisted of organochlorines such as DDT, the "drins" (aldrin, endrin, dieldrin), heptachlor, mirex, and toxaphene. Early commercial herbicides such as 2,4-D, 2,4,5-T were based on phenoxyacetic acids and were later joined by the triazines, all of which are still in widespread use. The herbicide Glyphosate which was introduced in the 1970s is probably the most widely used pesticide in the world. Fungicides were originally based on inorganic chemicals adapted from accidental discoveries that salt, lime, sulfur, and copper, mercury-, and arsenic-based compounds prevent fungal attack on grains or fruits. Later, the dithiocarbamate and phthalimide chemicals were developed which were easier to use and less phytotoxic than the inorganics. In the 1970s, new generations of pesticides began to emerge. Organophosphate and carbamate insecticides attacked the nervous systems of insects and other pests, as opposed to the stomach acting organochlorines. Herbicide technology was also becoming more diverse and selective with products also being designed to be less environmentally persistent and mobile. Fungicides were now extremely diverse and operating on a range of biological functions.

The Green Revolution of the 1960s revolutionized agricultural production in many countries allowing net importers of food such as Mexico and India to become net exporters in a short time. The strategies adopted by many countries broadly focused on selecting crop varieties that could produce high yields, accompanied by agrochemical inputs of fertilizers and pesticides. While previously agrochemicals were largely a tool of farmers in the more developed countries, now their use was becoming much more widespread globally.

By the 1960s, hundreds of chemicals were being marketed as pesticides. Information about the undesirable side effects of these chemicals was also beginning to emerge. Rachel Carson's seminal work of 1962 on the impact of pesticides on wildlife was published in *Silent Spring* which has been credited with the creation of the environmental movement. The book also triggered political interest in many countries which eventually led to the establishment of regulatory systems to control pesticides.

However, surprisingly, preceding the publication of *Silent Spring*, the 11th FAO Conference of November 1961 urged that "FAO provide leadership in achieving international understanding in the controversial aspects of the use of pesticides (e.g. pesticide residues, hazards to farm workers, operators and factory workers, insect resistance to insecticides, and marketing requirements) in furnishing guidance to governments" (FAO Conference, 1961). The Conference also highlighted the lack of uniformity of approach by governments and the agricultural chemicals industry and "other problems, such as the development of resistance to insecticides in insects, occupational hazards in connection with the production, handling and use of pesticides in agriculture, and the registration and marketing requirements" (FAO Conference, 1961). This early recognition of the problems associated with pesticide use and the need to assist countries with sustainable pest management solutions and in improving pesticide management was translated into the creation of a Committee on Pesticides in Agriculture, which evolved over the years into the current Expert

Panel on Pesticides Management. The 11th Conference also called for continued collaboration with the World Health Organization (WHO) and International Labor Organization which preceded it and is ongoing today.

Recognizing that some control needed to be exerted over which chemicals were being used, by whom, and under which circumstances, the first regulatory system for pesticides was probably the United States Insecticide Act of 1910 which focused mainly on pesticide quality. The Federal Insecticide, Fungicide and Rodenticide Act which was originally put in place in 1947, and has been amended several times since, changed the focus toward protecting users and gradually evolved to emphasize the protection of the public and the environment. The first US denials of registration for safety reasons occurred in 1964.

Other countries began with registration systems in the same period. A Benelux (an economic union consisting of Belgium, the Netherlands, and Luxemburg) Committee existed in the early 1950s, and a Council of Europe group in the late 1950s. The situation in the developing world, however, was somewhat different. Most developing countries had no pesticide regulatory system in place until the 1990s and many still have only rudimentary systems in place. But pesticide use has increased significantly in these regions with Asia and Latin America in particular emerging as the fastest growing pesticide markets in the world.

With a growing understanding of the need to regulate pesticides and their potential impacts on health and the environment, FAO members called upon the Organization to take measures to assist developing countries to manage pesticides safely and effectively.

One of the early examples is the Codex Alimentarius, which set up its Codex Committee on Pesticide Residues (CCPR) in 1964, with support from FAO and WHO through the FAO/WHO Joint Meeting on Pesticides Residues (JMPR) which started the same year.*

CODE OF CONDUCT

The *International Code of Conduct on the Distribution and Use of Pesticides*† (see Chapter 12 of this book) was one of the first voluntary Codes of Conduct adopted within a broad framework of support for increased food security, while at the same time protecting human health and the environment. It was adopted in 1985 by the FAO Conference at its 23rd Session, and was subsequently amended to include provisions for the Prior Informed Consent (PIC) procedure at the 25th Session of the FAO Conference in 1989. The Code established voluntary standards of conduct for all public and private entities engaged in, or associated with, the distribution and use of pesticides, and since its adoption has served as the globally accepted standard for pesticide management. It is at the heart of all FAO and WHO activities in the broader area of pest and pesticides management.

Collaboration between FAO and WHO has been in existence for as long as the Code of Conduct has been in place. The organizations have consistently worked

---

* http://www.codexalimentarius.net/web/jmpr.jsp

† *International Code of Conduct on the Distribution and Use of Pesticides*, FAO, 2002.

together to prepare pesticide hazard information sheets, recommend pesticide Maximum Residue Limits (MRLs) in produce, and to improve the management of pesticides.

In 2007 an agreement was signed between FAO and WHO for the joint development and implementation of the Code of Conduct. WHO established a panel of experts to support the Code of Conduct which operates jointly with the FAO panel of experts. The panels meet annually in the Joint Meeting on Pesticides Management (JMPM) to discuss policy and technical issues and offer guidance to FAO and WHO on more effective approaches to addressing pesticide management issues.

The Code and its supporting technical guidelines were instrumental in assisting countries to put in place or strengthen pesticide management systems. The number of countries without legislation to regulate the distribution and use of pesticides greatly decreased; awareness of the potential problems associated with pesticide use grew significantly; involvement in various aspects of pesticide management by nongovernmental organizations (NGOs) and the pesticide industry was strengthened; and many more successful IPM programs have been implemented in developing countries.

In 1998, the *Rotterdam Convention on the Prior Informed Consent Procedure for Certain Hazardous Chemicals and Pesticides in International Trade*[*] (see Chapter 14 of this book) was adopted and the article in the Code relating to PIC became redundant. The policy framework within which the Code was being supported and implemented had also changed significantly in the years since the code had been adopted. A revision of the Code was undertaken again which was finalized and approved in 2002. As well as the changes referred to above, a number of definitions were updated including the one for IPM, which still stands as the most widely supported definition of IPM in use among all stakeholders. Provisions were also made for mechanisms to monitor implementation of the code more effectively, including encouraging reports on noncompliance to FAO so that their causes could be addressed.

Today, the International Code of Conduct on the Distribution and Use of Pesticides provides a solid foundation to guide governments, pesticide manufacturers and traders, NGOs, pesticide users, policy makers, and legislators on the management of pesticides as a group of chemicals from their cradle, to their grave. It is a carefully and comprehensively considered document that in 12 articles every aspect of pesticide management is addressed.

The role of FAO and WHO has been to help its member countries to implement the Code effectively. This has largely been achieved by preparing technical guidelines that expand on the various articles and sections, and by building capacity through training, information exchange, demonstration projects, and similar means.

The central role of FAO in pesticides management was largely a reflection of the fact that most pesticides by far were being used in agriculture. Nevertheless, pesticides have always been used in the health sector to control disease vectors such as mosquitoes carrying malaria and dengue fever, sand flies transmitting leishmaniasis, and plague-transmitting rodents. Similarly, it is the health sector that is concerned with the potential and actual toxic effects of pesticides to those that have been

[*] http://www.pic.int

directly exposed to them, as well as those indirectly exposed through contaminated food or water.

SPECIFICATIONS/QUALITY CONTROL

If pesticides are to do their job effectively, they must be provided in standard and stable formulations that can be relied upon by farmers and other end users consistently. Chemicals vary in quality and may contain impurities. There are over 900 different pesticide-active ingredients in use which are in turn are formulated into tens of thousands of pesticidal products. Each product contains one or more active ingredient mixed with various carriers and inert ingredients that assist in the preservation and effective delivery of the active ingredient. Changes in the formulation and degradation or contamination of any of the ingredients in the formulation may alter the way in which the product behaves.

Pesticide manufacturers have their own quality control systems—some better than others. However, it is important for pesticide users to be able to check that the products they are buying are of good quality.

To help countries with this issue, FAO and WHO started collaborating in 2001 to determine standardized pesticide specifications.* WHO focuses on pesticides used in public health, while FAO focuses on agricultural pesticides. These specifications apply to specific formulations and are put in the public domain for anyone to refer to and use. The intention is to help countries that import pesticides to test the quality of the products they are buying to ensure that they comply with the standard specifications.

To date around 300 specifications have been developed and published by FAO and WHO. The specifications apply to agricultural as well as public health pesticides and are specific to a particular formulation rather than an active ingredient.

It is of course a significant limitation that many developing counties do not have adequate laboratories that can accurately analyze pesticide formulations and their contaminants. Without such facilities it is impossible to accurately test compliance of products with their JMPS specification. At the same time, establishing such laboratories needs a major investment in infrastructure and equipment, a stable and well-trained staff team, funds for an ongoing supply of materials and maintenance of equipment, and a cost recovery mechanism to pay for the process of analysis. Many countries are unlikely to have enough samples annually to justify the establishment of a laboratory, yet regional solutions are hard to come by and the logistics of transporting chemicals samples internationally for analysis are complicated. As a result, few countries actually check whether the pesticides they import comply with accepted standards.

In 2001 FAO and WHO carried out a survey of pesticide quality and found that on average, 30% of all pesticides traded in developing countries are substandard, that is, do not comply with JMPS specifications of any other accepted quality standard. As a consequence, it can be assumed that these substandard products are not as effective against the pests they are being used to control, and may be more toxic to

* http://www.fao.org/agriculture/crops/core-themes/theme/pests/pm/jmps/ps/en/

humans and the environment than compliant pesticides. Strengthening capacity to regulate, enforce, and monitor pesticides in all countries is necessary to overcome this type of situations.

RESIDUES IN FOOD—CODEX, JMPR

Pesticide residues in agricultural produce have been a matter of concern for decades. For more than 40 years, FAO and WHO have collaborated to determine MRLs for pesticides in the crops on which those pesticides are used according to their legal registration. In turn, those MRLs that are recommended by the Joint (FAO/WHO) Meeting on Pesticides Residues (JMPR) are transmitted to the Codex Alimentarius (FAO/WHO) that adopts them as its standard that is then recognized by WTO as an acceptable international trading standard. Some (wealthier) countries determine their own MRLs based on crop production and dietary regimes that may be specific to that country, but the JMPR MRLs are considered to be a good level at which the majority of countries, which do not necessarily have the capacity to determine their own MRLs, can set their national standards.

The current JMPR comprises the WHO Core Assessment Group and the FAO Panel of Experts on Pesticide Residues in Food and the Environment. It is recognized as a successful model on the collaboration with WHO. The JMPR consists of experts drawn from governments and academic circles, who attend as independent internationally recognized specialists who act in a personal capacity and not as representatives of national governments.

The WHO Core Assessment Group is responsible for reviewing pesticides toxicological and related data and estimating no-observed-adverse-effect-levels of pesticides and Acceptable Daily Intakes (ADI) of their residues in food for humans. In addition, as data and circumstances dictate, the Group estimates acute reference doses (ARfDs) and characterizes other toxicological criteria such as nondietary exposures.

The FAO Panel is responsible for reviewing pesticide use patterns, Good Agricultural Practice (GAPs), data on the chemistry and composition of pesticides, environmental fate, metabolism in farm animals and crops, methods of analysis for pesticide residues and processing studies and for estimating maximum residue levels, supervised trials median residue values, and highest residues in food and feed commodities. The toxicity of the active ingredient and its metabolites, evaluated by the WHO Core Assessment Group, is taken into consideration in deciding if residues may or may not give rise to problems of public health. The maximum residue levels are recommended to the Codex CCPR as suitable for consideration as Codex Maximum Residue Limits (Codex MRLs) to be adopted by the Codex Alimentarius Commission (CAC).

To date 42 sessions of JMPR have been conducted, and about 250 compounds with more than 2000 MRLs have been discussed and recommended to the Codex CCPR (FAO/WHO).

The output of JMPR, among others, constitutes the essential basis for the decisions of the CAC when it adopts MRLs for food and agricultural commodities circulating in international trade, its health-based guidance for pesticides (i.e., ADIs and

ARfDs) and when it recommends maximum residue levels, but also benefits to the governments of FAO and WHO member countries and regions.

The maximum residue levels proposed by the JMPR are considered by the CCPR. The CCPR, a subsidiary body of the CAC, is an intergovernmental meeting whose prime objective is to reach agreement between governments on MRLs for pesticides residues in food and feed commodities moving in international trade.

ROTTERDAM CONVENTION AND HIGHLY HAZARDOUS PESTICIDES

The Rotterdam Convention[*] (see Section III, Chapter 14 of this book) is designed to help countries to control the import of particularly hazardous chemicals. In recognition of the fact that the majority of chemicals addressed by the Convention are pesticides, and that FAO has the international mandate to help countries to manage pesticides more effectively, half of the Secretariat of the Rotterdam Convention is hosted by FAO. The second half is hosted by UNEP.

Despite the adoption of the Rotterdam Convention and its ratification by many countries, and despite the more widespread adoption of the Code of Conduct, there is high concern about the continued use of chemicals that are particularly toxic or hazardous, especially in developing countries. Many of these highly hazardous pesticides (HHPs) pesticides continue to be used because they are cheap, and because farmers are familiar with them and their effects on pests. Nevertheless, death, illness, and in some cases suspected intergenerational effects such as birth defects have been reported, especially from poorer countries. The Convention Secretariat has received more than 800 notifications of regulatory actions taken by countries on pesticides due to their health and/or environmental impacts, and these notifications are passed on to Convention Parties in order to support informed decision making on the import, registration and use of these chemicals.[†] Users of pesticides typically are untrained, are often illiterate, and either do not have access to or do not want to use protective equipment. Under such circumstances the advice of FAO has consistently been to avoid using pesticides that are classified as Class 1a or 1b under the WHO classification of pesticides by hazard in developing countries. FAO also advises against the use of WHO Class II pesticides or any products that require the use of specific protective equipment on the part of applicators.

In 2006, the FAO Council, debating FAO's endorsement of the Strategic Approach to International Chemicals Management (SAICM), called upon FAO to help its member countries to stop using HHPs, including the possibility of countries banning their use. The reduction of risks associated with pesticide use has since become the focus of FAO's Pesticides Management Programme. In 2007 and 2008, the JMPM discussed what constitutes an HHP. A list of criteria was drawn up that defined what might be an HHP, but recognition was also given to the fact that different countries

[*] http://www.pic.int

[†] http://www.pic.int/Reports/06-ICRs-Country-Parties.asp

may define different chemicals as HHPs, depending on conditions of use and related factors. FAO now works with pesticide regulators to help them to define HHPs in use in their countries, and develop strategies for their replacement with alternatives.

LINKS WITH OTHER INSTRUMENTS STOCKHOLM CONVENTION: REMOVAL OF OBSOLETE STOCKS

Stockholm Convention

The Stockholm Convention adopted in 2002 (see Chapter 15 of this book), seeks to end the production and use of Persistent Organic Pollutants. Of the 12 chemicals addressed by the Convention in its original version, nine were pesticides. None were still approved for use in agriculture, some had limited approved uses for termite or other specific pest control situations, and DDT was in use in some countries for malaria vector control. In terms of agricultural pesticides management and the role of FAO therefore, the relevance of the Stockholm Convention was fairly limited. The one area where POPS pesticides did feature in the work of FAO was among the ubiquitous stockpiles of obsolete pesticides that existed in countries.

Obsolete Stocks of Pesticides

FAO began working on the issue of obsolete pesticides in 1994. The main problem identified was in countries where locust control operations required the provision of large volumes of pesticides in emergency situations. A major campaign to deal with locust swarms in 1986–1987 provided large volumes of pesticides to countries across the Sahel belt in Africa as well as the countries of the Horn of Africa. North African countries, heavily dependent on intensive agricultural production and especially export crops, also stocked up on pesticides in order to repel locust swarms moving north from the Sahel as the rains shifted. Ultimately, the swarms of 1986–1987 ended their days by being blown out to the Atlantic. One of the consequences was a massive stock of pesticides, much of which consisted of dieldrin—a favorite control option of the time. Dieldrin is one of the original 12 POPS, but was in fact banned for agricultural use, including locust control, by most countries long before the Stockholm Convention came into force.

In an exercise to quantify the scale of the obsolete pesticide stocks in Africa, FAO estimated that there were about 50,000 tons of obsolete pesticides and heavily contaminated soils and other materials dispersed throughout the continent. No country claimed to be free of obsolete pesticides, but equally, no country had the infrastructure or resources to eliminate the problem. Those that tried burying the pesticides found that a few years later the problem had expanded many times by pesticides leaking into the soil and ground water. Those that stored their pesticides in buildings found that drums corroded and the chemicals leaked onto the floors and into the surroundings of the stores. The problem was not only substantial, but continued to grow as long as it was not addressed.

Surveys were carried out in other regions of the world and a similar picture of ubiquitous stockpiles emerged. Projects mounted to quantify the stocks in detail and try to eliminate them in Zambia, Seychelles, Ethiopia, Tanzania, and outside Africa in Yemen, Lebanon, and Colombia found that in reality, obsolete pesticide stocks were typically twice as big as initial estimates. FAO's efforts to eliminate the obsolete pesticides were always accompanied by a parallel program to strengthen national capacity in pesticides management in order to prevent any future buildup of obsolete stocks. The title of the FAO program was therefore "Prevention and Disposal of Obsolete Pesticides."

With the adoption of the Stockholm Convention and the creation of a financing mechanism for its implementation in the Global Environment Facility, the FAO program stimulated the creation of the Africa Stockpiles Programme (ASP). The ASP was designed to eliminate all obsolete pesticides and prevent their recurrence in all 53 African countries. The ASP was launched in 2005 in seven African countries considered to hold the largest stockpiles and representing a subregional distribution.

INTEGRATED PEST MANAGEMENT

The concept of crop production in an artificially controlled environment of chemical nutrition and pest control was anathema to many who saw agriculture as operating in the natural environment and dependent on biological processes. Healthy plants are more resistant to disease and pest attack, and natural ecosystems can generally keep themselves in balance. Fungal and bacterial diseases often result from poor cultural practices. Insect attack at low levels can be tolerated by healthy plants, and natural parasites and predators are generally capable of keeping pest populations at tolerable levels.

IPM is now mainstreamed into FAO policies and guidance and is defined succinctly in the pesticides Code of Conduct as

> *Integrated Pest Management (IPM)* means the careful consideration of all available pest control techniques and subsequent integration of appropriate measures that discourage the development of pest populations and keep pesticides and other interventions to levels that are economically justified and reduce or minimize risks to human health and the environment. IPM emphasizes the growth of a healthy crop with the least possible disruption to agro-ecosystems and encourages natural pest control mechanisms.

However, farmers that had become reliant on pesticides to kill pests were reluctant to stop using chemicals and allow natural processes to take their place, for fear of losing their crops. It was therefore an FAO program based in the Philippines in the 1970s that very practically demonstrated the viability of IPM. Special plots were established as farmer field schools where groups of farmers came weekly throughout the growing season to carry out guided experiments to see for themselves how biological processes functioned. They not only learned which were the good bugs and which the bad, but also learned that small amounts of damage to rice leaves could actually stimulate additional growth, and that insecticides did not kill the bugs that were the greatest problem for their crops, because they had developed resistance to the chemicals, but they did kill the natural enemies of the pests.

Eventually, 10 million farmers in the Philippines, Indonesia, and South East Asia learned that growing rice without the use of insecticides was more economical, more productive, and safer. IPM methodologies were developed for other crops including cotton—the crop using more pesticides than any other. IPM programs were developed in all regions of the world, and a wide range of agroecosystems and crops. A regional approach in the West African Sahelian countries is promoting IPM among farmers of rice, vegetables, fruits, and cash crops such as cotton. Building on methodologies and experience gained primarily in Asia, the Programme has high-level political support and is being adopted as national policy to support the holistic approach of sustainable crop production intensification (SCPI).

OUTLOOKS

Pesticide markets as defined by the OECD-based pesticide industry represented largely by the industry association CropLife International, expanded until the late 1990s, and have remained stable at around US$30 billion annually. Recent expansion to a value of around US$40 billion can be attributed to lower dollar values and higher oil prices that affected prices rather than volumes of pesticides sold. Meanwhile, however, emerging pesticide producers in Asia have steadily and dramatically expanded their markets worldwide so that China is now a bigger producer and exporter than the United States which has consistently led the market. Production in Latin America is also growing fast with Brazil in particular leading on the production and marketing of generic pesticides.

While more pesticides are being used, regulatory systems and institutional infrastructure such as laboratories, import controls, and law enforcement are not developing fast enough to cope with the expansion in pesticides use. The same problems that were identified in the 1970s in terms of health risks, environmental impacts, residues in food, pesticide misuse, lack of farmer knowledge, labels not being read, poorly maintained and calibrated application equipment, misuse of empty pesticide containers, and obsolete pesticide stockpiling continue largely unabated.

The chemicals conventions (Basel, Rotterdam, and Stockholm) address part of the problem associated with pesticide, but have not yet made substantial changes to the global scene. SAICM provides the most comprehensive framework to date to potentially bring about significant changes in the way chemicals are managed, but lack the finances to mobilize the programs and activities necessary to help governments implement the required developments.

FAO continues to focus the agricultural sector and on pest management and pesticides. Pesticides are part of the food production system. They are seen as a tool that requires careful management within the drive to intensify food production to meet the demands of growing populations, changing diets consumer demand for safe food, and adaptation to climate change.

A BROADER PERSPECTIVE ON FUTURE FAO PRIORITIES

In the wake of the spiralling food prices crises of 2008, the world finds itself with over a billion hungry people and a population set to expand by almost 50% in the coming four decades. The highest population growth will occur in countries where food security is most tenuous and where the impacts of climate change, encroaching desertification, water scarcity, and nutrient-depleted soils are often most acute.

Competition for arable land comes from urbanization and the lure of cash crops, be they food for export or industrial crops such as biofuels with uncertain futures. In some countries, highly productive land is being leased by expatriates to produce crops to feed the populations in their less productive homelands; thus, for example, African land is feeding populations of the Gulf States or South East Asia. Changing diets with more consumption of meat and animal products require higher production of feed crops, which are converted to protein with reduced efficiency, and hence more land is needed to feed fewer people.

Access to and sustainable provision of adequate suitable fertilizers and good quality, locally adapted seed is difficult in many countries where the necessary infrastructure is poorly developed. Access to mechanization is beyond the reach of many farmers. Extension services have been eroded to such an extent that in many countries they no longer exist. Farmers turn to the only available source of free advice, the agricultural suppliers, who are often untrained and whose main interest is to sell more inputs.

The livelihoods of small-scale farmers are extremely precarious when essential inputs are unavailable, of poor quality and overpriced; when environmental factors are stacked against successful crop production; and when market prices for produce are typically at their lowest at harvest time. Climate change will soon be creating additional apparently insurmountable pressures with rising ambient temperatures, unpredictable rains, transboundary pests and diseases, and receding watertables.

There exists a need to develop an extensive and comprehensive strategy to help farmers to produce more food sustainably.

This can be achieved with a holistic approach that encompasses ecological, agronomic, institutional, economic, and social factors. The concept proposed by FAO is SCPI which depends on knowledge about good agronomy and proper planning of the production cycles as well as the timely provision of high-quality inputs at the forefront of which must be good quality seeds and varieties that are adapted to maximize yields under prevalent local conditions. Other inputs would typically include fertilizers, land, water, pesticides, tools, and energy, as well as labor. However, it has long been recognized that inputs merely complement the natural processes supporting plant growth. Examples include nutrient cycles that degrade organic matter and allow plants to access key nutrients and maintain a healthy soil structure which promotes water retention and the recharge of groundwater resources; pollination services; and natural predation for pest control. Farmers who use information and knowledge about the supporting biological processes more effectively can help to maximize the efficient use of conventional inputs.

Some examples of agricultural production systems that manage ecosystem services to improve productivity and reduce environmental impact through an integrated ecosystem approach include IPNM, IPM, CA, participatory plant breeding approaches, crop–livestock systems, agroforestry systems, integrated weed management, and pollination management. However, these approaches are knowledge intensive, and to assist small-holder farmers to develop the necessary knowledge and skill sets locally will require a sustained investment for capital formation (physical and human).

With specific relation to pesticides therefore, FAO will be focusing on helping farmers to produce crops in ways that reduce the impact of pest and disease attack, supports the use of ecological approaches to pest prevention and management, and encourages the development of locally based tools for nonchemical pest management, such as biological, mechanical, and environmental controls. Accepting that pesticides will nevertheless remain in use, FAO aims to help countries to reduce risks to health and the environment from pesticides that are being used through regulatory, technical, and institutional mechanisms.

FAO will continue to support the promotion of pesticides control through vigorous implementation of the Code of Conduct and support to countries for its implementation.

REFERENCES

Carson, R. 2002 (1962). *Silent Spring*. New York, NY: Houghton Mifflin.

FAO Conference. 1961. Activities and programs of the organization, H. Agriculture: Plant production and protection. *Report of the Conference of FAO*, 4–24 November 1961, Eleventh Session, Section VIII, Paragraph 161, Rome. Available at: http://www.fao.org/docrep/x5572E/x5572e0a.htm#plant%20production%20and%20protection

19 International Chemicals Management within the Global Environmental Governance Context

*Achim Halpaap**

CONTENTS

INTRODUCTION

International chemicals management has emerged as an important thematic area within the increasingly complex global environmental governance regime. A milestone in this process was the adoption of the Strategic Approach to International Chemicals Management (SAICM) at the first International Conference on Chemicals Management (ICCM) in 2006. The Conference was cohosted by a number of international bodies representing a wide and diverse range of constituencies. Yet, it was the Governing Council of the United Nations Environment Programme (UNEP) that had initiated and led the process to develop SAICM through Decision SS.VII/3 of February 15, 2002, subsequently endorsed by the World Summit on Sustainable Development (WSSD) in 2002† (see also Section IV, Chapter 17 of this book).

The consideration of chemicals management as an "environmental issue" is a relatively new phenomenon. As early as 1921, for example, the International Labour Organization (ILO) adopted a *Convention Concerning the Use of White Lead in*

* The opinions expressed in this chapter are those of the author and do not necessarily represent the views of UNITAR.

† The Conference was cohosted by the Inter-Organization Programme for the Sound Management of Chemicals (IOMC), the Intergovernmental Forum on Chemical Safety (IFCS), the Global Environment Facility (GEF), the United Nations Development Programme (UNDP), and the World Bank.

Painting which can be considered one of the first international agreements concerned with chemicals. Other examples of international chemicals agreements that predate the 1972 Stockholm Conference on the Human Environment—which marked the birth of international environmental governance (IEG) and UNEP—include the International Maritime Dangerous Good Code under the 1960 International Convention for the Safety of Life at Sea, or the 1971 ILO Convention concerning Protection against Hazards of Poisoning Arising from Benzene. Equally important, the work of Food and Agriculture Organization (FAO) and World Health Organization (WHO) on pesticides since the early 1960s predates the consideration of chemicals management within an environmental context.

Notwithstanding the emerging and growing role of UNEP in international chemicals management, intergovernmental bodies that are not directly accountable to environmental constituencies, such as FAO, WHO, ILO, United Nations Industrial Development Organization, and Organization for Economic Cooperation and Development have played—and will continue to play—an important role in international chemical standard setting, implementation, and capacity development.

Given the range and multitude of international organizations engaged in chemicals management and the growing number of international agreements that they service, one could expect a fragmentation of the international chemicals regime. Yet, significant steps have been taken with the goal to avoid such fragmentation. This chapter summarizes these developments and suggests that valuable lessons can be learned from international chemicals management in further developing and reforming of the broader IEG regime. Vice versa, the international chemicals regime can learn from—and needs to become fully compatible with—the overall IEG regime.

This chapter proceeds as follows. Following this introduction, the second part introduces the IEG system, including some of its achievements and cited shortcomings. Next, a snapshot of cooperation and coordination within the UN systems in the area of environmental management is presented. The fourth part examines to what extent the international chemicals management is characterized by some of the cited shortcomings of the IEG system. This chapter concludes by outlining some lessons learned and providing a short outlook concerning the broader IEG reform process currently under way.

THE IEG REGIME

The IEG regime has significantly developed and expanded over the past two decades. Progress has been made in particular in strengthening international environmental law in the areas of climate change, biodiversity, desertification, and chemicals and waste management. Similarly, procedural environmental rights have been promulgated at the international level through *Principle* 10 of the Rio Declaration, adopted at the Rio Summit on Environment and Development in 1992, and the 1998 United Nations Economic Commission for Europe Convention on Access to Information, Public Participation in Decision-making and Access to Justice in Environmental Matters (Aarhus Convention). Despite these achievements, IEG has become subject to growing scrutiny and criticism. What are some of the main concerns raised?

First, critics point to a proliferation of separate international legal and soft law instruments concerned with the environment or certain aspects of it.* In a comprehensive study, Knigge et al. (2005) refer to about 500 multilateral environmental agreements (MEAs) registered with the UN, covering a wide range of thematic areas, such as climate change, biodiversity, or chemicals management. Critics point out that this "treaty congestion" is not only costly, but also promotes a single issue approach that prevents identification and examination of cross-cutting issues and potential synergies. Indeed, each of the various thematic environmental areas is usually governed by one, if not several international agreements that are initiated, developed, and guarded by separate policy communities. These communities usually originate from specialized institutions in Member States, with relatively few interactions taking place across thematic areas, both at the national level and at the international level. Even within a particular thematic area, such as biodiversity, coherency may not automatically be ensured since relevant international agreements are not necessarily intertwined with each other, for example, through joint meetings.

Second, and closely related to the first concern, coordination among the various multilateral agreements and their secretariats can be considered to be sporadic at best. Cited reasons include, for example, the legal independence of environmental treaties, location of convention secretariats in different organizations and across diverse geographic locations and the absence of institutionalized coordinating structures. The secretariat for the Biodiversity Convention is, for example, located in Montreal, the secretariats for the Desertification Convention and the Climate Change Convention are located in Bonn, while Geneva has become the principal location of secretariats in the area of chemicals and waste-related conventions. While this, in itself, may not be a problem, it creates challenges for Member States who need to follow the work of various secretariats through geographically dispersed permanent missions that need to build up technical and policy expertise. This is a particular challenge for developing countries that usually have limited human and financial resources to maintain representational offices.

Third, international fragmentation is often mirrored at the implementation level in Member States. National focal points for international environmental agreements are usually scattered across various ministries concerned with environmental issues, such as ministries of environment, agriculture, and health, to name only a few. While some countries have developed national environment councils and environment strategies, integrated planning that addresses environmental management as a cross-sectoral theme is only slowly gaining currency. Recognizing this problem, the UN Joint Inspection Unit (UN JIU) has recommended the development of guidelines by the UN to support the establishment of country-owned national coordinating platforms that would support the integrated implementation of international environmental agreements (UN JIU, 2008).

* A recent UNEP report (2010) outlines at least nine thematic areas which are addressed through international environmental agreements including: (1) air pollution and air quality; (2) biodiversity; (3) chemicals and waste; (4) climate change; (5) energy; (6) forests; (7) freshwater; (8) oceans and seas; and (9) soil, land use, land degradation, and desertification, with environmental governance serving as a tenth and cross-cutting area.

The above discussion suggests that environment cannot be considered as a single policy area, as implied by referring to "environment" as one of three pillars of sustainable development, or using the term "international environmental governance." Rather, the environmental governance domain is made up of diverse policy communities and institutions that often work in isolation, rather than in collaboration. Thus, when examining the coherency of the IEG regime, one may ask three questions. First, does information exchange, coordination, and synergy development take place within a particular thematic area (e.g., within chemicals and waste management)? Second, are linkages and synergies explored across thematic areas, for example, climate change and biodiversity? And, third, how are linkages between specific thematic areas—or the environment as a whole—and other development sectors (e.g., agriculture and industry) explored and implemented?

ENVIRONMENTAL GOVERNANCE COORDINATION AND COOPERATION IN THE UN SYSTEM

When established at the 1972 Stockholm Conference on the Human Environment, UNEP was equipped with a system-wide governance framework, backed by the Environment Coordination Board and a common planning instrument—the System-wide Medium-Term Environment Programme. Yet, the 1992 Rio UN Conference on Environment and Sustainable Development (UNCED) created dynamics which emphasized the interface of development and environment. This new paradigm created a proliferation of international institutions and organizations engaged in environmental matters. New bodies were established and existing organizations concerned with development took on board environmental considerations. New institutions established at, or following, the UNCED Conference include, for example, the Commission on Sustainable Development, new MEAs, such as the UN Framework Convention on Climate Change (UNFCCC), and the GEF. In general, separate secretariats were established for these bodies in different regions of the world. These developments undermined the original mandate of UNEP as being the sole lead and coordinating body for environmental matters in the UN System.

In an attempt to fill the emerging coordination gap in environmental matters within the UN system, the UN General Assembly established the Environment Management Group (EMG) in 1999. EMG was established with the "purpose of enhancing inter-agency coordination in the field of environment and human settlements."* Members of EMG comprise specialized agencies, programs, and organs of the UN system, including the secretariats of MEAs, as well as the Breton Woods institutions and the World Trade Organization. EMG is a flexible mechanism that is tasked to identify emerging issues and brings together relevant knowledge available in the UN system. By taking an issue-based approach, EMG seeks to contribute to synergy development, ensure complementarity, and add value to the existing UN interagency cooperation. Given its focus on specific issues and tasks EMG has, however, not emerged as a mechanism that fosters coordination in a systematic manner, as some expected it to become.

* Resolution A/RES/53/242 of the UN General Assembly: Report of the Secretary-General on Environment and Human Settlements (1999).

Independently from the EMG, several coordinating mechanisms, both formal and informal, have been established around thematic areas of environmental management. In the area of biodiversity, for example, the Biodiversity Liaison Group (BLG) was established following a request by the Conference of the Parties of the Convention on Biodiversity. In the area of chemicals management, the IOMC is a formal coordinating mechanism that is endorsed by the heads of participating organization through the signing of an Memorandum of Understanding (MOA).

The overall coordination in the UN in matters of sustainable development is undertaken through the United Nations System Chief Executives Board for Coordination (CEB). CEB is responsible for coordinating the UN system's follow up to the WSSD and seeks to identify linkages with other relevant UN conferences, for example, in the area of water and energy. To address this challenge, CEB established interagency collaborative arrangements that deal with water and sanitation (UN-Water), energy (UN-Energy), oceans and coastal areas (UN-Oceans), and consumption and production. CEB members comprise the Executive Heads of the Specialized Agencies, Funds, Programmes, International Monetary Fund, and the World Bank. The secretariat functions for these mechanisms are provided by the UN Department for Economic and Social Affairs.

The United Nations Development Group (UNDG) is linked to the CEB and is one of four Executive Committees established by the Secretary General in 1997 to improve the effectiveness of UN and coordinate UN work in Member States. It also provides guidance and support to the Resident Coordinator system and UN country teams (UNCTs). Members of the UNDG include Funds and Programmes of the UN, the UN Secretariat Departments as well as, on a voluntary basis, specialized agencies. Its membership has grown to 28, plus five observers.* An important aspect of UNDG's work is the implementation of the Millennium Development Goals (MDGs) adopted in 2000 by the General Assembly which include an MDG 7 on environmental sustainability and several other MDGs closely related to environment. While environment is not the focus of UNDG, the group increasingly engages in environmental matters through its Task Team on Environment. The team has recently issued, or is working on, a number of guidelines concerning environmental management, climate change, and the green economy.

At the national level, UNCTs facilitate coordinated development support of the UN system. The teams are composed of the heads of UN agencies and other institutions represented in the respective country. County teams are responsible for monitoring the implementation of the UN Development Assistance Framework (UNDAF)—national planning frameworks that outline the common objectives and strategies shared by UN organizations providing development assistance. UNCTs also monitor the linkages between UNDAF and the formulation of programs and projects by each of the resident UN agencies, as well as the development of any joint initiatives. Recently, "Delivering as One" pilot projects were initiated in eight countries.† These pilot projects explore how the UN family—with its many and diverse

* For further details, see http://www.undg.org

† Pilot countries include Albania, Cape Verde, Mozambique, Pakistan, Rwanda, Tanzania, Uruguay, and Vietnam.

agencies—can deliver in a more coordinated way at the country level. The overall goal of the "Delivering as One" initiative is to ensure faster and more effective development operations and accelerate progress to achieve the MDGs.

A review of UN coordinating structures in the area of environment and sustainable development reveals different origins, mandates, and reporting structures. Some coordinating bodies, such as the EMG or the BLG, were established by Member States. In other cases, mechanisms were created through initiatives from within the UN system (e.g., in the case of the subsidiary bodies of CEB) or the IOMC. The analysis also shows that in light of the complexity and decentralization of "environmental" management activities within the UN system, an overarching and comprehensive coordinating and planning structure has not emerged. This is perhaps not surprising. A centralized top-down structure would be complex and costly and potentially infer with thematic arrangements which operate through decentralized networks that work on specific issues.

COHERENCY OF THE INTERNATIONAL CHEMICALS REGIME

Given that chemicals management cuts across various sectors, such as agriculture, transport, labor, and so on, a significant number of international institutions are engaged in international chemicals management. How do these bodies coordinate and to what extent is the international chemicals management regime characterized by coherency?

A milestone which set the path to a more cross-cutting approach to international chemicals management was the adoption of Chapter 19 of *Agenda 21* in 1992 at the Rio Summit (see Chapter 3 of this book). The chapter is titled *Environmentally Sound Management of Toxic Chemicals, Including Prevention of Illegal International Traffic in Toxic & Dangerous Products.* An innovative feature of Chapter 19 was its promotion of a life-cycle approach to the management to chemicals, covering different stages of the chemical life cycle from production, to transport and use, to disposal. This paradigm complemented the traditional approach of dealing with chemicals through sectoral approaches, for example, in the areas of agriculture, transport, water management, and so on.

The life-cycle approach of Chapter 19 of *Agenda 21* set the stage for the establishment of new, coherency-enhancing international institutions and coordinating structures in the 1990s. Chapter 19 called, for example, for strengthening the International Programme on Chemicals Safety—a collaborative Programme among WHO, UNEP, and ILO (see Chapter 33 of this book)—as a nucleus to foster institutional coordination at the international level. Building on the collaboration within in the IPCS, the IOMC (Chapter 23 of this book) was established in 1995 with the goal to coordinate the efforts of international organizations in the area of chemicals management. As discussed in other chapters, IOMC members consult on the planning, programming, implementation, and monitoring of activities undertaken jointly or individually, and help ensure that activities are mutually supportive, complementary, and avoid duplication of efforts.

Chapter 19 of *Agenda 21* also set the stage for the establishment of an intergovernmental body that promoted a cross-cutting approach to chemicals management—the Intergovernmental Forum on Chemical Safety in 1994. IFCS (see Chapter 21 of this book), drew up recommendations for governments and for international organizations concerning the implementation of Chapter 19 of *Agenda 21* in an integrated way. A milestone achieved by the IFCS was the adoption of the Bahia Declaration and Priorities for Action in 2000 which subsequently provided the foundation of SAICM. IFCS also developed the "famous" 2020 target to achieve the sound management of chemicals, subsequently taken up and endorsed by WSSD in 2002 (see Section II, Chapter 4 of this book) and through SAICM in 2006 (see Section IV, Chapter 17 of this book).

Notwithstanding the important role that IFCS fulfilled in the 1990s, governments realized that a more formal policy and institutional approach to international chemicals management was warranted. This led to the initiation of SAICM by UNEP in 2002 and the first International Conference on Chemical Management in Dubai in 2006 (ICCM-1, see Section II, Chapter 5 of this book), which was cohosted by UNEP, the IOMC, IFCS, UNDP, and the World Bank. ICCM-1 adopted SAICM as an international policy framework to achieve the sound management of chemicals throughout their life cycle so that, by 2020, chemicals are produced and used in ways that minimize significant adverse impacts on human health and the environment. This "2020 goal" had been adopted by the WSSD in 2002 as part of the Johannesburg Plan of Implementation. Yet, no guidance and specific targets were adopted by WSSD to support implementation and monitoring of progress at all levels in achieving the 2020 goal.

SAICM filled this void by developing a more detailed plan and targets, covering specific areas and setting precise interim milestones in particular through the Global Plan of Action (GPA).* Building upon the five main themes addressed by SAICM (i.e., risk reduction, knowledge and information, governance, capacity-building and technical cooperation, and illegal international traffic), the GPA elaborates for 39 specific work areas proposed activities, actors, targets, time lines, and indicators. The GPA includes, for example, a time target that by 2015 all countries should have established a national Pollutant Releases and Transfer Register.

Activities in the GPA are implemented, as appropriate, by governments and stakeholders, according to their applicability. However, relatively little funding is available through SAICM to support developing countries in implementation which can be considered as one of its weaknesses. However, the point made here is that SAICM established at least clear targets and indicators that allow monitoring of measures taken by countries and stakeholders and, through this, a review of progress made against internationally set targets.

An important dimension of SAICM is that it fosters coherency at the national level through a policy recommendation (and some related funding) to support the development of Integrated National Chemicals Management Programmes. These programs bring together concerned government sectors as well as stakeholders at the national level with the goal to translate the 2020 goal into national programs of action. Important aspects of this "strategic approach" to chemicals management at the national level include the preparation of a National Chemicals Management

* Other important SAICM elements comprise the Dubai Declaration on International Chemicals Management, expressing high-level political commitment to SAICM, and an Overarching Policy Strategy which sets out its scope, needs, objectives, financial considerations underlying principles and approaches and implementation and review arrangements. See Chapter 17 of this book.

Profile, development of a national SAICM capacity assessment, and identification of priorities of action which are taken up as part of an integrated and coordinated national program that pursues implementation of the 2020 goal.

SAICM did not only create driving forces for fostering coherency at the national level. It also provided an impetus for international organizations, via the IOMC, to align their work programs in a way that ensures complementarity and avoids duplication. For some specific work areas, for example, the development of guidance for national SAICM implementation or policy choices in chemicals management, IOMC is working jointly and develops publications endorsed by all participating organizations. This type of cooperation can be considered exemplary within the UN System.

Another important development, catalyzed by UNEP, is the development of synergies among the three key UNEP conventions dealing with chemicals and waste, that is, the Rotterdam Convention, the Stockholm Convention, and the Basel Convention. Agreed synergy enhancing measures include, for example, the colocation of the secretariats in Geneva (Rotterdam also has part of its secretariat with FAO in Rome), the appointment of a common Head of Secretariat, and the implementation of joint outreach and capacity development activities, such as a series of regional "synergy" workshops. These efforts culminated in, and generated further momentum, at the first *Simultaneous Extraordinary Meetings of the Conferences of the Parties to the Basel, Rotterdam and Stockholm Conventions in Bali* in early 2010 (Chapter 8 of this book).

Suggesting that these measures were a success does not imply that the international community is on track in achieving the commonly agreed 2020 target. Just as for other international policy domains concerned with protecting and advancing public interest issues, such as ensuring universal health protection or sound social standards, full protection from chemicals, in particular for vulnerable populations, has not yet been accomplished. In this context, it is important to note some remaining gaps in the international institutional regime to manage chemicals. The secretariats of the various chemicals-related conventions, for example, are not member organizations of the IOMC and discussions how to engage these entities more effectively have only recently started. Similarly, the negotiations of a new international agreement on mercury may result in another separate convention with an independent secretariat. Other conventions dealing with specific heavy metals may follow.

Similarly, there are unreaped opportunities to create linkages between international chemicals management and the broader IEG regime and its conventions. Chemicals affect, for example, biodiversity and contribute to climate change. However, structured discussions in the UN System on how to explore linkages have not taken place to date. Pollutant Release and Transfer Registers provide, for example, an opportunity to measure and report chemical and greenhouse gas emissions at the national level through one single inventory system. However, these types of synergy enhancing environmental management tools are not easily identified through fragmented negotiation processes.

LESSONS LEARNED FOR IEG REFORM

The steps taken within the international chemicals regime to foster coherency provide valuable lessons learned as the international policy community seeks to strengthen the broader IEG regime. The IEG reform process dates back, *inter alia*, to the decision of the UNEP Governing Council on IEG in 2002* the 2006 Report of the Secretary-General's High-Level Panel on United Nations System-wide Coherence in the Areas of Development, Humanitarian Assistance and the Environment,† and the Management Review of Environmental Governance within the United Nations System, prepared by the JIU (UN JIU, 2008). All of these initiatives identified the need to strengthen the IEG system. IEG reform also received major political attention through the 2005 World Summit Outcomes.‡ The General Assembly recognized, *inter alia*, the need for more efficient environmental activities in the UN System, and better integration of environmental activities in the broader sustainable development framework at the operational level.

Despite these initiatives, relatively little progress has been made to date. In 2007 and 2008, reform discussions were dealt with by the General Assembly, but in light of a lack of progress made, UNEP was mandated to take up the issue, culminating with the establishment of a Consultative Group of Ministers or High-level Representatives on Broader IEG Reform in 2010. The group completed its work by the end of 2010, and its outcomes were submitted by UNEP as a contribution to the preparatory committee of the 2012 Rio plus 20 Conference on Sustainable Development.§

One of the stumbling blocks in the IEG reform discussion is the possible establishment of a United Nations Environment Organization (UNEG) to address environmental matters with more resources and political weight. While this chapter is not the place to review the merits of such an organization in a comprehensive way, it raises the question how a new UN environment organization would integrate, or link up with existing structures concerned with thematic areas, for example, in the area of chemicals management. Would a UNEG take over normative functions from the various policy bodies engaged in chemical policy-making? Would it integrate chemicals management capacity development activities into one single international program? If so, how can the expected resistance be overcome from bodies that currently deal with chemicals management? Questions such as these are not only relevant for the area of chemicals management, but would need to be addressed for all other thematic environmental areas within an overall IEG reform process.

The case of reforming the international chemicals regime suggests that coherency of the broader IEG regime is unlikely to be achieved, unless coherency is advanced at the level of thematic environmental areas through a step-by-step and integrative process. The establishment of the IOMC and the adoption of an the overarching SAICM policy framework in 2006, for example, were all measures that fostered coherency through collaborative processes that involved all concerned parties, emphasized incremental rather than paradigmatic reform, and built on, rather than replaced existing institutions concerned with chemicals management.

* See UNEP Governing Council Decision SS.VII, February 15, 2002.

† See UN General Assembly A/61/583, November 20, 2006.

‡ See UN General Assembly A/60/L.1 A/60/L.1, September 15, 2005.

§ See UNEP UNEP/GCSS.XI/11, March 3, 2010.

Equally important, goal-oriented benchmarks, such as those established through the SAICM, need to be developed for all thematic environmental areas. This would not only allow better monitoring of progress made, but also create driving forces to foster institutionalized coordination and goal-oriented action both at the national and international level. In some areas other than chemicals management, progress is made to move in this direction, for example, through the 2010 Biodiversity Targets and the 2010–2020 Strategic Plan under the Biodiversity Convention. However, such goal-oriented frameworks as well as supporting coordinating structures in the UN are absent in a number of thematic areas. As a possible first step to take a more strategic approach, UNEP issued a comprehensive report in 2010 which outlines international agreed goals and objectives for key environmental areas (UNEP, 2010). This initiative could set the stage for identifying gaps and setting goal-oriented priorities at the thematic level. The main point made here is that unless progress can be made for specific environmental areas, the IEG regimes as a whole is likely to remain ineffective.

It remains to be seen how the above questions and challenges will be addressed through a possible new UNEG or other reform measures. Yet, governments seem committed to address the issues. The inclusion of the topic of sustainable development governance by the General Assembly as one of the main topics to be discussed at the Rio plus 20 Conference on Sustainable Development in Brazil in 2012 is an indication of the commitment and could provide a milestone in strengthening the IEG regime. Hopefully, negotiators engaged in the Rio plus 20 process have sufficient time to address these issues and take into account existing experiences, such as lessons learned from international chemicals management, in tackling the broader sustainable development governance reform challenge. Equally important, decision makers engaged with chemicals management need to engage fully in the Rio plus 20 process to ensure that international chemicals management is fully embedded within a future international governance regime that seeks to advance environmental sustainability.

REFERENCES

Knigge, M., J. Herweg, and D. Huberman. 2005. *Geographical Aspects of International Environmental Governance: Illustrating Decentralisation*. Berlin: Ecologic Institute for International and European Environmental Policy.

United Nations Joint Inspection Unit (UN JIU). 2008. *Management Review of Environmental Governance within the United Nations System*. Geneva. JIU/REP/2008/3.

United Nations Environment Programme (UNEP). 2010. *Existing International Agreed Environmental Goals and Objectives*. UNEP/GEG/1/2, February 12, 2010.

20 Managing Chemical Risks
The Role of the Chemical Industry

Michael Walls

CONTENTS

To say that accidents are due to human failing or human error is not so much untrue as unhelpful. It does not lead to effective action, only to advice to take more care. Instead of asking what is the cause of an accident we should ask what we can do to prevent it happening again. We may then think of ways of improving the training, supervision, design, and so on.

– Trevor Kletz
Critical Aspects of Safety and Loss Prevention, 1990.
Cited in *Common Sense, Scientific Method, and Accident Prevention*,
by Professor Cyril Domb, 1995*

INTRODUCTION: AGE-OLD CONCERNS AND NEW-AGE APPROACHES

In 2002, the International Council of Chemical Associations (ICCA) and its members joined with other stakeholders at the Johannesburg World Summit on Sustainable Development (WSSD) to establish a goal that by the year 2020, chemicals are used and produced in ways that lead to the minimization of significant adverse effects on human health and the environment. ICCA's decision reflected industry's long-term commitment to risk reduction and the safe management of chemicals.

It was the companies, engineers, and scientists themselves who pioneered the safe production, transportation, and use of chemicals. Accidents were and still are expensive and no company wanted to see its personnel hurt, facilities damaged, or products lost. Their record was imperfect, but their work in the fields of process safety and the safe production of chemicals set the foundation for future regulatory and voluntary actions to reduce risks associated with chemicals in the modern world.

PROCESS SAFETY AND CHEMICALS MANAGEMENT

Chemical companies seek to reduce the risk of accidents. Today, we recognize that "inherent safety" is a modern term for an age-old concept: to eliminate hazards rather than accept and manage them.† This concept goes back to prehistoric times—for example, building villages near a river on high ground, rather than managing flood risk with dikes and walls.

In chemistry, these efforts were evident throughout the rise of industrial society—for example, in 1866, following a series of explosions involving the shipment of nitroglycerine, and a state ban on its transport, the Central Pacific Railroad approved the idea of making the explosive onsite where it was building a line through the mountains of California. "This is an early example of an inherently safer design principle—*minimize* the transport of a hazardous material by *in situ* manufacture at the point of use" (*Inherently Safer Chemical Processes*, 2009, p. 5). Of course, the safety of nitroglycerine was taken to a new level one year later when Alfred Nobel invented dynamite, which reduced the risk of storing, shipping, and using this explosive (see *Inherently Safer Chemical Processes*, 2009, pp. 5–6).

---

* http://www.darchenoam.org/ethics/PREVENT/domb1.htm

† Cited on page 1, Dr. Trevor Kletz, an ICI Petrochemicals safety specialist, is widely regarded as the "father of inherently safer design."

Especially in the years since World War II, the chemical industry, its scientists, and engineers have disciplined and quantified process safety with the understanding of chemicals management throughout the value chain. At the same time, governments, intergovernmental organizations (IGOs) and nongovernmental organizations (NGOs) have broadened their efforts to help reduce risk, and protect communities, workers, and customers. At times these efforts and those of the industry seem to have operated at cross-purposes, fueled by misunderstanding and mistrust. Fortunately, the relationship between the industry and its stakeholders has evolved and continues to improve, as confrontation gives way to cooperation.

Today, industry strongly believes that the best approach to chemicals management is through a combination of scientific, risk-based regulations and strong, industry-driven initiatives.

This chapter will explain the voyage upon which industry is embarked, what it has accomplished, and what must still be done to arrive at the WSSD-2020 harbor.

But First: A Word About Society's Need for Chemicals

Chemicals management is an issue because society needs chemicals to achieve virtually all of its economic, health, agricultural, technological, and sustainability objectives.

For example, during his inaugural speech in January 2009, U.S. President Barack Obama declared: "The state of the economy calls for action, bold and swift, and we will act—not only to create new jobs, but to lay a new foundation for growth. We will build the roads and bridges, the electric grids and digital lines that feed our commerce and bind us together .... We will harness the sun and the winds and the soil to fuel our cars and run our factories" (Obama, 2009).

The challenges of the twenty-first century—including the quest for cleaner energy, greater efficiency, and a smaller environmental footprint—continue to *drive demand for chemical products*. From lighter automotive components, compact fluorescent light bulbs, building insulation, solar cells panels, wind generation turbines, and more, chemistry facilitates the green revolution sought by consumers and policy makers.

The role of chemistry was confirmed in a recent life-cycle analysis of the chemical industry's impact on greenhouse gas emissions which found that for every unit of greenhouse gas emitted by the chemical industry, the industry enabled 2–3 units of emission savings via its product and technologies. Some dramatic examples of the greenhouse gas (GHG) emissions savings enabled by chemistry include (ratio of emissions savings to emissions)*

- Building insulation foam: 233:1
- Foam coating in district heating: 231:1
- Synthetic diesel additives—fuel efficiency improvements: 111:1
- Glass and carbon fiber for wind turbines: 123:1

* The ICCA-commissioned study, "Innovations for Greenhouse Gas Reductions: A Life Cycle Quantification of Carbon Abatement Solutions Enabled by the Chemical Industry," was performed by McKinsey and Company in 2009 and reviewed by the Öko Institute. (Copies of the study are available at: http://www.icca-chem.org)

- Compact fluorescent lighting: 20:1
- Marine fuel reduction owing to the use of antifouling coating: 20:1
- Low-temperature detergents: 9:1

The bottom line is that the roads that lead to the sustainable future envisioned by President Obama and so many others across the world are paved with the products of chemistry. And, for industry to continue to develop innovative products that meet world needs, it recognizes—as do its critics—the importance of effective chemicals management and product stewardship.

THE JOURNEY SO FAR

Before WSSD: Stockholm, Rio, Responsible Care®, and ICCA

Long before the Johannesburg summit, the global chemical industry was developing the process safety knowledge and discipline required to promote safety and reduce risk. In 1962, Rachel Carson gave voice to a new fear in her classic opus, *Silent Spring*: "For the first time in the history of the world, every human being is now subjected to contact with dangerous chemicals, from the moment of conception until death" (p. 15). The global community, including the chemical industry, slowly awakened to a new concern.

Beginning in the 1960s, and certainly, by the mid-1980s, chemical companies knew they were being judged by stakeholders on a wide range of factors such as employee health, safety, and overall environmental stewardship. The business of chemistry now included awareness of, and accountability for, the effects of chemical products, processes, and commerce as governments worldwide launched new efforts to regulate chemicals.

Stockholm and Rio: International Environmental Awareness Grows

In 1972, the United Nations (UN) convened its first major conference on international environmental issues in Stockholm. The Stockholm Conference produced a Declaration, Action Plan and a Resolution to preserve and enhance the environment. The Stockholm Conference also led to the creation of the United Nations Environmental Programme (UNEP).

Twenty years later, in 1992, the UN hosted the "Earth Summit" in Rio de Janeiro. See also Section II, Chapter 3. The Rio Declaration contained principles which would help shape public policy affecting the manufacture and use of chemicals, especially by drawing a connection between sustainable development and the environment.

The Rio summit also adopted *Agenda 21* which addressed the environmentally sound management of chemicals (among other issues). Specific areas included assessing and reducing chemical risks and sharing risk-related information, the harmonization of classification and labeling of chemicals and improving chemicals management at the national level.

Responsible Care: Industry Expands Its Role in Chemicals Management

In 1985, armed with the growing understanding of process, facility, and chemical handling safety, and aware of worldwide public concerns about the use of certain chemicals, the industry launched the Responsible Care code of management practices. More than just a set of practices, Responsible Care reflects industry's understanding of its ethical responsibility to continuously improve health, safety, and environmental (HSE) performance. Buoyed by this ethic, chemistry became the first industry to embark on global performance reporting, and since its inception in Canada, Responsible Care has grown to include the chemical industries of 53 economies. In recent years, Responsible Care has begun expanding to emerging chemical economies including Eastern Europe, China, the Middle East, and Africa.

The principles of Responsible Care are the foundation for ICCA's efforts to lead the global chemical industry to a more sustainable future. Today, no national association may join ICCA unless it agrees to verifiably implement Responsible Care.

How does Responsible Care Translate into Practical Measures?

Every chemical association that subscribes to Responsible Care must complete an annual questionnaire that defines progress made in implementing the initiative. Progress within the 53 associations is measured using a scale of three status levels: "no progress to date," "developing plan," and "fully implemented," including the use of third-party monitoring.

Every Responsible Care programme has eight fundamental features:

- A formal commitment by each company to adhere to a set of guiding principles—usually signed by the CEO
- A series of codes, guidance, and checklists to help companies fulfill their obligations
- The development of indices against which performance can be measured
- Open communication about HSE performance
- Exchange of best practice experience among companies
- Expansion of Responsible Care to all companies within the industry
- A title and logo which clearly identify national programs as consistent with and part of the Responsible Care initiative (Figure 20.1)
- Procedures for verifying that members have implemented the measurable or practical elements of Responsible Care

FIGURE 20.1 Responsible Care logo.

In addition, the Global Product Strategy (GPS) addresses the product stewardship components of Responsible Care (see the section "The Global Product Strategy and Product Stewardship"). To fully understand the industry's performance, Responsible Care evaluates environmental, health, and safety (EHS) performance based on common definitions. For the most recent report (the 2008 report, published in 2009 and covering the years 2000–2007),* national associations reported on

- *Safety*: number of fatalities; lost time injury rate
- *Environment*: air emissions of sulfur dioxide, and nitrogen oxides; discharges to water; chemical oxygen demand; GHG emissions (direct and indirect CO_2 emissions); GHG intensity (emissions per tons of production)
- *Resources*: energy consumption (expressed as fuel oil equivalents); energy intensity (consumption per tons of production); and water consumption
- *Distribution*: number of transport incidents

The findings tell a powerful story of progress on all fronts, though there is still more to be done. For example, from 2000 to 2007, worldwide chemical production nearly doubled from 388 million metric tons to almost 760 million metric tons. Global employment rose from nearly 3.8 million to nearly 4.8 million, and Responsible Care employees rose from 1.7 million to 2.07 million, an increase of 21%. Yet, during this time frame, Responsible Care associations reported, for example,

- Worker fatalities fell by one-third
- Lost workdays fell by nearly 49%
- Emissions of nitrogen oxides and sulfur dioxides fell 25% and 40%, respectively, between 2003 and 2007
- Carbon dioxide emissions rose to a mere 5.3% despite the doubling in production, as CO_2 intensity fell by nearly 46%
- Energy intensity fell by 19%
- Water consumption fell by 11%

While not solely responsible for these improvements, Responsible Care is a driving force that requires associations and companies to track and report their EHS performance.

Responsible Care companies also work with their customers and suppliers to extend safety and stewardship throughout the chemistry value chain.

ICCA: The Global Chemical Industry's Platform for Progress

Few industries are as truly global as the chemical industry in production, sourcing, marketing, and sales. There has always been a strong connection between global

* http://www.responsiblecare.org/filebank/Status%20Report%2001_05.pdf

industrial development, commerce, and the chemical industry. Production of synthetic dyes for textiles was crucial during the early industrial revolution, and from this "colorful" start, the modern chemical and pharmaceutical industries were born. Innovation further drove chemical production, in an increasingly urban society. Since World War II, the growth of plastics, materials sciences, modern agriculture, and communications and information technology gave additional impetus to the development of global chemistry.

Amidst the growing awareness that industry needed a global voice, ICCA was created in 1989, four years after the introduction of Responsible Care, to coordinate the work of chemical companies and associations on issues and programs of international interest.

Comprised of national and regional trade associations representing companies involved in all aspects of the chemical industry, ICCA promotes the exchange of information, the use of best practices, and capacity building activity for newer producers and emerging economies. In addition, ICCA seeks to develop common positions on policy issues of international significance, especially chemicals policy and management, climate change and energy, and Responsible Care.

ICCA also serves as the main channel of communication between the industry and various international entities, such as IGOs and NGOs concerned with these global issues. For example, ICCA has developed strong relationships with the United Nations Environment Programme (UNEP), the United Nations Institute for Training and Research (UNITAR), and the Organisation for Economic Cooperation and Development (OECD).

ICCA and United Nations Environment Programme

United Nations Environment Programme (UNEP), the main United Nations environmental body, provides general policy guidance for direction and coordination of environmental programs, and reviews and approves such programs. ICCA, its member associations, and their member companies have joint projects with several UNEP branches including the chemicals branch.

ICCA efforts with UNEP include a wide range of capacity-building activities, several of which support UNEP projects such as the Awareness and Preparedness for Emergencies at Local Level program, as well as Safer Production and Corporate Social Responsibility at the site level which promote product stewardship.

ICCA and United Nations Training and Research Programme

The United Nations Training and Research Programme (UNITAR), jointly with ILO, is responsible for promoting training and research in developing countries and countries with economies in transition on issues of chemicals safety.

In collaboration with ICCA, UNITAR/ILO is developing a broad guidance document on implementation of the Globally Harmonized System (GHS) of Classification and Labelling of Chemicals mainly focused on helping small- and medium-size enterprises (SMEs) and developing countries.

ICCA, the OECD, and OECD's Business and Industry Advisory Committee

The OECD's Environment, Health and Safety programme deals with chemical issues such as classification and labeling of chemicals, mutual acceptance of notification, existing chemicals in nanomaterials, biotechnology, and more (see Chapter 26 of this book).

The chemical industry, via Business and Industry Advisory Committee (BIAC), is closely involved in the nanomaterials issue, currently a high priority on the OECD and Strategic Approach to International Chemical Management (SAICM) agendas. BIAC contributes to various OECD nanomaterials projects. Through BIAC, ICCA members also contribute to various OECD projects related to chemicals such as the test guidelines program, global data portal, and exposure assessment.

In addition, through the High Production Volume Chemicals (HPV) programme, ICCA via BIAC has been working in partnership with the OECD Environment, Health and Safety Division since 1998 (see the discussion in the section "Risk Assessment: Crucial for Chemicals Management and Sustainability").

As a result of these efforts, ICCA has been a participant in every major global forum dealing with crucial industry issues, including the WSSD, the SAICM, the initial International Conference on Chemicals Management (ICCM-1), and in 2009, ICCM-2. ICCA member associations and companies worked hard to establish the organization as the voice of the global industry through the development of long-term relationships with other stakeholders.

How ICCA's Relationships Contribute to Effective Chemicals Management: Chemical Weapons Convention, Persistent Organic Pollutants, and GHS

The industry is able to share its technical expertise with policy makers and others, working closely with stakeholders through SAICM and other global programs. The result is an approach to policy that reflects scientific realities. Three examples are the Chemical Weapons Convention (CWC), the Stockholm Convention on Persistent Organic Pollutants (POPS), and the Globally Harmonized System of Classification and Labelling (GHS).

ICCA and its members were active partners and leaders in transforming the CWC from concept to reality. ICCA supported the goal of eradicating the illegal uses of chemistry as a weapon, and helped promote government participation in the agreement. A total of 188 governments have joined the CWC, which bans the production, use, and storage of chemical weapons. The CWC is the first arms control treaty to directly engage the private sector in support of the Convention objectives, through the establishment of a clear reporting and inspection process for industrial facilities and opportunities for onsite inspections by an international inspection body. ICCA members not only cooperated in negotiating the text of the CWC, but also helped "road test" the treaty's verification provisions to ensure a workable solution that recognized the legitimate role of commercial chemical production. The success of this collaborative effort can be measured based on the fact that even after thousands of onsite inspections all over the world, no privately owned facility has ever been accused of violating the CWC strictures.

ICCA members similarly joined with governments and other stakeholders in negotiating the Stockholm Convention on POPS (see Chapter 15 of this book). The POPS convention reflects a risk-based approach to managing certain chemicals. By working closely with governments, ICCA's World Chlorine Council (WCC) helped assure that high-priority products receive greater attention and that candidate chemicals meet Stockholm treaty criteria. WCC provided financial and technical support for a UNEP program that helped 12 countries develop their national implementation plans for chemicals management under the POPS convention.

A third example is the GHS of Classification and Labelling of Chemicals. The GHS helps achieve the WSSD 2020 goal by providing a platform for harmonizing rules and regulations on chemicals at national, regional and worldwide levels (see Section III, Chapter 11 of this book). It reinforces SAICM creating a common regulatory language and specifying the information to be included on product labels and safety data sheets, which are crucial for effective chemicals management. This in turn facilitates safe international trade in chemicals. ICCA strongly supports GHS.

Moreover, ICCA's Responsible Care network has turned support for GHS into concrete action. New Zealand was the first nation to fully implement GHS and there, the industry worked closely with government officials to develop implementing legislation. South Africa's association has developed GHS guidance for SMEs and has helped drive GHS implementation in that nation.

INDUSTRY EMBRACES SAICM AND THE WSSD OBJECTIVES

The 2002 Johannesburg Summit led to the creation of the UN Strategic Approach to International Chemicals Management (SAICM) at the 2006 ICCM-1 in Dubai. SAICM's goal is to promote chemical safety around the world with the overall objective of achieving the WSSD goal. Industry, through the ICCA and its member associations, and at the company level, is working hard to achieve this goal. In fact, SAICM's core policy objectives—relating to *risk reduction, knowledge and information, governance, capacity-building and curtailing illegal international traffic*—have directly shaped the industry's efforts. ICCA introduced two voluntary initiatives in Dubai, *The Responsible Care Global Charter* and *The Global Product Strategy*, to help achieve the objectives of SAICM and to improve the sound management of chemicals. These programs align closely with SAICM's Overarching Policy Strategy and have seen substantial progress since 2006, as we note below.

THE RESPONSIBLE CARE GLOBAL CHARTER

The Responsible Care Global Charter addresses sustainable development, public health issues relating to the use of chemical products, and the need for greater industry transparency. The Charter builds on the original Responsible Care initiative, the industry's voluntary initiative under which companies, through their national associations, work to continuously improve their EHS performance.

Under the Global Charter, the global chemical industry is making progress in improving knowledge and information about chemicals, reducing risks, building capacity around the world, and extending Responsible Care along the industry's value chain. This effort reflects the energies of national associations and individual companies, and requires a considerable investment in human and information resources as well as financial commitment.

Examples of how the Global Charter helps the industry fulfill the goals of SAICM include

- *Risk Reduction*: An essential part of chemical management, risk reduction is the key element of the SAICM effort. Under the RCGC, ICCA members are making significant strides
  - The European industry association's Safety and Quality Assessment Systems evaluate the safety, security, quality, and environmental practices of logistics providers. The system is now used in Brazil, South Africa, and, most recently, China.
  - Ireland's association introduced a system for circulating process safety alerts to rapidly share information that could help avoid repeat incidents at other facilities.
  - Australia's and India's industry developed process safety guidance to promote health and safety performance including a step-by-step guide and formal process safety training.
  - South Africa's chemical industry association, working with government and labor, has created a web-based information system to promote safety and provide links to worldwide chemical safety sites.
  - Close cooperation in Chile has led to the development of EHS regulations and helped companies fulfill obligations under international conventions (such as CWC and Stockholm POPS).
- *Capacity Building and Sharing of Knowledge and Information*: The sharing of knowledge and information is a crucial element of capacity and constitutes an important aspect of overall industry capacity building. A prime example is the work that led to the Russian Chemist's Union membership in the Responsible Care Leadership Group. China too has been the focus of numerous workshops and programs to help the growing industry in that nation improve its EHS performance. Some other examples include
  - In Russia, with support from the Russian Chemical Worker's Union, 33 of the nation's leading chemical companies have joined the Responsible Care initiative.
  - In China, the Hong Kong chemical association is working with the national association to implement Responsible Care. In 2008, 24 leading companies signed the "Beijing Manifesto," committing them to implement Responsible Care in China. The industry hosted Responsible Care and product stewardship conferences attended by hundreds of participants including representatives of government and SMEs.

    - ICCA and its member associations sponsor product stewardship workshops in Colombia, Bulgaria, Turkey, Croatia, the Slovak Republics, Thailand, Singapore, Indonesia, Russia, Argentina, and Japan. These workshops provide best practice sharing opportunities for industry and government officials from neighboring nations. They also spread the word among local companies to help them implement crucial risk reduction and stewardship programming.
    - The Japanese industry works closely with government officials in Tokyo to build capacity throughout the ASEAN region, including the introduction of Responsible Care in Cambodia, Myanmar, and Vietnam and GHS work in Indonesia.
    - In Brazil, the chemical industry association helps to build capacity among key chemical industry customers by helping the paint and coatings industry comply with the Coatings Care initiative and reaching out to the automotive and electronics sectors.
    - In March 2009, ICCA signed an agreement to partner with the International Council of Chemical Trade Associations to jointly promote Responsible Care and Responsible Distribution worldwide, especially within emerging economies.
- *Governance and Verification*: Fifty-three associations are signatories to the Responsible Care Global Charter. Of these, 50 responded to the survey for the most recent report. Findings included:
    - Seventy-six percent have published all required guidelines for implementation.
    - Eighty-eight percent have modified their Responsible Care Guiding Principles to be consistent with the Global Charter.
    - Nearly 50% have now included chemical plant security in their Responsible Care initiatives.
    - Nearly half have a process in place to remove a member company for failing to meeting Responsible Care commitments.

In addition, the member associations strengthen verification and make Responsible Care a condition of membership. The following are some examples:

- The Canadian association works with the supply chain to enhance product stewardship by forming coalitions with industries such as: raw materials, energy suppliers, distributors, and chemical users.
- In the United States, association members and Partners must undergo third-party verification of Responsible Care and use the Responsible Care Management System® to drive EHS and security performance. The association and ISO14001 developed an integrated approach known as RC 14001, which is also subject to third-party verification.
- Starting in 2007, the South African association fully implemented third-party verification, with reverification required every 2 years.

Their protocols are also aligned with ISO14001. Others have taken similar approaches.
- The New Zealand industry association operates a self- and third-party assessment and accreditation system that makes it far easier for SMEs to comply with that nation's performance-based EHS legislation. More than 350 sites participate, including chemical manufacturers, distributors, retailers, and laboratories.
- India's chemical association introduced a compulsory audit by a team that includes association and company experts. This is required for use of the Responsible Care logo.

The examples noted here focus on association and industry-wide activity. Chemical companies too have initiated results-driven EHS activities to help other companies and value chain partners reduce risk, build capacity, and verify their actions under Responsible Care.*

- *Illegal International Traffic*: The global industry's efforts in fighting the manufacture of illicit drugs have earned respect from governments and law enforcement agencies. For example, in the Slovak Republic, Australia, New Zealand, and the United States, the associations work closely with the drug enforcement agencies to monitor trade in certain chemicals.

The Global Product Strategy and Product Stewardship

ICCA's Responsible Care initiative also focuses on product stewardship, the managing of EHS performance throughout the life cycle of products. One way industry accomplishes this is through the GPS, the second major ICCA initiative launched at Dubai (ICCM-1) in 2006.

Working within the context of the RCGC, the GPS requires the industry to adopt a new, more structured and comprehensive process of chemical product management, both within the industry and the value chain. "The ultimate purpose of the GPS is to increase public and stakeholder awareness of, and confidence in, the safe management of chemicals throughout their lifecycle by noticeably increasing the chemical industry's performance and transparency, and promoting the safe handling of its products in downstream applications."†

The Nine Strategic Elements of the GPS

- Guidelines for product stewardship, including ways to make relevant product stewardship information, are more transparent.

* (For a good introduction to the actions undertaken by individual companies, see ICCA's *Progress Report to the Second Session of the International Conference on Chemicals Management* (May 2009) available at: http://www.icca-chem.org)

† Id. at p. 4.

- A tiered approach for completing risk characterizations and recommending risk-management actions for chemicals in commerce.
- Improved product stewardship cooperation with industry groups and companies that are customers and suppliers to the chemical industry.
- Tracking industry performance and reporting to the public.
- Exploration of potential partnerships with IGOs and other stakeholders to enhance global product stewardship.
- Outreach and dialogue with customers, the public, and stakeholders.
- Participation in scientific inquiry to address health, environmental, and risk concerns.
- Constructive industry engagement in the public policy process.

How GPS Promotes Product Stewardship: Guidelines and Hazard and Exposure Info

By 2008, two-thirds of ICCA's associations were beginning to implement the GPS through which members agree to constantly improve their management of chemicals. To help achieve this goal, ICCA and its member associations have developed guidelines that companies can use to better characterize chemical hazards and risks.*

The GPS works hand-in-hand with the Responsible Care Global Charter in the improvement of product stewardship throughout the supply chain. While product stewardship has always been a discipline under Responsible Care, the GPS provides detailed focus and specific actions to improve industry performance in this critical area in the future. Under GPS, the industry has

- Adopted best practices for an initial set of hazard and exposure information adequate for conducting chemical safety assessments and a mechanism to promote data sharing to enhance risk assessment and management on a global basis.
- Developed a set of global product stewardship guidelines for use by member associations and companies to accelerate the implementation of their chemical management programs. These guidelines cover crucial disciplines such as: risk characterization, the gathering of relevant hazard and potential exposure information, and risk management—including the possible refusal to sell a product under certain circumstances, transparent public communications about the product, and close liaison with customers and the value chain.
- Under the SAICM Modalities of Reporting Project, industry proposed a mechanism for monitoring and reporting the chemical industry's progress in meeting the objectives of SAICM. (At the time of this writing, however, there is no agreement on the format for reporting.)
- Adopted a global Responsible Care Governance Process to assure greater accountability for performance and the upholding of the Responsible Care ethic.

---

* http://www.ICCA-Chem.org

- Provided capacity building projects in a number of developing countries in Africa, Asia-Pacific, Latin America, and in countries with economies in transition.

At the second UN International Conference on Chemicals Management in May 2009, industry reported on its progress toward the 2020 WSSD objective under the Responsible Care and GPS initiatives.

Risk Assessment: Crucial for Chemicals Management and Sustainability

The objective of global industry cooperation with governments, IGOs, and NGOs, as stated by Christian Jourquin, Chief Executive Officer of Solvay and former ICCA President, is to "demonstrate to governments, IGOs and the public that [the chemical] industry is a reliable, willing and responsible partner in meeting global sustainability objectives."*

The chemical industry's commercial success and public reputation rely heavily on an ability to apply best scientific practices. Indeed, risk assessment, the integration of hazard data with exposure information, is at the heart of the industry's ability to produce chemicals safely, and to assure their transportation and distribution in ways that protect human health and the environment.

HPV Chemicals Programme

One of ICCA's earliest risk assessment/risk management initiatives included voluntary efforts to increase the availability of information to governments and to the public on existing chemicals in commerce under the HPV Chemicals Programme. Established in 1998 by ICCA in cooperation with the OECD, the ICCA HPV program (and the US EPA HPV Challenge) delivers harmonized, internationally agreed hazard information and robust study summaries that meet national and regional program standards. ICCA members gather a Screening Information Data Set (SIDS) Dossier, which includes a robust summary of health and environment information required for making an initial hazard assessment of HPV chemicals by the member countries of the OECD.

SIDS Dossiers contain information that identifies: (1) possible hazards upon all major organ systems from both acute and repeated exposures; (2) potential hazards arising from *in utero* exposures; (3) the potential of a substance to affect reproduction; (4) the potential of a substance to damage DNA; and (5) dose levels that are without adverse effects. The acute toxicity studies are most critical to assure correct packaging, labeling, and handling to prevent poisoning incidents. Developmental and reproductive toxicity studies are most relevant to prevent exposures that could affect normal prenatal and postnatal growth, development, and maturation of children and reproduction in adults. SIDS Dossiers are used to screen chemicals for safety, to establish safe management practices, and are also used to set priorities for

* ICCA Review 2008–2009, p. 1.

further testing and to define what specific additional testing is needed to increase scientific certainty of the safety assessment.

Transportation

The industry also supported measures to minimize the potential effects of transportation incidents involving chemical products. One of the best known of these programs is the American Chemistry Council's (ACC) Chemical Transportation Emergency Response Center, or CHEMTREC®. The Center provides a 24/7 response capability to assist officials responding to chemical transportation emergencies, and has partnered with organizations like the US Departments of State and Defense to expand the reach of the Center's activities throughout the Americas.

Long-Range Research Initiative

Another ICCA collaborative program is the Long-Range Research Initiative (LRI), the purpose of which is to strengthen the scientific foundation for public policy and commercial decisions through quality research to protect human health and the environment. The LRI is financed by three association members of ICCA, and, since the program began in 1999, more than US$ 200 million has been invested. LRI projects sponsored by the European Chemical Industry Council (Cefic) include contributions to the European Commission and European Partnership for Alternative Approaches to Animal Testing. In the United States, LRI projects sponsored by the ACC have contributed to improved methods to characterize exposures to chemicals and their potential health risks. In Japan, through the Japan Chemical Industry Association, LRI has contributed to developing several new test methods and systems for chemical safety.

One area of focus for LRI has been research in conjunction with governmental agencies and academia to establish standardized, validated, and reproducible testing methods. Such methods, once established, can be used broadly to obtain consistent test results across geographies. In this regard, industry has been a significant participant in developing methods and sponsoring validation studies as part of OECD's program to adopt valid test methods that can be used globally to screen substances for their potential to interact with the endocrine system.

Most recently, a key global focus of the LRI has been on placing the results of human biomonitoring in a health risk context. For example, ICCA convened a multi-stakeholder workshop in July 2006 that generated the report *Interpretation of Human Biomonitoring Data: A Research Strategy* (International Council of Chemical Associations Long-Range Research Initiative, 2006). Similar workshops have been held annually to enhance the global efforts on addressing complex environmental health issues. Results will be used to guide regulatory efforts by governments, as well as further research on the role of biomonitoring.

The LRI fosters collaborations among industry, government, and other stakeholders on research to address new and emerging health and environmental concerns. For example, LRI projects are currently underway to improve interpretation and understanding of the data emerging from the new technologies. These new approaches can assess the potential for chemicals to produce effects at the cellular level, by using high-throughput technologies to quickly screen hundreds of chemicals. Industry hopes to work closely with government and other scientists to understand how people might be exposed to chemicals and the impact such exposure might have.

LOOKING AHEAD

ICCA members recognize that there is still much to be done to improve risk assessment and management practices, including improvements in collaborative approaches. ICCA members are confident, however, that they will build on the significant progress achieved to date in implementing the industry's policy advocacy, program, and research commitments.

ICCA's chemicals management and product stewardship efforts will become increasingly transparent. For example, the GPS includes a 2018 target date for companies to have assessed the safety of all their chemicals in commerce, with an interim 2012 target to report on progress in making these assessments. Under the GPS, of course, participating companies will make safety summaries available on chemicals that have been assessed.*

ICCA members are working on the following objectives as contributions to meet the WSSD goals at the latest by 2020:

- Establish a base set of hazard and exposure information adequate to conduct safety assessments for chemicals in commerce.
- Share relevant product safety information for all products with coproducers, governments, and the public.
- Work across the value chain so that suppliers and customers can effectively evaluate the safety of their products and enhance their performance.
- Make product safety summaries on chemicals publicly available.
- Extend their monitoring and reporting structure by including additional metrics to quantitatively track progress and support continuous improvement in the sound global management of chemicals.

ICCA is also in the process of developing principles and elements to foster greater consistency and transparency in chemical regulatory programs and to promote regulatory convergence. Industry hopes to provide enough flexibility to accommodate existing and anticipated national or regional laws and regulations. These principles and elements will be shared with governments and other stakeholders to promote cooperation and collaboration in reducing and, where possible, eliminating chemical risks.

CONCLUSION

Since ICCA declared its support for the WSSD in 2002, the business of chemistry has been seriously engaged in meeting sustainability goals. Governmental regulatory

* (For example, the ACC's product stewardship summaries are available at: http://reporting.responsiblecare-us.com/Search/PSSummarySearch.aspx)

programs such as REACH in Europe and the Toxic Substances Control Act in the United States may capture more of the headlines, but the global chemical industry has been an active partner in revising and enhancing those programs, and in initiating its own chemicals management and product stewardship initiatives.

Industry knows it can provide sustainable solutions to global problems. But it must be allowed to innovate and develop new products. The key to winning society's support in building a better future is transparent, open communications and strong, measurable performance. The Responsible Care Global Charter and the GPS are two crucial parts of the foundation upon which industry hopes to demonstrate such performance as it continues its journey toward the WSSD 2020 goal.

REFERENCES

American Institute of Chemical Engineers, Center for Chemical Process Safety. *Inherently Safer Chemical Processes*, 2nd Ed., 2009, pp. 5–6, John Wiley & Sons, Inc., Hoboken, NJ.

International Council of Chemical Associations Long-Range Research Initiative. 2006. *Interpretation of Human Biomonitoring Data: A Research Strategy*, Available at: http://www.americanchemistry.com/s_acc/bin.asp?CID=1389@DID

Obama, B. H. 2009. *Inaugural Address.* Presidential Inaugural, January 20, 2009, Capitol Building, Washington, D.C. Available at: http://www.whitehouse.gov/blog/inaugural-address/

21 The Intergovernmental Forum on Chemical Safety

Jack Weinberg

CONTENTS

INTRODUCTION

The Intergovernmental Forum on Chemical Safety (IFCS) was established in 1994 and has played an important role in building support among both governmental and nongovernmental actors for advancing chemical safety objectives, especially in developing countries and countries with economies in transition. IFCS was founded at an International Conference on Chemical Safety (ICCS) in Stockholm, Sweden that was convened by three United Nations (UN) organizations: the United Nations Environment Programme (UNEP), the World Health Organization (WHO), and the International Labor Organization (ILO). Representatives from 114 governments participated in the

Conference as did numerous representatives from intergovernmental organizations, nongovernmental organizations (NGOs), and chemical industry trade associations.*

The formal mandate of IFCS is to serve as a Forum where government representatives and stakeholders can consider issues relating to the environmentally sound management of chemicals and can provide advice and recommendations. According to its founding documents,† IFCS is neither an organization nor a program, but rather, it is a "noninstitutional arrangement." IFCS was set up to provide advice and recommendations. Although it is not empowered to play any explicit role in implementation, IFCS does evaluate and review progress.

AN INTERNATIONAL FORUM FOR NATIONAL CHEMICALS MANAGERS

Very few national governments have any single agency that is specifically responsible for chemical safety. Rather, most governments distribute the responsibility for ensuring that human health and the environment are protected from harms caused by exposure to toxic chemicals among numerous different agencies, such as Environment Ministries, Health Ministries, Labor Ministries, Agriculture Ministries, Industry Ministries, and others. Similarly, at the intergovernmental level, responsibilities for various aspects of chemical safety have been given to UNEP, WHO, ILO, the Food and Agriculture Organization (FAO), the United Nations Industrial Development Organization (UNIDO), the United Nations Development Programme (UNDP), and others. For each of these ministries and for each of these intergovernmental organizations, chemical safety is just one responsibility among many, and in most cases, it is not considered to be among the ministry's or the agency's highest priorities.

One consequence of the way chemical safety responsibilities are distributed among numerous ministries and agencies is that it suggests chemical safety itself is not an important governmental concern. Another consequence is frequent failures in effective cooperation and coordination between the various responsible ministries and agencies leading to significant coverage gaps and inefficiencies. While these consequences can be significant to countries at any level of development, they are of particular importance to many developing countries and countries with economies in transition, especially those with weak chemicals management policies and infrastructures.

The establishment of IFCS in 1994 helped in many countries to elevate the perceived importance of chemical safety at the national political level: it was a signal that the world community now considers the sound management of chemicals, in all its aspects, to be an important responsibility that national governments should assume. According to IFCS rules, each government is to be represented at Forum meetings by "a senior official concerned with chemical safety." As such, IFCS meetings are not environmental gatherings; nor are they public health gatherings; nor are

* Information about the International Conference on Chemical Safety, including the meeting report, is at: http://www.who.int/ifcs/forums/one/en/index.html

† The initial IFCS Terms of Reference and other founding documents are included in the meeting report referenced above.

they occupational safety gatherings, and so on. Rather, the IFCS is a high-level international forum for national chemicals managers.

The IFCS enables national chemicals managers to meet with their peers and with stakeholders to explore and address policy issues and solutions to problems of mutual concern. The very fact that every government is invited to send a senior official with chemicals management responsibilities to the meetings of IFCS helps elevate the status within national governments of chemical safety concerns and it also helps to raise the status of those individuals within government who are responsible for ensuring that chemicals are soundly managed.

Through their participation in IFCS, national chemicals managers are able to share views, knowledge, skills, and experiences both globally and regionally with other government officials from a range of sectors and also with NGOs and industry representatives. This has frequently been of greatest value to those from countries where the officials with responsibility for national chemicals management tend to lack authority and often feel weak and isolated. As a result, IFCS has had consistent strong support from many participating national chemicals managers from developing countries and countries with economies in transition. On the other hand, although highly industrial country governments were originally instrumental in establishing IFCS, over time, several of them became less supportive of the Forum. This has been a problem since it led to a decline in support for IFCS from a number of wealthy governments that had previously paid much of IFCS's operational costs.

IFCS AND CHAPTER 19 OF *AGENDA 21*

IFCS was established as a direct result of the 1992 Rio Earth Summit and the program of action adopted there, was *Agenda 21*. Chapter 19 of *Agenda 21* is titled the *Environmentally Sound Management of Toxic Chemicals Including Prevention of Illegal International Traffic in Toxic and Dangerous Products*.* Chapter 19 acknowledges that chemical contamination causes grave damage to human health, genetic structures, and reproductive outcomes, and also harms the environment. It notes that the impacts of this pollution and their importance are only recently becoming understood, and it calls for significantly strengthening both national and international efforts to achieve the sound management of chemicals.

Chapter 19 explicitly called upon UNEP, WHO, and ILO to convene a conference to establish an IFCS, and it defined six program areas for it to address:

a. Expanding and accelerating international assessment of chemical risks
b. Harmonization of classification and labeling of chemicals
c. Information exchange on toxic chemicals and chemical risks
d. Establishment of risk reduction programs
e. Strengthening of national capabilities and capacities for management of chemicals
f. Prevention of illegal international traffic in toxic and dangerous products

* The text of Chapter 19 can be found at: http://www.unep.org/Documents.Multilingual/Default.asp?DocumentID=52&ArticleID=67

IFCS TERMS OF REFERENCE

According to the IFCS terms of reference (TOR) the Forum is to seek consensus amongst government representatives on priorities and strategies for the implementation of Chapter 19 and related issues. It is to provide policy guidance with particular emphasis on regional and subregional cooperation and help foster an understanding of the issues. It is to identify priorities for cooperative action and recommend concerted international strategies for the environmentally sound management of chemicals. One particularly important function of the Forum is to promote establishing stronger national mechanisms to coordinate the activities of the various national ministries and agencies that share responsibility for chemical safety and to help strengthen national capabilities and capacities for chemicals management, especially infrastructure building, training, education, research, monitoring, and the provision of information.

Other IFCS functions include promoting information exchange and scientific and technical cooperation, including training, and education and technology transfer. IFCS is to advise governments in their work on chemical safety; promote cooperation among governmental, intergovernmental and NGOs; and promote stronger programs to prevent and respond to chemical poisonings. On a periodic basis, IFCS is to review the effectiveness of efforts to implement international chemical safety strategies; recommend further activities; and, where necessary, provide advice on how to strengthen or establish necessary follow-up mechanisms.

NGO INVOLVEMENT

According to its TOR, international NGOs involved in chemical safety issues are invited to participate in the IFCS but without the right to vote. In its practice, the IFCS recognizes four categories of NGOs: public interest NGOs (environmental organizations, health advocacy organizations, consumer organizations, etc.); industry trade associations; trade unions; and scientific/academic organizations.

NGOs have actively engaged in the IFCS from the start, and the participation rights afforded to them have been virtually complete. Although the IFCS TOR limits participation to international NGOs, this has not been a practical barrier to participation by relevant NGOs whose work is primarily at the national or subnational level. Such NGOs have grouped themselves together into international networks, associations, and other groupings, and IFCS has welcomed their participation through these arrangements.

Chemical and related industries participate in IFCS through international trade associations. Public interest NGOs from all regions have also been able to actively participate. In general, the large, well-recognized, international environmental organizations have not maintained an ongoing interest in IFCS. On the other hand, work in support of various aspects of chemical safety is important to many of the small and medium-sized, nationally based NGOs that are active in most countries. Representatives of such NGOs from all regions actively participate in all aspects of the IFCS through international networks, and the IFCS Secretariat has cooperated with these networks to raise the travel funds needed to ensure full, regionally balanced participation by public interest NGOs.

NGOs participate not only at full Forum meetings, but also in all other IFCS bodies. They have the same right to speak as other participants, can table proposals, and have sometimes even been selected to chair IFCS working groups. One consequence of these arrangements has been to foster increased collegiality between participating government chemicals managers and NGO stakeholders.

The broad NGO participation rights fostered by the IFCS were initially a significant departure from the practices of other intergovernmental fora. Over time, however, similar practices and the spirit of stakeholder engagement pioneered by IFCS have spread to other international and regional fora where chemical safety issues are being considered. Even though industry trade associations and public interest NGOs frequently represent highly divergent interests and often take opposed positions, in the context of IFCS they and government representatives have been able to successfully engage in productive dialogue. Based on the IFCS experience, this kind of stakeholder engagement has spread to other intergovernmental fora where chemicals policies are debated and this practice is now widely considered to be essential for successful formulation and implementation of chemical safety programs and policies.

FORUM MEETINGS AND BODIES

Full meetings of the Forum have been held approximately every three years since the IFCS was established. Under the terms of its initial TOR, an IFCS Intersessional Group (ISG) meets in the off years. However, the IFCS TOR was modified and updated in 2000 at a Forum meeting in Bahia Brazil,* and the ISG was eliminated. It was replaced by a smaller Forum Standing Committee (FSC) whose main responsibility is to organize the full Forum meetings.

The FSC prepares the agenda, the meeting program, and working documents, including draft resolutions. FSC members include the IFCS President, five regionally elected IFCS Vice-Presidents, 12 regionally elected government representatives, the chairperson of the Inter-Organization Coordinating Committee of the Inter-Organization Programme for the Sound Management of Chemicals (IOMC), and one representative each for the four categories of NGOs that IFCS recognizes.

In addition to the FSC, the IFCS's rules also permit the formation of *ad hoc* working groups to address specific topics. In contrast to most intergovernmental organizations, where staff and consultants generally prepare reports and working documents, FSC members and/or members of *ad hoc* working groups themselves generally write and revise IFCS reports and documents in open and transparent processes. This approach fosters dialogue on potentially controversial chemical safety issues by regionally diverse groups of government and stakeholder representatives, and it has frequently resulted in agreements with surprisingly broad support.

After IFCS eliminated ISG meetings, it put greater emphasis on the organization of regular regional Forum meetings in Africa, Asia, the Central and Eastern Europe (CEE) Region, and the Latin America and the Caribbean (GRULAC) Region. All important matters that are explored and debated within the IFCS are first and mainly

* The revised IFCS TOR can be found at: http://www.who.int/ifcs/documents/forums/forum3/en/index2.html

discussed regionally. As a result, IFCS has helped foster regional thinking and regional approaches in the field of sound chemicals management, and it has also provided a framework in which important regional leaders have emerged.

IFCS AND THE GLOBAL TREATY ON PERSISTENT ORGANIC POLLUTANTS

In 1995, the UNEP Governing Council (GC) took a decision* in which it asked IFCS to consider assessments on the health and environmental impacts of a class of chemicals termed persistent organic pollutants (POPS), and to develop "recommendations and information on international action, including such information as would be needed for a possible decision regarding an appropriate international legal mechanism." IFCS was asked to complete its work in time to submit recommendations to the February 2007 meeting of the UNEP GC.

In response to this request, the IFCS, at its second ISG meeting held in 1996 in Canberra, Australia, established an *ad hoc* IFCS POPS Working Group to develop the recommendations. The group met and decided that to fulfill its mandate it would organize two meetings to be held back-to-back in Manila, Philippines: an open-ended international experts meeting and then an open-ended meeting of the Working Group where recommendations could be debated and adopted.

Working Group members, themselves, did the work of preparing documents and materials for these meetings with financial and technical support from several governments. Representatives of governments, intergovernmental organizations, chemical industry trade associations and activist environmental organizations all participated in the Working Group.

In June, 1996 the Working Group convened an open forum meeting in Manila attended by representatives from 32 governments (with all regions represented), seven NGOs (including both environmental groups and chemical industry trade associations) and seven intergovernmental organizations. After considerable debate, this meeting was able to finalize and approve by consensus the *Final Report of the IFCS* ad hoc *Working Group on POPS*.†

The general conclusion of the report is that available information is sufficient to demonstrate the need for international action on POPS including the establishment of a global legally binding instrument to control them. The report also included detailed conclusions spelling out a proposed framework for that instrument and an outline of the provisions it should contain. Its final recommendations included an invitation to UNEP to convene an intergovernmental negotiating committee (INC), with a mandate to prepare an international legally binding instrument for implementing international action on POPS.

This report was considered by the 1997 meeting of the UNEP GC, and its recommendations were fully approved.‡ Following the decision, the IFCS Working Group

---

* UNEP GC Decision 18/32 http://www.chem.unep.ch/pops/indxhtms/gc1832en.html

† The report is on the Web at: http://www.chem.unep.ch/pops/indxhtms/manwgrp.html

‡ See UNEP decision 19/13 C. http://www.unep.org/Documents.multilingual/Default.asp?DocumentID=96&ArticleID=1470&l=en

co-organized with UNEP a series of workshops in all regions to provide information about POPS to relevant government officials in order to help prepare them for POPS treaty negotiations. The Working Group completed its work and was disbanded prior to the first meeting of the POPS INC, June 1998. The POPS negotiations were successful. In May 2001, governments adopted the Stockholm Convention on POPS, a global treaty whose terms very much followed the outline contained in the IFCS *ad hoc* Working Group report.

FULL FORUM MEETINGS FROM 1994 THROUGH 2003

There have been six full meetings of IFCS. This section will briefly describe major outcomes of the first four of them: the 1994 founding Conference in Stockholm, Sweden; Forum II in 1997 in Ottawa, Canada; Forum III in 2000 in Bahia, Brazil; and Forum IV in 2003 in Bangkok, Thailand.

Forum I

At its founding meeting, in Stockholm in 1994,* IFCS was established and the WHO agreed to host the secretariat. The meeting additionally adopted a resolution on *Priorities for Action in Implementing Environmentally Sound Management of Chemicals*. The resolution recognized and adopted the six program areas set forth in Chapter 19 of *Agenda 21* and established priorities for immediate actions as well as longer term goals to be achieved. Most of the proposals were for action to be taken by individual governments including the elaboration of the basic foundation for chemicals management at the national level. This included preparing National Chemicals Management Profiles,† setting priorities, preparing action plans, and establishing interministerial coordinating mechanisms. However, several proposals also included suggestions to international bodies to develop effective tools for use by governments. All Forum participants were urged to cooperate in the effective implementation of these recommendations with particular emphasis on cooperation at the regional level.

Forum II

Forum II, in Ottawa in February 1997, was organized on the theme *In Partnership for Global Chemical Safety*.‡ The meeting reviewed the progress made in the six program areas, noted accomplishments, and noted also areas where progress had been slow or nonexistent. Based on this review, it charted a further course

* The Final Report and the resolutions adopted at Forum I is on the Web at: http://www.who.int/ifcs/forums/one/en/index.html

† National Chemicals Profiles are national documents that provide an overview and assessment of the existing national legal, institutional, administrative, and technical infrastructure related to the sound management of chemicals. See National Profile Homepage: http://www2.unitar.org/cwm/nphomepage/index.html

‡ The Final Report and the resolutions adopted at Forum II is on the Web at: http://www.who.int/ifcs/forums/two/en/index.html

for the next three years. The meeting also reviewed the activities and work product of the *ad hoc* Working Group on POPS and agreed to extend the tenure of the group through the June 1998 the date of the first meeting of the POPS Intergovernmental Negotiating Committee, with a mandate to help governments prepare for negotiations. The Forum also recommended that the globally harmonized system (GHS) of classification and labeling move forward as a nonbinding international instrument and provided guidance for the work including a definition of its scope.

Forum II additionally took up the topic of endocrine disrupting substances as an emerging issue. An information session was held on the topic with presentations from leading scientific experts in the field. For many IFCS participants, this was their first, substantive introduction to endocrine system disruption as an important chemical safety concern. The meeting agreed that there exists a rapidly growing body of scientific research indicating that a number of chemicals have the potential to interfere with normal functions of the body that are governed by the endocrine system. It agreed further on the need for investigating, in depth, the human, environmental, and ecotoxicological aspects of endocrine disrupting substances; it recommended specific actions to facilitate and enhance research; and it also proposed that IOMC should facilitate information exchange on the issue.

Forum III

Forum III, in Bahia, Brazil in 2000 was also organized on the theme *In Partnership for Global Chemical Safety.*[*] The meeting undertook a full review of the Priorities for Action that had been adopted by the Forum in 1994. It revised and updated them to produce the widely circulated IFCS document: *Priorities for Action Beyond 2000.* The priorities are targeted at governments, international organizations, and other stakeholders and emphasize efficient coordination, the participation of employers and workers, and the strengthening of community's "right to know."

The meeting also adopted the *Bahia Declaration* which many consider to be the single most important output of Forum III. The declaration represents a commitment by IFCS participants to strengthen efforts and build partnerships to accomplish specific targets and recognizes the importance of technical and financial assistance and technology transfer to enable developing countries to meet the chemical safety challenges set out in Chapter 19 of *Agenda 21.*

Forum III additionally conducted a full review of the IFCS. It noted that the Forum had undergone a gradual evolution and it modified the IFCS TOR to improve its work and better meet the needs of its participants. Among the changes adopted were: the establishment of an independent IFCS President with a clear definition of the President's roles and responsibilities; the election of five IFCS Vice-Presidents, one from each region, with clearly specified roles and responsibilities; guidelines for IFCS National Focal Points; TOR for the FSC; and the discontinuation of ISG meetings.

---

[*] The full Forum III Final Report including *Priorities for Action Beyond 2000, and the Bahia Declaration* can be downloaded at: http://www.who.int/ifcs/documents/forums/forum3/en/index.html

Forum IV

Forum IV, in Bangkok in 2000 was organized on the theme *Chemical Safety in a Vulnerable World*.* The meeting identified the issue of pesticide poisonings as a priority with a focus on acute poisonings resulting from both occupational and community exposures to acutely toxic pesticides. It considered an assessment of acutely toxic pesticides prepared by an IFCS Working Group which concluded that these pesticides may pose significant public health problems for developing countries and countries with economies in transition including severe effects such as fatalities and permanent or temporary impairments, as well as mild effects such as skin irritations. Based on the assessment, the Forum adopted a full suite of recommendations addressing policy, regulations, and communications. Among these was a recommendation that countries prohibit or restrict the availability of acutely toxic pesticides including the possibility of controlling their import and/or export if this is desirable.

Another important topic considered by Forum IV was *Children and Chemical Safety*. One conclusion of this discussion was that when assessing how to protect children from toxic chemical exposure, chemicals managers should take into account that exposures can occur during preconception, throughout gestation, as well as during infancy, childhood, and adolescence. The Forum called upon governments to prepare national assessments of children's environmental health and chemical safety, identify priority concerns, and use these as a basis for developing action plans to address the concerns. The Forum additionally asked WHO to develop guidance tools and to assist at least three countries at different stages of economic development in each region to prepare assessments and action plans by 2006.

Forum IV also developed policy recommendations under the headings *Occupational Safety and Health, Hazard Data Generation and Availability, The GHS for Classification and Labelling of Chemicals,* and *Capacity Building,* including the provision of capacity building assistance to governments and also the widening gap among countries in their ability to implement chemicals safety policies and practices.

IFCS AND THE STRATEGIC APPROACH TO INTERNATIONAL CHEMICALS MANAGEMENT

In 2002, the Governing Council of UNEP decided that there was a need to further develop a strategic approach to international chemicals management (SAICM). It endorsed the IFCS's *Bahia Declaration* and *Priorities for Action Beyond 2000* as the foundation of this approach and it instructed UNEP to work with the IFCS and other relevant organizations and stakeholders to identify any gaps in the *Bahia Declaration* and the *Priorities for Action Beyond 2000* and also gaps in the implementation of these priorities, and to suggest remedies for any identified gaps. Finally, the UNEP

* The Final Report on Forum IV is at: http://www.who.int/ifcs/documents/forums/forum4/final_report/en/index.html

GC instructed UNEP to work with IFCS to convene an open-ended consultative process that involves all stakeholder groups to further develop the SAICM.*

Also in 2002, the UN convened the World Summit on Sustainable Development (WSSD) in Johannesburg, South Africa. In the Plan of Implementation adopted at WSSD,† paragraph 23 renewed the commitment of the world community to Chapter 19 of *Agenda 21* for the sound management of chemicals throughout their life cycle and of hazardous wastes. It established an aim to achieve, by 2020 that chemicals used and produced in ways that lead to the minimization of significant adverse effects on human health and the environment. As a means to achieve this aim, it endorsed the earlier decision of the UNEP GC to develop a SAICM based on the IFCS's *Bahia Declaration* and its *Priorities for Action Beyond 2000.* And it called upon UNEP, the IFCS, and other relevant international organizations to closely cooperate in this effort.

UNEP, IFCS, IOMC, and other relevant intergovernmental organizations decided to hold three Preparatory Committee meetings to prepare the SAICM. These would be followed by an International Conference on Chemicals Management (ICCM) where the SAICM would be adopted. The first of these meetings (PrepCom1) was held in 2003 in Bangkok immediately following the conclusion of IFCS Forum IV. At Forum IV, in addition to the other business, SAICM was extensively debated. The meeting adopted a document for consideration by PrepCom1: *Forum IV Thought Starter Report to SAICM PrepCom1.*‡ The Forum also produced a compilation of issues and views on gaps in present global chemicals policy and identified obstacles, opportunities, and enablers for change, financing, and coordination. These provided an analytical framework for further discussions about the nature and content of SAICM.

Three Preparatory Committee meetings were held and the SAICM was finally adopted at the ICCM held in February, 2006 in Dubai, United Arab Emirates.§ The ICCM adopted two core SAICM texts: the SAICM Dubai Declaration on International Chemicals Management, and the SAICM Overarching Policy Strategy, and recommended the SAICM Global Plan of Action as a tool and guidance document for meeting commitments to chemicals management. The ICCM further adopted a resolution on SAICM implementation arrangements which calls upon governments to designate SAICM Focal Points and which assigns to UNEP the responsibility of establishing a SAICM Secretariat colocated with UNEP's other chemical-related programs in Geneva. These and other ICCM decisions raised a question about what might be the institutional relationship between SAICM and IFCS.

In order to begin addressing the issue of this relationship, the ICCM adopted a resolution on IFCS. The resolution recognized the significant role IFCS has played;

* See Final Report of the Seventh Special Session of the UNEP Governing Council, decision SS.VII/3: http://www.unep.org/GC/GCSS-VII/Reports.htm

† The WSSD Plan of Implementation is at: http://www.un.org/esa/sustdev/documents/WSSD_POI_PD/English/WSSD_PlanImpl.pdf

‡ The Thought Starter can be downloaded at: http://www.who.int/ifcs/documents/forums/forum4/final_report/en/index.html

§ The core SAICM Texts and resolutions adopted at the ICCM are on the Web at: http://www.saicm.org/index.php?menuid=3&pageid=187

it invited the Forum to continue its important role in providing an open, transparent, and inclusive forum for discussing issues of common interest and also new and emerging issues, and to continue to contribute through this to the implementation of the SAICM; and it requested that the SAICM secretariat establish and maintain a working relationship with IFCS to draw upon its expertise. Nonetheless, there remained important outstanding questions about the relationship between IFCS and SAICM.

IFCS FORUM V

Six months after SAICM was adopted, IFCS Forum V was held in September 2006 in Budapest, Hungary. It was a substantive and dynamic meeting with participation from 81 governments, 12 intergovernmental organizations, and 64 NGOs. Forum V took up a number of priority topics on the international agenda. It had a major session which addressed applying the precautionary approach in the context of chemical safety. This session identified a number of tools and approaches that can be used in the application of the precautionary approach to domestic chemicals management activities with special emphasis on the needs of chemical managers in developing countries and countries with economies in transition.

Forum V also addressed the need for further global action on heavy metals and adopted *The Budapest Statement on Mercury, Lead, and Cadmium.* This statement recognized the worldwide harmful health effects of mercury, lead, and cadmium and it urged governments and other IFCS participants to take actions to address these public and occupational health effects and also environmental impacts. It further invited the UNEP GC to give high priority consideration to further measures to address mercury, and also as appropriate, to address lead and cadmium. It proposed that among other measures, consideration should be given to establishing a legally binding instrument. (In 2009, the UNEP GC adopted *Decision 25/5* which addresses lead and cadmium and which establishes an intergovernmental negotiating process to establish a legally binding instrument on mercury.[*])

Another major topic addressed at Forum V was *Toys and Chemical Safety.* The meeting heard presentations about toys that can cause harmful chemical exposures; it considered the risk associated with such exposures; and it identified actions that governments and others can take to protect children from these harms.

Forum V included an information session on the topic of the *Sound Management of Chemicals and Poverty Reduction,* which addressed, among other issues, how governments can approach multilateral and bilateral development agencies for financial support to strengthen national chemicals management infrastructures. The meeting also discussed topics to be addressed at the next Forum meeting. While many proposals were put forward, two topics garnered the greatest interest: substitution and alternatives, and manufactured nanomaterials.

The most intense debate at Forum V, however, was on the topic *The Future of IFCS.* A resolution was adopted that recognized the desirability of continuing the

[*] See Proceedings of UNEP GC at its 25th Session: http://www.unep.org/GC/GC25/Docs/Proceedings-English.pdf

Forum but that also indicated a regard for the need to efficiently use human and financial resources and avoid duplication of functions. The resolution established a working group to prepare a draft decision on the future role and functions of IFCS including the consideration that IFCS and SAICM should share a joint secretariat. It was further decided that Forum VI would consider the working group's product and, if it so decides, propose a draft decision for possible consideration by the SAICM's second International Conference (ICCM-2) scheduled for May, 2009.

IFCS FORUM VI

Forum VI was held in Dakar, Senegal, in September, 2008 with participation from 71 governments, 12 intergovernmental organizations, and 39 NGOs.* Its theme was *Global Partnerships for Chemical Safety Contributing to the 2020 Goal.* It was agreed that the outcomes of Forum VI would be submitted to ICCM-2 and to other relevant entities and organizations for consideration and further action.

One major outcome of Forum VI was the *Dakar Statement on Manufactured Nanomaterials.* The statement recognizes potential benefits, new opportunities, challenges, hazards, risks, ethical and social issues associated with manufactured nanomaterials and nanotechnologies. However, it focuses only on safety aspects of nanomaterials. The statement agrees that current efforts to identify potential environmental, health, and safety risks of manufactured nanomaterials have not yet been fully conclusive and efforts need to be expanded and supported globally. The statement notes many countries lack comprehensive policy frameworks to address nanomaterials and there is also no inclusive global policy framework. It recognizes that groups like children, pregnant women, and elderly people are especially vulnerable to manufactured nanomaterials; it emphasizes the need to take appropriate safety measures to protect their health; and it notes that developing countries and countries with economies in transition have special needs in addressing this issue. In order to minimize risks associated with manufactured nanomaterials, the statement recognizes that countries should have the right to accept or reject manufactured nanomaterials. The statement concludes with a comprehensive list of recommendations to governments, industry, and other stakeholders, and it calls upon ICCM-2 to consider these recommendations for further action.

Forum VI also adopted the *Dakar Recommendations on Substitution and Alternatives.* This statement defines substitution as the replacement or reduction of hazardous substances in products and processes by less hazardous or nonhazardous substances, or by achieving an equivalent functionality via technological or organizational measures, including the use of traditional low- and nonchemical practices. It recognizes that many multilateral environmental agreements and many national regulatory policies in the chemicals policy area advocate or mandate substitution and the use of alternatives. While it acknowledges that there is a need to promote and support the development and implementation of environmentally sound and safer alternatives, it also recognizes existing economic policies and other incentives that

* The Final Report of Forum VI is at: http://www.who.int/ifcs/documents/forums/forum6/report/en/index.html

work against substitution and support continued use of dangerous materials. The statement concludes with a comprehensive list of proposed actions for governments, intergovernmental organizations, and stakeholders.

Another important decision adopted at Forum VI was the *Dakar Resolution for Eliminating Lead in Paints*. The resolution recognizes that lead in paints poses serious risks to human health and the environment, especially to the health of children. It also recognizes that the greatest childhood exposure to lead is in developing countries and countries with economies in transition. It recognizes that household paints sold in many developing countries still contain lead even though safer and affordable alternatives exist, and it affirms that many consumers are still unaware of the dangers posed by lead in paints. The resolution recognizes that the 2002 World Summit on Sustainable Development in Johannesburg called for phasing out lead in lead-based paints and in other sources of human exposure, and it called for work to prevent children's exposure to lead and to strengthen monitoring and surveillance efforts and the treatment of lead poisoning. The resolution therefore decides that there should be a global partnership to support phasing out of lead in lead-based paints, especially for developing countries and countries with economies in transition. It establishes an IFCS working group to prepare draft TOR for such a partnership to be submitted to ICCM-2 for it to consider taking action. The resolution finally invites UNEP and WHO to support and participate in this initiative.

Forum VI additionally adopted the *Dakar Recommendations on Ecologically Based Integrated Pest Management and Integrated Vector Management*. These recommendations put emphasis on the contributions of ecologically based integrated pest management (IPM) and integrated vector management (IVM) as key elements of pesticide risk-reduction strategies. They call upon governments and other stakeholders to adopt a pesticide use-reduction strategy as the first step in risk reduction, and to consider IPM and IVM as the preferred options in responding to challenges posed by crop pests and vector-borne disease transmission. They call upon governments to develop the necessary regulatory and institutional framework in order to facilitate ecologically based IPM and IVM, and they also call upon governments, international organizations and NGOs to ensure the sustainability of IPM and IVM by implementing participatory approaches that aim at community empowerment. Other recommendations address the need for governments to develop national strategies; the need for technical and financial support; capacity building; awareness-raising; and additional research and development programs. The resolution calls upon WHO and FAO to strengthen their policy basis in support of national IPM and IVM programs, and it invites ICCM-2 and other relevant organizations to consider its recommendations for further action.

Finally, Forum VI devoted considerable time and effort debating the future of IFCS,

THE FUTURE OF IFCS

Forum VI adopted a resolution on the future of IFCS which recalls the decision adopted at ICCM-1 inviting the Forum to continue and contribute to the implementation of SAICM and the work of other chemicals-related international organizations and institutions. The IFCS resolution recognizes that it is desirable for the Forum to continue, and it decides that the role of IFCS will be to provide an open, transparent, and inclusive forum for enhancing knowledge and common understanding about current, new, and emerging issues related to sound chemicals management. The proposed future functions of IFCS will be: to provide stakeholders, especially developing countries and countries with economies in transition, an opportunity to share and acquire information through open discussion and debate; to provide an independent, objective source of synthesized information about chemicals management issues, including potential health, environmental, and socioeconomic impacts and possible response actions; and to disseminate reports that reflect a state-of-the-art understanding of key subjects that may stimulate action, particularly for ICCM.

The resolution invites ICCM-2 to decide to integrate the Forum into the ICCM by establishing IFCS as an ICCM advisory body. It proposes some key elements describing how the Forum will operate in the future, and it invites ICCM-2 to include these elements in new TOR and rules of procedure for the Forum. It recognizes that successful integration of the Forum into ICCM will require sufficient human, financial and in-kind resources to enable it to effectively serve its functions and it urges governments and others to provide such resources. Finally, the resolution decides that until ICCM integrates the Forum into ICCM, the Forum will continue to operate under its current TOR.

However, when ICCM-2 met in May 2009, it decided not to integrate IFCS into the Conference as an advisory body at this time. In response, the FSC met to consider the future work and arrangements for IFCS.* It was noted that only a session of the Forum itself can modify the IFCS TOR or take a decision on the future of the IFCS. In light of the lack of sustainable financing for the IFCS, the FSC agreed to suspend its work for the foreseeable future and by doing so the result is to also "suspend" the work of the IFCS. The work of the FSC and IFCS may resume in the future if stakeholders so desire and if adequate resources are made available.

* Information based on correspondence with the IFCS Executive Secretary on November 16, 2009.

22 International Labour Organization's Activities in the Area of Chemical Safety

*Pavan Baichoo**

CONTENTS

INTRODUCTION

Chemicals have become a part of our life, sustaining many of our activities, preventing and controlling many diseases, and increasing agricultural productivity. However, one cannot ignore that many of these chemicals may, especially if not properly used, endanger our health and poison our environment.

It has been estimated that approximately 1000 new chemicals come into the market every year, and about 100,000 chemical substances are used on a global scale. These chemicals are usually found as mixtures in commercial products. One to two million such products or trade names exist in most industrialized countries.

The growing number of substances and rising production means a rise in the storage, transport, handling, use, and disposal of chemicals.

* The opinions expressed in this chapter are those of the author and do not necessarily represent the views of ILO or of the governments of member countries.

The International Labour Organization (ILO) has occupational health as part of its mandate. In this respect, hazardous chemicals are one of the important issues.

Many substances that are used regularly at work will contain chemicals which, if not handled correctly, can cause harm. ILO estimates that of the 2.3 million occupational fatalities, 439,000 per year are caused by chemicals and of the 160 million cases of work-related diseases per year 35 million are because of chemicals. In addition, ILO also estimates that 10% of all skin cancers are attributable to workplace exposure to chemicals. There is widespread concern over the increase in fatalities and work-related diseases because of the rapidly increasing inventory of chemicals in commercial use, especially in developing countries where adequate control measures are often unavailable. Table 22.1 shows the estimated annual average number of deaths attributable to occupational exposure to hazardous substances by condition in the world.

MAJOR ACCIDENTS

Most chemical accidents have a limited effect and are not usually reported. Unfortunately, in certain cases, chemical accidents can have disastrous effects with great loss of life and extreme damage to the environment, and more than 20 years has passed since, probably, the worst chemical industrial disaster occurred, namely Bhopal. On the night of December 2, 1984, a gas leak caused a deadly cloud to spread over the city of Bhopal, in central India, leaving 2500 people dead and injuring over 200,000 in the space of a few hours. The accident occurred because of a runaway reaction in one of the tanks of methyl isocyanate (MIC). The concrete storage tank containing some 42 tonnes of this compound, which was used to manufacture pesticides, burst open and vented MIC and other breakdown chemicals into the air. To this day, the effects of this disaster are still being felt. However, although tragedies like Bhopal occur, most chemical accidents and diseases occur from activities that are far less dramatic and in fact routine. As an example, a plantation worker spraying pesticides without protection in a field is prone to accidental poisoning.

ILO INSTRUMENTS IN THE FIELD OF CHEMICAL SAFETY

ILO has been active in the area of safety in the use of chemicals at work since the year of its creation in 1919, through activities such as the development of international treaties and other technical instruments, the provision of technical assistance to its member states, and the development of chemical safety information systems. Other instruments, such as those dealing with agriculture, construction, and mining also include provisions dealing with chemical safety. Another important standard-setting activity related to chemical safety which is specific to ILO is the development of the ILO list of occupational diseases which includes a section dealing with diseases resulting from exposure to chemicals.

The key ILO labor standards in the area of chemical safety are the Chemicals Convention, 1990 (No. 170) and the Prevention of Major Industrial Accidents Convention, 1993 (No. 174).

The Chemicals Convention, 1990 (No. 170) entered into force in 1992 and has 17 ratifications to date, though more than a 100 member states have used it as the basis

TABLE 22.1
Estimated Annual Average Number of Deaths Attributable to Occupational Exposure to Hazardous Substances by Condition in the World[a,b]

Causes of Death	No. of Deaths per Year		Estimated Percentage Attributed to Hazardous Substances at Work		No. of Work-Related Deaths Attributed to Hazardous Substances
	Men	Women	Men (%)	Women (%)	
Cancer (Total)					314,939
Lung cancer and mesothelioma	996,000	333,000	15	5	166,050
Liver cancer	509,000	188,000	4	1	22,240
Bladder cancer	128,000	42,000	10	5	14,900
Leukemia	117,000	98,000	10	5	16,600
Prostate cancer	253,000		1		2530
Cancer of mouth	250,000	127,000	1	0.5	3135
Cancer of esophagus	336,000	157,000	1	0.5	3517
Stomach cancer	649,000	360,000	1	0.5	8290
Colorectal cancer	308,000	282,000	1	0.5	4490
Skin cancer	30,000	28,000	10	2	3560
Pancreas cancer	129,000	99,000	1	0.5	1785
Other and unspecified cancer	819,000	1,350,000	6.8	1.2	71,892
Cardiovascular disease, 15–60 years	3,074,000		1	1	**30,740**
Nervous system disorders, 15+ years	658,000		1	1	**6580**
Renal disorders, 15+ years	710,000		1	1	**7100**
Chronic respiratory disease, 15+ years	3,550,000		1	1	**35,500**
Pneumoconiosis estimate	36,000		100	100	**36,000**
Asthma 15+ years	179,000		2	2	**3580**
Total	**438,489**				

[a] *African Newsletter on Occupational Health and Safety*, Volume 16, number 3, December 2006. See http://www.ttl.fi/en/publications/electronic_journals/african_newsletter/african_archives/Documents/african_newsletter3_2006.pdf

[b] Safety in Numbers: Pointers for a global safety culture at work, ILO, Geneva 2003. See http://www.ilo.org/safework/info/publications/lang–en/docName–WCMS_142840/index.htm

for their chemicals legislation. It was also during its conception that the idea of the globally harmonized system for the classification and labeling of chemicals (GHS) was developed and has been recognized in the Strategic Approach to International Chemicals Management (SAICM) as the main convention dealing with the management of chemicals in the workplace.

The Prevention of Major Industrial Accidents Convention, 1993 (No. 174) came into force in 1996 and has 15 ratifications to date.

The Chemicals Convention, 1990 (No. 170) aims to encourage ratifying members to formulate, implement, and periodically review a coherent policy on safety and the use of chemicals at work. It has the broad purpose of protecting the environment and the public, and the specific objective of protecting workers from the harmful effects of chemicals. It applies to all branches of economic activity in which chemicals are used, and it covers all chemicals, with particular measures concerning hazardous chemicals.

The general provisions for a national policy on safety in the use of chemicals at work should be undertaken through a national system for the management of chemicals, which should cover areas, such as the classification of chemicals, and the responsibility of the supplier for labeling and the provision of chemical safety data sheets (SDSs). It should also provide for the identification of chemicals by the user, and the responsibility of employers for labeling and providing information to workers' training, as well as operational prevention and control measures (including the monitoring of exposure to chemicals at the workplace).

The objective of the Prevention of Major Industrial Accidents Convention, 1993 (No. 174), is the prevention of major accidents involving hazardous substances and the limitation of the consequences of such accidents. It seeks to protect workers, the public, and the environment against risks of major industrial accidents.

The Convention requires ratifying states, in consultation with other interested parties in their country, to formulate a coherent national policy to be implemented through preventive and protective measures for major hazard installations and, where practicable, promote the use of the best available safety technologies.

The Convention includes the following provisions:

- Each Party is to formulate, implement, and periodically review a coherent national policy concerning the protection of workers, the public, and the environment against the risk of major accidents. Competent authorities are to establish a system for the identification of major hazard installations (based on a list of hazardous substances, or categories of substances, with their threshold quantities).
- Employers are to identify any major hazard installations within their control and notify such installations to the competent authorities.
- Employers are to establish and maintain a system of major hazard control at major hazard installations, prepare and update safety reports, and submit the reports to the competent authorities.
- Employers must report major accidents.
- Competent authorities are to establish emergency plans and procedures.
- Competent authorities are to disseminate information to the public on safety measures and correct behavior in the event of an accident.
- Competent authorities are to establish a siting policy with appropriate separation of proposed major hazard installations from areas frequented by the public and appropriate measures for existing installations.

- Workers and their representatives are to be consulted in order to ensure a safe work system.
- Workers are to be informed of, for example, hazards associated with the major hazard installations and to receive relevant instructions and training.
- Workers are to comply with practices and procedures relating to the prevention of major accidents and the control of developments likely to lead to a major accident with the installations, and to comply with all emergency procedures should a major accident occur.
- Exporting Parties are to provide any importing country with information concerning the use of hazardous substances, technologies, or processes as a potential source of a major accident and the reasons for it.

More information may be found at: http://www.ilo.org/safework/areasofwork/lang–en/WCMS_DOC_SAF_ARE_CHEM_EN/index.htm

OTHER ACTION AREAS

Children and Chemical Safety

Pesticides are a major problem especially as 70% of child labor worldwide is in agriculture. The ILO is tackling child labor pesticides exposure in two ways:

a. Direct action through project work at national level. More information may be found at: http://www.ilo.org/ipec/lang–en/index.htm
b. Legal: An important instrument that policy makers can use as part of their strategy to tackle hazardous child labor is a legally binding list of hazardous work activities and sectors that are prohibited for children. Countries that have ratified ILO Convention No. 182 are obligated to do this under Article 4.

In drawing up a national list, countries must also identify where such hazardous work is found and devise measures to implement the prohibitions or restrictions included in their list. Because this list is critical to subsequent efforts to eliminate hazardous child labor, the Convention emphasizes the importance of a proper consultative process, especially with workers and employers organizations, in drawing up, implementing it, and periodically revising it.

Advice for governments and the social partners on some hazardous child labor activities which should be prohibited is given in the ILO Worst Forms of Child Labour Recommendation, 1999 (No. 190), which accompanies Convention No. 182:

"Recommendation 190, Paragraph 3. In determining the types of work referred to under Article 3(d) of the Convention, and in identifying where they exist, consideration should be given, inter alia, to:

a. work which exposes children to physical, psychological, or sexual abuse;
b. work underground, under water, at dangerous heights, or in confined spaces;

c. work with dangerous machinery, equipment and tools, or which involves the manual handling or transport of heavy loads;
d. **work in an unhealthy environment which may, for example, expose children to hazardous substances, agents or processes**, or to temperatures, noise levels, or vibrations damaging to their health;
e. work under particularly difficult conditions such as work for long hours or during the night or work where the child is unreasonably confined to the premises of the employer."

The Worst Forms of Child Labour Convention, 1999 (No. 182) came into force in 1999 and has 172 ratifications in 2010.

The Green Jobs Initiative

Green Jobs have become an emblem of a more sustainable economy and society that preserves the environment for present and future generations and is more equitable and inclusive of all people and all countries.

Green jobs reduce the environmental impact of enterprises and economic sectors, ultimately to levels that are sustainable. Specifically, but not exclusively, this includes jobs that help to protect ecosystems and biodiversity; reduce energy, materials, and water consumption through high-efficiency strategies; de-carbonize the economy; and minimize or altogether avoid generation of all forms of waste and pollution.

Green jobs in emerging economies and developing countries include opportunities for managers, scientists and technicians, but the bulk can benefit a broad cross-section of the population who needs those most: youth, women, farmers, rural populations, and slum dwellers.

However, many jobs which are green in theory are not green in practice because of the environmental damage caused by inappropriate practices. The notion of a green job is thus not absolute, but there are "shades" of green and the notion will evolve over time. Moreover, the evidence shows that so called green jobs do not automatically constitute decent work. Many of these jobs are "dirty, dangerous, and difficult." Employment in industries such as recycling and waste management, biomass energy, and construction tends to be precarious and incomes are low. If green jobs are to be a bridge to a truly sustainable future, this needs to change. Green jobs therefore, need to comprise decent work. Decent, green jobs effectively link Millennium Development Goal 1 (poverty reduction) and Millennium Development Goal 7 (protecting the environment) and make them mutually supportive rather than conflicting.

The Green Jobs Initiative is a joint initiative by the United Nations Environment Programme (UNEP), the International Labour Organization (ILO), the International Employers Organization (IOE), and the International Trade Union Confederation (ITUC), which has been launched to assess, analyze, and promote the creation of decent jobs as a consequence of the needed environmental policies. It supports a concerted effort by governments, employers, and trade unions to promote environmentally sustainable jobs and development in a climate-challenged world.

Work under the Green Jobs Initiative so far has focused on collecting evidence and different examples of green jobs creation, resulting in a major comprehensive study on the impact of an emerging green economy on the world of work.

UNEP, ILO, IOE, and ITUC are planning a second phase of the Green Jobs Initiative. The project will move from information gathering and analysis in the green jobs report to assistance in policy formulation and implementation through active macro-economic and sectoral assessment of potential green jobs creation.

More information may be found at: http://www.ilo.org/integration/greenjobs/lang–en/index.htm

INTERNATIONAL COOPERATION ON CHEMICAL SAFETY

A major part of ILO work in the field of chemical safety takes place within the context of established mechanisms for interagency cooperation. The International Programme on Chemical Safety (IPCS), a partnership between ILO, UNEP, and the World Health Organization (WHO) was established in 1980 with a mission to develop and disseminate internationally peer-reviewed chemical risk assessments and other activities related to chemical safety. Following the UN Conference on Environment and Development (UNCED) in 1992, the Inter-Organization for Sound Management of Chemicals (IOMC) was set up in 1995. The IOMC coordinates chemical safety activities of UNEP, FAO, ILO, WHO, United Nations Institute for Training and Research (UNITAR), UNIDO, and Organisation for Economic Cooperation and Development (OECD), and has United Nations Development Programme (UNDP) and the World Bank as observers. See Chapter 23 of this book on IOMC.

THE GHS

At the 76th Session of the International Labour Conference (1989), ILO adopted a Resolution concerning the harmonization of systems of classification and labeling for the use of hazardous chemicals at work. Work on the GHS started as a follow-up to the adoption of the Chemicals Convention, 1990 (No. 170). The work was coordinated and managed under the auspices of the IOMC and the technical focal points were the ILO, OECD, and the United Nations Economic and Social Council's Sub Committee of Experts on the Transport of Dangerous Goods (UN SCETDG). The GHS has been designed to cover all chemicals, including pure substances and mixtures and to provide for the chemical hazard communication requirements of the workplace, the transport of dangerous goods, of consumers, and the environment. As such it is a truly harmonized and universal technical standard that should have a far-reaching impact on all national and international chemical safety regulations.

Recognizing that unprecedented capacity building efforts would be required to enable countries, especially developing countries and countries with economies in transition, to implement the GHS, ILO, and UNITAR established the UNITAR/ILO GHS Global Capacity Building Programme. The UNITAR/ILO Global GHS capacity building program provides guidance documents, educational, awareness raising, resource and training materials regarding the GHS. Relevant topics include the development of national GHS implementation strategies, legislation, situation/gap analyses, chemical hazards, labeling, SDSs, as well as related support measures, such as comprehensibility testing. UNITAR and ILO are the designated focal point for capacity building in the UN ECOSOC Subcommittee of Experts on the GHS (SCEGHS). See Chapter 11 of this book for GHS.*

The IPCS International Chemical Safety Cards

The International Chemical Safety Cards (ICSC) summarize essential health and safety information on chemicals and are intended for use by workers and employers in the workplace and in education and training activities. The information on the cards is expressed as far as possible using standard phrases thereby enabling the use of computer-aided translation into various languages. The cards include data on

- Fire and explosion hazards and their prevention
- Symptoms and first-aid treatment following acute exposure
- Health effects of short-term or repeated exposure
- Occupational exposure limits
- Physical and chemical dangers
- Spillage disposal and emergency response
- Safe storage
- Packaging and labeling requirements
- Environmental data

The ICSCs offer a basic tool to supply workers with information on the properties of the chemicals that they use and may often be the principal source of chemical safety information in less developed areas or in small and medium-sized enterprises (SMEs).

The ICSC project is an undertaking of the joint WHO/ILO/UNEP IPCS and is being developed in the context of the cooperation between the IPCS and the Commission of the European Communities. The cards are prepared by participating institutions in various countries and go through several steps of consultation and editing before being peer reviewed by a group of international experts. This last step represents a significant asset of the ICSC as opposed to other packages of information prepared at national, local, or professional levels.

There are currently over 1400 cards. They are available to view and download on various Internet sites, including that of the ILO International Occupational Safety and Health Information Centre (CIS).†

Searches may be made by chemical name or synonym, Chemical Abstracts Service (CAS) number or ICSC reference number. The Cards are also available on the CIS SafeWork Bookshelf CD. The standard phrases used and criteria for their use are available in the Compiler's Guide.‡

* For more information, please visit: http://www.unitar.org/cwg/ghs/index.html and http://www.unitar.org/cwg/ghs_partnership/index.htm

† http://www.ilo.org/public/english/protection/safework/cis/products/icsc/index.htm

‡ http://www.ilo.org/public/english/protection/safework/cis/products/icsc/compguide.pdf

Translations of the cards into 15 languages (including Chinese, Japanese, Korean, Swahili, Thai, Urdu, Vietnamese as well as many European languages) are available through links on the CIS site and the NIOSH (U.S. National Institute for Occupational Safety and Health) site.*

A recent development of the cards has been the addition of the GHS classification. Thus, the cards provide a potential way of assisting in the worldwide implementation of GHS.

A planned development within the project is the design and implementation of an Internet-based system to allow for easier international editing of the cards and faster worldwide availability of the updated information. This development will include a change in the format of the data from standard phrases to standard sentences in order to facilitate translation.

SAICM

ILO, as part of the IOMC, was an active member in the development of the Strategic Approach to International Chemicals Management (SAICM). The SAICM is a policy framework for international action to advance the sound management of chemicals, adopted by the International Conference on Chemicals Management (ICCM) on February 6, 2006 in Dubai, United Arab Emirates.

See Section IV, Chapter 17 of this book on SAICM.

SAICM aims to encourage governments and other stakeholders to address chemical safety more effectively in all relevant sectors such as agriculture, environment, health, industry, and labor. The strategic approach will support the achievement of the goal agreed at the 2002 Johannesburg World Summit on Sustainable Development that, by the year 2020, chemicals will be produced and used in ways that minimize significant adverse impacts on the environment and human health. To this end, SAICM will promote capacity building for developing countries and countries with economies in transition and better coordination of international efforts to improve chemicals management.

The ILO Governing Body endorsed SAICM at its 297th Session (November 2006) and approved the follow-up activities proposed by the office to implement SAICM objectives. This included active involvement by the ILO in the operations of the SAICM Quick Start Programme Trust Fund Implementation Committee, as well as supporting ILO-related activities in the SAICM Global Plan of Action.

Furthermore, from December 10–13, 2007, ILO held a Meeting of Experts to Examine Instruments, Knowledge, Advocacy, Technical Cooperation and International Collaboration as Tools with a view to developing a policy framework for hazardous substances. ILO action in the field of chemicals was discussed and a roadmap was adopted for ILO's future work in the area of hazardous substances. The fulfillment of SAICM objectives featured prominently in the discussions.

---

* http://www.cdc.gov/niosh/ipcs/icstart.html

The Meeting of Experts recommended a plan of action* based on the following fundamental pillars: information and knowledge; international cooperation; preventative and protective systems aimed at reduction of risks; capacity building; social dialogue; and good governance. The plan of action is being implemented through a variety of instruments, including ILO standards and joint actions, and based on the principles of the 2003 *Global Strategy on Occupational Safety and Health* and SAICM, and in partnership with workers, employers, and governments, addressing the following areas:

1. Social dialogue: The joint support from employers and workers and their participation is essential for successfully achieving the goals of the organization with regard to the global management of hazardous substances.
2. Information and knowledge: The acquisition, management, and dissemination of information and knowledge related to hazardous substances need to be continuous and integrated in the process of developing and marketing chemicals. Universal access to this information and knowledge is essential to the development of prevention and protection tools.
3. International cooperation: In order to contribute fully to the implementation of the SAICM, the ILO should continue to actively collaborate with other IOMC members as this is an effective mechanism for policy coordination for chemical management, in particular, strengthen the technical collaboration with the UNITAR in developing chemical safety training tools for the GHS and guidance for the implementation of national chemical safety programs.
4. Awareness raising and capacity building: In order to promote an effective implementation of relevant instruments on the sound management of chemicals, the ILO should, in the context of the Decent Work Country Programmes (DWCPs), mobilize internal and external resources to include chemical safety components in its technical cooperation projects related to the building and strengthening of national occupational safety and health (OSH) systems and programs. In doing so, ILO, in collaboration with other members of the IOMC, should cooperate closely with employers, workers, and governments with a view to improving the sound management of chemicals at national and global levels, particularly within SMEs.
5. Good governance and knowledge dissemination: As provided by the Dubai Declaration, sound management of hazardous substances requires effective and efficient governance through transparency, public participation, and accountability involving all stakeholders. The application of a systems approach to the sound management of chemicals is essential, both at national and enterprise levels, particularly for SMEs. Thus, the ILO should promote, in collaboration with other members of the IOMC, and in the implementation

* The full meeting of experts report, as well as background paper and ILO plan of action in the field of hazardous substances may be found at the following URL: http://www.ilo.org/public/english/dialogue/sector/techmeet/mepfhs07/index.htm

of the SAICM, the ratification of Convention Nos. 170, 174, and 187, and the establishment of national OSH systems, programs and profiles.

6. Preventative and protective systems aimed at risk reduction: Prevention entails implementation of preventative and protective systems. In this context the ILO should focus its action in cooperation with employers, workers, and governments, and other IOMC members, on promoting the implementation of preventative and protective measures according to the hierarchy of controls as contained within Section 3.10 of ILO–OSH 2001 and applying appropriately the precautionary approach, as set out in *Principle* 15 of the Rio Declaration on Environment and Development, while aiming to achieve that chemicals are used and produced in ways that lead to the minimization of adverse effects on the health of workers.

FUTURE WORK

ILO work in the field of chemicals (and for most IOMC organizations) is under the framework of SAICM, with its mandate until 2020. ILO acts as the IOMC SAICM focal point for Central and Eastern Europe, and also assists member states in developing their projects for the SAICM Quick Start Programme.

23 Inter-Organization Programme for the Sound Management of Chemicals

Jan van der Kolk

CONTENTS

INTRODUCTION

Following the UN Conference on Environment and Development (UNCED) in 1992 (see Chapter 3 of this book), the Inter-Organization Programme for the Sound Management of Chemicals (IOMC) was established to foster increased collaboration between international organizations in the area of sound chemicals management.

The seven participating organizations of the IOMC are: the Food and Agriculture Organization (FAO), of the United Nations, the International Labour Organization (ILO), the Organisation for Economic Cooperation and Development (OECD), the United Nations Environment Programme (UNEP), the United Nations Industrial Development Organization (UNIDO), the United Nations Institute for Training and Research (UNITAR), and the World Health Organization (WHO). In addition the

United Nations Development Programme (UNDP) and the World Bank participate in the IOMC as observers.

The IOMC has its executive body, the Inter-Organization Coordinating Committee (IOCC). Administrative services for IOMC and IOCC are provided by WHO.

HISTORY

The IOMC was established in 1995 following recommendations made by the 1992 UNCED (also known as the Earth Summit; see Chapter 3 of this book) in Rio de Janeiro and in particular those in Chapter 19 of the conference report (*Agenda 21*) about toxic chemicals. FAO, ILO, UNEP, UNIDO, WHO, and OECD initially signed a memorandum of understanding; UNITAR joined the IOMC in 1997. In 2006, the International Conference on Chemicals Management (ICCM) reasserted the coordination function of the IOMC.

IOMC VISION STATEMENT

The IOMC is the preeminent mechanism for initiating, facilitating, and coordinating international action to achieve the WSSD 2020 goal for sound management of chemicals.

OBJECTIVE

The objective of the IOMC is to strengthen international cooperation in the field of chemicals and to increase the effectiveness of the organizations' international chemicals programs. It promotes coordination of policies and activities, pursued jointly or separately, to achieve the sound management of chemicals in relation to human health and the environment.

OPERATIONS

The IOMC organizations coordinate their activities on chemicals management through the IOCC. The IOCC is composed of representatives of the participating organizations who meet twice a year. Observer organizations may also attend the meetings. The Chair of the IOCC serves for one year on a rotational basis.

WHO is the current administering organization for the IOMC and provides secretariat services to the IOCC.

The IOCC coordinates the planning, programming, funding, implementation, and monitoring of activities undertaken jointly or individually by the IOMC organizations. In full consultation among all those involved, it helps identify gaps or overlaps in international activities, and makes recommendations on common policies.

The IOCC fosters information exchange and joint planning with the aim of ensuring effective implementation without duplication. Because intergovernmental organizations are mandated by their respective governing bodies and funded by governments, the latter will benefit directly from the IOMC through efficiencies which can be obtained through optimal coordination of the work of the IOMC participants.

TECHNICAL COORDINATING GROUPS

At the technical level, specific coordinating groups are established from time to time and when necessary. These groups provide a means for all interested bodies working in the respective areas to consult with each other on program plans and activities, and to discuss ways and means of ensuring that the activities are mutually supportive. They are periodically reviewed and are discontinued when their work is done. Past examples include: POPS, the GHS, and assessment of industrial chemicals.

Membership of the Coordinating Groups is not necessarily limited to intergovernmental bodies, but may involve nongovernmental organizations and appropriate national institutions.

SUBJECTS COVERED BY THE IOMC

The IOMC works on subjects related to chemical safety. For example, the IOMC addresses key areas such as those previously elaborated in Chapter 19 of *Agenda 21* and which are now covered by the SAICM (Strategic Approach to International Chemicals Management) Global Plan of Action:

1. Risk reduction
2. Knowledge and information
3. Governance
4. Capacity building and technical cooperation
5. Illegal traffic

CURRENT ACTIVITIES

Databases

The IOCC has established an IOMC database that includes

- A calendar of events, listing the past and future meetings, workshops, seminars, and so on. of the IOMC organizations.
- An inventory of activities, providing a summary table of all the past and current activities of the IOMC organizations. The IOCC keeps an inventory of relevant activities of its member organizations.

The inventory provides the title of each activity, the name of the IOMC participating organization responsible for implementation, any partners involved, programme area, outputs of the work, duration of activity, resources allocated, geographical coverage, and the relevant contact point.

The IOMC database has been developed and is hosted by the OECD.

Guidance Material

IOMC has published several guidance documents that assist countries in their activities to strengthen their chemicals management. Examples of this are:

- Developing and Sustaining an Integrated National Programme for Sound Chemicals Management (2004).*
- Developing a Capacity Assessment for the Sound Management of Chemicals and National SAICM Implementation (2007).†
- A Guide to Resource, Guidance, and Training Materials of IOMC Participating Organizations (2008).‡

SAICM IMPLEMENTATION

IOMC actively contributed to the development of SAICM§ (see Chapter 17 of this book). IOMC was a co-convenor of the first ICCM held in Dubai in 2006 (see Chapter 5 of this book) and actively contributed to preparations for ICCM2 (see Chapter 6 of this book). IOMC plays a key role in the implementation of government-mandated priorities agreed for SAICM.

For ICCM2, IOMC prepared a report on "activities of the IOMC and its participating and observer organizations for implementation of the SAICM" along with a number of documents for consideration by ICCM2.¶

IOMC PUBLICATIONS

The IOMC develops documents on issues of common interest to stakeholders working in chemical management. In particular, the IOMC has issued documents of direct relevance to SAICM (see Chapter 17 of this book). They are prepared in cooperation with inputs from all IOMC organizations and other stakeholders. See http://www.who.int/iomc/publications/publications/en/index.html.

Documents include those related to SAICM implementation, to capacity strengthening for the sound management of chemicals in the broader sense, implementation of relevant instruments such as GHS (see Chapter 11 of this book), and new technologies, such as nanomaterials.

OUTLOOK

The aims of the UN is to strengthen its implementation capacity and service to countries, delivering more and more as "one UN" underlines the importance of arrangements such as the IOMC. It is to be expected that during the implementation of SAICM and the many efforts to achieve the "2020 goal" (see Chapter 4 of this book) the role of IOMC will be further strengthened.

---

* http://www.pic.int/secEdoc/Developing%20NAP%20for%20Sound%20Chemicals%20Management%20Guidance%20doc-04.pdf

† http://www.pic.int/secEdoc/Developing%20Capacity%20Assessment%20for%20chemicals%20and%20SAICM%20implementation.pdf

‡ http://www.who.int/iomc/saicm/resource_guide.pdf

§ http://www.saicm.org

¶ http://www.iomc.info/saicm

24 The International Panel on Chemical Pollution

Martin Scheringer, Åke Bergman, and Heidelore Fiedler

CONTENTS

IMPACTS OF ANTHROPOGENIC CHEMICALS ON HUMAN AND ENVIRONMENTAL HEALTH: A NEED FOR IMPROVED COMMUNICATION BETWEEN SCIENTISTS AND POLICYMAKERS

There are more than 100,000 commercially relevant chemicals on the market worldwide with some 40,000 of significant production volumes (JRC, 2009; US EPA, 2009). These chemicals can largely be grouped into pharmaceuticals, pesticides for agricultural and nonagricultural uses, and industrial chemicals. In addition, there are unintentionally generated chemicals such as polycyclic aromatic hydrocarbons (PAHs), polychlorinated dibenzodioxins (PCDDs), and polychlorinated dibenzofurans (PCDFs). They are not intentionally manufactured for any reason other than laboratory purposes but are unintentionally formed in industrial–chemical processes, such as chemical manufacture, and thermal processes, such as waste incineration (Neilson, 1998; Fiedler, 2003; United Nations Environment Programme (UNEP), 2007). Pharmaceuticals and pesticides have to undergo a detailed investigation of their properties, the effects that they can cause in organisms and, on this basis, a risk assessment before they are admitted to the market. These groups comprise several hundred active ingredients or active substances, which is only a small fraction of the entire set of chemicals on the market. Industrial chemicals constitute, by far, the largest segment of the chemicals market. They include a wide variety of chemicals used for a multitude of purposes, for example, solvents, plastic monomers, plastic additives (e.g., flame retardants, antioxidants, and plasticizers), process chemicals, surface coatings, fragrances, surfactants, dyes, and many more. These chemicals do not have to undergo a testing procedure as detailed as for pharmaceuticals and pesticides/biocides, and for many of them, their physicochemical properties and toxicity are poorly known; often, no information is available at all (EEA, 1998; Allanou et al., 1999). To improve this situation is one of the main goals of the new chemicals legislation of the EU, the regulation on the Registration, Evaluation, Authorisation and Restriction of Chemicals (REACH) (REACH, 2007).

Threats to human and environmental health can be caused by all of the above categories of chemicals. Pesticides have been relatively well investigated, but many users in developing countries do not have access to the information that is, in principle, available. In many developing countries, there are pesticide stockpiles whose size and composition is unknown and which represent a risk for releases of highly potent toxic substances (FAO, 2009). For pharmaceuticals, the focus of the assessment, for example, in preclinical and clinical trials, is on their performance in humans whereas environmental risks are not directly relevant to whether they are admitted to the market. In recent decades, significant volumes of some pharmaceuticals have been identified to be released to the environment (Kümmerer, 2008). Industrial chemicals are in most cases less well characterized than pesticides, and several thousand of them are used in amounts of more than 1000 tonnes/year (high production volume chemicals). Many industrial chemicals are presently detected in all environmental media and also in human tissue. Also, unwanted by-products such as PAHs and PCDDs/PCDFs are released to the environment in significant quantities in many parts of the world, are distributed globally, and contribute to chemical pollution problems. A particular concern is that, in addition to the specific effects that may be caused by individual chemicals, the presence of many chemicals in the human body may impair the intellectual and sexual development of humans (Jacobson and Jacobson, 1996; Guillette et al., 1998; Hood, 2005).

It should be noted that under both national legislation and multilateral international environmental agreements such as the Stockholm Convention on Persistent Organic Pollutants (POPS), chemicals are evaluated on a substance-by-substance basis and not as mixtures. The only exemption for risk assessment so far is the toxicity–equivalency approach for mixtures of dioxin-like compounds (PCDD/PCDF and dioxin-like PCB) were the contributions of 29 individual congeners to the overall toxicity is measured (van den Berg et al., 2006). Otherwise, the combined risk caused by the simultaneous presence of mixtures of chemicals or the combination of toxic mechanisms is rarely addressed because of inherent difficulties in evaluation.

Because of the multitude of chemicals and their use in technical applications or consumer products and because of the different cultural, environmental, economic, and regulatory conditions in different countries, chemical pollution problems are extremely diverse. It is noteworthy that management of chemicals is relatively new on the global political agenda and typically is not a priority issue. However, first steps are underway and it is hoped that more information and better coverage will be obtained in the future. Chemical pollution occurs at scales from local to regional and global, and the management of this pollution is addressed by a wide range of national and international activities, approaches, and programs. It is not our intention to provide a comprehensive review of these activities regarding chemicals assessment and management. Instead we want to point out the need for more extensive communication from scientists to stakeholders regarding the state of the science related to chemicals, health, and environmental effects. Available scientific information about emissions of a chemical, the chemical's properties, environmental fate, and toxic effects; and about the most important pathways of human exposure is frequently incomplete, inconsistent, fraught with uncertainty, and difficult to access. The fact that a huge amount of information is in principle available on the Internet does not solve this problem, because the information needs to be gathered, evaluated, and interpreted before it can be used in chemicals management. This type of synthesis needs to be done by experts from the scientific side. Hence, there is a need for interface institutions bridging the gap between science, on the one hand, and politics, decision making, and chemicals management, on the other hand. To some extent, this need is met by existing procedures for including scientific expertise in decision-making processes, for example, from members of the roster of experts of the Review Committee of the Stockholm Convention on Persistent Organic Pollutants and similar mechanisms for other international agreements. However, there is a potential for more comprehensive input from the scientific side. Scientists may join their efforts when they are called in as experts or even create an institutional framework to offer scientific input on chemicals-related issues to various national and international institutions, the public, and other interested actors.

The International Panel on Chemical Pollution (IPCP) (Scheringer et al., 2006; IPCP, 2009) is intended to function as such an interface institution. The focus of the IPCP is primarily on large-scale (global) pollution problems, but also on emerging pollution issues, and on regional environmental-pollution problems for which international exchange and collaboration is desirable. IPCP is presented in further detail below.

A SELECTION OF EXISTING INSTITUTIONS AND APPROACHES IN THE FIELD OF INTERNATIONAL CHEMICALS ASSESSMENT

There are several institutions and approaches that work in various ways at the interface between science and chemicals management. It is not possible here to provide a review of their scope and activity, but our goal is to illustrate that there is still a need for interface institutions such as IPCP. In other words, our point is not that existing institutions and approaches are not successful and needed. Our point is that they do highly important and effective work, but at the same time the need for this kind of work at the interface between science and chemicals management is so multifaceted and pressing (Hansson and Rudén, 2006) that the IPCP, as a new institution, still can make a useful contribution.

The selection of institutions and approaches, below, is not comprehensive but is intended to show some examples of interface institutions. It includes institutions that mainly deal with chemicals in the environment and chemical safety; note that several of these are described more fully elsewhere in this book.

The Arctic Monitoring and Assessment Programme

The Arctic Monitoring and Assessment (AMAP) (http://amap.no/) was established in 1991 and has the mandate of providing "reliable and sufficient information on the status of, and threats to, the Arctic environment, and providing scientific advice on actions to be taken in order to support Arctic governments in their efforts to take remedial and preventive actions relating to contaminants" (AMAP, 2009). AMAP is one of five working groups of the Arctic Council; AMAP member countries are Canada, Denmark/Greenland, Finland, Iceland, Norway, Russia, Sweden, and the United States. AMAP runs its own monitoring program of priority chemicals in the Arctic, including POPS, heavy metals, radioactivity, and others. In addition, it carries out assessments of the status of the Arctic environment and publishes AMAP Assessment Reports, which are comprehensive presentations of scientific information, and also State of the Arctic Environment Reports, which are shorter and directed toward politicians as audience.

The European Monitoring and Evaluation Programme

The European Monitoring and Evaluation Programme (EMEP) (http://www.emep.int/), in full the cooperative program for monitoring and evaluation of the long-range transmissions of air pollutants in Europe, was established under the Convention on Long-Range Transboundary Air Pollution (LRTAP Convention). Its main objective is to "regularly provide governments and subsidiary bodies under the LRTAP Convention with qualified scientific information to support the development and further evaluation of the international protocols on emission reductions negotiated within the Convention" (EMEP, 2009). The activities of EMEP include three elements, namely collection of emission data, measurements in air and precipitation, and modeling of atmospheric transport and deposition. EMEP works in collaboration with a broad network of scientists, mainly from the countries that are parties to the LRTAP Convention; as of 2009, there are 51 parties to the convention, which are mainly European countries and Canada and the United States. The structure of EMEP includes several task forces and scientific centers.

The International Programme on Chemical Safety

The International Programme on Chemical Safety (IPCS) (http://www.who.int/ipcs/en/) was established in 1980 as a cooperative program of World Health Organization (WHO), International Labour Organization (ILO), and UNEP; WHO is the executing agency of the IPCS. The objectives of the IPCS are to provide a scientific basis for chemical risk assessments and to contribute to capacity building in developing countries and countries with economies in transition. The IPCS publishes a wide range of documents on risk assessment of individual chemicals and on guidance and method development for chemicals assessment. These documents include, for example, more than 70 Concise International Chemical Assessment Documents (CICADs), more than 200 Environmental Health Criteria (EHC), and various publications on risk-assessment methodology for different topics. All of these documents are available from the IPCS Web site. The IPCS works in cooperation with national governments, nongovernmental organizations (NGOs), nominated national focal points, and many scientific institutions worldwide.

Two other activities under IPCS are the Joint FAO/WHO Expert Committee on Food Additives (JECFA) and the Joint FAO/WHO Meeting on Pesticide Residues (JMPR). Both JECFA and JMPR are international scientific expert committees that are administered jointly by the Food and Agriculture Organization of the United Nations (FAO) and the World Health Organization (WHO). JECFA was established in 1956 and initially evaluated the safety of food additives; today it also deals with the evaluation of contaminants, naturally occurring toxicants, and residues of veterinary drugs in food. To date, JECFA has evaluated more than 1500 food additives, approximately 40 contaminants and naturally occurring toxicants, and residues of approximately 90 veterinary drugs. The Committee has also developed principles for the safety assessment of chemicals in food that are consistent with current thinking on risk assessment and take account of recent developments in toxicology and other relevant sciences (JECFA, 2009).

JMPR consists of the FAO Panel of Experts on Pesticide Residues in Food and the Environment and the WHO Core Assessment Group; it has held meetings regularly since 1963. JMPR provides scientific advice to FAO, WHO, the governments of their member countries, and to the Codex Alimentarius Commission. It has evaluated about 230 pesticides and establishes acceptable daily intakes (ADIs) and acute reference doses (ARfDs) on the basis of toxicological data; taking into account the ADIs and ARfDs in combination with typical food intake and pesticide applications, it also recommends maximum residue limits for pesticides found in food commodities. In addition to the evaluation of individual pesticides, JMPR develops principles for assessing the safety of chemicals in food with continuous efforts to review and update the evaluation procedures (JMPR, 2009). JMPR publications are available from the IPCS Web site (http://www.who.int/ipcs/publications/jmpr/en).

Scientific Societies

Scientific societies, such as the environmental chemistry branches of the American Chemical Society (ACS) and the European counterpart, the European Association for Chemical and Molecular Sciences (EuCheMS), the Society of Toxicology (SOT), the Society of Environmental Toxicology and Chemistry (SETAC), and International Society for Environmental Epidemiology (ISEE) play important roles in the dissemination of novel research results and the investigation of chemical risks in general. Important goals of scientific societies are their services to scientists, such as publication of peer-reviewed scientific journals and books and the organization of scientific meetings. In addition, they are important professional networks. Recently, for example, SETAC has decided to disseminate scientific knowledge about chemical risks more actively and to address the needs of decision makers. Constitutive to the work of SETAC is the tripartite structure based on members from academic institutions, government, and industry. This is very useful for the interaction and collaboration of members with different backgrounds, but may also lead to some presence of industry perspectives when statements about controversial topics are made.

Collaboration with various scientific societies dealing with environmental research is essential for the work of the IPCP. The IPCP has established links to ACS, SETAC, SOT, the Swedish Society for Toxicology, and EuCheMS, and intends to develop similar links also with other scientific societies in the fields of chemicals, environment, and health.

The Role of IPCP in Relation to Existing Organizations

The goals of the IPCP are to some extent related to the purposes and activities of the organizations and approaches summarized above. However, there are still some differences that define the specific role of the IPCP. Compared to AMAP and EMEP, the IPCP is not planning to run its own monitoring programs; its geographical scope, on the other hand, is wider than that of AMAP and EMEP.

The IPCS delivers results such as the CICADs and/or EHC that are similar to the intended outcomes from the work of the IPCP. The IPCP certainly does not want to duplicate work that has been done or is being done by the IPCS, but intends to make its own contribution that will complement the work done by the IPCS. For example, compared to the chemicals covered by the IPCS, there are additional chemicals for which summaries of the available knowledge are needed; there are also more topics than the ones covered by IPCS for which guidance documents and state-of-the-science reports should be prepared. In addition, several of the CICADs and EHCs were prepared by the IPCS in the 1990s or even earlier and may need to be updated and amended. Finally, in contrast to the IPCS, the IPCP is not a UN institution; the IPCP has been established by scientists from all parts of the world, because scientists want to introduce results from their research in a more effective way into decision-making processes than in the past. Accordingly, the IPCP wants to create an interface between academic science and policy in the field of chemical pollution problems, see the next section.

THE IPCP INITIATIVE

Motivation

The IPCP is an initiative that is driven mainly by academic scientists. It is motivated by the realization of many scientists that there is potential for more comprehensive, systematic, and timely scientific contributions to decision-making processes dealing with chemical-related problems. "More comprehensive" means that all scientific fields that are relevant to the issue under discussion should be included and that knowledge from the various fields—chemistry, toxicology, engineering, epidemiology, to name a few—needs to be synthesized in an interdisciplinary effort. But even within single fields, there may be different scientific positions that need to be compared and balanced in an assessment of the state-of-the-science. "More systematic and timely" means that more scientists want to be informed in more detail about the needs of decision makers and about the procedures leading to decisions about chemicals in, for example, the context of international agreements, and about the different stages and the schedules and deadlines of these procedures.

Moreover, an institutional framework addressing the transfer of scientific knowledge into applications may be helpful for academic scientists, because such a framework can give their activity a more well-defined status. Currently, the academic environment often does not reward efforts spent on the interpretation and transfer of scientific results into application contexts. An established international institution working on the science–politics interface in the field of chemicals may help scientists to justify that they spend some time on the interpretation, dissemination, and application of their knowledge.

Finally, it is important for academic scientists to foster international collaboration. Within the scientific system, this is done by scientific societies, see above, but international collaboration is also needed in the dissemination and application of scientific knowledge in collaboration with other, nonscientific actors. This is one of the goals of IPCP, as more detailed below. By joining their efforts, scientists also want to create awareness of chemical pollution problems among decision makers in all parts of the world.

Organization of the IPCP

The IPCP initiative was started in 2006 at the Dioxin Conference in Oslo, Norway. A growing group of scientists expressed their interest in such an initiative, and as a first step a declaration was formulated that summarizes goals of the IPCP initiative.* The initiative is based on a mutual understanding that scientists need a platform of their own to present the state-of-the-science in relation to chemicals, health, and environment and to disseminate the available scientific knowledge.

This declaration, which is still open for signature by individuals who support the IPCP initiative, quickly found support from individuals from all parts of the world. In this way, a first network of scientists interested in the initiative could be established; as of 2011, individuals from 42 countries have signed the IPCP declaration. In a next step, the IPCP initiative was made known to a broader audience by several activities, including an overview article (Scheringer et al., 2006) and a full session at the Dioxin 2007 Conference in Tokyo, Japan (Suzuki and Morita, 2007).

In 2008, the IPCP was established as an association based in Zurich, Switzerland, which gives a legal status to the IPCP. Members of the association can be individuals working on or interested in any aspects of chemical pollution problems. An important aspect is that the IPCP is defined as an association of independent scientists, that is, individuals who can pursue their research interests without direct influence of

* http://www.ipcp.ch/IPCP_Declaration.html

conflicting interests. These are mainly academic scientists or scientists employed by governmental research institutes, whereas scientists working for industry, authorities, or environmental activist groups are not in the focus of the IPCP. This does not at all imply that the IPCP holds the view that these stakeholders are not needed in developing solutions to chemical pollution problems. However, the intention of the IPCP is to focus on the scientific understanding of a chemical pollution problem, without constraints that may be caused by business or other interests. Once the scientific understanding has been evaluated and consolidated as far as this is possible at the present time, this will serve as an input to the chemicals management process. (The members of the IPCP are aware that even academic scientists are not completely independent, and the resolution of possible conflicts of interests will be an important responsibility for all members of the IPCP.)

With the legal status of an association, the IPCP is an NGO and can as such attend conferences organized by UN institutions such as the International Conference on Chemicals Management (ICCM) (ENB, 2009). In addition to the general assembly of the IPCP members and the executive board of the IPCP, important organs of the IPCP will be working groups. The purpose of these working groups is to address priority topics related to chemical-pollution problems and to summarize the available scientific knowledge on a topic in such a language that is understandable to nonscientists.

Work Approach of the IPCP

As stated in its bylaws, the IPCP has three goals: (1) to initiate, prepare, and disseminate condensed state-of-the-science documentation on all aspects of environmentally relevant chemicals; (2) to act internationally and in countries with particular needs for improving knowledge regarding chemicals for them to manage issues related to chemicals; and (3) to offer the scientific expertise accumulated within IPCP to international organizations, national governments, and other parties for discussions and review of all aspects of the scientific basis for regional and/or global management of chemicals.

The first goal is motivated by the fact that even in industrialized countries transitioning to "knowledge-based societies" the knowledge that is needed for successful chemicals management is not easily accessible for decision makers, but needs to be "managed" in itself: past experience shows that the simple "transfer model" assuming that the relevant scientific information "trickles down" from science to its application needs to be revised. Therefore, the IPCP needs to go beyond the normal endpoint of scientific research, that is, publications in scientific journals. To this end, the IPCP selects priority topics, for example, single chemicals of high importance, such as DDT or polyfluorinated alkyl substances, or important broader subjects such as the assessment of human exposure to chemicals. In working on these priority topics, it is not the intention of the IPCP to carry out new research, nor is it the IPCP's intention to repeat work done by other institutions. The IPCP aims to prepare condensed reports of the currently available scientific knowledge by drawing on the expertise of scientists working in all fields of environmental research. These reports will be written in nontechnical language and will include executive summaries of a few pages. Depending on the needs of the target audience, language and format of the reports may be adjusted.

To work on individual topics, the IPCP plans to establish working groups. The specific tasks of such a working group are

- To review the available knowledge about a topic
- To identify the consensus and the disagreement present in the available knowledge
- To discuss and evaluate uncertainties
- To identify knowledge gaps and needs for future research
- To make this review and interpretation of the state of the science available to national governments, international institutions, and the public

In the work flow of IPCP working groups it will be important that the working group members also learn about the decisions to be made by the users of the scientific knowledge. In other words, the work of IPCP working groups will not be a one-way process that just delivers a summary of scientific knowledge to users of this knowledge, but it will involve the selection and interpretation of scientific knowledge (and lack of knowledge) in the light of practical problems to be handled by the policy makers.

Topics envisaged for first IPCP working groups are the assessment of effects caused by mixtures of chemicals and the assessment of human exposure and methods of biomonitoring. In addition, the IPCP is planning to compile a global inventory of emerging issues related to chemical pollution that are relevant in different parts of the world. Emerging issues are of high priority and require particular attention and also technical and scientific collaboration; examples are exposure to flame retardants and many other chemicals from the dismantling of electronic waste or mercury exposure in small-scale gold mining. However, what is relevant as an emerging issue depends strongly on the situation in a country or region. The IPCP wants to address this question by compiling information from its members all over the world. This inventory of emerging issues may also lead into the selection of topics for IPCP working groups.

A challenge for the work of the IPCP is that it requires scientists to take on a double role: they need to be active in their field of research and they need to be willing to put the knowledge from their field, including the tacit knowledge that is not present at the "surface" of journal publications, into a larger perspective, to relate it to knowledge from other fields, and to translate it into nontechnical language. This requires additional time and effort and is often not rewarded by the academic environment. On the other hand, this process will make scientific knowledge better available for a range of nonscientific users, which is an intention shared by many scientists working on chemical-pollution problems.

The second goal of the IPCP is addressed by the international network that is formed by scientists working with the IPCP. On the one hand, the IPCP supports transfer of knowledge, data, methods, and experience from industrialized countries to developing countries and countries with economies in transition. On the other hand, experience and results from research into chemical pollution problems in developing countries will be delivered to the international community and to scientists in industrialized countries so that they can incorporate these results and insights into the further development and adjustment of methods and tools.

A first activity in this field is that the IPCP coordinated a successful application to the Quick Start Programme (QSP) of the Strategic Approach to International Chemicals Management (SAICM). The QSP provides funding to support "initial capacity building activities for the implementation of SAICM objectives" (QSP, 2009). The application coordinated by the IPCP was submitted as a multicountry project by the governments of Armenia, Chile, and Ghana; the IPCP acts as the executing agency of the project. The project will run for 1.5 years and deals with the application of tools and methods for chemical hazard and risk assessment in the different contexts of the three participating countries. It is intended to contribute to capacity building by providing training on risk-assessment methods on the basis of country-specific applications and case studies. On this basis, also a better understanding of the chemical-pollution problems, their type, origin, and severity, in the three countries will be achieved.

The third goal of IPCP is dissemination. The work conducted within the IPCP will be delivered to various stakeholders, including the public, intergovernmental organizations, and national governments. It is important to stress that the input given by IPCP will be based on science to establish a basis for chemicals management. Beyond this scientific contribution, management of chemicals-related environmental issues needs the input from several other sources.

OUTLOOK: TOWARD A NEW UN PANEL ON CHEMICALS?

The main goals of the IPCP, namely to evaluate and interpret available scientific knowledge, to disseminate the knowledge in a form that is suitable for nonscientists, and to foster international collaboration on research into chemical risks, are to some extent analogous to the goals of the Intergovernmental Panel on Climate Change (IPCC) (IPCC, 2009). The goals of the IPCC are stated as follows (IPCC, 2009):

> The IPCC was established to provide the decision makers and others interested in climate change with an objective source of information about climate change. The IPCC does not conduct any research nor does it monitor climate related data or parameters. Its role is to assess on a comprehensive, objective, open and transparent basis the latest scientific, technical and socio-economic literature produced worldwide relevant to the understanding of the risk of human-induced climate change, its observed and projected impacts and options for adaptation and mitigation. IPCC reports should be neutral with respect to policy, although they need to deal objectively with policy relevant scientific, technical, and socio-economic factors. They should be of high scientific and technical standards, and aim to reflect a range of views, expertise and wide geographical coverage.

However, there is an important difference between the IPCC and the IPCP. The IPCC was created in 1988 as a UN body on the basis of a decision of the Executive Council of the World Meteorological Organization (WMO), and with support by the

UNEP Governing Council. Members of the IPCC are countries; the IPCC is open to all member countries of WMO and UNEP. Scientists working for the IPCC are nominated by IPCC member countries or may be invited in their own right by the IPCC to contribute to the work of the IPCC.

The IPCP, in contrast, is an association of individual scientists and has the legal status of an NGO; it is not affiliated with the United Nations. Independent scientists working on problems related to chemical pollution can apply for IPCP membership; members of the IPCP come from all continents of the world.

Recently, an intriguing development has taken place that suggests a potential relationship between the IPCP and UN organizations. Two Swedish politicians—M. Wallström, and V. Bohn—recently have independently proposed that a new UN panel on chemicals be established (ECHA, 2009; ICCM2, 2009), but it is not yet clear in what way this proposal will be carried on and what the relationship of such a new panel and the IPCP will be. The IPCP supports the proposal for such a new UN panel on chemicals and may, by its work, prepare the ground for its establishment. It is, however, important to keep in mind that the IPCP is an initiative from academia and will also in the future be based on individual scientists, whereas a UN chemicals panel is an initiative at the level of intergovernmental organizations. This does not mean that these lines cannot merge in the future. Much more work is to be done in this context, work beyond the present IPCP responsibility.

REFERENCES

Allanou, R., B. G. Hansen, and Y. van der Bilt. 1999. *Public Availability of Data on EU High Production Volume Chemicals*. Ispra, Italy: European Commission, Joint Research Centre.

AMAP. 2009. *Arctic Monitoring and Assessment Programme*. Oslo, Norway. Available: http://amap.no

ECHA. 2009. Wallström opens ECHA's conference centre and calls for a UN chemicals panel. European Chemicals Agency, ECHA/PR/09/04. Available at: http://echa.europa.eu/doc/press/pr_09_04_inauguration_conf_centre_20090406.pdf

EEA. 1998. *Chemicals in the European Environment: Low Doses, High Stakes? The EEA and UNEP Annual Message 2 on the State of Europe's Environment*. Copenhagen, Denmark: European Environment Agency.

EMEP. 2009. European Monitoring and Evaluation Programme. About EMEP. Available: http://www.emep.int

ENB. 2009. ICCM2 Final, Earth Negotiations Bulletin 15, No. 175. Available at: http://www.iisd.ca/chemical/iccm2/

FAO. 2009. http://www.fao.org/ag/AGP/AGPP/Pesticid/Disposal/en/103401/index.html

Fiedler, H. 2003. Dioxins and furans. In. *The Handbook of Environmental Chemistry* (Vol. 3), Part O, *Persistent Organic Pollutants*. Berlin: Springer Verlag, pp. 125–201.

Guillette, E. A., M. M. Meza, M. G. Aquilar, A.D. Soto, and I. E. Garcia. 1998. An anthropological approach to the evaluation of preschool children exposed to pesticides in Mexico. *Environmental Health Perspectives*, 106: 347–353.

Hansson, S. O. and C. Rudén. 2006. Priority setting in the REACH system. *Toxicological Sciences*, 90: 304–308.

Hood, E. 2005. Are EDCs blurring issues of gender? *Environmental Health Perspectives*, 113, A671–A677.

ICCM2. 2009. Advance Report of the International Conference on Chemicals Management on the work of its second session, Geneva, May 11–15, 2009. Available at: http://www.saicm.org/index.php?content=meeting&mid=42&def=1&menuid=9

IPCC. 2009. Intergovernmental Panel on Climate Change, Geneva, Switzerland. http://www.ipcc.ch

IPCP. 2009. International Panel on Chemical Pollution, Zürich, Switzerland, http://www.ipcp.ch

Jacobson, J. L. and S. W. Jacobson. 1996. Intellectual impairment in children exposed to polychlorinated biphenyls *in utero*. *The New England Journal of Medicine*, 335, 783–789.

JECFA. 2009. Joint Expert Committee on Food Additives. Available at: http://www.who.int/ipcs/food/jecfa/en/

JMPR. 2009. Joint Meeting on Pesticide Residues, available at: http://www.who.int/ipcs/food/jmpr/about/en/index.html

JRC. 2009. EINECS online information system. Joint Research Centre, European Commission, Ispra, Italy. Available at: http://ecb.jrc.ec.europa.eu/esis/index.php?PGM=ein

Kümmerer, K., ed. 2008. *Pharmaceuticals in the Environment*. Berlin, Germany: Springer, 650pp.

Neilson, A., ed. 1998. *PAHs and Related Compounds. The Handbook of Environmental Chemistry* (Vol. 3) Anthropogenic Compounds, Part 3I. Heidelberg: Springer, 412pp.

QSP. 2009. *Strategic Approach to International Chemicals Management—Quick Start Programme*. Available at: http://www.saicm.org/index.php?menuid=22&pageid=252

REACH. 2007. Registration, evaluation, authorisation and restriction of chemicals. *Official Journal of the European Union*. http://www.reach-compliance.eu/english/legislation/docs/launchers/launch-2006–1907-EC-06.html

Scheringer, M., H. Fiedler, N. Suzuki, I. Holoubek, C. Zetzsch, and Å. Bergman. 2006. Initiative for an International Panel on chemical pollution. *Environmental Science and Pollution Research*, 13: 432–434.

Suzuki, N. and M. Morita. 2007. Potential needs on the global framework for the control of chemical pollution: Existing international framework and expected perspectives for the IPCP—International Panel on Chemical Pollution. *Organohalogen Compounds*, 69: 532–535.

UNEP. 2007. *Guidance for Analysis of Persistent Organic Pollutants (POPS)*. UNEP Chemicals Branch, DTIE, Geneva, Switzerland, March 2007. Available at: http://www.chem.unep.ch/pops/laboratory/analytical_guidance_en.pdf

US EPA. 2009. Chemical assessment and management program, TSCA inventory reset. US Environmental Protection Agency, Washington DC. Available at: http://www.epa.gov/champ/pubs/hpv/tsca.html

van den Berg, M., L. S. Birnbaum, M. Denison, M. De Vito, W. Farland, M. Feeley, H. Fiedler, et al. 2006. The 2005 World Health Organization re-evaluation of human and mammalian toxic equivalency factors for dioxins and dioxin-like compounds. *Toxicological Sciences*, 93: 223–241.

25 The Role of the International POPS Elimination Network

Mariann Lloyd-Smith

CONTENTS

INTRODUCTION

The International POPS Elimination Network (IPEN) is a unique organization: a public interest NGO promoting chemical safety that emerged in response to the global recognition of the need to eliminate POPS.

Since its inception in 1998 IPEN has grown to a global network of over 800 public interest nongovernmental organizations (NGOs) from more than 100 countries united in support of the common goal of a "toxic free future." IPEN facilitates the engagement of public interest NGOs in efforts to eliminate POPS and other persistent toxic substances (PTS), and to work for a world where exposure to chemicals is no longer a significant source of harm to public health and the environment. IPEN has emerged as a broad-based international chemical safety network with a global reach and the ability to translate chemical policy into concrete action on the ground. IPEN provides a bridge to ensure that international policy discussions are relevant to the concerns of local and national NGOs and their communities in developing and transitional countries, and that the issues highlighted by developing country NGOs are given a voice at international forums.

SETTING THE SCENE

Since World War II, approximately 80,000 new synthetic chemicals have been manufactured and released into the environment, with an estimated 1500 new chemicals being introduced each year. This growth in synthetic chemicals has been matched by growing concerns over their toxic impacts on humans and the environment. The vast majority of pesticides and industrial chemicals have not been adequately tested for their long-term health and environmental impacts, particularly in terms of emerging concerns such as endocrine disruption and the impacts of mixtures of chemicals, occurring in the environment. The little information that does exist is often not available to workers and exposed communities, particularly in developing countries and countries with economies in transition. Highly publicized incidents and issues such as mercury contamination in Japan's Minamata Bay, dichloro-diphenyl-trichloroethane (DDT) contaminated meat, urban air pollution, and the Seveso factory dioxin leak in Italy served to highlight the risks and hazards of the new chemical age.

Community concerns over chemical hazards galvanized with the 1984 Bhopal disaster, when over 3000 people died due to the toxic gas leak (methyl isocyanate) from the Union Carbide factory in India. Many more were to die later from the impacts of this chemical disaster and people today are still dealing with the toxic aftermath.

By 1992, governments meeting at the Rio Earth Summit had little choice but to acknowledge that chemical contamination could be a source of "grave damage to human health, genetic structures and reproductive outcomes, and the environment."[*] The subsequent Chapter 19 of *Agenda 21* focused on Environmentally Sound Management of Toxic Chemicals, and in particular, the needs of developing countries when faced with the chemical hazards of their rapidly industrializing economies.

Agenda 21 also provided clear acknowledgement that it is in the public interest for the community to be informed, to exercise their right to understand, to make informed choices, and to participate in informed decision making.[†] It acknowledged that environmentally sound management of chemicals and waste is reliant on effective stakeholder participation. As a result, in 1994 the Intergovernmental Forum on Chemical Safety (IFCS) was established. IFCS actively supported multisectoral, multistakeholder engagement in information exchange and capacity building initiatives related to

---

[*] *Agenda 21*, Chapter 19, Environmentally Sound Management of Toxic Chemicals, Including Prevention of Illegal International Traffic in Toxic & Dangerous Products, Section 19.2. Available at: http://www.un.org/esa/dsd/agenda21/res_agenda21_19.shtml

[†] *Agenda 21: Programme for Action for Sustainable Development Rio Declaration on Environmental Development.* United Nations Conference on Environment and Development (UNCED), June 3–14, 1992, Rio de Janeiro, Brazil; Also see *Principle* 10 Annex I, Rio Declaration on Environment and Development, *Principle* 10. Available at: http://www.un.org/documents/ga/conf151/aconf15126-1annex1.htm

chemical safety. Both public interest NGOs and industry associations received full participation rights along side national governments.

A large number of local, national and regional environmental and public health NGOs emerged around the world in response to the harmful environmental and human health impacts caused by toxic chemicals' production, use and disposal. International chemical campaigns by Greenpeace, Pesticide Action Network (PAN), and WWF* stimulated awareness and engagement on the part of domestic NGOs working on health and environmental issues and set the stage for the emergence of many new global networks. These included the Basel Action Network (BAN); Health Care without Harm (HCWH); the Global Alliance for Incinerator Alternatives (GAIA), and the IPEN.

In 1996, IFCS recommended that the international community respond to the growing problem of POPS. Capable of traveling the globe on air and water currents, toxic POPS contamination showed no respect for territorial borders, and it was obvious that no country acting alone would be able to deal with them effectively; rather a global response was necessary and urgent. In 1998, negotiations commenced for a global treaty that aimed to control and/or eliminate an initial 12 POPS chemicals, which became widely known as the "dirty dozen."†

In accordance with *Principle* 10 of the Rio Declaration, in 2000, at the third meeting of the IFCS, where IPEN participating organizations (IPEN POs) were active negotiators, it was recognized that communities had a right to participate meaningfully in decisions about chemical safety that affect them.‡ The subsequent *Stockholm Convention on Persistent Organic Pollutants* 2001 reflected this and specifically highlighted in Article 10 the important role of public participation for the national implementation of the Convention.

IPEN AND THE STOCKHOLM CONVENTION NEGOTIATIONS

IPEN was founded in June 1998 at the first session of the UNEP Intergovernmental Negotiating Committee (INC1) for a global POPS Convention. Throughout the three years of negotiations, IPEN working in alliance with PAN, GAIA, Greenpeace, WWF, and Arctic indigenous peoples effectively mobilized and coordinated the participation of more than 350 environmental, public health, and consumer NGOs from 40 countries, including many from developing countries and those with economies in transition. This meant that at the commencement of the second negotiating session (INC2), country negotiators were aware of the more than 100 representatives of civil society present in the conference room. IPEN's contribution to the successful conclusion of POPS negotiations was recognized in 2003 when the

---

* Founded as the World Wildlife Fund.

† The initial list of 12 POPS include industrial chemicals like polychlorinated biphenyls (PCBs) used in transformer oils; pesticides like DDT, endrin, dieldrin, aldrin, chlordane, toxaphene, heptachlor, mirex, hexachlorobenzene (HCB); and unwanted wastes like dioxins and furans.

‡ Intergovernmental Forum on Chemical Safety, Bahia Declaration on Chemical Safety, Third Session—Forum III Final Report of Intergovernmental Forum on Chemical Safety (IFCS/FORUM III/23w) Brazil, October 2000 at para 11/6.

governments participating in IFCS presented IPEN with a Special Recognition Award for their work.*

During the POPS negotiations, the IPEN POs prepared and distributed briefing papers, collated case studies and collectively built the capacity of the many national NGOs involved in chemical issues. The cooperative development of the IPEN Stockholm Handbook saw the consolidation of these and preferred NGO convention text, which not only informed the NGO network but also provided suggested text on which smaller developing nations could draw (Lloyd-Smith, 2003).

IPEN also worked closely with the indigenous peoples, especially those of the Arctic, involving them in the development of NGO policy and positions. Arctic peoples are most at risk from POPS due to the global distillation of these contaminants in the colder regions of the world. As a result, the special vulnerability of this group was recognized in the preamble to the Stockholm Convention.

IPEN helped NGOs working on the ground to understand the importance and potential impacts of the international negotiations on national policy decisions and design. Many NGOs addressing waste issues, pesticide stockpiles, and/or pesticide poisonings in their communities began to appreciate the connection between their issues and POPS negotiations. As difficult issues emerged, IPEN coordinated NGO discussions and sourced technical and scientific expertise from among the networks. This created a learning environment for NGOs around the complex dynamics of the issues and their policy implications.

IPEN provided a bridge to ensure that POPS negotiations were relevant to the concerns of local and national NGOs in developing countries, and that the issues highlighted by developing country NGOs were given a voice. IPEN POs have direct knowledge of the impacts of chemical exposure on their communities, local environments and countries, as well as the capacity for their national governments to implement or enforce national and/or local regulation. By involving these groups in intergovernmental policy negotiations, the practical challenges of designing international policies that are to be applied at the community or national level, can be more effectively addressed. For example, case studies provided by Indian NGOs highlighted the need for technical assistance to ensure the elimination of POPS pesticides in agriculture. Similarly, the exemption from confidentiality provision for information on the health and safety of humans and the environment was a direct result of lobbying by IPEN POs cognizant of the restrictions imposed on much needed chemical information by commercial interests.

Thereafter, when the chemicals policies are internationally agreed, IPEN POs are well positioned to promote the implementation of these policies in their country and their local communities. The IPEN network has regional and subregional components that operate in all areas of the globe and in all six UN languages, as well as other languages.

The majority of IPEN POs focus on public and environmental health issues with many having specialized and scientific expertise in areas such as agriculture, waste management, clean production, consumer issues, and workers' rights. Members of

* See IFCS 2003 Special Recognition Award. Available at: http://www.who.int/ifcs/documents/forums/forum4/sp_award/en/index.html

IPEN POs are also drawn from academia and the medical profession. IPEN has helped to elevate the national profile of its POs, assisting them in gaining access to their national decision makers. The relationship built between IPEN POs and their national governments continued to grow throughout the Stockholm Convention implementation process and in other chemical forums. IPEN is now a global network of over 800 NGOs from more than 100 countries. The majority of them are from developing countries and those with economies in transition and many also participate in other networks such as GAIA and PAN. All are united in support of a common POPS elimination goal, that is, to eliminate POPS and other PTS, and work for a world where exposure to chemicals is no longer a significant source of harm to public health and the environment, in line with the World Summit on Sustainable Development (Johannesburg, 2002) goal for 2020.

IPEN POST 2001, IMPLEMENTING THE STOCKHOLM CONVENTION

After the POPS negotiations concluded in 2001 IPEN continued to work with NGOs to promote national ratification of the Stockholm Convention and to encourage speedy and effective implementation planning in their countries. IPEN coordinates NGO participation in all the Conferences of the Parties (COPs) of the Stockholm Conventions, as well as the numerous intersessional processes. This requires the production of technical papers and information packages,* while the ongoing internal NGO discussions about the evolving issues assists in strengthening and fostering NGO capacity building efforts at the country and regional level.

IPEN POs participate in the ongoing intersessional work of the Convention by contributing to the BAT/BEP Expert Group† to reduce dioxin emissions, the DDT Expert Group to promote alternatives to combat malaria, the POPS Review Committee (POPRC) to add new POPS to the Convention and other regional and national awareness raising activities. Public interest NGOs have an essential role to play balancing the self-interest of industry, as well as making available an extensive range of expertise, information and data.

INTERNATIONAL POPS ELIMINATION PROJECT

Between 2004 and 2006, IPEN managed and coordinated a global project known as the International POPS Elimination Project (IPEP). IPEP was mainly funded by the Global Environment Facility, and conducted in partnership with UNIDO and UNEP.‡ The broad goals of IPEP were to enhance the skills and knowledge of NGOs to help build their capacity as effective stakeholders in the Stockholm Convention implementation process and to help establish regional and national NGO capacity in support of longer term efforts to achieve chemical safety.

* See http://www.ipen.org

† BAT—best available techniques, BEP—best environmental practice.

‡ See IPEP Projects and Report. Available at: http://www.ipen.org/ipepweb1

IPEP exceeded expected outcomes on many levels. At the conclusion of the two-year project, IPEP had funded more than 290 projects implemented by over 350 NGOs in 65 developing and transitional countries. One reason for its success was that it was based on NGO/country-driven project development, as well as collaboration across developing countries. IPEP was managed by IPEN through the establishment of regional hubs housed in well-established NGOs in eight different regions.* NGO regional hubs identified the project-supported activities they wanted to undertake, received and administered funds and assistance, and provided the necessary oversight to ensure they succeeded. IPEP demonstrated how IPEN efficiently enabled NGOs to transform a global treaty on POPS into local and national activities on the ground.

NGO MONITORING: "THE IPEN EGG REPORT"

IPEN also plays an important role in providing monitoring data to support implementation activities. In 2005, IPEN's Dioxin, PCBs, and Waste Working Group completed the Egg Report case study, which reported on the contamination of domestic chicken eggs.† The study sampled backyard and free-range chicken eggs as a useful bio-indicator of food and environmental contamination. It investigated dioxin, furan, PCB, and HCB contamination in eggs from 17 countries on five continents. The eggs were collected near waste incinerators, cement kilns, metallurgical industries, waste dumps, and chemical production facilities in Belarus, Bulgaria, Czech Republic, Egypt, India, Kenya, Mexico, Mozambique, Pakistan, Philippines, Russia, Senegal, Slovakia, Tanzania, Turkey, Uruguay, and the United States.

The vast majority of samples exceeded European Union (EU) limits for contaminants in eggs with some containing the highest dioxin levels ever measured in chicken eggs. The results demonstrated a possible link between pollution sources and exposure patterns and, most importantly, indicated priority areas for action. For many of the countries involved, the egg study provided their first datasets about unintentional POPS. The egg study was an example of a low cost, global monitoring and cooperative data collection exercise. It clearly demonstrated the role and benefits of an international NGO network in the generation of data pertinent for effective chemical management.

STRATEGIC APPROACH TO INTERNATIONAL CHEMICALS MANAGEMENT

In 2003, the IPEN Steering Committee acknowledged that POPS are part of a larger chemical safety problem and recommended that IPEN expand its policy operations to include the global effort to develop a Strategic Approach to International

* Regional hubs are based in Franco Africa, English Africa, Latin America, Asia-Pacific, Asia, Eastern Europe, Caucasus & Central Asia, and the Middle East.

† The Egg Report, Contamination of chicken eggs from 17 countries by dioxins, PCBs, and hexachlorobenzene, "Keep the Promise, Eliminate POPS!" Campaign and Dioxin, PCBs and Waste Working Group of the International POPS Elimination Network (IPEN) Report. Available at: http://www.ipen.org

Chemicals Management (SAICM). SAICM is the action plan to implement the 2020 goal of the World Summit on Sustainable Development that is "to achieve by 2020 that chemicals are used and produced in ways that lead to the minimization of significant adverse effects on human health and the environment" (see Chapter 6 of this book).

SAICM also recognized that chemical safety was an essential component to the sustainable development agenda. In February 2006, Ministers of more than 140 governments endorsed the SAICM high-level declaration* (see Chapter 17 of this book), which states:

> The sound management of chemicals is essential if we are to achieve sustainable development, including the eradication of poverty and disease, the improvement of human health and the environment and the elevation and maintenance of the standard of living in countries at all levels of development.

IPEN coordinated NGO engagement in the SAICM development process including the Preparatory Conferences, International Conferences on Chemical Management, regional meetings and in the Extended Bureau. In 2006, IPEN took the formal decision to expand its mission beyond POPS and endorsed the IPEN Dubai Declaration for a Toxic Free Future at the International Conference on Chemical Management (ICCM1). IPEN POs committed:†

> ... to work for and achieve by the year 2020 a Toxics-Free Future, in which all chemicals are produced and used in ways that eliminate significant adverse effects on human health and the environment, and where persistent organic pollutants (POPS) and chemicals of equivalent concern no longer pollute our local and global environments, and no longer contaminate our communities, our food, our bodies, or the bodies of our children and future generations.

The decision to broaden IPEN's remit came at a time when there was growing awareness of the threat posed by other persistent, bioaccumulative or toxic chemicals (PBTs) such as endocrine disruptors. Many are used in domestic products, such as cookware, clothing, and electronic goods. Some of the most toxic PBTs are regularly detected throughout the global environment and a growing number had been identified in the umbilical cord of newborn babies (Dallaire et al., 2002).‡ Most worrying, the number of new PBTs being detected in the Arctic appeared to be growing and some like the common stain repellant, perfluorooctane sulfonate (PFOS) demonstrated no degradation or break down under any environmental conditions.§ As well, IPEN POs particularly from the developing and transitional countries were facing

* http://www.saicm.org/documents/saicm%20texts/SAICM_publication_ENG.pdf

† http://www.ipen.org/ipenweb/saicm/dubai.html

‡ Also see *Body Burden—The Pollution in Newborns*, A benchmark investigation of industrial chemicals, pollutants and pesticides in umbilical cord blood. Environmental Working Group, July 14, 2005.

§ "PFOS is extremely persistent. It has not shown any degradation in tests of hydrolysis, photolysis or biodegradation in any environmental condition tested." Risk Management Evaluation for Perfluorooctane Sulfonate 2008, POPS Review Committee. Available at: http://chm.pops.int/Convention/POPSReviewCommittee/AboutPOPRC/tabid/221/language/en-US/Default.aspx

the escalating threats of the ever increasing waste streams as well as illegal dumping by developed countries. In particular, the quantity of hazardous electronic waste containing PBTs finding its way to developing countries was and is still growing exponentially.

GLOBAL NGO SAICM OUTREACH CAMPAIGN

In 2007, IPEN collaborated with UNEP Chemical and the SAICM Secretariat, and secured resources to initiate and coordinate the Global NGO SAICM Outreach campaign.* The purpose of the campaign was to raise awareness about SAICM and to secure commitments from NGOs in all regions to undertake efforts to elevate the threats posed by toxic chemicals. The Outreach Planning Committee consisted of representatives from IPEN, HCWH, International Society of Doctors for the Environment (ISDE), PAN, Women in Europe for a Common Future (WECF), and the World Federation of Public Health Associations (WFPHA) The campaign targeted not only environmental NGOs, but also organizations from other sectors including health, agriculture, and labor. As a result of the campaign, more than 1000 NGOs in over 100† countries endorsed a civil society statement supporting SAICM and its objectives, committing themselves to contribute to the SAICM implementation. The campaign spread the message for the need for chemical management to ensure the protection of human health and the environment but also human rights and national development.

The NGO SAICM Outreach Campaign produced educational and outreach materials that were translated into UN languages and distributed in all regions. It also supported numerous NGO projects aimed at advancing SAICM objectives. In 2009, in preparation for the second SAICM meeting, the International Conference on Chemical Management (ICCM2) in May 2009, IPEN published the Citizen's Report‡ detailing NGO SAICM-related activities and their findings regarding current chemical management and identified priorities for action. The report described many of the difficulties of implementing good chemical management in the developing world and many of the failings in the developed world.

IPEN INTO THE FUTURE

In response to the needs of its POs, IPEN continues to expand and grow. It has become increasingly engaged on heavy metal issues including collaborating with the German Government to undertake a global mercury survey, participating in the mercury INC meetings and initiating the global Mercury Free campaign.§ IPEN has also expanded is efforts to address lead. Despite the phase out of lead in paint in highly developed countries in the 1970s and 1980s, an IPEN PO, Toxics Link of India, found high levels of lead in paint in the local market. This prompted a joint collaboration

* See Global SAICM Outreach Campaign. Available at: http://www.ipen.org/campaign

† See http://www.ipen.org/campaign/signed.html

‡ See Citizen's Report. Available at: http://www.ipen.org/ipenweb/documents/ipen%20documents/citzreport_09.pdf

§ See http://www.ipen.org/hgfree

to conduct a global sampling activity, where IPEN POs sampled paint in 10 countries. The 2009 outcomes of this monitoring found that paint sold in many developing countries often contains high concentrations of lead. IPEN brought a resolution to IFCS calling for the establishment of a global partnership to promote phasing out the use of lead in paints. The resolution was adopted and the proposal was subsequently approved at ICCM2.

In July 2009, IPEN launched the Heavy Metals Working Group, co-chaired by the Indian NGO Toxics Link and the Czech NGO Arnika. The group is tasked with coordinating the efforts of IPEN POs relating to the mercury negotiations, which started in June 2010, the lead in paint partnership, and other relevant heavy metal initiatives that may arise.

In preparation for ICCM2 held in May 2009, IPEN POs collaborated with Friends of the Earth, the European Environment Bureau and others, to form the Nanotechnology (nano) Working Group. The group tracks and provides NGO input into the international initiatives surrounding nanotechnology and nanomaterials.

POPS REVIEW COMMITTEE

Other ongoing activities such as the public interest NGO input into the POPRC, has required significant commitment and in-kind contributions from the NGO delegation drawn from IPEN and PAN POs. The POPRC, the scientific committee set up by the Stockholm Convention to assess nominated POPS, has completed nine assessments, which were adopted by the Fourth Conference of Parties in May 2009 (COP4). However, the listing of four of the new POPS (PFOS, OctaBDE, PentaBDE, and Lindane) included substantial exemptions to the use and waste obligations of the treaty. IPEN has committed to monitoring and tracking these exemptions and continues to campaign for adequate public information to ensure individuals can make informed decisions in order to avoid exposure to these new POPS.

Throughout the POPRC assessments, the NGO delegation has provided expert comment and information. Members of the POPRC NGO delegation developed "alternatives assessment" guidelines to assist countries in choosing appropriate alternatives to POPS chemicals. Researchers have now identified over 800 chemicals that match the structural profile of known Arctic contaminants such as POPS (Brown and Wania, 2008). Many of these chemicals have already been detected in Arctic biota and marine mammals and require substantial actions to restrict or eliminate their contamination. Needless to say, currently with only 21 POPS listed in the Stockholm Convention, IPEN POs have much more work to do to eliminate POPS, and to curb the build up of their toxic legacy.

CONCLUSION: THE FUTURE OF CHEMICAL MANAGEMENT AND IPEN's ROLE

The future of chemical management is in alternative "cleaner and greener" products, processes, and approaches to life. No amount of capacity building, training, or negotiations will be enough to address the chemical onslaught of this century unless there is radical change in the way we manufacture products, in the way we consume them, and in the way we address their waste or recycling phase. This is all the more pertinent in the age of climate change.

Chemical management and safety in the twenty-first century cannot be just a matter of more risk management and the ongoing justification of old ways and old chemicals. The core principles of chemical reform need to underpin all decision making about chemicals. It is only through "Right to Know," that society will have the information to make informed decisions; only through the principles of "No data/No market" and "Precaution" that we will be able to protect ourselves and our families from toxic threats, and only through the "Substitution Principle" will we be able to make far better choices into the future.

As acknowledged by the international community, chemical safety is an essential component to the sustainable development agenda, and the reduction of chemical hazards can only be achieved through proactive and effective chemical reform. Such reform must also acknowledge the interaction of climate on chemical toxicity and exposure (see Patra et al., 2007; UNEP, 2010; and also Noyes et al., 2009)* as well as waste management activities.

IPEN will continue to collaborate with NGOs across the globe, learning from their experiences while doing its best to address the capacity building needs of IPEN POs and civil society in general. IPEN will continue to foster an organizational structure based on principles of inclusiveness and responsiveness, gender equity, equitable participation, multilingual access, and information provision.

IPEN will continue its involvement in SAICM and strive to improve chemical safety globally and importantly, to fully integrate this issue into the development agenda. While it will participate in new negotiations, such as the preparations for a global instrument for mercury, it aims to maintain its work on the scope and implementation of the current chemical conventions, particularly in the POPRC and the identification of new POPS.

IPEN POs are not naïve and they do not see international environmental governance as the only way to a toxic-free future but recognize it as a valuable tool that can support national activities and grass roots actions. IPEN has learnt that the active engagement of public interest NGOs from all regions plays an essential role in achieving good outcomes in intergovernmental negotiating processes.

As was recognized at the Rio Earth Summit, the benefits of NGO involvement in global and national chemical management are numerous. Not only do NGOs provide a wide range of expertise, they also provide pertinent local information and data to measure policy options against. IPEN POs have instigated valuable monitoring and data collection as well as initiating community outreach, public education, and awareness raising activities. As an effective conduit to broader civil society, networks like IPEN make possible so many of the "on the ground" cost-effective management activities and drive behavioral change within civil society.

IPEN POs will continue to play a significant role in the protection of communities and the environment from toxic threats, particularly for those most affected by chemical policy and management decisions. IPEN continues to help empower civil

* For example, research has demonstrated exposure to some pesticides reduces a fish species tolerance to increased temperatures.

society to address POPS and other chemical safety issues, and by doing so is an effective leader in the movement for a toxic free future.

REFERENCES

Brown, T. N. and F. Wania. 2008. Screening chemicals for the potential to be persistent organic pollutants: A case study of Arctic contaminants. *Environ Sci Technol.* 42(14): 5202–5209.

Dallaire, F., E. Dewailly, C. Laliberté, G. Muckle and P. Ayotte. 2002. Temporal trends of organochlorine concentrations in umbilical cord blood of newborns from the lower north shore of the St. Lawrence river (Québec, Canada). *Environ Health Perspect.* August, 110(8): 835–838.

Lloyd-Smith, M. 2003. *The Role of Technical Information Delivery in Environmental Disputes, Faculty of Law.* University of Technology, Sydney, Unpublished PhD Thesis.

Noyes, P. D., M. K. McElwee, H. D. Miller, B. W. Clark, L. A. Van Tiem, K. C. Walcott, K. N. Erwin, and E. D. Levin. 2009. The toxicology of climate change: Environmental contaminants in a warming world. *Environ Int.* 35: 971–986.

Patra, R. et al. 2007. The effects of three organic chemicals on the upper thermal tolerances of four freshwater fishes. *Environ Toxicol Chem J.* 26(7): 1454–1459.

UNEP. 2010. *Climate Change and POPS: Predicting the Impacts.* Report of the UNEP/AMAP Expert Group, December 2010. Available at: http://www.pops.int

26 Organisation for Economic Cooperation and Development

*Richard Sigman**

CONTENTS

INTRODUCTION

The Organisation for Economic Cooperation and Development (OECD)—an intergovernmental organization made up of 34 member countries from Europe, North America, and Australasia†—has been working on chemical safety issues, as part of its

* The opinions expressed in this chapter are those of the author and do not necessarily represent the views of the OECD or of the governments of member countries.

† Australia, Austria, Belgium, Canada, Chile, Czech Republic, Denmark, Estonia, Finland, France, Germany, Greece, Hungary, Iceland, Ireland, Israel, Italy, Japan, Korea, Luxembourg, Mexico, the Netherlands, New Zealand, Norway, Poland, Portugal, Slovak Republic, Slovenia, Spain, Sweden, Switzerland, Turkey, United Kingdom, United States.

Chemicals Programme, for almost 40 years. Its aims are to develop and harmonize chemical safety tools and policies so that governments and industry can achieve cost-effective approaches for protecting people and the environment from the risks posed by chemicals, without creating barriers to trade.

This chapter describes the history of the Programme, the role of OECD as a leader in the global management of chemicals, how work is carried out and the major areas of that work, and finally, how the Programme will continue to evolve to meet future challenges.

HISTORY OF THE OECD WORK ON CHEMICALS

The creation of the Chemicals Programme in 1971 was prompted by the growing concern in member countries over the widespread exposure to certain harmful chemicals, and the fact that transboundary pollution, and global trade in chemicals, necessitated an international response. In the early years, OECD's activities focused on specific industrial chemicals known to pose health or environmental problems such as polychlorinated biphenyls (PCBs), mercury, and chlorofluorocarbons (CFCs). Efforts were aimed at sharing information about the risks posed by these chemicals, and ways governments could act jointly to reduce these risks. An example of such action was the 1973 OECD Council Decision to restrict the use of PCBs. Shortly thereafter, a Council Recommendation was adopted on measures to reduce emissions of mercury to the environment, and work was undertaken in OECD on CFCs and lead that resulted in concerted risk management actions among countries.

As the Programme evolved, governments recognized that with tens of thousands of chemicals on the market, and many new chemicals being added each year, concentrating on just a few at a time would not be sufficient, and a more comprehensive approach was necessary. This approach would help countries anticipate, identify, prevent, and manage risks posed by chemical products, and do so in a way which optimized the use of government and industry resources, and avoided unnecessary nontariff distortions in trade. This led to work on a key foundation of the Chemicals Programme: The development of harmonized Test Guidelines and Good Laboratory Practice (GLP) Principles. The 1981 Council Decision on the Mutual Acceptance of Data (MAD) requires OECD governments to accept test data developed in another country, provided these data were generated according to OECD Test Guidelines and GLP Principles. MAD reduces duplicative testing which, in turn, reduces the cost of testing for companies, and limits the number of animals required for testing. The development of harmonized Test Guidelines and GLP Principles made possible the initiation of new work in the 1980s on the systematic investigation of chemicals (i.e., governments sharing the work of assessing existing chemicals). In parallel to this work, OECD also began new projects to develop methods for risk assessment, approaches to risk management, and principles for chemical accident prevention, preparedness, and response. In the years that followed, the Chemicals Programme, now called the Environment Health and Safety Programme, began new activities that would address the safety of pesticides, biocides, and, most recently, products of modern biotechnology and nanomaterials.

ROLE OF OECD

OECD has played and continues to play an important role in the global management of chemicals. As OECD countries account for about 75% of total world production of and trade in chemical products (OECD, 2008, p. 379), they have a large responsibility for ensuring the safe production and use of such products, and they have played an important role in promoting chemical safety domestically and internationally. Further, as OECD countries have a long history of managing chemicals, they have gained invaluable experience in what has worked, what has not, and what needs to change. By working together through OECD and tapping into this expertise and experience, governments can reduce the cost to governments and industry associated with the management of chemicals, while maintaining a high level of health and environmental protection. In a new publication (OECD, 2010), OECD estimates that governments and industry save approximately 150 million Euros per year as a result of OECD activities. This figure only accounts for quantifiable benefits and not the significant but unquantifiable benefits such as the large reduction in animals which are needed for testing, because of the elimination of duplicative testing across OECD countries.

The focus of OECD work is not limited to assisting member governments, but is also aimed at facilitating cooperation between OECD countries and nonmembers, assisting member countries in their implementation of the Strategic Approach to International Chemicals Management (SAICM),* and, through participation of the OECD Secretariat in the Inter-Organization Programme for the Sound Management of Chemicals (IOMC), coordinating the work on chemicals in nine international organizations.† Finally, OECD also aims to make the outputs of the Programme as accessible and useful to nonmember countries as possible, by, among other things, distributing such output, via the Internet, free of charge.

HOW OECD CARRIES OUT ITS WORK

In order to help governments reduce barriers to trade, optimize the use of their resources, and support the sound management of chemicals, OECD's work focuses on three broad objectives:

- *Harmonization*—by working through OECD to harmonize national approaches to the management of chemicals, governments are provided with a common infrastructure which allows them to work together on issues or chemicals of concern, and chemical companies are not faced with conflicting or duplicative requirements.

* SAICM, is a policy framework aimed at fostering the sound management of chemicals globally. It was adopted by the International Conference on Chemicals Management (ICCM) on February 6, 2006 in Dubai, United Arab Emirates.

† OECD, United Nations: Food and Agriculture Organization (FAO), International Labour Office (ILO), UN Environment Programme (UNEP), UN Industrial Development Organization (UNIDO), UN Institute for Training and Research (UNITAR), World Health Organization (WHO) as participating organizations, and the World Bank and UN Development Programme (UNDP) as observers.

- *Sharing the burden*—by working together on such issues or chemicals, countries can leverage their resources and accomplish more than what might be possible if they worked alone.
- *Sharing technical and policy information*—by exchanging such information, countries gain greater confidence in each other's approaches, which facilitates work sharing, and, as some of the best and most experienced experts in countries participate in this work, participants gain access to high quality material and technical expertise that no individual country could match.

In general, OECD's work falls into two categories: (1) Development of general tools and methodologies that can support the assessment and management of a wide range of chemicals, and (2) development of approaches for countries to work together on specific types of chemicals.

MAIN AREAS OF WORK: GENERAL TOOLS*

Test Guidelines

OECD Test Guidelines are methods for use in regulatory safety testing to identify the health and environmental hazards of chemical substances, mixtures, and preparations. These Guidelines cover a range of areas, including physical and chemical properties, effects on human health and wildlife, persistence and degradation in the environment, and presence of pesticide residues on crops and in livestock. To date, over 100 Test Guidelines have been published, and new ones are periodically developed, and existing ones updated. As a companion to the Test Guidelines, OECD also publishes Guidance Documents which provide additional material on, for example, the interpretation of results from tests conducted according to OECD Test Guidelines. The Test Guideline Programme is also working to develop "alternative" test methods which can either replace existing tests which use animals, or reduce the number of animals that are needed for tests. Since the adoption of the first set of Test Guidelines in 1981, many tests (e.g., short- and long-term toxicity and genetic toxicity) have been developed or revised to reduce the number of animals needed for such testing. In response to the concern about the link between reproductive and developmental effects and certain chemicals in wildlife (i.e., "endocrine disrupting substances"), the Programme has also been working on the development and validation of test methods for these effects on human health and wildlife.

GLP and Compliance Monitoring

The OECD Principles of GLP is a management tool covering the organizational process and the conditions under which nonclinical studies—conducted in the laboratory and in the field—are designed, performed, monitored, recorded, and reported. They are an important complement to OECD Test Guidelines relating to the quality,

* More information on each of the tools described in this section, can be found on OECD's Web site: http://www.oecd.org/ehs

rigor, and reproducibility of studies, and thus are an integral part of the MAD data system. The Principles were first published in 1981 and later updated in 1997. In 1989, an OECD Council Decision was adopted which requires governments to establish and maintain procedures for ensuring that test facilities have complied with the GLP Principles through inspections and study audits and for international liaison among GLP compliance monitoring programs. In 1997, the MAD system was opened to non-OECD economies who can participate, once they have implemented the relevant Council Decisions. Brazil, India, Singapore, and South Africa currently participate as *full* adherents to the Council Decisions, and Argentina, Malaysia, and Thailand participate as *provisional* adherents (observers).

Hazard and Exposure Assessment

Chemical specific data generated from animal testing or by other means are used to evaluate the hazards of chemicals, and, when used in conjunction with exposure information, are used to evaluate the risks. OECD works to support such assessments through the development of guidance documents on the latest scientific techniques, as well as the development of other tools that can help governments conduct and share assessments. For example, for hazard assessments, OECD has developed a (Quantitative) Structure-Activity Relationships [(Q)SARs] "Toolbox"*; (Q)SARs are methods for estimating properties of a chemical from its molecular structure (i.e., without actual testing). The "Toolbox" is a software program which incorporates information and tools from various sources and allows a user to systematically group chemicals according to the presence or potency of a particular effect for all members of a category as well as evaluate all members of a category for common toxicological behavior or consistent trends among results related to regulatory endpoints. For exposure assessments, OECD develops Emission Scenario Documents (ESD) which describe the sources, production processes, pathways, and use patterns associated with specific industries, and products (e.g., metal finishing, plastic additives, kraft pulp mills) and are used in risk assessment of chemicals.

Risk Management Tools and Experiences

Once a chemical risk assessment has been completed, a government can take a risk management decision as to whether or not a new substance can be placed on the market, or an existing substance removed or controlled. OECD's work in this area focuses on supporting member countries' efforts to develop national policies and actions, and, where appropriate, develop international responses based on national risk management measures. OECD guidance has been developed, for example, on ways to conduct socioeconomic analyses for risk management decisions and how governments can communicate about risks posed by chemical products. In addition, OECD's work aims to promote the design, manufacture and use of environmentally benign chemicals through sustainable chemistry (or "Green Chemistry"). This work focuses on identifying activities that can further the development and use of sustainable chemistry

* http://www.oecd.org/env/existingchemicals/qsar

programs including, *inter alia*, recognizing and rewarding sustainable chemistry accomplishments; promoting the research, discovery and development of innovative sustainable chemistry technologies; and promoting the incorporation of sustainable chemistry principles into various levels of chemical education.

Harmonization of Classification and Labeling

An important element in protecting the safe use and handling of chemicals is the identification of any hazard associated with a chemical, and the presentation of that hazard in a way which is easily understood by people who may use that chemical. Along with the International Labour Organization (ILO) and the UN Committee of Experts on Transport of Dangerous Goods (UN-CETDG), OECD has been one of the key actors in the development of the Globally Harmonized System of Classification and Labelling of Chemicals (GHS). OECD had the lead in developing classification criteria for all human health and environmental hazards, UN CETDG had the lead developing the criteria for physical hazards, and ILO had the lead in harmonizing hazard communication. The GHS system is comprised of harmonized criteria for classifying substances and mixtures according to their health, environmental and physical hazards, as well as harmonized hazard communication elements for labeling products.

MAIN AREAS OF WORK: DEVELOPMENT OF APPROACHES FOR SPECIFIC TYPES OF CHEMICALS

Existing Industrial Chemicals

Prior to the introduction of regulations in governments aimed at assessing new industrial chemicals before they could be been placed on the market, thousands of industrial chemicals were being produced and used with very little or no information concerning their potential risks to human health and the environment. It is estimated that from 20,000 to 70,000 of these "existing" chemicals are currently on the market (EEA/UNEP, 1998). Given the difficulty of conducting comprehensive testing on each of these existing chemicals, in 1990, OECD member countries agreed to work together to assess High Production Volume (HPV) chemicals (i.e., those chemicals which are produced in volumes of at least 1000 tonnes per year in any one OECD country or in the EU region). Under the existing chemicals program, OECD governments share the work, with each "sponsoring" a chemical for which it is responsible for gathering the necessary data, filling data gaps, and writing an initial assessment, which, if agreed by the other countries, becomes an agreed OECD hazard assessment. To date, almost 900 chemicals have been assessed in OECD, and these assessments are made available worldwide through UNEP Chemicals. The existing chemicals program also develops and maintains a number of databases and information technology tools to assist countries; most recently it has posted an "eChemPortal" on OECD's public Web site* which provides free access, via simultaneous search capabilities of multiple databases, to data

* http://webnet3.oecd.org/echemportal/Home.aspx

submitted to governments on physical/chemical properties, environmental fate and behavior, ecotoxicity, and toxicity of chemicals.

New Industrial Chemicals

OECD's work also focuses on "new" industrial chemicals (i.e., those which have not yet been placed on the market). As many new chemicals will eventually be marketed in more than one country, OECD's efforts aim at providing the infrastructure within which two or more governments can work together to review a notification on the same new substance. By doing so, not only can governments reduce the time and resources they spend on evaluating new chemicals, but companies can reduce the resources they spend in preparing and submitting data about these chemicals to governments. To support this work, OECD has provided, among other things, guidance on a process by which companies can notify multiple jurisdictions and governments can share information when conducting their reviews, and working definitions of key terms used in such a process.

Pesticides

OECD's Pesticides Programme aims at improving the efficiency and effectiveness of pesticide regulation by harmonizing testing and assessment methods, promoting work sharing and promoting risk reduction of agricultural pesticides. It has developed internationally agreed formats used in registering and reregistering agricultural chemical and biological pesticides, and is creating additional formats, guidance, and pesticide-specific test guidelines that will help governments jointly review data submissions for new pesticide-active ingredients. The Programme also works to promote pesticide risk reduction by facilitating information exchange and giving status and credibility to certain risk reduction goals and tools.

Biocides

Biocides are a diverse group of chemical products (e.g., sterilizers, disinfectants) that are designed to control unwanted organisms in nonagricultural settings. Similar to the Pesticides Programme, OECD's Biocides Programme focuses on tools and guidance that can help governments improve their management of these products. In particular, it focuses on harmonizing the testing of product efficacy, producing emission scenario documents on biocidal products and facilitating the sharing of data and reviews among governments.

Safety of Nanomaterials

With respect to the field of nanotechnology—the engineering of materials at the atomic or molecular level—OECD countries have begun to work together to ensure that governments share safety information with one another, examine whether existing risk assessment methodologies and testing schemes are adequate to address nanomaterials, and coordinate on regulatory issues.

Safety in Biotechnology and Safety of Novel Foods and Feeds

Finally, these two programs deal with environmental safety and human food and animal feed safety associated with products derived from modern biotechnology. Together, they help OECD countries evaluate the potential risks of these products, support greater communication and mutual understanding of the regulatory processes in different countries, and reduce the potential for nontariff barriers to trade.

FUTURE PLANS

While OECD will continue to focus on the main areas of work described above, as in the past, this work will evolve to reflect the latest scientific approaches and new regulatory policies in governments and address new environment health and safety concerns as they arise. For instance, new activities on the development and evaluation of information technologies and scientific techniques which predict health and environmental effects from exposure to chemicals, and the coordination of development of IT tools to store, organize, and disseminate data, are expected to play an increasing role in the future.

In 2008, OECD published a report, *OECD Environmental Outlook to 2030*, which included the projections of economic and environmental trends to 2030, and policy actions which can address the key challenges. For chemicals, the report noted that over the next two decades, chemicals production in OECD countries as a percentage of world production would drop from almost 75% today, to 63% in 2030 (OECD, 2008). This highlighted the need for OECD and its member countries to work more closely with nonmember countries to share information, build capacity, and promote good chemicals safety policies and the convergence of these policies. As mentioned above, OECD and its member countries are expected to play a large role in the implementation of SAICM. In addition, OECD has an active outreach program with a number of nonmember countries, and, following a 2007 Decision of OECD's Council to invite five countries* to open discussions for membership in OECD, and offer "enhanced engagement," with a view to possible membership to five more,† OECD is broadening its scope to meet the challenges of tomorrow.

REFERENCES

EEA/UNEP (European Environment Agency/ United Nations Environment Programme). 1998. *Chemicals in the European Environment: Low Doses, High Stakes?*, Geneva: UNEP.

OECD. 2008. *OECD Environment Outlook to 2030*. Paris: OECD.

OECD. 2010. *Cutting Costs in Chemicals Management: How OECD Helps Governments and Industry*. Available at: http://www.oecd.org/document/23/0,3343,en_2649_34365_44983063_1_1_1_1,00.html

* Chile, Estonia, Israel, Russia, and Slovenia. Chile, Estonia, Israel, and Slovenia became OECD members in 2010.

† Brazil, China, India, Indonesia, and South Africa.

27 The Chemical Weapons Convention and the Work of the Organisation for the Prohibition of Chemical Weapons

*Boitumelo V. Kgarebe and Cristina B. Rodrigues**

CONTENTS

THE DEVELOPMENT OF CHEMISTRY AND CHEMICAL WEAPONS

Chemistry is central to human existence, with its applications making life easier, safer, and more delightful. The properties of chemicals have been exploited in many different ways to benefit humankind; from new drugs based on nanotechnology

* The opinions expressed in this chapter are those of the authors and do not necessarily represent the views of the OPCW or of the governments of Member States.

systems to traditional herbal medicines; from collagen creams to almond oil for skin treatments; from new electronic devices to fireworks; from expensive cognacs to simple grape juices.

The properties of chemicals, though, have not always been exploited for their beneficial purposes. From time immemorial, chemicals have found uses as methods of aggression. Examples can be found in the 431–404 BC use of arsenic smoke during the sieges of Plataea and Delium by the Spartans during the Peloponnesian War, around 500 BC. "The Greek fire," a burning mixture of wood, pitch, and sulphur was used to incapacitate a beleaguered Athenian force prior to assault. The purgative hellebore was used ca 6000 BC during the siege of Cirrha, the port of Delphi, resulting in violent diarrhoea for the defenders of Cirrha and consequently in their defeat (Mayor, 2003). The San people in Southern Africa still use an extract from beetle larvae to disable their prey when hunting. More recent examples include the sarin gas attack in a Tokyo subway in 1995 (Tu, 2002) and the chlorine gas release in Iraq in 2008. These are all examples of chemical action on life processes which can cause death, temporary incapacitation, or permanent harm to humans or animals.

The exploitation of the lethal properties of chemicals started to grow in the nineteenth century. Advances in chemistry saw many more chemical compounds being synthesized and tested for their activity; many of which found utility for war purposes. The abhorrent effects of these chemicals drew countries together in an effort to eliminate forever their development and their use in situations of conflict. One early effort was the First Hague Convention in 1899 prohibiting "the use of projectiles, the sole object of which is the diffusion of asphyxiating or deleterious gases" (Croddy et al., 2005). But these attempts did not stem the use of chemicals as weapons. 1914 saw the use of tear gas by French troops against German positions. However, it was the use of chlorine at Ypres on April 22, 1915 during World War I (Croddy et al., 2005; Matousek, 2008) that defined the momentum in the modern use of chemicals as warfare agents. By the armistice in 1918, chemical warfare had claimed more than 90,000 dead and 1 million casualties. Chemical weapons created a trail of horror and social revulsion inspired by the indiscriminate and horrifying nature of this class of warfare. By 1925, the Geneva Protocol (Croddy et al., 2005; Matousek, 2008) barring the use in war of asphyxiating, poisonous or other gases, and bacteriological methods of warfare was signed by 30 countries. This protocol presented a number of shortcomings; namely, the prohibition against use only during times of war and the possibility of retaliation in kind (response with the use of similar methods if attacked), as well as the inclusion of reservations by the signatory states. These exemptions presented a major constraint as did the fact that the protocol was not binding on nonsignatory states. The existence of the Protocol did not prevent Italy using poisonous gas against Abyssinia in 1935, and Japan, a nonsignatory to the Protocol, used poisonous gas against China from 1937 to 1942 (Kenyon and Feakes, 2007).

After World War II, the biggest concern in the agenda on the nonproliferation of weapons of mass destruction became nuclear weapons. The International Atomic Energy Agency (IAEA) (Croddy et al., 2005) came into being in 1957 and the treaty on the nonproliferation of nuclear weapons (nuclear nonproliferation treaty, NPT) opened for signature in 1968 and came into force in 1970.

The Convention on the Prohibition of the Development, Production, and Stockpiling of Bacteriological (Biological) and Toxin Weapons and of their Destruction (Croddy et al., 2005), also known as the Biological and Toxin Weapons Convention (BTWC), prohibiting the development, production, stockpiling, and acquisition of these weapons, and thus supplementing the prohibition on use of biological weapons contained in the 1925 Geneva Protocol, opened for signature in 1972 and came in to force in 1975.

The end of the Cold War ushered in new developments in the debate on chemical weapons. The Wyoming memorandum (Kenyon and Feakes, 2007) in 1989 and the work of the Australia Group (Kenyon and Feakes, 2007) opened up the way for the final discussions that would lead up to the Chemical Weapons Convention (CWC).

THE CWC AND THE ORGANISATION FOR THE PROHIBITION OF CHEMICAL WEAPONS

After extended negotiations at numerous fora, the Convention for the prohibition of the development, production, stockpiling, and use of chemical weapons and their destruction, the CWC* was signed on January 13, 1993 in Paris by 130 countries. This gave birth to the only international treaty that totally eradicates a category of weapons of mass destruction, and includes a verification of nonproliferation regime. The Convention entered into force on April 29, 1997, with the creation of its implementing body the Organisation for the Prohibition of Chemical Weapons (OPCW), located in The Hague, the Netherlands.

The CWC comprises 24 articles and three annexes on chemicals, verification, and confidentiality. For verification purposes, this Convention sets out three Schedules with the most important toxic chemicals and their precursors. These schedules are arranged in accordance with the risk the chemicals pose within the framework the Convention and the extent of their industrial use. Schedule 1 lists toxic lethal chemicals and key precursors that have no peaceful uses; Schedule 2 contains less dangerous toxic chemicals and precursors that are produced in small quantities; and Schedule 3 lists toxic industrial chemicals that had previously been used as chemical warfare agents and precursors produced on a large scale. The Schedules do not constitute a definition of chemical weapons, neither do they provide an exhaustive list of toxic chemicals; rather they are to be used within the context and definitions provided in the Convention (Matousek, 2007).

In Article I, the CWC forbids any State Party to the Convention under any circumstances, to develop, produce, and use chemical weapons, or to engage in any military preparations that use chemical weapons. It is under the definitions and criteria provided in Article II, that this Convention becomes more embracing by extending the prohibition not only to the conventional chemical warfare agents, but also to any toxic chemical and its precursors which through their chemical action on life processes can cause death, temporary incapacitation, or permanent harm to humans or animals. This includes all such chemicals, regardless of their origin or of their method of pro-

* Convention on the Prohibition of the Development, Production, Stockpiling and Use of Chemical Weapons and on Their Destruction. UN, New York 1993.

duction, and regardless of whether they are produced in facilities, in munitions, or elsewhere (Article II, paragraph 2). This excludes defoliants and fumigants.

The Convention covers four main objectives (often referred to as the four pillars):

1. The destruction of chemical weapons and the elimination of the capacity to develop them.
2. The nonproliferation of chemical weapons through the verification of chemical industry.
3. Assistance and protection in the event of the use or threat of use of chemical weapons.
4. International cooperation in the peaceful uses of chemistry.

All States party to the CWC are members of the OPCW. In order to execute its mandate, the Organisation comprises: The Conference of the States Parties (the Conference), the Executive Council, and the Technical Secretariat, which is the implementing body of the CWC and acts under the supervision of the Director General. To render expert advice, the Technical Secretariat relies on subsidiary organs, namely, the Scientific Advisory Board, the Confidentiality Commission, and the Advisory Board on Administrative and Financial Matters.

The last 14 years have witnessed progress and accomplishments in the implementation of the CWC. Indicators of the progress include an increased membership of the Organisation (universality), the destruction of chemical weapons arsenals, the establishment of an effective verification regime, the improvement of protective capacity, and the development of international cooperation initiatives in the peaceful application of chemistry.

The universality of this Convention spells out the universal conviction to eradicate an entire class of weapons of mass destruction. In the First Conference of the States Party 128 countries attended, (May 2010), the OPCW boasts of a membership of 188 Member States, representing 98% of the global population and landmass. Two countries, Israel and Myanmar, have signed but have not yet ratified the Convention, whilst 5, Angola, the Democratic People's Republic of Korea, Egypt, Somalia, and the Syrian Arab Republic, have not signed the treaty (Organization for the Prohibition of Chemical Weapons, 2010). By comparison to the number of States Party to other multilateral agreements, that is, BTWC or NPT, the CWC has achieved universal recognition that no other nonproliferation treaty has accomplished.

The establishment of a national body, the National Authority, which creates a direct point of contact with the Technical Secretariat, is a requirement of the CWC. The support for National Authorities is an activity of the Technical Secretariat reflecting the necessity to build national capacities for the effective implementation of the provisions of the Convention. At the same time, however, such implementation support has a positive bearing on other fields. It can facilitate integrated solutions in such areas as training of national personnel, the organization of exchanges, and workshops between National Authorities to share their experience and discuss common issues, on-site assistance in areas such as declarations, legislation, chemical emergency response, scientific and technical infrastructure, and databases. Many activities are tailored to the needs of individual States Parties or (sub) regions and participation of representatives from developing countries is facilitated by the OPCW.

By 2007, with the recognition that a number of factors have hindered the development of national capacities in most African States Parties, the OPCW Director-General expressed readiness "to explore ways in which a programme might be developed by the Secretariat, to respond to the particular needs of Africa" (OPCW, 2007). The OPCW developed the Programme to Strengthen Cooperation with Africa on the CWC, also known as the Africa Programme, with the express view to accelerate and strengthen efforts both to achieve universality, ensure full implementation of the CWC in Africa and promote increased participation by African States Parties in programmes and activities designed to promote the peaceful uses of chemistry and enhance national protective capacities (OPCW, 2007).

Destruction of all Chemical Weapons and Chemical Weapons Facilities

Article I of the CWC stipulates that

a. Each State Party undertakes to destroy chemical weapons it owns or possesses, or that are located in any place under its jurisdiction or control, in accordance with the provisions of this Convention.
b. Each State Party undertakes to destroy all chemical weapons it abandoned on the territory of another State Party, in accordance with the provisions of this Convention.
c. Each State Party undertakes to destroy any chemical weapons production facilities it owns or possesses, or that are located in any place under its jurisdiction or control, in accordance with the provisions of this Convention.
d. Each State Party is obliged to declare and destroy all chemical weapons it owns, possesses in its territory, or it had abandoned in another State Party as well as any chemical weapon production facility (Article I, Paragraphs 2–4).

The Convention further establishes step-wise timelines for the total destruction of chemical weapons. The complete destruction of chemical weapons arsenals has to be achieved by April 29, 2012. To date, 40,514 metric tonnes or 56.91% of the worlds declared stockpile of 71,194 metric tonnes of chemical agents, as well as 3.93 million or 45.33% of the 8.67 million chemical munitions and containers have been destroyed. (Organization for the Prohibition of Chemical Weapons, 2010).

Verification of Nonproliferation

The Convention establishes a verification regime for the nonproliferation of these kinds of weapons of mass destruction. Article VI of the Convention states that:

> Each State Party has the right, subject to the provisions of the Convention, to develop, produce, otherwise acquire, retain, transfer, and use toxic chemicals and their

precursors for purposes not prohibited under this Convention. In addition, each State Party shall adopt the necessary measures to ensure that toxic chemicals and their precursors are only developed, produced, otherwise acquired, retained, transferred, or used within its territory or in any other place under its jurisdiction or control for purposes not prohibited under this Convention, as described under the Verification Annex.

The verification regime is composed of declarations and inspections to industrial sites that develop, produce, acquire, retain, transfer and use for normal industrial processes, chemicals that have been utilized or are precursors of chemical weapons. Worldwide, 4997 industrial facilities are liable to inspection. In the last 14 years since entry into force, 81 States Parties have received 3964 inspections at 1103 industrial sites. The OPCW inspectors verified the destruction as well as the nonproduction of new chemical weapons at 195 chemical weapons related sites (OPCW, 2009).

Assistance and Protection

Given the history of chemical warfare attacks, the threat posed by chemical weapons is of concern to all countries. Thus, preparedness to respond to a chemical emergency and readiness to protect against a chemical incident are necessary assets that all countries should have at their disposal. Through Article X, the CWC recognizes this need and proposes to provide support to Member States if requested.

Article X establishes the obligations and rights of a State Party concerning the Assistance and Protection against Chemical Weapons, and affords each State Party the right to receive from the Technical Secretariat, expert advice and assistance in identifying its programmes for the development and improvement of protective capacity against chemical weapons as well as the right to request and to receive assistance and protection against the use or threat of use of chemical weapons if it considers that:

a. Chemical weapons have been used against it.
b. Riot control agents have been used against it as a method of warfare.
c. It is threatened by the actions or activities of any State that are prohibited by Article I.

By the same token, State Parties undertake to provide annually to the Technical Secretariat, information on their national programmes related to protective purposes. Conversely, each State Party undertakes to provide assistance through the Organisation by electing to take one or more of the following measures:

a. Contributing to the voluntary fund for assistance.
b. To conclude agreements with the Organisation concerning the procurement, upon demand, of assistance.
c. To declare the kind of assistance that may be provided in response to an appeal by the Organisation.

Furthermore, each State Party undertakes to facilitate, and shall have the right to participate in, the fullest possible exchange of equipment, material and scientific and technological information concerning means of protection against chemical weapons.

The Technical Secretariat should provide emergency, humanitarian and supplementary assistance upon request by a State Party that has been attacked or faces threat of attack with chemical weapons or riot control agents as method of warfare. Also, considering the needs to develop or improve their protective capacity against chemical weapons, or in a broader sense, toxic chemicals, a State Party has the right to request and receive expert advice from the Technical Secretariat or bilaterally from another State Party, if so required.

International Cooperation

International cooperation for the development of chemistry as described under Article XI of the CWC supports the scientific community in an effort to build national scientific capacity in chemistry, chemical engineering and allied disciplines. Article XI recognizes that education and knowledge are pivotal in developing peaceful and ethical applications of chemistry that will contribute to the socioeconomic progress of Member States.

In order to achieve the above, Article XI affords States Parties to

a. Have the right, individually or collectively, to conduct research with, to develop, produce, acquire, retain, transfer, and use chemicals.
b. Undertake to facilitate, and have the right to participate in, the fullest possible exchange of chemicals, equipment and scientific and technical information relating to the development and application of chemistry for purposes not prohibited under this Convention.
c. Not maintain among themselves any restrictions, including those in any international agreements, incompatible with the obligations undertaken under this Convention, which would restrict or impede trade and the development and promotion of scientific and technological knowledge in the field of chemistry for industrial, agricultural, research, medical, pharmaceutical or other peaceful purposes.

In effect, Article XI supports the economic and technological development through international cooperation initiatives in the field of chemical activities for purposes not prohibited under the Convention. Such purposes include industrial, agricultural, research, medical, pharmaceutical, and other peaceful purposes.

THE CWC AND ARTICLES X AND XI: TOOLS TO MANAGE CHEMICAL AND ENVIRONMENTAL RISKS

For developing countries, and countries that are nonpossessor states, namely, states that do not have chemical weapons arsenals, and/or have a small chemical industry base, by ratification of the Convention, Articles X and XI present attractive possibilities for them to increase their civil protection (including appropriate response to chemical incidents/accidents), as well as their economic, scientific, technical, and technological bases. Articles X and XI provide scope for the exchange of capacity between States Parties bilaterally or multilaterally through the Technical Secretariat.

In the event of a chemical incident, an appropriate response substantially decreases not only the number (and severity) of casualties, but also the extent of any environmental harm caused by a chemical spill. The Technical Secretariat is obliged to provide emergency, humanitarian, and supplementary assistance upon request by a State Party that has been attacked or faces a threat of attack with chemical or riot control agents as methods of warfare. Implementation of Article X addresses the needs of Member States to develop or improve their protective capacity against chemical weapons, or in a broader sense, chemicals incidents.

In order to respond appropriately, States Parties must have the requisite technical expertise at their disposal. Through the implementation of Article XI and its associated programmes, the Technical Secretariat is able to assist Member States in building and maintaining effective and sustainable national and regional teams of scientific and technical capacity in areas of chemistry or chemistry related disciplines.

Taken together, implementation of Articles X and XI provides the OPCW with synergistic and effective means with which Member States can develop their capacities in all areas related to chemistry or involving chemicals products.

Assistance and Protection

Assistance is defined by Article X of the Convention as coordination and delivery to State Parties of, protection against chemical weapons, including, *inter alia*, the following: detection equipment and alarm systems; protective equipment, contamination equipment, and decontaminants, medical antidotes and treatments and advice on any of these protective measures (Convention on the Prohibition of the Development, Production, Stockpiling and Use of Chemical Weapons and on Their Destruction).

Article X has two main areas of focus:

a. The provision of expert advice, coordinated by the Technical Secretariat, upon request from a Member State to develop and improve its protective capacity programmes against chemical weapons attacks.
b. To respond and provide assistance and protection at the request of a State Party when chemical weapons have been used against it or if there is a threat of use.

The OPCW has provided training and development of protective capacities in 60 Member States upon request, as well as creating national and regional networks of expertise and at the same time developing procedures and instructions to respond to chemical incidents. In countries where there are no chemical weapons but a chemical or chemicals-related industry exists, the major concern of the country is the response to an accident involving chemicals. The trainings and expert advice provided by the OPCW can stand these Member States in good stead in developing a coordinated chemical emergency response system that will decrease the deleterious effects of any chemical spills to life and to the environment. Tailor-made activities have been developed in countries or regions with the express aim to improve and create national and regional protective capacity and expertise. Since 1997, the OPCW, with the support of Member States like Brazil, China, Czech Republic, Finland, Iran, Republic of Korea, Russian Federation, Serbia, Singapore, Slovakia, South Africa, Spain, Sweden, and Switzerland, has trained more than 2000 nationals from Member States creating a pool of expertise that can be shared among Member States.

Though the OPCW has never had to be deployed in response to a request from a Member State under attack or threat of chemical weapons or riot control agents used as a method of chemical warfare, procedures have been developed and are continuously tested during national and regional training exercises. Joint exercises with other organizations, for example, United Nations Office for the Coordination of Humanitarian Affairs (UN OCHA), North Atlantic Treaty Organisation–Euro Atlantic Disaster Response Coordination Centre (NATO–EADRCC), and International Humanitarian Partnership (IHP) involved in emergency and humanitarian assistance, are a part of the work of the Technical Secretariat to prepare the organization for a request of assistance from a Member State. Examples of such exercises are ASSISTEX 1 held in 2002 in Croatia and Joint Assistance 2005 in Ukraine. ASSISTEX 3, which was held in Tunisia in October 2010, was the first joint exercise developed by the OPCW in Africa. This event involved 15 national emergency response teams and subregional teams from a total of 32 Member States. Three of the subregional teams were from the Africa region and these had been trained by the OPCW under its Africa Programme.

International Cooperation for the Peaceful use of Chemistry

Paragraph 2 of Article XI of the Convention provides the framework for the OPCW's activities in the area of international cooperation. It recognizes the States Parties' desire to "undertake, facilitate, and have the right to participate in the fullest possible exchange of chemicals, equipment and scientific and technical information relating to the development and application of chemistry for purposes not prohibited under [the] Convention."

In order to achieve this, the OPCW has established a number of projects and programmes in the areas of

a. Assistance in the development of the scientific and technical infrastructure in Member States particularly in areas related to the implementation of the Convention.
b. Capacity building for peaceful uses of chemistry.

Effective and sustainable infrastructure building is an important aspect, not just of the implementation of the CWC, but also of other regulatory instruments in the field of the sound management of chemicals. Within the context of the Convention, two areas of science and technology are of particular relevance: destruction of toxic chemicals, and the analysis of chemicals. Whilst a few OPCW Member States are involved in the destruction of stocks of chemical weapons, most are currently addressing the issue of how to treat other toxic chemicals and wastes.

In the area of chemical analysis, the Technical Secretariat has embarked on an integrated approach toward assisting Member States in the development of their technical capabilities. The analytical skills development training courses supported by Finland, Spain, and South Africa are particularly popular with the Member States because of their singular focus on specific and specialized chemical analytical techniques related to scheduled chemicals and their precursors and degradation products. This approach recognizes that only very few countries can afford to develop and maintain highly specialized laboratories that deal with CW agents, but that many other Member States need to be able to conduct other types of chemical analysis for regulatory, environmental, and occupational safety reasons, including the analysis of the chemicals listed in the three Schedules of the Convention. Many of these chemicals are constituents of common commercial products such as dyes, surfactants, agricultural fertilizers, pesticides, and other commodities, which require analysis for a variety of regulatory purposes such as product licensing, chemical safety audits, import controls, or industry audits. Examples of OPCW cooperation are the joint training courses prepared by VERIFIN (Finland) and Protechnik Laboratories (South Africa) to build capacity in analytical skills within the framework of the Africa Programme, and the Spanish-government sponsored course at LAVEMA (Spain) designed for the Latin America and Caribbean region.

The Role of Science

Through Article XI, the Convention implicitly recognizes that meaningful implementation of the Convention relies heavily on the application of sound scientific knowledge. For this to happen, the pivotal role played by academic and research institutions in building a strong and sustainable national and regional scientific knowledge base is realized through programmes that provide financial support to practicing scientists and engineers from developing countries and countries whose economies are in transition to prepare and disseminate scientific information and educational materials that raise awareness of the Convention for scientists, policy makers, and the broader public. In addition, together with strategic partners such as the International Foundation for Science, financial support to undertake small-scale research projects that address issues of relevance to both science and society and to the Convention is also provided. The topics of the research projects have included environmental aspects of toxic chemicals, chemical analysis and detoxification, chemistry aspects and applications of natural resources, and other relevant peaceful applications of chemistry. Article XI further accommodates the exchanges of scientific hardware between laboratories from Member States including shipments and any requisite equipment installations and users' trainings. The Technical Secretariat also operates a free service for institutions in Member States which provides information on specific requests relating to technical know-how, equipment, and trade possibilities as well as on the properties and uses of scheduled and nonscheduled chemicals.

The flagship programme in international cooperation is the Associate Programme, designed for chemists and chemical engineers from developing countries and countries whose economies are in transition to familiarize themselves with the work of the OPCW and the requirements for the implementation of the Convention with regard to the chemical industry. This programme equips scientists and engineers with an integrated experience in modern safety, production, and management practices in the chemical industry.

THE FUTURE AND CHALLENGES FACING THE CWC

A requirement of the CWC is that States Parties review its implementation every five years. Two Review Conferences have taken place; the first in 2003 (OPCW, 2003), and the second in 2008 (OPCW, 2008). The purposes of the Conferences are to identify any necessary changes in the way the CWC is being implemented, and if necessary, to review the implementation so as to ensure that the CWC remains effective. The review conferences maintain the Convention in a state of responsiveness and proactivity to the potential global challenges.

The existence of the CWC does not rule out the discovery of new toxic chemicals which are not contained in the Schedules or their appearance and use on battlefields by non-States Parties or by States Parties being in breach of the CWC, or by non-State actors. For this reason, with a dynamic global scientific community living in an ever-changing world, the developments in science and technology have to be very carefully watched, international verification measures extended, national authorities and operation systems established, and respective legislation adopted in order to enable prevention, adequate response and remediation in real time in cases of emergency (Thakur and Haru, 2006). Containing, preventing, protecting against and rescuing from the adverse effects of incidents will always prove to be a challenge.

An increasing concern for Member States is the aspect of safety and security in the chemical industry. Industrial sectors in developing countries, particularly in Africa where the development of industry is based on a strong foreign-investment component need to be in line with globally accepted best practice, hence to adopt, develop, and inculcate transparent science-based risk assessment and risk-management procedures. Such an undertaking requires the provision of support to developing countries in strengthening their capacity for the sound management of chemicals and hazardous materials by providing technical and financial assistance.

During the Second Review Conference one of the concerns of the Member States was the evaluation and the measurement of the impact of the capacity building projects under Articles X and XI. The thrust of their arguments were the level of responsiveness to the need of the recipient States Parties and the level of optimization and effective use of the allocated resources. The Technical Secretariat was called upon to measure the quality and impact of all the OPCW international cooperation and assistance programmes (OPCW, 2008).

Assistance and Protection

States Parties should undertake to provide assistance through the OPCW and undertake to facilitate the fullest possible exchange of equipment, materials, and scientific and technological information concerning means of protection against chemical weapons and in a broader sense, toxic chemicals. Considering Technical

Secretariat's resource constraints and the increase in the requests to provide protective capacity building programmes, the OPCW has to improve the coordination of bilateral donations between countries in order to respond to all the needs especially in the case of provision of protective equipment and other materials for training purposes.

Significant progress that had been made on Article X over the last 14 years, there is still scope for additional efforts both by Member States and the Secretariat in order to achieve and maintain the highest level of readiness of the OPCW. Response measures and exercises have been developed to address a request for assistance from a Member State in case of use or threat of use of chemicals as weapons under Article X, but there are still several challenges that the Organisation needs to consider. One of them relates to the time taken to respond. The Technical Secretariat is situated in Europe as are the majority of the resources that the Organisation has at its immediate disposal to respond to this request. Will the response time for coordination and deployment of the assistance be swift enough to afford the requisite protection for and minimization of harm to human life and the environment? OPCW contingency plans take the time factor into account and procedures have been developed to minimize the effects of the chemical incident.

Another is the organizational preparedness to lead a big operation of assistance and coordinate an investigation of use of chemical weapons. The OPCW will have to coordinate the delivery of assistance whilst simultaneously undertaking an investigation of alleged use of chemical weapons. In order to save lives and protect the environment, the emergency response will most likely involve the decontamination of affected areas. The investigation of alleged use, on the other hand, would require the conservation of the affected area so that the integrity of the chemical is maintained to allow sampling, analysis and identification can be effectively undertaken. The importance of the smooth execution of these two activities has been realized and has been addressed through practical exercises and updating of procedures based on the lessons learned.

The Technical Secretariat has initiated discussions with UN agencies which are involved in the delivery of assistance, for example, UN OCHA, World Food Programme (WFP), and Department of Field Support (DFS), in order to increase cooperation and avoid duplication of efforts. A direct result of these negotiations will be the participation of UN OCHA in the third exercise for delivery of assistance (ASSISTEX 3) which will be organized by the OPCW in Tunisia in October 2010. Continuous efforts to increase partnerships with other international organizations involved in the delivery of assistance, for example, NATO–EARDCC, International Committee of the Red Cross (ICRC), and European Union Monitoring and Information Centre (EU MIC) are in the planning stages.

The OPCW Scientific Advisory Board has discussed in some detail, the current state of affairs regarding detection and field analysis, medical countermeasures, and decontamination (OPCW note by the Director-General on the Report of the Scientific Advisory Board on Developments in Science and Technology, Conference of States Parties, Second Review Conference). The Technical Secretariat addressed this concern by reviving the OPCW Protection Network, which consists of a pool of experts that provide advice on the implementation of Article X, in order to improve the impact of the OPCW efforts in national protective capacity building and in the delivery of assistance operations.

In 2008, the Second Review Conference (OPCW, 2008) noted the possibility of the use of chemical weapons or toxic chemicals or even attack on chemical facilities, by nonstate actors such as terrorists, and underscored the importance of the implementation of Article X as a platform for cooperation and coordination to mitigate the risks, in this regard by the States Parties and the Technical Secretariat.

International Cooperation for the Peaceful Use of Chemistry

Article XI programs relate to the economic and technological development of States Parties, and to international cooperation in relation to the production, processing, and use of chemicals for purposes not prohibited under the Convention. The development of science and in particular chemistry during the twentieth century has created global concerns regarding the use of toxic properties of chemicals as offensive agents.

The classical boundaries between the disciplines of chemistry, biology and physics are fast disappearing, and are being replaced by complementary areas which see more chemical compounds being synthesized and exploited for their biological activity; advances in particle engineering and nanotechnology lead to more effective delivery systems in medicine, food technology, in the fine chemicals industries. The wide use of batch processes for multipurpose production, the development of small-scale of production equipment, and the consequent ease of concealment mean that indicators of possible noncompliance will be more difficult to identify. This has consequences for both other chemical production facilities (OCPFs) and challenge inspections. As the International Union of Pure and Applied Chemistry (IUPAC) has noted (Pearson et al., 2006), OPCW inspectors must be fully familiar with the new technologies and the associated production changes they are likely to encounter. This was will assume greater importance with the ever-increasing pace of technological change.

The OPCW, through its international cooperation programs has to be able to keep up with technological trends and advances in order to provide state-of-the-art capabilities to Member States upon request. As science and technology advance, and in response to emerging challenges, some of the international cooperation programmes currently under implementation require modification, or new ones have to be introduced.

A beneficial and strategic approach entails fostering closer cooperation between industrial and business ventures in the States Parties. In response to the recommendations of the Second Review Conference (OPCW, 2008) regarding closer collaboration with other international organizations and chemical-industry associations, the OPCW, in cooperation with other regional and international organizations, chemical industry associations, and academia, has developed a new initiative to promote technical collaboration and commercial exchanges for the production, processing, or use of chemicals for purposes not prohibited under the Convention. In 2009, 2010, and 2011 initial efforts of this Industry Outreach Programme have essentially involved arranging regional meetings and seminars at which individual contacts between the concerned parties are made. Such seminars have been held in Japan, Germany, the

Netherlands, India, and latterly South Africa. The meetings also facilitate the sharing and dissemination of information and experience on safety technology, accident prevention, risk management, and safety-related research and development programmes. In an integrated approach, the OPCW is thus leveraging its close relationship with its Member States to promote global ideals and programmes that can be adopted to address chemical safety management issues and enhance State Parties' expertise and skills in the safe handling of chemicals and hazardous materials.

Many Member States have an emerging chemical industry with a large number of small and medium enterprises. These enterprises have special demands and require capacity to deal with issues of safety in production, handling, and transportation of these chemicals. Therefore, as part of the international cooperation activities, the Industry Outreach Programme has been initiated to promote the accepted best practices like the Responsible Care Programme and facilitate international collaboration amongst parties involved in chemical industry safety management issues.

Part of the demand for training in engineering processes, quality, and environmental management systems received by the Technical Secretariat remains unattended due to a limited number of places available in the international cooperation training courses with an industry component. The offer for these courses strongly depends on the training opportunities identified and supported in the chemical sectors of the industrialized Member States under International Cooperation programmes such as the Associate Programme. The Technical Secretariat's resource constraints and the increase in the requests to provide effective programmes, requires the OPCW to mobilize more voluntary contributions in order to fulfil the requests. Closer collaboration with other international organizations, chemical-industry associations and academia could assist in matching demand and supply.

CONCLUSION

The last 14 years that the Convention has been in existence have witnessed an increase in the membership of the OPCW, an indication that progress toward universality is steady, and as a consequence, the primary success of the CWC.

The main function of the CWC today and until 2012 is to secure elimination of existing stocks of chemical weapons and chemical weapons production facilities and the nonproliferation of scheduled chemicals as outlined in the Convention.

The chemical warfare agents of the twentieth century were born out of research in academic or industrial rather than military laboratories. Even when disarmament has been completed, the function of protection against future hostile application of chemicals and chemistry will remain paramount. The logical steps to assume a truly chemical weapons free world necessitate the intensification of measures that establish and maintain the security and oversight of chemicals, as well as accepting universal codes of conduct for scientists and industry.

As Member States exhibit different economic and technological growth rates, Articles X and XI will continue to facilitate the balance between the developed countries and the developing ones. Full implementation of Articles X and XI of the Convention will reduce the developmental gap between Member States by creating synergies and strengthening subregional, regional, and global collaboration.

Truly integrated and harmonized approach can be achieved through the recognition of the OPCW and the CWC as an important partner together with other international and regional treaties that regulate chemicals and their harmful effects (Stockholm, Basel, Rotterdam, Montreal, Seveso II, etc.). This will serve to avoid duplication of effort and maximization of efficacy.

REFERENCES

Croddy, E. A., J. J. Wirtz, and J. A. Larsen, Eds. 2005. *Weapons of Mass Destruction, an Encyclopedia of Worldwide Policy, Technology and History*, Vol. I. Santa Barbara, USA: ABC- Clio.

Kenyon, I. R. and D. Feakes, Eds. 2007. *The Creation of the Organisation for the Prohibition of Chemical Weapons. A Case Study in Birth of an Intergovernmental Organisation*. The Hague: TMC Asser Press.

Matousek, J. 2007. Status of implementing the Chemical Weapons Convention ten years after entry into force, and the way ahead. Paper presented at *26th Meeting of the Pugwash Study Group on the implementation of the Chemical and Biological Weapons Conventions: 10 Years of the OPCW: Taking Stock and Looking Forward*, Noordwijk, The Netherlands, March, 17–18.

Matousek, J. 2008. *Chemical Weapons Chemical Warfare Agents*. Kleinwachter: Frydek-Mistek.

Mayor, A. 2003. *Greek Fire, Poison Arrows and Scorpion Bombs: Biological and Chemical Warfare in the Ancient World*. Woodstock, NY: Overlook-Duckworth.

OPCW. Note by the Director-General on the *Report of the Scientific Advisory Board on Developments in Science and Technology, Conference of States Parties, Second Review Conference*, The Hague, the Netherlands, April 7–18, 2008. RC-2/4, April 18, 2008.

OPCW. *Report of the First Special Session of the Conference of States Parties to Review the Operation of the Chemical Weapons Convention (First Review Conference). Conference of States Parties, First Review Conference*, The Hague, the Netherlands, April 28–May 9, 2003. RC-1/5, May 9, 2003.

OPCW. Programme to Strengthen Cooperation with Africa on the Chemical Weapons Convention. *Fiftieth Executive Council*, The Hague, the Netherlands, September 25–28, 2007. EC-50/DG.17, September 26, 2007.

OPCW. *Report of the Second Special Session of the Conference of States Parties to Review the Operation of the Chemical Weapons Convention (Second Review Conference, RC-2/4). Conference of States Parties, Second Review Conference*, The Hague, the Netherlands, April 7–18, 2008. RC-2/4. April 18, 2008.

OPCW. Opening Statement by the Director-General to the Conference of the States Parties at its Fourteenth Session; *Fourteenth Session of the Conference of States Parties*, The Hague, the Netherlands, November 30–December 4, 2009. C-14/DG.13. November 30, 2009.

Pearson, G. S. and P. Mahaffy. 2006. Education, outreach, and codes of conduct to further the norms and obligations of the Chemical Weapons Convention. *Pure and Appl. Chem.*, 78(11): 2169–2192.

Thakur R. T. and E. Haru, Eds. 2006. *The Chemical Weapons Convention: Implementation, Challenges and Opportunities*. Hong Kong: United Nations University Press.

Tu, A. T. 2002. *Chemical Terrorism: Horrors in Tokyo Subway and Matsumoto City*. Colorado: Alken, Inc.

28 National and International Scientific Societies' Role in Global Collaboration in Chemicals Management

John Duffus

CONTENTS

INTRODUCTION

Historically, the first organizations to be concerned with chemicals management were national scientific organizations and related chemical societies. This chapter will consider the evolution of such societies, and especially of the two national chemical societies which are currently the most influential, the oldest, the U.K. Royal Society of Chemistry, and the largest, The American Chemical Society (ACS).

At the beginning of the twentieth century, as a result of the growing economic importance of the chemical industry, it became clear that national chemical societies were not adequate to deal with problems arising from the internationalization of chemistry. In particular, there were variations in chemical nomenclature which were causing confusion in exchange of knowledge and preventing effective patenting of processes and new substances. This situation made it essential to establish an international organization to create a coherent system of nomenclature. The leaders in academic science and in the chemical industry came together to found such an organization, known as the International Union of Pure and Applied Chemistry (IUPAC).

Inevitably, the safe production, use, and disposal of chemicals became a matter of concern for chemists at both national and international level and, as will be seen below, action has been taken in response to this concern.

NATIONAL CHEMICAL SOCIETIES

The oldest scientific society in existence is probably the Accademia dei Lincei (now the Italian Accademia Nazionale dei Lincei) founded in Rome, on August 17, 1603, by four young men, Federico Cesi, son of the first Duke of Acquasparta, who designated himself Consessus princeps et institutor (Prince and Founder of the Association), a Dutchman, Johannes Eck (called Ecchio in Italian) who had graduated in medicine at Perugia, and Francesco Stelluti and Count Anastasio De Filiis who were both Italian: Stelluti was an expert in natural science and also translator of the Latin poet Persius; and De Filiis was a kinsman of Federico Cesi. Cesi was 18, and the others only eight years older. Their chief aim was to see into the secrets of nature with a perception as acute as that of the lynx. Hence, the arms and the name Lincei attached to their association. Undoubtedly, chemistry and possibly alchemy were amongst their interests but their views on safe use of chemicals have not been recorded.

The second national scientific society to be founded, and which still exists, is the Royal Society of London. The Society started as in an "invisible college" of natural philosophers who began meeting in the mid-1640s to discuss the ideas of Francis Bacon. It was founded officially on November 28, 1660, when 12 members of the "invisible college" met at Gresham College after a lecture by Christopher Wren, the Gresham Professor of Astronomy, and decided to found "a Colledge for the Promoting of Physico-Mathematicall Experimentall Learning." This group included Wren himself, Robert Boyle, John Wilkins, Sir Robert Moray, and William, Viscount Brouncker. The Society was to meet weekly to witness experiments and discuss scientific topics. The first Curator of Experiments was Robert Hooke. It was Moray who first told King Charles II, of this venture and secured his approval and encouragement. The name, The Royal Society, first appears in print in 1661, and in its Royal Charter of 1663 the Society is referred to as "The Royal Society of London for Improving Natural Knowledge." Again, there is little evidence of great concern with safe use of chemicals during the first two centuries of the Royal Society of London's existence. However, when specialist chemical societies developed with the coming of the industrial revolution, such societies had a vested interest in maintaining an acceptable level of safety for their members in their laboratories and workplaces.

Perhaps the oldest society specializing in chemistry is the Royal Society of Chemistry that traces its origin to the establishment of the Chemical Society of London in 1841. The importance of the Chemical Society, and of the developing science of chemistry, was recognized seven years later with the award of a Royal Charter. Under the Charter, the Chemical Society's role extended beyond the advancement of the science of chemistry to encompass the proper development of the applications of chemistry, both in industry and in the community. Thus, the importance of good chemical management was recognized even at this early stage in the development of applied chemistry.

THE ROYAL SOCIETY OF CHEMISTRY

The Royal Society of Chemistry today has a global membership of over 46,000 and the longest continuous tradition of any chemical society in the world. It is the sole heir and successor to four long-established bodies:

The Chemical Society (founded in 1841)
The Society for Analytical Chemistry (founded in 1874)
The Royal Institute of Chemistry (founded in 1877)
The Faraday Society (founded in 1903)

The Royal Society of Chemistry fulfills the roles previously undertaken by all four of these bodies. In accordance with its first Royal Charter, granted in 1848, the RSC continues to pursue the aims of the advancement of chemistry as a science, the dissemination of chemical knowledge, and the development of chemical applications. However, over the years its responsibilities have broadened and its activities have become more extensive.

Today, the RSC's work spans a wide range of activities connected with the science and profession of chemistry. It is actively involved in the spheres of education, qualifications, and professional conduct. It runs conferences and meetings at both national and local level. It is a major publisher, and is internationally regarded as a provider of chemical databases. In all its work, the RSC aims to be objective and impartial. It is independent of government, trade associations and trade unions. It is recognized throughout the world as an authoritative voice of chemistry and chemists and makes many contributions to ensuring chemical safety at all levels of production, use and disposal.

Apart from facilitation of appropriate education, specific concern for chemical safety within the Royal Society of Chemistry is devolved to the Environment, Health, and Safety Committee (EHSC). EHSC is a long-established committee that has overseen the RSC's work in chemical safety and legislation for over 25 years. Most recently EHSC has led the RSC's work on REACH. EHSC produces a range of publications covering all aspects of health, safety, and environmental issues. These include the Notes and Professional Briefs, which are intended to provide background information for members on subjects which are not their specialty. Recent publications have included: "Contaminated Land," "Integrated Pollution Prevention and Control," "The Precautionary Principle," "Safety of Laboratory Workers with Disabilities," and "Occupational Health and Safety Management Systems." In addition, EHSC reviews and responds to all proposed legislation related to chemical safety from both the U.K. Government and the European Union legislators. EHSC also provides a representative for the RSC on government consultative bodies such as the current Chemical Stakeholders Forum.

THE AMERICAN CHEMICAL SOCIETY

Thirty-five chemists met at the College of Pharmacy of the City of New York on April 6, 1876, to found the ACS. From its inception, the ACS was committed to sharing its professional work with a public audience. Now, with more than 154,000 members, the ACS is the world's largest scientific society and one of the world's leading sources of authoritative scientific information. In particular, it is responsible for the Chemical Abstracts Service (CAS), which produces the CAS REGISTRY℠, the largest single collection of substance information available worldwide, as well as indexed references from more than 10,000 major scientific journals and 59 patent authorities around the world. The CAS databases are updated daily, built and quality controlled by CAS scientists, readily accessible through search and analysis tools provided by ACS, and form a single resource for tracing both patents and journal articles.

A nonprofit organization, chartered by the United States Congress, ACS is at the forefront of evolving worldwide chemical activity. It is committed to "improving people's lives through the transforming power of chemistry" and to "be a global leader in enlisting the world's scientific professionals to address, through chemistry, the challenges facing our world." Some of the challenges ACS has identified include providing sufficient energy, protecting the environment, assuring the availability of safe food and water for all people, and improving global healthcare.

Like the Royal Society of Chemistry, ACS has a committee, created in 1963, dedicated to chemical safety, appropriately named the Committee on Chemical Safety. The Committee on Chemical Safety has as its stated prime responsibility the encouragement of safe practices in chemical activities. The committee aims to serve as a resource to chemical professionals in providing advice on the safe handling of chemicals, and seeks to ensure safe facilities, designs, and operations by calling attention to potential hazards and stimulating education in safe chemical practices. The Committee also provides advice to other ACS units on matters related to chemical safety and health. This advice is made available in a range of publications, including *Chemical Safety for Teachers and Their Supervisors*, Grades 7–12 (ACS, 2001), *Chemical Safety Manual for Small Businesses*, 3rd ed., 2009 (ACS, 2009), and a *Handbook of Chemical Health and Safety* (ACS, 2001).

OTHER NATIONAL CHEMICAL SOCIETIES

Similar activities have been adopted by other national chemical societies and their collaboration is facilitated by the IUPAC as described below. For example, The Gesellschaft Deutscher Chemiker (GDCh) is the largest chemical society in continental Europe with members from academia, industry, and other areas. The GDCh supports chemistry in teaching, research, and application and promotes the understanding of chemistry by the public. The society, a registered charity, was founded in 1949 but builds on a long tradition that began in 1867 when its first predecessor organization, the Deutsche Chemische Gesellschaft was founded in Berlin. The GDCh defines its mission as to unite those people associated with the chemical and molecular sciences and support them in their responsible and sustainable endeavors for the good of the public and of the environment. This is achieved mainly through its conferences and educational activities, including many related to chemical safety.

In Asia, the Japanese Chemical Society was founded in 1878 by approximately 20 young scholars wishing to advance research in chemistry. Later, it was renamed The Tokyo Chemical Society, and eventually given the present English name of "The Chemical Society of Japan" (CSJ). In 1948, it merged with the Society of Chemical Industry, founded in 1898. The CSJ has a history encompassing 130 years, with a current membership exceeding 34,000, and covers most areas of pure and applied chemistry. Like the other national chemical societies, the prime mission of the CSJ has been defined as to promote chemistry for science and industry in collaboration with other domestic and global societies, with the overriding purpose to contribute to the betterment of human life. In 1998 the CSJ organized the Special Committee for Environment and Chemistry. It was renamed two years later as the Environment and Safety Promotion Committee. The Committee has examined how chemists and chemical engineers should strive to solve environmental and safety issues and the actions the CSJ should take as the core organization. These considerations have resulted in various activities including environmental symposia/seminars for the general public, and publications that include environment/safety-related pamphlets, and the following books: *Notes on Chemical Safety*, *Guide for Risk Communications in Handling Chemicals*, *Safety Guide for Chemical Experiments*, and *Dioxin and Environmental Hormones*. The CSJ, as a contractor for government agencies such as the Ministry of Environment and METI (Ministry of Economics, Trade and Industries), carries out studies in areas such as risk communication and handling of chemical materials, and the training of environment and safety personnel. In 2002, the CSJ established the Risk Research Group to examine risk-related issues and the role of chemists in sustainable development.

THE INTERNATIONAL UNION OF PURE AND APPLIED CHEMISTRY

The IUPAC is the organization that sets the appropriate scientific foundations for safe use of chemicals. IUPAC makes its journal content and Gold book of terminology available free of charge on the Internet, together with other information and educational materials, for example, on toxicology. Projects such as a Safety Training

Programme and associated Workshops, dissemination of the chemical education materials and laboratory equipment, the Flying Chemists Programme, and coordination with the Pan African Network, are examples of successful efforts led by IUPAC toward capacity building around the world. The Flying Chemists Programme has the aim of providing emerging countries with the means to improve the teaching and learning of chemistry at primary, secondary, and tertiary levels. Under the program, the host country provides local costs (boarding and lodging), while the IUPAC Committee on Chemical Education provides the air fares.

A major safety management activity was initiated when IUPAC was approached by the Organization for the Prohibition of Chemical Weapons (OPCW) in 2002 and again in 2007, and asked to organize the work of establishing the knowledge basis on which the revisions of the Chemical Weapons Convention would be based.

The IUPAC mission is to advance the worldwide aspects of the chemical sciences and to support the application of chemistry in the service of mankind. As a scientific, international, nongovernmental, and objective body, IUPAC addresses many global issues involving the chemical sciences. IUPAC was formed in 1919 by chemists from academia and industry. Over nine decades, the Union has developed worldwide communications in the chemical sciences and has united academic, industrial, and public sector chemistry with a common scientific language. IUPAC is recognized as the world authority on chemical nomenclature, terminology, standardized methods for measurement, atomic weights, and many other critically evaluated data. The Union sponsors major international meetings ranging from specialized scientific symposia to CHEMRAWN (CHEMical Research Applied to World Needs) meetings with broad societal impact. CHEMRAWN topics have included "Chemistry, Sustainable Agriculture and Human Well-Being in sub-Saharan Africa," "Chemistry for Clean Energy," "Toward Environmentally Benign Products," and "Processes, and Greenhouse Gases Mitigation and Utilization."

IUPAC is an association of various bodies, National Adhering Organizations, which represent the chemists of the member countries. There are 53 National Adhering Organizations, and 11 Associate National Adhering Organizations. Many chemists throughout the world are engaged on a voluntary basis in the scientific work of IUPAC, primarily through projects, which are components of eight Divisions and several Committees. The Divisions of IUPAC are as follows: I Physical and biophysical chemistry; II Inorganic chemistry; III Organic and biomolecular chemistry; IV Macromolecular chemistry; V Analytical chemistry; VI Chemistry and the environment; VII Chemistry and human health; VIII Chemical nomenclature and structure representation.

IUPAC and Terminology

IUPAC's terminology is used by professional chemists in academia, government, and the chemical industry throughout the world. Much of it has been published in books commonly identified by the colors of their covers: Gold—Chemical terminology; Green—Quantities, units, and symbols in physical chemistry; Red—Nomenclature of inorganic chemistry; Blue—Nomenclature of organic compounds; Purple—Macromolecular nomenclature; Orange—Analytical nomenclature;

Silver—Nomenclature and symbols in clinical chemistry. These are supplemented by papers published in the journal, Pure and Applied Chemistry, which is freely available to read and download from the IUPAC Web site (http://www.iupac.org/).

Chemical Standards

IUPAC publishes definitive and up-to-date data on atomic weights and isotopic abundances. It also publishes a wide variety of other chemical data: international thermodynamic tables of the fluid state; solubility data; metal–complex stability constants, available on disk; enthalpies of vaporization of organic compounds; thermodynamic and transport properties of alkali metals; recommended reference materials for achievement of specific physicochemical properties; and evaluated kinetic and photochemical data for atmospheric chemistry.

IUPAC is widely involved in establishing standard methods for use in analytical, clinical, quality control, and research laboratories. Some examples are: standard methods for the analysis of oils, fats, and derivatives; harmonization of international quality assurance schemes for analytical laboratories; protocol for self-auditing of analytical laboratories for ISO 9000 certification, quality assurance, and sampling; standardization of immunoassay determinations; standard methods for the determination of trace elements in body fluids; JCAMP-DX, a standard format for the exchange of spectra in computer readable form; standards for experimental thermodynamics; and standards for measurement of the transport properties of fluids; and standards for solution calorimetry.

Environment

The various Commissions and Committees of IUPAC have undertaken an extensive array of environmental projects. Some examples are: environmental analytical chemistry; study of environmental particles; polymer recycling; determination of trace elements in the environment; gas kinetic data for atmospheric chemistry; and glossaries of terms used in atmospheric chemistry, toxicology, and ecotoxicology.

Chemistry and Human Health: IUPAC Division VII

Toxicology is one of the concerns of this division. The main activities of the Division are covered by three Subcommittees: Nomenclature, Properties, and Units; Medicinal Chemistry and Drug Development; and Toxicology and Risk Assessment. All of these contribute to relevant areas of chemical safety management.

Subcommittee on Nomenclature, Properties, and Units

In 1995, the predecessor of the present Subcommittee, the Commission on Nomenclature, Properties and Units (C-NPU of International Federation of Clinical Chemistry (IFCC) and IUPAC) started publishing a series of papers on a coding system (i.e., a structure or a framework for pairs of codes and meaning) and a coding scheme (i.e., the pairs of codes and meaning), for properties in clinical chemistry. The codes offer unique and sufficient information about properties, and are designed to facilitate the transfer of information between laboratories and the end users of laboratory information. The codes make it possible to translate the data into any language automatically. The meanings have been tested for translation into 18 languages, including many of the European languages, Arabic, and Cantonese. The coding scheme is accessible on the IFCC Web site and now includes clinical pharmacology and toxicology, and environmental toxicology.

Subcommittee on Medicinal Chemistry and Drug Development

This subcommittee arranges two meetings a year, publishes books, most recently the IUPAC Handbook of Pharmaceutically Acceptable Salts, and prepares glossaries to aid communication between chemists working in this area.

Training of Medicinal Chemists, a series of papers on appropriate education, has been published. A syllabus for a short course on medicinal chemistry has also been published, and courses have been initiated in some Latin American countries. An IUPAC Recommendation entitled *Preservation and Utilization of Natural Biodiversity in Context of the Search for Economically Valuable Medicinal Biota* was published in 1996 in addition to a technical report intended to encourage collaboration in development of these resources (Andrews et al., 1996; Cragg and Newman, 2005).

SUBCOMMITTEE ON TOXICOLOGY AND RISK ASSESSMENT

In relation to safe global management of chemicals, the IUPAC Subcommittee on Toxicology and Risk Assessment has the following mission objectives:

1. Coordination of projects that have been approved by the Division VII Committee and which relate to toxicology and risk assessment.
2. Provision of links with other organizations concerned with toxicology such as the International Union of Toxicology (IUTOX), the International Programme for Chemical Safety (IPCS), the World Health Organization (WHO), the Organization for Economic Cooperation and Development (OECD), and the Strategic Approach to International Chemicals Management (SAICM).
3. Facilitation of IUPAC interaction with chemists in the chemical industry worldwide in the field of toxicology and risk assessment.

One of the main concerns of the Subcommittee on Toxicology and Risk Assessment and its predecessor, the Commission on Toxicology, has been the education of chemists in fundamental principles of toxicology. Activities here have involved the compilation of glossaries, educational modules, and reviews of matters of current concern. These can be found on the IUPAC Web site at: http://www.iupac.org/divisions/VII/VII.C.2/index.html.

The UN International Year of Chemistry 2011, Chemistry: Our Life, Our Future

The Year of Chemistry is being held to celebrate the achievements of chemistry and its contributions to the well-being of humankind, and to increase the public understanding of chemistry. One of the key outcomes will be to promote the role of chemistry in contributing to solutions to global challenges to health and to the environment. IUPAC will hold keynote events in Paris, San Juan, Puerto Rico, and Brussels during 2011, celebrating the achievements of chemistry and encouraging continued contributions from chemistry to the benefit of humankind. Individual organizations affiliated to IUPAC at national, regional, and local levels will hold events that involve students, chemical professionals, industry, the press, government officials and policy-makers, and the public in promoting understanding of chemistry. IUPAC will be particularly active in regions of the developing world in building beneficial collaborations. Further information will appear on the Web site at: http://www.chemistry2011.org.

THE INTERNATIONAL UNION OF TOXICOLOGY

The International Union of Toxicology (IUTOX) was founded in 1980. It has 53 national/regional Society members, eight institutional members and represents over 20,000 toxicologists from academia, industry, and government worldwide.

IUTOX has the stated aim in its "Vision Statement" of improving human health through the science and practice of toxicology worldwide. IUTOX pursues its vision by fostering international scientific cooperation for the global acquisition and utilization of knowledge in toxicology for the improvement of human health. Recognizing the value of a multi-national, multicultural organization, IUTOX has defined the following objectives:

1. Provision of an international platform and leadership to promote scientific cooperation and information exchange in toxicology.
2. Organization with member societies of triennial international toxicology congresses (the International Congress of Toxicology and the Congress on Toxicology in Developing Countries) in order to facilitate and encourage scientific exchange and leadership, exploring additional formats such as blogs, Web sites, conference calls, and focused workshops, and facilitating meetings of national leaders in toxicology during major scientific meetings.
3. Enhancement of opportunities for educational development and exchange in toxicology by supporting courses and lectures in developing countries and elsewhere, promoting Risk Assessment Summer Schools (RASS, held mainly in Europe), and Risk Assessment Workshops (RAW, held mainly in developing countries). The provision of international educational opportunities worldwide has involved raising funds in order to award fellowships for international training and exchange programs.
4. Broadening the geographic base of toxicology as a discipline by encouraging diversity of toxicological disciplines throughout the world by providing a means for toxicologists in areas with limited numbers and resources to identify with the science, increasing the number of member societies in a region, and building local capacity for toxicology.

5. Communication of the value of toxicology for improving human health by identifying partners to assist in achieving this objective, working with member societies, and facilitating discussion among regulators, academics, stakeholders, private sector, and public at large regarding toxicological issues.
6. Strengthening organizational effectiveness by reaching out to other partners to provide for adequate funding of the organization and its activities and improving dialogue between member societies.

IUTOX Congresses

One of the central activities of IUTOX is the organization of its flagship international congresses to spread knowledge of toxicology around the world. Thus, while the most recent was the twelfth International Congress of Toxicology (ICT XII), hosted by Spanish Society of Toxicology, incorporating the 47th European Society of Toxicology (EUROTOX) Congress, in Barcelona, Spain, it will be followed by the thirteenth International Congress of Toxicology (ICT XIII) to be hosted by the Korean Society of Toxicology Seoul, Korea, in June 2013.

In addition, the sequence of Congresses of Toxicology in Developing Countries will continue. These have already been held in Argentina (1987), India (1991), Egypt (1995), Turkey (1999), China (2003), Croatia (2006), and South Africa (2009).

IUTOX Growth Areas

IUTOX intends to expand RAW and to look for other opportunities to serve the developing and least-developed countries in capacity building and in other ways. There will be emphasis on programs in Africa and Latin America. This will require finding new sources of funding and identifying and recruiting new member societies in regions with the greatest need. It will also necessitate improving communications with member societies and other international societies such as IUPAC and SETAC through greater use of broadcast e-mails, and improved Web site, and other technology. Attention will be paid to the need to inform policymakers, regulators, and scientists in developing countries about the potential contribution of toxicology to solving the many issues associated with the environment, health, and chemical safety.

THE SOCIETY OF ENVIRONMENTAL TOXICOLOGY AND CHEMISTRY

The Society of Environmental Toxicology and Chemistry (SETAC) is a nonprofit, worldwide professional society comprised of individuals and institutions engaged in the study, analysis, and solution of environmental problems, the sustainable management and regulation of natural resources, environmental education, and research and development.

SETAC's mission is to support the development of principles and practices for the protection, enhancement and management of sustainable environmental quality and ecosystem integrity. SETAC promotes the advancement and application of scientific research related to contaminants and other stressors in the environment, education in the environmental sciences, and the use of science in environmental policy and decision making. The Society provides a forum where scientists, managers, and other professionals exchange information and ideas for the development and use of multidisciplinary scientific principles and practices leading to sustainable environmental quality.

The principles fundamental to any SETAC activity are as follows. Firstly, multidisciplinary approaches are essential to solving environmental problems. In applying these approaches, a balanced contribution from academia, business, and government is also essential. Last but not least, strict objectivity must be maintained in assessing each situation on its merits and objective conclusions must be science and evidence based.

SETAC is moving to join other leading global scientific organizations in working to make the world's environment safer and more sustainable. As a nongovernmental organization drawing on participation from scientists and environmental experts in academia, business, government, and other NGOs, SETAC offers a model for promoting environmental dialogue not only among those experts, but also between those experts and the public in countries around the world.

Building on its roots in North America and Europe, SETAC seeks to achieve a global impact through a concerted effort to connect with scientists and partners in developing countries. Establishing global networks of science and scientists is critical to successfully address global environmental problems. Creating national communities of committed scientists and officials across sectors is similarly crucial to solving national problems and building national consensus on the environment. Such networking also provides developing countries with needed links to people and resources. SETAC's programme to build membership around the world, supported effectively by its Global Partners, is integral to achieving its objective of "Environmental Quality Through Science®."

Action in International Chemical Management

One example of SETAC's engagement in global decision making relates to persistent organic pollutants. Several SETAC members are members of the review committee of the Stockholm Convention on Persistent Organic Pollutants (POPS).

Another example is the group of recent meetings organized by SETAC in 2008 on "Emerging Issues in Chemicals Management in Developing Countries (EMERCHEM)" to identify emerging, high-priority, science-based issues related to chemical contaminants, chemicals management, and chemical safety. The meetings were held during the SETAC Annual Meetings in Warsaw, Sydney, and Tampa. About 100 scientists from government, business, and academia identified and discussed such key issues as e-waste, nanomaterials and nanotechnology, linkages between chemicals and climate change, and a need for capacity building and improved chemicals management legislation. Learning from current transition economies, sharing knowledge with developed countries, and actively stimulating collaborations and expertise centers constitute first steps to addressing these problems.

SETAC has also contributed to the Strategic Approach to International Chemicals Management (SAICM), by participating in regional meetings in France, Panama, and Tanzania, and working to prepare an effective capacity-building program under the SAICM Quick Start Programme (QSP), in order to highlight relevant risk-assessment tools from SETAC, international governmental organizations, national governments, and business associations, particularly related to the whole life cycle of chemicals.

At the same time, SETAC continues to explore partnerships with a number of international and national scientific organizations, such as IUPAC and IUTOX, in order to generate synergies for more effective scientific activity in all the related disciplines around the globe.

CONCLUSION

National chemical societies such as the RSC, ACS, GDCh, and JCS have long recognized their responsibilities to promote good management for the safe use of chemicals within their relevant national boundaries. Subsequently, they responded to the growing international dimensions of chemistry by establishing IUPAC to harmonize terminology and good chemical practice. More recently, national toxicology societies have developed with the same aim. At international level, IUTOX, and SETAC have evolved from the national societies. IUPAC, IUTOX, and SETAC have a common interest in promoting safe and sustainable management of chemicals in the interest of human progress and environmental protection, based on sound science, and they are likely to collaborate further in pursuing this objective. Together with the national societies, their activities help to provide the solid intellectual foundation on which other international developments, such as SAICM, must be based in order to ensure success in preventing further deterioration of our environment and in the maintenance and improvement of human health.

REFERENCES

ACS Board-Council Committee on Chemical Safety. 2001. *Chemical Safety for Teachers and Their Supervisors, Grades 7–12*. Washington, DC: American Chemical Society.

ACS Committee on Chemical Safety. 2009. *Chemical Safety Manual for Small Businesses*, 3rd ed. Washington, DC: American Chemical Society.

Alaimo, R. J., ed. 2001. *Handbook of Chemical Health and Safety*. Washington, DC: American Chemical Society.

Andrews, P. R., R. Borris, E. Dagne, M. P. Gupta, L. A. Mitscher, A. Monge, N. J. de Souza, and J. G. Topliss. 1996. Preservation and utilization of natural biodiversity in context of search for economically valuable medicinal biota (Technical Report). *Pure and Applied Chemistry* 68(12):2325–2332.

Cragg, G. M. and D. J. Newman. 2005. International collaboration in drug discovery and development from natural sources. *Pure and Applied Chemistry* 77(11):1923–1942.

29 The Role of the United Nations Development Programme in Sound Chemicals Management

Jan van der Kolk

CONTENTS

INTRODUCTION

The sound management of chemicals (SMC) is integral to sustainable human development; it is a precondition for healthy environments for human settlement and physical well-being, including those with respect to provision of safe drinking water, air and food, and ecosystem health. Being a cross-cutting issue, the United Nations Development Programme (UNDP) works to ensure that, as a part of the overall poverty reduction efforts at country level, the SMC is incorporated.

The world's poorest routinely face the highest risk of exposure to toxic and hazardous chemicals, because of their occupations, living conditions, lack of knowledge related to unsafe handling practices, limited access to sources of uncontaminated food and drinking water, and the fact that they often live in countries where regulatory, health, and education systems are weakest.

Coordinating the UN development network, UNDP has since its inception been involved in chemicals work. The involvement in chemicals/chemical pollution management has been mainly as a part of projects supporting host governments' overall development goals. Much of the work during the 1970–1990s was co-implemented together with UN specialized agencies such as the Food and Agricultural Organization (FAO), the United Nations Industrial Development Organization (UNIDO), the United Nations Educational, Scientific, and Cultural Organization (UNESCO), and the United Nations Office for Projects Services (UNOPS). Today, UNDP is an observer at the Interagency Organisation for the management of chemicals (IOMC) (see Chapter 23 of this book) and works closely together with its members, but is in the process of becoming a full member.

With increased attention to global environmental issues and the adoption of a series of global environmental treaties in the 1980–1990s, UNDP was invited to assist developing countries and countries with economies in transition to fulfill their chemical management commitments under these treaties. Initially this work concentrated on countries obligations under the Montreal Protocol on Substances that Deplete the Ozone Layer and the Regional Seas Conventions, and has during the past 10 years been extended to cover activities under the Stockholm Convention on Persistent Organic Pollutants (POPS) as well as nonbinding instruments such as the Strategic Approach to International Chemicals Management (SAICM).

UNDP MAJOR AREAS OF WORK AND MAJOR INSTRUMENTS

The UNDP promotes the SMC as an important component of the global poverty reduction effort. Proper chemicals management links to the overall Millennium Development Goals (MDGs) and UNDP advocates for the importance of addressing issues related to chemicals management and chemically linked pollution in developing countries by integrating rigorous chemicals-management schemes into national development policies and plans. This "Mainstreaming of Sound Management of Chemicals" is one of the core approaches of UNDP's work in the international chemicals area.

In addition, UNDP actively supports governments in their efforts in implementing provisions and compliance with global environmental agreements. UNDP supports projects that aim to achieve SMC and help countries strengthen national capacities in the following areas:

- Incorporate the SMC into MDG-based national development policies and plans in support of the SAICM.
- Reduce and eliminate the release of POPS as specified under the Stockholm Convention on POPS.
- Phase-out Ozone-Depleting Substances (ODS) and achieve compliance with the Montreal Protocol on Substances that Deplete the Ozone Layer.
- Reduce and prevent chemical pollution of lakes, rivers, groundwater, coasts, and oceans in support of international water agreements.

ROLE OF UNDP

UNDP is on the ground in 166 countries, working with countries on their own solutions to global and national development challenges. As they develop local capacity, they draw on the people of UNDP and their wide range of partners, including government institutions, UN Agencies and international financial institutions, industry, representative organizations such as technical associations, agricultural institutes, academia, and civil society.

As an implementing agency of the Multilateral Fund (MLF) for the Implementation of the Montreal Protocol, the Global Environment Facility (GEF), the SAICM Quick Start Programme (QSP) Trust Fund, and a multitude of bilateral donors, UNDP is working with a broad range of partners, to help developing countries and countries with economies in transition to adopt and implement strategies that target the preservation of the environment and sustainable development and aim to achieve the objectives of chemical related Multilateral Environmental Agreements (MEAs).

How UNDP Carries Out Its Work

UNDP's chemicals programs and projects are implemented on the ground directly by host governments supported by UNDP's country-based Environment and Energy team on technical aspects as well as the country offices' operations team for financial and administrative project matters. The country offices are further technically supported and backstopped by regionally-based technical specialists and program coordinators. In addition to the above in-house expertise, UNDP relies on a panel of external experts for a range of topics, who are deployed to assist country-based projects on the ground.

Main Areas of Work

SAICM

Since the adoption of SAICM in 2006 (see Chapter 17 of this book), UNDP has been assisting national governments and UN country teams in beginning to integrate SMC more systematically into MDG-based national development planning processes and strategies. In this work, UNDP is drawing on its own experience and expertise gained in providing support to developing country partners in implementing other chemicals related multilateral environment agreements. UNDP's core expertise lies in elaborating national MDG-based development plans and poverty reduction strategies. UNDP interventions in this area include:

- Advancing countries' SMC capacity at the local and national level, based on UNDP's country-level expertise and presence on the ground.
- Providing strategic policy and economic guidance related to the interaction of SMC and the MDGs, as well as other national development priorities.
- Identifying specific areas of chemicals management that are likely to result in concrete environmental, health, and economic benefits as a result of introducing sound management practices, and supporting plans to begin addressing identified national priorities.
- Building capacity to assess national development and budgeting planning processes, and identifying opportunities to integrate national priorities for the SMC into these plans.

- Offering guidance on integrating chemicals management priorities into national discussions, development processes, policies, and plans, with the objective of fostering the national budget commitments.

To this end, UNDP has developed a technical guide to assist governments and UN Country Teams in recognizing and assessing opportunities for incorporating SMC into national development planning processes: the UNDP Guide on Integrating the SMC into Development Planning.* The guide is one of several tools that UNDP's Environment and Energy Group has developed to enhance assistance to partner countries through a comprehensive approach to sustainable development. The methodology for integrating SMC in MDG-based policies, as put forward in this technical guide, follows the steps that countries typically go through to advance their national chemicals management regimes.

Persistent Organic Pollutants

See Chapter 15 of this book for the Stockholm Convention on POPS.

UNDP, in its role as an implementing agency of the GEF, has been supporting developing countries, and countries with economies in transition, in their efforts to reduce and eliminate POPS and meet the objectives of the Stockholm Convention. Many of the challenges and priorities relating to the reduction and elimination of POPS require enhancement of national capacities with respect to human resources development and institutional strengthening, as well as increased availability of technical knowledge and training opportunities.

UNDP assists countries in meeting their commitments under the Stockholm Convention, including through:

- Meeting reporting obligations, sharing lessons learned and adopting global best practices.
- Building the necessary capacity to implement POPS risk reduction measures.
- Reducing the effects of POPS on human health and the environment.
- Demonstrating effective alternative technologies and practices that avoid POPS releases.

During the early years of the Convention's implementation, much of the focus was on national planning, as well as building necessary national capacity, meeting countries' reporting obligations, and compiling National Implementation Plans (NIPs). Through the implementation of "Enabling Activities" projects, UNDP has been instrumental in providing support to 29 countries in developing their NIPs. Since for most countries the preparation of POPS NIPs has now been completed, or is at a very advanced stage, country level action has been shifting gradually from preparation of NIPs to implementation of activities to address POPS priorities included in those plans.

* http://www.undp.org/chemicals/mainstreamingsmc.htm

In addition, UNDP is supporting 20 "post-NIP" country activities, as well as three global programs, with a combined portfolio of projects amounting to over US$ 215 million (including US$ 78 million of grants programmed through the Global Environment Fund (GEF). UNDP-supported country projects and global programs address a variety of national and Stockholm Convention priorities, as well as GEF Strategic Objectives. Through the implementation of "post-NIP" projects, UNDP supports the reduction and elimination of all types of POPS contaminants included under the Stockholm Convention, covering a multitude of sectors and activities.

These sectors and activities range from POPS-free agricultural practices to reduction of unintentional POPS releases related to medical waste disposal, and from sound management of PCBs contained in equipment to minimization of the exposure levels of communities living close to contaminated areas.

In addition, UNDP has supported capacity development with respect to POPS management in a large number of countries. This has been accomplished through GEF-funded global skills development projects, as well as through UNDP-implemented projects that aim to integrate the SMC into national development planning processes in support of the SAICM. Wherever possible and appropriate, UNDP POPS activities are undertaken within a country's broader framework for SMC, to ensure national coordination among chemicals related activities both in support of regional or global conventions and agreements on chemicals and in support of other national goals.

UNDP's key approaches to helping countries advance the SMC include:

- Campaigning and mobilization—Advocacy and awareness building among stakeholders about POPS management and SMC.
- Analysis and capacity building—Identification of innovative practices, policies, and institutional reforms to help countries put in place effective POPS and chemicals management structures that are informed by strategic needs assessments and financial evaluations.
- Technical assistance—Specific impact-driven technical assistance for addressing national challenges and constraints affecting the management of POPS and other chemicals.
- Monitoring and integration—Assistance to countries in tracking progress on mainstreaming of POPS priorities and sound chemicals management into broader national MDG-based development strategies.

Management of Ozone-Depleting Substances

UNDP Montreal Protocol Programme (see Chapter 9 of this book) implements activities supporting compliance with the Montreal Protocol, with financial support from the MLF, the GEF and bilateral donors. UNDP is working with a broad range of partners, in helping developing countries and countries with economies in transition to adopt and implement strategies that preserve the ozone layer while safeguarding the global climate.

To date, UNDP has been managing a global program of over US$ 500 million to provide financial and technical assistance to more than 100 countries, enabling them to phase out the use of ODS in activities such as foam production, refrigeration and air-conditioning manufacturing and servicing, aerosol and solvents applications, fire protection, and crop fumigation. In total these projects will prevent over 63,000 tons of ODS from being released into the earth's atmosphere.

UNDP assists its partners in complying with Montreal Protocol targets through:

- Capacity Development—Assisting governments to develop more effective national policies and programs to meet ODS elimination targets, including development of country programs related to ODS, institutional strengthening, and performance-based national phase-out management plans.
- Technical Assistance, Training, and Demonstration Programmes—Providing technical support and information dissemination regarding ozone and climate-friendly technologies and alternative substances to ODS. This is done through practical, hands-on training sessions and in-field demonstrations designed to build technical and economic confidence in alternative substances and processes.
- Technology Transfer—Facilitating access to the best available technologies and related technical assistance to allow governments and enterprises to adopt alternative production processes and ozone/climate friendly technologies.
- Increased Access to Funding—Supporting countries in securing financial support from the MLF to meet compliance with the Montreal Protocol, and seeking opportunities for ODS measures that also address other priorities, such as climate benefits. This may also result in additional sources of funding becoming available.

After successfully completing the first stage of ODS management, the emphasis of the Montreal Protocol has shifted toward phasing out hydrochlorofluorocarbons (HCFCs).

Many HCFCs have high global warming potential, some up to 2000 times that of CO_2, and thus contribute significantly to climate change.

In order to achieve agreed reductions, UNDP is, with financial assistance from the MLF, helping countries to prepare and implement their HCFC Phase-out Management Plans (HPMPs). In the global fight against climate change, all economically and technically feasible measures to reduce emissions of greenhouse gases are being actively pursued. Several instruments, including ODS banks, support the goals. They assist in preventing emissions of significant quantities of ODS from existing stockpiles, and from products that are discarded because they are no longer useful, or because they are replaced in connection with energy efficiency programs.

UNDP's current role as Lead Agency for a significant number of countries seeking to phase out HCFCs puts it in a unique position to help countries identify and develop appropriate greenhouse gas emission reduction projects while building market credibility and managing risks appropriately. Thus, achieving goals for this group of substances under the Montreal Protocol and under the United Nations Framework Convention on Climate Change (UNFCCC; see Chapter 16 of this book) are mutually supportive.

International Waters

Access to clean water plays a pivotal role in achieving sustainable human development, including poverty reduction. However, chemical pollution of water resources is one of the major threats to the achievement of sustainable water resources development and management.

Chemical pollution can be caused by: poorly treated or untreated municipal and industrial wastewater; pesticide and fertilizer run-off from agriculture; spills and other ship-related releases; mining; and other sources. It is one of the contributing factors to the current global crisis in which nearly a billion people lack access to safe drinking water. UNDP's response to this water crisis has been to emphasize an integrated approach to water resource management through effective water governance, referring to the range of political, social, economic, and administrative systems to develop and manage water resources and the delivery of water services at different levels of society. An integrated water governance system balances the mechanisms, processes, and institutions through which all involved stakeholders, including citizens and interest groups, articulate their priorities, exercise their legal rights, meet their obligations, and mediate their differences. UNDP's strategy in strengthening water governance—and thereby boosting progress toward the MDGs—includes:

- Incorporating water management, water supply, and sanitation into national development and poverty reduction strategies.
- Catalyzing financing for improved water governance.
- Supporting and participating in global, regional, national, and local dialogues on water governance.
- Building capacity to manage water resources effectively.
- Promoting women's empowerment and human rights as essential components of effective water governance.

As one of the implementing agencies of the GEF, UNDP administers and implements an important program on International Waters, assisting developing countries which share important water bodies—lakes, river basins, aquifers, and marine ecosystems—to improve their joint management of these transboundary resources through analysis and priority setting, and by developing and implementing joint action programs. A major portion of UNDP-GEF's International Waters funding is used to prevent or reduce chemical pollution originating from land-based human activities—including agriculture, industry, mining, oil and gas exploitation, and wastewater management—that place ecological stress on marine and freshwater systems and degrade them, often affecting their use by another country or community that shares the resources.

Examples of types of UNDP supported projects that aim to reduce chemical pollution of international waters include

- Projects implementing stress reduction measures in major transboundary water bodies that result in measurable reductions in pollution loads and evidence of ecosystem recovery, through the introduction of cleaner production

technologies, transfer of environmentally sustainable technologies and practices, sustainable financing and business models, harmonized legislation and improved environmental monitoring.
- Projects leading to a reduction in the release of mercury into the environment from artisanal gold mining, by supporting mining policy reforms, transferring sustainable mining technologies and practices, and introducing sustainable livelihood options.
- Creation of artificially engineered wetlands treating municipal wastewater, for national and regional replication.

CHALLENGES

Many challenges and additional work remains in assisting countries in their chemical management efforts. Much work is still needed on capacity building as well as guaranteeing access to mature, affordable, and environmentally friendly technologies.

Integrating SMC into the broader sustainable development agenda has been gathering pace in recent years much due to international recognition of the need of such a wider integration. SMC has too long been considered as a technical issue that could be addressed in a technical context. Although the importance of chemicals and of SMC for economic development, for health and for ecosystems is widely accepted, they have received very little attention on the broader development agenda and their financial mechanisms and the economic case for preventive rather than remediating action has been ignored

Separate financial mechanisms have been established for part of the agenda (see Section VIII, Chapter 49 of this book on Global Financial Instruments).

It is therefore essential, that these aspects of SMC and its importance for sustainable development, are translated in approaches that address aspects such as governance, industrial development, natural resource management, including mining, and education.

Linking SMC to the MDGs has been a first step. Achieving the WSSD 2020 goals (see Chapter 4 of this book) will only be possible, however, if SMC becomes an explicit goal in sustainable development.

30 United Nations Environment Programme and UNEP Chemicals

Jan van der Kolk

CONTENTS

INTRODUCTION

Following a recommendation of the Conference on the Human Environment in Stockholm in 1972 (see Section II, Chapter 2 of this book), the United Nations Environment Programme* (UNEP) was created.

Environment was an upcoming concern in several parts of the world, and the creation of UNEP as a program was an early response of the international community to this concern.

* http://www.unep.org

The Stockholm Conference had been organized at the initiative of Sweden, motivated by the visibly increased ecosystem disturbances that occurred when human activities started to affect the entire planet beginning more significantly in mid-twentieth century. Several factors converged in the 1960s that laid the basis for national and international political responses to environmental problems. They included strong public reaction in industrialized countries, influential publications such as Rachel Carson's *Silent Spring* in 1962, and large-scale environment-related accidents.

The choice of Nairobi as the city where UNEP would be located was also a sign of changing times. It was the first time a UN body was located outside the United States or Europe, in a country that had recently gained independence. Several industrialized countries fiercely opposed this location. But the times were receptive to change and hope, and to new initiatives to safeguard the future of humankind and its living environment. These would be essential to maintain or improve living conditions for a steadily increasing world population and its environment, which was showing various kinds of signs of degradation, inimical to healthy life. Programs within UNEP, from the very start, reflected these concerns of degradation of life-support systems, and included major themes such as desertification and quality of the oceans. During the Conference, the link between severe environmental degradation and poverty was clearly made by Indira Gandhi, the Prime Minister of India.

One of the changes that the Stockholm Conference has brought about, and that has continued to play a major role in UNEP over the years is the active involvement of nongovernmental organizations (NGOs). Against the resistance of several parties, NGOs have found their way to participating in many UNEP events, and have gradually gained more acceptance as a key stakeholder in many international gatherings.

The recommendation to establish UNEP proposed the following four entities:

a. An intergovernmental committee (*Governing Council*) under the General Assembly with the task of providing general policy guidance for direction and coordination of environmental programs within the UN system and to keep the world environmental situation under review.
b. A small *Environment Secretariat* to serve as a the focal point for environmental action and coordination within the UN system led by an Executive Director (ED), elected by the UN General Assembly on the nomination of the Secretary-General. This placed the ED on the same level as the heads of the specialized agencies. The ED was further mandated to advise UN intergovernmental bodies, a unique role for an international civil servant.
c. A voluntary *Environment Fund* to provide additional financing for environmental programs in order to finance, either wholly or partly, the costs of the new environmental initiatives undertaken within the UN system. The underlying rationale was that the specialized agencies would also increase their own financial resources in the field of the environment.
d. An *Environment Coordinating Board* (ECB) under the chairmanship of the ED in order to have maximum efficiency in coordinating UN

environmental programs. Governments were called upon to ensure their own coordination of environmental action, both national and international. The ECB was abolished in 1977.

Maurice Strong, whose role for the success of the Conference had been essential, was appointed as the first ED of UNEP.

THE INTERNATIONAL REGISTER OF POTENTIALLY TOXIC CHEMICALS

As environmental awareness necessarily encompassed the effects of chemicals on humans or ecosystems, chemicals were on UNEP's agenda from the very beginning.

After the Stockholm Conference, a larger meeting in Bilthoven, the Netherlands, in 1974 and a small meeting in Nairobi, Kenya in 1975 resulted in the creation of the International Register of Potentially Toxic Chemicals (IRPTC) which was to be located in Geneva. Proximity to the International Agency for Research on Cancer (IARC) (see Section V, Chapter 33 of this book, World Health Organization, WHO) in Lyon, France had also been considered, but a more narrow focus on cancer was considered less adequate, and Geneva offered a broader range of related organizations, such as WHO and the International Labour Organization (ILO) (see Section V, Chapter 22 of this book). Together they have been the nucleus of the later focus on Geneva as the location of many international chemical related activities. This was later confirmed by the creation of the International Programme on Chemical Safety, a collaborative program between WHO, ILO and UNEP (See Section V, Chapter 33 of this book on WHO).

The IRPTC Databank*

IRPTC was set up as a databank, being a register for all relevant information on potentially hazardous substances.

There were two sides to IRPTC's databank: one was data collection and validation; the other, management of this data by computer, including dissemination to users.

IRPTC aimed to maintain a storehouse of information adequate for an understanding of the hazards to health and the environment associated with toxic chemicals. Pharmaceuticals and radioactive substances were not considered within its scope. It offered concise overviews of selected data on chemicals for evaluation purposes, called chemical data profiles. It also contained a Register Index that offered a "profile of the profiles," an indication of the substances covered, their current status, and the subject areas where information existed or was absent. IRPTC also functioned as a network organization, in which many partners around the world provided relevant data.

* http://www.chem.unep.ch/irptc/irptc/databank.html

Data profiles were prepared by contributing network partners and IRPTC staff. All data were verified by IRPTC staff. The steps involved were

- To select the chemicals on which information was to be gathered.
- To search and acquire available scientific information.
- To select and validate the information received.

The IRPTC databank was accessible and distributed via print publications, telephone, and personal computer. As times and techniques evolved, and email and Internet became more widespread, IRPTC gradually developed into a networked organization, where validated information from a large variety of sources was made available to those searching for such data.

Over time, this led not only to major operational changes, but also to a name change—IRPTC became UNEP Chemicals.*

Other Functions of IRPTC

IRPTC also contributed to the gradual development of management instruments, such as the London Guidelines† which were the predecessor of the later Rotterdam Convention (see Section III, Chapter 14 of this book).

IRPTC also provided training in data assessment, preparing data profiles, and seeking the best ways to make use of the data for chemicals management. This has been an important contribution to the increased understanding of the potential role of management instruments, both at national and at international levels.

ENVIRONMENTAL GOVERNANCE

From its very beginning, governance issues have played a major role within UNEP's Governing Council and continue to do so, even more broadly, today.

A key issue has always been whether UNEP was to continue as a program of related environmental activities, with a mandate limited to certain areas within the environmental agenda, or was to develop into an Agency, comparable to WHO for health or FAO for agriculture. As environment is one of the three pillars of sustainable development (together with economy and social development), some maintained, even in Stockholm in 1972, that environmental issues needed to be at the heart of the programs and activities of most specialized UN agencies. This stance was used as an argument against having a single environmental agency. Others felt that a strong agency was needed to safeguard the inclusion of environmental considerations across the board. One of the consequences is that several key activities such as the Basel, Rotterdam and Stockholm Conventions and SAICM (Strategic Approach to International Chemicals Management) Secretariats (see respective chapters of this book) are not a formal part of UNEP Chemicals, but form separate entities. The

* http://www.chem.unep.ch/

† http://www.chem.unep.ch/ethics/english/longuien.htm

Conventions and SAICM have separate governing bodies and this has resulted in less explicit coordination and steering from UNEP.

For a more detailed discussion on this, see Section V, Chapter 19 of this book on Environmental Governance.

UNEP'S PROGRAMS AND UNEP CHEMICALS

The areas in which UNEP has played a major role over the decades are desertification, biodiversity, quality of the oceans and chemicals. For other areas, such as climate change and the decreasing ozone layer, mechanisms have been created separate from UNEP. Today, and especially since the UN Conference on Environment and Development (UNCED) in 1992 in Rio de Janeiro (see Section II, Chapter 3 of this book) international instruments, conventions and others, have been developed in at least nine thematic areas. A UNEP report of 2010 (UNEP, 2010) outlines the following areas addressed through international environmental agreements: (1) air pollution and air quality; (2) biodiversity; (3) chemicals and waste; (4) climate change; (5) energy; (6) forests; (7) freshwater; (8) oceans and seas; and (9) soil, land use, land degradation, and desertification.

UNEP has played a significant role in the development of these areas within the international agenda.

Examples for the chemicals management agenda are multifold.

The role of UNEP in SAICM (see Section IV, Chapter 17 of this book), as an example, has been crucial. The UNEP Governing Council, through decision 21/7 in its 1999 meeting, was the real starting point of the development of SAICM as a strategic approach to international chemicals management.* UNEP Chemicals, with support from WHO, continues to play a major role hosting the SAICM Secretariat.

UNEP also continues to play a key role in initiating a number of instruments, such as the recent development of a global instrument for the management of mercury (see Section VIII, Chapter 48 of this book).

UNEP CHEMICALS SUPPORT TO REGIONAL AND NATIONAL ACTIVITIES

One of the roles of UNEP Chemicals is to carry out regional activities and to support countries and regions to coordinate their efforts both during the elaboration of instruments and during their implementation. Immediately following the adoption of SAICM in 2006, a regional meeting took place in Cairo in September 2006.† This was followed by a series of regional meetings in Africa and other continents, coordinating implementation at regional levels.

Other roles include regional mechanisms for the exchange of information, such as the Chemicals Information Exchange Network‡ (CIEN).

* http://www.unep.org/gc/gc21/documents/K0100275-E-GC21.doc

† http://www.saicm.org/index.php?content=meeting&mid=12&def=4&menuid=16

‡ http://jp1.estis.net/communities/cien/

Information systems also support countries in their efforts to phase out certain hazardous chemicals such as Persistent Organic Chemicals (POPS),* replacing them with less hazardous chemicals or alternative approaches.

THE RESPONSE TO THE CHALLENGES OF RIO 1992 AND JOHANNESBURG 2002

THE FOLLOW-UP OF UNCED, RIO 1992

During the UNCED† in Rio de Janeiro in 1992, UNEP was entrusted with many follow-up activities, including the implementation of Chapter 19 on Chemicals‡ and Chapter 20 on Waste§ of *Agenda 21* (see Section II, Chapter 3 of this book). This has resulted in a major role of UNEP in the elaboration and adoption of the Rotterdam and Stockholm Conventions and its full support to the Secretariats of these Conventions (for the Rotterdam Convention shared with FAO). UNEP has also engaged in regional and sometimes national activities in support of these instruments.

More recently, UNEP has played a key role in the development of SAICM (see Section IV, Chapter 17 of this book).

THE FOLLOW-UP TO THE WORLD SUMMIT ON SUSTAINABLE DEVELOPMENT, JOHANNESBURG 2002

The World Summit on Sustainable Development (WSSD)¶ developed new ideas of promoting sustainable development, including the development of partnerships between different types of stakeholders. The WSSD + 10 (is Rio + 20) Conference in 2012 will evaluate to what extent this has contributed to reaching the goals of the WSSD. In the area of chemicals, the main "2020 goal" aims "to achieve, by 2020, that chemicals are used and produced in ways that lead to the minimization of significant adverse effects on human health and the environment, using transparent science-based risk assessment procedures and science-based risk management procedures, taking into account the precautionary approach, as set out in *Principle* 15 of the Rio Declaration on Environment and Development, and support developing countries in strengthening their capacity for the sound management of chemicals and hazardous wastes by providing technical and financial assistance."** The latter part of this declaration is often not quoted when reference is made to "the 2020 goal." UNEP created opportunities for ICCM-1 and its preparation process, which set the stage for the adoption of SAICM (see Section II, Chapter 5 of this book).

* http://www.chem.unep.ch/Pops/newlayout/infpopsalt.htm

† http://www.un.org/geninfo/bp/enviro.html

‡ Environmentally sound management of toxic chemicals including prevention of illegal international; traffic in toxic and dangerous products; available at: http://www.un.org/esa/dsd/agenda21/res_agenda21_19.shtml

§ Environmentally Sound Management of Hazardous Wastes, Including Prevention of Illegal International Traffic in Hazardous Wastes; available at: http://www.un.org/esa/dsd/agenda21/res_agenda21_20.shtml

¶ http://www.un.org/events/wssd/

** http://www.un.org/esa/sustdev/documents/WSSD_POI_PD/English/POIChapter3.htm

The Role of the MDGs

In broader UN framework, the Millennium Development Goals* (MDGs) were adopted in 2000, giving key directions for the main development goal of significantly reducing poverty. Although this has not been fully achieved, these MDGs show several interesting linkages to the SAICM agenda and environmental management, including management of chemicals and waste. Both the positive role of industrial development and the positive and negative roles that chemicals can play for sustainable development and eradication of extreme poverty, including access to sufficient food and clean drinking water and appropriate sanitation, are closely linked to sound management of chemicals. UNEP is one of the players with a key role in helping reach the MDGs.

BROADER UNEP AGENDA

UNEP's current agenda includes, besides chemicals, the main priority areas of work:

a. Climate Change
b. Disasters and Conflicts
c. Ecosystem Management
d. Environmental Governance
e. Resource Efficiency

Fact sheets and overviews of activities are available for each of these thematic areas.†

UNEP also has a range of regional activities and offices in all continents, supporting implementation and coordination at regional and national levels.

UNEP AND OTHER PLAYERS IN CHEMICALS MANAGEMENT

A number of international organizations are involved in sound management of chemicals. Together, they form the Inter-Organization Programme for the Sound Management of Chemicals (IOMC) and their steering body Inter-Organization Coordinating Committee (IOCC) (see Section V, Chapter 23 of this book). Programs and activities are being discussed and a certain amount of coordination achieved, and common guidance developed to support the implementation of international instruments. However, there tends to be fragmentation in the general environmental policy agenda and the important role of sound chemicals and waste management is not always adequately addressed. See also Section V, Chapter 19 of this book on Governance. For specifics on the concrete programs and activities of the respective organizations, see the chapters in Section V.

---

* http://www.un.org/millenniumgoals/

† http://www.unep.org/publications/contents/pub_details_search.asp?ID=4150

MAIN CHALLENGES

Increased Coordination within the UN System

The UN currently has a number of different mechanisms where coordination in environmental areas takes place.

Sustainable development is development that "meets the needs of the present without compromising the ability of future generations to meet their own needs" (United Nations, 1987).

As early as the 1970s "sustainability" was employed to describe an economy "in equilibrium with basic ecological support systems" (Stivers, 1976). The field of sustainable development is conceptually broken into three constituent parts: environmental sustainability, economic sustainability, and sociopolitical sustainability.

UNEP seems the most logical entity within the UN family to see to it that environmental sustainability considerations are adequately included in decision making at all relevant levels.

The current fragmentation in the environmental governance system, however, makes it extremely difficult to fully assume this responsibility.

Focus on the 2020 Goal

If the role of UNEP, notwithstanding hurdles mentioned above is to continue to be leading in the further development of the international chemicals agenda, a clear focus on the WSSD 2020 goal of achieving a high level of sound chemicals management in all parts of the world by that date needs to be the leading principle for all programs and activities in the area of chemicals and hazardous waste. As discussed in several other chapters of this book, chemicals management has made very significant progress on the international agenda, and should continue to do so. But limited resources both at national and international levels make it urgent to have a very clear focus. SAICM is the main mechanism to reach this goal.

REFERENCES

Carson, R., *Silent Spring*, Fawsett World Library, New York, reprinted in September 1997.

Stivers, R. 1976. *The Sustainable Society: Ethics and Economic Growth*. Philadelphia: Westminster Press.

United Nations. 1987. Report of the World Commission on Environment and Development. *General Assembly Resolution 42/187*, December 11.

United Nations Environment Programme (UNEP). 2010a. Existing International Agreed Environmental Goals and Objectives. UNEP/GEG/1/2, February 12.

31 United Nations Industrial Development Organization

Heinz Leuenberger and Elisa Tonda

CONTENTS

GREENING INDUSTRY: INNOVATIVE APPROACHES TO SOUND CHEMICALS MANAGEMENT

INTRODUCTION

The United Nations Industrial Development Organization (UNIDO) was established in 1966 and became a specialized agency of the United Nations (UN) in 1985. It promotes the creation of wealth and the alleviation of poverty through manufacturing industry. The Organization focuses on three interrelated thematic priorities:

- *Poverty Reduction through Productive Activities*: UNIDO seeks to enable the poor to earn a living through productive activities and thus find a way out of poverty. The Organization provides a comprehensive range of services customized for developing countries and transition economies, ranging from advice on industrial policy to entrepreneurship and the development of small and medium enterprises (SME), and from technology diffusion to sustainable production and the provision of rural energy for productive uses.

- *Building Trading Capacity*: Developing countries are benefiting from increased participation in the global trading system. Strengthening their capacity to participate in global trade is therefore critical for their future economic growth. UNIDO is one of the largest providers of trade-related development services, offering customer-focused advice and integrated technical assistance in the areas of competitiveness, trade policies, industrial modernization and upgrading, and compliance with trade standards, testing methods, and metrology.
- *Energy and Environment*: Energy is a prerequisite for poverty reduction. However, fundamental changes in the way societies produce and consume are indispensable for achieving global sustainable development. UNIDO therefore promotes sustainable patterns of industrial consumption and production. As a leading provider of services for improved industrial energy efficiency and sustainability, UNIDO assists developing countries and transition economies in implementing multilateral environmental agreements and in simultaneously reaching their economic and environmental goals.

UNIDO and Chemicals Management

In response to the international context shaped by the challenges of climate change and other global environmental threats, environmental impacts on trade agreements, and financial and economic crises, UNIDO developed the *Green Industries Initiative*. It focuses on promoting greater efficiency in the use of resources by industries to take advantage of cost reductions, better images, and less technical barriers to trade. Industries would benefit from the following aspects as indirect effects: the more efficient use of scarce, expensive natural resources, which would free up capital for more job-creating investments and would ameliorate balances of trade, and the creation of new enterprises (and thus new jobs) in the environmental services sector to assist "main-line" industries in becoming more efficient and clean.

To contribute toward the implementation of the Green Industries Initiative, UNIDO implements a number of programs that promote clean technologies and/or the preventive approach: the National Cleaner Production Centres (NCPCs) Programme, the Montreal Protocol Programme, the Transfer of Environmentally Sound Technologies approach, Chemical Leasing, and Corporate Social Responsibility based on the implementation of a Triple-Bottom Line approach, Environmental Management Systems, the elimination of persistent organic pollutants (POPS), sector-specific programs for the reduction of process wastes and pollution from the leather and textile sectors as well as other sectors, and so on.

Linkages with Other International Organizations

UNIDO has made a tangible contribution to the United Nation's system-wide coherence. It goes beyond process improvements and reduced transaction costs that make for the provision of more effective services to Member States. UNIDO's contribution is geared to overcoming the challenges of global development in unison with other agencies and organizations in the UN system. Important contributions in this direction have been made by the Organization in the area of trade and productive capacity and through its role in UN-Energy.

UNIDO's longstanding commitment to collaborate in the area of chemicals management has been expressed through its active participation in the Inter-Organization Programme for the Sound Management of Chemicals, which operates to strengthen international cooperation in the field of chemicals and to increase the effectiveness of the chemicals programs of its participating organizations (Food and Agriculture Organization, Food and Agriculture Organization, United Nations Environment Programme, UNIDO, United Nations Institute for Training and Research, World Health Organization, and Organization for Economic Cooperation and Development) and observers (United Nations Development Programme and World Bank).

One of UNIDO's specific roles is that it has been playing a very crucial role in closing the gaps in the industrial sector of the developing and transition countries through the adoption of the internationally recognized best practices in the area of chemicals management. This has been achieved with the capacity building in industries, and specifically SMEs; the dissemination of Environmentally Sound Technologies; and the fostering of public–private partnerships.

UNIDO Programs in the Area of Chemicals Management

The Stockholm Convention Unit

UNIDO plays a leading role in the implementation of the Stockholm Convention (SC) for the elimination of POPS. UNIDO is currently developing a number of large-scale thematic programs in the fields of energy/climate change and POPS. To date, over 50 countries have requested and received UNIDO's assistance in developing their SC National Implementation Plans. The assistance provided by the Organization in the implementation of the SC has addressed the strengthening of the institutional and technical capacity to ensure the environmentally sound management of polychlorinated biphenyls (PCBs), pesticides, and other POP wastes; the development and implementing of legislation on the environmentally sound management of POPS; the introduction of Best Available Techniques (BAT) and Best Environmental Practices (BEP) strategies in the industrial sector, the management of POP-contaminated sites, the preparation of new POPS inventories; and the establishment of noncombustion treatment facilities for the destruction of POP wastes.

For example, through the technical assistance for the environmentally sustainable management of PCBs and other POPS waste in Armenia UNIDO developed a detailed analysis of Armenia's institutional and technical capacity to ensure the environmentally sound management of PCBs and other POPS waste. UNIDO also assists the country in developing and implementing legislation on the ESM of PCBs. Priority objectives for UNIDO and Armenia are interim storage sites for PCB-containing waste so as to minimize the adverse effects of PCB releases to humans and the environment. Similarly, such project have been carried out in Mongolia, Macedonia, Egypt, Nigeria, and Romania.

Resource-Efficient and Cleaner Production

Cleaner Production (CP) is commonly understood as the application of preventive environmental strategies to the processing of products and services. CP aims at reducing environmental and health risks, as well as improving efficient natural resource usage.

Recognizing that resource efficiency requires CP and vice-versa, UNIDO and UNEP move toward Resource-Efficient and Cleaner Production (RECP). RECP catalyzes production efficiency by optimizing the productive use of natural resources (materials, energy, and water) by enterprises and other organizations. It further promotes environmental conservation by preventing the generation of wastes and emissions, as well as promoting the management of environmentally sound chemicals (Figure 31.1).

UNIDO's primary CP delivery mechanism is the NCPCs' approach. Together with UNEP, UNIDO established more than 40 NCPCs, as well as a number of related regional and sectoral centers. The NCPCs set out to train industry professionals, undertake plant-level CP assessments, raise awareness, disseminate information, and provide advice on policy matters. Some NCPCs are involved in the implementation of the BAT and BEP projects related to the SC.

The outreach in terms of technical activities to be addressed within the framework of the UNIDO and UNEP RECP Programme has been further extended,

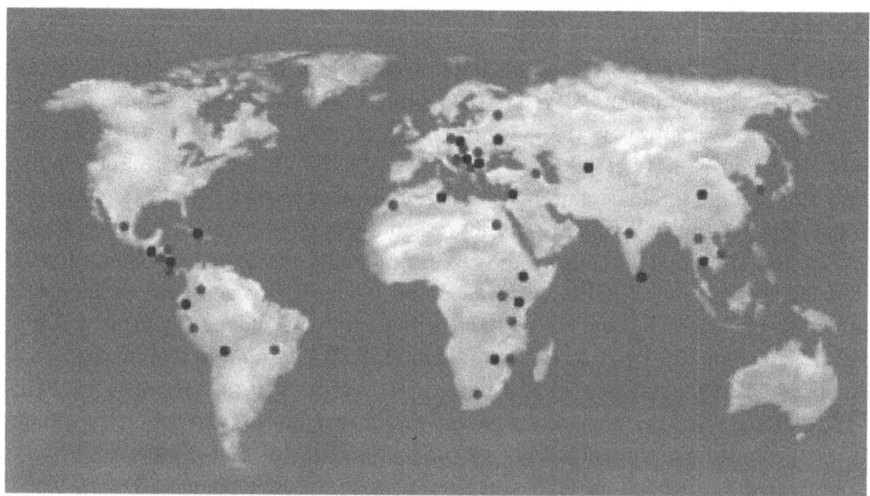

FIGURE 31.1 Worldwide distribution of NCPCs and Programmes. Asia: Cambodia, China, India, Lao People's Democratic Republic, Lebanon, Sri Lanka, Republic of Korea, Vietnam, and Uzbekistan. African and Arab countries: Egypt, Ethiopia, Kenya, Morocco, Mozambique, South Africa, Tunisia, Uganda, United Republic of Tanzania, and Zimbabwe. Central and Eastern Europe: Armenia, Bulgaria, Croatia, Czech Republic, Hungary, Romania, Russian Federation, Serbia, Slovakia, The former Yugoslav Republic of Macedonia, and Ukraine. Latin America: Bolivia, Brazil, Colombia, Costa Rica, Cuba, Ecuador, El Salvador, Guatemala, Honduras, Mexico, Nicaragua, and Peru.

based on the result of its independent evaluation carried out in 2007. One additional component of intervention is addressing the strengthening of national innovation systems aiming at improving national capacities for the widespread adoption of cleaner technologies and development of sustainable products, including much better health and environment properties.

Chemical Leasing

Chemical Leasing is an innovative service-oriented business model to increase the efficient and sustainable use of chemicals while reducing the risks and protecting human health. It builds on the preventive idea of CP and strives to improve the economic and environmental performance of the participating companies. Within Chemical Leasing, the responsibility of the producer is extended and may include the management of the entire life cycle of chemicals.

The Chemical Leasing model shifts the focus from increasing the sales volume of chemicals toward a value-added approach. The producer mainly sells the functions performed by the chemicals rather than the product itself, and functional units are the main basis for payment. Functions performed by a chemical might include the number of pieces cleaned, total area coated, and so on. In this manner, Chemical Leasing bundles the motivations of chemicals suppliers and users toward "less is more," as the value lies in increasing the delivery of services and not volume.

In 2004, UNIDO and the Government of Austria jointly launched a series of pilot projects to promote and test the Chemical Leasing concept at the global level. At their core, the projects build national capacity and develop Chemical Leasing demonstration projects in selected industries.

Initially, projects were implemented in Egypt, Mexico, and the Russian Federation in close cooperation with the respective NCPCs. The pilot projects demonstrated the applicability of the Chemical Leasing business model and contributed to further developing and promoting this innovative concept in developing countries and countries with economies in transition. As a result, four additional countries (Colombia, Morocco, Serbia, and Sri Lanka) joined UNIDO's Global Chemical Leasing Programme in 2008. In 2009 a global award for Chemical Leasing has been launched by UNIDO and the Austrian Minister of the Environment.

Chemical Leasing Tools

Based on the experience obtained, UNIDO developed its *Chemical Leasing guidelines*. These guidelines and respective worksheets cover the main steps to be undertaken to ensure smooth and efficient application and monitoring of Chemical Leasing in industries from different sectors and countries.

For more information, refer http://www.chemicalleasing.com/.

Mercury and Arsenic

At a global level, scientific, political, and civic communities have identified a need for the regulation and control of mercury. For over 15 years, UNIDO has been working to reduce the emissions of mercury from the largest intentional anthropogenic source of release: Artisanal and Small-Scale Gold Mining. One of the most successful interventions in the sector, the UNIDO Global Mercury Project, was implemented between 2002 and 2007 in six countries: Brazil, Indonesia, Lao People's Democratic Republic, Sudan, Tanzania, and Zimbabwe. The project raised global awareness of the issue, training over 10,000 miners, promoting the use of mercury-recycling devices and nonmercury alternatives to gold extraction. The project has resulted in a large reduction of mercury releases from the participating mining sites. It is also important to mention that mercury emissions from energy intensive sectors have been included in the BAT and BEP guidelines of the SC on POPS.

UNIDO's Support for SAICM Implementation

UNIDO is supporting developing and transition countries in the formulation and implementation of projects submitted to the Trust Fund Implementation Committee of the Strategic Approach to International Chemicals Management (SAICM) Quick Start Programme, as the executing agency and in partnership with other executing agencies. The projects target the implementation of life-cycle analysis for priority chemical products and substances in El Salvador, Peru, Sudan, and Uruguay. UNIDO also provides a supporting role in two regional projects in the area of artisanal gold mining, one in Cambodia and the Philippines and the other in Bolivia and Peru.

Challenges Encountered during Implementation

Despite the large number of activities implemented by the Organization in the area of chemicals management, similar challenges are faced in many countries, including

- Lack or ineffective enforcement of policy and regulatory frameworks for programs.
- Lack of information and research at national levels.
- Insufficient financial mechanisms to address the requirements of industries, and especially SMEs.
- Insufficient technical knowledge or access to up-to-date technologies.

Future Work of the Organization

As it recognizes the importance of disseminating sound chemicals management practices among developing and transition countries, UNIDO has prioritized intervention in the area of chemicals management in the joint UNIDO and UNEP RECP Programme. This will be achieved through the strengthening of the intervention of the NCPCs and the expansion of their network. The Centres will also strengthen their support to the countries in the implementation of the activities addressing the sound management of all chemicals included in the SC. The results achieved through the implementation of the joint program will provide an active contribution to the SAICM goals.

REFERENCES

United Nations Industrial Development Organization. 2008. *The Contribution of UNIDO to United Nations System-Wide Coherence: Synergy at Work*. Vienna.

United Nations Industrial Development Organization. 2009. *Annual Report 2008*. Vienna.

FURTHER READING

Jakl, T. and P. Schwager. (Eds). 2008. A win–win business model for environmental and industry. *Chemical Leasing Goes Global: Selling Services Instead of Barrels*. 245pp. New York: Springer.

32 United Nations Institute for Training and Research

Craig Boljkovac and Jan van der Kolk*

CONTENTS

INTRODUCTION

The United Nations Institute for Training and Research (UNITAR), was founded in 1965 following a recommendation of the Economic and Social Council to the General Assembly, which commissioned the Secretary-General with the establishment of UNITAR as an autonomous body within the UN system.† The creation of UNITAR occurred at the most opportune time in the history of the UN. Thirty-six states had joined the organization since 1960, including 28 African States. The training of diplomats representing newly independent countries was a success story for the UN. At the same time, however, it created a critical need for assistance, as many of the newly independent States lacked the capacity to train their new diplomats. The Institute, therefore, endeavored to satisfy that need in accordance with its Statute.

The Institute originally had its headquarters based in New York and a European Office in Geneva. In 1993, UNITAR's headquarters were transferred to Geneva.

* The opinions expressed in this chapter are those of the authors and do not necessarily represent the views of UNITAR.

† General Assembly resolution 1934 (XVIII) of December 11, 1963.

At the outset, the functions of UNITAR included the following: conducting training programs in multilateral diplomacy and international cooperation for diplomats accredited to the UN and national officials; carrying out a wide range of training programs in social and economic development; and ensuring liaison with UN organizations while strengthening cooperation with academic institutions. The Institute was equally entrusted to conduct research to improve outcomes of its training programs through the development of training tools and methodologies. The Institute's research program originally concentrated on three main areas: UN institutional issues, peace and security issues, and economic and social issues.

Since those days, UNITAR's mandate and purposes have been broadened but also more focussed: from environmental governance to management of intellectual property rights in trade negotiations, from peacekeeping in redeployment settings to UN reform, UNITAR is responding to pressing issues. The Institute strives to offer learning technologies that respond to life-long training needs of officials from UN Member States and national or nongovernmental institutions.

Over the past decades UNITAR has acquired the necessary expertise, accumulating experience, knowledge and capacities to design and implement a variety of training activities. The small size of the Institute and its independence within the UN system enables it to respond with a high degree of flexibility to new challenges in the area of training and research. It has a pragmatic and practice-oriented approach to training, based on sound and well proven methodology.

UNITAR has several programs in the environmental area, including environmental governance, climate change, biodiversity, and chemicals and waste.

THE CHEMICALS AND WASTE MANAGEMENT PROGRAMME

In 1990, UNITAR initiated its Chemicals and Waste Management Programme (CWM).

This took place shortly before the UN Conference on Environment and Development (UNCED) Conference in Rio de Janeiro in 1992, which through its *Agenda 21* encouraged countries and the international community to make serious efforts to improve the management of chemicals and waste. This also resulted in various international Conventions, such as the Rotterdam and the Stockholm Conventions, and other instruments, like the Globally Harmonized System for Classification and Labeling of Chemicals (GHS).

All activities in this program are based on methodologies that have been developed through practical experience and tested and adapted in field situations. The program includes training related to the full life cycle of chemicals and waste.

Today the program includes, for example, capacity building and training related to the following: National Profiles and National Priority Setting, training for the implementation of the Stockholm and Rotterdam Conventions, for the implementation of the GHS, for the implementation of Strategic Approach to International Chemicals Management (SAICM), and for the development of Pollution Release and Transfer Registers. It also has supporting activities like training on development of action plans in support of the various instruments.

Working in a multistakeholder setting is common to all activities of the program: although the entry point is always a request from a government structure (most commonly the Ministry entrusted with the Environment or the Ministry entrusted with Health), training and other activities always include other relevant stakeholders, such as other government agencies, representatives from business, nongovernmental organizations (NGOs), Universities, Labour Unions in as far they have an expressed interest.

National Profiles and National Priority Setting

Development of a National Profile of the current situation in a country regarding the management of chemicals and waste is the usual start of work on chemicals and waste management. The National Profile includes a full picture of the situation in a country, including general information, legislation, the various elements of the life cycle of chemicals and their management, the role of the various public and private actors, universities and civil society, resources and strengths and weaknesses.*

Obtaining clear insight in existing legislation, structures, and their coordination, the parts of the life cycle of chemicals that are relevant for a country (including industry and trade) and the steps already taken to address a number of challenges, and existing resources are essential for possible next steps.

Often, the development of a National Profile is followed by a selection of a limited number of priorities on which a country intends to focus. In addition, National Profiles have been endorsed by various Conventions and other agreements as a baseline document that can serve as a starting point for implementation of the specific commitments under these agreements.

UNITAR has developed extensive methodology to assist countries in developing their National Profiles. This is based on experience with now over 120 countries.

Rotterdam Convention

At the initiative of the Conference of Parties to the Rotterdam Convention on the Prior Informed Consent Procedure for Certain Hazardous Chemicals and Pesticides in International Trade (entered into force 2004, see Section III, Chapter 14 of this book) and together with the Secretariat of the Rotterdam Convention, UNITAR has been involved in about 10 countries in training for the implementation of the Convention.† For this training, specific methodology has been developed, which assists a country to investigate, in addition to the National Profile, how the various obligations of the Convention are currently being met or might be addressed in the context of the country. In a week long training all major obligations of a country regarding the Convention are systematically discussed, the roles and responsibilities of the different government or nongovernment structures defined, conclusions drawn with regard to a work plan and the available and necessary resources to implement this plan. As a result of the training and the workshop, all parties together come to common conclusions with

---

* http://www2.unitar.org/cwm/nphomepage/index.html

† http://www.unitar.org/cwm/pic

regard to the next steps for the full implementation of the Convention. Further cooperation between UNITAR and the Convention Secretariat (particularly in the field of capacity building for industrial chemicals management) is planned.

Stockholm Convention

The Stockholm Convention on Persistent Organic Pollutants (POPS) (see Section III, Chapter 15 of this book) entered into force in 2004.

The overall goal of UNITAR's POPS activities* is to provide support to developing countries and countries with economies in transition to take measures to eliminate or reduce the release of POPS into the environment. The POPS Programme Area, which is located within and draws upon methodologies developed through UNITAR's Chemicals and Waste Management Programme, aims to assist countries to strengthen the fundamentals of chemicals management in a number of areas; to provide technical assistance in specialized areas of POPS management; and to encourage integrated approaches involving all relevant ministries and stakeholders. An important part of the activities is to assist countries in developing a National Profile specifically oriented toward the obligations under the Stockholm Convention (National Implementation Plan, NIP), but related to chemicals and waste management in a wider sense.

UNITAR has assisted more than 80 countries in the development of NIPs, training for the development of action plans for the implementation, preparing submissions for funding of activities by the Global Environment Facility (GEF) (see Section VIII, Chapter 49 of this book) of activities in support of the implementation.

In one country, Ghana, UNITAR has worked with the Environmental Protection Agency and several other stakeholders in the development of a full sized US$ 6 million project for the elimination of polychlorinated biphenyls (PCBs) from electrical equipment and will continue to be involved during the whole implementation phase that started in 2009. UNITAR currently has a mandate from the Stockholm Convention Secretariat and GEF to update NIP guidance for the Convention, particularly in light of the nine new POPS that were recently added to the Convention.

Strategic Approach to International Chemicals Management

SAICM was adopted in 2006 in Dubai (see Section IV, Chapter 17 of this book).

Prior to the adoption of SAICM, UNITAR had worked with four countries to obtain valuable input into the final SAICM decisions and documents. Immediately after the adoption of SAICM, UNITAR initiated a pilot program with five countries to work on the implementation of SAICM. These countries are Belarus, Mongolia, Pakistan, Panama, and Tanzania, as a first concrete step. For this pilot, new methodology was gradually developed with the respective countries. SAICM as an overarching strategy has many different areas and selecting key priorities for a specific country and situation is essential.

After the necessary SAICM infrastructure had been put in place, many other countries also asked for collaboration from UNITAR in the implementation of

* http://www.unitar.org/cwm/pops

SAICM, and by mid-2010 more than 60 countries had been working with UNITAR on national SAICM implementation plans and activities. UNITAR is by far the most active agency assisting countries with such projects. The experience obtained in the five pilot countries has been of great help to develop the methodology and several guidance documents* and the approach for working with the many countries on their SAICM implementation and to assist them in having their proposals submitted to the SAICM Quick Start Fund (QSP) for funding, the large majority of them being approved for funding. As a result, many countries are now actively implementing several of the national priorities that are essential for improving their level of management of chemicals. In addition, UNITAR is facilitating the submission of a variety of other projects related to GHS (see below), the Rotterdam Convention, general chemicals management capacity building, Pollutant Release and Transfer Registers (PRTRs), and other issues to the QSP Trust Fund—therefore helping countries to use SAICM's resources to create capacities in many areas of chemicals management.

Globally Harmonized System for Classification and Labeling

The Plan of Implementation of the World Summit on Sustainable Development (WSSD), adopted in Johannesburg in 2002, encourages countries to implement the GHS as soon as possible with a view to having the system fully operational by 2008.

Although this date has not been met by all countries and regions, many countries are active in implementing GHS aiming at its full or sometimes partial implementation within their jurisdictions. GHS is not a Convention, but a global instrument, available for countries and (economic) regions, and its implementation will harmonize many diverging systems that have existed so far (see Section III, Chapter 11 of this book).

In response to growing requests from countries for capacity building to support GHS implementation, UNITAR and the International Labour Organization (ILO) initiated in 2001 the "UNITAR/ILO Global GHS Capacity Building Programme."† Building upon existing initiatives of international organizations, countries and others, the UNITAR/ILO program provides guidance documents, educational, awareness-raising, resource, and training materials regarding the GHS. Relevant topics include development of national GHS implementation strategies, legislation, situation/gap analyses, chemical hazards, labeling, safety data sheets (SDSs), as well as related support measures such as comprehensibility testing of GHS labels. UNITAR and the ILO are the designated focal point for capacity building in the UN ECOSOC Subcommittee of Experts on the GHS (SCEGHS).

Prior to the adoption of GHS, UNITAR had initiated a number of pilots in countries, obtaining thereby more insight in crucial elements for GHS implementation and for supporting countries in their national or regional endeavors to implement the GHS.

Since WSSD 2002 UNITAR and ILO have worked with more than 23 countries and 7 regions to assist in GHS implementation. A program advisory group (PAG)

---

* http://www.unitar.org/cwm/saicm

† http://www.unitar.org/cwm/ghs

has been established with representatives of countries involved in the implementation and international organizations that provide support.

A WSSD Global Partnership for Capacity Building to Implement the GHS has been initiated by UNITAR, ILO and OECD.*

Currently, an e-learning tool for GHS training is under development.

Pollutant Release and Transfer Registers

The goal of the PRTR Programme is to assist countries in the design of national PRTRs through multistakeholder processes. PRTRs are inventories of pollution from industry and other sources that have proven to be an effective tool for environmental management in many countries by providing government, industry, and the public with information on releases and transfers of toxic chemicals to air, water, and land. The design and use of a PRTR vary significantly with conditions prevailing in individual countries. Heavily industrialized countries will have a different type of PRTR than a nonindustrial country with mainly diffuse sources of pollution.

The PRTR Programme Area is implemented in cooperation with OECD and UN Environment Programme (UNEP) Chemicals.

PRTRs may be developed for different purposes, like the PRTR Design and Implementation under the Stockholm Convention.

UNITAR in close collaboration with UNEP has launched a GEF-supported Global PRTR Project on POPS monitoring, reporting, and information dissemination using PRTRs. The project will have duration of two years. Stakeholder involvement will be a critical project component.

This global pilot project will demonstrate the value of using PRTRs as a monitoring and reporting system for POPS at the country level in three countries in the Latin America and Caribbean region (Chile, Ecuador, and Peru), two countries in the Central and Eastern Europe region (Kazakhstan and Ukraine), and two countries in Asia (Cambodia and Thailand), providing a tool to address international requirements of Parties to the Stockholm Convention on POPS. The project will also include collaboration with the United States Environmental Protection Agency (US EPA) on the design of a regional PRTR in Central America, involving five countries: Costa Rica, Dominican Republic, El Salvador, Guatemala, Honduras, and Nicaragua.

UNITAR developed a Virtual Classroom on PRTRs. The Virtual Classroom supports information exchange and communication concerning the development of national PRTR systems in different regions and countries. The classroom provides a forum for sharing experience and knowledge on PRTRs. Registered participants benefit from access to an ongoing open forum, where queries can be registered, documents posted, and chats exchanged and maintained for future viewing. In addition, more limited discussion groups on specific issues can be formed.

Mercury

Mercury has been identified by the international community as a global chemical of concern. In 2002, the UNEP, in cooperation with the Inter-Organization Programme

* http://www2.unitar.org/cwm/ghs_partnership/index.htm

for the Sound Management of Chemicals (IOMC) developed the Global Mercury Assessment (GMA), indicating that mercury is a persistent chemical which cycles globally with serious health effects to humans and the environment. The GMA also pointed out that mercury-related problems are most challenging to developing countries.

In this context UNITAR initiated activities related to mercury in 2007, to assist countries in developing national strategies to reduce emissions and manage risks caused by mercury. An important aspect is the systematic collection of information concerning emissions from point (e.g., power plants) and diffuse sources (e.g., landfills, mercury containing products). PRTR are an important tool that can assist countries in identifying and reporting emissions and transfers of mercury on a sustained basis. Knowledge of mercury emission patterns and their magnitudes can later serve as a sound basis for targeting national reductions in mercury emissions through a national risk reduction strategy.

UNITAR's activities are in general closely related to UNITAR's long standing PRTR and Risk Management Decision Making specialized training and capacity building programs. Activities in this area take place through a country-driven and multistakeholder approach in collaboration with relevant international agencies, such as UNEP. In addition, with the support of Switzerland, UNITAR initiated a groundbreaking project with Kyrgyzstan (later joined by UNEP and the Government of the United States of America as partners) to address closure of the world's last remaining mercury mine, located in southern Kyrgyzstan. At a meeting in Bangkok in October 2009, the Kyrgyz Government agreed to its conditional closure.

Nanotechnology and Manufactured Nanomaterials

Nanotechnology/Manufactured Nanomaterials is an exciting new field that promises a broad array of benefits to humans and our environment. However, with these clear benefits come potential risks to the environment and human health—risks that, to-date, are not fully known.

UNITAR is embarking with its partner OECD, within the framework of the IOMC to raise awareness in countries about this new topic—including what the implications for developing and transition countries will be as nanobased or nano-containing products are traded across borders, into jurisdictions where there is little or no capacity to address them.

UNITAR's and OECD's activities commenced with a global series of regional awareness-raising workshops for all UN developing and transition countries. These workshops, the first of which was held in Beijing for Asia-Pacific countries, strove to brief participants on what is nanotechnology and manufactured nanomaterials, what some of the benefits and risks are, and what are the implications for them as government or other stakeholders.

This work takes its mandate from Resolution II/4 of the Second International Conference on Chemicals Management (ICCM-2), which was held in Geneva in May, 2009. The resolution was adopted by all governments and stakeholders present. The resolution comprises (in part) the following language: "[ICCM-2] ... encourages Governments and other stakeholders to assist developing countries and countries with economies in transition to enhance their capacity to use and manage nanotechnologies and manufactured nanomaterials responsibly, to maximize potential benefits and to minimize potential risks."

Further to the decision of ICCM-2, the June 2009 Joint Meeting of the OECD instructed UNITAR, in cooperation with the OECD Secretariat, to undertake awareness raising and other related activities in developing countries regarding the potential risks (e.g., to the environment or human health) and benefits (e.g., decreased costs of low-maintenance products, or use in environmental remediation) of nanotechnology and nanomaterials.

In addition to the series of awareness-raising workshops, UNITAR and OECD, with the support of the Government of Switzerland, will also undertake pilot projects to assist developing and transition countries to develop programmatic capacities to address nano issues at the national level.

INTER ORGANIZATION COOPERATION, IOMC

Since 1996, UNITAR has been a Participating Organization of the IOMC—a collaborative agreement among UNEP, World Health Organization (WHO), ILO, Food and Agricultural Organization (FAO), UN Industrial Development Organization (UNIDO), UNITAR, and OECD. UNITAR currently chairs IOMC. (See Section V, Chapter 23 of this book on IOMC.)

METHODOLOGY DEVELOPMENT

Objectives: CWM methodologies aim to assist countries and other stakeholders meet commitments and/or goals related to the protection of the environment and human health from the negative effects of chemicals. Such methodologies attempt to create a suggested framework that countries can refer to in order to take concrete actions toward fulfilling such commitments. Countries consistently approach UNITAR asking for assistance on how to start to address such commitments.

UNITAR's Chemicals and Waste Management Programme strives to address certain principles as its methodologies are developed. These include

- Responsiveness to country needs
- Provision of a framework of comprising *suggestions*—countries can take or leave the advice as appropriate
- Guidance is peer-reviewed by experts in the field and/or experts on methodologies
- Guidance can be given with minimal, key interventions from international experts/trainers
- Outputs based on the guidance are useful for reporting on and/or otherwise fulfilling international and national commitments

Methodologies under the Chemicals and Waste Management Programme are normally developed using the following typical approach: donor funding for initial pilot projects is sought for a specific subject area (e.g., National Profile development;

GHS, Nano, etc.); once secured, pilot countries are selected; a steering committee (formal or informal, depending upon the situation) participates in guidance development; draft guidance is circulated to pilot countries and other experts for comment; pilot workshops take place in the countries to test the guidance at a national setting; revisions are made as part of the UNITAR-pilot country agreement; and redrafted guidance is circulated for use by all countries and stakeholders (with, as appropriate, proposed endorsement from relevant bodies such as Conferences of the Parties, ICCM, etc.).

Currently, UNITAR CWM has a wide range of guidance in use and/or under development.

BUDGET AND STRATEGIC PARTNERS

UNITAR CWM's current biennial budget (2010–2011) is US$ 9.2 million per biennium. This is currently the largest program budget within UNITAR.

Strategic partners include the Secretariats of the Stockholm, Rotterdam, Basel, and Chemical Weapons Conventions. Key donors have included Switzerland, the United States, The Netherlands, the European Commission, Canada, Republic of Korea, UN Development Programme (UNDP), UNEP, and GEF.

OUTLOOK

WSSD in 2002 adopted the so-called 2020 goal: "the commitment, as advanced in *Agenda 21*, to sound management of chemicals throughout their life cycle and of hazardous wastes for sustainable development as well as for the protection of human health and the environment, *inter alia*, aiming to achieve, by 2020, that chemicals are used and produced in ways that lead to the minimization of significant adverse effects on human health and the environment" (see Chapter 4 of this book).

In order to achieve this goal, many challenges remain, often in the area of strengthening capacities for sound management of chemicals and waste. Together with SAICM, adopted in 2006, there remains a full agenda in support of the goal. UNITAR is committed to continue to work with countries, international organizations, and all stakeholders that work in support of this same goal.

UNITAR will continuously work to take new initiatives, improve existing methodologies, seize opportunities of strengthening collaboration with others, and seek synergies with other areas of work and other partners, both within the area of chemicals and waste and in other areas such as strengthening governance, mainstream chemicals and waste management in the general sustainable development agenda.

33 The Contributions of the World Health Organization to Sound Chemicals Management

*John A. Haines**

CONTENTS

* The opinions expressed in this chapter are those of the author and do not necessarily represent the views of the WHO or of the governments of member countries.

INTRODUCTION

The World Health Organization (WHO) Constitution was signed on July 22, 1946 and came into force on April 7, 1948, with the status of a specialized agency of the United Nations (UN). Its mandate is broad and recognizes that "The enjoyment of the highest attainable standard of health is one of the fundamental rights of every human being without distinction of race, religion, political belief, economic or social condition," defining health as "a state of complete physical, mental and social well-being and not merely the absence of disease or infirmity" (International Health Conference, 1946). It fulfills its objectives through its core functions which are reflected in its General Programme of Work*: providing leadership on matters critical to health and engaging in partnerships where joint action is needed; shaping the research agenda and stimulating the generation, translation, and dissemination of valuable knowledge; setting norms and standards and promoting and monitoring their implementation; articulating ethical and evidence-based policy options; providing technical support, catalyzing change, and building sustainable institutional capacity; and monitoring the health situation and assessing health trends.

WHO has 193 Member States, whose representatives meet annually at the World Health Assembly (WHA), the Organization's supreme decision-making body which determines its policies and elects its Director General. Further, an Executive Board, composed of 34 members technically qualified in the field of health, usually meeting twice a year, has as its main functions to give effect to the decisions and policies of the WHA, to advise it and generally to facilitate its work. The Organization undertakes its work at global, regional, and country levels and is one of the most decentralized of the UN specialized agencies. Members States are grouped according to regional distribution: Africa, the Americas, the Eastern Mediterranean, Europe, South-East Asia, and the Western Pacific, each region functioning with relative autonomy. A Regional Committee of the respective Member States formulates policies and supervises the activities of the corresponding Regional Office, which is headed by a Regional Director. Staff in 159 WHO country offices provide technical cooperation and advice, and work to implement WHO programs at national levels; many have technical expertise covering environmental health, food safety, chemical safety, and emergency response.

Several technical centers have been established in various WHO regions to support specific areas of work in the field of environmental health, including chemicals. Examples are

- The European Centre for Environment and Health (ECEH) which is part of WHO Regional Office for Europe and comprises two divisions, one in Rome and other in Bonn.†
- The Pan American Center for Sanitary Engineering and Environmental Sciences (CEPIS), which is the specialized center for environmental

* WHO's Eleventh programme of work, "Engaging for Health," covers the 10-year period from 2006 to 2015 (http://whqlibdoc.who.int/publications/2006/GPW_eng.pdf).

† http://www.euro.who.int/ecehrome and http://www.euro.who.int/ecehbonn

technology of the Pan American Health Organization (PAHO), Regional Office for the Americas of the WHO, established in 1968 and located in Lima, Peru.*
- The WHO Centre for Environmental Health Activities (CEHA), which is a specialized center for environmental health established in 1985 in Amman, Jordan, by the WHO Regional Office for the Eastern Mediterranean.†

Furthermore, there are more than 800 collaborating centers in 99 Member States working with WHO, which are designated by the Director General to carry out activities in support of the Organization's programs, of which some 40 are involved in activities related to the environment or chemicals. These centers are often research institutes, or parts of universities or academies.

The International Agency for Research on Cancer (IARC) was established in May 1965, as an agency of the World Health Organization, and is the WHO's leading source of information on cancer. The mission of IARC is to coordinate and conduct research on the causes of human cancer, the mechanisms of carcinogenesis, and to develop scientific strategies for cancer prevention and control. The activities of IARC are overseen by its own Governing Council with a membership of 21 countries. A series of monographs that identify the causes of human cancer have been published continuously since 1971. The series has reached the milestone of 100 volumes which review the accumulated evidence on the likely human carcinogenicity of selected pharmaceuticals, biological agents, metals, particles and fibers, radiation, lifestyle factors and chemical agents, and related occupations. This includes the evaluation of the carcinogenic risks of more than 850 chemicals or groups of chemicals.

HISTORY OF WHO ACTIVITIES RELATED TO CHEMICAL SAFETY

During the first two decades of its existence chemical safety-related activities were associated with WHO's continuing long-term policy priority areas for promotion and protection of public health: environmental sanitation, food safety, and health risks associated with the use of pesticides, occupational exposures to chemicals, and the standardization of pharmaceuticals. Environmental sanitation, a high priority from its inception, was defined by WHO in broad terms, and included chemicals used in the control of vectors of disease. The use of dichloro-diphenyl-trichloroethane (DDT), and subsequently other persistent organo-chorine pesticides, was encouraged in the control of mosquitoes responsible for transmitting malaria and other vectors of disease. However, by the early 1950s there was concern about insect resistance and vector control, as well as risks to health and damage to nontarget organisms from the use of pesticides. WHO contributed to the development of new pesticides and formulations and provided guidance on safe use of pesticides in public health (see the section "Thematic Areas of WHO's Work in

* http://www.bvsde.paho.org

† http://www.emro.who.int/ceha

Chemical Safety"). Currently, DDT is being reassessed for its use in relation to malaria control and health impacts (see the section "Chemicals Assessment").

Concern about the possible health implications of increasing use of chemicals in the food industry was referred by the WHA to a Joint FAO/WHO Expert Commission on Nutrition, established earlier in 1949, which lead to the establishment in 1956 of the FAO/WHO Joint Expert Committee on Food Additives (JECFA). Likewise, concern about the potential hazards to consumers of pesticide residues in food and feedstuffs lead to the establishment of the FAO/WHO Joint Meeting on Pesticide Residues (JMPR), which first met in 1961. At the same time the Joint FAO/WHO Food Standards Programme was established as the Codex Alimentarius in 1962 and the intergovernmental body, the Codex Alimentarius Commission first met in 1963. The Commission's work in harmonizing food standards is carried out through Codex Committees (on Food Additives and Contaminants, on Pesticide Residues, and on Residues of Veterinary Drugs in Food). The JECFA and JMPR provide the technical advice to the respective Codex Committees, the role of WHO being essentially to make the toxicological evaluations of the specific chemicals of concern (see the section "Chemicals in Food" and also Chapter 18 of this book).

The Joint ILO/WHO Committee on Occupational Health first met in 1950 and work has concerned, *inter alia*, diseases of chemical etiology and protection of workers from exposure to toxic chemicals. This work continued to be strengthened in the context of the International Programme on Chemical Safety (IPCS) (see below) and more recently with the development of the WHO Global Plan of Action for Workers Health 2008–2017, endorsed by the WHA at its 60th Session in 2007.*

ROLE OF WHO FOLLOWING THE STOCKHOLM CONFERENCE ON THE HUMAN ENVIRONMENT

Prior to the 1970s both national and international action on control of chemicals was piece-meal, being directed toward specific chemicals that caused health problems. The UN Conference on the Human Environment (Stockholm, 1972, see Chapter 2 of this book) provided a focus for chemicals viewed as pollutants in the environmental media and food chain, through which human beings could be exposed and the natural environment degraded. The main public concern was that of environmental degradation, particularly deteriorating air and water quality. The scientific community expressed concern about the rapidly growing numbers and types of chemicals, with the severe lack of information on their potential effects on human health (including long-term effects such as carcinogenicity and teratogenicity) and the environment; and their possible persistence in the environment, as well as of their possible long-range transport effects. The Stockholm Declaration and many of the Conference recommendations are directed toward the ways and means to control pollution.† Anticipating this, the 23rd WHA in 1970 recognized that WHO should continue its leading role in prevention and control of environmental

* Resolution WHA 60.26.

† Report of the United Nations Conference on the Human Environment A/Conf.48/14 Rev1.

factors adversely affecting human health,* and the Environmental Health Criteria (EHC) Programme, providing health and environmental risk assessments of potentially toxic chemicals, was formally launched in 1973.† WHO strengthened its program on vector control through the work of the WHO Expert Committee on Insecticides (subsequently on Safe Use of Pesticides) details of which are given below (see the section "Chemicals Assessment"). The joint work of FAO and WHO in the field of chemicals and food safety, through the JECFA and JMPR, was reinforced (see the section "Chemicals in Food").

WHO Regional Offices also started to develop chemicals-related activities. At the European regional level, WHO initiated in 1979 a program on environmental health aspects of the control of chemicals,‡ which promoted *inter alia* training, contingency planning for chemical emergencies, information exchange and development of methodologies, resulting in some 18 publications in the WHO/EURO Interim Documents "Health Aspects of Chemical Safety" series, and further publications in the WHO Regional Publications European Series, including for example "Assessing the Health Consequences of Major Chemical Incidents–Epidemiological Approaches" (WHO, 1997). This work led to projects on developing *Guidelines for Drinking Water Quality*, first published in 1993, and which is currently in its 3rd edition (2008),§ and *Air Quality Guidelines for Europe*. The recommendations concerning the chemical safety of drinking water are subject to a rolling cycle of revision. Air quality guidelines were published in 1987 and revised in 1997, a global update, including guidelines for particulate matter, ozone, nitrogen dioxide, and sulfur dioxide were published in 2005 and are applicable across all WHO regions.¶ Further, the ECEH (see the section "Introduction") was established.

Technical cooperation work on chemicals in the Americas through the PAHO was undertaken through its Centre for Human Ecology and Health (ECO) in Mexico from 1982, subsequently transferred in 1998 to CEPIS (see the section "Introduction") in Peru; and the Chemical Safety Programme for the Region of the Americas was approved in 1986 by the Pan American Sanitary Conference;** leading more recently to toxicology and environmental–epidemiologist networking and work in relation to surveillance and registration of pesticides in the context of a project on the occupational and environmental aspects of exposure to pesticides in Central America (known as the PLAGSALUD Project).††

Similarly, in the WHO Western Pacific Region technical cooperation work on toxic chemicals and hazardous waste was initiated, initially through a UNDP project, in the 1980s and undertaken through the Centre for the Promotion of Environmental Planning and Applied Studies (PEPAS) in Malaysia.

---

* Resolution WHA 23.60.

† Resolution WHA 26.58.

‡ Resolution EUR/RC29/R of the 29th session of the WHO Regional Committee for Europe.

§ http://www.who.int/water_sanitation_health/dwq/gdwq3rev/en/index.html

¶ http://www.who.int/phe/health_topics/outdoorair_aqg/en/

** XXII Pan American Sanitary Conference, September 1986, Washington, DC, USA.

†† http://www2.ops.org.sv/plagsalud/index.htm

Work in the other WHO regions developed from the mid-1980s and early 1990s, also initially with a parallel UNDP project on toxic chemicals and hazardous waste in the South-East Asian Region. In 1985 the WHO Eastern Mediterranean Region established its CEHA (see the section "Introduction") in Amman, Jordan, through which developed further activities on chemical safety. In the WHO African Region chemical-safety activities have developed on an *ad hoc* basis since the 1990s.

ESTABLISHMENT OF THE IPCS

One of the most significant outcomes of the Stockholm Conference was the establishment, in response to WHA Resolutions in 1977* and 1978,† of the International Programme on Chemical Safety (IPCS), in which the concept of chemical safety was enshrined. It was set up operationally in 1980 as a joint venture of WHO, the International Labour Organisation (ILO) and the United Nations Environment Programme (UNEP), through a Memorandum of Understanding, signed by the Executive Heads of the three Collaborating Organizations in April 1980. Administered by WHO, the IPCS was designated two main roles, that of providing the international scientific health and environmental risk-assessment basis on which governments could establish measures for safe use of chemicals; and of strengthening capabilities and capacities in countries for chemical safety, summarized in six main objectives: risk evaluation of priority chemicals, development of methodologies for health risk assessment, technical cooperation, management of chemical emergencies, prevention and treatment of chemical poisonings and human resource development. The concept of chemical safety was set out in 1986 in a paper to the IPCS Programme Advisory Committee:

Use of chemicals is essential for economic and social development. It is recognized that a sustainable development process requires that human health and the environment are protected from possible deleterious side effects and consequently all activities involving chemicals must be undertaken in such a way as to ensure the safety of human health and the environment from deleterious effects. This is commonly referred to as "Chemical Safety."

The current WHO work on chemical health and environment risk assessment along with the accompanying work on methodology became one of the cornerstones of the new thrust through IPCS; along with the promotion of effective international cooperation in emergencies and accidents involving chemicals, including poisoning prevention and management, and the promotion of training of human resources and technical cooperation among Member States.

Most of the scientific work of the IPCS is undertaken through committees and working groups of experts chosen independently for their competence in the field and endeavoring to ensure a balanced geographical participation. This work is supported through a network of IPCS Participating Institutions such as those participating in the preparation, review and updating of International Chemical Safety Cards (ICSCs) (see also Chapter 22 of this book). Further, work is often undertaken in

---

* Resolution WHA 30.47.

† Resolution WHA 31.28.

cooperation with the international scientific community (see Chapter 28 of this book) (such as, the International Union of Pure and Applied Chemistry (IUPAC) and the International Union of Toxicology (IUTOX) in areas of risk assessment) and with relevant professional bodies (such as the European Association of Poison Control and Clinical Toxicologists (EAPCCT) in areas of poisoned patient management), international workers federations in areas of safe use of chemicals in the workplace, and industrial associations, such as chemicals manufacturer associations and Crop Life International (formerly GIFAP) for capacity building for safe use of chemicals.

Promoting rapid access to internationally peer reviewed information on chemicals commonly used throughout the world, which may also occur as contaminants in the environment and food was considered as an important risk communication function of the IPCS. In partnership with the Canadian Centre for Occupational Health and Safety (CCOHS) a database was established consolidating current, internationally peer-reviewed chemical safety-related publications and database records from international bodies, for public access. Initially issued as a CD/ROM the IPCS INCHEM database is now available on the Web, offering a quick and easy electronic access to thousands of searchable full-text documents on chemical risks and the sound management of chemicals,* IPCS INCHEM contains the following:

- Concise International Chemical Assessment Documents (CICADs)
- Environmental Health Criteria (EHC) monographs
- Harmonization Project Publications
- Health and Safety Guides (HSGs)
- International Agency for Research on Cancer (IARC)—Summaries and Evaluations
- International Chemical Safety Cards (ICSCs)
- IPCS/CEC Evaluation of Antidotes Series
- Joint Expert Committee on Food Additives (JECFA)—Monographs and evaluations (part of WHO's Food Safety Programme)
- Joint Meeting on Pesticide Residues (JMPR)—Monographs and evaluations (part of WHO's Food Safety Programme)
- KEMI-Riskline
- Pesticide Data Sheets (PDSs)
- Poisons Information Monographs (PIMs)
- Screening Information Data Set (SIDS) for High Production Volume Chemicals

Similarly, a second database IPCS INTOX† was issued, directed toward support to the health sector in prevention and management of toxic exposures, which also has related database management software, details of which are described in the section "Poisoning Prevention and Management."

* http://www.inchem.org
† http://www.intox.org

PREPARATIONS FOR RIO AND UNCEDs IMPACT ON THE WORK OF WHO

The "Report of the World Commission on Environment and Development: Our Common Future"* in 1987 was a major stimulus in the convening of the United Nations Conference on Environment and Development (UNCED).† The resolution convening UNCED specifically refers to environmentally sound management of toxic chemicals as a main issue. An extensive preparatory process took place, which involved, *inter alia*, all the international organizations with chemical safety activities, including WHO/IPCS, and this cooperation was the precursor of the Inter-Organization Programme for the Sound Management of Chemicals (IOMC) (see Chapter 23 of this book). In response to a request by the UNCED Preparatory Committee, and by decision of the UNEP Governing Council, the executive heads of the three IPCS Cooperating Organisations convened in 1991 in London a meeting of government designated experts to draft proposals for an intergovernmental mechanism for chemical risk assessment and management (INCRAM). The meeting proposed the need for: a strengthened and expanded IPCS; enhanced coordination among international organizations; and the establishment of an intergovernmental forum on chemical safety. In the report of UNCED,‡ which met in June 1992 in Rio de Janeiro, Brazil, *Agenda 21*, Chapter 19 deals with environmentally sound management of chemicals, adopting an international strategy in six program areas and endorsing the proposals of the London meeting. Chemicals are also referred to in many of the other chapters of *Agenda 21*, such as Human Health (Chapter 6), Protection of the Atmosphere (Chapter 9), Sustainable Agriculture and Rural Development (Chapter 14), Protection of the Oceans (Chapter 17), Freshwater Resources (Chapter 18), Hazardous Wastes Management (Chapter 20), and Solid Wastes Management (Chapter 21) emphasizing that the use of chemicals and their effects influence other sectors of society.

By identifying sound management of chemicals as an essential element in sustainable development, UNCED gave a major political boost to chemical safety and stimulated activities both in countries and at the international level. Additionally, the momentum of the London meeting resulted in the convening of an International Conference on Chemical Safety in Stockholm, Sweden, April 25–29, 1994 which became the first meeting of the Intergovernmental Forum on Chemical Safety (IFCS),§ at which governments identified 42 priority actions, some for periods up to 1997 and 2000, others open ended, in order to implement the six program areas of UNCED *Agenda 21*, Chapter 19¶ (see Chapter 21 of this book). The WHO has been one of the initiators of IOMC (see Chapter 23 of this book). The WHO is the administering organization for the IFCS and the IOMC and provides secretariat services to the

* Transmitted to the General Assembly as an Annex to document A/42/427—Development and International Co-operation: Environment. See http://www.un-documents.net/wced-ocf.htm

† UN General Assembly Resolution A/RES/44/228, adapted December 22, 1989, New York, USA.

‡ http://www.un.org/esa/dsd/agenda21

§ http://www.who.int/ifcs

¶ With the further development of the Strategic Approach to International Chemicals Management (SAICM), through which much of the work of the IFCS may be undertaken, the Forum Standing Committee at its Teleconference on June 24, 2009, decided to indefinitely suspend the IFCS.

Inter-Organization Coordinating Committee (IOCC) of the IOMC, which coordinates and fosters joint planning of the seven Participating Organizations (POs) which contribute to the work of IOMC.

STRATEGIC APPROACH TO INTERNATIONAL CHEMICALS MANAGEMENT

The active participation of the health sector in the development of Strategic Approach to International Chemicals Management (SAICM) was facilitated by the WHA* in May 2003, which called for the participation of global health partners in SAICM and urged Member States to take full account of the health aspects of chemical safety in the development of SAICM. A set of health sector priorities for SAICM (see Chapter 17 of this book), identified through WHO consultations with its Member States, are reflected in the SAICM Overarching Policy Strategy and Global Plan of Action. These priorities may be summarized as

- Actions to improve ability to access interpret and apply scientific knowledge
- Filling of gaps in scientific knowledge
- Development of globally harmonized methods for chemical risk assessment
- Development of better methods to determine impacts of chemicals on health, to set priorities for action and to monitor progress of SAICM
- Building capacities of countries to deal with poisonings and chemical incidents
- Strategies directed specifically at the health of children and workers
- Work to promote alternatives to highly toxic and persistent chemicals
- Strategies aimed at prevention of ill-health and disease caused by chemicals

In May 2006, a specific resolution† of the WHA welcomed the completed SAICM, and urged Member States to take full account of the health aspects of chemical safety in national implementation of the SAICM, to participate in national, regional and international efforts to implement SAICM and to nominate a national Strategic Approach focal point from the health sector, where appropriate in order to maintain contact with WHO. The Resolution includes a request to the Director-General of WHO to facilitate implementation by the health-sector of the Strategic Approach, focusing on human health-related elements. The secretariat of SAICM is provided by UNEP and WHO.

Regular reports on planned and recent activities in relation to health-sector priority areas are coordinated by WHO headquarters and WHO Regional Offices. WHO also works to promote health sector involvement in the implementation of SAICM including those active in its global health-sector networks of poisons centers, emergency alert, and response operations for incidents of public health concern and networks of risk assessors (see the section "Poisoning Prevention and Management"). WHO also

* Resolution WHA 56.22.

† Resolution WHA 59.15.

leverages support for SAICM implementation though scientific bodies such as IUPAC and IUTOX and other bodies which have official relations with WHO.

The importance of the health aspects of chemicals management and the need to fully engage with the health sector was reflected in the adoption of a resolution at the second session of the International Conference on Chemicals Management, held in May 2009.* The resolution emphasizes the essential cross-sectoral responsibilities of national focal points, the importance of regional health and environmental interministerial processes as a springboard for effective intersectoral actions and underlines the need for all stakeholders to assist in the development of resources to permit a greater degree of sectoral balance in representation in SAICM fora and in implementation activities. The resolution calls on the WHO to intensify its activities in the sound management of chemicals in support of SAICM and requests that the outcomes of the Conference regarding human health be considered by the WHA. The Conference decided to develop a strategy for strengthening the engagement of the health sector in SAICM's implementation and to evaluate it at the third session of the International Conference on Chemicals Management to be held in 2012.

THEMATIC AREAS OF WHO'S WORK IN CHEMICAL SAFETY

As indicated earlier, the WHO activities related to managing chemical and environmental risks are carried out at global, regional, and country levels. At regional level WHO works with countries and partners not only on specific thematic areas, such as those listed below, but also on promoting the integration of chemicals issues into the broader development agenda and the ratification of relevant multilateral agreements. Activities may be carried out with other international bodies and countries, the PLAGSALUD project being a good example.† The related thematic areas of WHO's work are concerned with: risk assessment, including chemicals in food, methods for chemicals assessment, poisoning prevention and management, chemical incidents and emergencies and capacity building, which includes promoting the implementation of the relevant chemicals-related conventions (such as Stockholm, see Chapter 15 and Rotterdam, see Chapter 14 of this book) and multilateral agreements (such as SAICM; see Chapter 17 and GHS, see Chapter 11 of this book). Closely related is the area of children's environmental health. Furthermore, there are interlinkages among each of these thematic areas of WHO's work. For example each thematic area has a capacity building, including human resource development, element. Activities in several areas contribute to WHO's work in support of implementing Conventions and International Agreements. Tools developed in one area of work may be used in

* SAICM/ICCM.2/15. See http://www.saicm.org/index.php

† The "Occupational and Environmental Aspects of Exposure to Pesticides in the Central American Isthmus" (PLAGSALUD) project was implemented by PAHO in seven countries (Belize, Costa Rica, El Salvador, Guatemala, Honduras, Nicaragua, and Panama) between 1994 and 2003. This was a subregional project with financial resources provided by Danish International Development Assistance (DANIDA), and financial support from the governments in these countries, as well as PAHO, cooperation from the ILO, the Central American Commission for Environment and Development (CCAD), and the United States Environmental Protection Agency (US EPA).

other areas and in activities at different levels. Most chemicals-related activities of WHO are undertaken in the context of the IPCS.

Chemicals Assessment

The objective of chemicals assessment is to provide a consensus scientific description of the risks of chemical exposures. These descriptions are published in assessment reports and other related documents so that governments and international and national organizations can use them as the basis for taking preventive actions against adverse health and environmental impacts. For example the documents are often used as the basis for establishing guidelines and standards for the use of chemicals and for standards for drinking water and air and can assist with the implementation of international agreements such as the GHS and the Stockholm Convention on Persistent Organic Pollutants (POPS).

IPCS works cooperatively with other international organizations, such as the OECD, under the auspices of the IOMC avoiding duplication and thereby optimizing the use of assessment resources. Through this work several hundred chemicals have been evaluated for their risk to health and the environment and published as

- EHC monographs, which provide comprehensive data from scientific sources for the establishment of safety standards and regulations. EHC publications are monographs designed for scientists and administrators responsible for the establishment of safety standards and regulations and some 200 have been published on a wide range of chemicals and groups of chemicals.*
- CICADs are concise documents that provide summaries of the relevant scientific information concerning the potential effects of chemicals upon human health and/or the environment. They are based on selected national or regional evaluation documents or on existing EHCs. Before acceptance for publication as CICADs by IPCS, these documents have undergone extensive peer review by internationally selected experts to ensure their completeness, accuracy in the way in which the original data are represented, and the validity of the conclusions drawn. Mid-2010, 77 documents have been published.†
- HSGs, provide concise information in nontechnical language, for decision-makers on risks from exposure to chemicals, with practical advice on medical and administrative issues. Between 1987 and 1996, some 109 documents were published.‡
- ICSCs, developed cooperatively by the IPCS and published by the Commission of the European Union (EU), summarize essential health and safety information on chemical substances in a clear way, and are not only intended to be used at the "shop floor" level by workers, but also by other interested parties in factories, agriculture, construction, and other places of

* http://www.who.int/ipcs/publications/ehc/en/index.html
† http://www.who.int/ipcs/publications/cicad/en/index.html
‡ http://www.who.int/ipcs/publications/hsg/en/index.html

work for information on chemical substances. They are available in hard copy in 24 languages and on the Internet in 17 languages through the ILO site.[*] More than 1700 have been published in English.[†]

- PDS contain basic information for safe use of pesticides and are prepared by WHO in collaboration with FAO and give basic toxicological information on individual pesticides. Priority for issue of PDSs is given to substances having a wide use in public health programs and/or in agriculture, or having a high or an unusual toxicity record. The data sheets are prepared by scientific experts and peer reviewed. The comments of industry are provided through the industrial association, Crop Life International. The data sheets are revised from time to time as required. Some 94 have been published.
- The WHO Recommended Classification of Pesticides by Hazard was approved by the 28th WHA in 1975 and has since gained wide acceptance.[‡] Guidelines were first issued in 1978, and have since been revised and reissued at 2–3-year intervals. The classification will gradually be replaced with the GHS classification (see Chapter 11 of this book).
- The WHO Pesticides Evaluation Scheme (WHOPES) for evaluating and testing of pesticides for public health use, was established in 1960,[§] and the work continues in a broader framework on recommendations for application of pesticides.[¶]

WHO supports a number of other areas related to chemicals assessment including

- Global assessment of the state-of-the-science of endocrine disruptors, published in 2002.[**]
- Aircraft disinsection, published in 1995[††] reviews whether disinsection of aircraft is needed and how it should be implemented; which chemicals (pesticides, solvents, and propellants) and which methods can be recommended.

Further current and previous WHO activities relevant to POPS include

- The project for the reevaluation of human and mammalian toxic equivalency factors (TEFs) of dioxins and dioxin-like compounds.[‡‡]
- A joint project with the European Commission on Rapid Assays for Dioxins and Related Compounds in September 2001 was completed to further development of rapid assay methods for the screening of dioxins and related compounds in feed, food and environmental samples.[§§]

---

[*] http://www.ilo.org/public/english/protection/safework/cis/products/icsc/dtasht/index.htm
[†] http://www.who.int/ipcs/publications/icsc/en/index.html
[‡] Resolution WHA 28.62.
[§] The WHO Pesticides Evaluation Scheme, IPCS Document PCS/EC/90.16.
[¶] http://www.who.int/whopes/en/
[**] WHO/PCS/EDC/02.2
[††] Report of the Informal Consultation on Aircraft Disinsection, IPCS WHO/PCS/95.51
[‡‡] http://www.who.int/ipcs/assessment/tef_update/en/index.html
[§§] http://www.who.int/ipcs/assessment/dioxins/en/index.html

- A review of selected POPS: DDT, Aldrin, Dieldrin, Endrin, Chlordane, Heptachlor, Hexachlorobenzene, Mirex, Toxaphene, Polychlorinated biphenyls, Dioxins and Furans.*

Chemicals in Food

Part of WHO's work on chemicals risk assessment is directed toward the evaluation of the safety of food components, toxic natural constituents, and contaminants as well as food additives and residues of pesticides and veterinary drugs. These activities include providing the secretariats and scientific advice to the JECFA and the JMPR and carrying out international risk assessments of chemicals of concern such as acrylamide, produced as a by-product of food processing and cooking. For the assessment of chemicals in food, as with other chemicals assessment work, the development, harmonization and use of internationally accepted, scientifically sound and transparent principles and methods is vitally important. The development of the JECFA and JMPR is described in the section "History of WHO Activities Related to Chemical Safety" and their work is used by Codex Alimentarius as well as Member States to develop international food standards and to propose international food safety guidelines.

To date, JECFA has evaluated more than 1500 food additives, approximately 40 contaminants and naturally occurring toxicants, and residues of approximately 90 veterinary drugs. The Committee has also developed principles for the safety assessment of chemicals in food that are consistent with current thinking on risk assessment and take account of recent developments in toxicology and other relevant sciences. As a result of the JMPR work, 1041 monographs of toxicological evaluations of pesticide residues have been published. The JECFA and JMPR monographs are found on the IPCS INCHEM database (http://www.inchem.org/).

In light of the advances in the science of risk assessment and the recognition that the evaluations performed by JECFA and JMPR serve as the scientific foundation for international food standards that are of increasing importance within the Codex Alimentarius Commission and the World Trade Organization, FAO and WHO have initiated a joint Project to Update and Consolidate Principles and Methods for the Risk Assessment of Chemicals in Food. Further a document on "Principles for Modelling Dose-Response for the Risk Assessment of Chemicals" was published in 2009 as EHC.† Food is a feast of chemicals, most of them being an integral component of the food, important for nutrition and health. Examples are proteins, carbohydrates, fats, vitamins, trace elements, fiber, and antioxidants. However, some chemicals naturally present in certain foods, such as cyanogenic glycosides in cassava or solanine in potatoes, have toxic properties. Certain food processing techniques, including cooking, can produce toxic by-products, such as polycyclic aromatic hydrocarbons or acrylamide. In addition, environmental pollutants such as lead and chlorinated compounds, toxic elements such as cadmium from the earth's crust, and natural mould and algal toxins such as aflatoxins and shellfish poisons can be present at varying

---

* http://www.who.int/ipcs/assessment/en/pcs_95_39_2004_05_13.pdf

† Environmental health Criteria Document 239, http://whqlibdoc.who.int/publications/2009/9789241572392_eng.pdf

levels in the food supply. The objective of this work is to ensure that procedures are in place to estimate the health risk of these compounds in food, taking into account all other sources of exposure.

Methods for Chemicals Assessment

In the field of methodology, the work of WHO/IPCS aims at promoting the development, harmonization and use of generally acceptable, scientifically sound methodologies for the evaluation of risks to human health and the environment from exposure to chemicals. The results of such work enhance mutual acceptance of risk-assessment products, and includes the development of general principles of various areas of risk assessment, published as EHC (the "yellow series"), the Harmonization Project, addressing the risk-assessment uncertainties and challenges associated with new and emerging scientific issues, such as toxicogenomics, and the harmonization and update of the principles and methods for the risk assessment of chemicals in food.

As a response to meeting the priorities set out in *Agenda 21*, Chapter 19 and to ensure best use of limited resources, the IPCS and OECD developed a framework for cooperation in the field of risk assessment including work on methodologies. This ensures complementarity, mutual support and mutual involvement in the projects conducted by each organization. There are more than 40 methodology publications, and among the most important current projects are

i. The "IPCS Project on the Harmonization of Approaches to the Assessment of Risk from Exposure to Chemicals" (the Harmonization project)* has been a significant new approach to globally harmonize risk-assessment approaches by increasing understanding and developing basic principles and guidance on specific chemical risk-assessment issues. Harmonization enables risk assessments to be carried out using internationally accepted methods enabling the assessments to be shared more efficiently, avoiding duplication of effort. The project covers harmonization of the terminology used in hazard and risk assessment, exposure assessment, aggregate/cumulative risk assessment, physiologically based pharmacokinetic (PBPK) modelling, mode-of-action of both cancer and noncancer end points, mutagenicity, immunotoxicology, and dermal absorption. The work is relevant across various chemical-sectors including industrial chemicals, biocides, pesticides, veterinary products and pharmaceuticals, for different sources of exposure and in different regulatory and nonregulatory contexts, for example contributing to the GHS. The Harmonization project has been important in galvanizing the efforts of experts from national risk-assessment authorities from developed and developing countries, representatives of supra-national bodies such as the European Union (European Chemicals Agency, European Food Standards Agency and Joint Research Centre) and representatives of nongovernmental organizations in official relations with

* The Harmonization project Web site contains a list of publications and current activities: http://www.who.int/ipcs/methods/harmonization/en/index.html

WHO such as the European Centre for Ecotoxicology and Toxicology, the IUPAC and the International Life Science Institute. Given the objectives of the project, take-up and use of its products by risk-assessment bodies is an important indicator of success. It provides a framework for evaluating a mode of action of chemical carcinogenesis, the guidance for use of chemical specific adjustment factors and the descriptions of selected key generic terms used in chemical/hazard risk assessment (IPCS Conceptual Framework for Evaluating a Mode of Action for Chemical Carcinogenesis, 2001). The latter, being jointly prepared with the OECD, is among the best known and used products of the project.

ii. Work in relation to new and emerging scientific issues includes:
   - The use of toxicogenomics in risk assessment, a joint project with OECD, aiming to promote existing observational human data, involving collaboration among clinicians, toxicologists and risk assessors.
   - Developing an internationally applicable science-based approach, or model, to identify upper levels of intake for nutrient substances.
   - Establishing an integrated, holistic approach to risk assessment that addresses real life situations of multichemical, multimedia, multiroute, and multispecies exposures. The International Programme on Chemical Safety (IPCS) compiled a draft document on "Principles for Modelling Dose-Response for the Risk Assessment of Chemicals"* which was made available for public comment. The final version of this document is under preparation and will be published on the WHO/IPCS website as soon as it is available.

Despite the long-standing availability of guidance on risk-assessment methodologies and tools, on the Internet and in print, capacity is still lacking among risk-assessment bodies in developing countries and countries with economies in transition to locate and use these tools in various applications in countries and in support of obligations under global international agreements on chemicals. WHO has initiated development of a "toolkit" to make existing international tools on chemical risk assessment more readily available and to develop priority new tools and training materials for their use. While the toolkit is being prepared with developing countries and countries with economies in transition in mind, it is intended also to be of use to risk assessors worldwide.

Poisoning Prevention and Management

The IPCS poison prevention and treatment activities and the IPCS INTOX Programme were developed following a consultation in 1985 with the scientific and professional bodies in the field.† Poisoning is a significant global public health problem. According to WHO data,‡ in 2002 an estimated 350,000 people died worldwide from unintentional poisoning. In 2000, unintentional poisoning was the ninth most common cause

---

* http://www.who.int/ipcs/methods/harmonization/areas/aggregate/en/index.html

† Joint WHO/EU/World Federation of Association of Poisons Control and Clinical Toxicology Centres Consultation, held at WHO October 7–9, 1985, Geneva, Switzerland.

‡ http://www.who.int/healthinfo/statistics/mortestimatesofdeathbycause/en/index.html

of death globally in young adults (15–29 years), and in this age group it was the sixth most common cause of death in India and the ninth most common in China. More than 94% of fatal poisonings occurred in low- and middle-income countries. Snakebite is a largely unrecognized public health problem that presents significant challenges for medical management. While reliable data are hard to obtain, it has been estimated that about 2.5 million people are envenomed per year, and more than 125,000 die from this cause (WHO Bulletin, 1998). The IPCS seeks to build capacity in countries to deal with these problems. An important area of activity is promoting the establishment and strengthening of poisons centers, work carried out under the IPCS INTOX Programme. A world directory of poisons centers (YellowTox)* is maintained, which includes the contact information for poison centers in more than 90 countries. Other activities include the provision of information on chemicals, the provision of information management tools, and the development of internationally peer-reviewed guidelines concerning the prevention and clinical management of poisoning.

This work has resulted in the publication of guidance documents on setting up and operating poison control and related information, clinical and analytical facilities, as well as new series of internationally peer reviewed documents on antidotes (published with the EU), monographs on diagnosis and treatment of exposures to specific chemicals, pharmaceuticals, and toxins of biological origin, and guides on treatment of specific clinical features of toxic exposures, forming part of an information package for the health professionals in countries. This IPCS INTOX package (available in several languages) originally issued on CD/ROM and now available on the Web,† also provides database management software for harmonized recording of data in countries on toxic exposures and their management, and lays the foundations for collection of harmonized human toxicological data. The Package is supplied with a set of Authority Lists, which provide a controlled vocabulary for data entry into the system and a comprehensive and internationally agreed harmonized terminology, with definitions for each of the terms. The provision of a controlled, defined terminology within the INTOX Data Management System means that different poisons centers and staff within the same poisons center use the same terms to describe the same concept. This facilitates the comparison of case data collected by different centers, for example, for multicenter studies. It also makes it possible to pool data from different centers, for example, to compile national statistics.

Harmonized data recording is the basis for the IPCS International Project on the Epidemiology of Human Pesticide Exposures, through which data on exposures to pesticides are collected and analyzed. The Pesticides Databank on CD-ROM and Internet (in development) provides a collection of internationally peer-reviewed risk-assessment documents about pesticides, relevant Poisons Information Monographs and an Updated IPCS Manual on Diagnosis and Treatment of Pesticide Poisonings;‡ and there is a distance learning module on prevention, diagnosis and management of pesticide poisoning, aimed at three levels of user: community physicians, agricultural workers, and the general public. The tools of the Project, especially the Pesticide Exposure

* http://www.who.int/ipcs/poisons/centre/directory/en/index.html

† http://www.ipcs.intox.org

‡ http://www.who.int/entity/whopes/recommendations/IPCSPesticide_ok.pd

Record (PER) can assist countries in identification of hazardous pesticide formulations. The documents on pesticides in the Pesticide Data Management System and Databank assist developing countries in decision making as well as risk assessment.*

Analytical toxicology can assist in the diagnosis, management, prognosis, and prevention of poisoning. In addition analytical toxicology laboratories may be involved in a range of other activities such as the assessment of exposure following chemical incidents, therapeutic drug monitoring, forensic analyses, and monitoring for drugs of abuse. They may also be involved in research, for example in determining the pharmacokinetic and toxicokinetic properties of substances or the efficacy of new treatment regimens. The IPCS has developed a manual, describing simple analytical techniques for the identification of more than 100 substances commonly involved in acute poisoning incidents (Flanagan et al., 1995). These techniques do not need sophisticated equipment or expensive reagents, or even a continuous supply of electricity, and can be carried out in the basic laboratories that are available to most hospitals and health facilities. Also a guidance document advises on developing analytical toxicology services (Flanagan, 2005).

Chemical Incidents and Emergencies

Work of WHO in the field of chemical incidents and emergencies had its inception in the 1985 consultation referred to in the previous section. Subsequently, a number of activities were developed in cooperation with other international organizations, particularly UNEP/APELL and OECD chemicals accident activities, which resulted in joint publications.† Work was also initiated on developing database management software for harmonized recording of data in countries on chemical incidents and for preparedness for such incidents. See also Chapter 47 of this book.

Recognizing that many countries have limited capacity to respond to chemical incidents, and that chemical incidents occurring in one country could potentially be of international significance, the need to strengthen both national and global chemical incident preparedness and response through the development of an early warning system and a program of capacity strengthening in Member States, the 55th WHA in May 2002, agreed a resolution‡ expressing concern about the global public health implications of a possible release or deliberate use of biological, chemical, or radionuclear agents, and urging Member States to strengthen systems for surveillance, emergency preparedness and response. In a further development (May 2003), the 56th WHA agreed a resolution§ to revise the International Health Regulations (IHR) to cover not just cholera, plague, and yellow fever, but also biological, chemical, or

* http://www.who.int/ipcs/capacity_building/stockholm_rotterdam/en/

† For example, Health Aspects of Chemical Accident Awareness, Preparedness and Response for Health Professional and Emergency Responders, IPCS, OECD, UNEP-IE/PAC, WHO-ECEH (1994), *WHO/EURO Guidelines for Assessing the Health Consequence of Major Chemical Incidents—Epidemiological Approaches* (1997). WHO/IPCS Guidelines for the Public Health and Chemical Incidents for National & Regional Policy Makers in the Public/Environmental Health Roles (1999). See http://www.who.int/ipcs/emergencies/providing/en/index.html

‡ WHA 55 16.

§ WHA 56 28.

radiological events of "international concern." The WHO Chemical Alert and Response Team identifies, alerts, tracks, and where appropriate coordinates a response to chemical incidents and emergencies on a global basis, with the aim of strengthening capacity in countries, particularly developing countries and those in economic transition, to deal with chemical incidents and emergencies. In response to these developments, IPCS started to build upon previous activities for providing guidance for preparedness and response to chemical accidents and emergencies to include:

- Piloting of a Global Chemical Incident Alert, Surveillance, and Response System.
- Compiling a database of global chemical incidents of public health significance in order to improve the knowledge base.
- Establishing a joint operation center with the existing WHO Global Alert and Response (GAR) team for infectious diseases.* Everyday, the outbreak verification team screens information about disease outbreaks of potential international concern received from a wide range of sources, and carries out a risk assessment to determine whether there is a need to alert the government concerned and whether assistance should be offered in response to the outbreak.
- Compilation of a database of global chemical incidents. This database is compiled from various sources and includes details of: the date the incident occurred; the location and type of incident; the chemical(s) released; the public health impact of the incident; the public health action taken; and whether the event met the revised IHR criteria for an event of potential international public health concern. The database of global chemical incidents can serve to identify sentinel events and provide alerts, describing the public health consequences resulting from acute incidents and provide a mechanism for capacity strengthening. Further, a global network (called Cheminet) of partners has been established an essential first step in strengthening international cooperation for alert and response to chemical events.† A series of guidance documents have been produced to assist countries in developing their own public health response plans for both accidental and deliberate chemical events.‡ The latest WHO Manual for the Public Health

* http://www.who.int/csr/en/

† ipcsalert@who.int

‡ WHO Guidelines for the public health response to biological and chemical weapons (2004). WHO Guidelines to Assess National Health Preparedness and Response Programmes to Deliberate Disease from Biological and Chemical Agents (Draft, 2003). WHO Guidelines on crisis communication (Draft, 2003). WHO/IPCS Guidelines for the Public Health and Chemical Incidents for National & Regional Policy Makers in the Public/Environmental Health Roles (1999). WHO/EURO Guidelines for assessing the health consequence of major chemical incidents—Epidemiological approaches (1997). WHO Health Assessment Protocols For Emergencies (1999). Health Aspects of Chemical Accident Awareness, Preparedness and Response for Health Professional and Emergency Responders, IPCS, OECD, UNEP-IE/PAC, WHO-ECEH (1994). See http://www.who.int/ipcs/emergencies/providing/en/index.html

Management of Chemical Incidents (2009) provides a comprehensive overview of the principles and roles of public health in the management of chemical incidents and emergencies.*

Children's Environmental Health

Environmental quality is one of the key factors in determining whether a child survives the first years of life and strongly influences the child's subsequent physical and mental development. Children are at a greater risk from environmental hazards than adults. The special vulnerabilities of children were recognized at the World Summit on Sustainable Development (WSSD) particularly in relation to poverty eradication, health, and sustainable development. The Healthy Environments for Children Alliance was launched at WSSD to raise awareness and support policy maker and community action on children, health, and environment issues. The Children's Environmental Health Indicators (CEHI) initiative was also launched at the WSSD to help improve the better monitoring and reporting of key childhood environmental health indicators.†

WHO established a global plan of action for children's health and the environment.‡ The priority of children's environmental health has been reflected in a number of ministerial level initiatives linking health and environment. Most notably these include those of the Health and the Environment Ministers of the Americas (HEMA) in its Declaration of Mar del Plata and the European Health and Environment Ministerial Conference. Two of the three priority issues identified in the Declaration of Mar del Plata are the sound management of chemicals and children's environmental health. The fourth European Ministerial Conference on Environment and Health held in Budapest, June 23–25, 2004§ resulted in the adoption of the European Children's Environmental Health Action Plan (CEHAPE)¶ Regional Priority Goal IV of the Children's Environmental Action Plan covers chemicals safety. Progress with the CEHAPE has been reviewed at the fifth European Ministerial Conference on Environment and Health held in Parma, Italy, February 24–26, 2010.**

Chemical hazards that are recognized to be of particular relevance to children's environmental health include developmental toxins such as lead and mercury, persistent, bioaccumulative, and toxic chemicals such as those falling under the remit of the Stockholm Convention on POPS and pesticides, particularly in developing countries and chemicals responsible for acute poisoning injuries.

WHO activities in children's environmental health are focused in the following activity areas: national profiles on the status of children's environmental health such

* http://www.who.int/environmental_health_emergencies/publications/Manual_Chemical_Incidents/en/index.html

† http://www.who.int/ceh/indicators/en/

‡ http://www.who.int/ceh/en/

§ http://www.euro.who.int/budapest2004

¶ http://www.euro.who.int/childhealthenv/policy/20020724_2

** http://www.euro.who.int/__data/assets/pdf_file/0011/78608/E93618.pdf

as those developed in South America, Latin America and the Caribbean,* CEHI; capacity building, including the development of a training modules for pediatricians and other health professionals that can be used as the basis for training in both developed and developing countries; guidelines, good practice and tools; and research, including the development of materials supporting the development of longitudinal cohort studies on children's environmental health and collaborative research studies such as on asthma in children and the effects of arsenic exposure during pregnancy on children.

Capacity Building

Many developing countries are poorly equipped to respond to existing and emerging chemical safety issues. Strengthening the capacity of countries to soundly manage the chemicals they use is a theme that underpins most of WHO activities in this area. For example a major part of IPCS work on poisons information, prevention and management responds to the fact that a large number of countries do not have access to poisons centers.† Chemicals assessment;‡ and chemical incidents and emergencies§ are other important areas where IPCS supports capacity building. Most of the activities of WHO at regional and country levels are in support of capacity building. The following areas highlight recent activities undertaken and also the training and guidance documents that are available:

Chemicals Assessment

IPCS continues to apply its policy of regional balance in selecting experts to participate in committees convened to develop internationally agreed assessments of the risks of chemicals, and to develop internationally agreed methods for undertaking such assessments. This results in the selection of experts from developing countries and countries with economies in transition, thereby facilitating application of the methods and assessments in those countries. Meetings are also held in developing countries.

---

* http://www.who.int/ceh/profiles/amroprofiles/en/index.html

† Sound Management of Pesticides and Diagnosis and Treatment of Pesticide Poisoning, WHO-UNEP 2006, English; *Guidelines on the Prevention of Toxic Exposures*; *The Clinical Management of Snake Bites in the South East Asian Region*; *Guidelines for Poison Control*; *Analytical Toxicology*; *Management of Poisoning: A Handbook for Health Care Workers.*

‡ *Hazardous Chemicals in Human and Environmental Health: A Resource Book for School, College and University Students* (WHO/PCS/00.1); IPCS Training module No. 1: *Chemical Safety—Fundamentals of Applied Toxicology*: *The Nature of Chemical Hazards* (2nd rev. ed.) Section 1.1—Physical form of chemicals; Section 1.2—Health effects of chemicals (WHO/PCS/97.14); IPCS Training module no 2: Laboratory handling of mutagenic and carcinogenic products (WHO/PCS/98.9) 1998; IPCS Training module No. 3: Chemical risk assessment—Human risk assessment, environmental risk assessment and ecological risk assessment; IPCS Training module No. 4: General scientific principles of chemical safety (WHO/PCS/00.8); IPCS Training module No. 4 (document 2): General scientific principles of chemical safety.

§ Public health and chemicals incidents: Guidance for national and regional policy makers in the public environmental health roles; Public health response to biological and chemical weapons: WHO guidance (2004); Implementing the Stockholm and Rotterdam Conventions.

Implementation of the Globally Harmonized System for the Classification and Labelling of Chemicals*

WHO is committed to promoting implementation of this system, and IPCS is actively involved in engaging its Participating Institutions, networks of health professionals and scientific experts to identify activities and processes that will assist countries to have the system fully operational. The WHO Recommended Classification of Pesticides by Hazard has incorporated the GHS classification criteria in 2010.

Implementing the Stockholm (see Chapter 15 of this book) and Rotterdam (see Chapter 14 of this book) Conventions: WHO assists in mobilizing awareness in countries and regions about how chemicals subject to the conventions are used, through its global network of poisons centers and the work of its Regional Offices. Poisons centers can help to identify chemicals and the formulations which may be of concern. The WHO Environmental Health Emergencies team assists Member States in dealing with disease outbreaks of potential international concern caused by chemicals.

Pesticide poisoning represents a major concern worldwide, particularly in developing countries. The IPCS Project on Epidemiology of Pesticide Poisoning, described above, provides tools for developing countries to identify hazardous pesticide formulations for listing under the Rotterdam Convention (see also Chapter 14 of this book).

Emergency Preparedness and Response: WHO/IPCS has established a WHO Collaborating Centre for the Public Health Management of Chemical Incidents, the role of which is to develop training materials and to run training courses for chemical incident and emergency preparedness and response. In collaboration with the WHO Communicable Disease Cluster, WHO/IPCS has developed guidance and training materials to assist countries in enhancing preparedness and response to the potential use of chemical and biological weapons to cause harm. In addition, WHO has responded to requests by countries to provide technical assistance on preparedness and response in this field.

Training material on the health aspects of chemical releases caused by deliberate acts has been developed by WHO/IPCS, in collaboration with the WHO Department for Emergency and Humanitarian Action, for the UN Disaster Management Training Programme with funding from UNDP. The material is targeted at emergency managers in the field. The objectives are to raise awareness of the risks posed by chemical weapons, to improve the responds to the public health needs, and to strengthen coordination of these activities.

Women's Health and Pesticides Safety

Through regional meetings on Women's Health and Exposure to Chemicals,† awareness is raised on exposure to chemicals, especially pesticides, throughout women's life cycle and on specific effects on women's health. Training courses are also organized targeted at health and safety personnel in workplaces dealing with pesticides, including state officials with health and safety responsibilities, primary care occupational health practitioners, policy makers, managers in occupational

---

* See Chapter 11 of this book.

† For example that held jointly with PAHO in Nicaragua. See http://www.who.int/ipcs/capacity_building/pesticides/en/index.html

and environmental health programs, such as organized in Tanzania in 2003 for SADC (Southern African Development Community).* The University of Michigan Fogarty International Centre and the University of Cape Town have become collaborating institutions to assist the further development of environmental and occupational health infrastructure and human resources in Southern Africa, with a particular emphasis on improving research training and capacities on pesticides (see footnote * of previous page).

CONCLUSIONS

The history of WHO activities related to chemical safety and this chapter's review of the thematic areas of WHO work in this area show that since its establishment as a United Nations Specialized Agency, the World Health Organization has played an important role both in providing the scientific and evidence basis for sound management of environmental risks, particularly as they relate to human health and the environment, and in promoting capacity in Member States to use chemicals safely in furtherance of a sound and sustainable development process and in the protection and promotion of human health in accordance with its charter.

This role is undertaken in collaboration with other UN bodies and Programmes, as well as other intergovernmental and nongovernmental organizations involved in managing chemical and environmental risks. It operates at global, regional, and country levels. WHO encourages innovative partnerships among governments, the private sector, and civil society for promoting sound management of chemical and environmental risks, as well as good governance.

The World Summit on Sustainable Development provided an unprecedented opportunity to strengthen the role of health in sustainable development, reflecting Principle One of the Rio Declaration that "Human beings are at the center of concerns for sustainable development."† Since WSSD, there has been a burgeoning of the number of global legally binding agreements that address chemicals‡ and an increased emphasis on the value of multilateral environmental agreements as a means of controlling pollution at source and promoting the principles of sustainable development. This has been at a time of increasing recognition of the role that the health sector plays in chemicals management through gathering evidence about chemicals risks, informing the public, preventing and managing chemicals emergencies, working to protect the most vulnerable sections of society, and assessing the impacts of chemicals risk management policies through monitoring and evaluation.

Since Rio and WSSD, WHO has been active in new intergovernmental bodies that have shaped the international chemicals management agenda, notably IOMC,

* http://www.who.int/ipcs/capacity_building/pesticides/en/index.html

† http://habitat.igc.org/agenda21/rio-dec.html

‡ Of 23 global agreements involving chemicals from 1923 to 2007, seven (almost 30%) have come into force after 2002. These include Stockholm and Rotterdam Conventions, the IHR, the Kyoto Protocol, and IMO instruments on pollution incidents by hazardous and noxious sustances. Further information can be obtained from the book by John Buccini on *Global Pursuit of Sound Management of Chemicals*, available at: http://siteresources.worldbank.org/INTPOPS/Publications/20486416/GlobalPursuitOfSoundManagementOfChemicals2004Pages1To67.pdf.

IFCS, and SAICM. Pressure needs to be maintained to ensure that governments continue to and increase investment in developing the scientific basis for decision making for sound management of chemicals, and to ensure the necessary capacity building in developing countries and those in economic transition to implement chemical safety. WHO needs to continue and strengthen its role, in collaboration with all relevant partners, in promoting protection and enhancement of human health in relation to sound management of chemicals to the benefit of all peoples of the world, which requires countries to make an effort, first to harmonize their own health and environmental protection policies in a multistakeholder process and then to ensure that these policies are implemented through their multilateral agreements and other treaty obligations.

ACKNOWLEDGMENTS

The author gratefully acknowledges the contributions from historical material of Dr. Rune Lönngren's "International Approaches to Chemicals Control," published by the National Chemicals Inspectorate of Sweden (KEMI) in 1992.* The valuable comments and inputs of Lesley Onyon and Carolyn Vickers of WHO Secretariat are also gratefully acknowledged. Much of the descriptive sections on WHO programs and activities are adapted from text found in the WHO Web site http://who.int/en/.

REFERENCES

J. P. Chippaux. 1998. *WHO Bulletin* 76(5): 515–524.

R. J. Flanagan et al. 1995. *Basic Analytical Toxicology*. Geneva: World Health Organization.

R. J. Flanagan. 2005. *Developing Analytical Toxicology Services: Principles and Guidance*, Geneva, Switzerland: World Health Organization.

International Health Conference. 1946. Official Records of the World Health Organization, no. 2, p. 100, 19 June–22 July 1946, New York.

C. Sonich-Mullin et al. 2001. IPCS conceptual framework for evaluating a mode of action (MOA) for chemical carcinogenesis. *Regulatory Toxicology and Pharmacology*, 34: 146–152.

WHO. 1997. *Assessing the Health Consequences of Major Chemical Incidents—Epidemiological Approaches*. WHO Regional Publications European Series, No. 79, Copenhagen, Denmark.

* This book was a publication of Kemi, Solna, Sweden, 1992.

34 The Intergovernmental Panel on Climate Change

Bert Metz

CONTENTS

INTRODUCTION

The Intergovernmental Panel on Climate Change (IPCC) is the leading body for the assessment of climate change, established by the United Nations Environment Programme (UNEP) and the World Meteorological Organization (WMO) to provide the world with a clear scientific view on the current state of climate change, its potential environmental and socio-economic consequences and the possibilities to address it.*

The IPCC is a scientific body. It reviews and assesses the most recent scientific, technical, and socioeconomic information produced worldwide relevant to climate change. It does not conduct any research nor does it monitor climate related data or parameters. Thousands of scientists from all over the world contribute to the work of the IPCC on a voluntary basis. Review is an essential part of the IPCC process, to

* http://www.ipcc.ch

ensure an objective and complete assessment of current information. Differing viewpoints existing within the scientific community are reflected in the IPCC reports.

The IPCC is an intergovernmental body, and it is open to all member countries of the UN and WMO. Governments are involved in the IPCC work as they can participate in the review process and in the IPCC plenary sessions, where main decisions about the IPCC work program are taken and reports are accepted, adopted, and approved. The IPCC Bureau and Chairperson are also elected in the plenary sessions.

Because of its scientific and intergovernmental nature, the IPCC embodies a unique opportunity to provide rigorous and balanced scientific information to decision makers. By endorsing the IPCC reports, governments acknowledge the authority of their scientific content. The work of the organization is therefore policy relevant and yet policy neutral, never policy prescriptive.

HISTORY

It is because of the need of broad and balanced information about climate change that the organization was created back in 1989. It was set up by WMO and UNEP as an effort by the UN to provide the governments of the world with a clear scientific view of what is happening to the world's climate. The initial task for the IPCC as outlined in the UN General Assembly Resolution 43/53 of December 6, 1988 was to prepare a comprehensive review with respect to the state of knowledge of the science of climate change; social and economic impact of climate change, possible response strategies and elements for inclusion in a possible future international convention on climate (Bolin, 2008).

The scientific evidence brought up by the first IPCC Assessment Report of 1990 unveiled the importance of climate change as a topic deserving a political platform among countries to tackle its consequences. It therefore played a decisive role in leading to the creation of the UN Framework Convention on Climate Change (UNFCCC), the key international treaty to reduce global warming and cope with the consequences of climate change.

Since then the IPCC has delivered on a regular basis its Assessment Reports, the most comprehensive scientific reports about climate change produced worldwide. It also continued to respond to the needs of the UNFCCC for information on scientific and technical matters.

The IPCC Second Assessment Report of 1995 provided key input in the process leading to the adoption of the Kyoto Protocol in 1997. The Third Assessment Report that came out in 2001 played a key role in making the Kyoto Protocol operational. The negotiations on the operational details of the Protocol benefited from it, leading to the formal entry into force in February 2005. The Fourth Assessment report, published in 2007, is playing an important role in the negotiations on a new agreement for the period after 2012. Since the Third Assessment report a separate Synthesis Report is produced, incorporating the main findings of the three Working Group reports (Lohan, 2006).

Along with the Assessment Reports, the IPCC has produced several Special Reports on various topics that the political discussions had to deal with, such as Aviation and

Climate Change, Technology Transfer, Fluorinated Gases and CO_2 Capture and Storage. It also prepared methodologies and guidelines to be used by Parties under the UNFCCC for preparing their national greenhouse gas (GHG) inventories.

The participation of the scientific community in the work of the IPCC has been growing greatly, both in terms of authors and contributors involved in the writing and the reviewing of the reports and in terms of geographic distribution and topics covered by the reports (Zillman, 2007).

For the achievements over its lifetime the organization was honored with the Nobel Peace Prize at the end of 2007.

ORGANIZATIONAL STRUCTURE

The foundation and the main machinery of the work of the IPCC is provided by the thousands of scientists who perform the assessment or review the draft versions of the assessment report. To organize their work and provide guidance to the assessment process, an organizational structure was created with the following elements:

Working Groups

The IPCC is currently organized in three Working Groups. Working Group I deals with "The Physical Science Basis of Climate Change," Working Group II with "Climate Change Impacts, Adaptation, and Vulnerability" and Working Group III with "Mitigation of Climate Change." Each of the Working Groups is assisted by a Technical Support Unit, which is hosted and financially supported by the government of the country that provides the elected developed country co-chair of the Working Group.

The IPCC has also a Task Force on National Greenhouse Gas Inventories. The main objective of the Task Force is to develop and refine a methodology for the calculation and reporting of national GHG emissions and removals. In addition to the Working Groups and Task Force, further Task Groups and Steering Groups may be established for a limited or longer duration to consider a specific topic or question.

Bureau

The IPCC Bureau is the day-to-day governing body, at present composed of 30 members: two co-chairs of each of the Working groups and the Task Force, six vice chairs for each of the Working Groups, three overall vice chairs and a chair. The Bureau is elected by the Plenary of the IPCC, the supreme governing body. Its members are to provide guidance to the authors' teams in their preparation of an IPCC Assessment Report. Therefore, their mandate normally corresponds to the duration of an Assessment cycle (5–6 years). Bureau members have to be experts in the field of climate change and all regions are to be represented in the IPCC Bureau. The Bureau of the Task Force on National Greenhouse Gas Inventories (TFB) oversees the National Greenhouse Gas Inventories Programme. It is composed of two co-chairs, which are also members of the IPCC Bureau, and 12 members.

Panel

The Panel, comprised of government delegations of all member countries is the supreme governing body of the organization. It meets approximately once a year at the plenary level. These Sessions are attended by hundreds of officials and experts from relevant ministries, agencies and research institutions from member countries and from international organizations. Major decisions such as the election of the IPCC Chair, IPCC Bureau and the Task Force Bureau, the structure and mandate of IPCC Working Groups and Task Forces, as well as on procedural matters, work-plan and budget are taken by the Panel in plenary session. The Panel decides also on scope and outline of IPCC reports and accepts the reports. The IPCC is open to all member countries of the UN and the WMO. There are at present 194 member countries.

Secretariat and Technical Support Units

The small central IPCC Secretariat (5–8 full time people) plans the budget, prepares the work plans, manages the finances, organizes IPCC meetings, publicizes and disseminates IPCC reports and liaises with member governments and the parent organizations WMO and UNEP. Each of the Working Groups has a small secretariat (Technical Support Unit, about five full-time people) that does the planning for the Working Group contribution to an assessment report, preparing for the selection of authors, organizing the review process of the assessment report drafts and support of the work of the authors of an assessment report.

Observer Organizations

Sessions of the IPCC and the IPCC Working Groups are also attended by representatives of observer organizations. Any nonprofit body or agency, whether national or international, governmental or intergovernmental, which is qualified in matters covered by the IPCC, may be admitted as an observer organization.

Figure 34.1 gives a schematic overview of the organization.

FUNDING OF THE IPCC

The IPCC is funded by regular contributions from its parents' organizations WMO and UNEP, the UNFCCC and voluntary contributions by its member countries. WMO also hosts the IPCC Secretariat and WMO and UNEP provide one staff member each for the IPCC Secretariat.

The contributions are put into the IPCC Trust Fund which is administered by WMO. The Trust Fund supports the IPCC activities, in particular the participation of developing country experts in the IPCC work, and publication and translation of IPCC reports.

Governments provide further substantial support for activities of the IPCC, in particular through hosting Technical Support Units, supporting the participation of experts from their country in IPCC activities, hosting meetings and so on.

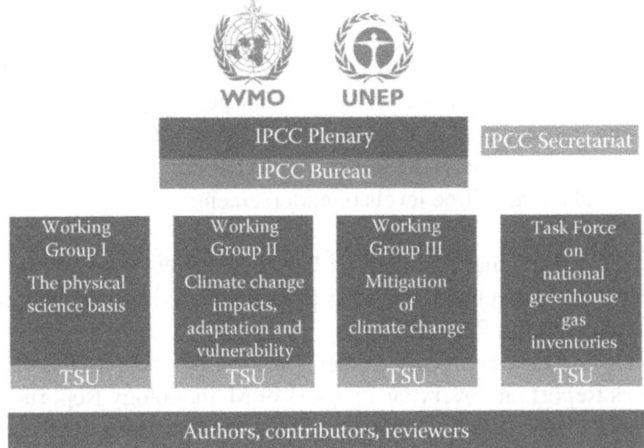

FIGURE 34.1 The organization of IPCC. (From IPCC. 2010. Brochure of the Intergovernmental Panel on Climate Change on Understanding Climate Change, 22 years of IPCC Assessment. With permission.)

THE IPCC AUTHORS AND THE PREPARATION OF THE IPCC REPORTS

Authors are selected by the Bureau of the Working Group from a list of nominations received from governments and international organizations. They can also be identified directly by the Bureau because of their special expertise reflected in their publications and work. The composition of lead author teams must reflect a range of views, expertise and geographical representation. All authors work on a voluntary basis and are not paid for their contributions, except for travel expenses of authors from developing countries.

The Coordinating Lead Authors (CLAs) have the role of coordinating the content of the chapter they are responsible for (there are usually two CLAs per chapter, one from a developing and one from an industrial country). The Lead Authors (LAs) work as a team to produce the content of the chapter they have been selected for. They are often supported by several Contributing Authors, who provide more technical information on specific subjects covered by the chapter and are usually not participating in the deliberations of the author team.

Review is an essential part of the IPCC process to ensure objective and complete assessment of the current information. The review is organized in two stages: a review of the first draft by independent experts and another review by the same experts plus governments. In the course of the multistage review process, both expert reviewers and governments are invited to comment on the accuracy and completeness of the scientific, technical, and socio-economic content and the overall balance of the drafts. The circulation among peer and government experts is very wide, with hundreds of scientists looking into the drafts to check the soundness of the scientific information contained in them. Authors are required to consider each and every review comment and record their considered opinion whether to accept the comment or reject it. These records are made public. To oversee the proper operation of the review process, Review Editors are selected for each chapter of the report (normally two or three per chapter). They have to make sure that all comments (normally hundreds of comments per chapter) are properly taken into account and they have to sign a formal statement at the end of the process.

All IPCC reports must be endorsed by the Panel during a Working Group or a Plenary session. There are three levels of endorsement:

1. "Approval" means that the material has been subjected to a "line-by-line" discussion and agreement. It is the procedure used for the Summary for Policymakers of the Reports.
2. "Adoption" is a process of endorsement "section-by-section." It is used for the Synthesis Report and overview chapters of Methodology Reports.
3. "Acceptance" signifies that the material has not been subject to line-by-line nor section-by-section discussion and agreement, but nevertheless presents a comprehensive, objective, and balanced view of the subject matter.

COMPREHENSIVENESS OF THE ASSESSMENTS

There has been an evolution in the comprehensiveness of the IPCC assessments over time. The first assessment report narrowly focused on climate change issues. Over time, however, the subject matter gradually broadened to include economic, social, and development issues. This was a reflection of the growing conviction in the scientific literature that these other dimensions form an integral part of the problem and the potential solutions. The last (fourth) assessment report explicitly considered dealing with climate change in the context of development policies, thereby for the first time looking at climate change adaptation and mitigation through the lens of other socioeconomic policies.

Many other interactions of climate change are being addressed in IPCC assessment reports, such as with biodiversity and availability of water. However, to keep IPCC assessments manageable, this was never done by fully integrating all related aspects. The focus has remained on highlighting the main interactions, without pretending to be doing also a full assessment of biodiversity, water, or development.

SCIENTIFIC INTEGRITY

Scientific integrity is one of the most precious characteristics of the IPCC assessments. It is guaranteed through a number of principles the IPCC applies to all its work:

- All peer-reviewed scientific literature should be included, no selective use of literature by the authors is allowed; this is where the review process fulfils a critical role in pointing out missing literature. The assessment process, however, critically evaluates the various publications about a certain issue and then gives an argued interpretation of its meaning. It is not a simple compilation, without a critical evaluation.

- "Dissidents" (i.e., people who argue climate change is not real) are invited as authors, provided they have a good scientific standing, and as reviewers; only strict scientific arguments are being honored in the review process.
- The review process overseen by review editors (see above) is a very strong mechanism to ensure scientific credibility, because authors are required to consider each comment and record their judgment in a publicly available record (all review comments of the Fourth Assessment report have been made public via the Web). The only aspect of the review process that might be improved is the change to an anonymous review process (currently names of reviewers are known to the authors).

INDEPENDENCE

Independence from political influences is another key characteristic of the IPCC assessments. It is ensured through the following processes and provisions:

- IPCC decides on its own products and terms of reference for each of its assessments. Very often suggestions and even requests are made through the political process of the UNFCCC, but IPCC is not held to honour those requests. It is fully independent.
- The full text of assessments and the Technical Summaries are strictly the authors' responsibility. The IPCC Panel "accepts" these texts without any change.
- In approving Summaries for Policy Makers (SPM) by the IPCC Panel there is of course the risk of political interference. The text is approved line-by-line and many amendments are made on the text submitted by the authors of the report. The way this risk is handled is by having the CLAs of all chapters in the room to judge if modifying the language of the SPM is still in conformity with the underlying text of the full chapters. If not, the amendment is rejected. This process allows the Panel to formulate conclusions in a more user friendly way without modifying the scientific conclusions of the report.

In practice there have been cases where IPCC Assessment reports whose SPMs were fully approved by the Panel, still met stiff resistance when they were brought forward in the framework of the UNFCCC discussions. This happened in particular with the Special Report on technology transfer that was published in the year 2000.

POLICY RELEVANCE

IPCC's charge is to make assessments to assist the policy process. Being policy relevant, but not policy prescriptive is one of the fundamental principles in IPCC's work. The way this is implemented is as follows:

- Assessment reports are written with a focus on policy relevant questions ("what does this mean for policy?"). On top of that the SPM of each of the

Working Group or Special Reports and the SPM of each Synthesis Report are particularly geared to the questions that policy makers have to deal with.

- Special Reports are often requested by the UNFCCC, because there are particular policy questions that countries feel could be better answered on the basis of an IPCC assessment report.
- The IPCC emission inventory work in the form of guidelines for making these inventories form a crucial input into the implementation of the UNFCCC and its Kyoto protocol.

There are, however, also some limits to the use of IPCC assessments in practical policy making. The rule that only peer reviewed literature (meaning material published in peer reviewed scientific journals and other publications that are considered to have undergone a comparable degree of scrutiny and review) can be used limits the use of more practical material that emerges from government analysis and private sector experience (Raes and Swart, 2007).

ACCESSIBILITY AND COMMUNICATION

Communication of the IPCC findings to policy users maybe the weakest aspect of the assessment process. A typical IPCC Working Group Report nowadays is about 1000 pages thick. Special Reports easily reach more than 300 pages. The language is technical and uses a lot of scientific and technical jargon (this is the way authors show they are not paraphrasing the scientific content of the publications they draw upon). This is somewhat compensated by having both a Technical Summary (of about 50–100 pages) and a Summary for Policy Makers (about 20 pages), but even these summaries are not jargon free and are often hard to read by people in practical policy making. IPCC has strong limitations in simplifying its reports to make them more readable. The fact that SPMs are approved by the Panel on a "line-by-line" basis means that the text is seen as balanced from a policy point of view. Rewriting such a summary in plain language can easily distort this balance which would upset the Panel. IPCC therefore consistently refrains from doing this. Other organizations however, such as UNEP, have produced simplified versions of IPCC reports for a broader audience.*

RECENT CONTROVERSIES

The IPCC came under pressure recently in the lead up to the Copenhagen Summit. Leaked emails from climate researchers at the University of East Anglia (House of Commons Science and Technology Committee, 2010) were picked up by climate skeptics and the media as an indication that prominent IPCC authors were trying to suppress unwanted information from others that would counter their own research, implying that IPCC assessments could no longer be seen as objective. Some errors were discovered around that time in the Working Group II part of the Fourth

* http://www.unep.org/dec/Information_Resources/Simplified_Guides.asp

Assessment report, further fueling media frenzy about the lack of credibility of the IPCC. Several investigations were performed into these allegations. The IPCC was completely vindicated, however. Not only was it made clear that the conduct of individual scientists had not influenced the quality and objectivity of the IPCC assessments (House of Commons Science and Technology Committee, 2010), it was also made clear that errors identified in parts of the IPCC Working Group II contribution to the Fourth Assessment report had not influenced the main conclusions about the seriousness of expected climate impacts (Netherlands Environmental Assessment Agency, 2010). A further investigation by a committee of the Interacademy Council (IAC), the umbrella organization of national Academies of Science, was set up to look into the organization and procedures of IPCC with the objective of improving these to avoid similar problems in the future. In August 2010, the IAC published its recommendations for improving organization and procedures (Interacademy Council, 2010).

The events prove that the public scrutiny of IPCC assessments has grown very significantly, as have the social and economic implications of climate change and action to prevent such change. This will require a further strengthening and professionalization of IPCC's organization, procedures, and its communication abilities.

REFERENCES

Bolin, B. 2008. *A History of the Science and Politics of Climate Change.* Cambridge: Cambridge University Press.

House of Commons Science and Technology Committee. 2010. The disclosure of climate data from the Climatic Research Unit at the University of East Anglia, London, available at http://www.publications.parliament.uk/pa/cm200910/cmselect/cmsctech/387/387i.pdf

Interacademy Council. 2010. Climate Change Assessments, Review of the Processes & Procedures of the IPCC, Amsterdam.

Lohan, D. 2006. Assessing the mechanisms for the input of scientific information into the UNFCCC. *Colorado Journal of International Environmental Policy*, 17(1): 249–308.

Netherlands Environmental Assessment Agency, Assessing an IPCC Assessment. 2010. An analysis of statements on projected regional impacts in the 2007 report, Bilthoven.

Raes, F. and R. Swart 2007. Climate assessment: What's next? *Science*, 318: 1386.

Zillman, J. 2007. Some observations on the IPCC assessment process 1988–2007. *Energy and Environment*, 18(7 and 8): 869–891.

Section VI

Representative Country Implementations

35 Chilean Approach to Chemical Safety and Management

*Sergio Peña Neira and Asish Mohapatra**

CONTENTS

* The opinions expressed in this chapter are those of the author and do not necessarily represent the views of any organization.

INTRODUCTION

Chile, a founding member of the United Nations and the Union of South American Nations, has been very active in the areas of chemical waste management, compliance and enforcement of regulations, and application of global chemical conventions and policies. The country has been identified as a significant depository of various minerals and chemicals (e.g., copper, lithium, gold, iron, silver, zinc, manganese, molybdenum, lead, iodine, etc.), and is undergoing increased resource exploitation, resulting in significant economic growth and at the same time generating hazardous wastes and imposing risks to human health and environment.

A BRIEF FOCUS ON THE CHILEAN CHEMICAL MINING SECTOR

At the time of the writing this chapter, Chile had completed a successful mining rescue of 33 miners from the San Jose copper gold mine (Empresa Minera San Esteban), which was cited 42 times for safety violations between 2004 and 2010 and was shut down temporarily in 2007 due to safety concerns and a human fatality incident.* Chilean engineers, with collaboration and support from the National Aeronautic and Space Administration (NASA), and The University of Texas Southwestern Medical Center at Dallas scientists, concluded a successful rescue event that made global headlines. However, the mine collapse incident also generated discussions related to mining workplace health and safety across the globe and chemical exposure and management issues associated with chemical and mining industries. These mining industries are pivotal to the economic growth of Chile and other South American countries and their specific contribution to global demand of gold, copper, lithium, and other minerals and chemicals in the twenty-first century. National industry developments in the 1940s and further developments in various industrial zones for Chile (e.g., mining industry—1st to the 6th Region; the petrochemical industry—5th, 8th, and 12th Region; and other industries in the Metropolitan Region) have provided a strategic advantage over other South American countries. After the end of the military rule (1973–1990), Chile became one of the most prosperous South American nations in the region. However, significant environmental issues, chemical contamination, and management issues remain a high priority for the country. In this chapter, a Chilean perspective of the regional and global chemical management initiatives is discussed.

The mining sector in Chile has played an important strategic role in the Chilean economy toward fiscal contributions and the generation of foreign exchange.

* http://en.wikipedia.org/wiki/Chile; http://en.wikipedia.org/wiki/2010_Copiap%C3%B3_mining_accident

Regionally, Chile is a key player in attracting external investments that tie to regional employment. However, in addition to these growth indicators, the mineral and mining industry model in Chile has also been designed as an export oriented sector due to its high levels of investment and advanced technological requirements. Therefore, the mining sector has also been responsible for a certain vulnerability in the economy and for distortions in the pattern of economic development. As the United Nations Environment Programme (UNEP) (1999) report indicated, these drawbacks are apparent in the unbalanced growth of mining to the detriment of other economic activities, a limited creation of external economies and irregularity in fiscal contributions to the Chilean economy (Borregaard, 1999). Furthermore, mining is concentrated mostly in the northern and central parts of Chile with rich deposits of copper, iron, and iodine. This area is characterized by mountains and deserts and historically precluded diversified economic activities. Therefore, mining activities have flourished in these areas (up to 60% of the economic activities of Chile). In southern Chile, mining activities are confined to coal, petrol, and gas. However, they are less important when compared to northern and central Chile. Overall, mining accounts for approximately 8% of economic activity. Because of its growing demand, copper leads the mining activities in Chile. However, iron, silver, zinc, manganese, molybdenum, and lead also contribute to the mining sector.

The free trade liberalization of Chilean industries and the economy as a whole have raised issues such as free exchange of hazardous and toxic chemicals associated with products. With differing regulations in different countries, free trade agreements may also reduce the incentive to establish stringent national environmental standards. This can occur when toxic wastes are accepted for disposal in countries with lower environmental standards (Johnstone, 1996).

NATIONAL INITIATIVES IN CHEMICAL MANAGEMENT IN CHILE*

Based on the United Nations Institute for Training and Research (UNITAR)/Inter-Organization Programme for the Sound Management of Chemicals (IOMC) guidance document, the national profile report to assess the Infrastructure for Management of Chemicals for Chile was updated in 2008 by taking into account updated and background information from the last two national profiles in 2000 and 2003. During this process of updating Chile's chemical management profile, monthly meetings with the National Coordination Team were held integrated by professionals from Comisión Nacional del Medio Ambiente (CONAMA), Ministry of Health, Ministry of Foreign Affairs, Ministry of Transports, Ministry of Labour, Servicio Agrícola y Ganadero (SAG), the Chilean Copper Commission (COCHILCO), Oficina Nacional de Emergencia del Ministerio del Interior (ONEMI), Asociación Nacional de Fabricantes e Importadores de Productos Fitosanitarios Agrícolas (AFIPA), Asociacion Gremial de Industriales Quimicos (ASIQUIM), Sociedad Nacional de Minería (SONAMI), Greenpeace, Rapal, and the University of Concepcion. Different public and private

* This section is an excerpt from the National Country Profile Report available from UNITAR. The Spanish language report was translated into English and relevant information was included in this section.

sources of information were used to obtain a general overview of the current management of chemicals in Chile. Geographical features of Chile and national industrial development policies furthered in the 1940s have resulted in the distribution of population and the formation of chemical industrial zones (e.g., mining industry from the 1st to the 6th Region; petrochemical industry in the 5th, 8th, and 12th Region; and a great diversity of industries in the Metropolitan Region).

The life-cycle analysis and management of chemicals in Chile is in the process of much improvement to necessitate the update of the national inventories of some chemicals, and to build registries of priority chemicals. Sources of chemical contaminants included in current systems that provide information to the Pollutant Release and Transfer Register (PRTR–RETC) are in the process of further refinement and improvement. The Registro de Emisiones y Transferencias de Contaminantes (RETC) system is a national effort in Chile similar to PRTR (currently as a pilot project) that groups and consolidates environmental information by integrating information from several sources. At the end of 2008, a second report was published with environmental information for 2006. As per the National Chemical Management profile for Chile, in 2009 the PRTR-RETC was going to be modified and the atmospheric emissions information was collected for inclusion in a Web platform, with the incorporation of new pollutants and information on their sources from industrial activities.*

In 2002 the government of Chile, through its National Environment Commission (CONAMA) and with the support of Environment Canada and UNITAR, initiated a process to promote the establishment of a national PRTR in Chile (2002–2005). A key outcome was the National PRTR Proposal with complete specifications. Gradually, CONAMA developed PRTR regulations that agreed on an institutional structure for the PRTR and defined its administrative role, and set the procedures for transferring information from various sectoral agencies. All national activities in Chile were also implemented with the active participation of civil society. Furthermore, this project included a parallel initiative to strengthen non-governmental organizations' capacities to enable their participation in various multistakeholder processes related to environmental conservation. At present, Chile has implemented its PRTR system and in 2009 entered into its second year of reporting. Chile's PRTR will be strengthened through the Global Environmental Facility-supported Global PRTR Project on Persistent Organic Pollutants (POPS) for monitoring, reporting, and information dissemination using PRTRs. Within this project, Chile will include the necessary elements in its PRTR to ensure POPS reporting through the existing system. The project will provide lessons learned and standards for capacity-building activities on POPS reporting, which may be replicated in other countries and regions. This project also involves Ecuador and Peru, which will design a national PRTR system.†

In addition to other Latin American countries such as Ecuador and Panama, Chile has developed national strategies for the integration of the data generated by the Mercury Emission Inventory into existing or future national PRTRs, which aims at institutionalization of the Mercury Inventory and ensures regular reporting of mercury. With these developments in a pollutant registry and information sharing, chemical

* http://www.unitar.org/cwm/prtr/, accessed Oct. 25, 2010.

† See http://www.unitar.org/cwm/prtr/

management priorities (e.g., atmospheric emissions, industrial liquid wastes, management of hazardous wastes, and potentially contaminated sites) have been set for various industrial sectors. In Chile, legal instruments for management of chemicals and wastes have been formulated and are in force (e.g., the regulation on hazardous substances, atmospheric emissions, and liquid wastes). The Chilean government has also proposed a specific Ministry of Environment that would provide directives on legal and regulatory aspects related to the protection of the environment and natural resources. Industries in Chile have also implemented nonregulatory instruments, which have been effective in risk reductions from industrial activities. The rising risk awareness associated with the deficient management of chemicals has pushed the public and private sectors to carry out training and risk communication activities, to effectively inform people and workers about chemical safety issues.

CHILEAN NATIONAL CHEMICAL SAFETY POLICY

On October 21, 2008, the National Policy on Chemical Safety was approved by the Board of Directors of CONAMA (Maximum Environmental Authority) of Chile. The safety policy has a set of strategic guidelines to address the life-cycle management of chemicals that are consistent and complementary to the principles and objective of Chilean environmental policy within the framework of the intersectoral coordination scheme of the 19,300 ruling policies on the general environment. In this policy (based on a precautionary approach), categories of chemicals that are used as raw materials in mining and agricultural industries and are classified as dangerous in the Chilean Official Standard No. 382.Of.2004 are included. Pesticide for agriculture are regulated by Decree 3557 of 1980 and supplementary resolution, the Farm Service Livestock.

Chemicals that are not included in this policy are drugs and psychotropic substances; radioactive materials; chemical weapons and their precursors; pharmaceuticals, including human and veterinary medicines; cosmetics; and chemical food additives. However, the scope of work included in this policy is chemical exposure issues and occupational health and safety that since has included relevant international instruments to protect workers.

The objective of this policy "is to reduce the risks associated with handling and/or life-cycle management of chemicals, including stages of import, export, production, use, transport, storage, and elimination to protect human health and the environment."

CHILEAN CHEMICAL MANAGEMENT PROGRAMS AND TOOLS FROM REGIONAL AGENCIES

CONAMA has developed the National Implementation Plan for POPS in the context of Chile's participation as a signatory to the Stockholm Convention-related activities. This is aimed toward gradual elimination of these toxic chemicals by the year 2025, which is why all the related research associated with the development of diagnostics and analytical technologies for treatment and disposal of these compounds made to date is based on guidelines that comply with specific objectives of the Stockholm Convention.

Activities to implement the Program Mercury World are an elaboration of the Risk Management Action Plan, which includes activities related to the minimization of health effects from exposure to mercury. The results in generation of specific management plans for storage and disposal of products containing mercury wastes.

The General Law on Waste and Extended Liability

A provider is in development in Chile that will aim to define which priority products have to be managed in order to reduce the hazards from quantities generated and disposed.

Inventories of Hazardous Waste Risk Assessment, Treatment, and Disposal of Chemicals in Contaminated Sites

The hazardous waste inventory can be generated from the statements obtained from the Web-based reporting system SIDREP (http://sidrep.minsal.gov.cl/sidrep/index.php). This may be supplemented with information from the Ministry of Health, which administers this site. Regarding contaminated sites, the Board of Directors of the Ministry of Health promotes specific management tools, including Power Sites Cadastre contaminants. CONAMA has this tool, which allows gathering information in the field to obtain a prioritized list of contaminated sites to assess preliminary risks. This is accomplished by collecting data from these sites for the source contaminant, the exposure routes, and the receptors involved. This information is then processed through qualitative risk assessment. Further analysis of the risk assessment provides information about the national situation of contaminated sites. Finally, steps to be implemented in the management system are recommended to enable future targeted assessments to account for and identify the real polluter existing in each of those identified and prioritized sites.

Dissemination of Scientific and Technical Information that Addresses the Various Aspects of Environmental Hazardous Waste

The National Environmental Commission, CONAMA, has a National System of Environmental Information (http://www.sinia.cl) where it shares information on hazardous wastes. The Web site of the Sanitary Authority and Ministry of Health also provides general information on the environmental health aspects of hazardous wastes.

SIDREP (Declaration and Monitoring System of Hazardous Waste) (sidrep.minsal.gov.cl) provides an alternative format to declare waste electronically, thus saving on paper and cost. There is also the project delivery Respelido (http://www.respel.cl), which contributes to the understanding of the processes to be followed for proper management of hazardous waste.

Notification Systems and Registries of Exposed Populations

The Ministry of Health is responsible for these notification systems and maintaining a registry of chemical exposure incidents and exposed populations. The Ministry of

Health also deals with the management of chemical exposure and health effects associated with international traffic of hazardous wastes.

Policies Aimed at Prevention and Minimization of Waste and Its Reusability and Recycling

With respect to policies aimed at prevention and minimization of waste already mentioned, there is a policy, regulations, and agreements in cleaner production that encourage these practices. However, the Waste General Policy is expected to have specific binding policies and to promote initiatives that are aimed at minimization of wastes under these regulatory instruments.

Establishment of Environmentally Sound Disposal Facilities, Including Technologies to Convert Waste into Energy

Using methane emissions from landfills can be a viable sustainable management practice of converting waste into reusable energy. It is important to note that Chile has standard incineration facilities. On the other hand, there are some landfills in the Metropolitan Region (Biobío regions) selling carbon credits through the Clean Development Mechanism (CDM). It is important to reiterate that in DS 189 on Sanitary and Security Policies, the basic requirements of landfills is that they should have gas collection systems.

Mechanisms for Financial Tools in Chemical Wastes Management

Policy on Integrated Waste Management has emphasized the importance of generating more efficient financing systems to enhance and facilitate more sustainable waste management practices, especially at the municipal level. The Undersecretary of Regional Development and Administration (SUBDERE) is funding projects related to the improvement of both waste facilities and health conditions across populations and is also funding projects related to the minimization and reuse of wastes. These financial tools and support are essential for improving sustainable waste management practices in Chile.

Chemical Management Example: Mercury

Chile participated in drafting the National Plan for Risk Management as part of the Global Mercury Project (a multilateral global initiative by UNEP). The objective is to implement priority actions to reduce risks associated with the consumption and emission of mercury, and to protect human health and the environment. This initiative has helped develop a toolkit for large-scale mining sectors (during 2009–2010) titled "Protocol for Measurement and Analysis for Determining Mercury in Different Procurement Processes of Copper," for which *in situ* analysis of geological territory will be undertaken. Several pilot projects are underway to test this toolkit in mining sectors.

INTERNATIONAL COLLABORATION AND REGIONAL IMPLEMENTATION OF CHEMICAL MANAGEMENT IN CHILE

A number of collaborative projects have been carried out in Chile in response to activities and action schedules derived from international conventions like the Rotterdam, Stockholm, Basel, and Montreal protocols. Chile signed the Organization for Economic Cooperation and Development (OECD) Convention on May 7, 2010, thereby pledging its full dedication to achieving the Organization's *fundamental aims*, and became the first South American country to join the OECD. As a result, Chile leads South America in supporting and contributing to the development of international cooperation programs in developing countries instead of receiving such support. This leads to more opportunities to invest in technological innovations and the development of applied science, and to improve legislation related to the sound management of chemicals.[*]

CANADA–CHILE ENVIRONMENTAL COOPERATION AGREEMENT[†]

These two countries signed the Canada–Chile Environmental Cooperation Agreement (CCECA) in 1997 in addition to the bilateral free trade agreement. Fourteen years later, the ongoing cooperation and collaboration between Canada and Chile shows that the two countries are strongly committed to the Agreement with increasing economic and environmental cooperation. The Council of the Canada–Chile Commission for Environmental Cooperation met for its tenth regular session on May 4, 2010 in Ottawa.

As per the agreement texts, except as otherwise provided in Annex 44.2, for purposes of Article 14(1) and Part Five:

a. "environmental law" means any statute or regulation of a party, or provision thereof, the primary purpose of which is the protection of the environment, or the prevention of a danger to human life or health, through
    i. the prevention, abatement or control of the release, discharge, or emission of pollutants or environmental contaminants,
    ii. the control of environmentally hazardous or toxic chemicals, substances, materials and wastes, and the dissemination of information related thereto, or
    iii. the protection of wild flora or fauna, including endangered species, their habitat, and specially protected natural areas.

---

[*] Chile's accession to OECD: http://www.oecd.org/about/0,3347,en_33873108_39418658_1_1_1_1_1,00.html. Accessed Oct. 25, 2010.

[†] http://ncrweb.ncr.ec.gc.ca/can-chil/default.asp?lang = En&n = 90612BD2-1; http://ncrweb.ncr.ec.gc.ca/can-chil/default.asp?lang = En&n = AF64227B-1

ANALYSIS OF CHILEAN COLLABORATION WITH MULTILATERAL CHEMICAL AGREEMENTS AND EXAMPLE OF BASEL CONVENTION RATIFICATION AND IMPLEMENTATION

Chile has developed numerous projects for international cooperation with respect to chemicals, particularly with regard to projects arising from signing international conventions such as Rotterdam and Stockholm, and ongoing work around the demands of the Basel and Montreal conventions. International organizations have provided ongoing support to Chile in terms of technical and financial help to Chilean government institutions such as CONAMA, Ministro de Agricultura (MINAGRI), and Ministerio de Economia Ministerio de Salud (MINECON MINSAL) among others, to implement programs, projects, and studies related to various aspects of chemical management. This support has been crucial and indispensable in order to comply with various chemical-related multilateral agreements signed by Chile.

Coordination regarding the implementation of international activities and agreements in the area of chemical management, developed through programs/agendas of committees or working groups, is mostly under the responsibility of CONAMA.

Examination of Chile's Adoption of the Basel Convention

One of the main problems in Chile has been the transboundary movement of chemical waste. Chile has been the depository of a large amount of hazardous waste. The "Basel Convention on the Control of Transboundary Movements of Hazardous Wastes and Their Disposal" (known as the "Basel Convention") is an international treaty that is part of, in Chilean terms, "the international chemical agenda." The objective of the Basel Convention is hazardous waste as defined in the Convention (Article 1) and annexes of the Convention and by every member state. It is based on obligations arising from the convention and the liability of every member state *vis-à-vis* the protection and preservation of the environment of such international obligations.*

As far as the Chilean legal system is concerned, the Basel Convention shall follow one specific procedure in order to be part of the national legislation. The Chilean constitution in Article 54 states that international treaties signed by the representative of the country (on a regular basis by the president of Chile) have to be approved by the senate of the parliament (one of the chambers of the parliament of Chile). Later, it has to be sent to the Constitutional Court of Chile and, following approval, the Parliament sends the treaty for publication. After publication and approval by the Chilean Constitutional Court, the treaty is considered to be approved and ratified by the president, following the procedure of the treaty itself. Chile ratified the Basel Convention on January 31, 1990 and implemented it on August 11, 1992.†

* Secretariat of the Basel Convention on the Control of Transboundary Movements of Hazardous Wastes and Their Disposal, *Text of the Basel Convention on the Control of Transboundary Movements of Hazardous Wastes and Their Disposal*, pp. 3–6.

† Secretariat of the Basel Convention on the Control of Transboundary Movements of Hazardous Wastes and Their Disposal, *Status of Ratification*, http://www.basel.int/ ratif/ratif.html (3.09.09).

The amendment to the Basel Convention (Decision III/1) has not been implemented in Chile.

In accordance with the explanations, it is possible to state that accomplishment of the obligations of the Basel Convention on the Control of Transboundary Movements of Hazardous Wastes and Their Disposal has taken a long time. It is related to the step of development of Chilean industry and Chilean economics. However, in the first step of evaluation, it is possible to find changes in the behavior of Chilean companies toward high standards of prevention against chemical contamination and transboundary movements of chemical waste. Their interest today is to work and export products as clean as possible because international markets are asking for clean and environmentally sustainable or greener products. Particularly in the agriculture and related agro-industry sectors (with products such as olive oil, wine, and fish), such an objective has been clearly stated. All wastes, even those that are not hazardous, require authorization from the National Sanitary Authority for sustainable management, including their transport.

CHILEAN RULES ON THE ENVIRONMENT

Since 1993, the Chilean Law of the Environment has been enacted under number 19.300. Article 2 includes a series of legal definitions, in which is included "chemical contamination." At the same time, the "primary rules of environmental quality" include the concept of chemical contamination, and the "industrial factories" include "chemical factories." Also the "primary rules of environmental quality" include a large number of "bylaws" on the contamination of various chemical products.*

In Chile, there are bylaws ruling the management of hazardous waste, and waste from chemical products and processes on a national scale.†

Chilean Rules on Chemical Waste

On February 18, 2005, the Chilean Sanitary Regulation on Management of Hazardous Wastes was integrated into its official communication to the Basel Convention Secretariat pursuing the accomplishment of international obligations arising from Articles 3 and 13 of the Basel Convention on the Control of Transboundary Movements of Hazardous Waste and Their Disposal.‡

* República de Chile, *Ministerio de Medio Ambiente*, http://www.sinia.cl/1292/propertyvalue-13572.html (5.09.09).

† República de Chile, *Decreto que regula a los establecimiento y fuentes emisora de anhídrido sulfuroso, material particulado o arsénico*, Diario Oficial 29 de septiembre de 1991; República de Chile, *Aprueba reglamento sanitario sobre manejo de residuos peligrosos*, Diario Oficial de 16 de junio de 2006; República de Chile, *Reglamento del Sistema de Evaluación de Impacto Ambiental*, Diario Oficial de 7 de diciembre de 2002.

‡ Secretariat of the Basel Convention on the Control of Transboundary Movements of Hazardous Wastes and Their Disposal, *Notification of national definition of hazardous wastes pursuant of article 3 of the Basel Convention Chile*, Geneva, Oct. 17, 2005.

The aforementioned legislation is a "bylaw that is considered as one of the most important Chilean rules on this specific topic under chemical management.[*] According to this bylaw, if chemical wastes affect public health and the environment and produce environmental toxicities (e.g., corrosion and other effects) in accordance with Article 11 of the "bylaw," that specific waste category will be declared as "risky" to the environment or to public health. In Article 12, the legal rule expresses that waste will have the qualification of strong "toxicity" when it is lethal to humans in low amounts or in an amount equal to or lower than 50 mg/kg body weight in an animal toxicology experiment (e.g., mouse toxicity studies). The article expresses other definitions of "strong toxicity." In order to define such toxicity, the "bylaw," in the same article, has expressed other forms of calculations for such toxicity. The ways in which chemical waste might be able to enter the body, in accordance with Article 12, is through the skin (dermal), in an oral (ingestion) form, or by inhalation. The maximum content of the contamination by waste differs, in accordance with the article, and depends on the exposure pathways of specific contaminants. A list of maximum concentrations in waste is included in Article 14. Article 15 focuses on the chemical reactions of components. Other articles explain other forms of contamination by different waste classifications. For example, Article 18 focuses on waste from hospitals (i.e., biomedical waste) but the "bylaw" includes the possibility of exemption based on noncontamination of the waste.[†] Articles 19 and 20 develop the possibility of exemptions based on noncontamination of the waste. Article 21 prevents reutilization of containers exposed to contaminated waste. Article 23 considers certain massive mineral residues not dangerous when their disposal has not been done along with residential solid waste. An important rule has been established in Article 24 of the "bylaw" in which packages that contain pesticide have been declared as contaminated. If the packages are treated, certain exemptions allow their use and, therefore, they are not considered contaminated waste. This article includes a procedure for cleaning and washing as well as treatment of these packages.[‡] The "bylaw" has also established a list of chemical products that are considered waste in various articles such as 88 and 90.

Chile is preparing a national rule on the final disposal and recovery of chemical waste with restrictions on transport of such waste. The Ministry of the Environment has included a training course on mercury and recently a course on the chemical industry. A national plan on POPS has been discussed and approved.[§] The National Policy on Chemical Security was approved in 2008 and implemented in 2009 (Jara, 2009).

---

[*] Secretariat of the Basel Convention on the Control of Transboundary Movements of Hazardous Wastes and Their Disposal, *Op. cit.*, pp. 2–3.

[†] Secretariat of the Basel Convention on the Control of Transboundary Movements of Hazardous Wastes and Their Disposal, *Op. cit.*, pp. 4–6.

[‡] Secretariat of the Basel Convention on the Control of Transboundary Movements of Hazardous Wastes and Their Disposal, *Op. cit.*, pp. 7–8.

[§] Comisión Nacional de Medio Ambiente, *Plan Nacional de Implementación para la Gestión de los Contaminantes Orgánicos Persistentes (COPs) en Chile*, Santiago de Chile, 27 de diciembre 2005, http://www.sinia.cl/1292/articles-37765_pdf_PNI.pdf (5.9.09).

RECOMMENDATIONS*

Because of the continuing migration of the rural population to urban centers and the lack of territorial planning, chemical industries that were initially installed in unpopulated areas of cities are now surrounded by residential areas, thus creating hazardous situations.

MANAGEMENT OF CHEMICALS

The information needed to make decisions and get an overview of the current situation on the handling of chemicals is still insufficient; additionally, the various sources of information differ in their periods of development and the purpose for which they were made.

The management of chemicals throughout their life cycle should be improved, for which work on the following should be completed: updating the inventory of some chemicals, developing cadastres of major chemical contaminants, and increasing information on pollution sources incorporated into existing information systems that deliver PRTR (Pollutant Release and Transfer of Pollutants).

A PRTR is an effort to group and strengthen a country's environmental information, which merges information from different sources and is in the phase of a trial run. It is expected that in 2011, when the results of emissions for 2009 are published, the PRTR would have reached a state where comparisons can be made to extract meaningful information for chemical emissions risk management.

Information is available on chemicals and waste, which has been used to establish priorities regarding management issues such as atmospheric, liquid industrial waste, and hazardous waste management sites that are potentially contaminated. Due to limited resources of agencies with expertise in the life-cycle management of chemicals, in financial terms and number of professionals, there is insufficient enforcement capacity of the existing rules.

There are chemical laboratories in universities and in private sectors and groups distributed throughout Chile. However, the analytical capacity in some parameters is deficient, such as in the determination of dioxins and furans.

REGIONAL LEGAL INSTRUMENTS FOR CHEMICAL MANAGEMENT

As discussed in this chapter, Chile has developed various legal instruments that directly affect the management of chemicals and their residues, which are fully in force as are regulations on hazardous waste, air emissions, and runoff. However, loopholes still exist in various matrices associated with certain environmental and chemical species that are deemed hazardous.

REGIONAL NONREGULATORY INSTRUMENTS

Some voluntary measures and nonregulatory mechanisms have been implemented with nationwide nonregulatory instruments by companies such as APL, ISO, and

* This section is excerpted from a translated version of the National Country Profile on Chemical Management for Chile (from the UNITAR Web site: http://www.unitar.org).

OSHA (among others), which have been effective in risk reduction and lower environmental impacts generated by industrial activities. Alongside this, there has been a growing awareness of the risks related to poor management of chemicals; hence, both the public and private sectors carry out activities periodically for training and dissemination to both community and workers on chemical safety issues.

NATIONAL COORDINATION

Public and private sectors in Chile can effectively work together toward implementing mechanisms, coordination, access to information sources, and capacity building around chemical management issues. However, it is evident that there is no unifying body to supervise the life-cycle management of chemicals, systematize information and statistics, and monitor and effectively manage chemical-related incidents and accidents. Avoiding the overlapping of functions and minimizing the possibility of voids in policy are essential.

CONCLUSIONS

Unfortunately, the possibility of ending the introduction of waste in Chilean environmental metrices (i.e., air, soil, water, food, and consumer products) is far from achieved. The shoreline of Chile is vast and the Chilean navy is doing its best to capture ships bringing illegal waste to the country, but it is very difficult to effectively manage various other waste categories. There are no specific population level health-assessment reports available at this time on the public-health consequences of chemical exposure. Environmental contamination could grow as a result of increased mining activities that can potentially impact environment and human health taking into account Chile's evolving economy without having capacity for implementation of effective clean up and waste management programs.

It is important to note that since the application of the Basel Convention in Chile, the transboundary movements of hazardous waste have diminished. The control and management of imported wastes have increased with the creation of a special division of the Chilean police that controls these waste movements, which is a positive outcome. Moreover, the risk awareness of citizens on chemical wastes and management has increased during the last decade of implementation of the convention. Today, Chilean citizens as well as the Chilean State have shown significant interest in the effective application of legal rules based on sanctions and control of chemicals as well as the creation of new environmental and chemical management instruments and institutions.

REFERENCES

Borregaard N., G. Volpi, H. Blanco, F. Wautiez, and A. Matte-Baker, 1999. UNEP report. *Environmental Impacts of Trade Liberalization and Policies for the Sustainable Management of Natural Resources—A Case Study on Chile's Mining Sector National Institution Leading the Study: Centro de Investigación y Planificación del Medio Ambiente (CIPMA) Santiago, Chile.* New York: United Nations. Available at http://www.unep.ch/etu/etp/acts/capbld/rdone/chile.pdf

Jara, C. 2009. *Política Nacional de Seguridad Química, Comisión Nacional de Medio Ambiente*, Sao Paulo, 2009. Available at: http://www.fundacentro.gov.br/dominios/CTN/anexos/Poltica%20Nacional%20de%20Seguridad%20Qumica%20espanhol%20.pdf

Johnston, N. 1996. International trade and environmental quality, in Swanson, T. (Ed.). *The Economics of Environmental Degradation: Tragedy for the Commons?* United Nations Environment Programme. Cheltenham, PA: Edward Elgar Publishing.

36 Chemical Management System in China
Past, Present, Future

DaeYoung Park

CONTENTS

INTRODUCTION: EVOLUTION OF CHEMICAL REGULATION IN CHINA

ENVIRONMENTAL PROTECTION AND CHEMICAL REGULATION FROM 1970S TO 2002

The Environmental Protection Law (Trial) of September 13, 1979 in China contained one short provision requiring the registration of toxic chemicals without specifying any registration procedure.* From the 1970s to the beginning of 2000, most chemical-related regulation in China covered the protection of workers involved in the operation and handling of chemicals.† In the 1980s and 1990s, however, there was a surge of media-specific environmental regulations, for example, the Marine Environmental Protection Law of August 23, 1982; the Water Pollution Prevention Law of May 11, 1984; the Air Pollution Prevention Law of September 5, 1987, the Environmental Protection Law of December 26, 1989, the Solid Waste Pollution Prevention Law of October 30, 1995, and the Environmental Noise Pollution Prevention Law of October 29, 1996.

Until China introduced a modern form of chemical regulation in 2003, that is, the Measures on the Environmental Management of New Chemical Substances of September 12, 2003,‡ the Regulation on the Safety Management of Dangerous Chemicals of February 17, 1987 and the Registration Measures for Environmental Management of First-Time Import and Export of Chemicals and Toxic Chemicals of January 1, 1990 were the two main pieces of legislation to protect health and safety of workers and the environment. See Figure 36.1 for a historical development of the chemical regulation in China from 1978 to 2010.

Chinese Chemical Regulation Since 2002

With the access to the World Trade Organization (WTO) membership in 2001 and the enactment of the Measures on the Environmental Management of New Chemical Substances on September 12, 2003, China adopted a series of implementing measures to meet internationally acceptable standards, for example, Technical Rules on Toxicity Testing of Chemicals of July 11, 2005, Guideline on Good Laboratory Practices (GLP) for Chemical Testing (HJ/T155–2004), Guideline for Chemical Testing (HJ/T153–2004), Guideline for the Hazard Evaluation of New Substances (HJ/T154–2004). For example, the Technical Rules of July 11, 2005 contains clear reference to the Organization for Economic Cooperation and Development (OECD) Guidelines for Testing of Chemicals (1981–2002) and the Health Effects Test Guidelines (1996–2000) issued by the United States Environmental Protection

---

* "Registration and control of toxic chemicals (You du hua xue pin) are required. Leakage of acute toxic chemicals (Ju du wu pin) shall be prevented during storage or transport," Environmental Protection Law (Trial) (promulgated by the Standing Comm. Nat'l People's Cong., September 13, 1979, effective September 13, 1979), art. 24, *translated by the author.*

† Amongst others, they included the following regulations: Rules on the Fire Prevention from Combustible Chemicals (promulgated by the Ministry of Public Security, April 1, 1961, effective April 1, 1961), Regulation on the Safety Management of Hazardous Chemicals (promulgated by the State Council, Feb. 17, 1987, effective Feb. 17, 1987).

‡ The date, that is, *12 September 2003*, in which China enacted the Measures on the Environmental Management of New Chemical Substances, is noteworthy in order to understand the current status of Chinese chemical control system. Compared to China, the Directive 67/548/EEC of *27 June 1967* on the approximation of laws, regulations and administrative provisions relating to the classification, packaging and labeling of dangerous substances, the Law on the Control of Examination and Manufacture of Chemical Substances (Law No.117 of *16 October 1973*) and the Toxic Substances Control Act (15 U.S.C. secs. 2601–2671, 1976) were introduced in the European Union, Japan, and the United States, respectively. Around 30-year experience gap on chemical management exists between China and other advanced countries.

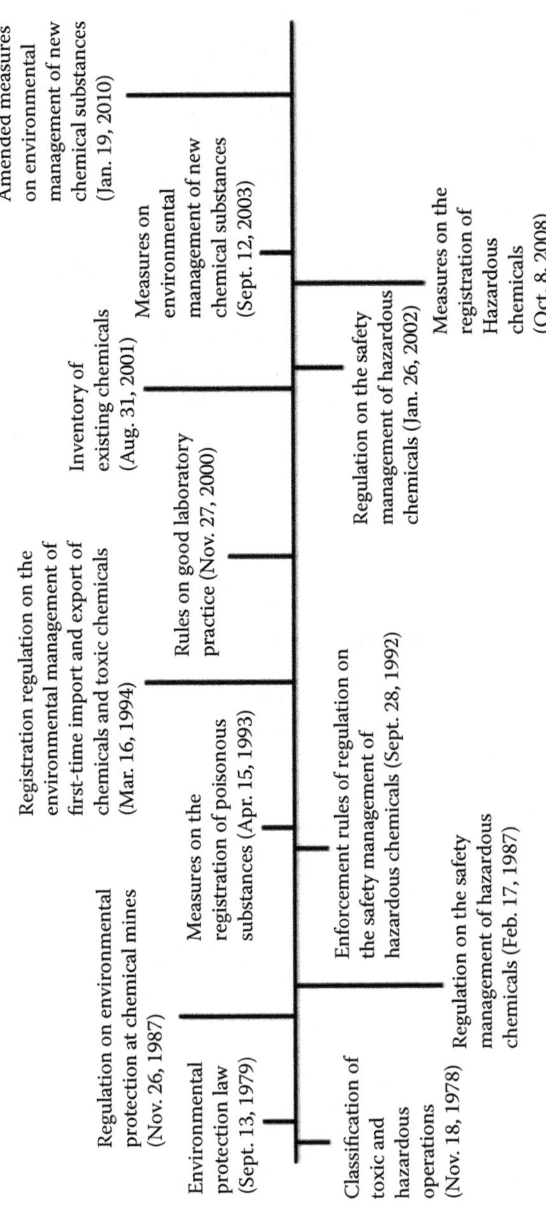

FIGURE 36.1 Brief historical overview of Chinese Chemical Regulation.

Agency's Office of Pollution Prevention and Toxics (OPPTS), which are also based on OECD guidelines.

Despite the effort to establish laboratories in China that can qualify for recognition under GLP by the Ministry of Science and Technology, the Ministry of Health and the State, Food and Drug Administration, still there are no laboratories which can generate OECD-acceptable chemical test data. So far, no specific experience exists on comprehensive risk assessment on existing chemicals, for example, High Production Volume (HPV) Program. It should be noted, however, that China has adopted more than 120 National Standards on Chemical Testing and GLP (including 65 Standards on Physico-chemical Tests, 58 Standards on Toxicity and Eco-toxicity) in 2008 and 2009. This indicates that China is aggressively improving scientific and technical infrastructure for stricter and sound management of chemicals, which may lead to further regulatory measures soon. Refer to Figure 36.2 for a general functional framework of the Chinese Chemical Regulation based on a multi-jurisdictional responsibility by multiple levels of governmental agencies in the area of occupational, public health and safety, and chemical management.

On the health and safety aspects of chemical management in China, the Occupational Health Law of October 27, 2001, the Regulation on the Safety Management of Hazardous Chemicals of January 26, 2002,* the Measures on the Registration of Hazardous Chemicals of October 8, 2002,† and the Regulation on the

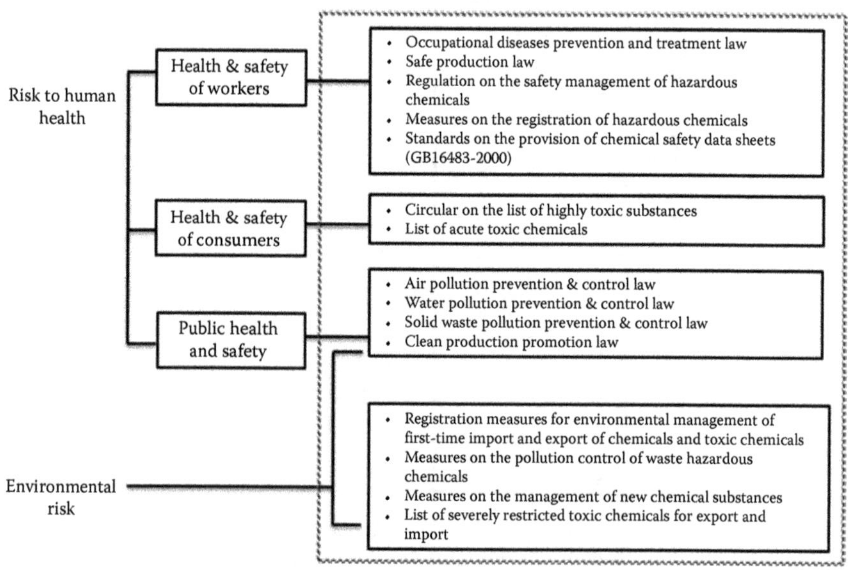

FIGURE 36.2 Structure of Chinese Chemical Regulation.

* Regulation on the Safety Management of Hazardous Chemicals (promulgated by the State Council, Jan. 26, 2002, effective Mar. 15, 2002).

† Measures on the Registration of Hazardous Chemicals (promulgated by State Economic and Trade Commission, Oct. 8, 2002, effective Nov. 15, 2002).

Labor Protection in Workplaces Handling Toxic Materials (RLPWHTM) provide a framework on the control of chemicals at the workplace.

The Regulation on the Management of Pesticides of May 8, 1997 provides requirements on production license, pesticide registration, and product quality standards. Between December 2007 and January 2008, the Ministry of Agriculture released six separate regulations on pesticides, including legislation to phase out highly toxic pesticides like the organophosphate methamidophos and a new list of laboratories was approved to carry out environmental toxicology tests.

CHEMICAL REGULATION OF CHINA IN 2010

On January 19, 2010, the Ministry of Environmental Protection released the amended Measures on the Environmental Management of New Chemical Substances (otherwise known as Decree or Order 7), which came into force on October 15, 2010.

The key points are:

- *Chemicals in products*: Article 2 of the 2010 Amendments provides a statement, "This Regulation is applicable to products which release new chemical substances in their normal use."
- *Globally harmonized system on classification and labeling of chemicals (GHS)*: The 2010 Amendments is one of the few Chinese regulations clearly referring to the GHS Standards which were issued on October 24, 2006. This is a positive move as it provides a clear link between law and standards.
- *Classification of chemicals*: There are three classifications under the 2010 Amendments, that is, general new chemicals, hazardous new chemicals, and priority hazardous chemicals.
- *Pollutant release and transfer register (PRTR) or toxic release inventory (TRI)*: Annual reporting scheme in the 2010 Amendments may provide a framework to implement PRTR requirements in China in the medium and long term. In particular, it has extended reporting requirements for producers or importers of hazardous new chemicals (including priority hazardous new chemicals).* With the entry into force of the 2010 Amendments in October 2010, such annual reporting is required for new chemical substances.
- *Phase-out of hazardous chemicals*: There are many promotional provisions for the development and use of environmentally friendly chemicals in the 2010 Amendments. The Ministry of Environmental Protection may also foresee natural phase-out of hazardous chemicals in China.

The amended Measures on the Environmental Management of New Chemical Substances are only applicable to new chemicals which are not included in the Chinese

* Although a chemical reporting and registration system exists under the Measures on the Registration of Hazardous Chemicals, it is not a version of the pollutant release and transfer register in the terms of the Organization for Economic Cooperation and Development (OECD). Currently, there is no capacity to track and provide presence of chemicals in the environment online like the European Pollutant Emissions Register.

Inventory of Existing Chemicals. However, companies industries should not misinterpret that the Measures on the Environmental Management of New Chemical Substances are not the only chemical regulation controlling new and existing chemicals in China. As mentioned in the introduction section of this chapter, the chemical control and management system in China has a complex jurisdictional structure controlled by multiple governmental ministries. This overwhelming jurisdictional complexity of chemical control may be prohibitive for integrated chemical control at the workplace and make it difficult to introduce a holistic chemical regulation like EU REACH (Regulation (EC) No. 1907/2006 of 18 December 2006 concerning the Registration, Evaluation, Authorisation and Restriction of Chemicals (REACH)).

REGULATORY FRAMEWORK AND INSTRUMENTS ON CHEMICALS IN CHINA

Classification and Control of New and Existing Chemicals

Risk Assessment of New Chemicals

Pursuant to the Measures on the Environmental Management of New Chemical Substances,* when a facility produces or imports a new chemical substance, it shall register the chemical substance with the Chemical Registration Center of the Ministry of Environmental Protection and obtain a "certificate of registration of new chemical substance" prior to the production or import of the substance.†‡

The following standards provide details for the risk assessment of new chemical substances in accordance with the Measures on the Environmental Management of New Chemical Substances:

- Standards on the Guidance for the Generic Name of New Chemical Substances (HJ/T 420–2008).
- Guidance for GLP for Chemical Testing (HJ/T 155–2004).
- Guidance for Hazard Evaluation of New Chemical Substances (HJ/T 154–2004).
- Guidance for Testing of Chemical Substances (HJ/T 153–2004).

* The original text of the Measures on the Environmental Management of New Chemical Substances are accessible online at: http://www.mep.gov.cn/gkml/hbb/bl/201002/t20100201_185231.htm

† A "new chemical substance" means a chemical that is not produced or imported at the time of the registration with the Ministry of Environmental Protection. The Ministry of Environmental Protection shall maintain and publicize the Inventory of Existing Chemicals in China (MEMNCS-Art.4). The Inventory of Existing Chemicals (2008 Version) covers all the chemical substances which are produced, processed, distributed, used, or imported for commercial uses between January 1, 1992 and October 15, 2003. In all, 45,290 chemical substances are covered in the Inventory (2008 Version). A dual-language (Chinese and English) version for the Inventory of Existing Chemical Substance database (2008 Version) is accessible online at: http://www.crc-mep.org.cn/iecscweb/default0.aspx

‡ The application has to include the information on a new chemical substance such as the name, testing methods, usage, annual quantities of production or import, physical and chemical characteristics, toxicological data, ecological data, accident prevention and emergency response measures, pollution prevention and remediation measures, waste disposal methods, and likely GHS compliant product labels.

In addition to the Measures on the Environmental Management of New Chemical Substances which regulate new chemicals from an environmental aspect, the Measures for Hazardous Chemicals Registration Management of October 8, 2002 regulate new chemicals from health and safety aspects. Pursuant to the Measures of October 8, 2002, when a facility produces a new chemical substance, it has to register it with the National Registration Center for Chemicals at least one year before the start of production. The testing of a new chemical substance has to be carried out by a chemical testing laboratory authorized by the State Administration of Work Safety.

Risk Assessment of Existing Chemicals

When a chemical substance is entered into a list of hazardous substances,* a facility has to register the substance with the National Registration Center for Chemicals under the State Administration of Work Safety within six months.

At present, there is no operational HPV Challenge Program to make companies disclose health and environmental effects data to the public on chemicals produced or imported in China. Since August 2009, the Chemical Registration Center and the Shanghai Environmental Protection Bureau has operated a trial project on TRI or PRTR. The Ministry of Environmental Protection initiated other pilot projects on notification of pollutant emissions in 2010. These activities would result in further scrutiny or regulatory actions to control existing chemicals in China.

Classification of Chemicals

The amended Measures on the Environmental Management of New Chemical Substances, the Measures for Hazardous Chemicals Registration Management of October 8, 2002 and the RLPWHTM of May 12, 2002 have the following classification of chemicals, and differentiated regulatory requirements are applied to each chemical classification. See Table 36.1 for a detailed account of different chemical classifications and their definitions, respective laws, and associated governmental ministries responsible for each chemical categories.

Chemical Packaging and Labeling

For the implementation of the Globally Harmonized System of Classification and Labeling of Chemicals (GHS), 26 Standards on Classification, Precautionary Labeling and Precautionary Statements of Chemicals, which have been implemented in China from January 1, 2008, and the Standard on Labeling of Chemicals based on GHS (GB/T 22234-2008) provide guidance on pictograms, warning

* In the Measures for Hazardous Chemicals Registration Management of October 8, 2002, "hazardous substances" means chemicals listed in the List of Dangerous Goods (GB12268-2005), the List of Hazardous Substances (2002) or the List of Poisonous and Deleterious Substances (2002). The List of Dangerous Goods, the List of Hazardous Substances and the List of Poisonous and Deleterious Substances are updated and maintained by the State Administration of Quality Supervision, Inspection and Quarantine and the State Administration of Work Safety, the Ministry of Environmental Protection, the Ministry of Health, the Ministry of Railways, the Ministry of Communication and the Civil Aviation Administration of China.

TABLE 36.1
Classification of Chemical Substances

Law	Classification	Definitions	Competent Ministries
Measures on the Environmental Management of New Chemical Substances	New Chemical Substances	Chemical substances that are not listed in the "Chinese Inventory of Existing Chemical Substances"	Ministry of Environmental Protection
	General New Chemicals	New chemical substances that do not have any hazard properties or, if any, lower than the threshold specified in relevant national standard for the hazard classification of chemical substances	
	Hazardous New Chemicals	New chemical substances having physico-chemical hazard, health hazard, or environmental hazard, and reach or exceed the threshold specified in relevant national standard for the hazard classification of chemical substances	
	Priority Hazardous Chemicals	Among hazardous new chemical substances, if the substances are persistent, bio-accumulative, hazardous to the environment and the human health, they are classified as priority hazardous new chemical substances	
Measures for Hazardous Chemicals Registration Management	Dangerous Goods	Chemicals included in the List of Dangerous Goods (GB-12268-2005)	State Administration of Work Safety
	Hazardous Substances	Chemicals included in the List of Hazardous Substances (2002)	State Administration of Quality Supervision, Inspection and Quarantine, State Administration of Standardization

	Poisonous and Deleterious Substances	Chemicals included in the List of Poisonous and Deleterious Substances (2002)	State Administration of Quality Supervision, Inspection and Quarantine, State Administration of Work Safety, Ministry of Environmental Protection, Ministry of Health, Ministry of Railways, Ministry of Communication, Civil Aviation Administration of China
Regulation on the Labor Protection in Workplaces Handling Toxic Materials	General Toxic Substances	Chemicals included in the List of General Toxic Substances (2002)	Ministry of Health
	Highly Toxic Substances	Chemicals included in the List of Highly Toxic Substances (2003)	

words, hazard statement, and precautionary measures. However, enforcement of these GHS standards is questionable, as the issuing authority (i.e., Standardization Administration of China) is not a directly competent authority having enforcement power and the standards are not clearly referred to in existing chemical regulations.

Among others, the Regulation on the Safety Management of Hazardous Chemicals of January 26, 2002 provides that if the facility manufactures hazardous chemicals, it has to affix appropriate chemical safety labels on the packaging of the hazardous chemicals. The RLPWHTM of May 12, 2002 requires toxic substances to be affixed with appropriate marks so as to indicate the information, such as product property, ingredients, and existing factors of occupational poisoning hazards, possible dangerous consequences, precautions for safe use, and emergency and first-aid measures.

Restrictions/Prohibitions on Use and Marketing of Chemicals

Pursuant to the Measures for Hazardous Chemicals Sales Permit Management of October 8, 2002, if the facility sells hazardous chemicals or has separate sales shops, it shall obtain a sales permit and an appropriate business license.* Under the Enforcement Measures on Safe Production Permit for Production of Hazardous Chemicals of May 17, 2004, when producing hazardous chemicals included in the List of Hazardous Substances (2002 version), the facility has to obtain a safe production permit from the local Safety Inspection Bureau.

Chemical Import and Export Restrictions

The amended Regulation on Environmental Management of the First-Time Import of Chemicals and Import and Export of Toxic Chemicals of October 8, 2007 and the Registration Measures for Environmental Management of the First-Time Import/Export of Chemicals and Import/Export of Toxic Chemicals of January 1, 1990 provide that, if the facility imports or exports banned or strictly restricted toxic chemicals, it shall obtain a registration certificate for environmental management of the import/export of toxic chemicals and a clearance notification for environmental management of the import/export of toxic chemicals.†

* Operating permits are divided into Type A and Type B permits. The facility having a Type A operating permit can handle toxic chemicals and other hazardous chemicals. The facility having a Type B operating permit can only handle hazardous chemicals other than toxic chemicals.

† "Chemical" means a chemical substance whether by itself or in a mixture or preparation, whether manufactured or obtained from nature and includes such substances used as industrial chemicals and pesticides. "Banned chemical" means a chemical which has, for health or environmental reasons, been prohibited for all uses. "Strictly restricted chemical" means a chemical for which, for health or environmental reasons, virtually all uses have been prohibited, but for which certain specific uses remain authorized. "Toxic chemical" means a chemical damaging to the health and environment through environmental accumulation, bioaccumulation and biotransformation or chemical reaction after entering the environment, or posing serious harm to the human health and posing potential hazard through contact. "Prior informed consent" means the principle that international shipment of a chemical that is banned or Strictly Restricted in order to protect human health or the environment shall not proceed without the agreement of the designated national authority in the importing country.

Prior Informed Consent Procedure for Hazardous Chemicals

The Rotterdam Convention on the Prior Informed Consent (PIC) Procedure for Certain Hazardous Chemicals and Pesticides in International Trade (PIC Convention) was ratified by the National People's Congress in December 2004 and China started implementing the requirements for the promotion of information exchange on hazardous chemical substances, the procedure and notification of imports and exports of the substances for the protection of the health of humans and the environment, and the promotion of joint responsibility and efforts among contracting parties in the trade of hazardous chemical substances.

Persistent Organic Pollutants

The Stockholm Convention on Persistent Organic Pollutants (POPS) Convention was ratified by the Standing Committee of the National People's Congress in June 2004. Amongst others, countries that signed and ratified the POPS Convention have to make efforts to reduce releases of POPS (e.g., dioxins and furans that are the by-products of combustion and industrial processes and other POPS as identified by the POPS convention) with the goal of their continuing minimization and, where feasible, ultimate elimination. However, China has made reservations regarding application of any amendment to Annex A, B, or C of the POPS Convention.

In the China's National Implementation Plan (NIP) for the Stockholm Convention on POPS of April 14, 2007, the State Council has also specified the implementation plan of the Chinese Central Government for the Stockholm Convention on POPS.

Restriction on Export and Import of Chemicals and Wastes in China

The amendment to the Basel Convention (Decision III/1) was implemented in China after it was approved by the 9th standing committee, the National Congress of the People's Republic of China on October 31, 1999. In line with the Basel Convention, the Solid Waste Pollution Prevention and Control Law only allows the import of waste which is either stipulated in the List of Waste that can be used as Raw Materials under Restricted Import Licensing Control or in the List of Waste that can be used as Raw Materials under Automatic Import Licensing Control.[*] At the same time, the Foreign Trade Law does not allow the import of waste which is listed in the List of Solid Waste which is prohibited from being imported.

The Measures on the Approval for the Export of Hazardous Waste[†] prohibits the export of hazardous waste to a country that is not a party to the Basel Convention.[‡] In case a facility exports hazardous waste to a country that is a party to the Basel Convention, it has to obtain an Approval for the Export of Hazardous Waste Export from the Ministry of Environmental Protection prior to the export.

---

[*] Imported solid waste has to comply with applicable Environmental Protection and Control Standards for the Import of Waste which can be Used as Raw Materials) and be inspected by the General Administration of Quality Supervision, Inspection and Quarantine.

[†] Measures on the Approval for the Export of Hazardous Waste (promulgated by the State Environmental Protection Administration, Jan. 25, 2008, effective Mar. 1, 2008).

[‡] "Hazardous waste" means waste included in the List of Hazardous Waste or waste meeting the Identification Standards of Hazardous Waste.

Regarding the transit of hazardous waste, the Radioactive Pollution Prevention and Control Law and the Solid Waste Pollution Prevention and Control Law prohibits the transit of hazardous waste, radioactive waste, or any materials contaminated by radiation via China. When a facility imports radioactive waste and contaminated materials by radiation or transfers them via China, it may be subject to an order from the General Administration of Customs to return radioactive waste or contaminated materials and a fine of between 500,000 and 1,000,000 Yuan. In cases where criminal charges can be applied, a facility may be investigated for criminal liabilities.

Material Safety Data Sheets

A material safety data sheet is an important component of product stewardship and workplace safety, as it provides workers with procedures for handling or working with chemicals substances in a safe manner. The Regulation on Safety Management of Hazardous Chemicals of January 26, 2002, the Measures on the Registration of Hazardous Chemicals of October 8, 2002 and the Regulation on Safe Use of Chemicals at Workplaces of December 20, 1996 require producers or importers of hazardous chemicals to provide chemical safety data sheets (CSDS) to its customers.* In case of industrial users of hazardous chemicals, they have to provide appropriate training on CSDS to the workers handling those chemicals.†

KEY CHALLENGING ISSUES IN CHINESE CHEMICAL CONTROL SYSTEM

Complex Institutional Governance on Chemicals

One of the major bottleneck issues in the chemical management in China is a complex control regime on use and marketing of chemicals. The Ministry of Health, the Ministry of Communication, the Ministry of Public Security, the State Administration of Quality Supervision, Inspection and Quarantine, the General Administration of Customs, the Ministry of Commerce, the State Food and Drug Administration, the State Administration of Work Safety, the State Administration of Standardization, and the National Development and Reform Commission all have some or overlapping control on chemicals. A good governance structure for chemical management may be a necessary prerequisite for sound chemical management in China in order to avoid potential jurisdictional conflict between agency rulemakings on chemical issues.

---

* Producers or importers are responsible for the accuracy for the information in CSDS. It is prohibited to sell hazardous chemicals without providing CSDS.

† The Standards on Safety Data Sheets for Chemical Products—Content and Order of Sections (GB/T 16483-2008) provide detailed guidance for the drafting and management of a CSDS 16 headings covered under the Standard GB/T 16483-2008 are provided as follows: chemical product and company identification, hazard summary (GHS: hazard(s) identification), composition/information on ingredients, first-aid measures, fire fighting measures, emergency response measures (GHS: accidental release measures), handling and storage, exposure control/personal protection, physical and chemical properties, stability and reactivity, toxicological information, ecological information, disposal (GHS: disposal considerations), transport information, regulatory information, other information.

Lack of Appropriate Technical Infrastructure

A sound chemical management system requires a well-structured and functioning chemical regulatory regime, scientific infrastructure (e.g., chemical testing laboratories), technical enforcement capacity, scientific data on chemicals, a technical expert pool, and a chemical tracking mechanism locally and nationally.

By building more sound capacities for chemical control through the upgradation of existing chemical regulations, China may introduce more advanced control systems on chemicals in a mid- and long-term legislative planning.

Enactment, Implementation, and Enforcement: Continued Action

Pursuant to the 11th 5-Year Plan on National Environmental Standards of February 6, 2006, the Ministry of Environmental Protection reviewed 1332 environmental standards between 2006 and 2010. This may be one example that demonstrates emphasis on legislative rulemaking of the Chinese government rather than implementation of existing laws and regulations. The European Union Network for the Implementation and Enforcement of Environmental Law would be a good benchmarking point for the Chinese government to develop initiatives to assess the implementation status of existing environmental regulations.

Transparency

Without involving broader stakeholders in the process of rulemaking, it would be unreasonably difficult to assess technical capabilities of industries or the necessity for a transitional period for compliance preparation. Invitation-only consultation on regulatory proposals, which is the current practice in China, may add more ambiguity, create confusion, bring significant difficulties in corporate risk management and cause significant unnecessary costs on businesses.

2011 and Beyond

Although there were some difficult challenges, what China accomplished between 2009 and 2010 (e.g., development of 120 National Standards on Chemical Testing and GLP and the adoption of the amended Measures on the Environmental Management of New Chemical Substances) should be considered as a significant progress for sound management of chemicals. These key activities and progress also provides a framework for the introduction of more advanced chemical control measures in further environmental regulatory changes. Several new environmental standards were already put into effect between January 2011 and April 2011. In 2011 and beyond, it is anticipated that the Chinese government may issue subsidiary legislation containing details on chemical exposure, risk assessment, classification of chemicals, and collection of hazard data after exposure analysis and confidentiality issues. Broader involvement of stakeholders in environmental rulemaking may have to be ensured while precisely assessing technical and enforcement capacities with a focus on the implementation of existing laws.

37 Chemicals in Egypt
A Generic Perspective

Mohamed Tawfic Ahmed and Naglaa M. Loutfi*

CONTENTS

* The opinions expressed in this chapter are those of the author and do not necessarily represent the views of any organization.

EGYPT, THE EARLY GLIMPSES OF HISTORY AND CHEMISTRY

Egyptian civilization is probably best known for the great pyramids of Giza, masterpieces of construction artistry and architectural distinction. Egypt is also known for the Sphinx, the mystical symbol of eternity, facing the sunrise every morning for thousands of years, narrating a seldom-occurring episode of serenity and immortality. But little is known about the capability behind those striking colors of the numerous engraved icons and plates depicted on the walls of old temples, and their remarkable persistence against the daunting aging and withering conditions of both time and environment. It is the secret of old Egypt, and their superior artistry of chemistry.

EGYPT: A GENERIC COUNTRY PROFILE

Egypt is located in the northeast tip of Africa at the junction between Africa, Asia, and Europe, covering an area of about 1 million km^2. Egypt's distinguished role in geopolitics is mostly based on its strategic geographical position, with the Suez Canal connecting the Red Sea to the Mediterranean. Egypt has always been an influential actor in the region, because of its strong ties with the Arab world and sheer size of its population, being the most populous country in the area, and the 16th worldwide with about 82 million people.

The closing decades of the ninteenth century have marked the early signs of industry. Machinery and technology were imported from Europe, and Egyptians were sent to Europe for training in various fields. Early industries were based on the agricultural products of Egypt. The Egyptian Salt and Soda Company was established in 1899, as the first industrial facility to produce edible oil, soap, and detergent.

EGYPT, INDUSTRY AND CHEMICAL USE

By World War I, the textile industry had become well established in Egypt. Cotton research and development have yielded a number of internationally renowned long staple varieties that allowed the prestigious ranking of Egyptian cotton, paving the way for a flourishing textile industry. Because of the high insect pest pressure, reliance on chemical control has been well established since the early 1950s. Wide use of synthetic organic pesticides in cotton has been extensively practised for decades (El-Sebae et al., 1993). Egypt has been one of the top 10 countries with highest use of both HCH and DDT, between 1948 and 2000 (Indian and Northern Affairs Canada, INAC, 2003).

The extensive application of organochlorine pesticides has had a heavy impact on environmental health and the quality of life in Egypt. El-Gamal (1983), reported cases of poisoning and number of fatalities each year from 1966 to 1982. His report indicated that after 1977, accidents of acute intoxication were decreased due to the introduction of some preventive measures. But, he added that over 60% of the workers engaged in pesticide applications suffered from chronic toxicity.

El-Sebae (1977) reported the incidents of delayed neurotoxicity in farm animals due to the large-scale application of leptophos (phosvel) on cotton fields during the period 1971–1974. In 1971, a mysterious epidemic of paralysis struck several hundreds of water buffaloes at some villages in the middle Delta, and eventually resulted in the death of 1300 animals. Evidence strongly pointed to the organophosphate leptophos as responsible for this delayed neurotoxic syndrome.

In 1996, a Ministry of Agriculture Decree banned the import and use of about 80 pesticides that included a large number of organochlorine pesticides. The list included DDT, heptachlor, chlordane, aldrin, dieldrin, chlordane, toxaphene, mirex, endosulfan, and some others (Table 37.1).

A number of elaborate reports have portrayed the impacts pesticides have caused on various environmental segments, food, and human health (Naglaa Loutfi et al., 2006; Mansour, 2004; Tawfic Ahmed et al., 2001, 2002; Sebae et al., 1993).

The development of resistance to toxaphene, in the early 1960s, and the catastrophic economic repercussion that followed were a turning point in cotton pest control strategies in Egypt. The use of organophosphates and carbamates insecticides, has gradually replaced the use of organochlorine, and eventually synthetic pyrethroids and chitin inhibitor compounds have been also used since the early 1980s, beside organophosphates and carbamates compounds. The total amount of pesticides used in Egypt within the last 15 years are portrayed in Figure 37.1. The sharp increase in pesticides used recorded in 2007 might be explained in view of the

TABLE 37.1
Total Active Ingredient Insecticides Used on Cotton in Egyptian Agriculture during the Period 1952–1990

Compound	Total Metric Tons (MT)	Years of Consumption
Toxaphene	54,000	1955–1961
Endrin	10,500	1961–1981
DDT	13,500	1952–1971
Lindane	11,300	1952–1978
Carbaryl	21,000	1961–1978
Trichlorfon	6500	1961–1970
Monocrotophos	8300	1967–1978
Leptophos	5500	1968–1978
Chlorpyrifos	13,500	1969–1990
Phosfolan	5500	1963–1983
Mephosfolan	7000	1968–1983
Methamidophos/azinphos-methyl	7500	1970–1990
Triazophos	8500	1977–1990
Profenofos	8000	1977–1990
Methomyl	9500	1975–1990
Fenvalerate	8500	1976–1990
Cypermethrin	6300	1976–1990
Deltamethrin	5400	1976–1990
Cyanophos	3000	1984–1990
Thiodicarb	5000	1984–1990

Source: Adapted from El-Sebae, A. H., M. Abou-Zeid, and M. A. Saleh. 1993. *Chemosphere*, 27(10): 2063–2072.

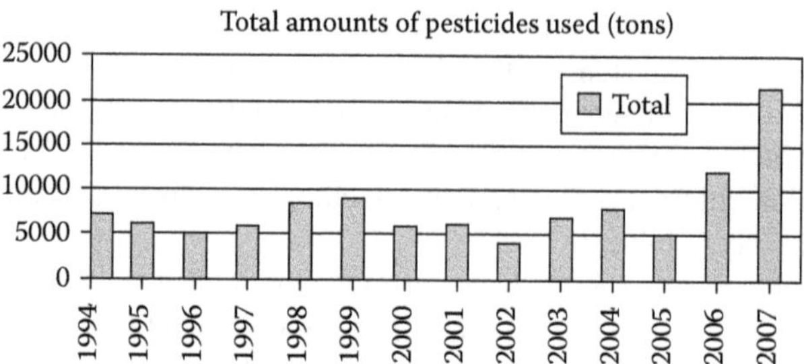

FIGURE 37.1 Total amount of pesticides used in Egypt, 1994–2007. (Adapted from Egypt State of the Environment Report, 2007.)

increased acreage of newly reclaimed land fed by the new fresh water canal extending below the Suez Canal to feed areas of Sinai (Shiekh Gaber Canal).

EGYPT AND THE INTERNATIONAL CODE OF CONDUCT ON THE DISTRIBUTION AND USE OF PESTICIDES

The International Code of Conduct on the Distribution and Use of Pesticides (FAO, URL), is a voluntary instrument that provides guidance on the management of pesticides. The Code, including technical guidelines that have been prepared to assist countries with implementation of specific measures, provides an overarching framework that helps ensure pesticides management in ways that minimize risks to public health and the environment. The code is supported by Egypt, along with Food and Agricultural Organization (FAO) member states, intergovernmental organizations, pesticide/crop protection industry, and nongovernmental organizations (NGOs)/civil society.

Within the framework of the Code, Egypt is laying special emphases on pesticide management as part of chemical management, as well as of sustainable agricultural development. In Egypt, new pesticides should be tested for three successive years, on both laboratory and field scale. A variety of tests are conducted to examine both the pesticidal efficacy, along with environmental toxicity and other impacts the new pesticide may cause. Testing is usually conducted in authoritative public research stations, including universities located in various ecological zones of the country. Trade and distribution of pesticides are also regulated according to special measures to ensure safety and reliability of handling pesticides at all levels of manufacture, transport, distribution, storage, and use. However, the use of life cycle analysis (LCA), one of the constituent of the Code, is not performed in Egypt.

EGYPTIAN INDUSTRY, SECOND PHASE

The mid-1950s marked a new stage of Egyptian industry that began with more diversified patterns of industrial works, with Egypt going through relentless efforts to widen its industrial profile. Beside Egypt's well-established industries such as textiles, and food processing, some new industries have emerged such as pharmaceutical, pesticides, fertilizers, paints and dyes, cement, paper and pulp, petrochemicals, car manufacture industries, and many others. A huge influx of chemicals and chemical related products has been reported, with considerable trade and finance activities. Demand for chemical imports within the last four years is shown in Figure 37.2. Increased demand for imported chemicals was recorded throughout, reflecting the growing chemical industry in Egypt.

Meanwhile, an integrated database was developed concurrently to determine types and amounts of hazardous chemical substances imported to Egypt via customs releases, in order to identify the current situation of chemicals imported into Egypt for coordination with concerned agencies on efficient use. In 2007, imported hazardous substances, according to a customs release, have accounted for a total of 8,947,221 tons.

CHEMICALS AND ENVIRONMENTAL SETTING

One of the major repercussions of the Rio Conference 1992 on the environment was raising environmental awareness and the emergence of new legislative bodies in many parts of the world that deal primarily with environment issues.

In Egypt, a state ministry for environmental affairs was first established in 1994, and a new law for the protection of the Egyptian environment (the environmental protection law, Law 4, 1994), was introduced as the first comprehensive law to focus on environmental issues. The new law endorsed a number of mandatory processes; welfare and the use, importation and hazardous waste of chemicals were some of the focal issues. Environmental impact assessment (EIA) was introduced in Egypt in 1994 as a legal requirement for licensing new projects and expansions in existing ones. EIA reports are required in the Central Department of the Egyptian Environmental Affairs Agency (EEAA) (EEAA, 1994). The review process may result in requiring additional information, requesting change of proposed technology, or in setting certain conditions for approval. The introduction of EIA, and the strict requirements it imposed have forced newly established chemical facilities to have their emissions of pollutants within permissible levels indicated in Law 4, 1994. Meanwhile, already-existing chemical facilities have had to introduce necessary measures in order to meet the new standards. Facilities failing to meet the new regulations were fined and closed.

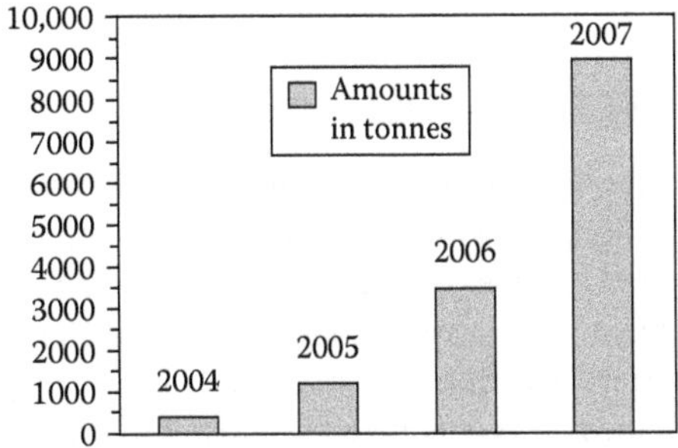

FIGURE 37.2 Amount of chemical substances identified from customs releases, 2004–2007. (Adapted from Egypt State of the Environment Report, 2007.)

EGYPT NATIONAL PROFILE FOR THE MANAGEMENT OF CHEMICALS

In the wake of the new environmental laws introduced in Egypt, a generic framework for the management of chemicals was put forward as the first, most comprehensive body to embrace guidelines that control the use of chemicals, along with related activities in the country (El Zarka, 1999, EEAA URLa). The development of the profile followed the recommendations of the Intergovernmental Forum on Chemical Safety (IFCS) which was established as a follow-up to the Rio Conference. At this conference, Heads of States or Governments adopted *Agenda 21*, a document outlining responsibilities toward the achievement of sustainable development. One main part of *Agenda 21* deals with environmentally sound management of chemicals as well as illegal international traffic in toxic and dangerous products.

CHEMICALS IN EGYPT, PROBLEMS

Despite the long history of using, importing, and exporting chemicals in Egypt, there seems a significant gap in awareness, and technical, and logistic support in the process of chemical management. The national chemical profile of Egypt has identified major hurdles that impede a sound chemical management in Egypt as the following issues:

- Inadequate capabilities to assess the potential toxicity and to control the nature and purity of imported or domestically produced chemicals.
- Handling of chemicals by inadequately informed or trained personnel, especially operators in small-scale enterprises.
- Shortage of management skills needed to deal safely with technology transfer and with the storage, transport, use or disposal of chemicals.
- Lack of effective mechanisms for coordinating the work of those responsible for different aspects of chemicals safety.
- Lack of means of coping with chemicals accidents, including the treatment of victims and the subsequent rehabilitation of the environment.
- Inadequate proper management of chemicals and enforcement of regulations.
- Lack of reliable information sources to establish properly coordinated infrastructures, controls, and procedures to deal properly with chemical safety.

CHEMICALS IN EGYPT: MEASURES

In recognition of the importance of establishing an information and management system for the identification, registration, categorization, and management of chemicals, EEAA was assigned to develop a comprehensive database for Hazardous substances including chemicals. Data were collected from different sources, including producers, users, importers, and distributors of chemicals. Categorization and specifications of these substances was conducted following international codes.

EEAA, in consultation with the competent ministries, has also been assigned to develop pollution prevention programs based on economic incentives and clean technologies that are compatible with Egyptian development after adaptation to the specific economic, cultural, social, and institutional context in Egypt in order to encourage the steady reduction of hazards resulting from mishandling chemicals. The highlights of the system include, among others, the following activities:

- Establishment of an adequate infrastructure monitoring and controlling systems for importation, manufacturing, transportation, storage, usage of chemicals.
- Establishment of information and database for handling hazardous wastes.
- Exchange of experience in the field of research and development, with the relevant national and international organization.

ENVIRONMENTAL MANAGEMENT SYSTEMS IN CHEMICAL INDUSTRY

Environmental management systems are a set of tools, procedures, and processes that allow an organization to analyze and reduce environmental impacts of its activities. Chemical managements include a variety of environmental management systems that all have the objective of minimizing risks related to the chemicals use.

Many organizations have come to realize both the business and environmental benefits of an EMS, such as:

- Improved environmental compliance
- Improved environmental performance in nonregulated areas, such as energy and water conservation
- Increased ability to identify pollution prevention opportunities
- Enhanced operational control and efficiency
- Reduced costs
- Improved relationships with regulators

CHEMICAL LEASING CASE STUDY

Industrial process: Cleaning with hydrocarbon solvent in automotive industry
Chemicals used: Hydrocarbon solvents
Service provided under ChL: Supplying the hydrocarbon solvent and supervising and giving recommendations for the application of the solvent in the process of cleaning equipment at GM Egypt. Furthermore, it takes back the solvent waste for recycling at its plant.

- *Economical benefits:*
  1. Cost reduction by 15% (saving of raw material with recycling)
  2. Reduction of solvent consumption from 1.5 L per vehicle to 0.85 L per vehicle
  3. Sharing of liability and benefits
  4. Ensuring long-term business relationship based on long contract
  5. Saving the cost of getting rid from the solvent waste

- *Environmental benefits:*
  1. Recycling of solvent waste and stopping dumping (Closing the Loop)
  2. Better hazardous waste management in accordance with environmental regulations and international environmental corporate policy

- *Organizational and management benefits:*
  1. More efficient cleaning process with hydrocarbon solvent by applying a procedure for cleaning by batch cleaning
  2. Enhancement of supply chain management and other environmental management systems
  3. Stopping of using the hydrocarbon solvent in purposes other than cleaning of equipment (e.g., stop washing worker hands, cloths, etc.)

Capacity building and awareness of operation staff.

Environmental management of the chemical sector in Egypt was a major concern of a number of ministries and administrative sections, in which some systems were implemented, mostly in cooperation with international donors, and/or bodies, in order to forestall any undesirable effects the use of chemicals may cause. Among these approaches are the following.

Chemical Leasing

Egypt's Centre of Cleaner Production was one of the few centers at a global level to apply the chemical leasing (ChL) concept as a viable tool for cleaner production and pollution prevention. ChL is a service-oriented business model that shifts the focus from increasing the sales volume of chemicals toward a value-added approach. The producer mainly sells the functions performed by the chemical and functional units, which are the main basis for payment. ChL involves new forms of payments for chemicals that direct the economic interests of all partners toward process optimization and the reduction of chemicals consumption. Within ChL business models, the responsibilities of the producer and service provider are extended, and, in some cases, may include the management of the entire life cycle.

Egyptian Pollution Abatement Program (EPAP)

Soft financing is provided by international bodies to industrial facilities to help develop plans and methods for industrial pollution abatement in industrial establishments, in both public and private sectors. The program was also meant to strengthen the inspection and enforcement capabilities of the environmental management units in a number of governorates and to promote awareness of the importance of industrial pollution abatement.

MINIMIZING SOLVENT EXPOSURE OF WORKERS IN A TIRE FACTORY, ALEXANDRIA, A SUCCESS STORY

During the manufacturing process, tires are sprayed inside and outside with a mixture of chemicals containing natural rubber, synthetic rubber, carbon, stearic acid, paraffin oil, and nonoxidizing agent all dissolved in organic solvents (heptane). Before the project, the process consumed 56 tons of spray annually. Approximately 77% of the mixture was aliphatic heptane. The spraying booth used for manual spraying is not equipped with an appropriate suction system and the heptane concentration in the occupational air exceeded the limits for heptane, 400 ppm and 500 ppm for average- and short-term exposure, respectively.

PROJECT OBJECTIVE

The objective was to reduce heptane concentrations in the work environment by performing automatic spraying in a closed area. In addition, the solvent-based mixture will be replaced by a water-based mixture in the internal tire spraying process.

ENVIRONMENTAL BENEFITS

Partial conversion into water-based spray leads to a significant reduction of the consumption of heptane. The automatic spray booth with efficient suction prevents the workers from hazardous exposure to heptane.

ECONOMIC BENEFITS

The new automatic spray booth resulted in replacing the solvent-based spray used for the inside spraying with water-based spray. Consequently, there was a significant decrease in the use of the solvent-based compounds by an annual amount of 50%, which resulted in saving 40,000 Egyptian Pounds annually.

CAIRO AIR IMPROVEMENT PROJECT, HARNESSING LEAD POLLUTION

Lead poisoning, sometime known as silent epidemic is one of the most hazardous heavy metal pollution, with proven effect on human health. Lead pollution has been one of the most controversial issues in Cairo with a number of studies on its impact on the wellbeing of some age groups in Cairo, especially children in areas adjacent to lead emission sources (Sharaf et al., 2008; Boseila et al., 2004).

The Cairo air improvement project is one of the major steps taken to improve Cairo air and to rid it of the high level of lead contamination.

Production of lead in the Great Cairo Area is estimated to be between 35,000 and 45,000 tons/year. The demand for lead for batteries and other products is expected to steadily increase over at least the next decade.

In Cairo, some secondary lead smelters that are located in densely populated areas have come under increased scrutiny. These smelters have been the source of lead in the environment for many years. With this in mind, the government of Egypt developed the Lead Smelter Action Plan (LSAP) to reduce the impact of these lead smelters on the environment. The LSAP focuses on reducing airborne and lead emissions from lead smelters and the relocation of smelters to industrial zones.

MULTILATERAL AGREEMENTS IN CHEMICALS MANAGEMENT

The international community has developed a number of agreements intended to manage chemicals, minimize their impacts on humans and their environment, regulate their use and promote safety. Egypt is an active member in these agreements, including the Stockholm Convention on Persistent Organic Pollutants (POPS), the Basel Convention on the Control of Transboundary Movement of Hazardous Wastes and their Disposal, the Rotterdam Convention, the Montreal Convention on Ozone Depleting Chemicals, and the Strategic Approach to International Chemical Management (SAICM).

STOCKHOLM CONVENTION ON POPS

Egypt has ratified the Stockholm Convention, and is in the process of implementing the convention's components. Egypt has already completed its national action plans, which include detailed information about emission inventories, stocks, contaminated sites, and legislation. Furthermore, the International POPS Elimination Network (IPEN) began a global NGO project, the International POPS Elimination Project (IPEP), in partnership with United Nations Industrial Development organization (UNIDO) and United Nations Environment Programme (UNEP). The Global Environmental Facility (GEF) provided core funding for the project. As a result of these efforts, RAED, a network of Arab NGOs, based in Cairo was formed to address the release of POPS in the region, with more than 200 NGOs from all the Arab countries (EEAA URLb).

CONVENTION IMPLEMENTATION, HIGHLIGHTS

Emission Inventories

A preliminary inventory of Dioxins and Furans in Egypt has been produced within the framework of the National Implementation Plan. Estimate of dioxins and furan release are estimated according to the equation:

Source Strength (dioxin emission per year) = Emission Factor × Activity Rate

A preliminary inventory of dioxin/furans release to the environment was assessed accordingly. Emission factors were produced for the following major activities:

- Municipal solid waste incantation
- Medical waste incineration

- Ferrous and nonferrous metal production: iron ore sintering, coke production, iron and steel production plants, copper plants, lead plants, aluminum plants, zinc production
- Power and heat generation plants
- Mineral Products, such as cement, glass, ceramics, asphalt mixing, and bricks
- Transport
- Uncontrolled combustion processes: waste burning and accidental fire
- Production and use of chemicals and consumer goods: pulp and paper production, textile, chemical industries, petrochemical industries, leather refining
- Others: tobacco smoking, dry cleaning residue, drying of biomass and others
- Disposal/landfill
- Hot spots, production sites of chlorine, chlorinated organics, and PCBs capacitors.

MINISTRY OF AGRICULTURE

1. Arrange monitoring programs on pesticides, toxics, POPS, and heavy metals in both imported goods and domestic products.
2. Establishing a system for monitoring imported and exported agricultural and nutritional products and making sure that they comply with the national and international chemical safety standards.
3. Controlling using of pesticides.

MINISTRY OF HEALTH AND POPULATION

1. Monitoring and developing pesticides for home use.
2. Establishing the chemical safety unit for enlisting imported chemical materials and analyze the toxicity level within them.
3. Establishing specific standards for water analysis to ensure the complete absence of POPS pesticides.

Ministry of Environment: It is the main ministry responsible for Environmental protection in Egypt. The ministry integrates all ministerial efforts and coordinates them as well as application of the Environmental decrees.

Ministry of Industry: Embracing the Cleaner Production direction and applying it in all industries in order to decrease the rate of these pollutants.

THE BASEL CONVENTION ON THE CONTROL OF TRANSBOUNDARY MOVEMENT OF HAZARDOUS WASTES AND THEIR DISPOSAL

The Basel Convention is the most comprehensive global environmental treaty on hazardous and other wastes. Egypt is a signatory to this convention, which aims to protect human health and the environment against the adverse effects of the generation, management, transboundary movement, and disposal of hazardous and other wastes.

Egypt is also hosting the Basel Convention Regional Centre for Training and Technology Transfer of Arab Countries. The Centre is conducting several training and capacity building programs in the field of hazardous waste and POPS to candidates from different Arab countries.

Despite the support for the convention, the definition of hazardous waste in Egypt and most Arab countries remains controversial. In Egypt, the terminology applied for hazardous materials, is defined in Law 4, 1994, as "Materials with dangerous properties that harm human health or have a harmful effect on the environment such as contagious, gas, explosive or flammable substances, or those with ionizing radiation." Meanwhile, hazardous waste is defined as "wastes of activities and processes or their ashes that maintain their harmful properties and have no subsequent original or substitutive uses, such as clinical wastes from medical treatment and wastes resulting from the manufacture of any pharmaceutical products, drugs, organic solvents, printing fluid, dyes and painting materials." Technically, both definitions are concerned with two states of matter, namely the solid and liquid forms, a fact that set the topics of air and water pollution as separate issues legally (El Dars, 2007).

STOCKPILING OBSOLETE PESTICIDES

One of the mandates of the Basel Convention is to help countries eliminate their stockpiles of obsolete pesticides. Stockpiles have accumulated largely because some products were banned for health or environmental reasons, but were never properly discarded. These pesticides contain some of the most dangerous insecticides produced—members of the POPS group. These dangerous chemicals threaten communities through the potential contamination of food, water, soil, and air. Poor communities are the most vulnerable to environmental degradation of this sort. According to FAO baseline information, Egypt harbors nearly 600 tons of obsolete pesticides (FAO URL) (Figure 37.3).

FIGURE 37.3 Containers of obsolete pesticides in one of the stores in Egypt. (Adapted from The International POPS Elimination Project, Country Situation Report for Egypt, http://www.fao.org/DOCREP/003/X8639E/x8639e05.htm#TopOfPage.)

CEMENT KILNS AND OBSOLETE PESTICIDES TREATMENT

Cement kilns are occasionally used to treat hazardous waste in a number of countries, with some pilot work in Egypt on obsolete pesticides. The high temperatures of the kilns are most suitable for the total breakdown of these pesticides and would also ensure no emission of dioxins and furans if performed correctly. The use of cement kilns offers a very competitive and cost-effective alternative to the elimination of obsolete pesticides, especially in countries like Egypt with very little waste management facilities. But, on the other hand there is some skepticism about the validity of the process as expressed by some stakeholders who believe that cement factories, especially in developing countries, may have neither the adequate technology nor the competence.

ELECTRONIC WASTE ANOTHER SECTION THAT CAN BE SIGNIFICANTLY SHORTER

Electronic Waste (E-waste) is an informal term for any electrical or electronic appliance that has reached its end-of-life (EoL). It contains hazardous elements like lead, mercury, arsenic, cadmium, selenium, hexavalent chromium, and flame retardants; therefore, it is classified as hazardous waste. Especially in the informal structures, few basic precautionary measures are applied to protect workers' health. Additionally, the methods applied, like the open incineration of E-waste, lead to severe environmental pollution. The prevalence of Information and Communication Technology (ICT) in Egypt is now below that of international average (CEDARE, 2008). Consequently, there is a huge growth potential for this waste stream.

CHEMICAL TRADE IN EGYPT

Egypt's chemical exports exceeded E£10 billion in first half of 2008 (*Egypt Daily News*, 2008) Egypt's ministry of Trade and Industry implements an integrated strategy to develop Egyptian industry, increase technological component and promote exports. The strategy is based on transforming activities depending on using natural resources to medium technology then to high technological industry. Egypt's chemical export during the first half of this year reached E£10.4 billion, an increase of 25% compared to the same period last year. Exports to the European Union accounted for 32% of the total volume. The number of factories stands at 3311. Thirty-two new factories have been established with investments totaling E£227 million. Chemical industries in Egypt should abide by the European regulatory system Registration, Evaluation, Authorisation and Restriction of Chemicals (REACH), (REACH URL), which is a new European community regulation on chemicals and their safe use. The system has been enforced on June 1, 2007 to deal with the REACH of chemical substances. They still have until the end of November 2008 to readjust and comply with the new European legislation, otherwise, their products will be banned from entering European markets.

REFERENCES

Boseila, S. A., A. A. Gabr, and I. A. Hakim. 2004. Blood lead levels in Egyptian children: Influence of social and environmental factors. *Am. J. Public Health*, 94(1): 47–49.

CEDARE. 2008. Centre for environment and development in the Arab region and Europe. *Mapping Study of E Waste Management in the Arab Region*, 48pp.

Indian and Northern Affairs Canada (INAC). 2003. Contaminants level, trends and effects in Biological Environment–Canadian Arctic Contaminants Assessment Report 1, 2003.

Egypt Daily News, September 1, 2008. URL available at: http://www.thedailynewsegypt.com/article.aspx?ArticleID=16151. Accessed September 6, 2009.

Egyptian Environmental Affairs Agency (EEAA). 1994. Environmental Promulgation Law (Law 4/1994), Cabinet of Ministers, Egypt.

Egyptian Environmental Affairs Agency URLb available at: http://www.eeaa.gov.eg/english/NIPP/NIPP_conv.asp

El Dars, F. 2007. Development of an industrial hazardous waste management system in Egypt: Constraints and proposed guidelines, environment international. *African J. Environ. Assess. Manage.*, 12: 13–27.

El-Gamal, A. 1983. Persistence of some pesticides in semiarid conditions in Egypt. In *Proceedings of the International Conference on Environmental Hazardous Agrochemicals*, Vol. I, Alexandria, Egypt, November 8–12, 1983, pp. 54–75.

El-Sebae, A. H. 1977. Incidents of local pesticide hazards and their toxicological interpretation. In *Proceedings of the Seminar/Workshop on Pesticide Management*, UC/AID University of Alexandria, Alexandria, Egypt, March 5–10, 1977, pp. 137–152.

El-Sebae, A. H., M. Abou-Zeid, and M. A. Saleh. 1993. Status and environmental impact of toxaphene in the Third World—A case study of African agriculture. *Chemosphere* 27(10): 2063–2072.

El Zarka, M. 1999. National profile for the Management of Chemicals in Egypt. Egyptian Environmental Affairs Agency, EEAA URL a, available at: http://www2.unitar.org/cwm/publications/cw/np/np_pdf/Egypt_National_Profile.pdfEEAA

FAO URL, available at: http://www.fao.org/docrep/005/Y4544E/y4544e00.HTM

http://www.fao.org/DOCREP/003/X8639E/x8639e05.htm#TopOfPageFAOFAO, URL 2

Mansour, S. 2004. Pesticides exposure, Egyptian scene. *Toxicology*, 198: 91–115.

Naglaa Loutfi, M., Fuerhacker, P. Tundo, S, Raccanelli, A. G. El Dien, and M. Tawfic Ahmed. 2006. Dietary intake of dioxins and dioxin-like PCBs, due to the consumption of dairy products, fish/seafood and meat from Ismailia city, Egypt. *Science of the Total Environment*, 370: 1–8.

Nevin E. Sharaf, Alia Abdel-Shakour, Nagat M. Amer, Mahmoud A. Abou-Donia, and Nevin Khatab. 2008. Evaluation of children's blood lead level in Cairo, Egypt. *American-Eurasian J. Agric. & Environ. Sci.*, 3(3): 414–419.

Registration, Evaluation, Authorisation and Restriction of Chemicals, REACH, available at: http://www.eu-delegation.org.eg/en/EU-Egypt_Trade_issues/REACH.asp.

State of The Environment Report, State Ministry of Enviornmental Affairs, Egypt, 2007.

Tawfic Ahmed, M., N. Loutfy, and E. El Shiekh. 2002. Residue levels of DDE and PCBs in blood serum of women in the Port Said region of Egypt. *J. Hazard. Mater.*, 89(1): 41–48.

Tawfic Ahmed, M., N. Loutfi, and Y. Mosleh. 2001. Residues of chlorinated hydrocarbons, polycyclic aromatic hydrocarbons and polychlorinated biphynel in some aquatic organisms in Lake Temsah around Ismailia. *J. Aquatic Ecosystem Health Manage.*, 4: 165–173.

38 Capacities for Chemicals and Pesticides Management in Ghana

*John A. Pwamang**

CONTENTS

INTRODUCTION

The use of pesticides is an essential means of achieving economic and social development, hence, pesticides such as insecticides, herbicides, fungicides, and nematicides, are widely used for pest control to ensure food security in agriculture and vector control in public health.

Despite their beneficial effects, pesticides also present risks to human health and the environment. These risks arise from the inherent toxicities of pesticides, and

* The views expressed in this chapter are those of the author and do not necessarily represent the views of the Environmental Protection Agency or the Government of Ghana.

from their misuse. Some growers sometimes harvest crops immediately after treatment with pesticides, ignoring the preharvest interval. Others are also of the view that any pesticide could be used on any crop and in some instances certain pesticides meant for cotton are used on vegetables and even on cocoa.

Some unscrupulous persons also go to the extent of using pesticides for fishing and hunting. These practices cause pesticides to enter the food chain and affect people who consume the catch both far and near, with the consequent health effects.

The Government of Ghana is very concerned about the abuse of pesticides, and has tasked Agencies such as the Environmental Protection Agency (EPA) and Plant Protection and Regulatory Services Directorate (PPRSD) of the Ministry of Food and Agriculture (MoFA) to assist in the control and regulatory efforts. This necessitated the promulgation of the Pesticides Control and Management Act (Act 528) in 1996 [which has now become Part Two of the EPA Act, 1994 (Act 490)] to provide the legal framework for the control and management of pesticides in the country.

Although the pesticides law was promulgated in 1996, it was not until 2004 that procedures and structures were finally put in place for full operation of the law. Prior to 2004, pesticides control was based on a provisional recommended list of pesticide products that were already in use in the country before the law came into force. Control was exercised by a technical committee comprising the PPRSD of the MoFA, the Chemicals Control and Management Centre of the EPA, the Chemistry Department of the Ghana Atomic Energy Commission (GAEC) and the Ghana Health Services of the Ministry of Health.

STRUCTURE FOR PESTICIDES REGISTRATION IN GHANA

The structure for registration of pesticides in Ghana is shown in Figure 38.1 and includes the following:

- Secretariat—Chemicals Control and Management Centre of the EPA receives all applications.

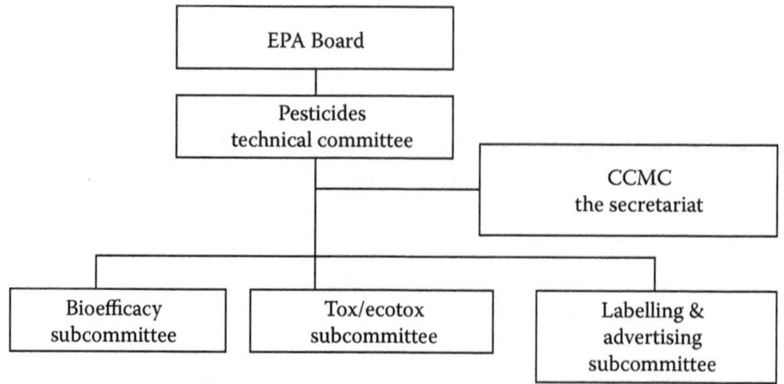

FIGURE 38.1 Organizational structure for pesticides registration in Ghana.

- Three subcommittees of Pesticides Technical Committee (PTC) evaluate applications and submit reports to the (PTC). The subcommittees are Toxicology/Ecotoxicology, Bioefficacy, and Labeling and Advertising.
- The thirteen-member intersectoral PTC considers the reports of the subcommittees and makes recommendations to the EPA Board.
- The EPA Board takes the final decision to register or deny registration of pesticides.

Composition of the PTC

The PTC provided for in Section 53 of the EPA Act, 1994 (Act 490) comprises the following:

1. Chairman appointed by the EPA Board.
2. Representative of Chemistry Department of the National Nuclear Research Institute of GAEC.
3. Representative of Cocoa Services Division of COCOBOD (Cocoa Board).
4. Representative of PPRSD of MoFA.
5. Representative of Veterinary Services Directorate of MoFA.
6. Representative of Ministry of Health.
7. Representative of Ghana Standards Board.
8. Representative of Customs, Excise, and Preventive Service.
9. Representative of Association of Ghana Industries.
10. Representative of Ghana National Association of Farmers and Fishermen.
11. Representative of Ministry of Land and Forestry.
12. Representative of Ministry Responsible for the Environment.
13. Representative of EPA (Secretary).

Composition of Bioefficacy Subcommittee

1. PPRSD/MoFA
2. Cocoa Research Institute of Ghana
3. Council for Scientific and Industrial Research (CSIR)/Crops Research Institute
4. EPA (two representatives)

Composition of Human Toxicology and Ecotoxicology Subcommittee

1. GAEC
2. Food and Drugs Board
3. Ghana Health Services
4. EPA (two representatives)

Composition of Advertising and Labeling Subcommittee

1. PPRSD/MoFA
2. Customs, Excise, and Preventive Service
3. Ghana Standards Board

4. Agrochemical Dealers Association
5. Ghana National Association of Farmers and Fishermen
6. EPA (two representatives)

PROCEDURES FOR THE REGISTRATION OF PESTICIDES IN GHANA

The purpose of the registration process is to determine before registration that the product can be used safely and effectively in accordance with its label directions. The pesticide product label and scientific data must be reviewed and found acceptable before the product can be registered.

To register a product, an applicant must submit a completed application form and required data (technical dossier) on the product for evaluation. The dossier is checked for completeness at the Secretariat and processed to extract relevant information needed to run exposure assessment models. The dossier is then evaluated by three subcommittees of the PTC and this involves the following.

- Scientific review of product chemistry, residue chemistry, toxicity to fish and wildlife, phyto-toxicity, and efficacy of the pesticide product.
- Risk assessment to determine potential risks of pesticide products to human health and the environment.
- Review of acute toxicological and chronic toxicological studies which are submitted in support of obtaining new product registration.
- Review of product labels to ensure that they meet EPA label requirements.

After considering the evaluation reports from the three subcommittees, the PTC may recommend that the EPA Board register the product and classify it for:

a. Full registration valid for three years either for General use or for Restricted use
b. Provisional clearance (maximum 1 year)
c. Suspended or banned

General use: If applied for the use for which it is registered, it will not have unreasonable adverse effects on human health and the environment.

Restricted use: If its use in accordance with widespread commonly recognized practice in the absence of additional regulatory restrictions may cause unreasonable adverse effect on people, animals, crops, or on the environment.

Banned or suspended: If the pesticide will have unreasonable adverse effects on human health and the environment in normal use, or in accordance with International Conventions, for example, Rotterdam Convention, Stockholm Convention, Montreal Protocol, and so on.

Provisional clearance: If adequate information has been provided and pesticide presents no toxicological risk to people, animals, crops, or the environment. Provisional clearance is temporary pending formal registration and is subject to conditions (e.g., limitations on quantities that can be placed on the market or analysis of samples). Provisional clearance is cancelled if application is refused.

LICENSING OF PESTICIDES DEALERS AND FACILITIES

Pesticides dealers are required to obtain licenses for the particular pesticides-related activity. The license is subject to conditions and the dealer's license can be acquired through an application. Licenses are issued based on satisfactory report of inspection of premises, facilities, and personnel. Pesticides activities which require licenses are: retail, import, distribution, commercial pest control, transportation, manufacturing, repackaging, formulation, warehousing, and any other related pesticides activity.

ENFORCEMENT

Members of the relevant subcommittee of the District Assembly may be appointed by EPA Board as pesticides inspectors to enforce provisions of the pesticides law including the following:

- Inspect any equipment used or to be used in applying pesticides.
- Inspect any storage or disposal facilities or areas used for the storage or disposal of pesticides.
- Inspect any land actually, or reported to be, exposed to pesticides.
- Investigate complaints or injury to human beings and animals, or damage to land and pollution of water bodies resulting from the use of pesticides.
- Take samples of pesticides applied or to be applied.
- Monitor the sale and use of pesticides.
- Examine and take copies of a license or other documents required by this Act or any regulations made under this Act.
- Declaration of office and evidence of authority must be produced before entry and search, and other cases on request.
- Issue written receipts, where practicable for items seized and reasons for seizure stated therein.
- Additional powers of entry/stoppage to search; seizure and arrest.

ACHIEVEMENTS

Since the operationalization of the registration scheme in August 2004, the EPA has received support from collaborators both nationally and internationally to strengthen the scheme. Some of the achievements to date include the following:

Register of Pesticides as of December 31, 2009

Category	Full Registration	Provisional Clearance	Banned	Total
Insecticides	72	3	25	100
Fungicides	23	1	0	24
Herbicides	57	19	0	76
Others	4	0	0	4
Total	156	23	25	204

Register of Licenses as of December 31, 2009

Category	New Licenses Issued	Licenses Renewed	Total
Formulators/Repackagers/Manufacturers	1	1	2
Importer/Distributors	4	7	11
Retailers	207	57	264
Commercial Operators (Pest controllers)	17	36	53
Warehouses	5	7	12
Transporters	3	3	6
Total	237	111	348

Other Achievements

- Developed a Pesticides Registration Manual outlining the data requirements, risk-assessment guidelines, and detailed description of the pesticides registration process.
- Prepared draft regulations under the pesticides law and these are being considered by the EPA Board and the Ministry of Environment, Science and Technology. The draft regulations are:
  - Draft Pesticides (Advertising) Regulations
  - Draft Pesticides (Labeling and Packaging) Regulations
  - Draft Pesticides (Licensing) Regulations
  - Draft Pesticides (Registration) Regulations
  - Draft Technical Guidelines for Pesticides Transportation, Storage and Disposal.
- Prepared Fertilizer and Plant Protection bills, which are currently being considered by Parliament.
- Trained 70 Pesticides Inspectors on sound management of Pesticides and on enforcement of pesticides law.
- Developed guidelines for conducting and reporting bioefficacy trials for the registration of pesticides for agricultural purposes.
- Developed guidelines for safe transport of hazardous materials, which include pesticides.
- Developed a reference guide on pesticides for the horticulture sector in Ghana.
- Collaborating with COCOBOD to train and monitor the activities of teams involved in the Cocoa Mass Spraying Exercise.
- Successfully withdrew two pesticide products from use on Cocoa since August 2004 when the registration process became operational. Gammalin 20EC (Lindane) was withdrawn due to international requirements of countries that import cocoa beans from Ghana while Carbamult 20EC (Promecarb) was withdrawn due to reports of nausea and vomiting by applicators in the field.

PREVIOUS AND ONGOING CAPACITY STRENGTHENING ACTIVITIES

Ghana has benefited and continues to benefit from a number of capacity building projects related to pesticides control and management including the following.

- Under the auspices of the Food and Agricultural Organisation (FAO) staff of EPA and the PPRSD/MoFA were trained in ecotoxicological/environmental risk assessment of pesticides.
- Under the auspices of the World Health Organisation (WHO), staff of EPA and PPRSD/MoFA were trained in human toxicological risk assessment.
- The EPA and PPRSD/MoFA are collaborating with the GAEC to run postgraduate courses on chemicals control and management at the School of Nuclear and Allied Sciences, University of Ghana—Atomic Campus. The programme commenced in the 2007/2008 academic year and the third batch of students is currently in their second semester of the two-year Masters of Philosophy Programme.
- Collaborating with the Ghana Agricultural Associations Business and Information Centre (GAABIC) on the certification of trained pesticides dealers under the Ghana Agro-Dealer Development Project. This three-year project is being implemented by the International Centre for Soil Fertility and Agricultural Development (IFDC) in collaboration with GAABIC with funding from the Alliance for a Green Revolution in Africa (AGRA). To date, more than 1000 pesticides dealers have been trained on safe handling of pesticides and provisions of pesticides law.
- Collaborating with CropLife International through CropLife-Ghana on the Clean Farms Initiative aimed at conducting inventory, collection and Safeguarding of Obsolete Pesticides and Empty Containers. Training on inventory and safeguarding has been conducted and an outreach campaign is underway requesting the public to register obsolete pesticides in their custody.
- Developed guidelines for the registration of biopesticides under the auspices of the Crop Protection Programme of the Department for International Development (DFID) of the United Kingdom.
- Developed Plant Protection Policy under the auspices of the FAO.
- Participating in the Economic Community Of West African States (ECOWAS) initiative to harmonize pesticides registration procedures in the subregion. Ghana drafted protocols for bioefficacy trials for crop–pest combination for the humid areas of the ECOWAS subregion.
- Participating in the subregional fruit fly control programme and have establishing national fruit fly control committee under the MoFA.
- Developing a national food safety policy through inter-agency collaboration.

CHALLENGES

There are a number of challenges to overcome including the following:

- Discrepancies in legal texts due to changes in mandates following ministerial rearrangements. This was effected without consultations with the key agencies and is hampering implementation of the pesticides law.
- Inadequate facilities for pesticides quality control analysis.
- Potentially high levels of pesticide residues in food and environmental media.
- There are inadequate funds for awareness creation programs for farmers and the general public and for Pesticides Inspectors to conduct post registration field surveillance and monitoring.
- Inadequate numbers of staff and inadequate logistics for monitoring.
- There is currently only one Poison Information Centre located at the Ridge Hospital in Accra.

OTHER AREAS OF CHEMICALS MANAGEMENT

EPA Ghana is involved in many areas of work regarding chemicals management. Of particular relevance is that of the implementation of the Stockholm Convention.

This activity is aimed at strengthening the capacities and capabilities of government officials and stakeholders outside of government to address polychlorinated biphenyl (PCB) identification, and manage existing sources of PCBs as well as their elimination/destruction, as identified as a priority in the National Implementation Plan for Persistent Organic Pollutants for the Republic of Ghana.

The project develops and implements a strategy, and the required steps, from the current unsustainable management of PCB-containing equipment to sound management and disposal practices. The strategy commences by strengthening the legal framework and the management capacity both within government institutions and among PCB holders. The project will also eliminate, as a first step, the PCB-containing equipment, mainly transformers, and in a second step start phasing out PCB-contaminated equipment. The project is the first major step to meet the obligations of Ghana under the Stockholm Convention.

Ghana is the first country in the subregion that has developed a Full Size Project (FSP) of this nature to eliminate PCBs as required under the Stockholm Convention. The experiences obtained during the implementation of the project will be shared with the other countries in the subregion that are currently developing concrete phase-out activities. The project is part of the general strategy of Ghana to significantly improve power production and distribution and to strengthen the management of the sector.

After three years of preparation with United Nations Development Programme (UNDP) and United Nations Institute for Training and Research (UNITAR), Ghana has succeeded in obtaining the necessary funding from Global Environment Facility (GEF), with co-financing from many, mainly national, sources, to start the implementation of this project that will take 5 years (2008–2012).

Many challenges are to be resolved, of a technical, financial, and organizational nature to successfully eliminate PCBs from this equipment in the country.

Ghana is, however, committed to making this a challenge a reality.

RECOMMENDATIONS/LOOKING INTO THE FUTURE

- Make presentation to the Attorney General to separate Act 528 from Act 490 and make amendments to Act 528 to create the Pesticides Management Fund (PMF). The proposed PMF under Act 528 could provide adequate funds from the Industry to support pesticides management programs.
- Ghana needs to establish a system to check the quality of imported pesticides to ensure that they conform to the data submitted for registration. Establish facilities for pesticides quality control analysis.
- Establish stronger collaborative arrangements with Academic and Research Institutions to monitor effects of pesticides on human health and the environment.
- Appoint additional pesticide inspectors and support pesticides inspectors with funds to conduct postregistration monitoring and surveillance and enforce provisions of the pesticides law.
- Establish Poison Information Centres in at least all Regional Hospitals in Ghana.

39 Chemicals Management and Safety in India

Ravi Agarwal

CONTENTS

CHEMICAL INDUSTRY IN INDIA

India's gross domestic product (GDP) has been consistent at around 8% annually over the past three years.* Alongside, industrial production is rising, to tap both domestic as well as international markets. Urbanization is increasing with more than 40% of India's population expected to live in urban settings by 2030 (MOH, 2009).

* GDP was 8.6% in June 2009 and 8.3% in March 2011 adjusted for inflation.

The associated demand for new goods and services is also leading to new environmental and chemical management needs.

India ranks 12th in the world for the production of chemicals by volume, with a 7% global share. India's chemical industry contributes about 6.7% to the nation's GDP, with a turnover of about US$ 30 billion and exports of about US$ 2 billion. It accounts for 17.6% of the overall manufacturing sector, 13–14% of total exports and 8–9% of total imports of the country. It is claimed to have a potential to grow to US$ 100 billion (KPMG, 2008). Chemicals users are both other industrial consumers who account for more than 33% of the total production, as well as individual consumers (MOEF, 2006) (see Table 39.1).

Structure

The Indian chemical industry has undergone many phases. From its early beginnings in the 1950s until around 1980, the industry was largely in the public sector, owned by the state. Subsequently, over the next 10 years, it grew quickly, though still as a primarily Indian owned and fragmented industry with small capacities and protected by the state through tariff and nontariff structures (such as more than 160 "reserved" products allowed for small-scale enterprise manufacturing only). During this time, it also diversified from manufacturing basic agrochemicals and pharmaceuticals in the 1950s to include paints, plastics, petrochemicals, consumer chemicals, and other chemicals, which exceed 70,000 products today. Since 1992, following the economic liberalization phase in India, the industry has had new investments through global multinational corporations, aided by a lowering of tariffs and a reduced role for the state-run public sector. Presently, the industry has become more globally conscious, with large investments in petrochemicals, creation of larger scale capacities, and trying to compete with the global industry such as in China (KPMG, 2009). However,

TABLE 39.1
Chemical Industry Sectors

Chemical Sectors	Types of Chemicals/Uses
Petrochemicals	Basic chemicals like ethylene, propylene, benzene, and so on, intermediates. Synthetic fibers like nylon. Polymers like LDPE/HDPE, PVC, polyester, PET, synthetic rubber, and so on.
Inorganic chemicals	Caustic, chlorine, sulfuric acid, and so on. Inorganic chemicals are mostly used in detergents, glass, soap, fertilizer, alkalis, and so on.
Organic chemicals	Methanol, formaldehyde, acetic acid, phenol, acetone, acetic anhydride, nitrobenzene, chloromethane, aniline, maleic anhydride, pentaerithitol, PNCB, MEK, citric acid, ONCB, iso-butyl alcohol, and so on.
Fine and specialty chemicals	Adhesives, additives, cleaning agents, and so on, used in textiles, leather, paper, detergent, rubber, paints, polyester, oil, gas industries, and others.
Agrochemical	Herbicides, insecticides, fungicides, fumigants, and so on.
Paints and dyes	Paints, inks, textiles, polymers, and so on.
Bulk drugs	Various.

the shift is still slow, as the domestic market continues to be a key focus, leading to smaller high-cost operations as compared to those international companies who have high capacity and much lower costs. Today, with falling costs, increased competition, and new requirements for accessing international markets, the current phase is a challenging one.

The industry as of now is still fragmented, with many small-scale capacities and some large ones. It is dominated by medium and scale units, distinguished by the scale of investments. More than 90% of the industry is in the Micro and Small and Medium Enterprise (SME) sector, which stretches down to tiny and household industry levels, and is concentrated in North India. There are fewer larger players, for example, the Indian pesticide industry has less than 10 large actors. As per the Third Census of the Small Scale Industries (SSI) sector, there are reportedly more than 110,000 chemical manufacturing units in India, of which 70% are illegal and unregistered (MSME, 2002). While larger companies produce bulk chemicals, specialty chemicals are produced by both large and small players, who also do formulations. Though the SMEs are a massive employment generation sector, with more than 6 million people employed, they typically have higher costs and are also very polluting and hazardous.

The Indian Chemical Council (ICC) is the main national body dedicated to the interests of the industry. It represents all segments of the chemical industry and has been in existence for more than 60 years, and has several smaller industry-specific associations. It has an active environment and safety division that also helps in capacity building with initiatives like ISO 9000 and 14001 certifications and Responsible Care initiatives and is a member of the International Council of Chemical Associations (ICCA). Initiatives with the Government include setting up of Waste Minimization Circles (WMC) to share waste as a resource among industries (http://www.wmc.nic.in).

Export Market Access

International regulations like the European Union's Registration, Evaluation, Authorisation and Restriction of Chemicals (REACH), Restriction of Hazardous Substances (RoHS), and Waste Electrical and Electronic Equipment (WEEE) directives have thrown up new challenges for accessing international markets. For example, as per the REACH Regulation, all chemical substances that are manufactured in or imported in the European Union (EU) in the quantity of more than 1 ton/year need to be registered with the EU Chemicals Agency (ECHA). The new law affects all manufacturers, importers, retailers, brand managers, and distributors of chemical substances or products in the EU, across industries as wide as electrical and electronics, aerospace and automotive, metals, plastics, textiles/garments, accessories, toys, and many other consumer goods industries. The Indian industry is still coping to understand the implications of this, and trying to build appropriate capacity.

Industry Response

The Industry response to the rising environmental and health safety demands on it has broadly ranged from downright resistance to cautiously adopting some change.

In several instances, such as waste management, the industry has been forced to change owing to the issue being brought to its door by media or the judiciary. There was early resistance to India ratifying the Stockholm Convention, for instance, since it was feared that this would lead to other chemicals being brought into its framework and currently the Industry is strongly resisting the addition of new chemicals like Endosulpan into the Treaty. Similarly highlighting problems like pesticides contamination of food and chemicals in soft drinks has led the industry to first confront and then comply.

POLICY AND LEGISLATIVE FRAMEWORK

Development of the Chemicals Management Regime

India's environment management efforts started quite early, after independence in 1947. The Insecticides Act to regulate the harmful impacts of pesticides was passed in 1968. After the Stockholm Conference in 1972, India enacted two major environmental laws—The Water (Prevention and Control of Pollution) Act in 1974, passed for the purpose of prevention and control of water pollution, followed by the Air (Prevention and Control of Pollution) Act in 1981. It also led to the institutionalization of the regulatory infrastructure through the establishment of the Central and State Pollution Control Boards (SPCBs) (see the section "Administrative and Institutional Framework").

However, it was after the Bhopal disaster in 1984 that chemical safety issues became specifically recognized through the passing of umbrella legislation, the Environmental Protection Act (EPA) (1986), under which specific rules on chemical safety have been framed from time to time. Subsequently after the Rio Summit in 1992, the World Summit on Sustainable Development (WSSD)—held in Johannesburg in 2002, which adopted the 2020 goal (see Section IV, Chapter 17 of this book), as well as the Strategic Approach to International Chemicals Management (SAICM) process, chemicals safety—has slowly become part of an overall development policy agenda. Even the Indian Planning Commission, the highest national planning body, has recognized this in its ongoing Eleventh 5-Year Plan, "Serious environmental health problems affect millions of people who suffer from respiratory and other diseases caused or exacerbated by biological and chemical agents, both indoors and outdoors. Millions are exposed to unnecessary chemical and physical hazards in their home, workplace, or wider environment" (Planning Commission, 2007).

In terms of legislation, India is well placed, even though there are gaps. Almost in all steps of chemical management from cradle to grave, some legislation has been laid down and consists of more than 15 Acts and 19 Rules (NCPM, 2006). Yet areas such as substitution of hazards in products and processes or the remediation of contaminated sites need attention.

Constitutional Provisions

The Indian Constitution, through its 42nd amendment, under Article 48 states, "The State shall endeavor to protect and improve the environment and to safeguard the forests and wildlife of the country." Through various judgments, the Indian Supreme

Court has also interpreted the provisions of Article 21, which enshrines a Right to Life, to hold that environmental degradation violates the fundamental right to life.

The Environment (Protection) Act (1986), Amended 1991

The EPA is a key and fundamental act enabling an umbrella legislation, under which the designated ministry in the Central Government (Ministry of Environment and Forests, MOEF) can frame rules and procedures to deal with chemical safety issues. It provides for both criminal as well as civil penalties and severe fines for violations as well as imparts wide-ranging powers for regulators to take action in the event of noncompliance to the Act, including ordering the closure of units. However, since these penalties have been ineffectual, this is now being reexamined to make this a real deterrant.

Chemicals-Related Rules Under the EPA

The Hazardous Waste Management and Handling Rules (1989), Amended 2000 and 2003

This law, enacted after the advent of the Basel Convention makes it mandatory for industries that generate hazardous waste to seek authorization from the SPCB before commencing operations. It mandates procedures for waste transport, storage, processing, and disposal. It also requires the maintenance of data and its reporting to the SPCB. The law follows the Basel list of wastes and largely conforms to the requirements of the Convention. Under it, all hazardous wastes are required to be treated and disposed off in the manner prescribed.

Manufacture, Storage, and Import of Hazardous Chemicals Rules (1989), Amended 1994 and 2000

This law mandates that all hazardous chemicals that fall under the criterion defined it must be stored and kept as stated. The purpose of the law is to ensure the proper handling of such chemicals and to ensure that chemical handlers appropriately warn employees, maintain a safe workplace, and are prepared if an emergency situation should occur. This law is applicable depending on both the type and quantity of a chemical that one is handling.

Chemical Accidents, Emergency Planning, Preparedness, and Response Rules (1996)

This law requires that the local administration draws up and implements plans to deal with such accidents. It has provisions for designating officers to be contacted in times of an accident and establishes a chain of command in such an event, through the constitution of a Central Crisis Alert System, and connected State and local level bodies. There are now 1464 Major Accident Hazard (MAH) Units in 19 States of the country, and 1395 on-site plans and 118 off-site plans have been prepared. All the states except two have constituted State Level Crisis Groups. In addition, a GIS-based Emergency Planning and Response System has been developed for four States.

Ozone-Depleting Substances Regulation and Control Rules (2000)

This law was enacted to implement the provision of the Montreal Protocol in India. It seeks to control the production, distribution, sale, import, and export of such substances, and has an associated monitoring and implementation mechanism.

Biomedical Waste Management and Handling Rules (1998)

This law regulates the management of wastes generated by health care establishments all over the country. Under these Rules, the wastes generated by them are divided into 10 categories and disposal methods for all the categories of wastes are also specified. The status of implementation of these rules is monitored, though not regularly.

Battery Management and Handling Rules (2000)

This law regulates the collection, channelization, and recycling as well as import of used lead acid batteries in the country. These rules make it mandatory for consumers to return used batteries. All manufacturers, assemblers, reconditioners, or importers of lead acid batteries are responsible for collecting used batteries against new ones sold as per a schedule defined in the rules. Such used lead acid batteries can be auctioned or sold only to recyclers registered with the Ministry of Environment on the basis of their possessing environmentally sound facilities for recycling and recovery.

Plastics Waste Management and Handling Rules (2011)

This law makes it illegal to use plastic bags less than 40 microns, and bans the use of plastic pouches for chewing tabacco. It also mandates setting up plastic waste collection systems in each municipality.

The Fly Ash Notification (1999), Amended 2003

This law makes it the responsibility of construction agencies to use fly ash-based bricks or products in a time bound manner, utilizing fly ash from thermal power plants.

E-Waste Management and Handling National Guidelines (2007)

These are to provide guidance for the collection, recycling, and disposal of electronic and electrical waste. They include Extended Producer Responsibility (EPR) as part of a life-cycle approach as well as the removal of hazardous substances from such products.

Other relevant guidelines are (1) Guidelines for Adoption of Cleaner Technologies, Hazardous Waste Management Guidelines and (2) Guidelines for Common Effluent Treatment Plants (CETPs).

Acts Impacting Chemicals Management and Safety

The Insecticides Act (1968) and Insecticide Rules (1971)

This Act (and Rules) regulates the import, manufacture, sale, transport, distribution, and use of insecticides to prevent risk to human beings and animals. It is also responsible for setting the body for the registration of new insecticides through a Central Insecticide Board (CIB). It is regulated by the Ministry of Chemicals. This Act is currently proposed to be replaced by a new Pesticides Management Act, and the Bill is pending in Parliament. It sets a maximum permissible pesticides food residue limit based on the Food and Food Safety Act (2006).

The Food and Food Safety Act (2006), Previously Known as the Prevention of Food Adulteration Act (1954)

This Act is where maximum tolerance limits for pesticide and chemical contamination in Food are laid down. These are now to be linked to the registration of new pesticides in the pending Pesticides (Management) Bill. The prescribed Acceptable Daily Intake (ADI) and Maximum Residue Limit (MRL) are determined based on the recommendation of the FAO Codex (Toxics Alert, 2009) (see also the section "Globally Harmonized System").

The Public Liability Insurance Act (1991), Amended 1992

This Act provides that owners of certain chemicals and substances must take out an insurance policy in favor of the workers. The Act was passed to provide public liability insurance for the purpose of providing immediate relief to the person affected by an accident occurring while handling any hazardous substance and for matters connected therewith or incidental thereto. It also provides for liability to give relief in certain cases on the principle of "no fault." It is the duty of the owner to take out insurance policies.

The Factories Act (1948)

This is the main Act that deals with all aspects of Factory operations including worker exposure and compensation in case of occupational injury. It lists 116 chemicals with their permissible limits, both in terms of short-term (15 min) as well as average (8 h) exposure in the work place. It is regulated by the Ministry of Labour in association with the Departments of Labour in each state.

Over time, awareness of Occupational Health and Safety (OH&S) has improved to some extent. The Bureau of Indian Standards (BIS) has formulated an Indian Standard on OH&S management systems, IS 18001:2000, for organizations.

Right to Information Act (2005)

In the absence of a public pollution registry of any sort, the newly passed Right to Information Act (2005) has been used extensively by citizens to access hitherto unavailable information and to help improve environmental governance in general but also related to chemicals pollution. The Act is to

> provide for setting out the practical regime of right to information for citizens to secure access to information under the control of public authorities, in order to promote transparency and accountability in the working of every public authority, the constitution of a Central Information Commission and State Information Commissions and for matters connected therewith or incidental thereto. (RTI, 2005)

Nonregulatory Instruments

Charter on Corporate Responsibility for Environmental Protection (2003)

The MOEF and the Government of India launched this initiative to go beyond the compliance of regulatory norms for prevention and control of pollution through various measures such as waste minimization, in-plant process control, and adoption of clean technologies. It is a voluntary initiative within the industry. The public Charter between the industry and the Government sets targets concerning conservation of water, energy, recovery of chemicals, reduction in pollution, elimination of toxic pollutants, and process and management of residues that are required to be disposed off in an environmentally sound manner. It also enlists the action points for pollution control for various categories of highly polluting industries. A Task Force has been constituted for monitoring the progress of implementation of CREP.

Ecomark Scheme

This voluntary eco-labeling scheme known as "Ecomark" was launched in 1991 by the MOEF, for easy identification of environment-friendly products. Consumer products, including detergents, soaps, oils, cosmetics, paints, textiles, and so on, which met certain criterion to significantly reduce the harm to the environment could be granted a logo. The scheme has been a nonstarter (http://envfor.nic.in/cpcb/ecomark/).

ADMINISTRATIVE AND INSTITUTIONAL FRAMEWORK

Environmental protection is a "concurrent" arrangement between the Center and State Governments. The Parliament has the power to make laws for the entire country, while State Legislatures can only do this for their respective territories. However, the States cannot dilute the provisions notified by the Center.

The Central MOEF is responsible for setting environmental policy and notifying rules as well as minimum standards for environmental regulation for the country. Several technical bodies support the MOEF, the prime one being the Central Pollution Control Board (CPCB), which draws up all technical standards (minimum national standards—MINAS) as well as gathers information on the environmental status in the country. The notified standards and laws are implemented by the SPCBs, which report to the State Governments.

Other key national institutes dealing with chemicals management include the Industrial Toxicological Research Institute (ITRC, http://www.itrcindia.org), The National Environment Engineering Institute (NEERI, http://www.neeri.res.in), and the National Institute of Occupational Health (NIOH, http://www.nioh.org) (see Table 39.2).

Laboratory Infrastructure

There is a wide network of laboratories that are certified by national bodies like the National Board for Laboratories (NABL), the Ministry of Science and

TABLE 39.2
Responsibility for Different Laws

Central Insecticides Board/Ministry of Agriculture	The Insecticides Act
Ministry of Chemicals and Fertilizers	Chemicals, Petrochemicals, and Pharmaceutical Industries
Ministry of Commerce and Industries	The Explosives Act
Ministry of Environment and Forests	Hazardous Chemicals Rules (1989), Hazardous Waste Rules (1989), Chemical Accidents Emergency Planning Preparedness and Response Rules (1996), and others
Central Board of Excise and Customs, Ministry of Finance	The Customs Act
DGFASLI, Ministry of Labor	The Factories Act and Rules
Ministry of Shipping, Road Transport, and Highways	Transportation of hazardous chemicals and wastes
Ministry of Railways	Transportation of hazardous chemicals through rail and storage at railway godowns
Ministry of Petroleum and Natural Gas	The Petroleum Act and related laws

Technology, the Council for Scientific and Industrial Research (CSIR), and the CPCB.

Labor Safety

The Ministry of Labour and Employment, along with two organizations under it, namely, The Directorate General of Mines Safety (DGMS) and the Directorate General of Factory Advice Service and Labour Institutes (DGFASLI) are responsible for implementing occupational safety and health in mines, factories, and ports. DGFASLI, located in Mumbai, functions as a technical arm of the Ministry in regard to matters concerned with safety, health, and welfare of workers in factories and ports/docks, while the DGMS is the designated regulatory agency for safety in mines and oil-fields. The implementation of the various schemes and laws falls under the purview of the State Governments.

INTERNATIONAL TREATIES AND COOPERATION

India is a signatory to all the multilateral, chemicals-related, environmental treaties thus far. These are the Montreal Protocol, The Rotterdam Convention, The Basel Convention, and the Stockholm Convention, besides signing on to SAICM. It has also actively participated in the Intergovernmental Forum for Chemical Safety (IFCS) as well as the United Nations Environmental Programme Governing Council

(UNEPGC) meetings, and is currently engaged in the negotiations for the upcoming global mercury treaty, mandated by the UNEPGC.

Basel Convention

India ratified the Basel Convention in 1992, and the national legislation is based on the Convention. The Ministry participates in various meetings of the Basel Convention regularly. India is also actively involved in the work relating to preparation of technical guidelines for environmentally sound management of ship breaking along with Norway and the Netherlands under this Convention. However, India still has to ratify the Basel Ban (Amendment II/XII, COP II).

Stockholm Convention

India ratified this Convention in 2003. The Convention seeks to eliminate production, use, import, and export of identified persistent organic pollutants (POPS) (currently 21). In India, except for dichloro-diphenyl-trichloroethane (DDT), all the other intentionally produced POPS in the original list of 12 have been discontinued, though dioxin and furans need attention as an unintended by-product of waste combustion and other activities. The status of the nine new POPS is still to be determined. Meanwhile, a Preliminary Enabling Activity Project to prepare a National Implementation Plan (NIP) for POPS has been assigned to ITRC in association with the United Nations Industrial Development Organization (UNIDO), but it is yet to be completed.

The problem of polychlorinated biphenyls (PCBs) has not been fully documented in India. Recently (January 2010), the UNIDO started a collaboration with the Government of India (US$ 14.5 million co-funded project by the Global Environment Facility [GEF]) for the phase-out of PCBs in the country. The project objective is to reduce or eliminate the use and release of PCBs into the environment through the disposal of PCB-containing equipment through Environmentally Sound management (ESM) procedures and to phase them out, evidently to comply with deadlines of the Convention.

Dioxins and Furans as unintended byproducts of medical waste disposal have also received attention through a new US$ 40 million project with UNIDO. The project will be implemented in five States with four large hospitals, eight medium-sized hospitals, and 16 small hospitals being benefited in each State. The project, which would help in reducing POPS, comes within the broad framework of the Country Programme of Technical Cooperation between India and the UNIDO signed in May 2008.

Another UNDP/GEF project, which has already commenced, will help reduce mercury, dioxins, and furans emissions from hospitals. The global, US$ 15 million project, is a 7-country initiative, with India being one of the countries. It focuses on protecting public health and the global environment and will demonstrate best environmental practices and best available technologies at healthcare facilities that have been selected to serve as models within seven countries including India, where it is being implemented in the States of Tamil Nadu and Uttar Pradesh.

Montreal Protocol

India ratified the Montreal Protocol in 1992. Further, Copenhagen, Montreal, and Beijing Amendments to the Montreal Protocol have also been ratified in 2003.

As per the reporting requirements under Article 7 of the Montreal Protocol, the data on production, import, export, and feedstock of CFCs, Halons, CTC, HCFC, and Methyl Bromide for each year is to be submitted to the Ozone Secretariat. Over the past 12 years, India has made significant progress in controlling the production and consumption of ODS. From a consumption level of 10,370 MT of ODS in 1991, the unconstrained demand was forecasted at about 96,000 MT by 2005, although the consumption of these substances by end-2004 was only about 9000 MT annually. These reductions were achieved with technical and financial assistance from the Multilateral Fund and implementing agencies and due to proactive policy actions. CFC has been completely phased out as of August 1, 2008.

Rotterdam Convention on Prior Informed Consent

India ratified this Convention in 2004. This Convention, which seeks to promote the exchange of information between parties for trade in the listed hazardous products, is relatively benign in its intent, and the implementation of it has been noncontroversial in India until recently. However, since 2008 India been opposing a proposal for the addition of the pesticide Endosulphan (India is the largest user and major producer at over 4500 mt per annum.) Earlier, India had successfully opposed the inclusion of chrysotile asbestos (India imports over 125,000 mt from countries like Canada) to the list. This led to the charge that India was protecting its commercial interests and diluting the Convention.

Mercury Treaty Intergovernmental Negotiations Committee

At its 25th session in February 2009, the UNEPGC agreed to elaborate a legally binding instrument on mercury, and asked UNEP to convene an intergovernmental negotiating committee (INC) with the mandate to prepare the legally binding instrument, commencing its work in 2010. The first session of the committee was held in June 2010, in Stockholm, Sweden and the second session was held in January 2011, in Chiba, Japan. In the interim, the Governing Council established an *ad hoc* open-ended working group to discuss the negotiating priorities, timetable, and organization of the INC. India participated in the meeting held in Bangkok in October 19–23, 2009.

Globally Harmonized System

The Globally Harmonized System (GHS) of Classification and Labeling of Chemicals was initiated at the Rio Summit (1992), adopted in 2002 by the United Nations Economic and Social Council's Subcommittee of Experts on the GHS (UNSCEGHS) and endorsed by United Nations Economic and Social Council (ECOSOC) in July 2003. The GHS provides a universal tool for chemical classification and hazard communication, to be made available to workers, consumers, and the public.

India is a signatory to the treaty concerning GHS. The MOEF and the Government of India have initiated some steps and established a dialogue with the industry, although it has not implemented this system fully as yet.

International Agencies

Several multilateral and bilateral agencies have been collaborating with the Government of India for projects relating to chemical management. Prime amongst them are the World Bank (WB, http://www.worldbank.org.in), German Agency for Technical Cooperation (GTZ, http://www.gtz.de), Department for International Development (DFID, http://www.dfid.gov.uk), the Asian Development Bank (ADB, http://www.adb.org), and the European Commission (EC). Many of them approach this as part of development assistance and poverty reduction country strategies. Some key initiatives include capacity building and pollution prevention programs (WB), hazardous and e-waste infrastructure development (GTZ), and most recently, examining the linkages between climate change and chemical safety issues (DIFID, GTZ, and World Health Organization [WHO]).

Besides UN agencies like the United Nations Development Programme (UNDP), United Nations Institute for Training and Research (UNITAR), UNIDO, and the WHO have contributed to this area. The work includes the Small Grants Programme (UNDP), participation in the preparation of the NIPS and a PCB management project (both UNIDO, see also the section "Stockholm Convention"), GEF–UNDP-related projects on POPS, developing the National Chemicals Management Profile (NCMP)(UNITAR), and working on mercury in the health care sector and on medical waste (WHO).

Other agencies have prepared reports such as on the ship breaking industry in India (Baily, 2000), and participated on issues related to food safety to strengthen the National Codex Committee under the information on the Codex Alimentarius Commission, Food and Agriculture Organization (FAO) (see also the section "The Food and Food Safety Act (2006), previously known as the Prevention of Food Adulteration Act (1954)"). India has also signed 40 International Labour Organization (ILO) Conventions dealing with workplace protection as well as safety issues.

National Chemicals Management Profile (NCMP, 2006)

This is the first of this type of report, developed by the MOEF with assistance from UNITAR and supported by the Canada–India Environmental Institutional Strengthening Project. The profile is a comprehensive information base on the chemicals in use in the country and their management framework. As a "first" version, it analyzes 10 major chemical sectors and potential impacts from production, use, and transport of the chemicals covered. It also examines the legislative, legal, and infrastructure aspects, besides identifying gaps and making recommendations (see also the sections "Implementation Issues" and "Data Gaps").

Bilateral Partnerships

India has entered into various bilateral environmental partnership agreements with some countries like the United States, Sweden, Norway, Denmark, and others, as also with the EU, which also include issues such as hazardous waste and mercury.

OTHER STAKEHOLDER ROLES

JUDICIARY

The Indian Judiciary has played a stellar role in directing implementation of laws relating to chemical management. There have been literally hundreds of judgments by the various High Courts as well as the Supreme Court of India, which have taken exceptional measures for the protection of the environment *per se*, but also in the area of chemical and hazardous substances, specifically. The judgments range from laying down principles, to ordering closures of polluting units, relocations, management of waste, examination of pesticides registration processes, and payment of worker compensation. The Courts have reiterated, amongst others, the "Polluter Pays" principle (Indian Council, 1996), the "Precautionary Principle" (Vellore, 1996), and asserted the state was only a "trustee of natural resources for public use and enjoyment" (Bangalore Medical Trust, AIR 1902).

The Indian Judiciary has allowed any person to bring in Public Interest Litigations (PIL) to the notice of the High Courts or the Supreme Court for the redressal of any fundamental rights and genuine violation of statutory provisions, provided it is not for personal gain, private profit, political motive, or any oblique consideration.* Over the past three decades, PILs have been used very effectively by nongovernmental organizations (NGOs) and individuals to obtain environmental orders, especially where there has been failure of the executive to protect the environment or human health.

An exhaustive overview is outside the scope of this chapter; however, the following examples illustrate the range of issues impacted.

In the well-known "Sriram Gas Leak Case" (Mehta, 1987), one of India's largest corporations was asked to relocate from the thickly populated city of Delhi. The Court also directed the Government to lay down a national policy for location of chemical and other hazardous industries, and established the principle of "absolute" liability. This led to the Indian Parliament enacting the Public Liability Insurance Act (1991) to give statutory recognition to no fault liability though limiting the maximum compensation to Rs. 25,000 (US$ 500). In another landmark case, the Kerala High Court stopped the operation of a large 10,000-ton ammonia storage tank that was operating in a giant public sector fertilizer company (Fertilizer and Chemicals Travencore Ltd, Udyogmandal, Kerala) in South India, as a preventive measure against accidental leakage in the city. The Court held that this was a "plain and clear violation of Article 21 of the Constitution of India" (Law Society, 1994).

The implementation of the Basel Convention and the management of hazardous waste has been driven by Courts. In 1996, the Delhi High Court issued the first order to ban the import of "toxic" wastes into India owing to rampant import of

* In 1981, Justice P. N. Bhagwati in *S. P. Gupta v. Union of India*, 1981 (Supp) SCC 87, articulated the concept of PIL.

scrap lead-acid batteries. Subsequently in 1996, the Supreme Court of India (Research Foundation, 1995) reaffirmed the ban, and set up an expert committee (HPCMHW, 2001) to investigate the situation in the country and recommend a way forward. This report is now a key guidance to the setting up of hazardous waste management infrastructure in the country.

Civil Society

Civil Society has played an important role in India to raise awareness of chemical safety issues as well as to help in policy formulation, even though issues relating to urban environment, waste, and industrial pollution are relatively new and need technical capacity. Many civil society organizations are involved in chemical safety issues including national organizations like Toxics Link (http://www.toxiclink.org), the Center for Science and Environment (http://www.cseindia.org), and Greenpeace (http://www.geenpeace.org/india). At the local State levels, NGOs like Thanal (http://www.thanal.org) in Kerala and Disha (http://www.dishaearth.org) in West Bengal have an important presence.

Civil Society International Networks

Many NGOs are also members of international civil society networks, which include IPEN (International POPS Elimination Network, http://www.ipen.org), HCWH (Health Care Without Harm, http://www.noharm.org), PAN (Pesticides Action Network, http://www.panap.net), BAN (Basel Action Network, http://www.ban.org), and GAIA (Global Alliance for Incinerator Alternatives, http://www.noburn.org). All the networks emerged after 1996, and rose mainly to respond to emerging needs such as participating in ongoing international multilateral treaty negotiations (mainly Basel Convention, Stockholm Convention), or to tackle the transboundary movement of polluting technologies and waste such as waste incinerators or hazardous waste. They have, over time, become established, and have developed capacities to raise issues and present contextual solutions. Similarly in response to the problem of mercury, which was recognized with the first UNEP Global Mercury Assessment (UNEP, 2003), Indian groups have come together under the banners of the Zero Mercury Working Group (ZMWG, http://www.zeromercury.org) and the IPEN Heavy Metal Working Group to participate in the upcoming mercury Intergovernmental Negotiations (INC) for a new international multilateral treaty on mercury scheduled to be completed in 2013. Another international alliance, which includes Indian NGOs, is the NGO Platform for Ship Breaking (http://www.shipbreakingplatform.org), which is spearheading a campaign on this issue.

Impacts

Civil society actors have been influential in creating national policy and are helping in opening up space for cleaner alternatives. Noteworthy are issues of hazardous and medical waste, immunization waste, food and water contamination by heavy metals and pesticides, electronic waste, POPS, mercury, and lead in toys and paints. The strategies employed include research, advocacy, awareness raising, training, capacity building, and also catalyzing the Courts. Global policy at forums such as at the

IFCS, SAICM, UNEP, and WHO has been informed as well. For example, "lead in paints" was adopted as an emerging issue at SAICM (May 2009), after it was proposed by Toxics Link and IFCS leading to a Global Partnership under UNEP and WHO (SAICM, 2009).

Challenges

However, the challenges faced by civil society cannot be overlooked. Often NGOs in India have been the targets of strategic lawsuits against public participation (SLAPP)* like court cases (both criminal as well as civil) filed against them by the industry, many of which are ongoing (Narain, 2006). In addition, financial support for such activities is relatively small, and sustaining the efforts over a longer period is challenging.

Labor/Trade Unions

In India, labor unions exist under the Trade Unions Act of 1926. Worker protection from chemical risks is one of the biggest under-recognized challenges in India. Even though labor unions have had a strong political presence in the sub-continent with a long history starting from 1890, their participation in occupations and chemical safety issues has been very sparse, despite their global counterparts having a history of this. These issues have been a low priority of trade unions here, which have been mostly engaged with more classical wage negotiation roles. The state too has desisted from enunciating its role in the various environmental laws. Legally, such an opportunity to participate has only been provided through various laws dealing with the workplace, such as the Factories Act (see the sections "The Factories Act (1948)" and "Labor Safety"), since India lacks a comprehensive occupational health and safety legislation such as Occupational Health and Safety Acts in some counties.

Absence

The reasons for this glaring absence are many, and some lie in the difficulty or reluctance to diagnose chemical effects on workers, especially from chronic exposures. Even the national, state-run Employee State Insurance (ESI) scheme, which provides mandatory health coverage to workers, is ill equipped, both in terms of doctors' abilities or diagnostic facilities, to identify such effects. This alone greatly reduces the perceptions of risks or impacts, and a possibility for labor unions to take up these issues. In addition, often no worker register of health effects exits. Another complicating factor is that less than 10% of the total work force is organized,† and the majority, which include women and children, work as migrant or agricultural labor, who are not unionized, are informal, and hence not covered through the ESI scheme. Their vulnerability is highest on many counts, as in cases of stone quarry workers suffering from silicosis, an "occupational" exposure listed under the Factories Act,

* A strategic lawsuit against public participation (SLAPP) is a lawsuit that is intended to censor, intimidate, and silence critics by burdening them with the cost of a legal defense until they abandon their criticism or opposition.

† The National Sample Survey (NSS, http://www.mospi.nic.in) conducted in 1999–2000 shows a total work force to be 406 million (as of January 1, 2000), with only 7% employed in the formal or organized sector while the remaining 93% worked in the informal or unorganized sector.

but misdiagnosed as tuberculosis, an "environmental" disease, which carries no liability or compensation. This leaves a majority of workers who are employed in hazardous industries such as textiles, dying, chemicals, and mining sectors very vulnerable to chemical exposures and risks, without recourse, and nowhere to turn to, since labor unions are unable to protect their interests.

Initiatives

Despite these handicaps, there have been some important efforts by key national trade unions to participate in conversations related to worker safety and chemical exposures. These include the Centre for Trade Unions (CITU), the All India Trade Union Congress (AITUC), and Mumbai Port Trust Dock and General Employees Union (MPTDGU). They have taken positions and even conducted joint activities along with environmental NGOs to address hazards in ship breaking, the recycling of hazardous wastes, asbestos and mercury-based products manufacturing, and stone quarrying. Other initiatives include improved attention to conditions of work at the workplace, and this has been supported by the ILO as well. As has been the case elsewhere, here too perceived conflicts between environmental protection and a loss of livelihood have been stumbling blocks in forming larger and more sustainable alliances.

Stakeholder Participation

Consultation, especially with public interest stakeholders, has had a mixed record in India, improving or deteriorating depending on the administration at any given time. There are no formally laid down procedures or institutionalized forums within the Government for seeking regular stakeholder inputs on chemicals safety and management issues *per se*, and this is more a matter of practice. For example, in the period 2004–2006, NGOs were literally barred from being invited for consultations in the MOEF, while the situation improved in 2009. Only in the event of a new set of Rules being proposed is it mandatory to seek public comments before the "draft" Rules can be finally notified. Public hearings are also mandated in case of certain new industrial projects that require undergoing an environmental impact assessment (EIA), as defined under the Environment Impact Assessment Notification (2006). However, it is not disclosed which concerns or comments have been taken on. On other occasions, NGOs have been selectively invited for consultations on a case-by-case basis depending on the issue and their area of work. Several high-level policy and technical committees constituted by the MOEF and the CPCB also have NGOs as members. The areas include national standards for waste disposal technologies, and various chemical- and waste-related policies and rules. Government representatives also participate in various meetings and consultations organized by better known public interest NGOs, especially those that have a track record in the area. However, it is rare for such consultation or participation to be held with local or smaller community groups.

On the other hand, Government and industry consultation and participation on various forums are far more visible and frequent. For example, industry has been part of the preparation of the National Chemical Profile, while public-interest NGOs were largely excluded. In some cases, the Government forms high-level interministerial commissions and coordinating mechanisms to deal with inter-sector issues arising from various chemical and waste treaties or areas dealing with pesticides and food safety regulation. Here the stakeholder ministries have included health, industry, chemicals, agriculture, and representatives of the chemical industry. Consultations with labor unions are nonexistent.

CURRENT STATUS

It is increasingly evident that poor environmental quality and chemical exposures have adversely affected human health. Environmental factors are estimated to being responsible for nearly 20% of the burden of disease in India (MOEF, 2006). Likewise, urban waste recycling is carried on largely in unorganized backyard operations, which have little capacity to deal with chemical hazards. Consumers are ill-informed about chemical safety issues, although some of them have only just begun to engage with issues of food and water contamination, mercury, and lead in products. Labeling for chemicals in products is missing as well.

Also "soil pollution from heavy metals due to improper disposal of industrial effluents, along with the excessive use of pesticides and mismanagement of domestic and municipal wastes, is becoming a major concern. Though no reliable estimates are available to depict the exact extent and degree of this type of land degradation, it is believed that the problem is extensive and its effects are significant" (MOEF, 2009a). In addition, several severely contaminated industrial hotspots exist, especially in more industrialized States (CPCB, 2009). Chemical effluents from industrial estates have also caused crops to be contaminated (Toxics Link, 2009).

IMPLEMENTATION ISSUES

Implementation of environmental laws has been India's Achilles' heel. The reasons are both a lack of capacity as well as structural inadequacies in environmental governance. The current system is inadequate to keep pace with rapid industrialization and its outfall. Several disputes relating to new industrial siting, violations of emission norms, contamination of vital resources like ground and river waters, use of toxics chemicals in products, or the continued use of chemicals banned in other countries have arisen in the past two decades.

The National Chemicals Profile (NCMP, 2006) suggests measures to improve implementation. These include mandating a chemicals registration system, strengthening of laboratory accreditation systems, an increase of resource allocation for developing infrastructure for chemical emergencies, improving public awareness in relation to chemical safety matters through educational programs, and providing information to the decision makers and the public at large concerning chemicals.

Significantly, the report recommends the setting up of a legally bound permanent mechanism for coordination and cooperation, with formal procedures for broad consultation among stakeholders laid down and encouraged, aside from creating an institutional monitoring system for improving compliance.

Encouragingly, the Government has taken this on board. Acknowledging that regulatory bodies do not "have the capacity or the resources to ensure compliance with various environmental regulations" (MOEF, 2009b), and the fact that the matter of disposal of toxics waste in Bhopal's erstwhile UCIL plant has not been resolved 25 years after the Gas Tragedy illustrates the point (Parliament of India, 2007). There is a new discussion to set up an independent National Environmental Protection Authority.

Data Gaps

Systematic and usable data for policy making are still unavailable in India. According to the NCMP (2006) report, it is recognized that "there is a general lack of health-related data, either in relation to public health or the workplace, where one-third of work-related deaths are due to chemicals. Exposure data, where very little exists, and location sensitivity is not well understood and needs attention for all categories. The use of available data for local risk assessment and other related activities still has limitations."

The report also outlines a need for improved data collection and its management. In particular, it states the need for (a) harmonized definitions even within sectors, (b) a standardization of data reporting, (c) the resolution of large differences between installed capacity and production of chemicals in some sectors, and (d) strengthening overall data on consumer chemicals. It calls for the development of methodologies for including accurate data on raw materials and intermediates, reduction of time lag for data collection, greater coordination on importation data, and the need for tracking certain key chemicals, for example, mercury.

Consumer and worker health-related data are also poor. This is complicated by the fact that over 70% of the workforce is agricultural labor, while the rest is a constantly mobile migrant labor working in urban sites. Less than 10% of the labor force works in the organized sectors. Owing to the nature of migration, documentation of either immediate or chronic effects from chemical exposures on workers does not exist.

Industrial Clusters and Contaminated Sites

The rapid industrialization in India has created chemical-based contamination in several places. Based on a Comprehensive Environmental Pollution Index (CEPI), 88 polluted industries clusters have been identified, in order of contamination. Out of the 88 clusters, 43 are "critically" polluted. Many of these industrial clusters have a mix of SMEs and manufacture pharmaceuticals, drugs, dye and dye intermediaries, pesticides, chlor alkali manufacturing, sponge iron manufacturing, etc. (CPCB, 2009). However, contaminated sites still need proper definition and do not have special remediation or preventive provisions to deal with them. Cleaning them up is an expensive and complex issue.

Industrial Hazardous Wastes

With growing industrialization, India's generation of hazardous wastes has been rising and is currently slated to be about 4.4 million mt per year (CPCB, 2008).

Petrochemical, pharmaceutical, pesticide, paints, dyes, fertilizer, and ship-breaking industries top the list of waste generators. Since more than 50% of the waste generated is from SMEs, like leather tanning, textile dying and printing, chemical production, electroplating, and so on, this is scattered all over the country. Further, heavy metal contamination from thermal power, tannery, and mining activities has occurred in several locations. The Central and SPCBs have recently identified 1532 "grossly polluting" industries in India, although almost none of the industries comply with the emission standards (HPCMHW, 2001).

Wastewater generation from industry has been estimated to be 13,500 million cubic meters per day, of which 5500 million cubic meters are dumped untreated directly into local rivers and streams without prior treatment (CPCB, 2009). It is often contaminated with highly toxic organic and inorganic substances, some of which are persistent pollutants. This has been one of the factors that has caused all of India's 14 major river systems to be heavily polluted. Though Common Effluent Treatment Facilities (CETFs) have been set up, they are plagued by problems such as inadequate power, lack of downstream landfills, and remixing of treated with untreated effluents. Most industries either are not connected to CETFs or only partially treat their wastewater before disposal.

Infrastructure for Hazardous Wastes

There has been progress made on this front. Twenty-two Common Treatment, Storage, and Disposal Facilities (TSDFs) have been set up in 10 states for a total waste handling capacity of about 1.5 million mta, leaving a gap of about 1.2 million mta, which has already been inventoried for land disposable hazardous wastes, and need, additional infrastructure. All these have been made on a Build Own Operate Model (CPCB, 2009).

Other Hazardous Wastes

Other hazardous waste streams are part of household municipal waste, and include household medical waste, mercury and Ni–Cd batteries, mercury-based thermometers, and florescent lamps, particularly owing to new shifts to CFLs being promoted to save energy. Currently, there are no separate collection and recycling or disposal systems for this portion of the waste stream.

E-Waste

Along with growing computerization with new users and rural computerization plans (about 40 million PCs until 2009) and a growing base of mobile phones (currently more than 500 million connections), there is mounting hazardous electronic and electric waste, currently at over 400,000 mt per year (MAIT–GTZ, 2009). This does not include other white goods like refrigerators and televisions sets. Over 95% of the recycling is carried out in the informal sector, employing women and children who are exposed to acid fumes, toxic compounds like dioxins, heavy metals, brominated flame retardants, and occupational hazards. They recover precious metals like gold, aluminum, palladium, and also copper, plastics, and glass (Toxics Link, 2003). There are new investments in recycling, and a new law is expected very soon to incorporate both EPR (extended producer responsibility) as well as RoHS-type provisions, but owing to the large grey markets and informal sector involvement, the problem involves issues of livelihoods as well and is a challenging one.

Biomedical Waste

Biomedical waste was brought under legislation in 1998. Although the waste collection and disposal is improving owing to the setting up of more than 160 Common Waste Treatment facilities under public-private partnerships, only about 50% of the waste generated is being collected (CAG, 2007). The waste rules discourage on-site incineration as a means of treatment, and prohibit the incineration of chlorine-based waste like PVC plastics to prevent the production of dioxins and furans. Primary health centers and the implementation of this law pose special challenges in rural areas. In 2007, the Ministry of Health included the issue of waste generated there as part of national policy (MOHFW, 2007).

Ship Breaking

India has one of the largest ship-breaking yards in the world at Alang, Gujarat, with over 170 beaching yards. Employing up to 40,000 workers, serious injuries and even death of workers have been reported constantly besides concerns about their health and safety from chemical contamination. The conditions of disposing of ships on the yards are controversial because they are beached (as opposed to docked) and cut open manually. Many of them contain hazardous substances like asbestos, PCB oils, and toxic paints containing lead and tributyltin (TBTs), making their safe disposal at site a near impossibility. Of late, though on the grounds, infrastructure and working conditions have improved with better health and safety measures as well as improved disposal of hazardous wastes. Several ships have drawn international attention, although the French navy's aircraft carrier, the Clemenceau, is the best known among them. Containing asbestos, it was sent to Alang, but later recalled, amid huge controversy, at the instance of the then French Premier (Greenpeace, 2008). The CPCB had issued guidelines for environmentally sound ship breaking as early as 1998 (CPCB, 1998); however, these have been violated several times and have been the subject of Supreme Court investigations. Evidently owing mainly to the yards' contribution to scrap steel, up to 4 million tonnes per year, the Government has been ambiguous about the continuing imports, tending to treat ships as "nonwaste."

Global Linkages

Outsourcing, shifting of production bases into India, and exports of products are growth opportunities, if taken responsibly. In the textile and apparel sector, for example, major international brands such as H&M, Ikea, Banana Republic, and Tommy Hilfiger are now outsourcing materials from India, putting pressure to green supply chains in this generally very polluting activity, which is also an important user of chemicals. The leather tannery industry is another export-based sector, which has been in the eye of the storm. Additionally, chemicals is one of the few high-priority sectors in which 100% Foreign Direct Investment (FDI) is allowed, and FDI in it grew at 227% in FY2009 (ASSOCHAM, 2009). On the other hand, under new Free Trade Agreements (FTA) being negotiated (e.g., with Japan and the EU), waste is being referred to as "non-new goods"* to sidestep legally bound definitions, and to permit its trade (DNA, 2010). Simultaneously, waste imports of electronics goods, plastics, scrap batteries, and waste oil have been permitted.

There is also a reported relocation of industry from other countries to India, such as bulk drugs, often after this has been restricted overseas (Reddy, 2007). For example, in 1984, the multinational company Ponds was permitted to set up a second-hand 100% export-based mercury thermometer manufacturing plant in Kodikanal, South India. Subsequently acquired by Unilever in 1997, it was found to have caused mercury contamination hundreds of times more than the permissible limits (Greenpeace, 2003). The giant MNCs Pepsi and Coke were found to contain high levels of pesticides in their soft drinks despite this not being present in their European and U.S. products (CSE, 2003). Lead in paints was found in India, in brands of international majors again despite their overseas products not containing this toxic metal (Toxics Link, 2007).

International Treaties

Despite India having ratified international treaties, its position in them has changed along with its economic interests. As an early supporter and leader of the G77 group of countries, India was one of the early supporters of the call for a ban on the international trade of hazardous waste. It helped negotiate a strong Basel Convention, and even supported a call for the Basel Ban in 1994. The Ban Amendment (decision II/XII, COP 2) called for a ban on the transboundary movement of hazardous waste either for dumping or for recycling. However, today India has become one of the main opponents of the Basel Ban Amendment and has steadfastly refused to ratify it. In fact, India has now openly supported the import of hazardous waste into the country, as a means to build India as an international recycling hub. For example, the Government of India for the first time issued an import license for electronic waste, which is a Basel-listed waste, to a private company to source materials from the United States and the United Kingdom. Under Article 11 of the Basel Convention, it is illegal for India to trade with the United States in the absence of a bilateral treaty (Toxics Link, 2009).

Also both in the Rotterdam Convention and the Stockholm Convention, India has been actively supporting its trade interests. It has blocked all efforts to place Asbestos and Endosulphan (IEP, 2008), two substances that have been cleared by the Convention's scientific committee for inclusion, and has been opposing the inclusion

* Article X-15 of the draft India–EU FTA talks about "non-new goods," and says that neither party shall apply to non-new goods, measures, including enforcement measures, which are more restrictive than to new goods. The treaty is expected to be finalized in early 2010.

of Endosulphan in the POP Review Committee of the Stockholm Convention. India has also not as yet ratified the ILO Chemical Convention C170 (1990) regarding safety in the use of chemicals at work.

FUTURE OUTLOOK

The chemicals sector in India, largely protected till recently, is slated to double its output by 2015 compared to 2009, to service both domestic and overseas markets. Domestic consumption of chemicals is growing quickly, but with the opening up of the economy, the competition will be from more efficient and lower cost global players. With increased investments from overseas companies, the industry would have to collaborate and consolidate capacity to survive. Simultaneously, domestic hazardous waste generation is rising fast with increased industrialization and urbanization, with newer waste streams like e-waste being added to the mix. The capacity to manage these is seriously wanting, and will need urgent infrastructure augmentation. Also, evidence of impacts on public health, rivers, groundwater, and land is growing rapidly. Areas such as data generation, information transparency, stakeholder participation, and regulatory compliance will be key, needing structural and attitudinal changes at the governmental level. It is also a time of opportunity for the chemical industry, but to partake in it will need a change from "business as usual." Till now, the emphasis has been on end-of-pipe measures. Cleaner production, toxic-free products, waste minimization, and stakeholder involvement have not been a focus. The burden will have to shift to upstream measures, in order to integrate chemical safety issues cost effectively, and to compete in an increasingly globalized environment and a more aware and demanding consumer. Not surprisingly though, in the immediate term, the 2020 MDG (Millennium Development Goals), which India committed itself to, seems to be a far cry from being achieved.

REFERENCES

ASSOCHAM—The Associated Chambers of Commerce and Industry of India. 2009. Press release, July 12, 2009.

Baily, P. J. 2000. Is there a decent way to break ships, a discussion paper, International Labour Organization. Available at: http://www.ilo.org/public/english/dialogue/sector/papers/shpbreak/

Bangalore Medical Trust v B.S. Muddappa AIR SC 1902, 1911, 1924 and other cases, p. 42. In *Environmental Law and Policy in India.* D. Shyam and A. Rosencranz, eds., Oxford University Press, India.

Center for Science and Environment. 2003. Colonisation's Dirty Dozen. Available at: http://www.cseindia.org/node/491

Central Pollution Control Board. 1998. Environmental Guidelines for Ship-breaking Industry. Available at: http://cpcbenvis.nic.in/newsletter/hwastemanagjun1998/jun98vi.htm

Central Pollution Control Board. 2008. National Inventory of Hazardous Waste Generating Industries and Hazardous Waste Management in India.

Central Pollution Control Board. 2009. Comprehensive Environmental Assessment of Industrial Clusters. Available at: http://www.cpcb.nic.in

Central Pollution Control Board. 2009. Wastewater Generation and Treatment. Available at: http://www.cpcb.nic.in/oldwebsite/News%20Letters/Latest/ch2–0205.html

Comptroller and Auditor General of India. 2007. Performance Audit of Management of Waste in India (PG).

DNA, January 12, 2010. Available at: http://www.dnaindia.com/money/report_free-trade-agreements-may-aid-toxic-waste-trade_1333702

Greenpeace. 2003. Kodaikanal, Tamil Nadu. http://www.greenpeace.org/india/en/about/Success-Stories/

Greenpeace. 2008. Available at: http://www.greenpeace.org/india/campaigns/toxics-free-future/ship-breaking/deathship

HPCMHW. 2001. Report of the High Powered Committee on the Management of Hazardous Wastes, Menon, M.G.K (chairperson), Supreme Court of India.

Indian Council for Enviro Legal Action v Union of India (Bichiri Case) AIR 1996 SC 1446, 1466, p. 42. In *Environmental Law and Policy in India*. D. Shyam and A. Rosencranz, eds. Oxford University Press.

Indian Environmental Portal. 2008. Asbestos, Endosulphan Escape Blacklist. Available at: http://www.indiaenvironmentportal.org.in/node/267090

KPMG. 2009. Indian Chemical Industry Report Vision, KPMG and CHEMTECH Foundation, Available at: http://www.in.kpmg.com/TL_Files/Pictures/KPMG_Chemtech_Report.pdf

Law Society of India v Fertilizer & Travencore Ltd., AIR 1994 KER 308, pp. 537–540.

MAIT—GTZ. 2009. E Waste Generation in India, GTZ-MAIT Study. Available at: http://www.mait.com/admin/press_images/press77-try.htm

Mehta, M. C. v Union of India, AIR 1987 SC 982, pp. 521–536. In *Environmental Law and Policy in India*. D. Shyam, and A. Rosencranz, eds. Oxford University Press.

Ministry of Environment and Forests. 2006. National Environmental Policy. Available at: http://www.envfor.nioc.in

Ministry of Environment and Forests. 2009a. State of India's Environment Report. Available at: envfor.nic.in

Ministry of Environment and Forests. 2009b. Proposal for a National Environmental Protection Authority, Discussion Paper. Available at: http://www.envfor.nic.in

MOH—Ministry of Housing and Urban Poverty Alleviation. 2009. Urban Poverty Report.

MOHFW—Ministry of Health and Family Welfare. 2007. Infection Management and Environmental Plan, Policy Framework, National Rural Health Mission. Available at: http://www.mohfw.nic.in/NRHM/IMEP/IMEPPolicyFramework.pdf

MSME—Ministry of Micro, Small and Medium Enterprises. 2002. Third Census, Government of India. Available at: http://www.dcmsme.gov.in

Narain, S. 2006. Strategic SLAPPs, Down to Earth, May 23, 2006.

NCMP. 2006. Ministry of Environment and Forests and UNITAR National Chemical Management Profile for India. Available at: http://cpcb.nic.in/upload/NewItems/NewItem_112_nationalchemicalmgmtprofileforindia.pdf

NEP. 2008. National Environment Policy, p6. Ministry of Environment and Forests, Government of India. Available at: http://www.envfor.nic.in

Parliament of India, Rajya Sabha. 2008. One Hundred and Ninety Second Report on Functioning of Central Pollution Control Board. Available at: http://rajyasabha.nic.in/

Planning Commission of India, Govt. of India. 2007. Eleventh Five Year Plan, Vol. II, p. 180.

Reddy, B. S. 2007. Globalisation and sustainable development, p 296. In *Ecology and Human Well-Being*. P. Kumar and B. S. Reddy, eds. New Delhi: Sage.

Research Foundation for Science, Technology and Natural Resource Policy v Union of India and others, writ petition (civil) no 657/1995.

RTI Act. 2005. Government of India. Available at: http://righttoinformation.gov.in/

SAICM. 2009. Resolution II/4 on Emerging Policy Issues, ICCM2, May 2009. Available at: http://www.saicm.org

Toxics Alert. 2009. How Safe is the Food you Eat? Toxics Link. Available at: http://enews.toxicslink.org/report-view.php?id=10. Food for Thought, Factsheet. Available at: http://www.toxicslink.org/ovrvw-int.php?prognum=4&intnum=12&area=2

Toxics Link. 2003. High Tech Scrap, E Waste in India.

Toxics Link. 2007. A Brush with Toxics. Available at: http://www.toxicslink.org/pub-view.php?pubnum=179.

Toxics Link. 2009. Government opens floodgates for e-waste dumping, Press release.

UNEP. 2003. *Global Mercury Assessment Report.* Available at: http://www.chem.unep.ch/mercury

Vellore Citizens Welfare Forum v Union of India AIR 1996 SC 2715, 2712 and other cases, p. 42. In *Environmental Law and Policy in India.* D. Shyam, and A. Rosencranz, eds. Oxford University Press.

40 São Tomé and Príncipe and the Issue of Persistent Organic Pollutants

*Arlindo Carvalho**

CONTENTS

INTRODUCTION

São Tomé and Príncipe is a small island state, consisting of the two islands of São Tomé and Príncipe. It is situated about 200 km west of the African continent on the equator. It has a population of about 160,000, the greater part living on the São Tomé Island. It has a mainly agricultural economy, based on the historical plantations of cocoa and coffee and is the smallest economy of Africa.

Given the agricultural activities, it is not surprising that certain amounts of obsolete pesticides, including some persistent organic pollutants (POPS), are present on the islands.

THE POPS

Essentially manmade in origin, persistent, bioaccumulating and prone to transport over long distances, far from the place of release, POPS are travelers without borders

* The opinions expressed in this chapter are those of the author and do not necessarily represent the views of any organization.

that do not escape to any nation in the world, contaminating more or less severely the environment in their biotic as well as their abiotic compartments. Among negative impacts, one can mention those on the genetic health, intellectual performance, and immunological defense. Several POPS are also considered carcinogenic and are suspected of causing birth defects.

São Tomé and Príncipe can therefore not escape the problems related to POPS: humans and their living environment are undoubtedly exposed, to a certain extent, to negative impacts of POPS utilized voluntarily (substances in Annexes A and B of the Stockholm Convention) or produced without intention during burning processes or in certain industrial processes (those on the list in Annex C of the Convention).

The quality of a state while having the problem of POPS has not been addressed in São Tomé and Príncipe with the level of priority that it deserves. This has close links to the health of the population. The issue was raised by the Stockholm Convention, aiming at protecting human health and the environment, is relatively new and poorly understood in most of its aspects in São Tomé and Príncipe by decision makers, workers, and the general public.

Having decided to take an active part together with the international community in its endeavors to fight the negative impacts of POPS at the global level, São Tomé and Príncipe has participated in the negotiations of the text of the Stockholm Convention that it signed on April 1, 2002, and ratified by presidential decree no 3/2006 on February 8, 2006.

In its quality of a state having signed the Convention, São Tomé and Príncipe has benefited from support of the Global Environment Fund (GEF) for the realization of habilitating action in the framework of a project entitled "Habilitating activities aiming at initial action for the implementation of the Stockholm Convention on persistent organic pollutants (POPS) in São Tomé e Príncipe."

The main goal of the project for habilitating action was to assist São Tomé and Príncipe in its preparation of its national plan of implementation of the Convention, in conformity with its Article 7. In addition, this project aimed at strengthening national capacities for the management of POPS at a global level, maximizing the involvement of the state, and to facilitate the ratification of the Convention.

São Tomé and Príncipe has chosen the United Nations Industrial Development Organisation (UNIDO) as the executing agency to assist in the implementation of the Convention.

At the national level, the Ministry of the Environment and Natural Resources, through the Directorate General of the Environment, is the national executing agency for the project of habilitating action.

This project has allowed, for the first time, to proceed with a full analysis of the situation in São Tomé and Príncipe regarding the issue of POPS. Initial inventories of sources and quantities of emissions have been made, and exposure of humans and the environment and impacts on their health have been examined. On top of this, an evaluation of national resources for the management of chemical products, including POPS, has been made through the preparation of a chemical profile with the financial and methodological assistance of UNITAR.

These inventories and evaluations now give sufficient insight into the problems related to POPS in the country as well as the nature and the quantities of emissions, the existing policies and legislative framework, socioeconomic, health and environmental impacts, and proposals for starting to work on solutions.

EVALUATION OF CHEMICAL PRODUCTS OF ANNEX A (PART I): POPS PESTICIDES

The substances of the first part of Annex A of the Convention are organochlorine pesticides of the first generation: aldrin, chlordane, dieldrin, endrin, heptachlor, hexachlorobenzene (HCB), mirex, and toxaphene.

The initial national inventory of these POP pesticides has been carried out on the whole national territory, according to the method of physical inspection, in conformity with the relevant Food and Agriculture Organization (FAO) guidelines. The data sought concern the identity of the POP pesticides, the amounts used or for sale, the types of use, the stocks of obsolete pesticides, and the practices for their disposal, the sites potentially contaminated by POP pesticides as well as the regulations, the level of awareness of stakeholders, the infrastructure and the national capacities for monitoring, available monitoring data, and vulnerable social groups.

São Tomé and Príncipe is neither a producer nor exporter of POP pesticides. But in the past, in colonial times (ending in 1975) POP pesticides have been imported (endrin, dieldrin, heptachlor, mirex, toxaphene, and aldrin), for agricultural and public health uses. All substances mentioned in Annex A (Part I) are no longer officially used in São Tomé and Príncipe, either in agriculture or in public health.

In agriculture, these products are currently replaced with biodegradable organic chemicals such as endosulfan, acephate, deltamethrin, fenitrothion, aldicarb, phostoxin, and so on.

For vector control, use is made of synthetic pyrethroids such as α-cypermethrin (interior spraying of houses) and deltamethrin and permethrin (impregnation of mosquito nets).

No obsolete stock of pesticides from Annex A of the Convention has been found during the inventory.

The study also reveals the absence of relevant legislation and the nonadherence of the country to certain important conventions, such as the Rotterdam (prior informed consent) Convention. The Basel Convention (transboundary transport of hazardous waste) was ratified in June 2010.

The country currently does not have sufficient human, material, or financial resources to cope with the sound management of POP pesticides. Moreover, the absence of a responsible body for the control of import, production, sale, and use of hazardous chemical products needs particular attention in the framework of the implementation of the Convention on POPS in the area of international collaboration.

EVALUATION CONCERNING CHEMICAL PRODUCTS OF ANNEX A (PART II): POLYCHLORINATED BIPHENYLS

PCBs have, of course, been imported into the country for use in production and distribution of electricity as components in dielectric fluids in transformers and condensers.

The results of the inventory carried out in 2005 are incomplete and do not allow us to make a projection of the evolution in time (histogram of age and histogram of quantities from 2005 to 2025) of the quantities of PCB dielectric fluids and their wastes for disposal.

PCBs have been reported in sites that dispose of transformers and other electric equipment such as condensers and circuit breakers.

For the whole of the reported equipment, 129 transformers and 17 condensers and circuit breakers have been reported, with a total estimate of 17.5 tons of dielectric fluid suspected of being PCBs. The inventory has shown one transformer in the "Fábrica de Confecções de Água Grande" that contains 235 kg of PCB dielectric fluid.

For 45 transformers (48% of the total investigated) information on the quantities of PCB suspected oils is available, but the names of the dielectric fluids and of the manufacturers are not exactly known. The others are considered as being PCB.

The amount of equipment out of use is 17 circuit breakers of 6–30 kV. For the electric equipment in use, the inventory has detected 127 pieces of equipment expected to be PCB contaminated.

Two pieces of equipment that are suspected of PCB content are in high poles of Conceição et de Alfândega.

Seventeen circuit breakers have been registered in the stock of EMAE (company for electricity production) dating from 1940 and 1950 that have been replaced in 2000 with interior circuit breakers of type SF6 ORTHOFLUOR type FP.

Among the 129 transformers in the inventory, 36 are indoors, 83 are on poles and two of them do present leakages. In total, 83 transformers on high poles have been identified in 123 sites visited. The equipment on poles for distribution of electricity consists mainly of transformers.

The completion of this inventory (physical inspection of electrical equipment, identification test for PCBs, labeling of PCB equipment) is a priority for the action plan of the management of PCBs and the equipment containing them, in the framework of the implementation of the National Implementation Plan.

EVALUATION OF DICHLORODIPHENYLTRICHLOROETHANE; ANNEX B

Like the pesticides of Annex A, DDT has never been produced industrially in São Tomé and Príncipe. Like all other organochlorine pesticides, it has been imported in the country for use in agriculture (culture of cocoa, coffee, and market gardening), animal health, and public health, in particular in fruitless attempts to eradicate malaria by large-scale airborne spraying during the colonial period (1954) and, much later, from 1981 to 1983. During this last period, it has been estimated that a total of 138 tons was used for aerial spraying, finally being abandoned because of the appearance of resistance of the mosquito.

Today, in the chemical vector, control synthetic pyrethrinoids are being used, such as: α-cypermethrin (indoor spraying) and deltamethrin and permethrin (impregnation of mosquito nets).

It has to be emphasized that the nonchemical methods of control, based on prevention and prophylaxis among pregnant women, raising awareness among the population, hygienic measures and sanitation, and the use of mechanic barriers (simple mosquito nets, impregnated mosquito nets, small-mesh wire-netting, etc.) form the cornerstone of the current strategy to control malaria in São Tomé and Príncipe.

The 2005 inventory has allowed noting that the biggest stock of POPS pesticides is about 500 kg of DDT identified in a storage building in Morro Carregado, district of Lobata in the northern region of São Tomé Island.

Next to this storage building is a container of unidentified pesticides that come from CIAT (International Center for Tropical Agriculture). Leaking of products, ruptured sacks, and beginning of exterior oxidation of metal barrels have been found. It is most probably also a contaminated site.

EVALUATION REGARDING THE EMISSIONS OF CHEMICAL SUBSTANCES FROM ANNEX C

Unlike POP pesticides and PCBs, explicitly produced by humans for different uses, polychlorodibenzodioxins (PCDD or dioxins) and polychlorodibenzofurans (PCDF or furans) have never intentionally been produced on a large scale for any specific use.

On the contrary, this category of POPS comprising 135 congeners of PCDD and 75 of PCDF occurs by accident as by-products of combustion processes (high temperature incineration, burning of biomass, burning of fossil fuels, etc.), of certain industrial processes (metallurgy, chemical production, paper production, mining, etc.), and also certain biological processes (biomethanization, compositing, etc.).

Next to PCDD/PCDF, Annex C contains equally hexacholorobenzene (HCB) and PCBs that are also formed as unintentional by-products in similar conditions as those that generate dioxins and furans. Nonintentional POPS are essentially due to human activities, but natural sources (volcanic activity) are also known. These last products are not covered by the Convention.

The standardized methodology that is recommended by UNEP for the inventory of unintentional by-products (toolkit) for the moment only addresses the identification of different sources of emission of dioxins and furans and their quantification. In accordance with this methodology, based on the source identification, the determination of their activity statistics and the use of emission factors, the initial national inventory of dioxins and furans was completed in São Tomé and Príncipe in 2005.

Of all POPS, dioxins and furans are the least known, even almost unknown in the country and very few people from São Tomé and Príncipe have had the opportunity to hear about dioxins and furans.

The inventory of these substances in São Tomé and Príncipe has allowed identifying the burning of waste (household and biomedical waste), uncontrolled processes of combustion, and transport as the main sources of emission of dioxins and furans.

As a conclusion of the inventory and the study of the data collected, the value of dioxin and furan missions in São Tomé has been estimated at 16.58 gTEQ per year in 2005.

INFORMATION REGARDING OBSOLETE STOCKS AND CONTAMINATED SITES

POPS Pesticides

The storage building and site of Morro Carregado contains the biggest stock of POPS pesticides and has about 500 kg of DDT and other dangerous products. This storage is located in the Praia das Conchas zone, district of Lobata in the Northern region of São Tomé Island. Next to the storage stands a container with chemical products/pesticides from CIAT.

The poor condition of the storage of these products makes them a real threat to the health of humans and for the environment and could form a secondary source of emissions of POP pesticides.

Technical and financial assistance in the framework of the common project *Africa Stockpile Program* between the African Union and the World Bank aiming at removing all stocks of obsolete pesticides from Africa is an opportunity for São Tomé and Príncipe. It is now necessary to create awareness among the main stakeholders to this threat to avoid the implementation of environmentally unsound elimination practices such as uncontrolled landfill (a source of contamination of soils, water, and groundwater) or combustion in open air, and so on.

The presence of a significant stock of 500 kg DDT, of waste, and of a potentially (most probably) contaminated site is a risk for human health and the environment. This stock and its wastes need to be secured urgently and their immediate ecologically sound elimination is a major challenge to the country.

In the short term, securing the stocks should be envisaged through collecting and bringing together all small individual stocks in a storage site in conformity with international standards.

All stakeholders will have important responsibilities in this activity according to their respective mandates and their specific tasks.

In the absence of a full evaluation of the nature of the contamination and its size and of a full analysis according to up-to-date standards of health and environmental risks, the sites that may be considered to pose a problem are called "potentially contaminated sites."

There is currently no regulation concerning stocks and POPS-contaminated sites in São Tomé and Príncipe.

THE FUTURE OF SOUND POPS MANAGEMENT IN SÃO TOMÉ

São Tomé and Príncipe is neither a producing nor an exporting country of POPS pesticides. All substances of Annex A (Part I) are no longer officially used in São Tomé and Príncipe, neither in agriculture nor in public health. Nevertheless, illegal import and trade of POPS products in the country is probable. These products are utilized by the population in the absence of knowledge about the potential danger for their health.

Strengthening the sound management of chemicals, including POPS, is one of the priorities for São Tomé and Príncipe. However, given the very limited resources available in the small country, São Tomé needs expertise and resources from abroad to work together to achieve the goals of the Stockholm Convention and, in a broader sense, of SAICM.

In particular, for the implementation of the National Implementation Plan for the Stockholm Convention and for the elimination of the stocks of obsolete pesticides and other chemicals, external support is essential. Cofinancing requirements such as those from GEF funding for the elimination of PCBs will make it almost impossible to fully implement the obligations under the Stockholm Convention.

Partnership with industry may allow solving the problem of obsolete stocks.

41 Activities, Challenges, and Accomplishments of the Republic of Slovenia in the Implementation of Chemical Management Instruments

Marta Ciraj

CONTENTS

INTRODUCTION

Slovenia is a small but beautiful country in the middle of Europe, picturesque and rich in biodiversity, with a very sensitive environment because of the varied natural conditions (karst, water sources, natural parks), and thus it is very important to prevent the spillage and intrusion of chemicals into the soil and environment. Most of the Slovenian area is very vulnerable and as such the development and maintenance of chemical safety are very important. Moreover, Slovenia has been concerned about people's health in connection with chemicals, since there have been certain negative experiences in some regions concerning chemical production. This is the main reason that Slovenia has made great strides as far as developing chemical safety is concerned. Slovenia took the first steps in developing chemical safety within the framework of the former Socialist Federal Republic of Yugoslavia (SFRY) by having its experts participate in the then Federal Commission for Toxins. In addition to relevant rules applied throughout the SFRY, Slovenia also had its own rules in place when it came to toxic substances and Plant Protection Products (PPP). After gaining independence, Slovenia made full use of the expertise gained in the former SFRY to start setting up a modern approach to chemical safety based on fresh impetus and associations with a number of other European Union (EU) Member States, through the Organisation for Economic Cooperation and Development (OECD and the country's efforts to join the EU, as well as through its active participation in international forums such as the Intergovernmental Forum on Chemical Safety (IFCS).

At the very beginning, it was necessary to make an inventory of which of the existing authorities were responsible for what stages of the chemical life cycle. Four main actors were found to be the most competent: the Ministry of Health (MH), the Ministry of the Environment and Spatial Planning (MESP), the Ministry of Agriculture, Forestry and Food (MAFF), and the Ministry of the Interior (MI). The MH was in general responsible for the assessment of poisons before putting them on the market. Since many PPPs are also harmful, the MH was co-responsible for the process of registration of these substances while the main responsibility prior to putting such products on the market was vested in the MAFF. The MESP was responsible for chemical waste, and the MI for transport on the road and other means. At that time, PPPs were the only chemicals for which regulation was already in place.

Slovenia very soon found out that intersectoral cooperation was vital for achieving chemical safety—that not only the above-mentioned ministries should cooperate, but that other ministries and nongovernmental stakeholders should participate as well. When it comes to the ratification of international legally binding instruments, the Ministry of Foreign Affairs should have a role, and in regard to the education of children on chemical safety, awareness raising, worker protection, consumer safety, and research, the relevant stakeholders should be involved. It is indispensable to cover the entire life cycle of chemicals.

Thus, in 1996 the Government of the Republic of Slovenia (RS) set up the Intersectoral Committee for Dangerous Substances (subsequently named the Intersectoral Committee for Chemical Safety (ICCS)), and in 1999 the National Chemicals Bureau at the MH was also charged with providing chemical safety in Slovenia.

INTERSECTORAL COMMITTEE FOR CHEMICAL SAFETY

An intersectoral coordinating mechanism for chemical safety matters in Slovenia has been operating since 1996 based on a governmental decision. After adoption of the first law on chemicals in 1999, it was re-established on the basis of this law. The ICCS is charged with coordinating with responsible ministries and other stakeholders in pursuing national policies, programs, and measures based on the Chemicals Act and other chemical regulations, as well as providing comprehensive and coherent development of chemical safety nationally. Through this mechanism, the preparation of the National Profile in 1997 and the National Chemical Safety Programme (NCSP), adopted by the Government of the RS, were coordinated. During the preparatory period for the Strategic Approach to International Chemicals Management (SAICM), all stakeholders were kept informed of the process and the Committee provided a mechanism for channelling views and guidance to representatives of Slovenia. The governmental decision in 2008 designated the National Chemicals Bureau to coordinate the implementation of the SAICM in Slovenia and authorized the ICCS as the mechanism for ensuring full stakeholder involvement in the process. What is more, the ICCS is charged with providing annual reports on its performance to the Slovenian Government.

The ICCS systematically coordinates relevant line ministries' performance when it comes to prevention of and response to chemical accidents, as well as reflecting on the draft NCSP with a focus on the priority area of chemical accidents.

In December 2005, the ICCS set up the Intersectoral Committee for Coordination of Monitoring Water Pollution with Chemicals, and charged it with harmonizing water-monitoring programs of different stakeholders in Slovenia and a database linkage. The aim of coordinating these monitoring programs and integrating relevant databases is also to increase the quality and compatibility of data on pollution of drinking water sources (underground and surface water) so as to use these data for conducting risk assessments and for taking decisions on risk reduction measures.

The ICCS also has the responsibility of advising on initiatives that Slovenia may launch in the international arena, such as the UN Specialized Agencies and the Organization for the Prohibition of Chemical Weapons (OPCW). The ICCS made the proposal of appointing the Director of the Chemicals Office of the Republic of Slovenia (CORS) as the vice-chairperson of the IFCS for the Central and Eastern Europe (CEE) region.

One of the main tasks of the ICCS was preparation of the NCSP, which took almost 10 years to complete and to be finally adopted by the National Assembly in 2006 covering the period between 2006 and 2010. In parallel and in some cases prior to these efforts, there were also other national programs related to other phases in the chemical life cycle being drafted by other relevant line ministries. Such national programs also govern certain aspects of chemicals and refer to the NCSP. A recent development is the conversion of the Committee into the coordination authority for SAICM implementation in Slovenia, the adoption of the SAICM Implementation Action Plan and designating CORS as the authority responsible for coordinating SAICM implementation and reporting on progress achieved prior to meetings of the International Conference on Chemicals Management (ICCM) based on the decisions of the Government of the RS.

The Committee consists of relevant representatives of ministries, the economic sphere, social activities, nongovernmental organisations (NGOs), and other stakeholders engaged in any phase of the chemical life cycle:

- MI
- Ministry of Foreign Affairs

- Ministry of Defence
- MAFF
- Ministry of Labour, Family and Social Affairs
- Ministry of Public Administration
- Ministry of the Economy
- Ministry of Transport
- MESP
- MH
- Ministry of Higher Education, Science and Technology
- Ministry of Education and Sport
- National Statistical Office
- Slovenian Institute for Standardization
- Chamber of Commerce and Industry
- Chamber of Crafts
- Worker's organizations
- Slovenian Society of Toxicology
- NGO
- Council for Environmental Protection

The purpose of the ICCS is primarily to ensure the integrity and horizontal harmonization of intersectoral legislation in the area of chemical safety, to increase its feasibility and effectiveness and, according to needs, improve vertical harmonization with the European legal order. A further purpose of the ICCS is to ensure effective functioning of a broader complex of legislation which falls within the competence of different portfolios and which covers the complete life cycle of chemicals. To achieve as much ownership of chemical safety problems as possible, the ICCS has a rotating chairmanship and deputy appointed from among its members. The appointment requires a majority vote. Its first chairperson was the director of the CORS. Currently, the chairmanship is held by the NGO, Slovenian Society of Toxicology and was formerly the responsibility of the nongovernmental Consumer Sector.

The ICCS Secretariat, based within CORS, coordinates the work of the ICCS and its provisional and standing committees. The ICCS's secretary is not necessarily one of its members, but is appointed by the CORS director. The Secretariat is responsible for carrying out professional, administrative, and technical tasks.

To accelerate and increase the level of chemical safety, it is necessary to constantly adjust the activities that lead to reaching common goals. The Committee performs the following tasks:

- Monitors and adjusts the work of competent governmental and interested nongovernmental portfolios.
- Evaluates the state of chemical safety, defines key problems, gives suggestions, and derives common solutions, which also includes short-term, mid-term, and long-term national and global strategic goals.
- Cooperates in the implementation of the NCSP.

- Monitors planning and implementation of other national programs in those parts which relate to chemical safety.
- Deals with questions of effective acquisition of national and foreign project funds, especially at the EU level.
- Shapes project proposals, encourages and coordinates the financial participation of ministries and consequently monitors their implementation.
- Encourages the exchange of information connected with chemical safety in Slovenia as well as in the international environment.
- Proposes activities that are necessary for improving the protection of health and mitigating the negative effects of chemicals on the environment.
- Contributes to realizing the intersectoral strategy for handling chemicals (SAICM).

The inclusion of the various ministries in the ICCS is based on their particular role in chemical safety:

- The MESP is concerned with the direct and indirect effects of releasing chemicals into the environment as emissions and waste to air, water, and land, and with the prevention of chemical accidents.
- The Ministry of Agriculture is concerned with the use of agricultural chemicals for the benefit of securing food supplies and the protection of plants and animals of economic benefit.
- The MH is concerned with the short- and long-term health impacts of chemicals on the general public, the medical response to people exposed to toxic chemicals, regulation of industrial and consumer chemicals, as well as pesticides used in the health sector.
- The Ministry of Labour is concerned with occupational health and safety issues related to the use and handling of chemicals in the workplace, including accidents involving chemicals.
- The Ministry of the Economy is concerned with the production of chemicals and chemical products, and the introduction of cleaner production technologies.
- The Ministry of Interior is concerned with the safe transport and storage of chemicals during the distribution phase.
- The Ministry of Finance (Customs Authority) is responsible for ensuring that chemicals do not enter or leave the country in contravention of government regulations and international agreements.
- The Ministry of Defence is responsible for emergency services, including response to emergencies involving chemicals.
- The Ministry of Foreign Affairs coordinates all international aspects of chemical management, such as participation in relevant international agreements and conventions.
- The Ministry of Public Administration is responsible for coordination of all inspection authorities.

- The Ministry of Higher Education, Science and Technology and the Ministry of Education and Sport are responsible for education in respect of chemical safety issues and expert knowledge.

THE CORS (UNTIL 2008 THE NATIONAL CHEMICALS BUREAU)

When the Chemicals Act was adopted in 1999, one of the first tasks was the establishment of the National Chemicals Bureau (now the CORS) as a constitutive body of the MH for which the legal basis was provided in this act. It was established and began to function on August 29, 1999. The mission of CORS is to ensure chemical safety in Slovenia, and its basic task is to provide acceptable risks to the entire population of the country concerning their health and the environment when it comes to the production and use of chemicals. This is fulfilled by integrating and coordinating views of the relevant national authorities, professional institutions, and the public to provide a balanced policy on chemical safety, by raising awareness at all levels through the public and the educational system, as well as through the work of the highly qualified CORS staff.

Beginning with six employees, CORS has developed into an organization with 28 professional employees with a wide range of responsibility for all chemicals which have not been regulated since its establishment. In general, only pharmaceutical and medical products, agrochemicals including PPPs and chemicals used in food processing do not fall within the competence of CORS. CORS is responsible for cosmetics, chemical weapons and other strategic materials, drug precursors, industrial chemicals, biocides, good laboratory practice, international legally binding and nonlegally binding agreements such as Rotterdam Convention, Stockholm Convention, SAICM, and for cooperation with relevant international institutions in the chemical safety field.

The CORS has six internal organizational units as follows:

- Sector for Planning, Coordination, and Development
- Sector for Chemicals
- Sector for Pesticides and Monitoring
- Sector for Risk Assessment

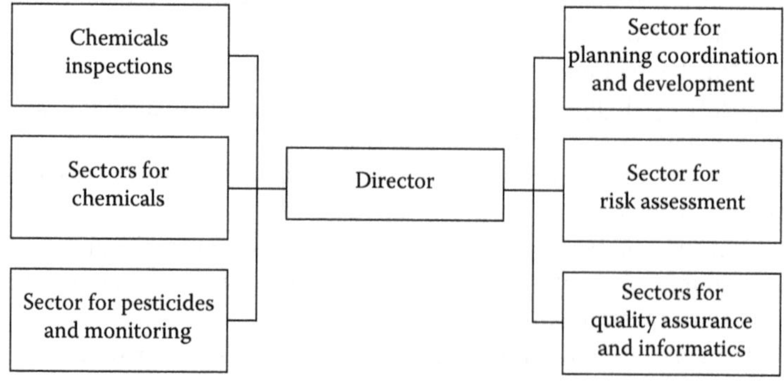

FIGURE 41.1 Structure of the CORS in 2009.

- Sector for Quality Assurance and Informatics
- Chemicals Inspection (see Figure 41.1)

After chemical safety, via the Chemicals Act, was embedded in one of the ministries and CORS was closely connected to international organizations, Slovenia joined in the preparation of international documents at the very beginning. Through such activities Slovenia was informed at the international level and could prepare for the ratification (if necessary) and then implementation of international instruments. For instance, Slovenia was the second country in the world to ratify the Rotterdam Convention.

For strengthening chemical safety as an important factor in the overall safety of life and the environment, it is also of great importance that chemical safety was incorporated into the Resolution on the National Healthcare Programme adopted by the National Assembly in 2008, and that chemical safety is part and parcel of each Consumer Protection Programme. In doing so, chemical safety has been mainstreamed into various national development programs; however, there is still room for improvement. What is more, it is also important that each program, while still in the pipeline, be examined for its possible consequences regarding chemical safety. This endeavor requires an active role on the part of CORS and good intersectoral cooperation, particularly through the ICCS.

INTERNATIONAL AGREEMENTS AND INTERNATIONAL ACTIVITIES OF THE RS IN THE CHEMICALS SAFETY AREA

As Slovenia had applied for the membership of the EU during the years 1999–May 2004, it, therefore, was required to incorporate into its legal system all regulations and directives of the EU in the area of chemicals safety, as was the case with other subject areas. This was achieved subsequently before entering to the EU on May 1, 2004. This meant that Slovenia accomplished full compliance in the chemicals safety field, since no transitional period was set.

After participation in the whole negotiation process, Slovenia ratified the *Rotterdam Convention* in 1999 with the Act Ratifying the Rotterdam Convention on the Prior Informed Consent (PIC) Procedure for Certain Hazardous Chemicals and Pesticides in International Trade and is implementing the European Community (EC) regulation concerning the export and import of certain hazardous chemicals. Slovenia is also a member of Chemicals Review Committee, which was chaired by a Slovenian expert at its fifth meeting in March 2009. CORS is the responsible authority for its implementation.

Slovenia is a party to the *Montreal Protocol.* As an independent state since 1992, Slovenia ratified the Montreal Protocol on substances that deplete the ozone layer by the Act on Accession and afterwards all listed Amendments to the Protocol.

As a result of the Global Environment Fund (GEF)-funded Ozone Depleting Substances (ODS) Phase-out Project and reestablishment of the institutional strengthening component in the period 1995–1998, Slovenia became a non-Article 5 (= developed country) party to the Montreal Protocol in 2000.

With membership in the EU, Regulation (EC) No. 2037/2000 on Substances that Deplete the Ozone Layer which implements the Montreal Protocol on substances that deplete the ozone layer and its relevant decisions, entered into force.

The EC has a leading role in global efforts to phase out ODS in order to preserve the ozone layer. The Regulation is broader in scope than the Protocol in several areas, such as its phase out schedules for ODS (especially hydrochlorofluorocarbons (HCFCs)) and its provisions on products and equipment. It prohibits the production of controlled substances, limits their import/export through issuing licences, sets rules of recovery, recycling and reclamation of used controlled substances, and establishes monitoring mechanisms for refrigerating and air conditioning equipment containing controlled substances.

Slovenia does not produce ODS. As a member of the National Experts Committee under the EC Regulation, the country is playing an active role in global efforts in the pre-phaseout of ODS at the meetings of the parties to the Montreal Protocol. The trend of abandoning consumption is also made clearly evident by the consumption quantities for Slovenia.

The Decree on the use of products and equipment containing ODS or fluorinated greenhouse gases (F gases) was adopted in 2008 (OG RS, No. 78/08). The Decree merged the area of use of ODS and of F gases and laid down leakage checking and recovery conditions for maintaining and installing equipment, conditions for the reclamation and disposal of recovered gases, conditions and methods of certifying fitters and undertakings, and the method of training fitters to perform leakage checking and recovery of gases, and the maintenance and installation of equipment containing ODS or F gases.

Slovenia ratified the *Stockholm Convention* on Persistent Organic Pollutants (POPS) in 2004 with the Act Ratifying the Stockholm Convention on POPS. The objective of the Convention is to protect human health and the environment from POPS by prohibiting, phasing out as soon as possible, or restricting the production, placing on the market and use, and by minimizing releases of POPS. As an EU Member State, Slovenia also implemented the Regulation on Implementation of Regulation 850/2004 of the EP and EC on POPS. On the basis of Article 7 of the Stockholm Convention, Slovenia also developed a national implementation plan for the management of POPS, which was recently adopted by the Government of the RS.

Slovenia ratified the *Convention for the Protection of the Marine Environment and the Coastal Region of the Mediterranean* (Barcelona Convention) in 1993 (by accession) and its amendments in 2003, as well as three protocols* of the Convention.

Slovenia is actively working on the Barcelona Convention and implementation of its protocols. It is also actively involved in the United Nations Environmental

* Protocol for the Prevention of Pollution in the Mediterranean Sea by Dumping from Ships and Aircraft (Dumping Protocol) from 1976, amended in 1995. Slovenia signed the amended Protocol in 2004. Protocol concerning cooperation in preventing pollution from ships and, in cases of emergency combating pollution of the Mediterranean Sea (Prevention and Emergency Protocol) from 2002. Slovenia ratified this Protocol in 2002. Protocol for the Protection of the Mediterranean Sea against Pollution from Land-Based Sources and Activities (LBS Protocol) from 1980 (amended in 1996). Slovenia ratified the Protocol in 1993 and its amendments in 2003.

Programme (UNEP) Mediterranean Action Plan, which was designated in 1975 by 16 Mediterranean countries and the EC as the first-ever Regional Seas Programme under the UNEP umbrella. The Contracting Parties now number 22, and they are determined to protect the Mediterranean marine and coastal environment while boosting regional and national plans to achieve sustainable development.

Slovenia ratified the *Basel Convention* on the Control of Transboundary Movements of Hazardous Wastes and Their Disposal in 1992 and its amendments in 1995.

The main objective of the Basel Convention is to reduce waste formation, to ensure environmentally sound waste management, to reduce the transboundary movements of wastes to a reasonable level, to prevent transboundary movements of wastes if presumed that these wastes will not be managed in an environmentally sound way, and to ensure suitable control to prevent waste management that would be in contradiction with the Convention.

Slovenia ratified the *Aarhus Convention* and is in process of ratification of the Protocol on *Pollutant Release and Transfers Registers (PRTR).* A system of reporting industrial emissions and transfers to the national authority was established through the European Regulation on Implementation of Regulation EC 166/2006. It contains all the main principles from the Annex and is provided for larger emission sources and defined pollutants. In addition, industrial emissions to air and water for all industrial sources that are subject to legal requirements for monitoring are available on the Web page of the MESP.

As mentioned above, Slovenia has actively participated in the negotiation or preparation process of instruments for chemical safety. Let us also mention the country's activity in the preparation of the EU's Registration, Evaluation, Authorisation and Restriction of Chemicals (REACH) regulation, in which Malta and Slovenia were involved in preparing a joint proposal as an alternative approach to the original Commission proposal for the regulation in the area of registration of low-volume chemicals. The idea behind this was based on the fact that although the 1–10 tons band accounts for only a very small percentage of total exposure to chemicals, it also includes the greatest number of potential registration dossiers. Effective management of this information is therefore critical in ensuring the workability and cost-effectiveness of REACH.

The key concept underpinning the Slovenian–Maltese proposal (SI–MT proposal) for this tonnage band was that it is possible and necessary to have a differentiated approach for chemicals in the lowest tonnage band. The solution was based on technical scientific knowledge and a pragmatic approach. The SI–MT proposal was supported first by the European Commission and by different EU presidencies (Netherlands, Luxembourg, and the United Kingdom) and key persons in the European Parliament. Then, it was also taken up by the European Parliament, where it was renamed. At the same time, it was also supported by all Member States in the Council and included in the Compromise Registration Package prepared by the UK presidency. The content of the SI–MT proposal was the key to finding a political compromise agreement in the Council (December 2005), which was the first step to final adoption of REACH in 2006. The result was that the SI–MT proposal was included in the final text of the adopted regulation and constituted a major change in the text of the original Commission proposal.

Slovenia was one of the most supportive parties of the *Strategic Approach to International Chemicals Safety*. The ministers of Health and Environment were signatories for the RS of the SAICM, adopted at the first meeting of the ICCM held in Dubai, United Arab Emirates, in February 2006. According to Article 23 of the Overarching Policy Strategy, the Strategic Approach national focal point has been designated by the Government (CORS), and ICCS was appointed to be the coordinating body for implementation of the SAICM. As the SAICM implementation body, CORS was authorized by the Government in February 2008, while also adopting the National Action Plan on SAICM implementation in Slovenia. All relevant stakeholders are required, within their financial and human resource constraints, to report progress to CORS for submission to the ICCM (in 2009, 2012, 2015, and 2020) concerning implementation of SAICM activities within their own areas of competence. Moreover, Slovenia is funding the Quick Start Programme for the period 2006–2011 and the SAICM Secretariat for the period 2006–2020. Additional funding of EUR 100,000 was provided in the year of the Slovenian Presidency of the EU, and the contribution will be kept at a high level in the following years. Slovenia was much honored and at the same time feels great responsibility to be part of the SAICM process, as the country was elected to the presidency of the second ICCM which took place in Geneva on May 11–15, 2009 and is now chairing the SAICM process in the interim period between two conferences until 2013.

It is necessary to emphasize that the IFCS was one of the first information sources where Slovenia gained all-important information and exchanged views with other countries and stakeholders on questions regarding chemical safety. Slovenia has been an active member of the IFCS since its establishment in Stockholm in 1994. Slovenia has actively participated in the Forum Standing Committee (from 2000 to 2006) since September 2006; from Forum V (September 2006) to Forum VI (September 2008) Slovenia was the IFCS Vice President for the CEE region and a member of the Forum Standing Committee.

Slovenia is of the opinion that the IFCS has greatly contributed to the development of the SAICM; in particular, the Global Plan of Action was prepared much on the basis of input on IFCS elaborated topics.

SLOVENIA AS AN EU MEMBER

As an EU country, Slovenia is consequently working in its preparation then in adopting and implementing of all the relevant legislation. In the preaccession period and after joining the EU Slovenia successfully applied for financial support through the Poland and Hungary Assistance for Reconstruction of the Economy (PHARE)* and other Programmes. The Phare Programme is one of three preaccession instruments financed by the EU to assist the applicant countries of (CEE) in their preparations for

* Initially founded to help Poland and Hungary in reconstruction of their economy and called Poland and Hungary Assistance for Reconstruction of the Economy. Afterwards, it was transformed into the Assistance for Economic Restructuring in the Countries of Central and Eastern Europe, but the original abbreviation remained unchanged, since it had become well known under this designation.

joining the EU. Later, Slovenia benefited from three rounds of the so-called EU-funded Transition Facility Twinning "Advanced Chemical Safety Project." These programs were very supportive in capacity building, for example, education, training and developing of legislation, including tackling new and timely issues such as nanomaterials and nanotechnology.

Nanoparticles have become a part of our everyday lives, since nanotechnologies have an extremely wide range of applications in new materials, food, communication and information technologies, medicine, environmental technologies, the leisure industry, cosmetics, and so on, to name just the most important ones. But besides huge benefits and great business opportunities, nanotechnologies have also raised completely new safety concerns. The big question is whether the consideration of safety aspects and the development of suitable safety practices and required regulations are keeping pace with the expanding development of nanotechnology.

For this reason, the MH—CORS and the Umweltbundesamt (Federal Environment Agency) in Austria, as partners in the EU-funded Twinning Project Chemical Safety three in cooperation with the OECD and European Commission, organized a CEE regional Conference on Nanosafety in Ljubljana from April 22–24, 2009. This was the third event on nanosafety in Slovenia in a period of three years. The main goal of the conference was to bring together interested stakeholders from (CEE), and from elsewhere, to exchange information on different policy approaches concerning the safety of nanomaterials as well as on new developments in this field. The benefits and risks of these new technologies must be properly communicated to the wider public. Slovenia is preparing its strategy on nanosafety to reflect the conclusions and recommendations at EU level.

SLOVENIA AS A CEE COUNTRY

Slovenia is a CEE country and as such, through active participation in the IFCS, the idea developed to share experience in chemical safety in the region. Slovenia therefore specified in its strategic plan the organizing of at least one regional conference a year at which different topics should be discussed in order to support the development of chemical safety in the region. Furthermore, Slovenian experts either actively participate in or host different bilateral meetings with other countries, particularly in the countries of the former SFRY, with the aim of sharing experience. In addition, Slovenia has hosted and provided many missions to help other countries in the region to develop chemicals safety and to share experiences with them.

THE MOST IMPORTANT INDICATORS AND PRACTICAL TOOLS ON CHEMICALS SAFETY IN SLOVENIA

When it comes to *domestic chemical safety*, Slovenia is particularly proud of very positive trends in lowering poisonings and of the fact that the national chemicals database comprises information on more than 45,000 dangerous chemicals including high-quality safety data sheets, the quality of which has been rising steadily and a biocides database comprising information on more then 1000 biocidal products. These databases are very important tools for inspectors and other officials in the legislative or enforcement process as well as for risk assessment and chemical management activities.

The number of poisoning cases reported to the National Poisons Centre of the RS between 2001 and 2008, shows a nearly 50% decline in recorded poisonings (see Figure 41.2).

Slovenia will continue its preventive activities to decrease poisonings and long-term damage as a consequence of exposure to chemicals with strict implementation of national and international law, whereby it is important to mention REACH and GHS regulations, the latter adopted during the Slovenian EU Presidency. These activities are among other measures in the Chemicals Strategic Plan for 2008–2012, adopted by the Minister of Health on the proposal of CORS.

One of the most important preconditions for rapid implementation of legislation in Slovenia was the training and education of the responsible persons, the so-called chemical advisers, in order to inform them of all legal obligations regarding chemical safety. This training is legally binding. In a period of 10 years, more then 2000 chemical advisers were qualified and certified for more then 1500 enterprises registered by CORS for the production, trading, or use of dangerous chemicals. What is more, they are obligated to report to CORS the quantities of chemicals that come into Slovenia.

For the future strategic planning of measures for the prevention of negative impacts of chemicals and also for evaluation of the efficacy of these measures, Slovenia provided the legal basis for biomonitoring and prepared a medium-term plan for implementation so that the whole area of the country will be covered by 2012.

For the above-mentioned activities, sufficient financial resources are necessary. In order to enable the execution of all necessary chemical safety activities, Slovenia allocated a budget of approximately € 2.5 million per year for CORS. This amount covers work and material costs, with about half of these funds being used for the various programs.

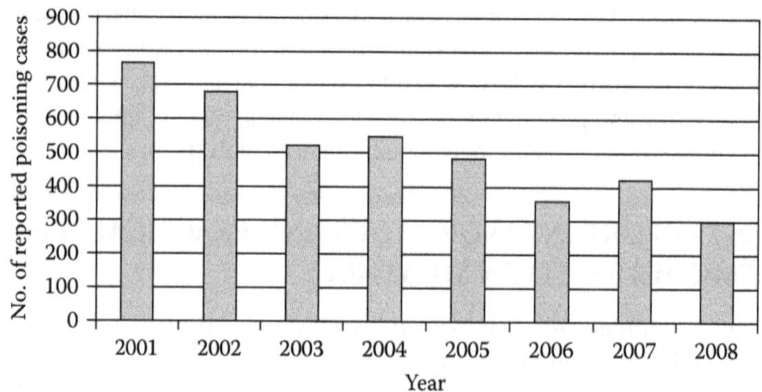

FIGURE 41.2 Poisonings reported to the poison centre of the Republic of Slovenia, 2001–2008.

CONCLUSION

It is necessary to emphasize that very little headway would have been made had there not been sufficient political will. This was first manifested in the establishment of the ICCS in 1996. Setting up such a mechanism required courage based on good arguments. This courage had first of all to be mustered by officials and experts familiar with chemical safety in Slovenia and beyond. What is more, they were insistent enough to place this issue high on the political agenda to be addressed by relevant state secretaries and finally by ministers. Further, a body of like-minded professional and political opinion was assembled, and then the right persons were employed in responsible posts to promote chemical safety. Without having the right people in the right places, there would have been no way forward. One of the key lessons that Slovenia learnt in the course of promoting chemical safety is that making great strides calls for major political will, the right people in the right places and tolerant and transparent communication and cooperation among relevant governmental and nongovernmental stakeholders.

The conjuncture of circumstances which led to the above-mentioned political will provided the opportunity for taking bold and innovative action, enabling Slovenia to take the governmental decision to prepare the Chemicals Act with a legal basis for important infrastructure and activities, such as the CORS, the Intersectoral Committee on Chemical Safety, the Integrated Programme for Sound Management of Chemicals through a life cycle approach, obligatory training of the responsible persons in industry and a chemical information database system. Slovenia recognized the essential role of chemicals in the development process, involving nearly all economic and social sectors, as well as the potential negative impacts on human health and the environment if chemicals were not managed properly. The need for a multisectoral approach which considered the use of chemicals in Slovenia in a holistic way was embraced and the decision taken to establish the NCSP for the RS with the necessary administrative infrastructure for its implementation.

Slovenia was able to benefit from the long-standing experience of the international community in the field of chemical safety and to launch activities to develop an integrated program for sound management of chemicals through a life cycle approach. The initial activities involved making a comprehensive assessment (National Profile) of the country's infrastructure and capacity relating to the legal, institutional, administrative, and technical aspects of chemicals management, along with an understanding of the nature and extent of availability and use of chemicals throughout their life cycle. These activities highlighted the need to involve all stakeholders, and a governmental decision established an intersectoral governmental coordinating mechanism for chemical safety during the period of preparing the NCSP with its legal mandate. Six national priority areas were chosen for action: (1) integrated chemical safety legislation; (2) integrated inspection for chemical safety; (3) safety and health at work with chemicals; (4) management of waste chemicals; (5) chemical accidents; and (6) integrated chemicals monitoring. Following an introduction regarding the background and formulation of the Programme, each of the six priority areas are described in individual chapters, divided into subchapters consisting of a number of pertinent issues ranging from an assessment of the situation to future prospects, objectives, indicators of progress, and program implementation requirements.

It was recognized that chemical safety issues relating to particular fields where chemicals and waste are being employed can only be resolved by introducing changes in the concerned sectors, such as agriculture, industry, and transport. Owing to the complexity of problems concerning chemicals, other relevant Slovenian national programs need to be strengthened to a greater or lesser degree regarding chemical safety. They generally comprise basic guidelines and directions for coordinating economic, technical, scientific, educational, organizational, and other measures, as well as measures for the implementation of international obligations. Some national programs are more directly concerned with chemicals than others, but nevertheless all of them include at least a brief description of potential chemical risks and associated hazards. These programs are: the National Environmental Action Programme, the National Programme for Protection against Natural and Other Disasters, the National Health Care Programme, the Agri-Environmental Programme, and the National Strategy in Occupational Safety and Health. Besides these, it is also important to take into consideration the National Security Strategy of the RS. A number of other national programs are relevant in a secondary sense. These programs are: the National Road Safety Programme, the National Motorway Construction Programme, the National Programme of Slovenian Railway Infrastructure Development, the National Research Programme, the National Programme for Higher Education, and the Strategy for Consumer Protection. The NCSP supplements these programs, providing guidelines, priorities, and strategies concerning chemical safety issues. The NCSP takes into consideration different perspectives from which chemicals can cause direct or indirect effects on people, animals, plants, property, and the wider environment. Besides previously implemented national obligations in the field of chemical safety, the NCSP also considers international instruments and initiatives. In this regard it derives from existing, established principles and requirements of related UN, EU, and OECD multilateral agreements and to those of subordinate organizations, specialized agencies, and programs.

The ministers of Health and Environment were signatories for the RS of the SAICM, adopted at the first meeting of the ICCM (Dubai, February 2006). Each of the 273 SAICM activities in the Global Plan of Action were examined as to their relevance to the situation in Slovenia, and revised priorities and timelines were established and will be incorporated into the forthcoming National Programme for the years beyond 2010, an agreement reached through the Intersectoral Committee on Chemical Safety. It assigned responsibilities for implementation of different activities from the SAICM Global Plan of Action to various stakeholders. The Government of the RS adopted this plan in February 2008, designating the CORS to coordinate implementation of the SAICM in Slovenia and requiring all stakeholders, within their areas of competence and responsibilities and in accordance with their financial and resource constraints, to report progress to the ICCM.

In view of the difficulties in obtaining full NGO-sector participation in the Programme, the Chemicals Office established a "platform" where NGOs could meet regularly and discuss and prepare contributions to the Programme, as well as their own roles in implementing the SAICM.

The RS has taken a huge step forward when it comes to chemical safety by adopting the NCSP and Action Plan for SAICM Implementation and by adopting the EU Regulation on REACH. Furthermore, another action plan is under preparation concerning children in relation to chemical safety. As for the impact of chemicals on the environment, annual biomonitoring has been put into place and scientific developments will be followed closely, as well as studies aimed at identifying any harmful consequences of chemicals on human health and the environment. Moreover, appropriate records will be set up and kept on causal relationships between certain chemicals and certain diseases. Safer chemical management will make a contribution to a decrease in the presence of chemicals in organisms and the environment.

All actors (producers, legal entities, and individuals placing chemicals on the market, users, and consumers) are encouraged to be involved in sound management of chemicals. These common endeavors will lead to an increase in sound, reliable, and safe use of chemicals and their management in all stages of their life cycle. Important activities increasing chemical safety in Slovenia are as follows: strengthening public awareness about chemicals, introduction of chemical safety in most curricula at all levels of education and establishing study programs related to chemical safety.

Because chemical safety has been globally introduced into healthcare and environmental protection, Slovenia will continue to play an active role in improving regional and global chemical safety, taking into consideration the interdependence of health and the environment, and the fact that leading a healthy life can only be done in a healthy environment, and therefore problems related to health and the environment need be addressed in a harmonized manner.

REFERENCE

Intersectoral Committee for Chemical Safety. 2009. *An Integrated Approach to Sound Management of Chemicals and Waste, Initiation of Implementation of the SAICM in Slovenia.* Editor: John Haines, Ministry of Health/Chemicals Office, TF project Chemical Safety.

42 Implementation of the Rotterdam Convention in Third World Countries
The Tanzania Experience

*Ernest Mashimba**

CONTENTS

---

* Late Chief Government Chemist, Dar es Salaam, Tanzania, http://www.gcla.go.tz, and former member of the Chemicals Review Committee of the Rotterdam Convention, died suddenly September 18, 2010.

INTRODUCTION

Tanzania was among the first countries that signed the Rotterdam Convention in 1998 and ratified it in 2002. After being signatory to the Convention, activities were initiated leading to fulfillment of the national obligations. Prior to signing the Convention, there was no comprehensive legislation on chemicals management in Tanzania. Presently, there are three laws that have provisions for the Rotterdam Convention. These are the Industrial and Consumers (Management and Control) Act No. 3 of 2003, the Plant Protection Act (1997), and the Environmental Management Act of 2004.

However, Tanzania also has policies that address some aspects of chemicals management. These policies include: the National Environment Policy of 1997, the National Agriculture Policy of 1997, the National Health Policy of 1990, revised in 2007, and the Mineral Policy of 2003. However, there is no specific policy on chemicals management. Under the initial three years of SAICM implementation, the government of Tanzania has agreed to establish a specific policy on chemicals management. Presently, a team of experts is working to develop a first draft of the policy. An annual budget is set aside for prior informed consent (PIC) activities at the Government Chemist Laboratory Agency (GCLA) through the Ministry of Health and Social Welfare. The political will to develop such a policy is also high.

Being a Party to the Convention, Tanzania has made progress in the implementation of the Rotterdam Convention through undertaking several actions in order to meet its obligations.

INFRASTRUCTURE

Focal Point

The Vice President's Office, Division of Environment is the focal point of all Multilateral Environmental Conventions including the Rotterdam Convention. This promotes supervision of implementation of this and other Conventions in the country.

Designated National Authorities

After being signatory to the Convention, Tanzania designated two designated national authorities (DNAs) (Mashimba, 2004, 2006; Mashimba and Katagira, 2007) for administrative functions; the Chief Government Chemist (for industrial and consumer chemicals) and the Department of Plant Health services in the Ministry of Agriculture and Food Security (for pesticides). The role of the DNA as per Article 4 of the Convention is to perform administrative functions obligatory by the Convention. The DNAs are responsible for communicating official decisions to the Secretariat and relevant authorities nationally like notifications of control actions, information exchange, training facilitation, conduction of PIC meetings, sensitization, and raising awareness on PIC procedures to stakeholders, and dissemination of information relevant to the Convention to all stakeholders.

To date, Tanzania has completed import decision responses on 28 pesticides (including Severely Hazardous Pesticide Formulations [SHPF]) and eight industrial chemicals listed in Annex III of the Convention; some of the responses are interim. Decisions on the remaining chemicals are pending assessment to ascertain their presence and use in the country as well as determination of their effects including those on environment and human health. These include interim responses. Tanzania has neither notified a final regulatory action nor submitted a proposal of SHPF to the secretariat due to the lack of scientific data and inability to perform risk evaluation for potential risks due to exposure to the environment or human health. Therefore, Tanzania resorts to the use of bridging information (see Chapter 14 of this book).

National PIC Committee

Tanzania has a PIC Committee whose responsibility it is to deliberate on imports and make decisions on future imports of PIC chemicals. It is also responsible for making recommendations and assessing whether additional banned or severely restricted chemicals should be made subject to the PIC procedure.

Accredited nongovernmental organizations and intergovernmental organizations are invited to participate as observers to the PIC Committee meetings. It comprises 13 members from both private and public sectors to ensure a multistakeholder participation in all decisions made regarding whether Tanzania shall allow imports of chemicals in Annex III. Members of the committee are as follows:

- GCLA—Chief Government Chemist DNA—Chairman
- The Department of Plant Health Services in the Ministry of Agriculture and Food Security for Pesticides DNA—Deputy Chairman
- Vice President's Office—Expert from Pollution Control Section
- National Environment Management Council—Expert from Environment Compliance and Enforcement Division
- GCLA—Expert from Chemicals Management Division
- Tropical Pesticide Research Institute (TPRI)—Expert from Pesticide Registration and Control Division
- Ministry of Industry and Trade—Expert from Department of Industry
- University of Dar es Salaam—Expert from Chemical and Process Engineering
- Agrochemicals Association of Tanzania—Member—Private Sector
- Confederation of Tanzania Industries—Member—Private Sector
- Occupational Safety and Health Agency (OSHA)—Occupational Health Doctor
- Ministry of Agriculture, Natural Resources, Environment and Cooperatives Department of Plant Protection (Zanzibar)—Member
- Ministry of Health—Government Chemist Laboratory (Zanzibar)—Member

The PIC committee meets and deliberates on the future status of the chemicals. Through this committee, all PIC issues are thoroughly discussed. The Committee has taken eight import decisions on industrial and consumer chemicals and 28 on pesticides. The problem encountered by this Committee is that it has no legal mandate. Also, the meetings of the committee are not regular due to lack of funds at DNAs. In order to resolve this problem, sustainable sources of funding for the Convention activities from all relevant institutions from which the committee is represented should be allocated in their budgets.

NATIONAL ACTION PLAN FOR PIC IMPLEMENTATION

A five-year (2008–2013) National Action Plan (Katagira, 2008) was developed and endorsed by stakeholders in July 2008. The Action Plan indicates the output, indicators, actors, and time frame for implementation of various activities. However, the implementation of some activities is lagging behind the time schedule because of financial constraints and availability of data to facilitate reporting. In order to resolve this problem, funds should be sought through mainstreaming of chemicals management activities under the national budget, development partners, and technical assistance from the PIC Secretariat and other international organizations such as UNEP, Swedish Chemical Agency (KEMI), German Society for Technical Cooperation (GTZ), and WHO.

REGULATORY INFRASTRUCTURE FOR IMPORT OF CHEMICALS

Chemicals management in Tanzania is a task performed by coordination, collaboration, and cooperation through a multisectoral approach realized under the existing sectoral laws in chemicals management. Specifically, there are three laws that have provisions for the implementation of the Rotterdam Convention, as indicated previously.

The enforcement of the three laws is done through assessment, registration, inspection, and import control. The achievements through the implementation of the three laws include the following:

- Import control is done through issuing of chemicals import permits prior to importation.
- National chemicals registers are in place with all information on importers and chemicals imported including availability of a Safety Data Sheet for each chemical.
- Tanzania receives and responds to Import Responses and Export Notifications.
- Through inspection, authorities encourage industries to promote chemical safety.
- The Harmonized System (HS) Custom Codes for all chemicals has been adopted and is in use.
- National labeling requirements with regard to risks/hazards to human health and the environment are in place.

ACCESS TO INFORMATION AND INFORMATION EXCHANGE

In developing countries like Tanzania, the flow of information is available but not adequate, and disseminated through workshops and seminars; the media, the public, and decision makers have access to information on chemical handling, accident management, and alternatives that are safer for human health and the environment.

Information exchange is done with international organizations such as UNEP, WHO, ILO, FAO, IPEN, KEMI, GTZ, and IFCS. Local NGOs involved include AGENDA for environment, the Tanzania Plantation Workers Union (TPAWU), and others.

A good and well-established communication infrastructure for DNAs, such as the Internet, is in place to communicate with other relevant groups including customs authorities, importers, users, and the general public about the decisions as published in the PIC circular.

Exchange of information with the PIC Secretariat and other nations takes place through the PIC Web site, aiming at PIC decision making, safe handling and other precautionary measures, hazard classification, or other detailed information, for example, nature of the risk of a chemical, relevant safety advice, and summary results of toxicological and ecotoxicological tests.

Awareness raising is done among relevant authorities and other stakeholders in the issues of safe handling and proper use of chemicals for smooth implementation by using workshops, seminars, meetings, media, TV, radio, newspapers, brochures and leaflets, and so on. An information system for networking and access to information and information exchange as mentioned is available through Web sites at GCLA and other institutions.

TRAINING

There is regional and international cooperation and collaboration in training and capacity building for PIC chemicals management. In Tanzania, staff has been trained on chemicals management. This was done when it was realized that for an effective implementation of the law, the human capacity needs to be strengthened by sharpening skills and encouraging expertise. Presently, any decision on whether to register a chemical is based on a risk assessment and safety data sheet submitted to the registrar and is reviewed by experts in chemicals management.

The training includes ecotoxicological training, which was conducted in 2006 with experts from the public and private sectors. The training focused on monitoring the impact of pesticides in the environment. Training on Multilateral Environmental Agreements (MEAs) including the Rotterdam Convention was conducted in 2004 for decision makers of relevant Ministries and other stakeholders and chemicals management experts. The training of DNAs and their support staff on submission of import responses and notification forms was sponsored by GTZ and the Chief Government Chemist in 2006. The two DNAs have extensively been trained on the Rotterdam Convention through the assistance of the PIC Secretariat.

LABORATORY CAPACITY

Tanzania has well-equipped laboratories at the GCLA and other institutions. This also includes qualified staff such as chemists and laboratory technologists. GCLA acts as a national toxicological and environmental laboratory for analysis of all poisoning cases. The GCLA carries out analytical tests and interprets laboratory results on occupational, toxicological, food (including pesticide residues), drugs, and environmental samples.

It is also involved in surveillance activities on populations at risks. Furthermore, the Agency collaborates with governmental and nongovernmental agencies (e.g., the police, NGOs, and environmental, food, and drugs authorities) on analytical methods. However, there are some problems that the Agency encounters, including the following:

- Poorly maintained equipment.
- Delay in procurement of appropriate consumables such as reagents, standards and prompt arrangement for maintenance of equipment.
- Insufficient practical analytical techniques that can provide timely results.

It is important to strengthen collaboration among the laboratories in the country to improve the quality of services and share resources. There is also a need to strengthen the laboratory in terms of prompt acquisition of consumables, spare parts, and maintenance of instruments in order to diagnose poisons and monitor chemical effects on humans and the ecosystem. In addition, it is important to establish networks at regional and international levels.

IMPORT RESPONSES AND USE OF INTERIM DECISIONS FOR PIC CHEMICALS IN TANZANIA

Tanzania has made import responses (Kalima, 2008), some of which are interim, that include consent to import with or without specified conditions or no consent to import during the interim period.

S/N	Name of the Chemical	Import Response
1	Poly Brominated Biphenyls (PBBs)	Interim; consent, subject to specified conditions
2	Poly Chlorinated Biphenyls (PCBs)	Interim; consent, subject to specified conditions
3	Poly Chlorinated Terphenyls (PCTs)	Interim; consent, subject to specified conditions
4	Tri (2,3 dibromopropyl) phosphate	Interim; consent, subject to specified conditions
5	Asbestos	
	• Actinolite	Interim; consent, subject to specified conditions
	• Athropyllite	Interim; consent, subject to specified conditions
	• Amosite	Interim; consent, subject to specified conditions
	• Tremolite	Interim; consent, subject to specified conditions
	• Crocidolite	Final; no consent to import

Reasons for Interim Decisions

- Absence of specific national PIC legislation and regulation.
- Poor capacity in risk assessment and risk evaluation. No proper assessments have been made for risk management of PIC chemicals.
- Poor documentation, collection, and reporting of poisoning and other chemical-related incidents.
- No comprehensive inventory of industrial chemicals; hence, difficulty in ascertaining the presence or absence of some PIC chemicals and their related health and environmental effects.
- Economic factors—alternatives to some banned or restricted chemicals are very expensive, and some are not readily available.
- Information available in the country is sometimes insufficient, particularly regarding human health.
- Only a weak poison center is available.
- Inadequate regional networking for DNAs to exchange information, for example, on incidences related to PIC chemicals.
- Irregular meeting of the PIC committee due to finance constraints at DNAs.
- Inadequate public education and awareness on chemical hazards/national control actions on PIC chemicals among stakeholders, particularly the informal sector which plays a notable role in the chemicals distribution network.
- Scanty and uncoordinated research efforts on chemical toxicology and safer alternatives to hazardous chemicals.
- Limited information exchange and networking among chemicals managers and institutions.
- Lack of financial capacity to conduct various activities related to implementation of the Convention.

These problems can be addressed by developing the capacity to conduct research or studies on the toxicological/environmental effects of the chemicals on human health and getting safer alternatives, thereby enabling Tanzania to make final import responses.

Export Notification

Since the notification of the Convention, Tanzania continues to receive and acknowledge export notifications. Typically, eight export notifications for industrial chemicals have been received and acknowledged to date as follows:

Year	Type of Chemical	Country of Origin
2002	Sodium chromate	United States
2004	Potassium chromate	United States
2005	Tanalith C3310 (containing arsenic compounds)	European Union
2005	Tanalith C3310 (containing arsenic compounds)	European Union
2005	Chloroform	European Union
2006	Benzene	European Union
2006	Sulfuric acid	South Africa
2006	(Phenol, 2-(2H-benzotriazol-2yl)-4,6-bis (1,1-dimethylethyl)	Japan

Export notification is vital as it can trigger national action to ban or restrict importation of such chemicals. Hence, the exports notification for the previously mentioned chemicals will help Tanzania to take national action in the future.

CHALLENGES IN THE IMPLEMENTATION OF THE ROTTERDAM CONVENTION IN TANZANIA

Incident Reporting

In Tanzania (Ndiyo, 2008; Mashimba and Katagira, 2007, 2009; Mashimba, 2004, 2007), pesticide dealers/distributors discourage incident reporting. In addition, it is difficult to link the health effects to a specific pesticide, due to multiple exposures to different types of pesticides. Therefore, detailed information on pesticide poisoning is mostly not available to DNAs. Some information is available at local administrations, but it is fragmented.

The poison center that exists in essence does not work. Establishment of an effective and efficient poison center in Tanzania could be useful in providing information on poisons and advice on poisoning cases as well as providing systematic education and prevention programs.

The recognition of poison symptoms of nontarget organisms, particularly invertebrates such as insects, is not easy and therefore such incidents are not reported. These circumstances, in conjunction with those mentioned in the section "Why Not Many Proposals for SHPF Are Submitted to Secretariat" in Chapter 14 of this book, discourage Tanzania from submitting any proposal for SHPFs to the secretariat.

In Tanzania, community health monitoring programs have started. This exercise provides a good flow system from farmers to DNAs to facilitate information collection and reporting to the DNAs (see also the section "Why Not Many Proposals for SHPF are Submitted to Secretariat" in Chapter 14 of this book). The system may result in identification of severely hazardous pesticide formulations and subsequently enable Tanzania to submit proposals for SHPFs in the future to the PIC secretariat.

In order to improve the reporting system, training and financial resources are needed for establishing national registers and databases, reporting systems, poison control centers, rural health centers, communication infrastructures, and awareness raising campaigns targeted at farmers, extension workers, health workers, researchers, agricultural suppliers, customs officers, and many others.

Strengthening Laboratories

There is a need to continuously strengthen national laboratories and networking to ascertain rapid identification of chemical ingredients used and the kind of effects that are expected from using those chemicals and pesticides. Strengthened analytical laboratories will result in effective and efficient analysis of samples from poison incidences.

Coordination, Cooperation, and Communication

At present, there is inadequate coordination, cooperation, and communication among the various organizations concerned with chemicals management at a national level. The key stakeholders need to have a common understanding of the necessary procedures involved. Many front line personnel lack adequate and accurate technical knowledge or information of how to implement the Convention effectively. This ends up in poor networking.

Limited Enforcement

There is inefficient and ineffective enforcement of the existing laws in chemicals management because of, among other reasons, limited human and financial capacity.

Illegal trade of chemicals across the borders is a major concern. There are ongoing efforts that include strengthening of inspection and regular surveillance.

Access to and Exchange of Information

At present, there is little knowledge and awareness of the Rotterdam Convention among policy makers, some regulatory bodies, and the public in general, especially for developing countries.

There is also no comprehensive inventory of industrial and pesticides chemicals. Hence, it is difficult to ascertain the presence or absence of some PIC chemicals and their related health and environmental effects.

Strengthening of access to and exchange of information with stakeholders such as importers and exporters of chemicals about the PIC Convention will facilitate its implementation.

Integration of Chemicals Management into National Strategies

Integration of chemicals management activities into national development plans and strategies is still a challenge. There are ongoing efforts to this end (see the section "Introduction").

Training

There is a need to train medical personnel including toxicologists and epidemiologists to enable them to diagnose and manage chemical-related diseases. Presently, this knowledge is scanty among medical personnel.

Noting this shortcoming, Tanzania has embarked on a process to include chemical management-related subjects in tertiary health training institutions. In addition, training of other stakeholders will be conducted continuously.

BENEFITS FROM ROTTERDAM CONVENTION IN TANZANIA

Tanzania, like other developing countries and being a Party to the Convention, has benefited especially for the proper use of chemicals and pesticides while guaranteeing that its development is environmentally sustainable. This is because developing countries can avoid most of the mistakes that happened in the developed world where misuse of chemicals in one way or another had negative impacts on people and the environment. Thus, knowledge on how chemicals are hazardous, the kind of hazards they cause, and the best way of handling them is an important step in minimizing risks associated with hazardous chemicals.

Other benefits gained by Tanzania from the Convention include the following:

- The training that the DNAs have attended at the Convention and the membership of Dr. Ernest Mashimba in the CRC from 2004 to 2009 have made Tanzania to have experts on the Rotterdam Convention. They have added to the pool of experts in chemicals management.
- There is coordination, collaboration, and cooperation among different stakeholders responsible for chemicals management.
- The Focal Point for the Rotterdam Convention is under the Vice President's Office, Division of the Environment. This helps in proper supervision of the Rotterdam Convention in the country's coordination.
- Through the Convention, there is access to information on hazardous chemicals and pesticides from different stakeholders and technical assistance is gained from the Secretariat of the Convention and other international organizations such as GTZ, UNITAR, and UNEP. The PIC Circular and the Decision Guidance Documents are good sources of information in this regard.
- Tanzania participates in the decision-making process, seminars, and workshops at the regional and international levels concerning hazardous chemicals and pesticides. This increases awareness on chemical issues and its management as a whole.
- It is possible to get information about any hazardous chemical through the DNA (networking of the DNAs), particularly those in countries with similar conditions. This is because the Convention guarantees the cooperation of other governments on known and existing risks through its provisions regarding information sharing through export notifications, information accompanying export, and the PIC procedure.
- Tanzania gets export notifications from some exporting countries especially the European Union and the United States. The information enables Tanzania to know the regulatory actions in the exporting countries and helps to understand the presence of unwanted chemicals in other countries.
- The knowledge of chemicals listed in Annex III of the Rotterdam Convention and the export notification provide an early warning on trade of chemicals that are banned or restricted elsewhere.
- Tanzania is making full use of the expertise gained by experts.

In summary, the Convention is a major tool for Tanzania to reduce the risks associated with pesticide use by accessing alerts on chemicals that are banned or severely restricted elsewhere. The Convention will promote sustainable agriculture in a safer environment, thereby contributing to an increase in agricultural production and supporting the battle against hunger, disease, and poverty through proper use of chemicals and pesticides. Proper use of pesticides and industrial chemicals helps to increase food production while minimizing negative impacts on the environment and human health.

The Convention promotes transparency in the sale of dangerous chemicals abroad and less vulnerablility to abuse by encouraging harmonized labeling of exported chemicals, as well as the provision of safety data and information on health and environmental risks. As such, it assists Tanzania to avoid using pesticides that are recognized as being harmful to human health and the environment and highly toxic pesticides that cannot be handled properly by small farmers.

INTEGRATING ROTTERDAM, STOCKHOLM, BASEL, AND SAICM IN TANZANIA

Tanzania ratified the Rotterdam Convention in 2002, the Stockholm Convention in 2004, and the Basel Convention in 1993. The structure of implementation of these conventions in Tanzania is coordinated under the Vice President Office as the Main Focal Point. Since ratification of the conventions, various activities related to their implementations have been undertaken through the national chemicals management committee before and sectoral committees after the establishment of three laws on chemicals management, which in most cases involves the same stakeholders. Some of these activities include:

- Development of the National Profile, which assessed the national capacities and set priorities for chemicals management.
- In 1997–1998, a national inventory of obsolete pesticides and veterinary wastes was carried out. The findings of this project led to the development of a National Implementation Plan (NIP) of the Stockholm Convention and an African Stockpiles (ASP) project was developed titled "Enabling activities to facilitate early actions on the implementation of the Stockholm Convention."
- The Industrial and Consumer Chemicals Act, the Environment Management Act, and Plant Protection Act have provisions for all the major MEAs in chemicals management.
- Inventory of hospital hazardous waste management, Dar es Salaam, 1996.
- Assessment of feasibility and viability of using cement kilns in incineration of hazardous wastes, February 1999. DNOC pesticides were incinerated.
- Training workshop organized by the GCLA on hazardous chemical wastes, 2001, at British Council Dar es Salaam, was sponsored by WHO.
- Training workshop for Executives and Experts on International Conventions to promote chemical safety, jointly organized by the GTZ, ILO, and GCLA, August 2004.

- Participation in the development and implementation of SAICM at the national level.
- Implementing the SAICM Pilot Project (2006–2009) in which all the activities conducted during the project involved all stakeholders in chemicals management. The activities included, among others, conducting of the national capacity on chemicals management, priority setting on chemicals management, and conducting of two priority partnership projects.

In further implementation of SAICM, Tanzania aims at linking all chemicals management activities through the prospective policy in chemicals management and an Interministerial Coordination Committee which will involve all ministries, agencies, and other stakeholders including those involved in the direct implementation of the Convention. Consequently, this approach will enable direct implementation of the MEAs and SAICM.

In facilitating the implementation of SAICM, the National SAICM focal point is the Chief Government Chemist in the Ministry of Health and Social Welfare, who is also the DNA for industrial chemicals for implementation of the Rotterdam Convention. This setup enables a smooth cooperation, coordination, and communication with other stakeholders in chemicals management.

The Industrial and Consumer Chemicals Management and Control Act of 2003 appoints the Chief Government Chemist as Registrar, whose functions among others include fostering cooperation between stakeholders and coordinating chemicals management policies and programs in the country.

The Tanzania Ministers for Health and Environment signed the Libreville Declaration of 2007 for the United Republic of Tanzania. Since then, joint activities between the two ministries that are dedicated to implementation of the declaration have started.

The DNA for industrial and consumer chemicals participated in the development of a draft strategy for strengthening engagement of the health sector in line with SAICM, February 2010 Ljubljana, Slovenia.* Upon endorsement of this strategy, Tanzania will benefit through direct engagement of the health sector in SAICM implementation.

The Ministry of Health and Social Welfare has also included chemicals management in its annual plans. In addition, chemicals management activities have been devolved to the local government to ensure sound management of chemicals at the local level.

LESSONS LEARNED: EXPERIENCE TO SHARE

Tanzania has made progress and has much experience in the implementation of the Rotterdam Convention. However, following ratification of the Convention, implementation needs to be reinforced as follows:

- Allocation of adequate resources for data collection and incident reporting should be considered by the government in order to have enough data for assessing effects.

* http://www.iisd.ca/media/chemical_management.htm

- Health personnel also need to be trained on how to correctly identify pesticide-poisoning symptoms.
- Networks of chemical regulators at regional and international levels should be established for efficient information exchange on effective management of chemicals.
- Coordination, cooperation, and communication of various stakeholders should be strengthened to implement the Rotterdam Convention and other MEAs at national and regional levels.
- Continuous training of DNAs, their support teams, and relevant authorities on the MEAs should be conducted.
- Sensitization and raising awareness on the Rotterdam Convention as well as dissemination of information about this Convention to relevant stakeholders, for example, importers and exporters of chemicals, should be conducted continuously.
- There should be preparation and enactment of regulations to domesticate international obligations into national legislation. Regulations specific for the Rotterdam Convention are important.
- There should be effective and efficient collecting and reporting of information on pesticide and chemicals poisoning, and using community health services should be established.
- Poison control centers should be established.
- Improved access to international literature, databases, and risk/hazard evaluations is essential.
- Inventories should be undertaken of PIC chemicals, including obsolete ones, and comprehensive databases of industrial and consumer chemicals should be established.
- Feedback should be provided and follow-up from delegations to international workshops and seminars.
- Tanzania and other developing countries should make use of "bridging information" to submit notifications for severely hazardous pesticides formulations.
- Policy should be established integrating the initiatives under the various conventions and to provide for mainstreaming of chemicals management in the National Plan.
- Making progress requires major political will to empower the right people in the right places, tolerant and transparent communication, cooperation, and collaboration among relevant government and nongovernment stakeholders (Haines, 2009; Mashimba, 2004).
- Considering the usual fluctuations and movement in personnel at all levels in any organization, training of staff should be carried out to ensure stability and continuity of activities at organizational, national, and international levels.

REFERENCES

Haines, J. 2009. Ed: *An Integrated Approach to Sound Management of Chemicals and Waste*, Initiation of Implementation of SAICM in Slovenia.

Kalima, J. 2008. *Import Responses and the Use of Interim Decisions for the PIC Chemicals in Tanzania*. Paper presented at Stakeholders' National Seminar on Review of National Action Plan for Implementation of the Rotterdam Convention.

Katagira, F. 2008. *Current Implementation Status of the Rotterdam Convention in Tanzania*. A paper presented during the National Seminar on the National Action Plan for the implementation of the Rotterdam Convention. Tanzania.

Mashimba, E.N.M. 2004. *The Need for Sustainable Chemicals Management*. A paper presented at a training workshop for Executives and Experts to Promote Chemical Safety, August 9–13, Bagamoyo, Tanzania.

Mashimba, E.N.M. 2004. *Implementation of the Rotterdam Convention in Tanzania*. A paper presented at a Training Workshop for Executives and Experts on International Conventions to Promote Chemical Safety, August 9–13, Bagamoyo, Tanzania.

Mashimba, E.N.M. 2006. *Implementation of Rotterdam Convention in Tanzania*. A paper presented to stakeholders, November 22, Mbezi Garden, Dar es Salaam.

Mashimba, E.N.M. and F. Katagira, 2009. *National Experience in Collecting Information of Severely Hazardous Pesticide Formulations (SHPF)*. A paper presented during the joint Workshop of the Rotterdam and Stockholm Convention for effective participation in the Chemical Review Committee's Work–CRC and POPRC, November 17–19, Cairo, Egypt.

Mashimba E.N.M and F.F Katagira, 2007. Sub-region meeting to foster cooperation among DNAs in the implementation of the Rotterdam Convention for English-speaking Africa countries, November 5–9, Pretoria, South Africa.

Ndiyo, D. 2008. *Notification of Final Regulatory Action for Implementation of the Rotterdam Convention*. Paper presented at stakeholders' National Seminar on Review of National Action Plan for Implementation of the Rotterdam Convention.

Section VII

Regional Activities

43 Tripartite Environmental Collaboration between China, Japan, and Korea in Chemical Management

DaeYoung Park

CONTENTS

INTRODUCTION

Tripartite Environment Ministers Meeting among China, Japan, and Korea

China—the most populated country in the world, Japan—the world's second economy, and Korea—one of the fastest growing economies, are not only closely linked by economic and political relations but geographically share causes and adverse impacts of air, marine, and other environmental pollution. While recognizing and discussing regional environmental problems in Asia through the Ministerial Conference on Environment and Development in Asia and the Pacific since 1985, China, Japan, and Korea have made respective bilateral agreements for environmental protection since the beginning of 1990* and the three countries finally came together to form the Tripartite Environment Ministers Meeting (TEMM) among China, Japan, and Korea in 1999. For the last 13 years, through TEMM, China, Japan, and Korea have cooperated on environmental awareness, information exchange, and research and development of environmental technology, policy measures for the prevention of air pollution, protection of the marine environment, and handling of global environmental issues such as biodiversity loss and climate change. TEMM has provided a high-level policy dialogue forum playing a leading role in local, regional, and global environmental management and to discuss current environmental conditions and outstanding environmental issues of each country and the Asian region. Table 43.1 highlights a brief historical timeline of TEMM and various other regional chemical regulations, agreements and important meetings between these three countries.

Tripartite Policy Dialogue on Chemicals Management in China, Japan, and Korea

Owing to its political nature and broad cooperation framework, TEMM has carried out various capacity building projects which include the Ecological Conservation in Northwest China, the Joint Environmental Education Project, the Tripartite Environmental Education Network, the Freshwater (Lake) Pollution Prevention Project, and the Environmental Industry Roundtable. Based on an increasing number of international and other activities and instruments on sound chemical management in 2000s, for example, the Globally Harmonized System of Classification and Labeling of Chemicals (GHS), the Strategic Approach to International Chemicals Management (SAICM), the Intergovernmental Forum on Chemical Safety (IFCS), and the European Union's Regulation on Registration, Evaluation, Authorisation and Restriction of Chemicals (REACH), TEMM needed to address more specific environmental policy and regulatory issues. In the 2007 Tripartite Policy Dialogue on Chemicals Management in China, Japan, and Korea, technical and policy experts of the three countries started to discuss the future challenges of environmentally sound chemicals management in areas such as implementing GHS, an information sharing system on chemicals used in products, and regional collaboration on global chemicals management.† In the 9th TEMM held in December 2007, the three ministers officially adopted the agreement for periodic expert-level policy dialogue on chemical management and illegal export of electronic waste. Between 2007 and 2009, the

* The Agreement on Environmental Cooperation between the People's Republic of China and South Korea and the Agreement between Japan and Korea on Cooperation in the Field of Environmental Protection were signed in October 1993 and June 1993, respectively. The Agreement between the People's Republic of China and Japan on Cooperation in the Field of Environmental Protection was signed in March 1994.

† Since 2007, TEMM has added a dedicated chemical regulatory information Web site. Available at: http://ncis.nier.go.kr/temm_cmp/index.html

TABLE 43.1
Historical Overview of TEMM, and Environmental Institutions and Chemical Regulations in China, Japan, and Korea

	TEMM	China, Japan, and Korea	
2011	13th TEMM	Major discussion topics included dust and sandstorms (DSS), climate change, biodiversity, transboundary movement of hazardous wastes, and other regional environmental issues in North East Asia (NEA) and more collaboration in environmental cooperation. Other discussions included South Korea's bid to host the 18th Conference of the Parties to the UNFCCC (COP18), Four Major Rivers Restoration Project and joint cooperative measures for disasters. A special agenda topic was also adopted to discuss the recent earthquake and tsunami in the northeastern part of Japan. This special topic was contained in a Joint Communique adopted on 29 April 2011 (excerpted from http://www.temm.org/sub05/view.jsp?id=20)	
2010	12th TEMM	(China) Measures on the Environmental Management of New Chemical Substances amended	
2009	11th TEMM		
2008		(China) Ministry of Environmental Protection established	
2003		(China) Measures on the Environmental Management of New Chemical Substances adopted	
2001	2nd TEMM		
1999	1st TEMM		
1998		(China) State Environmental Protection Administration established	
1995		Third Meeting of Ministerial Conference on Environment and Development in Asia and the Pacific	
1994		Agreement between China and Japan on Cooperation in the Field of Environmental Protection	(Korea) Ministry of the Environment established
1993		Agreement on Environmental Cooperation between China and Korea	Agreement between Japan and Korea on Cooperation in the Field of Environmental Protection
1992		First Meeting of the Northeast Asian Conference on Environmental Cooperation (NEAC)	
1991		First Meeting of Ministerial Conference on Environment & Development in Asia and the Pacific	
1990		(Japan) Ministry of the Environment established	(Korea) Toxic Chemicals Control Act adopted
1985		First Meeting of Ministerial Conference on Environment and Development in Asia and the Pacific held	
1984		(China) National Environmental Protection Agency established	
1980		(Korea) Environment Agency established	
1976		(Japan) Law on the Control of Evaluation and Manufacture of Chemicals adopted	
1971		(Japan) Environment Agency established	

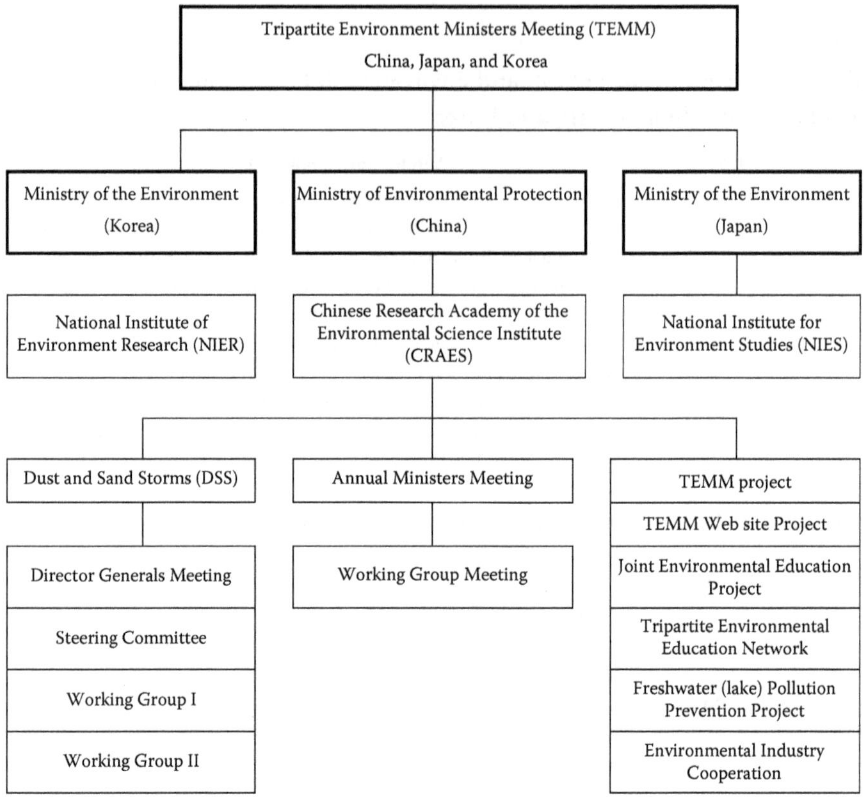

FIGURE 43.1 Organizational structure of TEMM.*

Tripartite Policy Dialogue on Chemicals Management discussed and partly implemented cooperation projects, amongst others, including information sharing on chemicals of high concern, cooperation on the investigation of existing chemicals, the response to REACH and the SAICM, harmonization of classification and labeling, pollutant release and transfer register (PRTR), and the notification and supervision of new chemical substances. Figure 43.1 shows overall organizational structure of TEMM.

LOCAL REGULATIONS VERSUS REGIONAL AND GLOBAL HARMONIZATION

Similarities and Differences in Chemical Regulations of China, Japan, and Korea

The industrial structures of the three countries are quite varied from agriculture (38.1% for China, 3.9% for Japan, 3% for Korea), and industry (27.8% for China,

* TEMM Home Page, http://www.temm.org

26.2% for Japan, 39.4% for Korea), to service (34.1% for China, 69.8% for Japan, 57.6% for Korea) (*CIA World Factbook*, 2011). The regulatory focus and legislative timing of each country has not been the same as a result of this difference in economic structure. In addition, China, Japan, and Korea have different legal cultures even though all three countries have a civil law system in place. When the People's Republic of China was established in 1949, the roots of the Chinese legal system were derived from the Soviet Union model of the socialist legal system (Chao, 2005). The Japanese Constitution, which was drafted in 1948, relied heavily on the legal systems of European countries (in particular, Germany). During the Japanese rule of Korea from 1910 to 1945 and after the independence of the Republic of Korea in 1948, the Korean legal system adopted many elements from the Japanese legal system.

The aforementioned fundamental differences appear in the chemical regulation of each country. Japan, Korea, and China have introduced the first advanced form of chemical regulation on new and existing chemicals in 1976, 1990, and 2003, respectively.* Compared to the Japanese Law on the Control of Examination and Manufacture of Chemical Substances, which has the most sophisticated provisions on the control of new and existing chemicals among the three countries, the Chinese Measures on the Management of New Chemical Substances regulate new chemical substances only and do not have built-in control system on existing chemicals.† Due to these diversities in experience, regulatory framework and economic structure, coordinating cooperation on chemical policy and regulation at TEMM has been a big challenge until the three countries met the common and overwhelming challenge, that is, REACH, in 2006. REACH has provided competent ministries of the three countries and TEMM with legitimate causes to work together closely and to amend existing chemical regulations in a better or advanced manner.

Globalization Impacts on the 2008–2009 Changes of Chemical Regulation in China, Japan, and Korea

From the beginning of 2008 to July 2009, there have been clear moves indicating that China, Japan, and Korea started incorporating REACH-like elements in their existing chemical regulations. At the same time, the three countries have actively implemented the GHS system (UN Globally Harmonized System on Classification and Labeling of Chemicals; see Section III, Chapter 11 of this book) into their existing chemical regulations in order to meet the implementation schedule agreed at the Asia Pacific Economic Cooperation (APEC) level and at the United Nations Economic Commission for Europe (UNECE).

* Japanese Law on the Control of Examination and Manufacture of Chemical Substances (Law No. 117 of October 16, 1973); Chinese Measures on the Management of New Chemical Substances (State Environmental Protection Administration Ordinance No. 2 of September 12, 2003); Korean Toxic Chemicals Control Act (Law No. 4261 of August 1, 1990).

† This does not imply that China does not have any regulations controlling existing chemicals. There are separate regulatory instruments on hazardous chemicals. Amongst others, they include the Decision on the Ratification of Rotterdam Convention on the Prior Informed Consent Procedure for Certain Hazardous Chemicals and Pesticides in International Trade (December 29, 2004, National People's Congress), Measures on the Management of Hazardous Factors in Workplaces with Chemical Operations (Hualaozi, 1990, p. 640 of October 13, 1990).

China

In May 2009, the Ministry of Environmental Protection of China proposed the amending draft to the Measures on the Environmental Management of New Chemical Substances. Pursuant to the amending draft of May 2009, amongst others, China would introduce a risk management concept for new chemicals in the Chinese chemical control system, that is, regulating both hazard and exposure of chemicals. It would also introduce three chemical classifications of new chemicals, that is, general chemicals, hazardous chemicals, and chemicals of environmental concern. The scope and limit of the amending draft of May 2009 is that the provisions would only be applicable to new chemicals which are not included in the Chinese Inventory of Existing Chemicals (http://www.crc-mep.org.cn/iecscweb/default.aspx).*

During 2008 and 2009, on the other hand, China has adopted more than 120 National Standards on Chemical Testing and good laboratory practice (GLP) which include 65 Standards on Physico-chemical Tests, 58 Standards on Toxicities and Ecotoxicities, mainly based on Organization for Economic Cooperation and Development (OECD) guidelines. This indicated that China is aggressively putting in place scientific and technical infrastructure for stricter or sound management of chemicals and this may lead to further regulatory instruments soon.

Japan

The outcome of the Kashinho Revision Working Group,† which was established by the Ministry of the Environment, the Ministry of Economy, Trade and Industry, and the Ministry of Health, Labor, and Welfare in January 2008, appeared as the adoption of the amended Law on the Control of Examination and Manufacture of Chemical Substances of May 20, 2009 in Japan.‡ In the amended Law of May 20, 2009 for the manufacture or import of chemicals, manufacturers, or importers are required to submit an annual report on the chemicals to the Ministry of Economy, Trade and Industry. Through screening of the reported existing chemicals, the Ministry of the Environment, the Ministry of Health, Labor and Welfare, and the Ministry of Economy, Trade and Industry will determine "priority chemicals for assessment" and may require further hazard and risk-assessment studies from industries. Also the amended Law of May 20, 2009 provides a framework for the stricter control of internationally controlled chemicals, in particular, chemicals in semiconductors (e.g., perfluoro compounds) or fire extinguishers. The key point of the amended Law of May 20, 2009 is that the Law will transform the Japanese chemical control system from a hazard-based into a risk-based approach.

In light of the recent earthquake and tsunami of March 11, 2011 and the Fukushima I and II nuclear plant disaster, Japan's Nuclear and Industrial Safety

* Available at: http://www.crc-mep.org.cn/iecscweb/default.aspx

† "Kashinho" is the Japanese acronym for the Law on the Control of Examination and Manufacture of Chemical Substances.

‡ The Enforcement Ordinance of 30 October 2009 of the Law on the Control of Examination and Manufacture of Chemical Substances (Government Ordinance No.257) set various entry into force dates, that is, hazard communication requirement by April 1, 2010, introduction of new chemical classification system by April 1, 2011.

Agency (NISA) responsible for regulating nuclear industries may be under global scrutiny. The nuclear plant operators were left to conduct their own risk assessments and emergency response planning (excerpted from http://www.brookings.edu/opinions/2011/0401_nuclear_meltdown_kaufmann.aspx).

Adequate oversight and strict regulatory structures may be required to avoid failure of containments of nuclear infrastructure that ultimately lead to radiation leaks.

Korea

By issuing its chemical policy action plan Green SHIFT (Safety, Health, Information, Friendship, Together) Plan of February 12, 2009, the Ministry of the Environment of Korea showed its clear signals to introduce an advanced chemical control system in Korea. In particular, the Green SHIFT Plan stipulated the introduction of more testing items for new chemicals from 2011 and preparation of testing guidelines from 2009, introduction of alternative chemical test methods between 2009 and 2012 and prevention of repetitive chemical testing, mandatory submission of hazard data and exposure information of existing chemicals, chemical data collection through the domestic HPV (High Production Volume) program between 2010 and 2012, introduction of chemical information communication in supply chains, and introduction of risk assessment (including consumer exposure assessment) and socioeconomic analysis between 2009 and 2012. The amending proposal of June 1, 2009 to the Toxic Chemicals Control Act clearly indicated that the Ministry of the Environment implements its Green SHIFT Plan. The amending proposal of June 1, 2009 would require manufacturers, importers or sellers of severely restricted or banned substances to communicate appropriate information (e.g., permitted activities, restriction or ban applied to substances) to its customers. In order to allow transitional period, the Ministry of the Environment would publish a candidate list of severely restricted or banned substances and a potential enforcement date before it designates toxic substances as severely restricted or banned.

On the matter of national legislative activities including chemical regulation, TEMM has limited competency to be involved, excluding communication among competent ministries. In coming years, however, TEMM would need to play a key role in harmonizing national chemical policies in order to prevent unnecessary regulatory barriers to free trade and ensure safe and sound chemical management in the region. Growing international chemical regulations and stricter environmental responsibilities on manufacturers or exporters of chemicals would provide legitimacy for the extended role of TEMM in regional chemical-policy coordination.

TEMM AS FACILITATOR AND COORDINATOR FOR BETTER REGIONAL CHEMICAL POLICYMAKING

The sound chemical management system requires a well-structured and functioning chemical regulatory regime, scientific infrastructure (e.g., testing laboratories with GLP), technical enforcement capacity, scientific data on chemicals, a technical expert pool, and a chemical tracking mechanism locally and nationally.

As members of the OECD, Japan and Korea have an internationally acceptable regulatory framework on the control of new and existing chemicals including PRTR and GLP.* Although a chemical reporting and registration system exists under the Measures on the Registration of Hazardous Chemicals, China, which is a nonOECD member, does not have a well-developed PRTR system on paper or electronic format. Less capacity exists to track and provide information on the presence of chemicals in the environment online like the European Pollutant Emissions Register.† Despite the effort to establish internationally recognizable Chinese GLP laboratories by the Ministry of Science and Technology, the Ministry of Health, and the State Food and Drug Administration, there is still no GLP laboratory which can generate internationally acceptable chemical test data. Furthermore, the Chinese chemical control system has complex jurisdictional structure controlled by multiple governmental ministries. This overwhelming jurisdictional complexity on chemical control may be prohibitive for integrated chemical control and make it difficult to introduce a holistic chemical management regulation.

At the international level, there is a converging trend of environmental regulation, which imposes enormous challenges on all parts of industry being linked through a global supply chains. Even though this trend is supposed to work for better global harmonization in chemical management, it may create more ambiguity, confusion, significant difficulties, and costs on industries when a country introduces an advanced environmental regulation because of domestic or international political reasons while lacking its own scientific and technical capacity (Deutsche Bank Research, 2005). Being a policy dialogue venue for China, Japan, and Korea, TEMM has provided a good mechanism to share and discuss local, regional, and international environmental policies and regulations. However, TEMM may need to find a better way to communicate and coordinate policy and regulatory developments on chemicals and environmental issues in a shorter period. In the coming years, policymakers and technical experts involved in TEMM may also have to discuss disadvantageous regulatory provisions of a particular country in a more open manner. Otherwise, they lose better opportunities of communication and coordination to the APEC‡ or the World Trade Organization (WTO).§

---

* The Ministry of Labor of Korea has introduced a precise certification scheme for good laboratory practice (GLP) in health impact testing and physico-chemical testing through the Public Notice of February 20, 2008 on Standards on Hazard and Risk Assessment of Industrial Chemical Substances. Regarding test methods of hazard and risk assessment, the Public Notice of February 20, 2008 provides 13 test methods for health impacts (e.g., acute oral toxicity test, acute skin irritation test, acute eye irritation test, carcinogenicity test) and 22 test methods on physico-chemical properties (e.g., melting point, boiling point, water solubility).

† The European Pollutant Emissions Register provides access to information on the annual emissions of around 9200 industrial facilities in the 15 Member States and around 12,000 facilities in the 25 Member States in the European Union. Available at: http://eper.ec.europa.eu/eper/

‡ The APEC Chemical Dialogue (CD) provides a forum to discuss trade and economic impacts of chemical policy and regulation between regulatory authorities and industry representatives. In addition to its active involvement in the UN Strategic Approach to International Chemical Management and GHS, the Chemical Dialogue has coordinated to communicate potential adverse trade impacts of REACH on APEC economics between APEC and the European Commission.

§ The WTO Technical Barriers to Trade (TBT) Committee provides a multilateral forum to address potential trade impact of technical regulations including environmental regulations.

CRITICAL REVIEW FOR FURTHER IMPROVEMENT

As a simple policy dialogue forum that does not have an established and independent institutional support, TEMM would have clear limits in playing a balancing and harmonizing role for the three countries. Following the recommendations of the Tripartite Joint Research of 2008 (Tripartite Joint Research on Environmental Management in Northeast Asia, 2008) TEMM may have to institutionalize an independent secretariat, and develop a good governance structure, for example, the establishment of basic principles and objectives, a financial mechanism, project implementation, and monitoring geographical distribution of projects. The Commission for Environmental Cooperation (CEC) which is established by Canada, Mexico, and the United States under the North American Agreement on Environmental Cooperation (NAAEC), may provide one of the good benchmarking mechanisms to address regional environmental concerns and to ensure prevention of potential trade and environmental conflicts and effective enforcement of environmental law.

What may have been problematic in previous trends of transboundary movement of global and European environmental regulations to Asia, in particular, China and Korea, was "free-riding justification for further enactment" (Park, 2010). Without carrying out systematic regulatory impact assessments, Asian environmental authorities had a tendency of borrowing all prejustification and political backup for further environmental regulations from studies of the EU. In those occasions, TEMM has not been able to monitor, supervise, and involve stakeholders to discuss emerging environmental regulations impacting local and global industries. The result was unnecessary administrative and business costs on market players in the Asian market even though a regulatory initiative itself has legitimate and positive benefits to society in a long term.*

Deutsche Bank forecasted that the Chinese chemical industry would become the second biggest producer after the United States as of 2015 (Deutsche Bank Research, 2005). China is already the world's second in the chemical shipment value of US$ 388 billion and Japan is the fourth in the total shipment rank after the United States of America, China, and Germany (Japan Chemical Industry Association, 2009). If the region grows without sound chemical management mechanisms and well-balanced coordination of chemical policies and regulations, the environment, the public, and the chemical industry would be at risk.

CONCLUSION

The Tripartite Policy Dialogue on Chemicals Management was an excellent initiative to communicate existing and emerging chemical policies and regulations among

---

* While lacking sufficient administrative and technical capacity, for example, China introduced the Chinese version of the European Union Directive 2002/95/EC on the Restriction of the Use of certain Hazardous Substances in Electrical and Electronic Equipment (RoHS) in February 2006. Although the enactment of the Chinese RoHS had some justification for the protection of the environment and human health, it has not been effectively implemented and enforced. If the same legislative action is taken by a Chinese authority on Regulation (EC) No 1907/2006 of the European Parliament and of the Council of December 18, 2006 concerning the Registration, Evaluation, Authorisation and Restriction of Chemicals (REACH), this may cause enormous operational interruption to all industries in China.

policymakers. China, Japan, and Korea may need to institutionalize their dialogue into an organization and incorporate good operational and governance structure so as to implement further regional and global cooperation.

REFERENCES

Chao, Xi. 2005. *Transforming Chinese Enterprises: Ideology, Efficiency and Instrumentation in the Process of Reform*, J. Gillespie, and P. Nicholson, (Eds.), *Asian Socialism and Legal Changes: The Dynamics of Vietnamese and Chinese Reform*. Canberra: ANU, pp. 91–114.

CIA World Factbook (China). 2011. Available at: https://www.cia.gov/library/publications/the-world-factbook/geos/ch.html

CIA World Factbook (Japan). 2011. Available at: https://www.cia.gov/library/publications/the-world-factbook/geos/ja.html

CIA World Factbook (Korea, South). 2011. Available at: https://www.cia.gov/library/publications/the-world-factbook/geos/ks.html

Deutsche Bank Research. 2005. Chemical Industry China: Overtaking the Competition (25 October 2005), Available at: http://www.dbresearch.de/PROD/DBR_INTERNET_DE-PROD/PROD0000000000191318.pdf

Japan Chemical Industry Association, Chemical Industry of Japan. August 2009. Available at: http://www.nikkakyo.org/upload/2661_3613.pdf

Park, DaeYoung. 2010. *REACHing Asia Continued*, Paper presented at Innovation for Sustainable Production 2010, Brugge, Belgium, 18–21 April 2010.

Tripartite Joint Research on Environmental Management in Northeast Asia (Summary for Policymakers). 2008. January, pp. 10–11. Available at: http://enviroscope.iges.or.jp/modules/envirolib/upload/2254/attach/summaryforpolicymakers_final.pdf

44 REACH
Next Step to a Sound Chemicals Management

*Arnold van der Wielen**

CONTENTS

INTRODUCTION

The publication of the REACH Regulation (EC, 2006) on December 30, 2006 marks the end of a long period of drafting and debating new chemicals management

* The opinions expressed in this chapter do not necessarily reflect the views of the Ministry of Infrastructure and the Environment, the Netherlands.

legislation in the European Union (EU*) involving all stakeholders, such as trade associations, non-governmental organizations, scientists, and authorities. After having been subject to heavy lobbying and debates for almost 10 years, there was finally an agreement and REACH came into effect in June 2007. Rules for manufacture, import, and use of chemical substances and preparations in the EU have been established within the framework of total harmonization of national legislations. The new legislation elaborates on the experiences and working practices from the previous EU chemicals management system. REACH is the culmination of a long effort to improve and integrate chemicals regulation in the EU and to replace more than 40 existing laws and regulations. The most important feature of REACH is that it seeks to reverse the burden of proof with regard to chemical safety so that industry, rather than public authorities, is responsible for identifying and controlling risks associated with chemicals manufactured, imported, and used in the EU.

At the end of the 1990s, the weaknesses of the previous system on chemicals management in the EU sparked discussion about ways of restructuring laws on chemicals. In response to criticism of the ineffective legislative system on chemicals in the past, the European Commission published its proposal for a Regulation in October 2003. A Regulation is directly applicable in EU member states, once it has been formally adopted. This proposed new regime involves processes of Registration, Evaluation, Authorisation and Restriction of Chemicals, known by the acronym REACH. The new strategy extends far beyond the previous legislation by building on the principles of (a) no-data-no-market, (b) one-substance-one registration, (c) shift in responsibility and burden of proof from public authorities to industry, and (d) supply chain responsibility. In addition, animal welfare became an important issue, for which procedures were established to prevent duplication of animal testing and to make effective use of available data from animal testing.

KEY OBJECTIVE OF REACH

The key objective of REACH is to ensure that chemicals, during their full life cycle, are managed in ways that no significant risks to humans and environment occur, similar to the WSSD 2020 goals (see last paragraphs of this chapter). In order to realize this objective, REACH has been designed to manage and control the potential hazards and risks to humans and the environment from the manufacture, import, and use of chemicals. The main elements of REACH are as follows:

- All substances within the scope of REACH and either manufactured or imported in 1 tonne or more per manufacturer/importer (M/I) per year will have to register; the no-data-no-market concept.
- New substances for the first time in the EU (so-called "nonphase-in" [new] substances) are to be registered before the manufacture or the import into the EU market. Existing substances already in the EU market (so-called "phase-in" [existing] substances) are to be registered by an interim

* In the abbreviation EU is understood the European Economic Area (EEA: 27 countries of the European Union and 3 countries from the European Free Trade Association).

procedure with deadlines defined by certain quantity thresholds (the so-called staggered registration procedure).

- For reasons of animal welfare, data generation by animal testing shall be reduced as much as possible by preventing duplication of animal testing, by promoting non-animal testing, by promoting use of alternative methods, and by sharing available animal test data. Based on the principle of *one-substance-one registration*, registrants of the same substance are participants of a *Substance Exchange Information Forum* (SIEF) and are obliged to cooperate in preparing a registration and in sharing core animal test data; sharing of nonanimal test data is encouraged.
- High volume substances and substances of concern will be evaluated by authorities to identify whether additional information is needed for completing missing knowledge, or whether sufficient information is available for initiating a community-wide action. In order to achieve a harmonized evaluation procedure within the EU, the task of evaluating registration dossiers is mandated to the European Chemicals Agency.
- The uses of substances of very high concern (SVHCs) may be subject to authorization; the SVHCs are well-defined CMRs (carcinogenic, mutagenic, or reprotoxic substances), PBTs (persistent, bioaccumulative or toxic substances), vPvBs (very persistent, very bioaccumulative substances), or substances of equal concern.
- Manufacture, import, or use of high-risk substances and even articles containing high-risk substances may be restricted in the case of serious concern for the health for humans or the environment.
- Reversal of burden of proof from authorities to industry, which is reflected in the obligation of industry in preparing a technical dossier as part of the registration, in assessing the safe use of a chemical, and communicating the results of the assessment up and down the supply chain.
- There is greater responsibility for all actors including distributors throughout the supply chain.

SCOPE OF REACH

REACH has a very wide scope. It applies to all chemical substances that are manufactured, imported, placed on the market, or used within the EU, either on their own, as a component in a preparation, or in an article.

The concept of "substance" is very broadly defined; it includes chemical elements and their compounds in either natural or manufactured forms, including process-related impurities and additives essential for the stability of the substance. Some substances are exempted from the overall scope of REACH (e.g., radioactive substances, waste, including hazardous waste, nonisolated intermediates, and substances in transit under custom supervision), while others are exempted from certain provisions of REACH, for example, substances and preparations already covered by specific legislation having comparable requirement of risk management. The regulation does not apply to the transport of dangerous substances or preparations. These subjects are covered by other pieces of EU legislation.

KEY POINTS OF REACH

General

REACH abolishes the previous distinction between "new" and "existing" substances in adopting a single legislative system, although new substances require registration prior to manufacture, import, or use, while existing substances may continue manufacture, import, and use until the deadline for submitting a registration. REACH requires manufacturers and importers to obtain relevant information on the substances they manufacture or import and to use that data to manage the substances safely.

It will ensure that substances currently on the EU market in 1 tonne/year or more will undergo at least basic health and safety testing. For planning purposes and for enabling data sharing, a pre-registration procedure has been introduced in which all potential registrants of existing substances are required to pre-register (simple administrative action) accompanied by the expected deadline for submitting a registration. The pre-registration exercise unexpectedly revealed that about 140,000 existing substances may be expected to be registered over the coming period instead of the presumed 30,000. REACH includes requirements to register all chemicals in a staggered time-schedule, giving priority to high volumes (for more details, see the section "Registration").

Any substance presenting a risk may be required to be evaluated and its manufacture or use can be subject to restrictions. The previously existing restrictions (EC, 1976) have been taken up in Annex XVII to the REACH Regulation that entered into force on June 1, 2009. Directives 67/548/EEC (dangerous substances directive: EC, 1967) and 1999/45/EC (dangerous preparation directive: EC, 1999) have been amended. Directive 76/769/EEC (limitation directive) has been withdrawn.

Registration

Registration (applicable to all chemicals manufactured or imported in quantities of 1 tonne/year (tonnes per annum: tpa) or more per M/I) is the centerpiece of REACH and consists of a registration procedure that is designed to provide systematic information about the properties of relevant substances. The aim of the registration dossiers is for manufacturers and importers in the EU to demonstrate that they manage their chemicals safely or that they can be used safely by downstream users within the supply chain. For that reason, a registration dossier may contain a chemical safety assessment (CSA) (required for substances ≥10 tpa per M/I), in which the registrant assesses the physicochemical and (eco)toxicological properties, assesses the emission and exposure from identified uses, and determines recommended risk control measures. The registration will be stored in a central database under the management of the Agency. Registration applies both to substances on their own and to components of preparations. Substances in articles are also subject to registration requirements when they are intended to be released from the articles. Examples of articles intended to release substances are toner cartridges, printer ribbons, pens and ballpoints, wet wipes, scented children's toys, and firecrackers.

The registration schedule will depend on whether the potential registrants are manufacturing or importing a "nonphase-in" (new) or a "phase-in" (existing) substance. For nonphase-in substances, registration is required since June 1, 2008 before manufacture, import, or placing on the market. Nonphase-in substances will primarily include all new substances, and any substance that does not meet the definition of phase-in as given in the Regulation.

More complicated is the process of phase-in (existing) substances, those that were listed in the European Inventory of Existing Commercial Chemical Substances (EINECS), or manufactured but not placed on the market before June 1, 2007, or known from previous legislation as "no-longer-polymer" substances. The registration of existing substances will be implemented progressively over a period of 11 years (see Table 44.1), beginning with the registration of existing substances produced in high tonnages (1000 tpa or more per M/I) and the most dangerous ones of 1 tpa or more; they had to be registered before December 1, 2010.

Existing substances of 100 tpa or more per M/I have to be registered before June 1, 2013. The remaining part of existing substances between 1 tpa and 100 tpa per M/I per year have to be registered before June 1, 2018. Substances for which these deadlines will not be met will have to be removed from the market and will be considered "no-phase-in substances." Therefore, after 2018, when all existing substances have been registered in REACH, a simple straightforward procedure of registration before manufacture/import will be the only procedure.

Information to be supplied includes physiochemical, toxicological, and ecotoxicological data on the substance. In addition, individual identified uses of downstream users throughout the supply chain, as well as assessments of the associated risks and safety measures derived from these uses, must be specified. Information

TABLE 44.1
For Phase-In Substances, the Following Deadlines Apply

Deadline	Categories of Phase-In Substances
Nov. 30, 2008	CMRs[a], Categories 1 and 2, ≥1 tonne/year
	N; R50–53[b] ≥100 tonne/year (screening criteria for PBTs[c] or vPvBs[d])
	≥1000 tonnes/year per M/I
June 1, 2010	≥100 tonnes/year per M/I
June 1, 2018	≥1 tonne/year per M/I
From June 1, 2011	Notification of substances of very high concern in articles (CMR, PBT, or vPvB if >0.1% weight by weight as component/impurity in an article); within 6 months after introduction in the EU market.

[a] Carcinogenic, mutagenic, reprotoxic substances.

[b] Substances classified as dangerous for the environment (symbol N), very toxic to the aquatic environment (risk phrase R50), and may cause long-term adverse effects in the aquatic environment (risk phrase R53).

[c] Persistent, bioaccumulative, toxic substances.

[d] Very persistent, very bioaccumulative substances.

to be supplied is progressively related with the yearly volume manufactured or imported. At volumes ≥100 tonnes/year per M/I, respectively, ≥1000 tonnes/year per M/I, additional information is required. For additional information involving animal testing, submission of one or more testing proposals for the missing additional information must be submitted to the Agency for approval of conducting such testing.

At volumes ≥10 tonnes/year per M/I, a detailed assessment of the safe handling CSA must be conducted and documented in a chemical safety report (CSR) that is included in the registration dossier.

Testing and Data Sharing

The testing requirements for collecting the information needed for the registration are not rigid. The aim of the registration procedure is that manufacturers or importers demonstrate that their substances can be used in a safe way. For this reason, manufacturers or importers may deviate from the standard testing requirements, provided that sound justification is given. Already available information or information derived by alternative methods can be used and tests can be waived.

In order to minimize testing and costs, REACH encourages sharing of data among potential registrants of the same substance on the basis of a fair, transparent, and non-discriminatory compensation of costs. Sharing of animal test data is even obligatory. A system of preregistration of phase-in substances had been applied as a tool to facilitate the sharing of core data. The preregistration period had a 6-month run: from June 1, 2008 through December 1, 2008. During this period, potential registrants had to submit elementary information of all existing substances intended to be registered to the Agency in the following 11-year period.

From this, the facility of a Substance Information Exchange Forum (SIEF) has been created for potential registrants of the same substance, with the aim of reaching agreement on the sameness of the substance identity and, if the same, on the sharing of tests and costs, and on who will play the role of lead registrant and on who will perform future tests on behalf of the SIEF members. REACH does not regulate how the cooperation between the members in the SIEF should be organized. A possibility is to form a consortium and enter into a consortium agreement between the members of the SIEF. A consortium is not a mandatory requirement by REACH but is considered to be a very effective way to organize the cooperation between the members of the SIEF and to fulfill their SIEF objectives and joint submission of data. Since data sharing under the SIEF usually regards the joint submission of data to the Agency, the members of a consortium usually are the potential registrants. However, participants of a consortium are not in any way restricted and a non-EU manufacturer or importer can also participate through its Only Representative (see the section "Consequences for Imports from Outside the EU") or directly by itself. Consortia may also be formed between different SIEFs.

A company that did not preregister a phase-in (existing) substance will not be able to make use of the staggered phase-in registration procedure. Without preregistration, their substance would have the status of a nonphase-in substance and be registered immediately.

For chemicals not raising special concerns or generally produced in amounts of less than 100 tonnes/year, registration will be the only step required. But all substances considered to present a risk will be subject to evaluation by the authorities on the basis of the registration data. This system aims to reduce unnecessary animal testing and to ensure total compliance with all recommended control measures.

The task of evaluating substances of potential concern is given to the authorities of the EU member states. Substances will be evaluated on a priority basis, to clarify suspicions of risks to human health or the environment. Substances selected for evaluation will be put on the Community Rolling Action Plan, from which member states will choose which substances to evaluate. Such evaluations may lead to the conclusion that additional information is needed or that action needs to be taken under the authorization or restriction procedure of REACH.

Communication Down the Supply Chain

All chemicals must be classified and labeled if they fulfill the classification criteria (see also the section "Information in the Supply Chain"), and all information relevant for the safe handling and use of substances must be communicated through the supply chain. The provisions related to the safety data sheets (SDSs) (EC, 2001) have been incorporated into the REACH Regulation. Their quality will be improved, thanks to the additional information deriving from the registration requirement in REACH.

Authorization

Authorization of substances raising the most concerns (substances of very high concern, SVHCs) constitutes the third pillar of REACH. It relates to "CMR" substances, "PBTs" and "vPvBs" which are considered to have no threshold below which there is no risk. It applies to manufacturers, importers, and downstream users who place a substance on the market or who use the substance themselves. The principle behind SVHCs being identified for authorization is "banned, unless authorized for a specific use." Companies requiring authorization need to demonstrate (a) essential need for which viable alternatives do not exist, and (b) risks from use are well controlled and fairly balanced with the economical benefit. If one or more viable alternatives exist, the company must present a plan for gradual substitution of the incriminated substance with a safe(r) alternative. The aim of authorization is to ensure that risks from SVHCs are properly controlled or that these substances are replaced by suitable alternatives or technologies. The use of SVHC will only be authorized if the company demonstrates that there is no safe alternative and that the socio-economic benefits of the chemical outweigh the risks.

The scope of the authorization regime does not fully overlap with the scope of REACH. There is no volume threshold for the application of authorization and substances subject to authorization do not necessarily have to be registered. Moreover, substances exempted from the registration requirements of REACH are not exempt from the scope of authorization.

As entry point for identifying substances for which the authorization regime is applied, a candidate's list of substances meeting the CMR, PBT, or vPvB criteria is drawn up. Substances on the candidates list shall be prioritized for being formally authorized substances on the basis of the hazardous properties, wide dispersive use, or if manufactured in high volumes. Substances adopted for inclusion in the list of authorized substances are specified with a starting date and a sunset date; the starting date is the entry into force (EIF) of the authorization status of the substance and the sunset date is the EIF of the "ban, unless authorized for a specific use." Between the starting and the sunset date, a company may submit a request for authorization of a specified use providing a technical dossier including a detailed CSR and a socioeconomical analysis. Adequate control shall be demonstrated in the applicant's CSR. All authorizations will be subject to a time-limited review.

The candidates list and the list of authorized substances will be updated regularly; updates of the candidates list is expected twice a year and of the authorization list once a year. The decisions on placing substances on the authorization list and on granting requests for authorization are taken by the European Commission in consultation with the EU member states.

Restrictions

A proposal to restrict the manufacture, marketing, and use of the substance can be made by the Commission or by an EU member state, and is intended to provide a safety net to manage risks that have not been addressed (properly) by other instruments of REACH. It allows for community-wide risk reduction measures to be introduced where there is an unacceptable risk to human health or the environment due to the manufacture, import, placing on the market, or use of a substance.

The decisions on restriction are also taken by the Commission in consultation with member states. Restriction can apply to a substance on its own, in a preparation, or in an article.

Management of the System: The Agency

The Agency (established in Helsinki, Finland) will manage the technical, scientific, and administrative aspects of the REACH system. The June 1, 2007 EIF has already triggered a cascade of deadlines affecting businesses (manufacturers, importers, and downstream users), public authorities of EU member states and the Agency, and will trigger further deadlines in the coming years. (Most of REACH's substantive provisions for business applied a year after, from June 1, 2008, with the start of a 6-month window for manufacturers and importers to pre-register all existing substances intended to be registered in the following period of 10 years (the last deadline is June 1, 2018 and refers to registration of substances between 1 and 100 tpa).

CHEMICAL SAFETY ASSESSMENT (CSA) DOCUMENTED IN A CHEMICAL SAFETY REPORT (CSR)

General Aspects

Key information in registration and authorization of substances is the risk assessment from identified uses, in REACH called the chemical safety assessment (CSA) to be documented in a chemical safety report (CSR). A CSA must be prepared by the registrant as part of the registration dossier for "hazardous" substances in volumes of 10 tonnes/year or more, or as part of a request for authorization for a specified use of an authorized substance. The CSA has to demonstrate that the risks are being adequately controlled. The results of the CSA must be documented in a CSR.

In brief, a CSA consists of:

- An assessment of the hazardous properties of the substance. If the substance should be classified in one of the hazards classes (receive a hazard label) or may pose risks to the environment on the basis of persistence, bioaccumulation, and toxicity (a so-called PBT or vPvB substance), the manufacturer or importer needs to add an assessment of the nature and extent of exposure and emission taking into account the control measures already to be taken (such as gloves). All relevant exposure and emission scenarios must be included in the assessment.
- The risk characterization on the basis of the hazard and emission/exposure assessments taking into account the desired risk control measures.
- Such risk control measures being implemented at the production location or recommended for the intended uses by downstream users that the substance can be used safely.

For the manufacturer, the CSA includes both the manufacture at its own site as well as all intended uses downstream. For the importer, the CSA only includes all intended uses by the importer or by downstream users.

When carrying out the CSA, first the intrinsic risks of the substance for humans and the environment are established, with special attention paid to persistent, bioaccumulative, and toxic. If the substance is hazardous, then the exposure for humans and the environment is evaluated, and finally the risks are charted by comparing hazard and the level of exposure. In the final step of the assessment, the exposure of humans and the environment is compared with the established toxicological limit values (no effect levels, NELs). In the end, the exposure must always be lower than the limit values: only then can the situation be declared safe with sufficient certainty. In situations where limit values cannot be established (for toxic effects having no threshold, such as carcinogenicity, mutagenicity, and reproductive toxicity), the applicant has to substantiate why no risks are expected. The result should preferably be that the use is safe for the identified uses, since if this is not the case, this fact will certainly be singled out for evaluation at the registration.

Identified Uses and Exposure Scenarios

The assessment of the exposure scenarios (ES) for humans and the environment is currently the most complex and less experienced part of the assessment (see Figure 44.1). In order to be able to prepare a solid CSA of a substance, it is important that the manufacturer or importer be fully aware of how the substance is used further down the supply chain and how the customers in the supply chain might be exposed to the substance. On the one hand, manufacturers and importers may only supply a substance if they are certain that the customer can or will use the substance in a safe

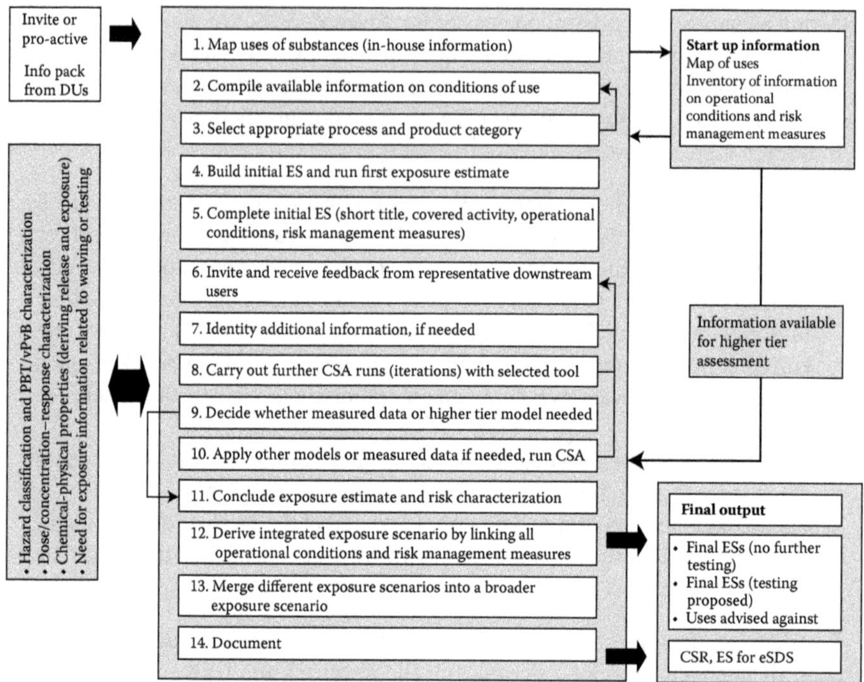

FIGURE 44.1 Steps for exposure scenario (ES) development related to downstream uses. (*Abbreviations:* CSA, chemical safety assessment; DU, downstream user; ES, exposure scenario; e-SDS, extended safety data sheet.)

manner. On the other hand, it is virtually impossible for a manufacturer or importer to carry out their own thorough CSA for every application. However, this is what REACH asks for. For downstream users, especially the formulators, it is crucially important to help the raw material suppliers in developing ESs for the applications of their customers and even for those of their customers' customers.

The desired/recommended risk control measures identified for the manufacture and each identified use lead to the development of ESs for substances and their uses. ESs are sets of conditions that describe how substances are manufactured or used during their life cycle and how the manufacturer or importer controls or recommends controlling exposures to humans and the environment.

An ES includes four elements:

- Substance properties (physicochemical and (eco)toxicological properties)
- Product properties (physicochemical properties of the final product including composition and release of substances)
- Operational conditions (use conditions, conditions at the workplace, in general public, etc.)

- Measures to control the risk (physical measures, protection measures, emission control, etc.).

ESs are developed to cover all identified uses, which are the M/I own use or uses that are made known to them by their downstream users. In this respect, the ESs are developed through intensive dialogue between the manufacturer or importer and their downstream users, as well as dialogue between the downstream users themselves.

Many risk assessments take a substance and look at the uses to which it is put: the so-called “substance approach.” For a professional market in which small volumes are traded, this substance approach is not only time-consuming, but also probably unfeasible. Industry is therefore developing “standard” scenarios based on the manner of application and the corresponding application method. Many categories of professional use can be unravelled into standard application scenarios. For that reason, a use-descriptor code system has been developed as a preferred tool for upstream and downstream communication of uses. The various uses are identified by “sector of use” (SU-code: main use category), “product category” (PC-code: type of product), “process category” (PROC-code: type of industrial process), “article code” if applicable (AC), and “environmental release category” (description of environmental release). Standard ESs are linked to these use descriptor codes; see the following for an example of an ES development related to downstream uses.

The ESs identified by the manufacturer or importer in the CSA are communicated to their downstream users. The downstream user must then decide whether the particular use of the substance is covered by the identified ES. If not already covered, the downstream user can identify its use to his manufacturer or importer, and the manufacturer or importer can choose to include this use in the registration and in the CSA. On the other hand, a downstream user may choose not to identify the use to the manufacturer or importer, for example, for confidentiality reasons. If a downstream user chooses this, then he must inform the Agency of his non-identified use, and prepare his own CSR. Overall, early and intensive dialogue is essential between M/Is and downstream users for getting an overview of identified uses and for including all identified uses and subsequent emissions/exposures in the registration dossier.

RISK ASSESSMENT

Risk assessment is an iterative process in a tiered approach (see Figure 44.2). The first step is making an initial assessment of the exposure for humans and the environment on the basis of the identified uses. The extent of the exposure or concentration in the environment can be determined with (proven) model estimations or on the basis of measured data that have been collected. This can be compared to the limit values below which the use is considered as safe. Initially, a rough estimate often suffices for exposure estimations (“Tier I”). If the estimated exposure is too high, an attempt can be made to make a better estimation in a second step (“Tier II”). The exposure itself can be determined more precisely by carrying out measurements or by using a better

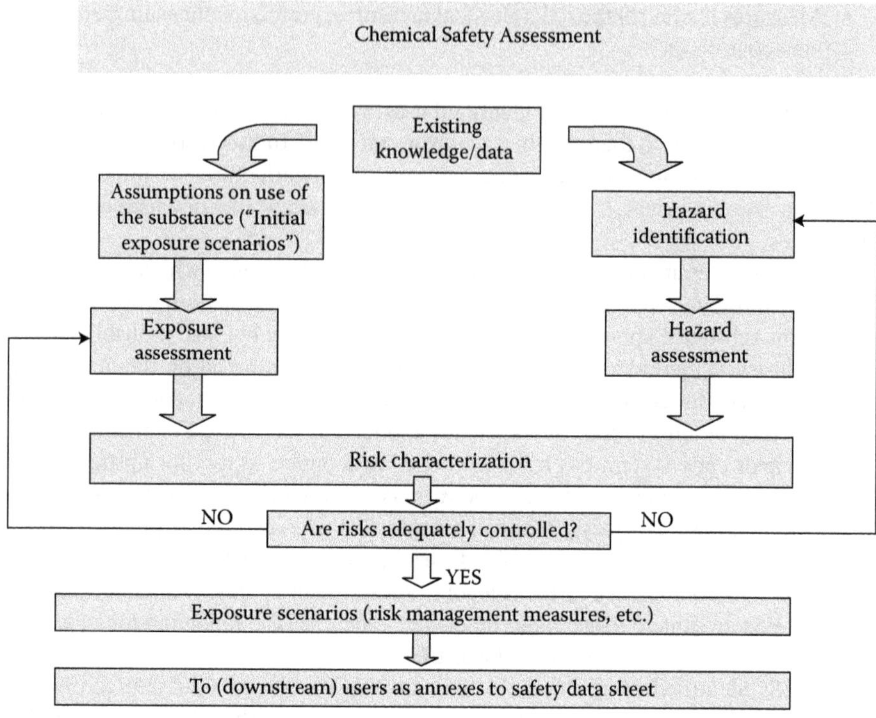

FIGURE 44.2 Process of risk assessment.

model. Better models often also need more precise information about the use and the conditions. This information needs to be gathered, and must finally be included again in the use scenario. An adjustment of the exposure estimation can also be made by adjusting the use process itself, by taking other control measures, or by making a more precise determination of the exposure. The limit values can be adjusted, among other things, by carrying out new tests with the substance or by involving the data of closely related substances. The limit values can often be decreased by reducing the uncertainty; for example, testing several different species instead of only one species. The use of several species may reduce the safety factor for extrapolating a limit value found in an experiment. After the adjustment of the level of exposure or the limit value, these two have to be compared again. If the exposure still exceeds the limit value, the cycle mentioned above must be followed again, for as long as it takes to reach an exposure below the limit value where the use can be called safe (Tier III, IV, etc.) or to conclude that safe use is not possible.

INFORMATION IN THE SUPPLY CHAIN: THE SAFETY DATA SHEET

Where a substance or a preparation is classified as hazardous in accordance with Directives 67/548/EEC or 1999/45/EC (since December 1, 2010 according to the

CLP regulation [EC, 2008] implementing the Globally Harmonized System for Classification and Labelling [GHS] in the European Community) or the substance is a PBT or a vPvB, then the supplier of that substance or preparation must provide the recipient with a safety data sheet (SDS). The format of the SDS is outlined in the regulation. The SDS will be the main tool for communication down the supply chain. Any person who is required to carry out a CSA will need to ensure that the information in both the SDS and the CSR are consistent with each other. The ESs and the applied or recommended risk control measures for all identified uses, documented in the CSR, are annexed to the SDS (so-called extended SDS or e-SDS). These can be used by downstream users when deciding on and implementing their risk control measures or when compiling their own SDSs for identified uses down to their customers. Downstream users receiving an e-SDS with ESs and recommended risk control measures are obliged to implement the recommended measures or equivalent measures provided that the equivalent safety can be justified.

PRACTICAL ASPECTS IN IMPLEMENTING REACH

To enable a smooth transition from the previous existing chemicals legislation to REACH, the European Commission developed an interim strategy to ensure that all stakeholders, especially industry and member states authorities, were adequately prepared for the practical application of the new system by the time REACH entered into force. Since the establishment of the ECHA, further development of additional guidance documents and tools has been taken over by the Agency. Within the interim strategy, a number of REACH Implementation Projects (RIPs) have been conducted covering the following tasks:

- The preparation of the IT formats and software (IUCLID-5 and REACH-IT) to enable technical dossier development by the industry and member states, and the submission of these dossiers to the Agency under REACH.
- Preparation of the detailed technical guidance documents to provide advice to industry and member states on the detailed requirements of the new system. Further details of the work on-going on the interim strategy can be found on the Web site of the European Chemicals Agency: http://echa.europa.eu.
- The practical arrangements for the establishment of the ECHA in Helsinki (since June 1, 2007).

A Web-based REACH-IT system has been developed as an interface to an IUCLID-5 database (International Uniform Chemical Information Database) in order to allow computerized exchange of information and submission of dossiers by uploading IUCLID-5 files into the REACH-IT system. IUCLID-5 must be used by companies to prepare their registration and authorization application dossiers before their submission to REACH-IT. IUCLID-5 software, including helpful tools, has been made available to companies (http://www.iuclid.eu).

The guidance documents for industry deal with issues such as substance identification, use of alternative methods for collecting information on intrinsic properties, preparing a CSA, classification and labeling, preparing a socioeconomic analysis, requirements for the downstream users, and requirements regarding substances in articles. The extensive guidance for industry is, next to subject-specific guidance documents, made accessible by a Web-based navigator tool that allows different stakeholders to quickly find and retrieve the pieces of information they need for understanding and fulfilling their obligations under REACH. The Web-based navigator tool and many detailed guidance documents are available in the final version and accessible on the ECHA Web site.

It has to be noted that guidance documents are not part of the legal REACH text. REACH allows industry to apply alternative approaches provided that they are justified. However, REACH requires the Agency to develop guidance for the implementation of REACH. The guidance will be applied by the Agency and the EU member states to check *inter alia* compliance of information submitted. Similar as under the previous EU legislation, guidance will become more authoritative depending on the use by authorities in the decision-making process.

CONSEQUENCES FOR IMPORTS FROM OUTSIDE THE EU

In general, the REACH regulation does not affect companies outside the EU, but their exports to the EU may be concerned by REACH. The importers established in the EU of a non-EU exporter are primarily responsible for registering imported substances as such, in preparations or in articles intended to release substances as part of normal use. In this respect, the importers are relying on their non-EU exporter's information about the chemical identity of imported substances as such or as components in preparations, composition of an imported preparation or article hazard data, and safe use of a substance, required for registration and assessment under REACH. Detailed information about chemical composition of preparations or about chemical identities of substances for many non-EU exporters may be considered as confidential business information that they do not want to disclose to EU importers. If preferred for confidentiality concerns, the non-EU exporter may appoint an "Only Representative" (OR) being a legal entity in the EU to fulfill the REACH tasks of an importer in the EU and to be responsible for the registration obligations on behalf of the EU importers. Only manufacturers, producers of preparations, and producers of articles outside the EU are allowed to establish an OR. In consequence, the other EU-importers shall be regarded as downstream users of the OR and do not have to register the same substance. For that reason, the OR needs to keep record of the listed importers including the imported volumes per importer and will be responsible for the information flow downstream by way of classifying and labeling the substance or preparation and providing SDSs for the professional users downstream.

By appointing an OR, a non-EU exporter may take part in the cooperation and data sharing under REACH by participating in the SIEF of an exported substance. The SIEFs have already been formed after the pre-registration period has ended, aiming to facilitate data sharing for the purposes of registration and to agree on the classification and labeling of the substance. In practice, many companies have organized in consortia for fulfilling the tasks of SIEFs. While a SIEF is only open for potential registrants established in the EU, non-EU exporters can voluntarily participate in consortia for sharing data and agreeing on the hazard characterization.

REACH will have a major impact on the trade of products on both sides of the EU border. EU importers and ORs will need more information on imported substances as such, on the composition of imported preparation, and on the composition of articles. Because of the costs involved, REACH may create serious problems to small importers (and exporters). For most of the small importers, the costs for preparing a registration will be high, certainly because of expertise needed for planning tests, deriving ESs, evaluating the data, and drafting a CSR for identified uses that may have to be hired externally. If non-EU exporters still have interest to be on the EU market, then they have to bear the costs.

NEXT STEP TO A SOUND CHEMICALS MANAGEMENT

REACH will create a broad information base that will be of great value to all stakeholders who are authorities, industry in the broadest sense, non-governmental organizations, scientists, and the general public. REACH defines clearly what information is considered to be non-confidential and will be disseminated; among it, all information about physical-chemical and (eco)toxicological properties, classification and labeling, ESs for intended uses, and e-SDSs. New findings will enable industrial manufacturers and producers to improve their products and production processes, and users of chemicals to improve their working procedures. In the regulatory sphere, the new findings will also have an impact on legal areas outside of the scope of REACH. Among others, the data obtained through REACH will facilitate authorities' actions for improving water and soil quality because regulations pertaining to water and soil often focus on hazard criteria. The situation is similar with legislation regarding consumer protection and occupational health and safety.

Next to the impact on the regulatory sphere, REACH will have a greater impact on the market forces worldwide. Downstream users being aware of the REACH requirements will look for (a) REACH compliant substances, and (b) less dangerous alternatives or technologies for substances being identified as being of very high concern (listed in the candidates list or in the authorization list). Non-EU companies exporting substances, preparations, and articles into the EU are forced to generate data in support of their EU importers for fulfilling the REACH requirements. However, in general, non-EU companies not having assessed their substances according to the REACH standard may be in a competitive, less profitable position in the world market because of the favorable position of their competitors marketing well-assessed substances.

In the plan of implementation of the World Summit on Sustainable Development (WSSD, 2002) in Johannesburg in 2002, the summit with the participating countries commit themselves to a wide range of actions; among them Action 23:

> Renew the commitment, as advanced in Agenda 21, to sound management of chemicals throughout their life cycle and of hazardous wastes for sustainable development as well as for the protection of human health and the environment, inter alia, aiming to achieve, by 2020, that chemicals are used and produced in ways that lead to the minimization of significant adverse effects on human health and the environment,

> using transparent science-based risk assessment procedures and science-based risk management procedures, taking into account the precautionary approach, as set out in principle 15 of the Rio Declaration on Environment and Development, and support developing countries in strengthening their capacity for the sound management of chemicals and hazardous wastes by providing technical and financial assistance.

With the implementation of REACH aiming to achieve by 2018 that all new and existing chemicals are used and produced in ways as committed in the WSSD implementation plan, the EU made a substantial move to the direction of a future society built on sustainable chemistry.

GLOSSARY AND DEFINITION OF TERMS

Agency: The European Chemicals Agency (ECHA) as established by REACH; established in Helsinki.

Article: An object which during production is given a special shape, surface, or design which determines its function to a greater degree than does its chemical composition.

Authorization: Decision of the European Commission on a request for granting an authorization for a specific use of a substance for which authorization is required. These substances are listed in Annex XIV of REACH and belong to the category of very high concern substances: carcinogenic, mutagenic or reprotoxic substances (CMR-substances) and substances dangerous for the environment (PBTs or vPvBs) and "substances of equal concern."

CMR substances: Substances being characterized as carcinogenic, mutagenic or toxic for reproduction.

CSA (Chemical Safety Assessment): An instrument to assess the intrinsic hazards of substances including determining the hazard classification, further characterizing hazards, including where possible derivation of no-effect-levels (Derived No-effect-Levels for human health, Predicted No-Effect-Concentrations for environment), and assessing properties relating to persistent, bioaccumulative or toxic chemicals (PBT). This includes generation of new information if needed. In addition, when the substance is classified as dangerous or assessed to have PBT or vPvB properties, to assess the emission/exposure of humans and the environment resulting from manufacture and uses throughout the life cycle of the substance (this includes the generation of sufficiently detailed information on uses, use conditions and emissions/exposures of the substance), to characterize risks following such emission/exposure, and ultimately to identify the conditions of manufacture and use which are needed for controlling the risks to human health and the environment. This includes operational conditions (OC) and risk management measures (RMM).

CSR (Chemical Safety Report): The CSA documented in a report.

Downstream user: Any natural or legal person established within the Community, other than the manufacturer or the importer, who uses a substance, either on its own or in a preparation, in the course of his industrial or professional activities. A distributor or a consumer is not a downstream user.

ES (Exposure Scenario): The set of conditions, including operational conditions and risk management measures, that describe how the substance is manufactured or used during its life cycle and how the manufacturer or importer controls, or recommends downstream users to control, exposures of humans and the environment. These ESs may cover one specific process or use or several processes or uses as appropriate.

GHS (Globally Harmonized System): A globally harmonized hazard classification and compatible labelling system recommended by the United Nations Economical and Social Council's Sub-Committee of Experts on the Globally Harmonized System of Classification, implemented in the EU by the Regulation on classification, labelling and packaging of substances and mixtures.

Identified use: A use of a substance on its own or in a preparation, or a use of a preparation, that is intended by an actor in the supply chain, including his own use, or that is made known to him in writing by an immediate downstream user.

Import: The physical introduction into the customs territory of the Community.

Importer: Any natural or legal person established within the Community who is responsible for import.

Manufacturer: Any natural or legal person established within the Community who manufactures a substance within the Community.

Manufacturing: Production or extraction of substances in the natural state.

Non-phase-in substance: A substance not fulfilling the criteria of a phase-in substance (practically considered as new substances).

PBT substances: Substances being characterized as persistent, bioaccumulative and toxic according to the criteria as defined in Annex XIII of REACH.

Phase-in substance: A substance that meets at least one of the following criteria:

a. It is listed in the European Inventory of Existing Commercial Chemical Substances (EINECS);
b. It was manufactured in the Community on May 1, 2004, or in the countries acceding to the European Union on January 1, 1995, on January 1, 2007, but not placed on the market by the manufacturer or importer, at least once in the 15 years before the entry into force of this Regulation, provided the manufacturer or importer has documentary evidence of this;
c. It was placed on the market in the Community on May 1, 2004, or in the countries acceding to the European Union on January 1, 1995, on January 1, 2007, by the manufacturer or importer before the entry into force of this Regulation and it was considered as having been notified in accordance with the first indent of Article 8(1) of Directive 67/548/EEC in the version of Article 8(1) resulting from the amendment effected by Directive 79/831/EEC, but it does not meet the definition of a polymer as set out in this Regulation, provided the manufacturer or importer has documentary evidence of this, including proof that the substance was

placed on the market by any manufacturer or importer between September 18, 1981 and October 31, 1993 inclusive.

OR (Only Representative): A natural or legal person in the EU appointed by mutual agreement by a natural or legal person outside the EU who manufactures a substance on its own, in preparations or in articles, formulates a preparation or produces an article that is imported into the EU in order to fulfill, as his only representative, the REACH obligations on importers.

Placing on the market: Supplying or making available, whether in return for payment or free of charge, to a third party. Import shall be deemed to be placing on the market.

Preparation: A mixture or solution composed of two or more substances.

Registration: A technical dossier to be submitted for a non-phase-in substance prior to manufacture or import, or for a phase-in substance before the deadline as defined in the staggered registration schedule.

Registrant: The manufacturer or the importer of a substance or the producer or importer of an article or the only representative submitting a registration for a substance.

Restriction: Any condition for or prohibition of the manufacture, use or placing on the market of a substance.

SIEF (Substance Information Exchange Forum): A forum participated by all potential registrants, downstream users and third parties who have information from the same phase-in substance, or by registrants who have submitted a registration for that phase-in substance with the aim to share the information and to agree on classification and labelling.

SME (Small- and Medium-size Enterprise): Small- and medium-sized enterprises as defined in the Commission Recommendation of May 6, 2003 concerning the definition of micro-, small-, and medium-sized enterprises.

Staggered registration procedure: A transitional registration procedure for phase-in substances with deadlines of submitting a registration based on manufactured or imported yearly volume or based on hazardous properties.

Substance: A chemical element and its compounds in the natural state or obtained by any manufacturing process, including any additive necessary to preserve its stability and any impurity deriving from the process used, but excluding any solvent which may be separated without affecting the stability of the substance or changing its composition.

SVHC (Substance of Very High Concern): Substances characterized as carcinogenic, mutagenic or reprotoxic substances (CMR-substances), substances dangerous for the environment (PBTs or vPvBs) and "substances of equal concern."

Use: Any processing, formulation, consumption, storage, keeping, treatment, filling into containers, transfer from one container to another, mixing, production of an article or any other utilisation.

Use and exposure category: An exposure scenario covering a wide range of processes or uses, where the processes or uses are communicated, as a minimum, in terms of the brief general description of use.

vPvB substances: Substance being characterised as very persistent and very bioaccumulative according to the criteria as defined in Annex XIII of REACH.

REFERENCES

EC Regulation (EC) No 1907/2006 of the European Parliament and of the Council of December 18, 2006 concerning the Registration, Evaluation, Authorisation and Restriction of Chemicals (REACH), establishing a European Chemicals Agency, amending Directive 1999/45/EC and repealing Council Regulation (EEC) No 793/93 and Commission Regulation (EC) No. 1488/94 as well as Council Directive 76/769/EEC and Commission Directives 91/155/EEC, 93/67/EEC, 93/105/EC and 2000/21/EC. *OJ L* 396 of 31/12/2006, p. 1.

EC Council Directive 67/548/EEC of June 27, 1967 on the approximation of the laws, regulations and administrative provisions relating to the classification, packaging and labelling of dangerous substances, *OJ* 196 of 16/8/1967, p. 1. Directive as last amended by Commission Directive 2004/73/EC (*OJ L* 152 of 30/4/2004, p. 1). Corrected in *OJ L* 216 of 16.6.2004, p. 3.

EC Regulation (EEC) No 793/93 of March 23, 1993 on the evaluation and control of the risks of existing substances, *OJ L* 84 of 5/4/1993, p. 1. Regulation as amended by Regulation (EC) No 1882/2003 of the European Parliament and of the Council (*OJ L* 284 of 31/10/2003, p. 1).

EC Directive 76/769/EEC of July 27, 1976 on the approximation of the laws, regulations and administrative provisions of the Member States relating to restrictions on the marketing and use of certain dangerous substances and preparations (*OJ L* 262 of 27/9/1976, p. 201). Directive as last amended by Directive 2005/90/EC of the European Parliament and of the Council (*OJ L* 33 of 4/2/2006, 1976, p. 28).

EC Directive 1999/45/EC of the European Parliament and of the Council of May 31, 1999 concerning the approximation of the laws, regulations and administrative provisions of the Member States relating to the classification, packaging and labelling of dangerous preparations (*OJ L* 200 of 30/7/1999, p. 1). Directive as last amended by Commission Directive 2006/8/EC (*OJ L* 19 of 24/1/2006, p. 12).

EC Directive 2001/58/EC of July 27, 2001 amending for the second time Directive 91/155/EEC defining and laying down the detailed arrangements for the system of specific information relating to dangerous preparations in implementation of Article 14 of European Parliament and Council Directive 1999/45/EC and relating to dangerous substances in implementation of Article 27 of Council Directive 67/548/EEC (safety data sheets) (*OJ L* 212 of 7/8/2001, p. 24).

EC Regulation (EC) No 1272/2008 of the European Parliament and of the Council of December 16, 2008 on classification, labelling and packaging of substances and mixtures, amending and repealing Directives 67/548/EEC and 1999/45/EC, and amending Regulation (EC) No. 1907/2006 (*OJ L* 353 of 31/12/2008, p. 1).

Plan of Implementation of the World Summit on Sustainable Development, Johannesburg, 2002 (WSSD, 2002).

45 North America Cooperation in Chemical Management

Asish Mohapatra and Philip Wexler

CONTENTS

INTRODUCTION

This chapter discusses management of chemicals in North America and regional cooperation and collaboration between Canada, United States, and Mexico. It introduces linkages, global initiatives, and agencies responsible for such chemical management initiatives, as well as the relationship of the regional scenarios. Further, it notes current and future collaboration scenarios within the regional context. Safe chemicals management is a local, national, and global concern, and significant changes in the production, storage, transportation and use of the chemicals, as well as an increased awareness of the potential harms from exposure to the chemicals are necessary. North American activities are an important contribution to the implementation of the Strategic Approach to International Chemicals Management (SAICM) (see Section IV of this book) and the realization of the World Summit on Sustainable

Development (WSSD) 2020 goal as adopted in Johannesburg in 2002 (see Section II, Chapter 4 of this book). The three countries are also active members of OECD and its chemicals program (see Section V, Chapter 26 of this book).

Chemical policy issues related to chemical risk management can include elements of information gathering, description of hazards, description of uses and potential for exposures, comparison of hazards and exposures (risk assessment), and actions to reduce risk. Chemical policy can be related to government actions, market forces, industry best practices, and product stewardship. Global and regional initiatives, as well as act and regulations identified in this chapter are moving forward or being revised to best manage chemical risk while protecting health and the environment, and allowing the three countries to remain competitive in the global marketplace.

NORTH AMERICA'S COLLABORATIVE CHEMICAL MANAGEMENT INITIATIVES

The Montebello Agreement (2007) (Cooper, 2007), between Canada, the United States, and Mexico, sets out a plan to coordinate risk assessment and risk management activities across North America, building on work done under the Canadian Chemicals Management Plan (CMP) and the U.S. Environment Protection Agency (US EPA) High Production Volume (HPV) Chemical Challenge. These efforts are being carried out to ensure the safe manufacture and use of industrial chemicals by entering into a regional partnership for assessing and managing potential chemical risks, with the goal of the agreement to enhance trade among the three countries, while ensuring protection of human health and the environment and retaining sovereignty.

The Montebello Agreement is considered by some as the North American response to the European Union's Registration, Evaluation, Authorisation and Restriction of Chemicals (REACH) program. Data generated under the agreement are intended to be used in jurisdictions outside of North America. For example, data generated in the HPV and Medium Production Volume (MPV) chemical programs are expected to be useful in meeting REACH registration requirements.

In 2008,* Canada, United States of America, and Mexico signed a joint Statement of Intent on North American Chemicals Cooperation affirming their commitment to developing a regional framework for the safe management of harmful chemicals, building on existing cooperative initiatives toward increased capacity building and improved chemicals management. The agreement will ensure that the three countries have access to consistent information that can be used to strengthen chemical management in North America. Canada agreed to share its analysis of more than 23,000 chemicals currently in use today and will gain access to critical research and expertise that are being gathered in the United States and Mexico to take action on harmful chemicals.

* Backgrounder: The Statement of Intent on North American Chemicals Cooperation, a joint statement of intent between Canada, Mexico, and the United States, can be accessed at: http://www.ecoaction.gc.ca/news-nouvelles/20081209-1-eng.cfm

The nature of those regional chemical management activities are described below:

By 2012:

- The United States will assess and initiate needed action on the over 6750 existing chemicals produced in amounts above 25,000 lbs per year (11,350 kg per year) in the United States.
- Canada will complete assessment of and take regulatory action on their highest priority substances as well as initiate assessment of medium-priority substances.
- Mexico will develop an information system for dangerous materials.
- The three countries will enhance appropriate coordination in areas such as testing, research, information gathering, assessment, and risk management actions on chemicals and nanoscale materials, and in facilitating pollution prevention approaches, improved environmental performance of industry sectors, and strengthened chemicals import safety.

By 2020, the trilateral cooperation intends to achieve the following:

- The establishment or updating of inventories of chemicals in commerce in all three countries.
- Enhanced capacity in Mexico to assess and manage chemicals.
- The sound management of chemicals (SMOC) in North America as articulated by the WSSD Johannesburg Plan of Implementation and reinforced by the SAICM.

The trilateral cooperation hopes to contribute toward the improvement of public health and environmental protection programs through a practical and focused approach, and to further strengthen chemicals management in North America over the long term.

NORTH AMERICA COMMISSION FOR ENVIRONMENTAL COOPERATION

The Commission for Environmental Cooperation (CEC)* is a regional organization created by Canada, Mexico, and the United States under the North American Agreement on Environmental Cooperation. The CEC was established to address regional environmental concerns, help prevent potential trade and environmental conflicts, and to promote the effective enforcement of environmental law. The Agreement complements the environmental provisions of the North American Free Trade Agreement (NAFTA).

The North American SMOC† initiative was established by a resolution of the Council of the CEC in October 1995, in Oaxaca, Mexico. SMOC is a trinational

* Commission of Environmental Cooperation (CEC): http://www.cec.org

† North American Sound Management of Chemicals (SMOC): http://www.cec.org/SMOC; http://www.cec.org/Page.asp?PageID=1225&SiteNodeID=237&BL_ExpandID=&AA_SiteLanguageID=1

initiative to reduce the risks of industrial chemicals to human health and the environment in North America. It promotes a regional approach to strengthen regional chemical management in collaboration with experts and the public. It has been active in helping to reduce or eliminate the use of substances of mutual concern such as DDT, chlordane, and mercury. It is important to note that in the context of Canada, the United States, and Mexico's contribution and participation in international chemical conventions such as Stockholm, Basel, and Rotterdam, these countries have their fair share of regional as well as global leadership and initiatives toward fulfilling several of those objectives identified under those conventions. However, while Canada and Mexico have ratified the Basel, Rotterdam, and Stockholm Conventions, the United States has not yet ratified these conventions.

Historically, Canada has worked to ensure that the global approach for the POPS Convention is effective. In March 2000, Canada became the first country that committed $20 million funding for POPS capacity building in developing countries and other emerging economies at that time. This funding helped those countries to find alternatives to the use of POPS, such as DDT. Further, the Government of Canada has also worked toward reduction and elimination of toxic substances under the Toxic Substances Management Policy, and regional action programs such as the Northern Contaminants Program (NCP). Furthermore, specific Canadian legislations, in particular, the Canadian Environmental Protection Act (CEPA), the Pest Control Products Act (PCPA), the Fisheries Act, and the Hazardous Products Act (HPA) were implemented to control and manage chemicals and protect ecosystems and public health. The Canadian leadership responsible for developing the convention has been recognized globally. Dr John Buccini chaired the negotiations and the first negotiating session was held in Montreal in 1998. In this session, Canadian science and research on the long range transport and the effects of POPS on humans and wildlife helped form the major scientific basis for the convention.

More recently, The trinational SMOC initiative provides a framework for "regional cooperation for the sound management of the full range of chemical substances of mutual concern throughout their life cycles, including by pollution prevention, source reduction and pollution control." A new direction was proposed in 2006 when the CEC Council directed SMOC to present the proposed realignment at the 2008 Council meeting for approval. This new direction focuses on strategies to catalyze cooperation in the following four areas:

1. Establish a foundation for chemicals management across North America—the development of comparable tools, data, and expertise for the assessment and management of chemical substances to enhance regional capacity for the safe production, transport, use, and disposal of chemicals. An example involves supporting Mexico's effort to develop an inventory of industrial chemicals.
2. Develop and implement a sustainable regional approach for monitoring toxic substances in humans and the environment—Data on the presence of chemicals in humans and the environment can be used to evaluate and assess the performance of the SMOC program's work and set priorities for future action, and help identify the impacts of chemical sources outside North

America. In addition Canada, the United States, and Mexico are also aggressively pursuing regional chemical management initiatives. SMOC also includes supporting Mexico in meeting its obligations under the Stockholm Convention on Persistent Organic Pollutants, and a sustainable approach for Mexico's Environmental Monitoring and Assessment Program (Hansen et al. 2006) (*Programa de Monitoreo y Evaluación Ambiental*—Proname).
3. Reduce or eliminate the risk from chemicals of mutual concern in North America. Completing implementation of action plans for mercury and lindane, a strategy to identify actions required addressing issues of mutual concern for dioxins, furans and hexachlorobenzene, and activities related to polybrominated diphenyl ethers (PBDEs) a group of chemicals primarily used as flame retardants in electronics, building materials and fabrics. Initial trilateral efforts focus on collaboration between Canada, the United States, and Mexico to assess the needs with regard to this chemical group.[*]
4. Improve environmental performance of various sectors—focuses on identifying and addressing challenges in the management of chemicals and opportunities in specific industrial sectors, such as current efforts in the health care and other sectors for mercury-related wastes and others.[†]

These areas are compatible with the Dubai Declaration on a SAICM and encompass the direction of the SMOC program along with an emphasis on: stronger outreach to stakeholders as partners; aligning North American priorities; and, the establishment of stronger linkages with key regional initiatives, such as the Security and Prosperity Partnership (SPP),[‡] and global initiatives such as SAICM, and meeting WSSD 2020 goals.

Further, the development of the Commission of Environmental Cooperation (CEC) and SMOC and SAICM share historical roots of the *Agenda 21*, WSSD 2020 goal and reflect the global will to improve global chemicals management. CEC, SMOC, and SAICM can be mutually supportive policy forums to address chemical issues of mutual concern. SAICM provides guidance on implementing a SMOC program through its priority elements of risk reduction; knowledge and information; governance; capacity building and technical cooperation; and eliminating illegal international traffic.

CEC SMOC provides examples of implementation of SAICM objectives through regional approaches to the management of chemicals of mutual concern. These current and proposed activities include:

- North American Regional Action plans (NARAPs)[§]-Minimized risks of chlordane, polychlorinated biphenyls (PCBs), DDT, NARAP model on mercury, lindane, dioxins, and furans, NARAPs Strategies for Catalyzing cooperation within North America, North American voluntary challenge programs within sectors or groups of Chemicals.

---

* http://cec.org/Page.asp?PageID=122&ContentID=2776&SiteNodeID=595

† http://www.cec.org/Page.asp?PageID=122&ContentID=2417&SiteNodeID=465

‡ http://www.spp.gov/pdf/spp_reg_coop_chemicals.pdf

§ North American Regional Action Plans (NARAP): http://www.cec.org/Page.asp?PageID=924&SiteNodeID=312

- Environmental Assessment and Monitoring of chemicals of mutual concern, Continued Environmental Assessment and Monitoring, Information road map, Biomonitoring, Inventory project.
- Effective regional mechanism for chemicals management on issues of mutual concern, increased involvement of stakeholders adding skills, knowledge, and commitment.
- Focus on capacity building and technical development in Mexico; Shared practices for DDT* management to other countries in the Americas; and Emphasis on building Mexican technical capacity and governance that will influence the region overall.
- Links to CEC enforcement area, providing training to customs officers on transboundary movement of mercury, SMOC linkages to work elsewhere in CEC, such as enforcement area, the Trade and Environment components dealing with Green Supply Chains and Electronics.

CHEMICAL MANAGEMENT IN CANADA

Canada's Approach to Chemicals Management revolves around various tools to control chemical substances to protect human health and the environment. These tools range from appropriate information provision about risks, proper use and disposal, to regulations that restrict or completely prohibit the use of specific chemical substances. These tools are supported by strong scientific backgrounds of monitoring and assessment, and by using a risk-based approach to safe chemical management.

All levels of the Canadian government play a part in protecting Canadians and their environment from the risks of chemical substances, including federal, provincial and territorial, and municipal governments. Regulations, guidelines, and objectives are made for the whole country. Canadians are leaders in conducting scientific research relating to chemical safety in humans and the environment, providing information to other countries and making agreements to ensure safe chemical management in a sustainable global context.

The Canadian federal government protects human health and environment through the establishment of more than 25 laws that govern chemicals in food, drugs, pesticides, and other products, as well as those that cover pollution release into air, water, and natural wildlife habitats. The *Canadian Environmental Protection Act, 1999* (CEPA, 1999), is one of the most important environmental laws in Canada that governs chemical substance management. The primary purpose of CEPA 1999 is to protect the environment and the health of Canadians and acts to address any pollution issues that are not covered by other federal laws. Under CEPA 1999, the CMP is Canada's governing safe chemical management program. Figure 45.1 outlines the laws that cover the environment and environmental health.

* NARAP for DDT under CEC: http://www.cec.org/Page.asp?PageID=122&ContentID=1262&SiteNodeID=312&BL_ExpandID=

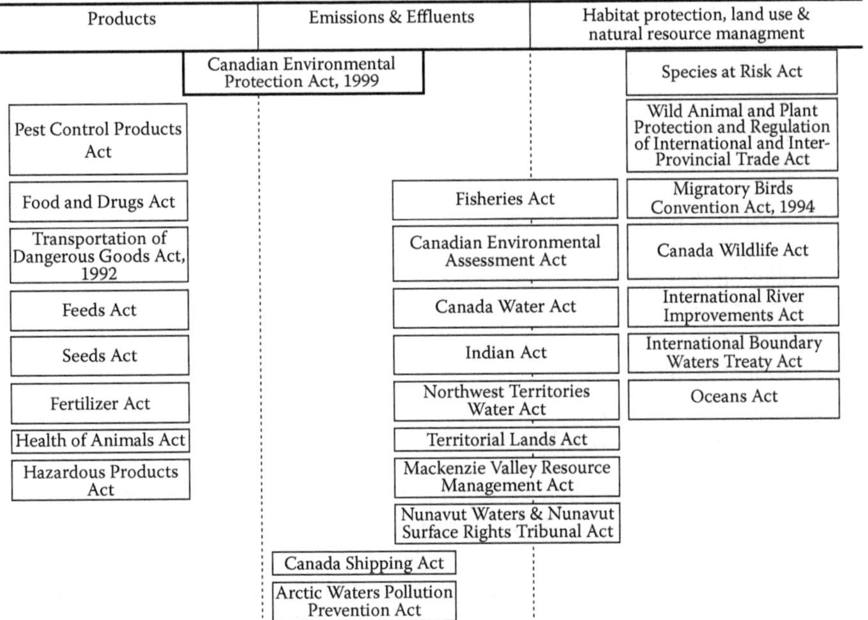

FIGURE 45.1 Major federal laws covering environment and environmental health issues. (Data from http://www.chemicalsubstanceschimiques.gc.ca/about-apropos/canada-eng.php)

The CMP*

The CEPA of 1999 required Environment Canada (EC) and Health Canada (HC) to conduct screening-level risk characterizations on all chemicals in the Canadian marketplace. This led to new approaches to risk evaluation, leading to a tiered, targeted, and risk-based approach to chemical risk assessment and risk management called the Canadian CMP. The initial tier makes use of conservative, predictive modeling and assimilates hazard data from chemicals with similar molecular structures to assist in the hazard characterization of the chemicals. Potential exposure to the substances was characterized employing general use categories and production/import volume. The Canadian CMP was announced on December 8, 2006. It is part of the federal Government's environmental agenda with a goal of positioning Canada as a global leader in the safe management of chemical substances and products. CMP has almost $300 million in funding between 2007 and 2011 from the Canadian government and is managed jointly by HC and EC.

The high-level outcomes of the CMP include:

- Identification, reduction, elimination, prevention, or better management of chemical substances and their use.

* Chemical Management Plan (CMP): http://www.chemicalsubstanceschimiques.gc.ca/plan/index_e.html

- Direction, collaboration, and coordination of science and management activities.
- Understanding of the relative risks of chemical substances and options to mitigate them.
- Risk assessment and risk management.
- Informed stakeholders and Canadian public.

CEPA guiding principals include:

- Pollution Prevention
- Precautionary Principle
- Intergovernmental Cooperation
- National Standards
- Polluter Pays Principle
- Science-based Decision Making

While the Ministers of Environment and Health are responsible for CEPA 1999, the management of chemical substances requires the collective efforts of the federal, provincial and municipal Governments, industry, health and environmental groups under a shared and collaborative model. CEPA 1999 involves the assessment and management of new and existing chemical substances. All new substances under the "New Substances Notification Program" are scientifically assessed for environmental and human health risks before being allowed to be used within Canada. Approximately 23,000 chemical substances already existed in Canada before the development of the testing program, and these are listed under the Domestic Substances List (DSL).* The DSL includes substances that were already used in Canadian industries, or imported into Canada in a quantity of 100 kg or more per year, during the time between January 1, 1984 and December 31, 1986.

Categorization and Screening

Under CEPA 1999, goals were set for the DSL to undergo a categorization and a screening level risk assessment. As required by the Act, this was completed by September 2006, making Canada the first country to take a systematic look at existing substances.

The criteria for categorization included (CEPA, 1999):

- *Inherent toxicity* in the environment (iTE); and
- Either:
  - *persistence* (P) of the chemical in the environment; or
  - *bioaccumulative* (B) potential in fish or other organisms in the food chain.
- *Greatest potential for exposure* (GPE) and
- *Inherent toxicity* for humans (iTH).

* Domestic Substances List (DSL) Categorization and Screening Program: http://www.ec.gc.ca/substances/ese/eng/dsl/dslprog.cfm

EC determined whether chemicals are P, B, or iTE, and HC categorized those chemicals for toxicity and human exposure (GPE and iTH). By categorizing the chemicals, the Government identified approximately 4300 chemical substances that meet criteria for further attention: 4000 were designated for further review by EC while 300 others warranted further attention from a human health perspective, thus analyzed by HC. Through categorization, it was determined that over 85% of the substances on the DSL do not need further action at this time. There is also more information than ever before on the remaining 4000. This information will be used by the Government to act quickly on high-priority chemical substances.

The screening level risk assessment considered whether the substance is "toxic" or capable of becoming "toxic." As per CEPA (1999) (http://laws.justice.gc.ca/PDF/Statute/C/C-15.31.pdf)*:

- A substance is toxic if it is entering or may enter the environment in a quantity or concentration or under conditions that
  - have or may have an immediate or long-term harmful effect on the environment or its biological diversity;
  - constitute or may constitute a danger to the environment on which life depends; or
  - constitute or may constitute a danger in Canada to human life or health.

Once this screening is applied, three possible consequences may occur: no further action taken, the chemical substance is identified as toxic and measures may be required for control, or it may be required to be placed on the Priority Substances List (PSL) where chemical substances are subject to an in-depth assessment. PSL chemicals are evaluated in great detail and are considered a high priority for the Government.

Challenge Program

The Government established an industry "Challenge Program" to address the approximately 200 top priority substances from the top 500 high-priority group. The intended action for the Challenge was published as a Notice in the *Canada Gazette*.† This was to provide information to enable the early completion of screening assessments and regulatory action of these substances. These included substances that

- Met the P,B, and iTE categories *and* believed to be in commerce in Canada, and/or
- Met the GPE or intermediate potential for exposure (IPE) *and* proving to have a high hazard to human health (carcinogenicity, mutagenicity, developmental toxicity, or reproductive toxicity).

* CEPA definition of "toxic": http://laws.justice.gc.ca/PDF/Statute/C/C-15.31.pdf

† Notice on Challenge substances in Canada Gazette: http://www.gazette.gc.ca/archives/p1/2006/2006-12-09/html/notice-avis-eng.html#i5

Under the conditions of the Challenge, the responsibility is placed on industry to effectively prove to the Government that these chemicals are being used safely. The Government, in turn, is reviewing these substances in "batches" or small groups, and providing results on a quarterly basis. Each batch is announced and then goes through a process that involves stakeholder comment and input, mandatory surveys and questionnaires. The timeline takes approximately 42 months and is depicted in* Figure 45.2.

Under CEPA 1999, chemical substances are assessed in a number of ways:

Screening Assessments: CEPA 1999 requires this type of assessment for all chemical substances on the DSL identified through categorization as requiring further attention.

Priority Substances: Chemical substances on the PSL are evaluated in depth.

Reviews of decisions of other jurisdictions: The federal Government works with other governments in Canada and around the world to share information on chemical substances.

Other Aspects of the CMP

One component of the CMP and CEPA 1999 is the "Significant New Activity" notice (SNAc).† These notices require Industry to provide relevant toxicological and environmental data for Government review before any new substances that are not

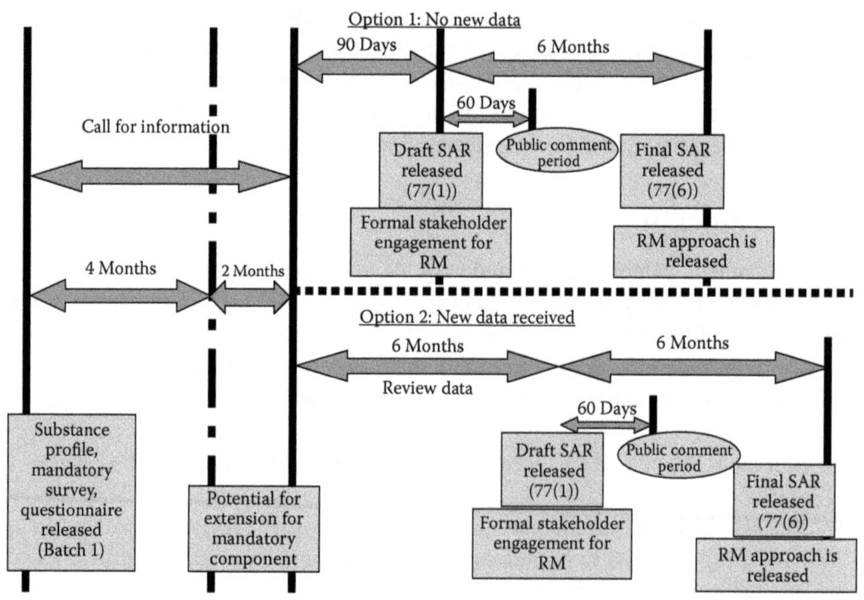

FIGURE 45.2 Timelines for the Challenge Program.*

* Timelines for the Challenge program: http://www.ec.gc.ca/substances/ese/eng/challenge/challenge_timeline_en.pdf

† Significant New Activity (SNAc) Approach: http://www.chemicalsubstanceschimiques.gc.ca/plan/approach-approche/snac-nac-eng.php

listed in the DSL (i.e., were no longer in commerce and are intended to be reintroduced). If there are concerns by HC or EC, they may delay, restrict, or prohibit the use of that chemical substance.

Other key programs under the CMP include:

- Regulatory prohibition of five chemical substance categories, including certain flame retardants, stain repellents, and nonstick coatings.
- Rapid screening of approximately 1200 chemical substances that met categorization criteria, but are low risk.
- Petroleum Sector Stream Approach (PSSA).*
- Mandatory ingredient labeling of cosmetic products.
- Enhanced management of environmental contaminants in food.
- Regulations to address environmental risks posed by the disposal of pharmaceutical and personal care products.
- Good stewardship of chemical substances.

CEPA 1999 has a "risk-based" method of making decisions. Risk is determined by evaluating harmful properties of the chemical substance and how much exposure there is for Canadians and the Canadian environment. The Act provides for the assessment of human health and environmental risks and management of chemical substances to prevent reduce or control environmental and human health impacts of:

- New and existing substances (including products of biotechnology)
- Marine pollution
- Emissions from vehicles, engines and equipment
- Fuels
- Hazardous wastes
- Environmental emergencies, including accidental spills

Since 1994, Canada has assessed all new chemical substances new to Canada (made there or brought in from other countries). Under Canada's New Substances Notification Program, more than 800 chemical substances are examined annually. The Government prohibits or puts restrictions on the use and disposal of those that could pose a risk to health and/or the environment. Further, the Government of Canada works with stakeholders such as the public, industry and health sector organizations, and environmental organizations to ensure that any decisions made on chemical substances are understood, and that the process used is transparent.

Globally, Canada has been and continues to be active in the SAICM development and implementation activities. Some Canadian activities include:

- Sponsored a project to develop draft indicators for reporting progress on SAICM implementation and a baseline report.
- Led in the development of a guidance document on the preparation of plans for the implementation of SAICM, with a view to achieving a consistent approach to preparing and reporting on such plans.

* Petroleum Sector Stream Approach (PSSA): http://www.chemicalsubstanceschimiques.gc.ca/plan/petroleum-petrole_e.html

- Participated in seminar in Tokyo, as part of Regional Focal point in Asia-Pacific on SAICM National Implementation Plans in the World, to disseminate information about SAICM and implementation activities in Canada, Sweden, and Thailand.

CHEMICAL MANAGEMENT IN THE UNITED STATES

In the United States, the key chemical management regulation is the Toxic Substances Control Act (TCSA).* Within the EPA, it is the responsibility of the Office of Pollution Prevention and Toxics (OPPT), formed in 1977, to manage and administer the law.† The law provides the EPA with the authority to ensure reporting, record-keeping, and testing requirements for new and existing chemicals, as well as the authority to restrict chemicals substances.

In 1990, the Pollution Prevention Act (PPA) was enacted, and extended the responsibilities of OPPT by making pollution prevention a national priority (http://www.epa.gov/lawsregs/laws/ppa.html). The objective of the PPA was to control industrial pollution at the source, therefore preventing pollutants from getting into the environment in the first place.

The TSCA addresses production, importation, use, and disposal of specific chemicals through requirements under various sections of the Act. The main requirement is Section 8(e) that requires manufacturers, importers, processors, and distributors to notify the EPA within 30 days of any new information regarding their chemicals that may impact human or environmental health (http://www.epa.gov/lawsregs/laws/tsca.html). Voluntary submissions also exist through the "For Your Information" submissions, which can provide the opportunity for reporting for those not captured within reporting requirements, for example, public interest groups, industry, and academics. (http://www.epa.gov/oppt/tsca8e/pubs/basicinformation.htm#fyi).

The EPA has worked to make information widely available to protect citizens and inform them of potential risks of existing chemicals. In 2008, these efforts were expanded to form the Chemical Assessment and Management Program (ChAMP).‡ As per the US EPA, "the Chemical Assessment and Management Program (ChAMP) was designed to develop screening-level hazard, exposure, and risk characterizations for chemicals produced or imported in quantities of 25,000 lbs (11339.8 kg) or greater a year. ChAMP has been superseded by the comprehensive approach to enhancing the Agency's current chemicals management programs."

Other Chemical Management Initiatives from OPPT, the United States

OPPT has a few other programs, including:

- New chemicals program§
- Existing chemicals program¶

* Toxic Substances Control Act (TSCA): http://www.epw.senate.gov/tsca.pdf
† See http://www.epa.gov/oppt/pubs/opptabt.htm
‡ See http://www.epa.gov/CHAMP/
§ See http://www.epa.gov/oppt/newchems/
¶ See http://www.epa.gov/oppt/chemtest/

- Specific chemical risk management programs, such as:
  - Lead[*]
  - Asbestos[†]
  - PCBs[‡]

There are various voluntary programs available through the OPPT that focus on pollution prevention and stewardship programs.[§] Examples of these are listed below, while a full list is available (see footnote[¶¶]).

- Design for the Environment (DfE):[¶] a voluntary partnership program that helps businesses design or redesign products, processes, and management systems that are cleaner, more cost-effective, and safer for workers and the public.
- Environmental Labeling[**] covers a broad range of activities from business-to-business transfers of product-specific environmental information to environmental labeling in retail markets and provides an opportunity to inform consumers about product characteristics that may not be readily apparent and guide their use in an environmentally beneficial manner.
- Environmentally Preferable Purchasing (EPP):[††] a federal government-wide program managed by OPPT that requires and assists Executive agencies in the purchasing of environmentally preferable products and services.
- Green Engineering:[‡‡] an initiative under the DfE program designed to promote the development and commercialization of environmentally beneficial design methods, risk-based design tools, and green technologies via education, outreach, and partnering with the academic, research, and industrial communities.
- Green Suppliers Network (GSN):[§§] a collaborative venture between industry and EPA that works with all levels of the manufacturing supply chain to achieve environmental and economic benefits; improve performance, minimize waste generation and remove institutional roadblocks through its innovative approach to leveraging a national network of manufacturing technical assistance resources.

Other statutes relating to chemicals management include:

- Federal Insecticide, Fungicide and Rodenticide Act (FIFRA);[¶¶]

---

* See http://www.epa.gov/oppt/lead/

† See http://www.epa.gov/oppt/asbestos/

‡ See http://www.epa.gov/epawaste/hazard/tsd/pcbs/index.htm

§ See http://www.epa.gov/oppt/pubs/oppt101-032008.pdf:

¶ See http://www.epa.gov/dfe/

** See http://www.epa.gov/oppt/epp/pubs/envlab/report.htm

†† See http://www.epa.gov/epp

‡‡ See http://www.epa.gov/oppt/greenengineering

§§ See http://www.epa.gov/greensuppliers/

¶¶ See http://www.epa.gov/pesticides/about/index.htm

- Federal Food, Drug and Cosmetic Act (FFDCA);* and
- Occupational Safety and Health Act (OSHA).†

Other programs include those aimed at providing chemical hazard and risk information to inform others of choices relating to chemical use, such as the HPV‡ Challenge Program and the Voluntary Children's Chemical Evaluation Program§ (VCCEP). The U.S. EPA had announced the HPV Challenge (1999), a voluntary program for toxicity and other testing on chemicals, produced or imported at amounts greater than 1 million lbs (453,592.37 kg) per year. Industry volunteered to provide information on over 2200 HPV chemicals, resulting in huge amount of hazard data. EPA had evaluated approximately 100 HPV chemicals and posted hazard assessments information on those chemicals. Under the HPV challenge, the US EPA was in the process of posting hazard information for another 200 chemical assessments.

The EPA intends to screen out the low toxicity chemicals first, and then use data from the 2006 Inventory Update Rule (IUR) reporting to characterize the risks of the remaining HPV chemicals that are categorized as moderately to highly toxic. Subsequently, EPA focuses on MPV chemicals (those produced or imported in amounts between 25,000 and 1,000,000 lbs [11339.8 kg and 453592.37 kg] annually). Existing data, conservative modeling, and structure-activity relationships are used to characterize the hazards of the MPVs where measured data do not exist. The hazard and risk characterizations for HPV and MPV chemicals will provide EPA with information necessary to conduct future risk assessment and risk management activities.

On September 29, 2009, the U.S. EPA Administrator Ms Lisa Jackson announced core principles that outlined the current administration's goals for legislative reform of U.S. chemical management law(s), the 1976 TSCA. The plans were also announced for a major effort to strengthen EPA's current chemical management program and increase the pace of the agency's efforts to address chemicals that pose a risk to the public.¶ The principles, listed below, present the current administration's goals for legislation that may provide EPA the mechanisms and authorities to expeditiously target chemicals of concern and promptly assess and regulate new and existing chemicals in commerce:

- Chemicals should be reviewed against risk-based safety standards based on sound science that protects human health and the environment.
- Manufacturers should provide EPA with the necessary information to conclude that new and existing chemicals are safe and do not endanger public health or the environment.
- EPA should have clear authority to take risk management actions when chemicals do not meet the proposed safety standard, with flexibility to take

---

* See http://www.epa.gov/lawsregs/laws/ffdca.html

† See http://www.epa.gov/lawsregs/laws/osha.html

‡ See http://www.epa.gov/oppt/chemrtk/

§ See http://www.epa.gov/oppt/vccep/index.html

¶ US EPA's current modified Chemical Assessment and Management program: http://www.epa.gov/oppt/existingchemicals/index.html

into account sensitive subpopulations, costs, social benefits, equity, and other relevant considerations.
- Manufacturers and US EPA should assess and act on priority chemicals, both existing and new, in a timely manner.
- Green Chemistry should be encouraged and provisions assuring Transparency and Public Access to Information should be strengthened.
- US EPA should be given a sustained source of funding for implementation.

Although legislative reform is necessary for an effective chemicals management program, US EPA is committed to strengthening the performance of the current program while Congress considers new legislation. This enhanced plan includes the development of chemical action plans which will outline the agency's risk management efforts on those chemicals of greatest concern. As per the revised chemical management program, US EPA may include initiation of regulatory action to label, restrict, or ban a chemical, or to require the submission of additional data needed to determine the risk. If US EPA determines that a chemical does not present a need for action, it will make the relevant information available to the public. As of August 18, 2010, US EPA has posted action plans for specific chemicals such as benzidine dyes; nonylphenol, and nonylphenol ethoxylates (NPEs); and hexabromocyclododecane (HBCD). Previously, US EPA posted action plans on bisphenol A (BPA), phthalates, perfluorinated chemicals (PFCs), PBDEs in products, and short-chain chlorinated paraffins. These action plans summarized available hazard, exposure, and use information; outlined the risk that each chemical may present; and identified the specific steps US EPA was taking to address those concerns.

Chemicals for which action plans have been posted on US EPA are*:

Benzidine Dyes (action plan released August 18, 2010); BPA (released March 29, 2010); HBCD (released August 18, 2010); Nonylphenol and NPEs (released August 18, 2010); PFCs (released December 30, 2009); PBDEs in products, (released December 30, 2009); Phthalates (released December 30, 2009); Short-chain chlorinated paraffins (released December 30, 2009). As of April 13, 2011, toluene diisocyanates and other related diisocyanates chemical action plans were released. Currently siloxanes are in the process of development of a chemical action plan. It will be released in the near future.

US EPA also supports the voluntary phase-out of industrial laundry detergents containing NPE. In this regard, the Textile Rental Services Association (TRSA) of the U.S. including all major industrial detergent manufacturers has pledged that members of their agency will phase out the use of industrial liquid detergents containing NPEs by December 31, 2013, and industrial powder detergents containing NPEs by December 31, 2014.

An additional focus of the chemical management program will be accelerating efforts to gather the critical information from industry that the agency needs to make chemical risk determinations. This will include filling the current gaps in health and safety data on HPV chemicals; enhanced, transparent, and more current reporting of use and exposure information; and a number of requirements for increased reporting

---

* Chemical Action Plan for priority chemicals: http://www.epa.gov/oppt/existingchemicals/pubs/ecactionpln.html

on nanoscale chemical materials. In addition, EPA is reviewing how nanoscale materials are managed under TSCA. EPA is also reviewing ways to increase public access to information about chemicals. Prioritizing chemicals for future risk management action is the final component of this effort and US EPA would formally engage stakeholders and the public.

The "Essential Principles for Reform of Chemicals Management Legislation"* can be found at: http://www.epa.gov/oppt/existingchemicals/pubs/principles.html.

Detailed information on EPA's enhanced chemical management program, including information on specific components of this effort, an initial list of chemicals under consideration for Action Plan development, new hazard characterization for 100 chemicals, and risk management actions recently announced on lead and EPA's plans for banning the use of mercury in certain products,† can be found at: http://www.epa.gov/oppt/existingchemicals/index.html.

United States supports SAICM as a mechanism to facilitate the efforts of countries and stakeholders to achieve chemicals management objectives, in particular the WSSD 2020 goal: "to achieve by 2020 that chemicals are used and produced in ways that lead to the minimization of significant adverse effects on human health and the environment, using transparent science-based risk assessment procedures and science-based risk management procedures" (see Section IV, Chapter 17 of this book for a detailed review of SAICM). Some of the actions of US EPA to contribute to meeting the objectives of SAICM include:

- Committed to complete risk characterizations and to initiate action, as appropriate, on 6750 HPV and MPV chemicals produced above 25,000 lbs (11339.8093 kg) per year by 2012. This information allows US EPA to generate a quantitative or semiquantitative description of the risks inherently possessed by a given chemical, and builds on the US EPA's High-Volume Production Challenge.

CHEMICAL MANAGEMENT IN MEXICO

When compared against Canada and the United States, Mexico does not have a very comprehensive chemical management program. However, it is an active participant in several chemical initiatives regionally and globally under the North America's SMOC. There is continuous and active collaboration between these three countries in the regional chemical management initiatives.‡

As per the Council for CEC, the "Capacity building" in the areas of chemical management is very critical for the regional SMOC initiatives and to the North American Regional Action Plans (NARAPs). A Capacity Building Task Force was

* Essential principles for Reform of Chemical Management Legislation: http://www.epa.gov/oppt/existingchemicals/pubs/principles.html

† US EPA's approach to ban Mercury in certain products can be found at: http://www.epa.gov/oppt/existingchemicals/index.html

‡ Various Chemical Management initiatives and related information on Mexico can be accessed at: http://www.cec.org/Storage/27/1802_CountryReport-Mexico-CHE_en.pdf;http://www.cec.org/Page.asp?PageID=1180&SiteNodeID=487; http://www.cec.org/Page.asp?PageID=1225&SiteNodeID=595

established in 1998 to develop the conceptual basis and strategy to help guide efforts in regional chemical management in North America. While Canada, the United States, and Mexico each have capacity building and coordination requirements related to the SMOC initiatives, however, the primary focus of capacity-building activities is in Mexico, and it will bear the major costs associated with implementation of the NARAPs. The major emphasis is on leveraging new or additional funds from international funding institutions to assist Mexico in implementing the SMOC program. As stated in this chapter (see the section "North America Commission for Environmental Cooperation"), collaborations and capacity building and reduction or elimination of chemicals (e.g., PCBs, Dioxins and Furans, Hexachlorobenzene, DDT, Lindane, Mercury, and emerging contaminants such as PBDEs, etc.) and development of a region specific Mexico chemical inventory with an emphasis on regional comparability and implementation of GHS, are some of the examples of initiatives of risk management of chemicals in Mexico.

Priorities for Mexico

Mexico promotes and ensures safe management of chemicals and hazardous materials by developing, strengthening, and improving the available regulatory instruments in the country and management and compliance with international commitments such as continuity of working toward Stockholm Convention objectives and priorities, establishment of effective linkages to UNIDO, GEF, capabilities inventory, and distribution of responsibilities, and POPS and other chemical waste inventory development. Mexico is also party to the SAICM Quick Start Program and has participated in internal chemical conferences such as ICCM2. It is also in the process of updating its National Profile of Chemicals. From the perspective of the Rotterdam Convention, Mexico is continuing to work toward effective implementation and participation in the Review Committee. Mexico continues in the path of cooperation and synergies in SMOC and CEC and participation in UNEP forums such as international mercury negotiations and Intergovernmental Forum of Chemical Safety.*

Challenges and Opportunities

Despite Mexico's effort in regional and global chemical risk management initiatives, there are various challenges. At the same time, these challenges can also be addressed as "realized new opportunities" as Mexico's various chemical management initiatives move forward. In this regard, institutional capacity, responsibility and effective governance may be required. Another challenge is the communication links and multistakeholder participation and involvement in a regional framework where affected parties can provide various perspective and input toward sustainable chemical management practices. Some of the challenges in this area can be seen as potential opportunities in capacity building among Canada, the United States, and Mexico.

* North American Regional Action Plan on Mercury Update including Mexico: http://www.cec.org/Page.asp?PageID=122&ContentID=1297&SiteNodeID=312&BL_ExpandID=#ANNEX4 Phase II NARAP on Mercury: http://www.cec.org/Storage/86/8204_Hgnarap.pdf

Several examples can be found in this area under CEC, SMOC, and NARAP.* In the areas of regional chemical management, intergovernmental collaboration, coordination and effective governance are necessary. Mexico's Chemical Inventory initiatives are such examples that encompass intergovernmental regional collaboration, coordination and knowledge sharing, and exchange (Development of National Chemical Substance Inventory for Mexico, 2009).

Furthermore, updating of the existing legal chemical management framework in the area of public health policy of chemicals would be necessary. Both regional and global sustainable approaches to maintain a balance between economic growths, economic impacts of chemical sectors, and optimization of use of chemicals and energy industry may pave the way to making Mexico a regional leader in chemical risk management.

ACKNOWLEDGMENTS

The authors acknowledge that Murali Ganapathy and Rajib Khettry of SENES Consultants Ltd. wrote the first draft of the chapter. Revisions of the organizational structure of the chapter were completed by Asish Mohapatra and Philip Wexler.

REFERENCES

Canadian Environmental Protection Act (CEPA). 1999: Available at: http://laws.justice.gc.ca/PDF/Statute/C/C-15.31.pdf (updated to September 19, 2010).

Cooper, J. 2007. Briefing Paper on the Montebello Agreement under the Security & Prosperity Partnership (SPP) of North America.

National Petrochemical & Refiners Association October 30, Available at: http://www.tcata.org/MontebelloAgreement103007.pdf

Development of National Chemical Substance Inventory for Mexico, April 2009 Update: Available at: http://www.cec.org/smoc/SanAntonio2009/Mexico%20chemical%20inventory-e.ppt

Hansen A.M., M. van Afferden, M.V. Canela, and L.F. S. Castañeda, 2006. Mexico's Environmental Monitoring and Assessment Program (Programa de Monitoreo y Evaluación Ambiental— Proname). Scoping study for the evaluation of the national program of monitoring and environmental assessment in Mexico Comisión de Cooperación Ambiental de América del Norte.

* Updates on Domestic Priorities in Mexico including International Efforts:http://www.cec.org/smoc/SanAntonio2009/Mexico%20update-e.ppt

46 Regional Cooperation among South Asian Association for Regional Cooperation Countries in the Areas of Chemical and Environmental Risk Management

Lakshmi Raghupathy, Asish Mohapatra, Ravi Agarwal, and Jan van der Kolk

CONTENTS

INTRODUCTION

The South Asian Association for Regional Cooperation (SAARC) was established on December 8, 1985 and includes eight countries from the region. Initially, the heads of State of seven countries (i.e., Bangladesh, Bhutan, India, Maldives, Nepal, Pakistan, and Sri Lanka) agreed to set up a body to facilitate the regional cooperation in this area. Later, in 2007, Afghanistan became the eighth member of SAARC. Amongst various issues such as regional security, trade and so on, environmental concerns have emerged as one of the important area is requiring cooperation amongst member countries. Though environmental challenges here are somewhat similar, the status of environmental programs is distinct, with each country having its set of focused priorities. The scope for better regional cooperation on chemical safety and management issues also remains a key requirement for the region.

As per UNEP (2002), the broader South Asian region (which also includes countries other than SAARC member countries), at least 21 chemicals or chemical clusters were identified as priority persistent and toxic chemicals based on their use and concentrations in the environment, and potential ecological, and human health effects. Those 21 chemicals include aldrin, chlordane, DDT, dieldrin, endrin, heptachlor, hexachlorobenzene, mirex, toxaphene, PCBs, dioxins and furans, atrazene, endosulfan, lindane, phthalates, PAHs, pentachlorophenol, organotin, organolead, and organomercury compounds. SAARC member countries such as India, Pakistan, Sri Lanka, Nepal, Bhutan, and Bangladesh were identified as hot spots for stockpiles of obsolete pesticides. From an industrial emission perspective, waste incinerators, pulp and paper industries, and chlorine-based manufacturing were identified in India and Pakistan as important sources of emission of dioxins and furans. Furthermore, PCB contaminations were also identified as major contaminants at the ship-dismantling areas in India, Bangladesh and Pakistan (UNEP, 2002).

It is important to note that in India, Pakistan, Sri Lanka, Nepal, Bhutan, and Bangladesh, persistent toxic chemical pesticides such as, aldrin, chlordane, dieldrin, endrin, heptachlor, HCB, toxaphene, mirex are either banned or not registered. The presence of these pesticides in the environment may be due to their legacy use in those countries (UNEP, 2002).

In view of the historical information on chemical contamination and management issues in the region, a brief physical geography and availability of natural resources of each SAARC member country are provided below:

Afghanistan is a landlocked country bordering Pakistan in the south and east, and Iran in the west. It has extensive deposits of minerals that include iron, copper, cobalt, gold, and other precious metals, precious and semiprecious stones and has a network of about 200 mines. Apart from this, Afghanistan also has a petroleum and natural gas industry. The flourishing drugs and narcotics industry of Afghanistan is a great concern for the security and development of the nation. Three decades of war and associated regional conflicts in the region have shown a rapid degradation of air, water, and soil ecosystems. As per UNEP, due to poor infrastructure and poor progress in the areas of environmental conservation, industrial manufacturing practices, exposure to chemical contaminants, biomedical wastes, and other pollutants remains a high concern. A phased approach toward environmental progress and sustainability by UNEP has shown some success in the areas of natural resource management, the establishment of legal and management frameworks, implementation of multilateral environmental agreements, and capacity building in the country.[*,†]

Bangladesh lies in the east of South Asia and on a major river delta. Industrialization and urbanization has led to the growth of industries engaged in the production of fertilizers, petrochemicals, cement, textile, leather, and mining. The country is vulnerable to natural disasters and is the subject of several global climate change concerns. The southern region of Bangladesh is the focus of several research projects on natural arsenic exposure from various drinking water sources and its human health effects. In 2001, UNEP helped Bangladesh in formulating the State of Environment (SoE) report as a response to the recommendations provided in the *Agenda 21* at the Earth Summit.[‡] The five key environmental issues identified in the 2001 report were land degradation, water pollution and scarcity, air pollution, biodiversity loss, and natural disasters. Those issues indicated that the environmental condition of Bangladesh is deteriorating, despite several policy measures undertaken under the different regulatory and nonregulatory instruments of the Government of Bangladesh. Regional issues such as rapid population growth, unsustainable land uses, poor practices in resource management, and poorly regulated discharge of pollutants from industries and vehicles were identified as major causes of degradation. These issues were also contributing to poor practices in regional chemical management associated with a lack of institutional capabilities, untrained human resources, lack of awareness, low community participation in resource management, and lack of research for enabling policy makers to take effective risk-management decisions. From a country-specific perspective, addressing these risk management deficiencies will enable Bangladesh in its progress toward attaining sustainable balance between environment and development.

Bhutan is a landlocked country in the Indian subcontinent. It has large mineral resources but they are not easily accessible. Of late, industrial development has been a driver and there is a major focus on developing wood-based, mineral-based, often also energy-based industries besides service industries.

India is the largest country in the region. India's economy is based on agriculture and industry. It is the most industrialized nation in the region and one of the fastest growing economies in the world.

Maldives is an island nation in the Indian Ocean with Sri Lanka on the east and India on the north. The country faces threat due to global climate change and sea level rise.

Nepal is a landlocked country in the central Himalayan Region bordering India, China, and Bhutan. It also has the world's highest mountain, Mount Everest. In all, 29% of total land area is forest and it has a varied climate and ecosystems.

---

[*] http://www.unep.org/Documents.Multilingual/Default.asp?ArticleID=3201&DocumentID=277

[†] http://www.unep.org/conflictsanddisasters/UNEPintheRegions/CurrentActivities/Afghanistan/tabid/287/language/en-US/Default.aspx

[‡] http://www.rrcap.unep.org/pub/soe/bangladesh_part1.pdf

Pakistan is bordered by Afghanistan and Iran in the west, India in the east and China in the far northeast. A total of 10.4% of the land area is protected area, with National Parks, Sanctuaries, and Game Reserves, and so on. Pakistan's industrial sector accounts for about 24% of its GDP. Cotton textile production and apparel manufacturing are Pakistan's largest industries and other major industries include mining and quarrying of minerals and exploration of petroleum and natural gas. Pakistan also has cement, fertilizer, edible oil, sugar, steel, tobacco, chemicals, machinery, and food processing industries.

Sri Lanka is an island strategically located in the Indian Ocean, on the southern tip of India. It is a major exporter of agricultural and industrial products, as well as minerals. Its major imports include consumer products, intermediate and investment goods. Sri Lanka has diverse group of industries, which have been classified into micro, small, medium and larger-scale industries. These include sectors like food and beverage, textile, apparel and leather products, wood and wood products, pulp, paper and paper products, chemical, petroleum, plastic and rubber, nonmetallic products, basic products and fabricated metal products.

REGIONAL COOPERATION IN CHEMICAL AND ENVIRONMENTAL RISK MANAGEMENT

South Asian Association for Regional Cooperation

The SAARC secretariat is located in Kathmandu, Nepal. Though the main objectives of SAARC are to promote economic growth, social progress, and cultural development and welfare of the people in South Asia in order to improve their quality of life in the region, it also promotes active collaboration and mutual assistance in the economic, social, cultural, technical, and scientific areas. Environment has become an important agenda in the region and its various summits have emphasized the need to intensify regional cooperation in the areas of environmental protection and disaster management. SAARC ministerial level meetings have been held each year with different thematic focuses on environmental issues, and have recommendations for cooperation amongst member countries. Alongside, a Technical Committee on Environment was formed in 1992 to coordinate such regional cooperation. Since then two studies have been carried out to provide a basis for regional initiatives.

South Asian Cooperative Environment Programme

South Asian Cooperative Environment Programme (SACEP) was established as an Intergovernmental Organization by the SAARC countries in 1982, with an objective to support the protection, management, and enhancement of the environment in the region. The SACEP Secretariat is located in Colombo, Sri Lanka. The focus of SACEP has been the management of rich biodiversity of the region and the problems and issues pertaining to coastal areas and protection of the seas. Activities have also been initiated in hazardous chemical management and the need to regulate such chemicals and wastes and to prevent the illegal traffic of hazardous chemicals and wastes from different industrialized nations.

CHEMICALS MANAGEMENT IN THE SAARC COUNTRIES

The use of industrial and consumer grade chemicals among SAARC member countries is exponentially increasing. There is extensive trade in chemicals within the region as well as imports from other countries, as these countries are developing from agricultural based economies to being industrially based ones. As discussed earlier in this chapter, the past legacy of indiscriminate use and application of agricultural chemicals (e.g., pesticides) has been one of the major causes of environmental degradation, especially from persistent organic pollutants (POPS). Improved cooperation and collaboration in regional chemical management can lead to effective management of ecosystems and human health among SAARC member countries.

MANUFACTURE, USE, AND IMPORT OF CHEMICALS

India has the maximum number of chemical manufacturing industries.* There is extensive use and trade in chemicals and related products between India and other countries, including member countries (see Section VI, Chapter 39 of this book on India). Other countries in the region like Pakistan also have chemical manufacturing units, which are primarily for agricultural chemicals comprising fertilizers and pesticides. Most of the chemical fertilizers are locally produced. Pesticides are generally imported as either branded/packed from other countries or formulated or packed locally from imported base materials. For these chemicals, there exists adequate control in as much as their production/import, transportation, storage and distribution are concerned. However, there is poor regulatory control or no control or management of expired pesticides and fertilizers. Bangladesh has been in the process of industrialization and urbanization. Industries dealing with the production, packaging, and transportation of fertilizer, petrochemical products, cement textile, leather, and mining have been set up. The improper usage and handling and indiscriminate disposal of chemical wastes have become a major environmental concern in Bangladesh. In Sri Lanka, the management of chemicals could broadly be classified under pesticides and industrial chemicals. Pesticides are regulated under the Pesticides Act while other chemicals are regulated based on concerns associated with the life cycle of use, environmental pollution, occupational health, consumer safety, and so on. Chemicals management is being adopted in accordance with multilateral instruments such as Stockholm Convention, Basal Convention, Rotterdam Convention, Montreal Protocol, and the Chemical Weapon Convention. There is virtually no management of hazardous chemicals in Afghanistan nor is there monitoring of pesticide residues in humans or in the environment. Water resources are still being polluted due to poor storage of these chemicals as well as indiscriminate disposal of untreated industrial effluents. In some aquifers, the concentration of hazardous chemicals exceeds hygienic standards, and in parts of Kabul city, pollutants make the water unsafe for consumption. Pesticides such as DDT and benzene hexachloride were used intensively for locust control in the northern agri-

* National Chemical Profiles of SAARC member countries: http://www2.unitar.org/cwm/nphomepage/np2.html

cultural regions of the country for several decades. Lack of proper management of these persistent organic chemicals represents a potential threat to the health of humans and wildlife in these regions. Bhutan, Maldives, and Nepal are largely dependent on imports of chemicals and in some cases there are formulations made from imported chemicals. Again problems occur due to the improper handling and disposal of hazardous chemical wastes.

Regulatory Controls and Enforcements

India has the most extensive regulatory system in the region (see Section VI, Chapter 39 of this book on India).

Bangladesh has been in the process of industrialization and urbanization, Industries dealing with the production, packaging, and transportation of fertilizer, petrochemical products, cement textile, leather and mining have been set up, though the improper usage and handling and indiscriminate disposal of chemical wastes is a major environmental concern. Legal instruments governing this situation are the Environment Protection Act, 1995 (EPA) and the Environment Protection Rules 1997, which focus on the control and prevention of industrial pollution and awareness generation. However, there is no comprehensive legislative or regulatory framework to deal with hazardous and toxic chemicals.

Bhutan's[*] pristine environment has several chemical management challenges, as it economy changes. Environmental Governance falls under Article 5 of The Constitution of the Kingdom of Bhutan (2008). An apex Environmental Commission, with multisector representations and with the Prime Minister as its Chair has been set up. Environmental legislations include the Waste Prevention and Management Act of Bhutan, 2009, The Pesticides Act of Bhutan, 2000, the National Environmental Protection Act, 2007, the Environmental Assessment Act, 2000, the Mines and Minerals Act, 1995, and the Forest and Nature Conservation Act, 1995. Chemical Management is distributed over several of these Acts as well as others, which deal with taxation, narcotics, medicines, and imports.

Maldives[†] has no specific legislation to regulate chemicals. The principal legislative act is on drugs, which was enacted in 1977 and deals with narcotic drugs and psychotropic substances.

> It is in the Strategic Plan of the Maldives Food and Drug Authority to regulate chemicals (imported, produced and exported) on its health based issues. It has been decided to form a technical committee to decide on the uncertainties and new chemicals and their impacts and so on arising and issues relating to it. This committee comprises of intersectional members (Defense, Customs, Agriculture, Fisheries, Environment and Health).[‡]

Nepal is predominantly an agricultural country and the focus is mainly on agrochemicals. The Pesticide Act was passed in 1991, to regulate the formulation, sales, distribution, use, and import of pesticides with an objective to minimize the

---

* http://www.env.go.jp/chemi/saicm/forum/100325/mat03-1.pdf
† http://www.who.int/ifcs/documents/forums/forum5/maldives.pdf
‡ http://www.who.int/ifcs/documents/forums/forum5/maldives.pdf

adverse effects to humans and other organisms. The Pesticide Regulation (2050)* was introduced in 1993 and became operative from July 1994. Under this Act, it is mandatory to register all pesticides before these are formulated or imported for use within the country and only notified pesticides are allowed to be imported or used.

Pakistan's Explosives Act, 1884 and the Factories Act, 1934 are the oldest legal instruments regulating chemical manufacturing and use. However, the most effective legal instruments in the field of chemical management are the Agricultural Pesticides Ordinance of 1971 and the Pakistan Agricultural Pesticides Rules, 1973 that control the manufacture and use of pesticides through their registration process and ban any pesticide which they considers harmful. It also allows for the introduction of new products but does not provide for destruction of date expired pesticides. No legal instrument is available for the registration or deregistration of fertilizers and other industrial or consumer chemicals. The Pakistan Environmental Protection Act 1997 has some provisions, but not encompassing, to cover manufacture, storage, and transportation of chemicals and so on. Recently, Pakistan–U.S. led support programs in Science and Technology (support program funding from the United States is worth $10 million) has added chemical management initiatives to the list of priority areas of science support.†

In Sri Lanka, pesticides are regulated under the Pesticides Act, 1980 while other chemicals are regulated based on concerns associated with the life cycle of use, environmental pollution, occupational health, consumer safety, and so on. Chemicals management in the country also includes compliance with international instruments such as Stockholm, Basal, and Rotterdam Conventions, the Montreal Protocol as well as the Chemical Weapon Convention. Chemical pollution issues are regulated by the National Environmental Act, 1980 and related regulations and standards notified. The Food Control Act is implemented by the Health Ministry and which regulates exposures to chemicals through food items while the Ministry of Agriculture mandates enforcement of the laws related to the pesticides. Afghanistan, Bhutan, Nepal, and Bangladesh are largely dependent on imports of chemicals and in some cases those chemical formulations are made from chemicals imported from elsewhere. All these countries have stringent monitoring requirements for all imports and exports of dangerous chemicals that are regulated using the PIC procedures (see Section III, Chapter 14 of this book on the Rotterdam Convention).

Chemical Risk Management: Challenges and Issues

The SAARC member countries today may be facing greater risks due to extensive past use and rapid increase of current use of some chemicals and various products, the improper handling of such chemicals, and an indiscriminate and poor management practice of disposal of chemical wastes generated thereof. Critical water resources in the region are being polluted as a result. Pesticides such as DDT, benzene hexachloride, Endosulphan, and so on, that are banned in many places, are still being used for pest control in some of these countries, and many stockpiles of such

* http://www.moacwto.gov.np/documents/ThepesticideRules.pdf

† http://sites.nationalacademies.org/PGA/dsc/pakistan/index.htm

pesticides could exist. The lack of proper management and use of such POPS in the countries of the region pose a potential threat to human health and the environment. Apart from this, pesticides, petrochemicals, and other hazardous chemicals are used extensively. In most of the countries here, adequate precautionary measures are not adopted during the transportation of hazardous chemicals and wastes. As a result, there are many cases of accidental spills due to hazardous transportation corridors in those countries, besides a lack of training and protective measures. Problems also arise due to the chemicals being handled by small-scale and microenterprises that operate in large numbers in these countries under a poor regulatory framework.

Generally, countries in the SAARC region have similar problems relating to chemicals safety issues. First the issue is often spread over several user sectors, and not integrated or coordinated adequately into development policy. Second, chemical legislation is a recent phenomenon and there are insufficient resources dedicated to it for infrastructure, testing facilities, technology availability, or human capacity building. Third, there is inadequate information and a general lack of awareness of the issues concerned and a poor database for information on toxicity and safety to refer to. Besides, illegal imports also pose a threat. The magnitude of problems is on the rise as industrialization is the new focus in this region.

SAARC AND INTERNATIONAL TREATIES

The three International Conventions dealing with chemicals and wastes include the Basel Convention on the Control of Transboundary Movements of Hazardous Wastes and their Disposal, the Rotterdam Convention on the Prior Informed Consent (PIC) Procedure for Certain Hazardous Chemicals and Pesticides in International Trade and the Stockholm Convention on POPS (see Section III, Chapters 8, 14, and 15 of this book on the Basel, Rotterdam, and Stockholm Conventions, respectively).

Status of SAARC Countries Party to the Three Chemicals Conventions

Countries	Basel Convention	Stockholm Convention	Rotterdam Convention
Afghanistan	No	No	No
Bangladesh	Yes	Yes	No
Bhutan	Yes	No	No
India	Yes	Yes	Yes
Maldives	Yes	Yes	Yes
Nepal	Yes	Yes	Yes
Pakistan	Yes	Yes	Yes
Sri Lanka	Yes	Yes	Yes

REGIONAL COOPERATION: SOME CASE EXAMPLES AMONG SAARC MEMBERS

Owing to the South Asian region still being largely agricultural, SAARC identified the need for cooperation in agriculture-related research, generation of agricultural-research statistics, and an exchange of scientific and technical information. The cooperation also focuses on food safety/standards, technology development and increased production and so on. In 2009, the projects developed included those concerning the balanced use of agricultural chemicals and fertilizers for the development of food products, in order to meet international standards. The ban on the use of POPS chemicals is one of the important issues being addressed. Other projects aim at helping the smaller nation states such as Bhutan, Nepal, and Bangladesh in the areas of chemicals and waste management. Many such projects will be implemented through India to facilitate the neighboring countries.

Other projects are:

A Regional Oil and Chemical Pollution Contingency Plan for South Asia was prepared and a Memorandum of Understanding (MoU) developed in association with the International Maritime Organization (IMO) for enhanced cooperation among five maritime countries of South Asia in the event of an oil spill. All member countries have consented to the Regional Plan and the process of signing the MoU has been initiated.

Blue Flag Beach Certification Programme for South Asia is an initiative of SACEP/SASP (South Asian Seas Programme) with the message, "A Clean beach—A Tourist Haven." It has the aim of promoting sustainable tourism, in collaboration with the Foundation for Environmental Education (FEE) Denmark, and UNEP. SACEP organized National Workshops to raise awareness in the SASP member states in February and March 2010 in collaboration with the National Focal Points.

The 12th meeting of the Governing Council of SACEP was held on 1–3 November 2010 at Colombo, Sri Lanka (http://www.sacep.org/).

HARMONIZING THE POLICIES FOR STRENGTHENING THE REGIONAL COOPERATIONS

Chemical Safety does not appear high enough in the issues of development priorities of individual countries. Another key problem in the region is a lack of harmonization of policies, and coordination in the activities between the member countries.

Although some of the countries in the region like India and Sri Lanka are far ahead in chemicals' management, others are yet to be aware of the consequences of mismanagement. While there are already some programs with International bodies such as UNEP, United Nations Institute for Training and Research (UNITAR), World Bank, and so on, to provide technical and financial support for activities, much more can be done. A critical evaluation of the policies of the member countries is required to streamline them and integrate them into a regional policy as well as norms to be adopted by member countries.

The following steps have been identified to achieve this, though this needs to be implemented.

1. *Policy Review:* Study and review of existing policies, strategies, and action plans of all member countries and evaluation of their adequacy to deal with transboundary issues.
2. *Assessment of chemical risks:* Knowledge about the manufacture, use and import of chemicals in each of the countries in the region to assess the chemical risks posed by the countries individually and the regional consequences and to develop a regional assessment of these to understand the regional requirements to control under worst case scenarios.
3. *Information Sharing:* One of the important issues is to be well informed about the hazardous and toxic nature chemicals and the risks posed by them. An Information Exchange on toxic chemicals and chemical risks is required.
4. *Planning and executing risk reduction programmes:* As a preventive approach a risk reduction program needs to be worked out for the region. Base line information on the chemicals being manufactured, imported, and used in the country and the characteristics of these chemicals should be included.
5. *Strengthening of national capabilities and capacities:* Mutual cooperation, capacity building, and information exchange on chemicals and wastes and compilation of a database on chemicals used in the region and chemical wastes generated.
6. *Prevention of illegal transboundary movement:* Any transboundary movement of the chemical should be legal and with the consent of the importing country. The common transboundary issue should be identified and resolved among the SAARC member counties.
7. *Capacity building:* Technology transfer and sharing of the knowledge and experiences with other countries.
8. *Addressing climate change issues:* Promoting energy-efficient technologies, introducing energy-efficient products and conservation of natural resources through encouraging reuse and recycling practices and calculation of carbon foot prints.
9. *Governance in SAARC region:* Institutional framework for monitoring and evaluation system should be provided for the same and the member countries should abide by the norms.
10. *Relevant Regional Indicators and benchmark successful regional initiatives.*

CONCLUSIONS

Regional cooperation has been the primary objective for establishing SAARC and SACEP. However, these have yet to fully explored for its potential. Chemicals management participation of some member countries in the international bodies has been poor, and there is no regional representation in such fora. Specifically, there is scope for many new initiatives and concrete action in the area of regional chemical management, since the region shares water and air resources and a significant trade in chemicals and products use.

REFERENCES

United Nations Environment Programme; UNEP. 2002c. Regionally based assessment of persistent toxic substances: Indian Ocean regional report. UNEP Chemicals, Geneva, Switzerland. http://www.chem.unep.ch/pts/regreports/IndianOcean.pdf

FURTHER READING

Needs Assessment for the Proposed Regional Training Centre for the SAARC Region under the Basel Convention, 2001, APCTT, New Delhi, India October 4–5, 2001.

Needs Assessment for Training of Technology in Environmentally Sound Management of Hazardous Wastes in the SAARC Region Asia and Pacific Center for Transfer of Technology. 2001. New Delhi, November 2001.

Regional Cooperation (SAARC) countries (Afghanistan, Bangladesh, Bhutan, India, Maldives, Nepal, Pakistan, SriLanka). Available at: http://www.saarc-sec.org

Report on the Needs Assessment for Training of Technology in Environmentally Sound Management of Hazardous Wastes in SAARC Region, 2001.

SACEP News Vol. 24, No. 1, March 2006.

SACEP News Vol. 24, No. 2, June 2006.

SACEP News Vol. 24, No. 3, September 2006.

SACEP News Vol. 24, No. 4, December 2006.

SACEP News Vol. 25, No. 1, March 2007.

SACEP News Vol. 25, No. 2, June 2007.

SACEP News Vol. 25, No. 3, September 2007.

SACEP News Vol. 25, No. 4, December 2007.

The Seventh GINC Tokyo Meeting April 18–20, 2001, Tokyo, Japan.

South Asia Cooperative Environment Programme (SACEP) Web Resource, Available at: http://www.sacep.org/

State of the Environment, South Asia, UNEP-SACEP, NORAD and DA 2001.

Workshop Report of South Asia Workshop for MEA Negotiations, SACEP, Colombo, Sri Lanka, October 5–7, 2005.

Section VIII

Invited Essays

47 A Global Approach to Environmental Emergencies

*Chris Dijkens and Peter Westerbeek**

CONTENTS

* The opinions expressed in this chapter are those of the authors and do not necessarily represent the views of the Inspectorate or the Ministry.

INTRODUCTION

There is a great variety of disasters. The fact is that all of them have negative impacts on various elements of the environment and can have devastating primary or secondary humanitarian impacts. More generally speaking in terms of providing environmental assistance to countries in disaster situations, it can be stated that "humanitarian aid includes environmental assistance."

The Global Risk Forum (Davos forum*) stated in 2008 that new principles, policies, and strategies, innovative mechanisms, and methods have to be designed to address the variety of risks that communities face, and a collaborative global risk reduction management process becomes increasingly important. Whereas this statement can be underlined, it is equally true that a collaborative global crisis management approach is of importance, especially if an adequate response requires a level of technical expertise, specialized information, or equipment that are beyond the capacity of individual states. This is the case when a disaster also includes an environmental emergency, caused by chemical compounds, which pose threats to human health, life-support systems, and nature. In this chapter, the global approach to an adequate response to environmental emergencies will be explored. The first section serves as a general introduction and addresses the safety chain as an instrument for policy making in terms of risk, with emphasis on elements preparation and response. The second section focuses on chemical hazards and how they can turn into an environmental emergency. The third section sets out the specialized response required in case of an environmental emergency and takes a closer look at the character of environmental emergencies. The fourth section explores the existing international instruments and models used during environmental emergencies response. The final section discusses the international system for environmental emergency response and maps out a strategy for strengthening the system.

THE SAFETY CHAIN: A MODEL FOR POLICY MAKING AND EVALUATION

The chain approach is widely used for policy making and evaluation in terms of risk management. This safety chain consists of five phases of risk mitigation and crisis management: proaction, prevention, preparation, response, and recovery.

* DAVOS-forum, IDCR Davos 2008; http://www.idrc.info/pages_new.php/Davos-2008/565/1/527/

Mitigating Measures: Risk Management

Proaction and prevention phases' attempts are aimed at preventing hazards from developing into accidents or, in the worst case, into disasters. Proaction stands for elimination of structural causes of incidents and prevention of incidents. For example, spatial planning and land use can reduce risk by excluding the development of a hazardous industrial complex near residential areas, thus eliminating the effects on human health if an accident occurs. Nevertheless, there are many instances where operations close to populated areas cannot be avoided and pose health risks. The prevention phase is—given this current situation—focused on reducing the effect of an accident that is likely to happen. This mitigation phase differs from the other phases because it focuses on long-term measures for reducing or eliminating risk (Haddow and Bullock, 2004).

Mitigating measures can be structural or nonstructural. Structural measures use technological solutions, like flood banks or spatial planning. Nonstructural measures include legislation, land-use planning (e.g., the designation of nonessential land like parks to be used as flood zones), and insurance. Mitigation is the most cost-efficient method for reducing the impact of hazards. Mitigation does include providing regulations regarding evacuation, sanctions against those who refuse to obey the regulations (such as mandatory evacuations), and communication of potential risks to the public (Lindell et al., 2006).

Preparation and Response: Crisis Management

However, mitigation is not always enough or suitable. Besides risk management, considerable effort is necessary to ensure an effective crisis management (preparation and response) to prevent matters from getting out of control.

In the preparation phase, planning, education, training, and practicing are some of the measures to prepare for the response to incidents. Common preparedness measures include:

- Risk evaluation, accident scenarios, and casualty prediction.
- Communication plans for the public and emergency workers.
- Maintenance and training of emergency services, such as emergency response teams.
- Development and exercise of assessment methods and evacuation plans.
- Development and stocking of inventory, disaster supplies, and equipment.
- Institutionalizing professional emergency agencies, networks, and coordination centers.

Response is the actual response to serious accidents and disasters. The response phase includes the mobilization of the first responders in the disaster area. This is likely to include a first wave of core emergency services. They may be supported by a number of secondary emergency services, such as environmental assessment teams. A rehearsed emergency plan developed as part of the preparedness phase must enable an effective coordination and employment of these services. In addition,

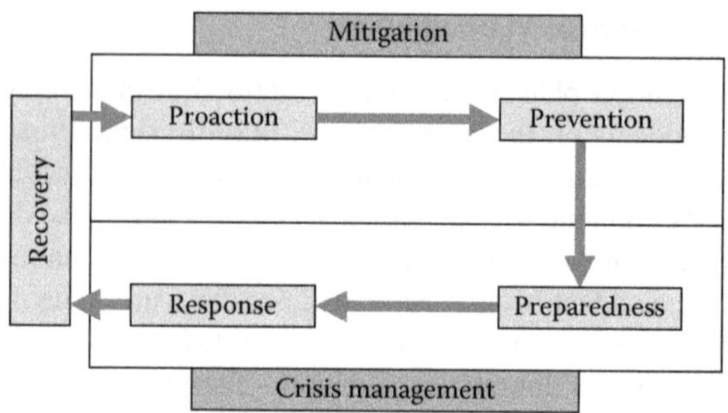

FIGURE 47.1 Continuous evaluation of accidents and disasters.

an international organizational response to any disaster should be based on existing and trained emergency management organizational systems and processes.

Recovery

After the initial response, the recovery phase is needed to return to a "normal" situation after an incident or disaster. The aim of the recovery phase is to restore, if possible, the area to its previous state. It differs from the response phase in focus; recovery efforts deal with issues and decisions that must be made after the immediate needs are taken care of. Recovery actions include rebuilding destroyed property and the repair of vital infrastructural works. An important aspect of recovery is evaluation. The evaluation should not only be limited to the cause or the response, but in addition, efforts should be made for evaluation of the state-of-affairs in the proaction and prevention phase, building up to the incident or disaster. The results of this should be brought back into the proaction or prevention phase quickly, as the aftermath of a disaster gives a "window of opportunity" for legislation and implementation of proactive and preventive measures and strengthening of emergency responses. In this the implementation of mitigation strategies can be considered a part of the recovery process if applied after a disaster has occurred (Haddow and Bullock, 2004).

The elements in the safety chain are interlinked. Continuous evaluation will lead to optimal risk and crisis management of accidents and disasters (see Figure 47.1).

CHEMICAL HAZARDS AND ENVIRONMENTAL EMERGENCIES

In general, the combination of the world's expanding urbanization and generalized globalization has aggravated the risks to all communities and nations. While risks worldwide are well identified and known, the number of disasters is increasing, striking all parts of society in all regions of the world. There is a wide variety of risks from natural causes to technical and biological hazards, from pandemics to terrorism. In addition, climate change can worsen the overall global vulnerability. In this, the character of chemical risks takes its own place.

Chemical emergencies substantially differ from natural disasters. The dynamics of an accident can be extremely fast. In opposition to some natural disasters like earthquakes, there is mostly no warning or alerting sign. From a policy view, chemical risks further complicate a common approach caused by difficulty of communicating at the multidisciplinary table with the representatives of natural risks.* There are different ways to look at chemical hazards. Common is the approach to base oneself on chemicals with hazard properties based on their belonging to hazard groups or classes. It goes no further than the connotation that a chemical is a toxic gas, explosive, flammable, and combustible or a liquid that is toxic to human health or fish species or a substance that is persistent and accumulating. Building on this basic subdivision, a more detailed subdivision can be made of main hazard types.

Direct Impact and Long-Term Impact

Of course, there are also chemical substances that pose more than one type of hazard. However, we prefer a more operational approach to chemical hazards, in which disasters with chemicals involved can be divided into two groups direct impact and long-term impact.†

Direct impact on human health includes the following:

- Immediate death.
- Immediate adverse health effects.

Direct impact on life-support functions and nature includes the following:

- Humans are impacted through effects on their life-support functions, for example, direct impact on crops, fish resources, agricultural land, and water supply.
- The same direct impacts as those on life-support functions can threaten the intrinsic value of nature, biodiversity, and certain species or ecosystems.

Long-term impact on life-support functions, nature, and humans includes:

- Toxic persistent substances entering the food chain and natural ecosystems and substances with long-term effects, such as carcinogens.

However, this division still does not explain why chemical hazards (risk) turn into environmental emergencies. For that, consideration of more factors is needed.

Main Factors that Determine the Impact of a Disaster

The impact of a disaster on human health, life-support functions, and nature can originate both from chemical hazards that are caused by (1) chemical spills, and

* Treu M. C. (coordinator), QUATER PROJECT, Territorial Risk Management Systems of Municipality, INTERREG IIIB MEDOC, Research group of Politecnico di Milano, 2004.

† Dutch National Institute for Public Health and the Environment (RIVM), FEAT-report 609000001/2009, p. 25.

(2) from chemical risks caused by physical hazards, such as landslides, waves of water, and saltwater intrusion. There are three main factors that together determine the magnitude of the impact of a disaster where chemicals are involved: hazard, exposure, and quantity. In other words, the impact of a (chemical) disaster is determined by three variables* and the three factors together determine the risk:

- The (intrinsic) toxicity of a chemical compound for a receptor (hazard).
- The presence of a pathway from source to receptor (exposure).
- The quantity of the substance released by the spill and to which the receptor is exposed (quantity).

In a disaster situation, hazardous substances will have a significant impact only if the hazard, the exposure, and the quantity are all significant. This means that the substance that is spilled must be toxic (significant hazard), that contact between the chemical and humans needs to be present or possible (pathway), and that a significant amount must be spilled (quantity). If one of these factors is not relevant or substantial, then it is unlikely that there is substantial impact in such a way that a hazard will turn into a disaster. Such situations are of low priority as compared to situations where all aspects are relevant. The magnitude of the impact is always determined by the proportional contributions of each of these three main factors. A large quantity of highly toxic material has a small impact if minimal exposure takes place, whereas a smaller quantity of material with moderate toxicity has a high impact if people, life-support systems, and nature are exposed at a higher level.

Exposure Route

The main difference that determines the distinction between intrinsic toxicity (potential impact of a chemical compound) and impact (actual exposure and associated effects) is considered to be mainly caused by the difference in the *exposure route* (the pathway through a medium). In other words, a combination of a gas and an aquatic endpoint will not turn into a disaster because the gas must first dissolve in water (the pathway) before it can result in damage to fish and the food chain. *Exposure* is only possible if a receptor and a transport mechanism or transport medium is present. Possible receptors of concern are humans (direct exposure through air), elements of the environment that play an important role in the life-support function (e.g., fish, cattle, water, crops, and fertile soil) and valuable ecosystems. Pathways are air, (drinking) water, and soil. Next to the pathway, many factors are known to influence the transport mechanisms and the vulnerability of the receptor. Local conditions are often important, and specific expertise would be required to evaluate the local situation. However, the quantity, the toxicity, the dispersion, and the vulnerability of the receptor contribute to the determination of whether a risk will turn into a disaster.

* Dutch National Institute for Public Health and the Environment (RIVM), FEAT-report 609000001/2009, pp. 25–27.

ENVIRONMENTAL EMERGENCIES AND RESPONSE

The response to environmental emergencies in which chemicals are involved differs from the traditional first response aimed at saving people and providing relief. Given the possible effects of chemical compounds, the main concern of the response must be a scan on the impact of the eventual hazards.

Assessment and Detection

The first concern is to get information on the scale of the environmental contamination and related health effects due to exposure to chemical compounds. For this purpose, the response should consist of a first assessment of pathways and quantities, the environmental and possible health implications, and a sampling, detection, and analysis strategy. This type of response requires the following: sampling (blood, urine, and food), detection, and (on site) analyses of (toxic) compounds.

- The sampling, detection, and analysis instrumentation should cover a broad spectrum of environmental matrices like air, water, soil, and sediment (the pathways). In addition, waste, liquids, solids, and a large number of compounds also should be included.
- Integrating and scientifically interpreting data gathered by sampling, detection, and analysis.
- Performing of model calculations: dispersion of compounds in air, dispersion of compounds in water, exposure modeling.
- Information about physical–chemical properties of compounds.
- Toxicological information.
- Health and ecological risk assessment.
- Specific expertise about air, water, soil, or food contamination.
- External safety risks.

Expertise

The above-mentioned tasks require expertise like academically graduated chemists or physicists, highly qualified analysts, and technically skilled staff members with experience in environmental sampling. In addition to this expertise, appropriate equipment is necessary.

This equipment should include instruments and facilities to determine concentrations of hazardous compounds in air, both gaseous compounds and particles, such as sensors to measure the concentration of several inorganic gases like sulfur dioxide, hydrogen sulfide, nitrogen oxides, chlorine, and ammonia. In addition, a photo-ionization detector (PID) can be used to measure the total concentration of volatile organic components. Canisters and Tedlar bags can be used to take samples of air. Tools for sampling surface water, groundwater, or drinking water are available as well. These samples can be analyzed using a special toolkit containing chemicals and a spectrophotometer. Similar to water samples, the equipment must also consist of several tools to take samples from soil or sediment or to sample deposited particles (wipe sampling). The samples must then be analyzed with, for example, an x-ray fluorescence (XRF) analyzer or a gas chromatograph–mass spectrometer (GC–MS) to identify and quantify organic compounds. In most cases, the samples need to be prepared for analysis by extraction and other techniques.

The data gathered with the above-mentioned equipment, expertise, and skills will provide information such as an estimation of the actual and potential exposure of humans and the environment, the assessment of the possible toxicological, medical, and ecological effects of the exposure, and an assessment of the potential future risks that may exist as a result of persistent contaminants dispersed in the environment. Following this information, the competent authorities can take appropriate measures.

We notice that assessing the impacts of disasters with chemicals involved requires a high level of technical expertise, specialized information, and equipment. This type of response is often beyond the capacity of individual states, especially developing countries. Therefore, the next sections address the availability of and the need for international mechanisms to address this situation.

Environmental Emergencies: A Closer Look

For providing assistance, if requested, in practice a compartmentalization between humanitarian and environmental disasters and emergencies reflects how the disaster management community often responds to disasters. Many organizations are capable of providing humanitarian relief, but just a few have the mandate and experience to respond to environmental disasters. Another problem is that integrated response capacity is limited. A more concrete definition can be useful to make the distinction between both types of disasters.

A definition according to United Nations Environment Programme's (UNEP's) Governing Council (2002)* is as follows:

> An environmental emergency can be defined as a sudden onset disaster or accident resulting from natural, technological or human-induced factors, or a combination of these, that cause or threaten to cause severe environmental damage as well as harm to human health and/or livelihoods.

In general, environmental emergencies distinguish between three scenarios: (1) classic environmental emergency (also referred to as technological or industrial accidents); (2) sudden onset natural disaster with major negative impacts on human life and welfare and on the environment; and (3) complex emergencies with major environmental impacts (more commonly known as conflict or war). These scenarios are described in more detail as follows.

* UNEP/GC.22/inf/5, November 13, 2002.

Classic Environmental Emergencies

Environmental emergencies in the narrow sense are technological or industrial accidents and usually involve some type of hazardous material and can occur at any location where these materials are produced, used, or transported. Common locations include pesticide storage sites, hospitals, railway marshalling yards, and numerous different industrial-manufacturing sites.

While radiological, nuclear and biological emergencies are typically excluded as they require a different specialized response, large forest and wild land fires are generally included in this definition as they are often "human-induced" and can create serious humanitarian suffering. The haze covering southeast Asia in 1997 and 1998 can be identified as a clear case in that perspective.

Natural Disasters

Large sudden-onset natural disasters, such as floods, hurricanes, earthquakes, and volcano eruptions can have big negative impacts on human health and the environment. Floods causing landslides, earthquakes resulting in fires in refineries, volcano eruptions leading to petrol station explosions, gas exploration works resulting in mud volcanoes, and typhoons sinking ferries containing large quantities of hazardous pesticides or a tsunami flooding, a nuclear reactor are examples of how secondary impacts have caused further death and suffering.

Complex Emergencies

In situations of complex emergencies, such as have happened in Somalia, Darfur (Sudan), and Iraq, the breakdown of authority, looting, and attacks on strategic industrial installations can all cause environmental emergencies. For example, the bombing of the fuel storage tanks of a power station in Lebanon led to a severe spill of heavy fuel oil in the Mediterranean Sea.

INTERNATIONAL INSTRUMENTS AND MODELS FOR ENVIRONMENTAL EMERGENCY RESPONSE

Requests for International Support

The international response to, and management of, environmental risks and impacts of disasters often require a level of technical expertise, specialized information, and equipment that is beyond the capacity of individual states. For these reasons, governments in affected countries often must seek expertise and resources available from the international community to address such crises. In addition, needs of countries for unbiased assessments can inspire countries to request international assistance and support. Situations of transboundary impacts of a disaster that have to be assessed and that have the need for overall coordination can also lead to requests for international support. Such support may be provided bilaterally, directly from one assisting country to the affected country, or multilaterally, through international organizations.

A problem in the past has been that countries suffering from environmental emergencies had little or no knowledge of what services were available to them internationally. Even today, many potential recipients of assistance are unfamiliar with the procedures already in place to request and receive assistance, which may result in delays, thereby having negative impacts on the population at risk and the environment. International organizations like the United Nations (UN) have recognized this problem and are preparing plans to improve awareness and knowledge. Training courses, seminars, capacity building activities, and contingency planning are important ingredients for building a robust and sustainable situation.

International Coordination of Environmental Emergencies

Coordinating bodies are needed for the worldwide management of the international response. At the international level, the Office for the Coordination of Humanitarian Affairs (OCHA) and UNEP founded the Joint UNEP/OCHA Environment Unit in 1994 to address this. The Monitoring and Information Center (MIC), operated by the European Commission (EC) in Brussels, is the operational heart of the Community Mechanism for Civil Protection in Europe.

Joint Environment Unit

The Joint UNEP/OCHA Environment Unit* (JEU) is the UN mechanism to mobilize and coordinate emergency assistance to countries affected by environmental emergencies and natural disasters with significant environmental impact.

In 1993, the UN General Assembly acknowledged the serious threats posed by environmental emergencies. Governments had come to recognize the connections between environmental conditions, human health, and the success of development efforts. Member states also determined there were needs to improve the international response to environmental emergencies and dedicated the UN to that role. The JEU is the result of cooperation between the UNEP's technical expertise and the humanitarian response coordination structure of the OCHA and is housed in OCHA's Emergency Services Branch.† Over the last 15 years, the JEU has been involved in the response to more than 100 environmental emergencies.

The JEU has a single purpose in dealing with environmental emergencies: to mobilize and coordinate international assistance. This assistance can be used in two ways:

1. Responding to environmental emergencies.
2. Helping countries to improve their preparedness to respond to environmental emergencies.

The JEU works with affected countries to identify and mitigate acute negative impacts stemming from emergencies, providing independent, impartial advice, and practical solutions. It also works with organizations dedicated to medium- and long-term rehabilitation to ensure a seamless transition to the disaster recovery

* http://ochaonline.un.org/ToolsServices/EmergencyRelief/EnvironmentalEmergenciesandtheJEU
† http://ochaonline.un.org/AboutOCHA/Organigramme/EmergencyServicesBranchESB/tabid/1219/language/en-US/Default.aspx

process. In addition, the JEU bridges gaps between emergency prevention, preparedness, and response, and between disaster management stakeholders by supporting information sharing, facilitating collaborative activities between partners, and increasing the range of stakeholders involved in environmental emergencies management. The JEU also acts as the secretariat to the Advisory Group on Environmental Emergencies (AGEE). The AGEE brings together disaster managers and environmentalists from developed and developing countries to exchange ideas and experiences on global environmental emergency response issues.

European Commission: Monitoring and Information Centre

In Europe, the MIC,* operated by the EC in Brussels, is the operational heart of the Community Mechanism for Civil Protection (hereafter, the Mechanism). It is available on a 24/7 basis and is staffed by duty officers working on a shift basis. It gives countries access to the community civil protection platform. Any country affected by a major disaster—inside or outside the EU—can launch a request for assistance through the MIC.

During emergencies, the MIC plays three important roles. First, the MIC functions as a focal point for the exchange of requests and offers of assistance. This helps in cutting down on the participating states' administrative burden in liaising with the affected country. It provides a central forum for participating states to access and share information about the available resources and the assistance offered at any given point in time. Second, the MIC has the role of "information provision." The MIC disseminates information on civil protection preparedness and response to participating states as well as a wider audience of interested parties. As part of this role, the MIC disseminates early warning alerts (MIC Daily) on natural disasters and circulates the latest updates on ongoing emergencies and Mechanism interventions. Third, the MIC also has an important role in the support of coordination, by facilitating the provision of European assistance through the Mechanism. This takes place at two levels: at headquarters level, by matching offers to needs, identifying gaps in aid and searching for solutions, and facilitating the pooling of common resources where possible; and on the site of the disaster, through the appointment of EU field experts, when required.

International Practices and Environmental Assistance

Natural disasters often have secondary impacts, including damage to infrastructure and industrial installations. These environmental emergencies may pose a threat to the health, security, and welfare of the affected population and the emergency responders. Too often, these risks are neglected, resulting in preventable deaths and injuries. It is therefore essential that information on the location of the hazardous facilities and the potential impacts is made available to relevant authorities and emergency responders at a very early stage of the disaster response or even prior to the onset of a disaster.

Disaster response teams are faced with the difficult task of not only dealing with the disaster at hand, but also identifying and responding appropriately to these potential

* http://ec.europa.eu/environment/civil/prote/mechanism.htm

environmental impacts. However, thousands of toxic chemicals could be involved in any given disaster, each with its own toxicity profile, and with a multitude of exposure pathways (e.g., air, water, and soil) and receptors (e.g., humans, livestock, and fishing grounds). In such complex situations, it can be easy to overlook or misjudge important risks. At the same time, given the often overwhelming demands of disaster situations, complex and full-fledged environmental assessments would be inappropriate.

When natural and environmental disasters occur, in the aftermath it is essential to assess the consequences as quickly and as thoroughly as possible. Sampling, conducting measurements and analysis for in-depth assessments, done by specialists, are very important activities in that respect. Several valuable instruments have been developed for operating quickly and effectively in the aftermath of the disaster, like the following.

Hazard Identification Tool

The Hazard Identification Tool (HIT*) has been developed to assist in the desktop research based identification of potentially hazardous facilities. The objective of the HIT is to limit the consequences of natural disasters and technological accidents for the human population; hence, to reduce the number of victims. To this end, the HIT is applied to identify potential secondary risks of a natural disaster and to provide information that can subsequently feed into the decision-making process on requests for further specialized assistance and targeted mitigation measures.

The methodology of the HIT is based on the Flash Environmental Assessment Tool (FEAT). FEAT is described in more detail in the next section. The HIT is complementary to FEAT and is compiled based on a desktop research, and sent to first responders or national authorities for in-country verification. It is important to emphasize that the HIT is only a first step to reach the goal of limiting the consequences of secondary risks.

The HIT consists of a list of "big and obvious" facilities and objects that may pose a risk to human health and life, as well as the natural environment. The list includes indications of the hazardous substances that are expected to be present in these facilities, as well as the hazard types associated with these substances and related estimated impact types. The hazard types and estimated impact types in the HIT are based on the already mentioned FEAT to facilitate ease of complementary use.

Flash Environmental Assessment Tool

A practical, accurate, yet simple tool was developed to assist initial response teams such as United Nations Disaster Assessment and Coordination (UNDAC) teams. With these challenges in mind, FEAT† is a carefully balanced compromise between simplicity and scientific rigor, with emphasis on usefulness to response mechanisms

* Hazard Identification Tool (HIT)—*Identifying secondary environmental risks in the context of natural disasters;* http://ochaonline.un.org/ToolsServices/EmergencyRelief/EnvironmentalEmergenciesandtheJEU/Reportsonemergenciesandactivities/HazardIdentificationToolHIT/tabid/1465/language/en-US/Default.aspx

† http://ochaonline.un.org/ToolsServices/EmergencyRelief/EnvironmentalEmergenciesandtheJEU/ToolsandGuidelines/tabid/5094/language/en-US/Default.aspx

such as UNDAC teams. It provides quick answers in complex disaster situations, even in the absence of specialized technical resources or expertise.

In summary, FEAT is a "first aid" tool to identify environmental impacts by identifying the most acute hazards to human health and the environment after natural disasters, and support initial response actions in disaster contexts. It does not replace in-depth environmental assessments, which may be appropriate at later stages of the disaster response. Findings from use of the FEAT should be communicated quickly to appropriate organizations so that appropriate actions can be taken.

Environmental Assessment Module

Natural and environmental disasters can occur anywhere, and they cannot always be prevented. In the aftermath, it is essential to assess the consequences as quickly and as thoroughly as possible. That is why the Netherlands has developed the Environmental Assessment Module (EAM*), a mobile laboratory that can be used for disasters involving hazardous substances. It consists of three specialists and two fully equipped off-road vehicles, with sampling and laboratory instruments. The EAM will mainly be deployed in countries that lack the specialist knowledge or capacity needed to deal with environmental disasters. The EAM will be deployed at the request of the UN or the EC–MIC, or is available based on bilateral agreements between a country and the Netherlands.

The EAM was developed on the basis of the sophisticated concept of a small, flexible, and well-trained team that can be deployed quickly anywhere in the world. The equipment is perfectly tailored to the situations the team may encounter. This means that the equipment and the team's logistics meet the highest requirements. For rapid worldwide deployment, the team and its equipment are highly mobile. That is why the EAM involves two off-road vehicles. One vehicle is equipped as a mobile measurement and analysis unit, while the other contains materials for logistical support. The entire unit can be transported in a cargo aircraft. To ensure maximum flexibility, the equipment consists of modules that are packed in crates for transport. If only a small team and only part of the equipment are needed, the required material can be taken on board the plane separately.

Of course, communication and navigation equipment is standard: the vehicles are equipped with satellite telephones and GPS. The EAM is mainly used to assess the medium- and long-term effects of environmental disasters and is specialized in identifying a large number of chemical substances in polluted material, and advising on the nature of the pollution and the threat it poses. The EAM is backed up by the expertise of 10 Dutch government institutes and services working together in the Policy Support Team for environmental incidents.

Back Office Support by Experts

In the Netherlands, as in many countries, there is considerable expertise residing in national institutes that, if available to emergency organizations, can greatly enhance the effectiveness of the response in case of major chemical accidents. The Dutch BOT-mi (*Beleidsondersteunend Team Milieu Incidenten*, or Policy Support Team for

* http://www.unep.org/Documents.Multilingual/Default.asp?DocumentID=543&ArticleID=5897&l=en

Environmental Incidents) was created to make that expertise promptly available to crisis management organizations. One of the key challenges of BOT-mi, and one of the key factors of its success, is the development of a standardized framework for communications within the multiagency cooperation (virtual Integrated Crisis Advise Web site—ICAWEB). That framework allows BOT-mi members to share information in real time, to discuss the information, and to generate integral advice for the recipients. It also offers the possibility of a common information platform for all emergency organizations, BOT-mi members and its partners to exchange key data related to the accident. Furthermore, the framework offers extra functionality for alerting, notification, data management, and communication.

Examples where BOT-mi supported international missions are the floods in Surinam (2006), technical assistance for the identification of environmental impacts of the mudflow in Indonesia (2006), and the hazardous waste dumping in Ivory Coast (2006). BOT-mi functioned in these cases as back office support for the experts abroad. By making use of ICAWEB and the possibilities of BOT-mi, the experts were continuously provided with (tailored) information they requested. This mechanism proved to be very effective and efficient.

The added value of BOT-mi (for international environmental missions) can be summarized as:

- Scientific backing
- Quick screening of risks/hot spots
- Integrated versus individual experts
- 24/7 availability and accessible from the field
- Natural disasters: quick screening
- Hazardous materials: detailed modeling

The advantage of this organization is that specific expertise is combined and integrated to create tailored and integrated advice to the local authorities. The BOT-mi members cover the following expertise:

- Chemical analysis
- Measure and sampling strategies
- Risk and plume calculations
- Risk analysis health impact
- Meteorological information
- Risk assessment aquatic environment
- Assessment food safety
- Assessment CBRN exposures
- Managing emergency response measures
- Crisis management experience

Green Star Award

In 2009, the Dutch government received the Green Star award for its international environmental disaster strategy. The Green Star award is a joint initiative of Green Cross International, UNEP, and the UN OCHA. The prize is awarded to companies, organizations, or governments that have made remarkable efforts to prevent, prepare for, and respond to environmental disasters around the world. The jury chose the Netherlands for its role in raising quality standards for the international response to humanitarian and environmental crises around the world. The Netherlands was awarded the Green Star trophy for a number of innovative projects that were carried out jointly by the Ministries of Foreign Affairs and Housing, Spatial Planning and the Environment, and the National Institute for Public Health and the Environment (RIVM). This collaboration resulted in FEAT (see the section "Flash Environmental Assessment Tool"), the EAM concept (see the section "Environmental Assessment Module"), and the development of an environmental emergencies training program for specialists. The Dutch government handed over the FEAT and the training program in custody to the JEU of the UN. Both instruments are now used worldwide as a UN standard.

The Environmental Emergency Response Model

Based on the experience of the JEU with the given response in numerous environmental disasters, the following model illustrates the methodology of assistance. The model contains three distinct phases in environmental emergency response* activities. In each phase, one or more described instruments are used.

The model is based on a scenario of a major sudden-onset natural disaster for which a UNDAC team has been requested and deployed. Although this model is obviously only a simplified reflection of the very complex reality of environmental emergency response and its case-by-case specificities, it provides a logical breakdown and can help in understanding the different needs that exist for preparedness and response to environmental emergencies.

The model is also applicable for industrial accidents and complex emergencies, for which the response would start in Phase 2 of the model. The different phases are determined by the information available and type of assistance needed (see Figure 47.2).

In situations where international assistance is requested for a sudden-onset (natural) disaster, the situation in that area is often chaotic. Little information and data are available. Information like how many people have been affected, where the most affected zones are, who is in charge of the response, and what national capacity is available is mostly lacking. Numerous questions need to be addressed, and the urgency of the situation makes the taking of well-informed decisions all the more challenging. In emergency situations where environmental emergency response is needed, the situation is similar. However, based on response experiences to date, a number of roles and tools have been identified to provide answers and attempt to create order during an often overwhelming and chaotic disaster situation. The HIT, FEAT, and other tools—as already described—proved to be very helpful and needed in practice.

* http://ochaonline.un.org/OchaLinkClick.aspx?link=ocha&docId=1109501; EU/AG/51, Advisory Group on Environmental Emergencies, 8th Meeting co-organized with the Monitoring and Information Center of the European Commission, Brussels, May 6–8, 2009, "Expanding the Resource Network for Environmental Emergency Response."

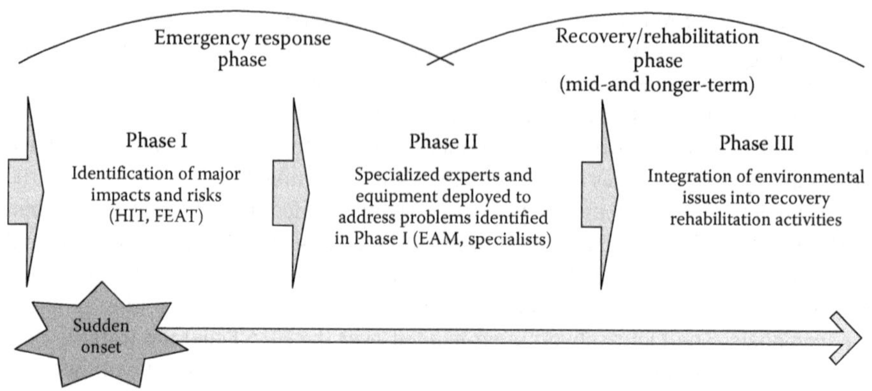

FIGURE 47.2 The different phases of an emergency and their responses.

Phase 1: Identification

When an UNDAC team is deployed to an affected country, the JEU will use the HIT to screen for any "big and obvious" secondary environmental hazards and impacts. The HIT will also help in deciding whether there is a need to deploy an UNDAC-trained environmental expert or to assign an environmental expert to the team to undertake an on-the-ground rapid environmental assessment, using the FEAT. During this phase of the emergency, a general background in environment, including some experience in dealing with hazardous substances, will be sufficient. In some instances, there may be a need for specific expertise, as with disaster waste management, for example.

Phase 2: Specialized Assistance

Once the situation following the onset of the natural disaster becomes clearer, there will be a shift from, for example, UNDAC teams, toward a need for specialized assistance in specific domains. It is in this second phase that specialists (e.g., chemists, disaster waste managers, engineers, toxicologists) are often needed, following the outcomes and findings of the FEAT. When necessary, organizations like the Dutch BOT-mi can provide the experts—at a distance—with in-depth information about the situation they will meet, information such as on hazardous materials which may have been used there, existence of drinking water wells, weather forecast information, scenario prediction and modeling, and so on. The BOT-mi specialists base the information on a rapid desktop research. A report with basic information can be provided at the earliest in such situations. The advantage is that the experts on the spot are already informed with some basic information and their attention is drawn to specific problems they will meet.

If there is a need for more detailed onsite sampling and analysis, the Dutch EAM can be deployed to determine which hazardous substances, and in what concentrations, have been released. Together with the backup systems such as BOT-mi, suggestions for prevention, mitigation, and clean up could be provided. In the event, for example, of an oil spill, a train accident involving toxic substances, or the bombing of an industrial facility in a conflict situation, the model is applicable from Phase 2, as the type of industry or substances involved are already known, and there is no need for a general assessment or generalist.

The type of expertise required for Phase 2 is very hard to define in advance, and will differ from case to case. Based on experiences in the recent past, experts in a broad array of fields (e.g., mining tailings, pesticides, cyanide, disaster waste management) were sent to affected countries.

Phase 3: Recovery and Rehabilitation

Once immediate emergency assistance has been provided, outstanding recovery needs are often still enormous. It is in this phase that a handover takes place to local authorities and international partners, who are best qualified and have developed specific expertise and methodologies to assist a country in getting back on its feet. During this phase, there is often a strong focus on building capacity of individuals and institutions to strengthen a country's coping mechanisms in case a similar disaster strikes again.

STRENGTHENING THE INTERNATIONAL SYSTEM FOR ENVIRONMENTAL EMERGENCY RESPONSE

Was the Bhopal chemical plant accident (1984) an environmental or a humanitarian disaster? What about volcanic eruptions, which had far-reaching consequences for the environment as well as a severe impact on local people? And any other disaster, which regardless of its origins, affects either people, the environment, or both? Too often, disasters get labeled as being one or the other without considering their consequences in a holistic and integrated fashion. Roy Brooke* (UNEP/OCHA) stated and answered these questions.†

He explained that, very often, natural disasters bring harm to people and simultaneously change the environment. As such, important volcanic eruptions may produce combined negative results: human victims, destruction of natural habitat, and heavy atmospheric pollution leading to climate change. Large-scale floods and landslides may damage soil and crops, affect important ecological areas, and kill people.

Brooke elaborates that,

> on the other hand, technological or industrial accidents may have devastating effects on both population and the environment. Toxic spills from dumping sites and obsolete dams pollute rivers, kill aquatic life and pose serious threats to human health. Fires at pesticide storage facilities may lead to serious environmental damage and have negative effects on humans. Chemical pollution is dangerous both to people and other elements of the environment. That is why various man-made accidents, such as explosions, fires, toxic leakages and pollution, are normally called by the single name 'environmental emergencies.' However, the application of this term would depend on the general approach to interaction between humanity and the environment.

---

* Programme Officer of the Joint UNEP/OCHA Environment Unit of the Office for the Coordination of Humanitarian Affairs (OCHA) and UNEP.

† http://www.grida.no/publications/et/ep3/page/2613.aspx; "Bridging the Gap between Human and Environmental Disasters."

Although there are certain differences in this great variety of disasters, from a practical point of view the causes of various emergencies are irrelevant. The fact is that all of them have negative impacts on various elements of the environment and can have devastating primary or secondary humanitarian impacts. More generally spoken in terms of providing environmental assistance to countries in disaster situations, it can be stated that "humanitarian aid includes environmental assistance."

John Holmes, Under Secretary General for Humanitarian Affairs and Emergency Relief Coordinator (OCHA) of the UN underlines this theory and says:

> disasters and conflicts can impact the environment in ways that threaten human life, health, livelihoods, and security. Disaster managers and humanitarian workers must therefore identify and address acute environmental risks quickly and consistently as an integral part of effective emergency response.*

A Global Approach toward a Strategy for Strengthening the System

It must be underscored that notable progress has been made in strengthening the worldwide system to prevent, prepare for, and respond to environmental disasters around the world. The effort and effect of individual countries and international organizations like the JEU and the EC–MIC are visible. However, the fact remains that the vast majority of this increase has been concentrated among a small number of traditional humanitarian donor nations. Although the seeds of collaboration have been planted and in some cases taken root in other areas of the world, such as Africa, the Asia/Pacific region, and Latin America, great strides have yet to be made. In light of certain global trends, the need for a better distribution of environmental emergency service providers has never been more important. For example, given ever-increasing industrialization worldwide, one must face the increasing likelihood of environmental emergencies occurring, often in nations where capacity to prepare for and respond to such disasters may not be sufficient. Also, given the ever-increasing links of the global community, environmental emergencies know no boundaries, and a disaster in one country can have direct and indirect impacts on and implications for other nations. Furthermore, the issue of climate change must also be faced. Increased numbers and intensity of weather-related natural disasters have been observed in recent years—most often affecting areas of the world that are least capable of responding to such emergencies—with much evidence linking this increase to global warming and climate change.

These challenges must be met by expanding the global network of readily available and deployable expertise and equipment to respond to environmental emergencies. Hence "humanitarian aid includes environmental assistance," countries need to provide environmental assistance to countries in disaster situations because:

- Humanitarian aid is a fundamental expression of the universal value of intrahuman solidarity.

---

* Environmental Emergencies: Learning from Multilateral Response to Disasters, p. 1. Available at: http://www.unocha.org/about-us/publications/other

- Multilaterism and international effort is a collective responsibility.
- Disasters know no boundaries.

This means that countries have to put effort into strengthening the system for environmental emergency response, including a readily available pool of diverse expertise, equipment to support operations, excellent bilateral and multilateral cooperation, and appropriate funding.

A strategy should be developed for a more equitable distribution of donor countries around the world that is beneficial to the strengthening of the system. Until now, capacity available for international deployments remained primarily within Europe. A next step would be creating regional informal networks, based on best practices with the objective to provide logistical support by the cooperation between the members to the countries in that specific region in sudden onset disasters. In addition, best practices like the EAM, BOT-mi, HIT, FEAT, and other instruments can be used. UN agencies and other international coordinating organizations can facilitate creating such networks. Response to a disaster in a country from within the same region is generally preferable, given that deployments could be expected to be faster and less expensive and there would be fewer linguistic and cultural issues to overcome.

It needs to be stressed that many other partners, such as NGOs, the private sector, and academia, have important roles to play in the provision of multilateral assistance.

In May 2009, the AGEE* recognized the fundamental importance of this strategy and invited a Steering Committee to develop strategies for establishing a more equitable distribution of donor networks.

REFERENCES

Haddow, G. D. and J. A. Bullock. 2004. *Introduction to Emergency Management*. Oxford: Butterworth-Heinemann.

Lindell, M., C. Prater, and R. Perry. 2006. *Emergency Planning*, 1st edition. Hoboken, NJ: Wiley.

* Report of the 8th Meeting of the Advisory Group on Environmental Emergencies.

48 Emerging Issues in Global Chemical Policy

Franz Perrez and Georg Karlaganis

CONTENTS

INTRODUCTION

Global chemical policy is a young and dynamic area of international environmental policy. Over the last two decades, several chemicals-related international conventions, protocols, and amendments to existing instruments have been negotiated and have entered into force both at the regional and the global levels. Moreover, with the Strategic Approach to International Chemicals Management, an institutional and political framework has been developed that promotes an integrated and comprehensive approach to chemicals management. Finally, by launching a process to strengthen institutional cooperation, coordination and synergies between the different relevant institutions and processes at the global level, the chemicals and waste area has taken the lead in strengthening international environmental governance. Over the last years, the international chemicals and waste regime has not only been able to address new challenges, it has also developed new approaches for effective policy making and implementation. Thus, the chemicals and waste area seems to be one of the most successful—if not the most successful—area of international environmental policy.

Chemistry is not a static science nor is the use of chemicals in industry, research, and our daily life a finite story. The rapid development of our society not only allows us to steadily increase the benefits from chemicals through new applications, new substances, or new uses of known substances. It also entails that the international community continues to be confronted with new and emerging challenges posed by

EVOLUTION OF THE INTERNATIONAL CHEMICALS AND WASTE REGIME

1979: UN-ECE Convention on Long-range Transboundary Air Pollution with several protocols on specific chemicals, such as the Heavy Metals Protocol (1998) and the POPS Protocol (1998) (regional, 51 parties)

1985: Vienna Convention for the Protection of the Ozone Layer (196 parties)

1987: Montreal Protocol on Substances That Deplete the Ozone Layer (196 parties)

1989: Basel Convention on the Control of Transboundary Movements of Hazardous Wastes and Their Disposal (172 parties)

1992: United Nations Conference on Environment and Development (UNCED, Rio Summit): Adoption of *Agenda 21* with specific chapters on chemicals and waste management (universal)

1998: Rotterdam Convention on the Prior Informed Consent Procedure for Certain Hazardous Chemicals and Pesticides in International Trade (131 parties)

2001: Stockholm Convention on Persistent Organic Pollutants (169 parties)

2002: World Summit on Sustainable Development (WSSD): Adoption of Johannesburg Plan of Action with specific provisions on chemicals, including the "2020 target" that, by 2020, chemicals are used and produced in ways that minimize significant adverse effects on human health and the environment (universal)

Globally Harmonized System of Classification and Labelling of Chemicals (GHS) (implemented by 67 countries)

2006: Strategic Approach to International Chemicals Management (universal)

Decisions by the Stockholm, Rotterdam, and Basel Convention to establish an *Ad Hoc* Joint Working group to launch a synergy process

2009: Decision of the UNEP Governing Council to launch negotiations on a legally binding instrument on mercury

2010: Decisions by the Stockholm, Rotterdam, and Basel Conventions to establish a joint head for the three secretariats, joint services and to further enhance cooperation and coordination between the three Conventions.

chemical substances for environmental safety and human health. In the light of these new and emerging issues, the international chemicals and waste regime is similarly not static but continues to evolve and becomes increasingly sophisticated. This contribution will summarize and describe some of the important emerging issues that shape and catalyze the global chemical regime. Thereby, it is important to note that there are two fundamentally different kinds of emerging issues that trigger the attention of policy makers: substances and policy-related emerging issues. Thus, the term "emerging issue" can relate to substances that emerge into the public attention. These can be new chemicals that have been newly invented by researchers, placed on the market and used by industry and consumers. These new chemicals can become a threat to the environment and human health and thus require a political response. Nanomaterials are an example of such a new and emerging issue. However, not only "new" chemicals, but also "old" and well-known chemicals that have been used several decades before (e.g., PCBs) can emerge into the public attention and require political response. The leaking of PCBs into a river from a waste disposal facility near Freiburg (Switzerland) (Schmid et al., 2009) is just one example of such an "emerging" issue that involves "old" chemicals. Second, the term "emerging issues" not only relates to substances that emerge as new issues in the global chemical policy, it can also relate to new policy approaches and new policy instruments to address the challenges posed by chemicals. Examples of such new international policies are public–private partnerships (PPPs), integrated approaches, or the institutional synergy process. All these new emerging issues will strongly shape the future international chemicals regime.

By discussing the challenges of heavy metals such as mercury, lead, and cadmium, of nanomaterials, endocrine disruptors (EDs) and perfluorinated compounds

(PFCs) and the possible policy responses to these challenges, the second part of this contribution will address both kinds of substances-related emerging issues. The experience gained from older policy responses and the better understanding of the complexity of chemicals management have helped to develop new policy approaches to chemicals management. In its third part, this contribution will address some of the most important, new and emerging instruments and approaches of international chemicals policy making, namely increased private actor involvement, integrated and life-cycle approaches, and strengthening institutional coordination and synergies. Referring to the identified new and emerging issues, we will then try to outline how the future international chemicals and waste regime could ensure a political and institutional framework that addresses new and emerging issues by building on and further developing the cooperative and synergetic approach developed over the last years. The contribution will conclude that the international chemicals regime has been impressively capable in addressing new and emerging challenges and in adapting and reforming itself by developing new and innovative policies in the past and that there is hope that it will continue to do so in a comprehensive, coherent, effective, and efficient manner.

SUBSTANCES-RELATED EMERGING ISSUES: EMERGING CHEMICALS IN THE GLOBAL CHEMICALS POLICY

Heavy Metals

The challenges posed by heavy metals are not new and emerging—they are well known. However, the recognition that the problems posed by heavy metals are not merely local but have a significant global dimension and therefore require an internationally coordinated response is rather new. While specific international rules—mainly at the regional level or with regard to specific areas such as maritime pollution or transportation of hazardous goods—exist,* so far no horizontal environmental agreement has been developed addressing heavy metals. Thus, at the World Summit on Sustainable Development (WSSD) in 2002, the international community decided to "promote the reduction of the risks posed by heavy metals that are harmful to human health and the environment."†

Mercury

Today, it is well established that the significant global adverse impacts from mercury require international action to reduce the risks to human health and the environment:‡ Mercury is a potent neurotoxin that interferes with brain functions and the nervous system. It accumulates in the food chain and is transported over

* See, for example, the 1998 Aarhus Protocol on Heavy Metals to the 1979 Geneva Convention on Long-range Transboundary Air Pollution, available at: http://www.unece.org/env/lrtap/hm_h1.htm

† World Summit on Sustainable Development, Johannesburg Plan of Implementation, available at: http://www.un.org/esa/sustdev/documents/WSSD_POI_PD/English/POIToc.htm , § 23(g).

‡ See generally: UNEP GC decision 22/4 of 7.2.2003; UNEP, Global Mercury Assessment (2002), available at: www.chem.unep.ch/mercury/publications/default.htm

long distances in water, the air, and in products. Mercury is a global pollutant that covers common considerations like POPS do, notably the characteristics of persistency, bio-accumulation, eco- and human toxicity, and the potential for long-range transport.

Mercury is a natural substance occurring in the environment. However, human-made releases of mercury have increased seriously over the last two centuries, giving thus considerable rise to levels of mercury in the global environment. Mercury is now present in various environmental media all over the globe at levels that adversely affect humans and wildlife.* Burning fossil fuels (mainly coal combustion) is the largest single anthropogenic source of mercury emissions, mining—particularly artisanal small-scale gold mining—being the second, and cement production the third primary anthropogenic source (UNEP, 2008). Secondary anthropogenic sources include industrial processes and emissions from unsound waste treatment.†

While mercury was addressed at the national and regional levels already in the last century,‡ a horizontal global approach to address the environmental risks posed by mercury was launched only this century: Recognizing that scientific studies have established that mercury cycles globally and referring to the calls from the UN/ECE region and the Arctic Council for a global assessment of mercury by UNEP, the UNEP Governing council mandated UNEP in 2001 to undertake a global assessment of mercury and its compounds.§ In 2003, UNEP's global mercury assessment concluded that mercury is transported globally, that therefore mercury is of global concern and that international action is needed to reduce the risks to human health and the environment from the release of mercury and its compounds to the environment.¶ On the basis of the clear results of these reports and taking the view that a legally binding approach provides the best framework for coordinated effective international action, Norway and Switzerland proposed in 2003 the development of a comprehensive legally binding instrument on mercury. However, the UNEP GC was not yet ready to agree on such an ambitious approach. It nevertheless confirmed "that there is sufficient evidence of significant global adverse impacts from mercury and its compounds to warrant further international action to reduce the risks to human health and the environment" and decided instead to launch a program for international action on mercury to facilitate and conduct technical assistance and capacity building activities to support the efforts of countries to take action

---

* UNEP, Global Mercury Assessment, supra note 5, at iii.

† Ibid., supra note at 12–14.

‡ For national approaches, see, for example, Swiss national legislation: Federal Law relating to the Protection of the Environment of October 7, 1983 (SR814.01); Ordinance on Substances of June 9, 1986, now replaced by Ordinance on Risk Reduction related to the Use of certain particularly dangerous Substances, Preparations and Articles (Ordinance on Risk Reduction related to Chemical Products, ORRChem) of May 18, 2005 (SR 814.81); Ordinance on Air Pollution Control (OAPC) of December 16, 1985 (SR814.318.142.1); Water Protection Ordinance (WPO) of October 28, 1998 (SR814.201); for regional approaches, see the 1998 Aarhus on Heavy Metals to the 1979 Convention on the Long-Range Transboundary Air Pollution, available at: http://www.unece.org/env/lrtap/hm_h1.htm

§ UNEP GC decision 21/5 of February 9, 2001.

¶ UNEP, Global Mercury Assessment, supra note 5, at iii–vii, 22 and 230.

regarding mercury pollution.[*] Namely, the Mercury Programme was aimed to assist countries in improving the scientific basis for mercury policies, exchanging information, developing and enhancing capacity for risk assessments, enhancing mercury-related risk communication, reducing demand for and uses of mercury, and reducing anthropogenic releases of mercury.[†]

Two years later, in 2005, the UNEP GC again did not accept Norway's and Switzerland's proposal for a legally binding instrument on mercury but requested UNEP to further develop the mercury program, including by launching voluntary partnerships for addressing the significant global adverse impacts of mercury.[‡] Subsequently, UNEP strengthened its Mercury Programme to become a main mechanism for the delivery of immediate actions on mercury.[§] The Partnership Programme areas include mercury management in artisanal and small-scale gold mining; mercury control from coal combustion; mercury reduction in the chlor-alkali Sector; mercury reduction in products; mercury air transport and fate research; mercury waste management; and mercury supply and storage. Today, the Partnership Programme has seven governmental partners, three international organizations, and numerous nongovernmental organizations (NGOs).[¶]

Over time, Norway and Switzerland have broadened the support for a legally binding instrument, and the EU and the African Group and countries like Japan, Uruguay, Nigeria, and Jamaica have become similarly key supporter of a convention. Moreover, important countries like Brazil have also become open to a legally binding approach. In 2006, following a full-day side event on mercury and other heavy metals organized by Switzerland, the Intergovernmental Forum on Chemicals Safety adopted at its 5th Forum Session its Budapest Statement which recognized that current efforts to minimize use and reduce release of mercury need to be expanded and invited UNEP to assess the need for further action, including the option of a legally binding instrument.[**] Subsequently, the UNEP GC, noting the Budapest Statement, not only decided to further strengthen its Mercury Programme but it also recognized "that current efforts to reduce risks from mercury are not sufficient to address the global challenges posed by mercury," concluded that further long-term international action is required, and agreed to establish an *Ad Hoc* Open-ended Working Group of Governments, regional economic integration organizations and stakeholder representatives (OEWG) to review and assess options for enhanced voluntary measures and new or existing international legal instruments.[††] During the work of the OEWG, the support for a legally binding approach has further increased. Moreover, it became clear that such an approach can best be realized through a new, freestanding legally binding instrument on mercury. Two reasons seem to have encouraged developing a new convention: first, there was a broad

---

[*] UNEP GC Decision 22/4 (07.02.2003), Section V, operative paragraphs 1 and 4 and annex.

[†] UNEP GC Decision 22/4, supra note 11, Annex.

[‡] UNEP GC Decision 23/9 (25.02.2005), operative paragraphs 22, 23 and 28–30.

[§] See generally: http://www.chem.unep.ch/mercury/partnerships/new_partnership.htm

[¶] See generally: http://www.chem.unep.ch/mercury/partnerships/new_partnership.htm

[**] Intergovernmental Forum on Chemicals Safety, Budapest Statement, available at http://www.who.int/ifcs/forums/five/en/index.html, preambular paragraph 7 and operative paragraph 6.

[††] UNEP GC Decision 24/3 (09.02.2007), preambular paragraph 4 and paragraphs 16, 17 and 28–30.

perception that using an existing instrument such as the Stockholm POPS Convention would require an amendment of that instrument; second, a new convention was seen as being able to become in the future also a framework for other chemicals of global concerns.[*] On the basis of the work of that open-ended working group, the UNEP GC decided in 2009 at its 25th session to launch negotiations for a legally binding instrument on mercury.[†]

The new convention on mercury is supposed to take a "comprehensive and suitable approach to mercury" that includes provisions to reduce the supply of mercury, the demand for mercury in products and processes, international trade in mercury, and atmospheric emissions of mercury.[‡] It should also address mercury-containing waste and contaminated sites, increase knowledge and information, specify arrangements for capacity building, technical and financial assistance, and address compliance.[§] Norway, Switzerland, and the other supporters of a mercury convention have always argued that a legally binding approach would not compete with or replace voluntary measures such as those undertaken within UNEP's Mercury Partnership Programme, but that it would, by providing a strong and committing framework, ideally complement and support such voluntary approaches. Thus, it can be expected that the mercury convention will also further encourage such voluntary initiatives. Moreover, the Intergovernmental Negotiation Committee is called to consider in the development of the new convention "flexibility in that some provisions could allow countries discretion in the implementation of their commitment."[¶] The Intergovernmental Negotiation Committee met the first time in June 2010 in Stockholm and discussed the main elements of the new convention. On the basis of this discussion, a substantive document with elements and options for the content of the convention has been prepared for the second meeting in January 2011 in Japan.[**] Whether the new instrument will be limited to address mercury only or whether it will include an "open door" to allow its further development to become also a framework for other chemicals of global concern remains to be seen (see Table 48.1).[††]

Lead and Cadmium

The Romans in the old Roman Empire were pioneers in distributing water in several Mediterranean counties through their famous aqueducts which were constructed with stone. In addition to these stone aqueducts, the Romans, however, also used lead water pipes which caused, as we know today, the poisoning of many. Several thousand years later, lead is still a major threat to human health and the environment and

[*] "Table with criteria for determining the best option," document from an informal meeting of like-minded countries organized by Switzerland in May 2008 in Glion, Switzerland, on file with the authors.

[†] UNEP GC Decision 25/5 (20.02.2009), paragraphs 26–31.

[‡] UNEP GC Decision 25/5, supra note 19, § 27. See also Final Report of the Ad Hoc Open-ended Working Group on Mercury to the UNEP Governing Council, Annex "Elements of a comprehensive mercury framework," UNEP Document DTIE/HG/OEWG.2/13 of October 16, 2008, at 19–23.

[§] Id.

[¶] UNEP GC Decision 25/5, supra note 19, § 28.

[**] UNEP(DTIE)/Hg/INC.2/3.

[††] See UNEP GC Decision 25/5, supra note 19, § 30.

TABLE 48.1
Possible Measures Addressing Mercury

Supply:

Phasing out primary mercury production
Export ban for mercury, with the exemption of exports for disposal and for essential uses

Products:

Ban of products containing mercury for which alternatives are available
Establish threshold limits for mercury in products (batteries, electronic components, *inter alia*)

Processes:

BAT/BEP (coal combustion, PVC vinyl chloride monomer production, recycling of batteries, recycling of electronic waste)
Ban certain uses of mercury
Regulate processes that use mercury (e.g., catalysts in chemical synthesis, artisanal small-scale mining)
Emission standards (into air for waste incineration, into water for chlor alkali electrolysis)

Disposal:

Standards for disposal according to special waste regulations

Public Awareness

Public information
Awareness raising campaigns (for the collection of batteries, fever thermometers

Note: These measures can be implemented through compulsory mechanisms, through voluntary approaches such as partnership initiatives, or through a combination of compulsory and voluntary approaches.

the problem still persists until today*: Lead is a heavy metal that is toxic at very low exposure levels. It bioaccumulates in most organisms and has acute and chronic effects on human health and is toxic to plants, animals, and microorganisms. It is released by various natural and anthropogenic sources into the environment, and it can move in the environment between water, air, and soil. Moreover, lead is used and traded globally as a metal in various products, for example, batteries, different compounds, lead sheets, ammunition, alloys, cable sheathing, and petrol additives.† Cadmium is a nonessential and toxic element for humans.‡ It is associated with kidney damage. Cadmium is released by various natural and anthropogenic sources to the atmosphere. It is used as a pigment, a stabilizer for plastics, and as a component of batteries. It is also widely distributed when it is present as a contaminant in fertilizers.

In 1921, the General Conference of the International Labor Organization (ILO) adopted the White Lead (Painting) Convention in order to prohibit the use of prod-

* See generally: UNEP (2008).
† Id. at 6 and 90–103.
‡ See generally: UNEP (2008).

ucts containing white lead pigments in the internal painting of buildings, with the exception of railway stations. This Convention number C13 came into force 1923 and was ratified by 63 countries.[*] In 1996, the OECD recognized the risks from lead exposure to human health, in particular for children and other high-risk and sensitive populations, and to the environment and it stressed the need for cooperative commitments to reduce any transboundary exposure (OECD, 1996). In 2005, the UNEP Governing Council requested UNEP to undertake a review of scientific information, focusing especially on long-range environmental transport, to inform future discussions on the need for global action in relation to lead and cadmium. UNEP's analysis concluded that while lead is transported in the atmosphere mainly over local, national, or regional distances, intercontinental transport occurs as well and at certain locations, contribution of intercontinental transport may be significantly higher than from local, national, and regional transport.[†] UNEP's assessment also concluded that cadmium may similarly be transported on local, national, regional, or intercontinental scales.[‡] In 2006, following a full-day side event on heavy metals organized by Switzerland, the Intergovernmental Forum on Chemicals Safety adopted the Budapest Statement on Mercury, Lead, and Cadmium, which recognized that the risks from lead and cadmium need to be addressed by further global action.[§]

So far, measures to reduce emissions of lead and cadmium have been adopted at the regional level.[¶] However, it appears that in the light of the economic importance of lead and cadmium—it seems that especially those countries oppose a stringent and legally binding international response to the challenges posed by lead and cadmium who are involved in primary mining of these heave metals—developing an effective approach at the global level to address the challenges posed by these chemicals seems to be quite difficult. In contrast to the old POPS such as DDT or PCB, the economic interest in lead and cadmium is much bigger. Thus, it was so far not possible to reach consensus for a mandate to start negotiations of a global legally binding instrument on lead and cadmium and the UNEP Governing Council has so far only taken note of the key findings of UNEP's review of the scientific information on lead and cadmium, noted the effects on human health and the environment in Africa of the trade in products containing lead and cadmium, noted that further action is needed, and requested UNEP to continue to address remaining data and information gaps and to finish its scientific review of lead and cadmium.[**]

---

[*] C13 White Lead (Painting) Convention, 1921, coming into force 31.08.1923, ILO Geneva, http://www.ilo.org/ilolex/cgi-lex/convde.pl?C013

[†] UNEP, Draft final review of scientific information on lead, supra note 26. at 3-4 and 103–124.

[‡] UNEP, Draft final review of Scientific information on cadmium, supra note 28, at 3 and 99–122.

[§] Budapest Statement on Mercury, Lead and Cadmium, preamblar paragraph 6, available at: http://www.who.int/ifcs/documents/forums/forum5/final_report_no_pl.pdf

[¶] See e.g., the UN-ECE 1998 Aarhus Protocol on Heavy Metals, supra note 3.

[**] UNEP GC decision 25/5, §§ 9-§7, available at: http://www.unep.org/GC/GC25/Docs/GC25-DRAFTDECISION.pdf

TABLE 48.2
Possible Measures Addressing Lead and Cadmium

Products:

Ban of products containing lead/cadmium for which alternatives are available (e.g., lead in paints, cadmium pigments in paints, lead in ammunition for hunting; lead in fish sinkers, cadmium in plastic materials, e.g., stabilizers in PVC)

Establish threshold limits for lead/cadmium in products (lead and cadmium in batteries, lead in electronic components,* cadmium in fertilizers)

Processes:

BAT/BEP†

BAT/BEP for lead mining

Occupational health standards for lead mining ban certain intentional uses of lead/cadmium

Regulate processes that use of lead/cadmium; production of lead and cadmium from both primary and secondary raw materials; surface treatment of metals and plastics

Emission Standards; emission limits for lead (into air) for waste incineration and for installations for the production of lead accumulators.

Disposal:

Standards for disposal for lead and cadmium

Public awareness

Public information

Awareness raising campaigns (for the collection of batteries)

Developing additional scientific information about risks of heavy metals

Development of models to predict the global distribution of lead and cadmium via air and sea

Note: These measures can be implemented through compulsory mechanisms, through voluntary approaches such as partnership initiatives, or through a combination of compulsory and voluntary approaches.

Other Heavy Metals

Other heavy metals that may be of concern include Antimony, Arsenic, Gallium, and Indium.

Antimony: The primary use of antimony is flame retardation, followed by uses such as ammunition and lead batteries (Mathys et al., 2007). Antimony trioxide is a flame-retardant synergist used in plastics like PVC, rubber, and textiles. In the EU, Antimony trioxide is classified as carcinogenic category 3 with limited evidence of carcinogenic effects.

Arsenic: It is used as alloy in metallurgy for hardening metals, in electronics for semiconductors (galliumarsenide) but also as pesticide, herbicide, insecticide, and for wood preservation. Most forms of arsenic are toxic (Regulation, 2008). Arsenic is a famous poison. It was used to kill people in the middle ages and can be detected in the hair of poisoned victims. It is banned in biocidal products, and it is also regulated for drinking water standards.

TABLE 48.3
Possible Measures Addressing Other Heavy Metals, for example, Antimony, Arsenic, Gallium, and Indium

Products:

- Ban of products containing heavy metals for which alternatives are available
- Establish threshold limits for heavy metals in products

Processes:

- BAT/BEP
- Ban certain intentional uses of heavy metals
- Regulate processes that use heavy metals
- Emission Standards

Disposal:

- Standards for disposal

Public Awareness

- Public information

Developing additional scientific information about risks of heavy metals

- There are many information gaps for Antimony, Arsenic, Indium, and Gallium. It is desirable to close these gaps in an international context in order to avoid duplication and to use synergies. Scientific information could cover hazard assessment, substance flow analysis, risk assessment, test guidelines, socioeconomic analysis, and strategies for risk management.

Note: These measures can be implemented through compulsory mechanisms, through voluntary approaches such as partnership initiatives, or through a combination of compulsory and voluntary approaches.

Gallium: The most important application for gallium is the production of semiconductors in electronic industry (gallium arsenide and gallium nitride). There are only little data about the toxicity of gallium compounds.

Indium: The most important application of Indium, a by-product of zinc mining, is in the electronic industry for the production of liquid crystal displays and touch screens.

Like mercury, lead, and cadmium, the other heavy metals are persistent and accumulate, and may have negative impacts on human health and the environment. However, while at this stage it is not yet entirely clear whether they involve long-range transboundary atmospheric transportation, it is evident that they are transported all over the globe in products, especially in new applications in electronic goods. Over time, these products become waste and need special attention. Thus, it may well be that a closer look at heavy metals other than mercury, lead and cadmium may conclude that these chemicals are similarly substances of global concern that need a coordinated and cooperative international response.

Nanotechnology

Background

Nanoparticles are very small particles, sized between 1 and 100 nanometer (μm). Nanoparticles may exhibit size-related properties that differ significantly from those observed in fine particles or bulk materials. Owing to the new physicochemical properties of manufactured nanomaterials (MN), the particle size might be more important than other parameters such as persistence or bioaccumulation. The most important physicochemical properties are size and size distribution of free particles, specific surface area, stability in relevant media including the ability to aggregate and disaggregate, surface adsorption properties, water solubility, photoactivation and potential to generate active oxygen. When nanomaterials come into contact with a biological fluid they may become coated with proteins and other biomolecules, which may change the nanomaterials properties considerably.*

Nanotechnology is an enabling technology that is expected to result in major changes in many economic sectors from medicine to energy. It will contribute to the production of many novel materials, devices and products. Depending on the area of application under consideration there are different timelines for the beginning of industrial prototyping and nanotechnology commercialization. First generation products are already on the market in products such as paints, coatings and cosmetics, medical appliances and diagnostics tools, clothing, household appliances, food packaging, plastics, and fuel catalysts. More sophisticated products such as pharmaceuticals, diagnostics and applications in energy storage, and production are under development.†

In considering the commercial introduction of MN to achieve potential environmental benefits, countries should also give due consideration to potential health or environmental implications of such use of nanomaterials during their whole life cycle. This includes the potential effects of production of the nanoscale materials, as well as the disposition of nanomaterials that may, for example, require new hazard communication programs to recyclers or new concerns for disposal.

Human Health and Ecological Risks

The lungs are the primary target site for inhaled nanoparticles. Lungs have an enormous exposed area, and some inhaled and deposited nanoparticles can get into the bloodstream through the air–blood–tissue barrier. In addition to the lungs, the skin provides a potential uptake surface following dermal exposures (such as for cosmetics, sunscreens, and nanoparticles-impregnated clothing and in the workplace). Another exposure route is oral ingestion of nanomaterials. Once in the bloodstream, studies have shown that nanoparticles can be transported around the body and are

* European Commission, Directorate-General for Health & Consumers, SCENIHR (Scientific Committee on Emerging and Newly Identified Health Risks), Risk assessment of products of nanotechnologies, January 19, 2009. http://ec.europa.eu/health/ph_risk/risk_en.htm

† SAICM/ICCM.2/INF/34, March 25, 2009, Background information in relation to the emerging policy issue of nanotechnology and manufactured nanomaterials.

taken up by organs and tissues including the liver, spleen, kidneys, bone marrow, and heart.

Finally, a growing number of studies highlight the specific vulnerability of fetuses (through the pregnant mother) and infants to many types of toxics and chemicals, which may strongly impact their future health. More research into potential toxicity of nanoparticles on this vulnerable population should also be undertaken so as to avoid any disruption of this critical window of development.*

For some nanomaterials, toxic effects on environmental organisms have been demonstrated, as well as the potential to transfer across environmental species, indicating a potential for bioaccumulation in species at the end of that part of the food chain.† Concerning risk assessment of nanomaterials SCENIHR1 concludes in its recent opinion: "The highest risk, and thus concern, is considered to be associated with the presence or occurrence of free (non bound) insoluble nanoparticles either in a liquid dispersion or airborne dusts."

Occupational Safety and Health

The workplace is of key importance when considering human safety and health with respect to MN. There is the potential for relatively high exposure. According to our present limited knowledge, worker exposure to nanoparticles occurs primarily through handling nanoparticles in the making of products. In China seven young female workers were exposed to polyacrylate nanoparticles and organic solvents for 5–13 months and showed serious damage of their lungs (pleural effusion, pulmonary fibrosis, and granuloma) (Song et al., 2009).

Activities of International and Intergovernmental Organizations

In 2006, OECD established a Working Party on Manufactured Nanomaterials (WPMN) as a subsidiary body of its Chemicals Committee.‡ As of 2010, the following areas are covered by the work plan of the WPMN:

- Development of an OECD Database on Human Health and Environmental Safety (EHS) Research.
- EHS Research Strategies on MN (including Occupational Health and Safety).
- Safety Testing of a Representative Set of MN.
- MN and Test Guidelines.
- Cooperation on Voluntary Schemes and Regulatory Programmes.
- Cooperation on Risk Assessment.
- The Role of Alternative Methods in Nano Toxicology.
- Exposure Measurement and Exposure Mitigation.

---

* CEH 2009, The 3rd WHO International Conference on Children's Health and the Environment, June 7–10, 2009, Busan, Korea, Busan Pledge for Action on Children's Health and the Environment.

† SCENIHR (Scientific Committee on Emerging and Newly Identified Health Risks), Risk assessment of products of nanotechnologies, January 19, 2009.

‡ Organization for Economic Cooperation and Development (OECD) Working Party on Manufactured Nanomaterials (WPMN): http://www.oecd.org/env/nanosafety

In addition, in 2007 OECD's Committee for Scientific and Technological Policy established a Working Party on Nanotechnology (WPN).* The following projects are in the work plan of the WPN:

- Statistical framework for nanotechnology.
- Monitoring and benchmarking nanotechnology developments.
- Addressing challenges in the business environment specific to nanotechnology.
- Fostering nanotechnology to address global challenges.
- Fostering international scientific cooperation in nanotechnology.
- Policy roundtables on key policy issues related to nanotechnology.

ISO† has established Technical Committee 229–Nanotechnologies. Presently, the following four working groups have been established: Terminology and nomenclature; Measurement and characterization; Health, Safety, and Environmental Aspects of Nanotechnologies; and Material specifications. The following two documents have been published: ISO/TR 12885:2008 Nanotechnologies—Health and safety practices in occupational settings relevant to nanotechnologies; and: ISO/TS 27687:2008 Nanotechnologies—Terminology and definitions for nano-objects—Nanoparticle, nanofibre, and nanoplate. About 30 work items are currently in development by the four working groups.

COMEST (The World Commission on the Ethics of Scientific Knowledge and Technology) proposed at the 6th Ordinary Session in June 2009 in Kuala Lumpur a web-based platform to serve as a clearing house in order to improve the dialogue on nanotechnology between philosophy (ethics) and natural sciences.‡

A plenary session was held on nanotechnology and MN during the sixth session of the Intergovernmental Forum on Chemical Safety (IFCS Forum VI, September 15–19, 2008, Dakar, Senegal). The Forum VI adopted unanimously the Dakar Statement consisting of 21 recommendations for further actions.§

At Strategic Approach to Chemicals Management (SAICM) ICCM2 nanotechnologies and MN were treated as an emerging policy issue. Governments and Industry were requested to promote appropriate action to safeguard human health and the environment. The conference agreed to refer the issue to the contact group established to discuss emerging policy issues and to include adding the issue on nanotechnology and MN to the Global Plan of Action on the agenda for the third session of the Conference.¶ As a next step, several countries, including

* OECD Working Party on Nanotechnology, http://www.oecd.org/sti/nano

† International Organization for Standardization - ISO/Technical Committee 229 – Nanotechnologies, http://www.iso.org/iso/iso_technical_committee?commid=381983.

‡ The World Commission on the Ethics of Scientific Knowledge and Technology, Ethics of Nanotechnology: Status Review, Ref: SHS/EST/COMEST2009/pub-9.

§ Final report of IFCS Forum VI, Sixth Session of the Intergovernmental Forum on Chemical Safety, Dakar, Senegal, September 15–19, 2008, Dakar Statement on Manufactured Nanomaterials, http://www.who.int/ifcs/documents/forums/forum6/report/en/index.html

¶ SAICM/ICCM.2/15, Geneva May 11–15, 2009, Report of the International Conference on Chemicals Management on the work of its second session.

Switzerland and the United States, have agreed with UNITAR and the SAICM secretariat to launch a series of regional workshops with the aim to enhance information on and provide capacity building on risk assessment and risk management of nanotechnology, based on the work of the OECD. The workshops should catalyze processes at the national level to develop nanotechnology-related policies (see Table 48.4).

TABLE 48.4
Possible Measures Addressing Manufactured Nanomaterials

Products:

- Prohibit use of MN in certain products
- Establish threshold limits for MN in products

Processes: Creating regulatory framework conditions for the responsible handling of MN

- Measures of occupational health protection
- Voluntary and stewardship schemes, self-supervision
- Risk assessment schemes
- Provision of safety information to the processing industry
- Informing consumers about MN, product declaration
- Reporting obligation for products containing MN
- Prohibitions and use restrictions for selected MN (e.g., carbon nanotubes) in selected applications
- Specific workplace limits for MN
- Emission limits into air and water for MN
- Threshold values of MN for accident prevention for storage and transport

Disposal:

- Standards for disposal for different categories of nano waste. Distinction between dangerous and nondangerous nano waste.;
- Recommendations and/or regulations for the disposal of industrial waste and of products containing MN

Public Awareness

- Communication on the national, regional and global levels
- Dialogue platforms
- Technology assessment
- Public information

Developing additional scientific information about risks of MN: Creating scientific and methodological conditions to recognize and prevent possible harmful impacts of MN on health and the environment

- Increased support for independent risk research in nanotechnology
- Terminology, standards, methods of measuring and testing

Note: These measures can be implemented through compulsory mechanisms, through voluntary approaches such as partnership initiatives, or through a combination of compulsory and voluntary approaches.

Endocrine Disruptors

Endocrine disruptors (EDs) are hormonally active chemicals that change the natural circle and cascades of hormonal reactions. Typical examples of EDs are brominated flame retardants (BFRs) in plastic materials, certain UV Filters in sun screens, phthalates in plastic drinking bottles, but also the old POPS dioxins, PCB and a metabolite of DDT. Some of them are regulated in the Stockholm Convention.

There are EDs which are not regulated by the Stockholm Convention, because they do not fulfill the POPS criteria. Such an example is the sunscreen UV filter 4-Methylbenzylidene camphor. They are so-called stealth chemicals, because they are not visible "on the radar" of the normal toxicity tests.*

The presence of such hormonally active chemicals in the biosphere has become a worldwide environmental concern. After 1990, it was generally agreed that there was a lack of knowledge about these substances (Roger et al., 1999). However, there are strong indications that EDs may have harmful effects on the environment and human health (Carson, 2002). The mode of action of EDs can be receptor dependent or receptor independent. Hormonal effects can be activated or blocked. EDs can interfere with estrogens, androgens, and other hormones. EDs can interfere with the synthesis, metabolism, and transport of hormones.

Several national and regional research programs have been started mainly in industrial countries. The Swiss Federal Government decided in 2001 on the National Research Programme No. 50 (NRP 50) (Felix et al., 2008). Similar programs have been started in other countries and regions. OECD developed in the last decade within the mutual acceptance of data program (MAD) a new test guideline for EDs.† Thereby, different metabolic pathways with different types of hormones in men and animals have to be taken into account. Substances having endocrine disrupting properties may be subject to authorization and may be included in Annex XIV according to Articles 57 and 58 of the REACH Regulation (EC) No. 1907/2006 (see Table 48.5).

Perfluorinated Compounds

Perfluorinated compounds (PFCs) refer to a class of organofluorine compounds that contain at least one carbon chain moiety in which all carbons are saturated with fluorine atoms. PFCs exhibit unique properties similar to fluorocarbons. They make materials stain, oil, and water resistant. There is a large number and structural variety of PFCs. However, two families of compounds among the PFCs have been most intensively studied and recognized as hazardous environmental pollutants: perfluoroalkyl sulfonates (PFAS), for example, perfluorooctane sulfonate (PFOS), and perfluoroalkyl carboxylates (PFCAs), for example, and perfluorooctanoic acid (PFOA). PFCs also include—among many others—the substances perfluorooctanesulfonyl fluoride (POSF), perfluorooctanesulfonamide (PFOSA), and fluorotelomer alcohols (FTOH). PFCs are used in a wide range of industrial applications such as polymers, surfactants,

* http://www.nrp50.ch/final-products/final-programme-reports.html; NRP50, endocrine disruptors: Relevance to Humans animals and ecosystems. June 2008.

† OECD Guidelines for testing of chemicals; Full list of test guidelines; March 2009.

TABLE 48.5
Possible Measures Addressing Endocrine disruptors

Supply/Production of EDs:

- Import and export restrictions, for example, for DDT

Products:

- Ban of products containing EDs
- Establish threshold limits for EDs in products

Processes:

- Harmonized national, regional or global authorization procedures
- Regulate processes that use of EDs
- BAT/BEP
- Emission Standards

Disposal:

- Standards for disposal as special waste in selected cases

Public Awareness

- Public information

Developing additional scientific information about risks of EDs

- Identification of chemicals and groups of chemicals as potential EDs (e.g., UV filters)
- Monitoring of endocrine disruptor exposure in humans
- Monitoring of endocrine disruptor exposure in the environment. Identification of hot spots. Use of mass flow models as a basis for risk assessment.
- Monitoring of data of sperm quality of young men from various geographic regions
- Use of new assays for EDs:
- OECD test No. 440 Uterotrophic Bioassay in Rodents: A short-term screening test for estrogenic properties
- OECD 21-day fish androgenized female stickleback endocrine screening assay
- OECD Hershberger Assay (review package) to detect androgen agonists, antagonists and 5-alpha-reductase inhibitors*
- OECD Stably Transfected Transcriptional Activation (TA) Assay (review package) to detect estrogenic activity
- OECD 21 day Fish Endocrine Screening Assay

Note: These measures can be implemented through compulsory mechanisms, through voluntary approaches such as partnership initiatives, or through a combination of compulsory and voluntary approaches.

fire-fighting foams, hard metal plating, manufacturing of printed circuit boards, oil exploitation, textile and carpet coatings, hydraulic fluids and lubricants, and pesticides. Unfortunately the strength of the carbon–fluorine bond makes these chemicals extremely resistant to abiotic and biological degradation. The OECD hazard assessment concluded that PFOS is persistent in the environment and bioconcentrates

* Website of OECD Environment Directorate: Home/Chemicals testing/Guidelines/Endocrine Disruptor Testing and Assessment.

in fish and it is persistent, bioaccumulative, and toxic to mammalian species.[*] PFOS, PFOA, and some other PFCs have been detected globally in humans and biota (Calafat et al., 2007; Giesy and Kannan, 2001; Tao et al., 2006; Houde et al., 2006), rivers and oceans (Wei et al., 2007; Yamashita et al., 2008; McLachlan et al., 2007) and are estimated to be emitted to a large extent directly to water during manufacturing and use (Prevedouros et al., 2006; Paul et al., 2009). These substances have been measured even in remote areas and in polar fauna (Kellyn, 2007; Tao et al., 2006).[†] PFCs can have significant harmful effects on the environment and human health.[‡]

Risk management of PFOS and PFOA is a good example to show that different regions can tackle the same problem with different approaches: In 2002 OECD adopted an assessment of PFOS within their existing chemicals program and concluded that PFOS is a candidate for further work.[§] OECD continued their work with the help of questionnaires among member countries and producing industry and published in 2005 and 2006 comprehensive lists of congeners.[¶] In January 2006, US EPA invited eight major companies in the fluoropolymer and fluorotelomer industries to commit to a voluntary program with global goals: 95% emission reduction to all media and product content of PFOA (perfluorooctanoic acid) by 2010 compared to 2000 baseline, elimination of PFOA and PFOA precursors by 2015.[**] The program is very efficient, since it covers the major producing companies. The European Union regulated PFOS in December 2006 with a ban with specific exemptions such as for photoresistant or antireflective coatings for photolithography processes, photographic coatings applied to films, papers, or printing plates, and mist suppressants for non-decorative hard chromium (VI) plating and wetting agents for use in controlled electroplating systems, and hydraulic fluids for aviation.[††]

The Conference of the Parties to the Stockholm POPS Convention decided in May 2009 to add PFOS, its salts and PFOSF to Annex B of the Convention. The parties are invited to provide relevant information of these substances in articles and in processes by July 2010.[‡‡] Listing PFOs, its salts and PFOSF in the POPS-list of the

---

[*] Hazard Assessment of Perfluorooctane sulfonat (PFOS) and its salts. Organisation for Economic Co-operation and Development (OECD), Environment Directorate, Joint Meeting of the chemicals committee and the working party on chemicals, pesticides and biotechnology, ENV/JM/RD(2002)17/FINAL, November 21, 2002.available at http://www.oecd.org/dataoecd/23/18/2382880.pdf

[†] Schadstoffe für die Ewigkeit: Perfluorierte Tenside – ein globales Umweltproblem; NZZ Online, 27.09.2006.

[‡] http://www.oecd.org/document/58/0,3343,en_2649_34375_2384378_1_1_1_1,00&&en-USS_01DBC.html#3

[§] Co-operation on existing chemicals; draft assessment of perfluorooctane sulfonate (PFOS) an its salts: conclusions and recommendations, ENV/JM(2002)29 September 16, 2002; 34th Joint Meeting of the chemicals committee and the working party on chemicals, pesticides and biotechnology.

[¶] List of PFOS, PFAS and other related compounds, OECD ENV/JM/RD(2005)7, amended in OECD ENV/JM/RD(2006)8.

[**] http://www.epa.gov/oppt/pfoa/pubs/stewardship/index.html

[††] Directive 2006/122/EC of the European Parliament and the Council of December 12, 2006 amending for the 30th time Council Directive 76/769/EEC on the approximation of the laws, regulations and administrative provisions of the Member States relating to restrictions on the marketing and use of certain dangerous substances and preparations (perfluorooctane sulfonates); OJ L 372, 27.12.2006, pp. 32–34.

[‡‡] UNEP SAICM/ICCM.2/15, May 27, 2009, Report of the International Conference on Chemicals Management on the work of its second session.

TABLE 48.6
Possible Measures Addressing PFCs and their Substitutes

Products:

- Ban of products containing PFOS, such as fire fighting foams
- Establish threshold limits for certain PFCs, for example, PFOS and PFOA in products

Processes:

BAT/BEP

Ban certain uses of PFCs, with the exception of essential uses such as photolithography processes, mist suppressants for plating and certain medical devices

Regulate processes that use of PFCs

Emission Standards into water, emission standards in drinking water

Disposal:

Standards for disposal as special waste

Public awareness

Public information

Developing additional scientific information about risks of PFCs

Monitoring of PFCs in environmental compartments and biota

Monitoring of PFCs in food and drinking water

Monitoring of PFCs human blood and mother's milk

Application of global exposure models to predict transport and accumulation

Note: These measures can be implemented through compulsory mechanisms, through voluntary approaches such as partnership initiatives, or through a combination of compulsory and voluntary approaches.

Stockholm Convention is a significant first step toward addressing PFCs at the global level. However, the current listing includes still many exemptions. Moreover, there are other PFCs that need international attention. And there is concern that whenever one perfluorinated compound that may cause a risk to human health or environment is regulated, substitutes will be developed and used that may be equally harmful for the environment or human health. Thus, it seems that PFCs will remain upcoming emerging substances that will require the attention of international decision makers (see Table 48.6).

POLICY-RELATED EMERGING ISSUES: EMERGING POLICIES IN THE GLOBAL CHEMICALS REGIME

As indicated, the development of international chemicals and waste policy has not only involved addressing new and emerging challenges as described in the previous section of this chapter; it has also been stimulated by the emergence of new policies. This chapter will focus on three new policy developments that will probably strongly shape the further evolution of the international chemicals and waste regime: the increasing involvement of nonstate actors in policy development and implementation, the shift from specific toward integrated policy approaches, and effort to increase efficiency and effectivity through synergy, coordination, and cooperation within the international chemicals and waste regime.

Private Sector/NGO Involvement and PPPs

Like general international environmental law and policy, international chemicals management policy and cooperation have traditionally focused on states and governments as main actors and subjects. And similar to the evolution within broader international environmental policy and law,* private actors became increasingly important also in the area of international chemicals and waste management. This stronger and more direct involvement of private actors not only as "object" but also as "subject" of international chemicals policy reflects the fact that they are directly responsible for taking and implementing the relevant decisions with regard to concrete chemicals management on the ground and that they have much of the expertise relevant for sound policy development.

There are several forms of how private actors are involved in international chemicals management policy. Thus, nonstate actors are fulfilling increasingly important roles as engines of international chemicals policy-making, setting agendas for international policy processes and development, providing knowledge and scientific information, monitoring implementation and lobbying state actors, and they are also involved in partnership initiatives with governments, as well as the development of voluntary standards, and implementation of environment and development programs.† This section will focus on the involvement of private actors (a) as participants in the development of new policies by governments, (b) as partners in PPPs, or (c) as actors who directly develop and implement sound chemicals management policies.

Private Actor Involvement in Policy Development

While traditionally international rules and policies were negotiated between governments, nonstate actors have always played a more prominent role in the area of environment. Today, NGOs, including industry representatives, regularly participate as observers at meetings of international environmental institutions and processes and make effective use of their right to intervene and submit their views and proposals.‡ However, in the context of chemicals management, this involvement of private actors in the development of policies has been significantly further developed beyond the normal "observer" status of NGOs: The Intergovernmental Forum on Chemicals Safety has always allowed nonstate actors to participate in its deliberation with equal rights like state actors. This has allowed a fruitful, open and very constructive direct dialogue and facilitated mutual understanding. Building on this positive experience,

* See, for example, Peter H. Sand (2007), describes this evolution as a process from a traditional era of coexistence towards an era of cooperation, then a modern era and finally a postmodern era.

† See generally: Franz Perrez (2008)

‡ For a critical view of traditional stake-holder involvement, see Franz Xaver Perrez (2009a).

it was decided to develop SAICM in a similarly participatory process.* Thus, nonstate actors could participate fully in the negotiations of SAICM and attend as equal partners in both the formal sessions and the informal negotiations, and nonstate actor involvement became one of the key characteristics of SAICM.† As a consequence, more than 60 NGOs from the agriculture, development, environment, health, industry, and labor sectors participated actively in the negotiations of SAICM with the full right to take the floor, express their views, and make concrete proposals (Gubb and Younes, 2006). This opportunity was used in a very active and constructive manner throughout the process and it can fairly be said that "SAICM is the result of intensive stakeholder participation including national governments, international organizations, trade unions, public-interest NGOs, industry, and other civil society organizations who share responsibility for its implementation."‡

SAICM clearly succeeded in bringing in the specific knowledge and expertise of the nonstate actors involved in chemicals management and in developing ownership by nonstate actors for both the process and the result. One could even argue SAICM set a precedent for leaving the traditional interstate paradigm and moving toward a collective concern or community interest approach.§

Public–Private Partnerships

At the international level, the use of PPPs is a relatively new phenomenon. Generally, PPPs are understood as formalized cooperation between public and private entities, where one or several government and one or several private actors form a nonhierarchical "partnership" or "joint venture" with the objective to implement a specific policy function.¶ PPPs are commended for enabling flexible, targeted, rapid, efficient, and effective concrete action on the ground. By involving nonstate actors, they engage additional expertise and resources. This can increase effectivity, ownership, and legitimacy.** At the same time, they still trigger political, institutional, and democracy concerns.†† And while PPPs are praised for attracting new and innovative funding support, they are in most of the cases not able to ensure sustainable financing.‡‡ Finally, PPP are also criticized for being primarily a way to evade responsibility of and real commitment by governments (Piest, 2003).§§ While PPPs have a

* See UNEP GC decision VIII/3, operative para. 6; UNEP GC decision 22/4, preambular para. 2, operative paras. 3 and 5; and UNEP GC decision 23/9, preambular para. 9 and operative para. 6.

† See generally: Franz Xaver Perrez (2006).

‡ Center for International Environmental Law (CIEL), The Strategic Approach on International Chemicals Management (undated), available at: http://www.ciel.org/Chemicals/Dubai_SAICM_Feb06.html.

§ Perrez, supra note 57, at 10. See generally: J. Brunnée.

¶ See generally: Börzel and Risse (2005), Karlaganis (2008), Linder and Rosenau (2000).

** See also: Karlaganis, supra note 84, at 47, referring to Börzel and Risse, supra note 84, at 195.

†† See generally: Franz Xaver Perrez (2009b).

‡‡ Id., with further references.

§§ Id., at 4 with further references. See also: Najam, A., M. Papa, and N. Taiyab, Global Environmental Governance: A Reform Agenda (Winnipeg: International Institute for Sustainable Development, 2006), at 67; Centre for Science and Environment (CSE), WSSD Turned Into Partnership Market (Press Release 31.8.2002), available at: http://www.cseindia.org/html/eyou/geg/press_20020831_1.htm

long-standing tradition at the national level at least in industrialized countries, PPPs have entered the international sphere only recently.[*] Namely, by offering a formal framework to launch PPPs to implement *Agenda 21* and the Johannesburg Plan of Implementation, the 2002 WSSD has created a strong momentum for PPPs in environmental policy.[†]

Partnerships have also gained popularity in the international chemicals management context. The Basel Convention was one of the first conventions to launch partnerships as an effective tool to catalyze the implementation of existing international instruments and goals. The Basel Declaration on Environmentally Sound Management,[‡] adopted in 1999, and the Strategic Plan of the Convention,[§] adopted in 2002, called for the development of PPPs between governments, industries and other NGOs to ensure practical application of environmentally sound management. Based on the initiative of Switzerland, the COP of the Basel Convention decided in 2002 to establish a Mobile Phone Partnership Initiative (MPPI).[¶] The MPPI was able to reach—in close cooperation with the world's most important mobile phone manufacturers—three targets: technically, it developed five guidelines and one overall guidance document which was adopted by the COP to the Basel Convention; institutionally, it demonstrated that PPPs can be a powerful and effective tool also within the framework of a Convention not only to implement specific activities but also to develop policies; and politically, it was an important incentive to bring the emerging problems of electronic scrap on the international—and in several countries also the national—political environment agenda.[**]

PPPs were not only initiated as instrument to strengthen implementation of existing instruments, they were also promoted as alternatives to legally binding approaches. For example, the United States, supported by Australia, Canada, China, and India, tried to avoid a legally binding framework on mercury by promoting a voluntary partnership approach.[††] However, it seems that experience shows that while PPPs do have the benefit of allowing for rapid concrete action on the ground, they benefit strongly from the presence of an overarching legally binding policy framework and that combining PPPs with a broader strong multilateral framework could help to address potential risks and shortcomings of purely voluntary initiatives.[‡‡] First, a broader framework could ensure that implementation is not isolated and *ad hoc* but part of a broader policy and comprehensive and global in its coverage. Second, the

[*] Karlaganis, supra note 63, at 46–47; Perrez, supra note 65, with further references.

[†] See Najam e.a., supra note 67, at 50; Lee Kimball, Franz Xaver Perrez, and Jacob Werksman, The Results of the World Summit on Sustainable Development: Targets, Institutions, and Trade Implications, in 13 Yearbook of International Environmental Law 3 (2002), at 16.

[‡] Available at: http://www.basel.int/meetings/cop/cop5/ministerfinal.pdf

[§] Strategic Plan for the implementation of the Basel Convention (to 2010), available at: http://www.basel.int/meetings/cop/cop6/StPlan.pdf

[¶] See Decision VI/31 of the Conference of Parties to the Basel Convention, available at: http://www.basel.int/meetings/cop/cop6/english/Report40e.pdf#vi31

[**] Basel Convention, Information Brief on the Mobile Phone Partnership Initiative (2007), available at: http://www.basel.int/industry/mppi/overview.doc

[††] Final Report of the Open-Ended Working Group, supra note 20, Annex II.V, Submission by the US.

[‡‡] Swiss Federal Office for the Environment, Experience of Switzerland with mercury Partnerships (Conference Room Paper for the Ad hoc Open-ended Working Group on Mercury of 13 November 2007), UNEP Document UNEP(DTIE)/Hg/OWEG.2/CRP.6, at 4. See also Perrez, supra note 65, at 10.

existence of a binding framework helps to ensure that PPPs are not a tool for evasion but part of a broader commitment and also help to ensure the political, financial, and institutional sustainability of partnership activities. Finally, a legally binding framework is able to provide an adequate institutional context for review, assessment, and further development of concrete activities. Nevertheless, PPPs will certainly continue to play an increasingly important role in international chemicals management within the existing and new legally binding frameworks.

Direct Private Sector Activities

Finally, private actors are also directly involved in developing and implementing international chemical policies. As discussed in separate chapters in this volume, both public interest and business NGOs are strongly engaged in such private activities.[*] This direct involvement is crucial, it does not only reflect the responsibility of private actors for sound chemicals management, it also is a proof of "good citizenship."

Integrated Approaches to Chemicals Management: Life-Cycle Approach, Integrated Chemicals Management and Chemicals Leasing

While traditional international chemical policy has addressed the challenges posed by chemicals for a long time by looking at the risks posed by a specific chemical in specific circumstances, over time the recognition has emerged that chemicals policy, in order to be effective and efficient, has to take a broader and a more integrated approach.[†] Moving in the chemicals context toward more integrated policies reflects not only the better understanding of chemicals and their interaction with each other and with the environment; it also reflects the general evolution of environmental policy and law from isolated *ad hoc* approaches toward comprehensive approaches that take into consideration the broader context of a specific challenge and that try to reflect the complex interdependencies in the natural environment.[‡]

By looking at chemicals throughout their entire life span and not only at the very moment when they become a risk to human health and the environment, the "life-cycle"—or "cradle-to-cradle" or "cradle-to-grave"—approach is the broadest reflection of such an integrated approach to chemicals policy making. Other integrated approaches to chemicals management include the concept of "Integrated Chemicals Management" developed by UNITAR and the Concept of "Chemicals Leasing" developed by UNIDO and the Austrian Ministry for the Environment.

Life-Cycle Approach

The "paradigm change" toward more integrated life-cycle policies in chemicals management can probably best be illustrated with the Basel Convention of 1989:[§] While the Convention as such covers a broad approach to waste management, it does

---

[*] See generally the contribution by IPEN (see Section V, Chapter 25) and ICCA (see Section V, Chapter 20) in this volume.

[†] See generally: Katharina Kummer Peiry (2009).

[‡] See generally: Franz Xaver Perrez (2000).

[§] Id., at 4.1.

not yet mention the term "life-cycle" at all. In the first period after the adoption of the Convention, the discussion focused mainly on the control and ban of transboundary movement of hazardous wastes. Only in 1999, at the Convention's 10th anniversary, the focus of the work under the Convention has really broadened toward preventing, minimizing, recycling, and recovery of hazardous wastes and their environmentally sound management, and three years later, the Strategic Plan of Implementation of the Basel Convention of 2002 specifically refers to an "integrated life-cycle management approach" and "life-cycle assessment."*

The *Agenda 21*, which was adopted three years after the Basel Convention at the Rio Conference on Environment and Development in 1992, explicitly requested that risk-reduction policies have to take into account the entire life cycle of chemicals, that a life-cycle analysis is needed and that a life-cycle approach should be taken to chemicals management.† Subsequent chemicals-related multilateral environmental agreements (MEAs) reflect the life-cycle approach: The Rotterdam Convention of 1998 requests exporting Parties to advise and assist importing parties to strengthen their capacities and capabilities to mange the chemicals listed in the Convention safely during their life-cycle and it calls Parties with more advanced programs to provide technical assistance to other Parties for this purpose.‡ And, the Stockholm Convention of 2001 stresses the need to take measures to prevent adverse effects caused by persistent organic pollutants at all stages of their life cycle.§ The WSSD renewed the general commitment to sound management of chemicals throughout their life cycle.¶

SAICM, adopted 2006, which includes over 30 references to the life-cycle approach, has definitively established the concept in international chemicals policy: the Head of States committed themselves to promoting the sound management of chemicals and hazardous wastes throughout their life cycle, to mobilize financial resources for the life-cycle management of chemicals, and to facilitate access to information on chemicals throughout their life-cycle.** SAICM's Overarching Policy Strategy, which covers chemicals at all stages of their life cycle, further concretizes the goal to "achieve sound management of chemicals throughout their life cycle" and underlines the necessity of taking a life-cycle approach in the context of risk reduction, risk assessment, knowledge and information, governance, and capacity building.†† And SAICM's Global Plan of Action contains a specific work area and 10 specific activities on the life-cycle approach.‡‡

* See generally: Basel Declaration on Environmentally Sound Management, supra note 70, § 6(a); Strategic Plan of Implementation of the Basel Convention, supra note 71. See also Kummer, supra note 77, at 4.1.

† *Agenda 21*, available at: http://www.un.org/esa/dsd/agenda21, §§ 19.44, 19.48 and 19.49.

‡ Stockholm PIC Convention, available at: http://www.pic.int, Art. 11.1(c)(ii) and Art. 16.

§ Stockholm POPS Convention, available at: http://www.pops.int, preambular paragraph 16.

¶ WSSD JPOI, supra note 4, Chapeau § 23.,

** Dubai Declaration Declaration on International Chemicals Management, available at: www.saicm.org/index.php?menuid=3&pageid=187, §§ 11, 16 and 21.

†† Overarching Policy Strategy (OPS), available at: http://www.saicm.org/index.php?menuid=3&pageid=187, §§ 3, 7, 13, 14(a) and (c), 15(a) and (b), 16(a) and 17(a).

‡‡ Global Plan of Action (GPA), available at: http://www.saicm.org/index.php?menuid=3&pageid=187, activities 23, 67, 71, 119–123, 190 and 229.

While it seems to be obvious that addressing chemicals in a manner that reflects and takes into account their whole life cycle is the best manner for developing effective, efficient, and sustainable policies, several reasons may explain why it took the international community so long to firmly accept this concept: Looking at chemicals in a holistic manner over their whole life span is more complex and needs more sophistication than looking at it at a specific moment in a specific situation. It may involve more uncertainties and knowledge gaps and makes information gathering more complicated. Moreover, addressing a more complex situation also requires more complex and mutually interdependent policies. This has also a strong impact on implementation and may challenge some of the institutional arrangements and competences that have emerged under the traditional *ad hoc* approach. This explains why SAICM and the Rotterdam Convention also explicitly refer to capacity building in the context of taking a life-cycle approach. However, today the concept is well accepted and used and also promoted by private actors.[*]

Integrated Chemicals Management/Integrated National Programmes

UNITAR, whose chemicals and waste program has the mandate to support governments and stakeholders to strengthen their institutional, technical and legal infrastructure, and capacity for sound chemicals management, has been a key catalyst in promoting an integrated life-cycle approach to chemicals management.[†] In the late 1990, UNITAR had developed, with the financial support of Switzerland, the concept of Integrated Chemicals Management/Integrated National Programmes. This concept, building on a life-cycle perspective and contributing to the implementation of SAICM, reflects the fact that in order to take an integrated and comprehensive approach to chemicals, national policies and institutions have similarly to be developed in a comprehensive, coordinated, and integrated manner. UNITAR's Integrated National Programmes thus help to establish and strengthen a collaborative framework at the national level to "provide a foundation for effective and coordinated action to address both national chemicals and waste management priorities as well as the implementation of international chemicals and wastes-related agreements and initiatives."[‡] Such a national coordinating framework facilitates interministerial coordination, access to and exchange of information, stakeholder participation, coordinated priority setting, and integration of chemicals management activities into national development planning processes.[§]

Chemicals Leasing

Chemicals leasing is another emerging policy of a more integrated approach to chemicals management. The concept has been developed by UNIDO in cooperation with the Austrian Ministry of the Environment with the goal to reduce ineffective and inefficient use of chemicals.[¶] The key element of chemicals leasing is a shift in paradigm away from selling chemicals as a normal product—an approach that

---

* See, for example, http://www.chemlifecycle.com
† See also Chapter 32 of this book.
‡ See generally: http://www.unitar.org/cwm/saicm
§ See http://unitar.org/cwm/inp
¶ See generally: Section V, Chapter 31 of this book on UNIDO.

focuses on increasing sales volume of chemicals—toward "a more service-oriented and value-added approach. The producer no longer sells the chemical but the associated merit and know-how. This relates to conditions of use, recycling concepts, and disposal. In addition, while in the traditional model the responsibility of the producer stops when the chemical is sold, in the current approach, the producer remains responsible throughout the use and treatment, disposal, and recycling phases."*

Synergies and the Emergence of an International Chemicals and Waste Cluster

The evolution from *ad hoc* solutions toward integrated policy responses is also reflected at the institutional level. Today's international environmental regime is generally characterized by a proliferation of institutions and processes.† Several hundreds of MEAs have been negotiated over time in a fragmented and *ad hoc* manner, each addressing specific environmental problems and each having its own "mini-institutional machinery," including a COP, a secretariat, and technical and legal subsidiary bodies (French, 2002; Churchill and Ulfstein, 2000).‡ The international chemicals and waste regime suffers from a similar *ad hoc* approach according to which, for each problem that has been identified, new institutions and instruments were developed.

Thus, there is broad recognition that, building on the existing instruments and respecting the need for the legal autonomy of the different treaties, there is a need for enhanced coordination and cooperation and a more coherent institutional framework and integrated structures.§ This need for enhanced synergies and linkages between MEAs was specifically recognized for the chemicals and waste context.¶

The idea of clustering MEAs was promoted at the political level already during the international environmental Governance (IEG) process: Switzerland strongly supported the goal to improve coordination among and effectiveness of MEAs, promoted the concept of clustering-related MEAs as an important tool to enhance synergies, linkages, coordination and cooperation, and called for a structural and organizational integration of related institutions as well as their geographic colocation where appropriate (Perrez, 2001). Based on the Swiss proposals, the Cartagena recommendations on strengthening IEG included several explicit references to the desirability of clustering related MEAs and to the further strengthening of the chemicals and waste cluster.** Namely, it referred specifically to chemicals in the context of requiring improved coherence in international environmental policy making,†† it

---

* See http://www.unido.org/index.php?id=5063

† See generally: Perrez and Ziegerer (2008).

‡ Perrez and Ziegerer, supra note 94, at 253–254. For an indepth analysis of the role of autonomous institutional arrangements see Robin R. Churchill and Geir Ulfstein, 2000.

§ General Assembly Resolution A/RES/60/1 of 16 September 2005, § 169, last bullet; UNEP decision SS.VII/1 on International environmental governance and Report of the open-ended Intergovernmental Group of Ministers or their representatives on international environmental governance (IEG Report), §§ 26-30, Appendix to UNEP decision SS.VII/1, available at: http://www.unep.org/gc/GCSS-VII

¶ IEG Report, supra note 96, § 27.

** IEG Report, supra note 96, §§ 8(n) and 27.

†† UNEP GC Decision SS.VII/1 on International Environmental Governance (15.02.2002), §§ 11(h)(ii).

called for improved coordination among and effectiveness of MEAs,* and it suggested enhancing synergies and linkages between MEAs with comparable areas of focus and "enhancing collaboration among MEA secretariats in specific areas where common issues arise, such as ... chemicals and waste."†

After the adoption of the Cartagena IEG decision, Switzerland successfully lobbied for the effective implementation of this decision. Within the process to develop a SAICM, Switzerland made several proposals on further concretizing and operationalizing the decision to develop an international chemicals and waste cluster.‡ Thus, SAICM reflects the need to improve synergies between the chemicals-related international instruments and processes, stresses the general determination to strengthen coherence and synergies in the international chemicals regime, and more specifically calls for increased cooperation and synergies in implementation of MEAs, considering developing common structures between the chemicals and waste conventions and organizing SAICM meetings back-to-back with meetings of the governing bodies of other relevant bodies.§

Underlining the necessity to enhance synergies, efficiencies, and effectiveness in the international chemicals and waste cluster, Switzerland further offered to colocate the secretariats of the Rotterdam PIC and Stockholm POPS Conventions within the emerging chemicals and waste cluster in Geneva (Perrez, 2003). After the successful colocation of the secretariats of the new chemicals conventions in Geneva, Switzerland called for a further integration of the secretariats of the Basel, Rotterdam, and Stockholm Conventions and convened an informal meeting in 2006 to present its idea about joint management for the three convention secretariats. The same year, Switzerland, supported by Norway and Senegal, presented, at the COP 2 of the POPS Convention a draft decision calling for a joint head of the three convention secretariats.¶ While the proposal for a joint head was not accepted by the COP, the Stockholm, Rotterdam, and Basel Conventions established subsequently a joint working group to explore further possibilities to enhance synergies between the three chemicals and waste conventions. The Joint Working Group concluded its work in March 2008 and submitted a comprehensive package of measures to enhance synergies and cooperation between the three conventions.** These measures were adopted by the Conferences

* Id., § 26–30.

† Id., § 27.

‡ See, for example, Swiss Submission to the SAICM Process of 18.7.2003, p. 2, available at: http://www.chem.unep.ch/saicm/SAICM/resprecvd/gov/SWISS.pdf

§ Dubai Declaration, supra note 85, § 8; OPS, supra note 86, §§ 5, 6(c) and 25; GPA, supra note 87, activities 170, 171 and 173. See generally: Perrez, supra note 60 at 247, 256 and 257.

¶ Draft Decision on Synergies submitted by Switzerland at POPS COP 2, on file with the author. See generally: International Institute for Sustainable Development, Earth Negotiation Bulletin on POPS COP 2, volume 15 No. 135, available at: http://www.iisd.ca/download/pdf/enb15135e.pdf, pp. 9 and 11.

** Report of the Ad hoc Joint Working Group on Enhancing Cooperation and Coordination Among the Basel, Rotterdam, and Stockholm Conventions on the work of its third meeting, Document UNEP/FAO/CHW/RC/POPS/JWG.3/3, available at: http://ahjwg.chem.unep.ch/index.php?option=com_content&task=section&id=16&Itemid=63. See also the Submission by Switzerland and Nigeria on joint managerial functions including joint head of secretariat, Document UNEP/FAO/CHW/RC/POPS/JWG.2/INF/8, available at: http://ahjwg.chem.unep.ch/documents/2ndmeeting/ahjwg02_inf08.pdf. See generally: http://ahjwg.chem.unep.ch/

of the Parties of all three conventions in 2008 and 2009. Thus, the decisions call for enhanced coordination at the national level, programmatic cooperation in the field, and coordinated use of regional offices and centers. They request the three convention secretariats to synchronize and streamline reporting; to strengthen cooperation on technical and scientific issues; to develop a common approach to awareness raising and information exchange; to act jointly in participating in other related processes; to establish, on a provisional basis, joint legal, joint financial, joint resource mobilization, joint information, and joint IT services; and to prepare proposals on possibilities for enhancing coordination on noncompliance once noncompliance mechanisms have been established for all three conventions. Moreover, the decisions request the UNEP ED to explore and assess the feasibility and cost implications of establishing joint coordination or a joint head of the secretariats of the Basel, Rotterdam, and Stockholm Conventions. Finally, they decide to convene a simultaneous extraordinary COP of the three conventions.*

The convening of three independent treaty conferences simultaneously in February 2010 was an unprecedented step and marked a historic departure for international environmental governance.† The simultaneous extraordinary COP have not only confirmed the decisions that have been adopted previously separately by the three conventions and adopted the joint secretariat services on a permanent basis, they have also decided to establish a joint head for the three convention secretariats. Moreover, they have agreed to review the new synergetic institutional arrangement in 2013. This will allow addressing remaining gaps and institutional weaknesses and to further deepen and broaden synergies between the three conventions. Through this, the Parties to the three conventions have adopted "a ground breaking framework for the achievement of coordination and cooperation at all levels between the legally separate instruments."‡ Not only does it reflect the need for enhanced coordination and synergies in the chemicals policy, at the same time—and for the first time in international environmental policy—three independent MEAs start to move together in a very concrete manner to form an international chemicals and waste cluster.

This process of strengthening coordination, cooperation, and synergies in the international chemicals and waste cluster will continue and further deepen. Thus, it will be interesting to see how the new convention on mercury will build on this synergy process and how it will be integrated in the existing joint structures. Moreover, it may well be that the growing recognition that it is inefficient and ineffective to establish legal and institutional frameworks with a narrow focus on a limited number of chemicals will further stimulate the upcoming negotiations on mercury.

* Decision IX/10 of the Basel Convention of 27.06.2008; Decision 4/11 of the Rotterdam Convention of 31.10.2008, and Decision 4/34 of the Stockholm Convention of 09.05.2009.

† See generally:. http://excops.unep.ch/

‡ Press Release of the three convention secretariats,. Bali, February 25, 2010, available at: http://excops.unep.ch/

THE FUTURE INTERNATIONAL CHEMICALS AND WASTE REGIME: PROVIDING A COMPREHENSIVE, COHERENT, EFFECTIVE, AND EFFICIENT FRAMEWORK FOR NEW AND EMERGING CHALLENGES

The second part of this chapter has described some of the substances and chemicals that emerge as new challenges which may have to be addressed by the international chemicals regime. Traditionally, policy responses to new and emerging problems have been issue specific, *ad hoc*, and without a broader overview and coherent strategy (French, 2002; Sands, 2003).* Over time, an international environmental regime has emerged that is characterized by a proliferation and fragmentation of frameworks, rules, and institutions (Weiss, 1993; Ivanova, 2010; Najam et al., 2006).† This is leading to a dilution of authority and competence; a lack of coordination, cooperation, and synergies among relevant international actors; to duplications, overlaps, inefficiencies, turf battles, inconsistencies, contradictions, and conflicts; to a lack of an overarching vision, of a common orientation and strategy, and of coherence and focus; and to a situation where limited financial resources are not always used efficiently.‡ Such a fragmented *ad hoc* approach makes it also difficult to ensure rapid policy responses to newly emerging challenges in a comprehensive and synergetic manner. Moreover, a fragmented approach makes it also more difficult to develop an adequate mechanism for financial support and capacity building. And, if the traditional issue-specific *ad hoc* approach is maintained instead of using an existing successful framework for similar new problems, new independent costly institutional and legal machineries will have to be developed.

Chemistry, the knowledge of potential new applications of chemicals, the use of chemicals in industry, production and in our daily live and thus the emergence of new challenges posed by chemicals to international policy are not static but evolving in a dynamic manner. In the light of the rapid evolution of the chemical reality and the dynamic progresses in the use of chemicals, the list of challenges that need international attention can never be closed. Therefore, a comprehensive, coherent, effective, and efficient framework is needed, that is able to address new and emerging chemicals-related problems (Roch and Perrez, 2005).§ The framework has to be dynamic as well, be able to respond to new challenges without reinventing the wheel each time and without building up unnecessary parallel structures and duplications. Developing full-fledged machineries for each chemical substance that emerges into the public attention is not a meaningful way forward.

* See generally: Franz Xaver Perrez (2000), *Cooperative Sovereignty: From Independence to Interdependence in the Structure of International Environmental Law*, The Hague: Kluwer Law International, at 272–277 with further references.

† Perrez and Ziegerer, supra note 94, at 254–255. See also International Law Commission, "Fragmentation of international law: difficulties arising from the diversification and expansion of international law" (Report of the Study Group of the International Law Commission; finalized by Martti Koskenniemi, April 13, 2006) A/CN.4/L.682.

‡ See, for example, Maria Ivanova and Jennifer Roy (2007); A. Najam, M. Papa, and N. Taiyab, (2006); Perrez and Ziegerer, supra note 94, at 254–258.

§ On the concept of a comprehensive, coherent, effective and efficient international regime ("double c/double e" approach), see generally: Franz Xaver Perrez (2001).

This chapter has described some of the most visible substance-related emerging topics that will require policy responses. Thereby, it is interesting to note that the possible measures that could be needed to address a new chemical substance that emerges into the public attention remain similar if not the same. In fact, the different new and emerging substances require regularly a similar set of measures such as bans with exceptions for specific products respectively for the use of a substance in certain products; prescription of thresholds for a substance in certain products; ban or control of trade; ban or control of certain uses in processes and process standards such as emission standards, best available technologies and best environmental practices (BAT/BEP); public information; and further research. In addition, the provision of adequate and effective financial, technological, and capacity support will be crucial for the success of any of these measures. Moreover, new emerging policy instruments identified in this chapter such as private actor involvement and taking integrated policy approaches are similarly important tools for addressing substances of global concern.

While measures that may be needed can be categorized into clusters or types of measures, the substances can similarly be categorized according to different criteria that trigger the need for international action:

- A first category includes persistent organic pollutants, that is, organic substances that persist in the environment and pose a risk to the environment and human health.
- Heavy metals—pollutants that are similarly persistent and toxic but inorganic—may be a second group of substances requiring international response.
- A third category could be the other chemicals of global concern where the risk is linked not to their persistence but to their size, human toxicity, or eco-toxicity. This category would include substances such as nanomaterials or nonpersistent EDs.

These three substances-based clusters could be complemented with two clusters addressing specific moments in the life of dangerous substances that require specific policy responses:

- The moment when a hazardous substance is traded internationally.
- Activities linked to its disposal.

A closer look at today's international chemicals and waste regime shows that it has emerged along the lines of these clusters: The Stockholm Convention deals with persistent organic pollutants; the UNEP Governing Council has agreed that a new convention will be developed to address the challenges posed by the heavy metal mercury; the international community has also started to look at nanomaterials; and nonpersistent EDs are similarly emerging into the international policy focus. Moreover, the Rotterdam Convention regulates the international trade of certain hazardous chemicals; and the Basel Convention addresses hazardous wastes. Gradually, a comprehensive regime is emerging.

However, as indicated, it is important that this regime emerges and further grows in an efficient manner that avoids fragmentation and ensures a coherent overview. As discussed previously,* the international chemicals regime was able to develop a new policy approach to address this challenge of fragmentation by launching a process to enhance coordination, cooperation, and synergies within chemicals and hazardous waste conventions. In order to make sure that the future international chemicals and waste regime is comprehensive, coherent, effective, and efficient, this synergetic approach will have to be further pursued. Thereby, different steps could be envisaged:

- First, it will be important to include new conventions such as the new convention on mercury into the developed framework of cooperation and coordination between the existing three conventions. Therefore, Switzerland has proposed that the secretariat of the mercury negotiation process should prepare an options paper on how to ensure the achievement of synergies and institutional cooperation and coordination between the mercury instrument to be negotiated and other relevant MEAs and policies.† This proposal was accepted‡ and the Negotiation Committee will thus have the opportunity to ensure that the new mercury convention will be embedded into the new synergetic institutional framework since its beginning, for example, by colocating the secretariat into the existing structure of the Basel, Rotterdam, and Stockholm Conventions and by integrating it fully into the joint secretariat and management structure.
- Second, the synergies process between the conventions should be further deepened and the coordinating framework should go beyond cooperation and coordination at secretariat level. Cooperation should also happen at the programmatic and policy levels. For example, simultaneous COP should be held on a regular basis to allow for joint decision making on joint issues. And ideas such as holding joint high-level segments together with the last COP of a given biennium instead of three separate high level segments at each separate COP should be further developed. By agreeing to review the existing cooperative framework in 2013, the simultaneous COP of the Basel, Rotterdam, and Stockholm Conventions have paved the way to such further progress.§
- Third, the synergies process between the conventions should be broadened and the coordinating framework should go beyond cooperating mechanisms between the conventions. Synergies should not only be ensured between the

* See the section on "Synergies and the Emergence of an International Chemicals and Waste Cluster."

† See Report of the ad hoc open-ended working group to prepare for the intergovernmental negotiating committee on mercury, § 75, available at http://www.chem.unep.ch/mercury/WGprep.1/documents/ii91)/English/WG_Prep_1_10_report_FINAL.doc.

‡ Report of the ad hoc open-ended working group, supra note 114, Annex II, § 2(o).

§ See e.g. Simultaneous Extraordinary Conference of the Parties of the Basel, Rotterdam, and Stockholm Conventions of February 2010, Conference Room Paper 4 on review, submitted by Switzerland on behalf of Côte d'Ivoire, Egypt, Norway, Switzerland, and Zambia; and Section VI of Conference Room Paper 2, submitted by the EU, both on file with the author.

chemicals and waste conventions, but also between the conventions and SAICM, UNEP Chemicals and other relevant international processes. And, synergy and effectivity could be enhanced significantly by adequate financial and other support that is well coordinated and linked to the three conventions. During the ministerial lunch organized by Indonesia and Switzerland at the Global Ministerial Environment Forum (GMEF) in 2010 in Bali, a general commitment toward taking such a broader look at the synergies process has emerged and it was concluded that this important process should be spread out in the future.* However, a proposal to ask UNEP to undertake additional analysis in that direction did not yet receive consensus at the GMEF.†

- Fourth, in order to avoid future proliferation of instruments and institutional fragmentation, the development of narrow, single-chemicals conventions should be avoided. So far, the international community has successfully avoided establishing single-chemicals instrument. For example, both the Rotterdam PIC and the Stockholm POPS conventions are dynamic and open to bring in additional substances. Following this example by broadening the scope of the new mercury convention to other heavy metals of global concern and by including the flexibility of adding other heavy metals into the new convention would be a significant additional step toward an effective and efficient regime. Thereby, it is important to note that allowing the future convention to also address heavy metals other than mercury would not preempt the decision whether such other heavy metals will indeed require a legally binding international response. It would simply make sure that instead of developing new independent machinery, an existing instrument could be used if the international community decides that other heavy metals also require a legally binding approach at the global level. Switzerland and others have highlighted this vision at the GMEF in Bali and called for a "future proof" mercury convention and for a discussion of this issue at the 26th UNEP Governing Council.‡
- Fifth, the international chemicals and waste regime should also provide a framework to address the third categories of substances of global concern identified above, namely those where the risk is linked not to their persistence but to their size, human toxicity, or eco-toxicity. This would allow ensuring a comprehensive international framework addressing all issues of global concern. The decision by the 2nd International Conference on Chemicals

* Summary of the Ministerial Working Lunch on Chemicals Management, Bali, 24.2.2010, and Swiss intervention on IEG during the Global Ministerial Environment Forum 2010, on file with the author and to be published i.a. on www.bafu.admin.ch/international/00692/00703/index.html.

† See report of the 11th special session of the UNEP Governing Council/Global Ministerial Environment Forum, forthcoming.

‡ See Earth Negotiation Bulletin, Summary of the Simultaneous Extraordinary COPs to the Basel, Rotterdam, and Stockholm Conventions and the 11th special session of the UNEP Governing Council/ Global Ministerial Environment Forum, available at http://www.iisd.ca/vol16/enb1684e.html and Summary of the Ministerial Working Lunch on Chemicals Management, supra note 117.

Management identifying nanotechnology as a new and emerging issue that needs to be addressed within SAICM is a step into this direction.*

- Finally, a comprehensive, coherent, effective, and efficient international chemicals and waste regime would not only have to provide for a broad framework for measures to address substances of global concern throughout their life cycle, it would also need an effective compliance mechanism and a financial mechanism. Thereby, it is important that the compliance mechanism is able to identify problems of compliance and to trigger support and response to the identified challenges. And the financial mechanism will not only have to be adequately funded but also close enough to the policy framework to be able to support the implementation of the international framework. The upcoming process to prepare a Rio + 20 Conference on Sustainable Development in 2012 could provide the framework for such a broader reform of the international chemicals and waste regime.† Thereby we have to bear in mind that the idea of a broad framework convention on chemicals management was discussed more than 20 years ago during the deliberations of the mandate for the negotiation of the Rotterdam Convention. A careful analysis would be needed to assess whether the same arguments that have led then to the rejection of the idea of a framework convention are still valid and whether approaches other than a framework convention are possible.

This vision seems to be very ambitious. However, the current policy developments within UNEP and within the existing conventions give promising signals that the international community has identified and understood the problem and that it is currently moving into the right direction.

CONCLUSIONS

This chapter has given an overview of some of the most important emerging issues of global chemicals and waste management. Noting that not only newly invented substances but also old and well-known chemicals can emerge into the public attention and require political response, it has presented some of the most pressing new challenges of today's international chemicals and waste policy, namely the challenges posed by heavy metals such as mercury, lead, and cadmium, nanomaterials, EDs and by PFCs. Moreover, it has also presented three new and emerging policy approaches that have been developed over the recent years based on the experience gained from older policy responses and the better understanding of the complexity of chemicals

* See supra, text accompanying note 44.

† See Earth Negotiation Bulletin, supra note 119, indicating that Switzerland highlighted the need for a strong framework to address chemicals, or Summary of the Ministerial Working Lunch on Chemicals Management, supra note 117, where it was indicated that synergies will have to be enhanced by financial support and that implementation needs not only a coherent policy framework, but also concrete and effective support.

management, namely increased private actor involvement, integrated and life-cycle approaches, and strengthening institutional coordination and synergies.

This overview of emerging issues—emerging substances and emerging policy approaches—of the global chemicals policy reveals the impressively dynamic and innovative character of the international chemicals and waste regime. It has not only been able to address new challenges, it has also developed and adopted new approaches for effective policy making and implementation. The international chemicals and waste regime seems to be able to respond to emerging issues and new challenges both by launching new conventions, furthering voluntary approaches, developing a comprehensive strategic framework, and by developing new policy approaches such as strengthened nonstate actor involvement, taking more integrated life-cycle approaches, and developing synergies for the traditionally fragmented conventions approach.

However, the list of emerging challenges that require policy attention cannot be closed. The rapid evolution of society and the progress of science will not only bring further benefits from already known and new substances, it will also create new challenges that have to be addressed. Thus, the international chemicals and waste regime has to remain dynamic as well and able to respond to new issues. Noting that there is a certain set of measures and policies that can be used to address new and emerging challenges and that there are significant synergies and efficiencies in using these policies in a coherent, complementary, and synergetic manner, the fourth part of this contribution has tried to outline how the future international chemicals and waste regime could provide a comprehensive, coherent, effective, and efficient framework. Most importantly, such a framework should make full benefit of the synergetic approach that is currently developed between and within the Basel, Rotterdam, and Stockholm Conventions.

With its process of enhancing effectivity and efficiency by strengthening institutional and political coordination, cooperation, and synergies, the global chemicals regime has even taken the lead in the broader efforts of strengthening international environmental governance. However, enhancing synergies retrospectively, that is, bringing institutions and processes together after having been developed and finalized as independent frameworks, is complicated and resource intensive. Moreover, addressing each specific new emerging issue in an *ad hoc* manner with separate full-fledged institutions is inefficient. Therefore, it is hoped that policy makers and technocrats will realize that new frameworks should be open and dynamic to be able to also address emerging issues that require future policy response. It would not be efficient to develop a new legally binding instrument for each substance. Mercury is only one of several heavy metals and chemicals of global concern that will require international cooperation for the protection of human health and the environment. Therefore, a new convention on mercury should be able to provide also a future home for other heavy metals of global concern. If successful, enhancing institutional synergies is a new and emerging issue that will not only help to further strengthen sound chemicals management efforts all over the world but that will also strongly impact future policy development in the broader international context.

ACKNOWLEDGMENTS

The authors would like to thank the comments and suggestions of Gabi Eigenmann, Bettina Hitzfeld, Josef Tremp, and Urs von Arx.

REFERENCES

Börzel, T. A. and T. Risse. 2005. Public–private partnerships: Effective and legitimate tools of international governance? In *Complex Sovereignty: Reconstituting Political Authority in the Twenty-First Century*. E. Grande and L.W. Pauly, eds. Toronto: Toronto University Press, p. 195.

Brunnée, J. 2006. International environmental law: Rising to the challenge of common concern? In *ASIL Proceedings*, 100:307–310 at 307.

Calafat, A. M., Z. Kuklenyik, J. A. Reidy, S. P. Caudill, J. S. Tully and L. L. Needham. 2007. Serum concentrations of 11 polyfluoroalkyl compounds in the US population: Data from the National Health and Nutrition Examination Survey (NHANES) 1999–2000. *Environ. Sci. Technol.*, 41: 2237–2242.

Carson, R. 1962. *Silent Spring*. Boston: Houghton Mifflin. Mariner Books, 2002, ISBN 0-618-24906-0; *Silent Spring* initially appeared serialized in three parts in June 16, June 23, and June 30, 1962 issues of *The New Yorker* magazine.

Churchill, R. R. and G. Ulfstein. 2000. Autonomous institutional arrangements in multilateral environmental agreements: A little-noticed phenomenon in international law. *Am. J. Int. Law*, 94: 623.

Felix, R. A., H. K. Hungerbühler, J. S. Jobling, U. Ruegg, A. Soto, and C. Studer. 2008. Endocrine disruptors: Relevance to humans, animals and ecosystems. *Chimia* 62: 316–317; special issue NRP50 318–438.

French, H. 2002. Reshaping global governance. In *State of the World*. L. Starke, ed., New York, NY: Norton, 174, pp. 176–177.

Giesy, J. P. and K. Kannan, 2001. Global distribution of perfluorooctane sulfonate in wildlife. *Environ. Sci. Technol.*, 35: 1339–1342.

Gubb, M. and M. Younes. 2006. SAICM—A new global strategy for chemicals. In *Environment House*. Available at: http://www.environmenthouse.ch/docspublications/newsletters/6a7b1dcbfde.pdf, Geneva: United Nations Environment Programme, p. 6.

Houde, M., J. W. Martin, R. J. Letcher, K. R. Solomon, and D. C. G. Muir, 2006. Biological monitoring of polyfluoroalkyl substances: A review. *Environ. Sci. Technol.* 40: 3463–3473.

Ivanova, M. 2010. UNEP in global environmental governance: Design, leadership, location. In *Global Environmental Politics* 10: 1, 30 at 46.

Ivanova, M. and J. Roy. 2007. The architecture of global environmental governance: Pros and cons of multiplicity. In *Global Environmental Governance—Perspectives on the Current Debate*. L. Swart and E. Perry, eds. New York, NY: Center for UN Reform Education, 48, pp. 5052.

Karlaganis, C. 2008. *Fair Trade Labels: A Case Study of the Max Havelaar Label in Switzerland* (Lizentiatsarbeit, on file with the author), at 46–49.

Kellyn, S. B. 2007. Perfluoroalkyl acids: What is the evidence telling us? *Environ. Health Persp.* 115(5): A250–A256.

Linder, S. H. and P. V. Rosenau. 2000. Mapping the terrain of the public–private policy partnership. In *Public–Private Policy Partnerships*. P. V. Rosenau, ed., Cambridge, MA: MIT Press, p. 5.

Mathys, R., J. Dittmar, and C. A. Johnson. 2007. Antimony in Switzerland: A substance flow analysis. Environmental studies no. 0724. Federal Office for the Environment, Bern, pp. 149.

McLachlan, M. S., K. E. Holmstrom, M. Reth, and U. Berger. 2007. Riverine discharge of perfluorinated carboxylates from the European continent. *Environ. Sci. Technol.* 41: 7260–7265.

Najam, A., M. Papa, and N. Taiyab. 2006. *Global Environmental Governance: A Reform Agenda.* Winnipeg: International Institute for Sustainable Development, pp. 14–16; 36–56; 67.

Paul, A. G., K. C. Jones, and A. J. A. Sweetman. 2009. First global production, emission, and environmental inventory for perfluorooctane sulfonate. *Environ. Sci. Technol.*, 43: 386–392.

Peiry, K. K. 2009. International chemicals and waste management. In *Research Handbook on International Environmental Law*. M. Fitzmaurice and D. Ong, eds. Celtenham: Edward Elgar Publishing at 3 and 4.1.

Perrez, F. X. 2000. Cooperative sovereignty: From independence to interdependence. In *The Structure of International Environmental Law*. The Hague: Kluwer Law International, pp. 272–277.

Perrez, F. X. 2001. Country Report: Switzerland. In *12th Yearbook of International Environmental Law* 452, Oxford: Oxford University Press, p. 454.

Perrez, F. X. 2003. Country Report: Switzerland. *14th Yearbook of International Environmental Law* 467, pp. 472–473.

Perrez, F. X. 2006. The strategic approach to international chemicals management—Lost opportunity or foundation for a brave new world? 15(3) *Review of European & International Environmental Law (RECIEL)* 245: 247–249.

Perrez, F. 2008. How to get beyond the Pareto optimum of stakeholder participation in environmental governance. In *Proceedings of the Conference on Environmental Governance and Democracy*, Yale, May 10–11. Available at: http://www.yale.edu/envirocenter/env-dem/documents.htm#track21, at 3-4.

Perrez, F. X. 2009a. How to get beyond the zero-sum game between state and non-state actors in international environmental governance. *Consilience: J. Sustainable Dev.* 2.

Perrez, F. X. 2009b. Public–private partnerships: A tool to evade or to live-up to commitment? Paper presented at the Conference Practical Legal Problems of International Organizations, University of Geneva, Faculty of Law, March 20–21. Available at: http://www.iilj.org/GAL/GALGeneva.papers.asp, pp. 4–5.

Perrez, F. X. and D. Ziegerer. 2008. A non-institutional proposal to strengthen international environmental governance. In *Environmental Policy and Law* 38/5 253, pp. 254–255.

Piest, U. 2003. A preliminary analysis of the inter-linkages within WSSD "Type II" Partnerships. In *17/1 Work in Progress* 25, Tokyo: United Nation University, p. 25.

Prevedouros, K., I. T. Cousins, R. C. Buck, and S. H. Korzeniowski. 2006. Sources, fate and transport of perfluorocarboxylates. *Environ. Sci. Technol.*, 40: 32–44.

Regulation (EC) No 1272/2008 of the European Parliament and of the Council of 16 December 2008 on classification, labelling and packaging of substances and mixtures. OJ L 353: 31.12.2008. pp. 1–1355.

Roch, P. and F. X. Perrez. 2005. International environmental governance: The strive towards a comprehensive, coherent, effective and efficient international environmental regime. *Colorado J Int. Environ. Law Policy*, 16: 1–25.

Roger, B., C. Studer, and K. Fent, 1999. Stoffe mit endokriner Wirkung in der Umwelt, Schriftenreihe Umwelt Nr. 308, Bundesamt für Umwelt, 3003 Bern, Schweiz.

Sand, P.H. 2007. The evolution of international environmental law. In *The Oxford Handbook of International Environmental Law*. D. Bodansky, J. Brunnée, and E. Hey, eds. Oxford: Oxford University Press, Vol. 29, pp. 30–31.

Sands, P. 2003. *Principles of International Environmental Law.* Cambridge: Cambridge University Press, pp. 73–74.

Schmid, P. et al. 2009. Polychlorierte Biphenyle (PCB) in Gewässern der Schweiz. Daten zur Belastung von Fischen und Gewässern mit PCB und Dioxinen, Situationsbeurteilung. Umwelt-Wissen Nr. 1002. Bundesamt für Umwelt, Bern, 2010.

Song, Y., X. Li, and X. Du. 2009. Exposure to nanoparticles is related to pleural effusion, pulmonary fibrosis and granuloma. *J. Eur. Respir.* 34: 559–567.

Tao, L., K. Kannan, N. Kajiwara, M. M. Costa, G. Fillmann, S. Takahashi, and S. Tanabe. 2006. Perfluorooctanesulfonate and related fluorochemicals in albatrosses, elephant seals, penguins, and Polar Skuas from the Southern Ocean. *Environ. Sci. Technol.*, 40: 7642–7648.

UNEP. 2008a. Draft final review of scientific information on lead (November 2008). Available at www.chem.unep.ch/Pb_and_Cd/SR/Draft_final_reviews_Nov2008.htmUNEP. 2008b. Draft final review of Scientific information on cadmium (November 2008). Available at www.chem.unep.ch/Pb_and_Cd/SR/Draft_final_reviews_Nov2008.htm

UNEP. 2008c. The Global Atmospheric Mercury Assessment: Sources, Emissions and Transport. Available at: http://www.chem.unep.ch/mercury/publications/default.htm at 11–12.

Wei, S., L. Q. Chen, S. Taniyasu, M. K. So, M. B. Murphy, N. Yamashita, L. W. Y. Yeung, and P. K. S. Lam. 2007. Distribution of perfluorinated compounds in surface seawaters between Asia and Antarctica. *Mar. Pollut. Bull.* 54: 1813–1818.

Weiss, E. B. 1993. International environmental law: Contemporary issues and the emergence of a new order. In *Georgetown Law Journal*, 81, pp. 675, 679.

Yamashita, N., S. Taniyasu, G. Petrick, S. Wei, T. Gamo, P. K. S. Lam, and K. Kannan. 2008. Perfluorinated acids as novel chemical tracers of global circulation of ocean waters. *Chemosphere*, 70: 1247–1255.

49 Financing Instruments Related to Chemicals and Waste Management

*Ibrahima Sow**

CONTENTS

INTRODUCTION

The existing financial mechanisms and instruments related to the management of chemicals and hazardous wastes include a variety of arrangements that may be tied to compliance with conventions obligations, or funding of activities generating global environmental benefits (GEB). Other arrangements are channeled through the development and implementation of national programs supported by development banks such as the World Bank, Asian Development Bank, African Development Bank, and other multilateral institutions, such as UNDP, UNIDO, UNEP, FAO just to mention a few.

Regarding chemicals and waste management, the Stockholm Convention on Persistent Organic Pollutants (POPS) and the Montreal Protocol on Substances that Deplete the Ozone Layer are the only legally binding instruments that rely on dedicated financial mechanisms, namely the Global Environmental Facility (GEF) for the Stockholm Convention and the GEF for Countries with Economies In Transition

* The views expressed in this chapter do not necessarily reflect the views of GEF or its member countries.

(CEIT) and the Multilateral Fund (MLF) for the Montreal Protocol for developing countries. Other instruments, whether legally binding or not such as the Basel Convention on the Control of Transboundary Movements of Hazardous Wastes and their Disposal, the Rotterdam Convention on the Prior Informed Consent Procedure for Certain Hazardous Chemicals and Pesticides in International Trade and the Strategic approach to International Chemicals Management (SAICM) are supported by specific Trust Funds of Voluntary nature.

This chapter provides an overview of existing sources and mechanisms supporting the management of chemicals and wastes issues, including some flexible mechanisms related to private sector initiatives, public private partnerships and to some economic instruments.

THE GLOBAL ENVIRONMENT FACILITY

The GEF, established in 1992 in Rio de Janeiro as a network organization is the world's largest funder of projects in developing countries to protect the global environment while supporting sustainable development. It is a financial mechanism structured as a Trust Fund. The replenishment of the Trust Fund takes place every four years based on donor pledges that are funded over a four-year period. The GEF Trust Fund has 39 donors that have committed funds. They are Argentina, Australia, Austria, Bangladesh, Belgium, Brazil, Canada, China, Côte d'Ivoire, Czech Republic, Denmark, Egypt, Finland, France, Germany, Greece, India, Indonesia, Ireland, Italy, Japan, Republic of Korea, Luxembourg, Mexico, Netherlands, New Zealand, Nigeria, Norway, Pakistan, Portugal, Slovak Republic, Slovenia, South Africa, Spain, Sweden, Switzerland, Turkey, United Kingdom, and the United States.

It operates as a mechanism for international cooperation for the purpose of providing new and additional grants and concessional funding to meet the agreed GEF Council incremental costs of measures to achieve agreed GEB in the following areas:

- Biological diversity
- Climate change
- International waters
- Ozone layer protection
- Land degradation
- POPS

The GEF is also the designated financial mechanism for a number of Multilateral Environmental Agreements (MEAs), starting with the United Nations Framework Convention on Climate. Change (UNFCCC) and the Convention on Biological Diversity (CBD), the Stockholm Convention on POPS, the United Nations Convention to Combat Desertification (UNCCD), and the Montreal Protocol on substances that deplete the ozone layer (for CEIT). As such, the GEF assists countries in meeting their obligations under the conventions that they have signed and ratified. The GEF also added two cross-cutting areas, one of which is sound chemicals management. The objective of this cross-cutting work is to promote sound management of chemicals practices in all relevant aspects of GEF programs, and to contribute to the overall objective of SAICM of achieving the sound management of chemicals throughout their life-cycle so that by 2020 chemicals are used and produced in ways that lead to the minimization of significant adverse effects on human health and the environment.

The GEF's governing structure includes primarily two levels. The Assembly, which is effectively the Conference of the Parties of the institution, meets every four years. Among its major functions, the Assembly reviews the general policies of the Facility and considers, for approval by consensus, the amendment to the Instrument (its charter) on the basis of recommendations of the Council. The other level of decision making of the GEF is the Council which meets twice a year. Among other things, the Council is responsible for developing, adopting, and evaluating the operational policies and programs for GEF-finance activities; reviewing and approving work programs; directing the utilization of funds; and considering and approving cooperative arrangements or agreements with the Conference of the Parties to the conventions, and ensuring conformity of GEF approved activities with the policies, priorities, and eligibility criteria of the conventions.

In exercising its responsibility for considering and approving cooperative arrangement/agreements with the Conference of the Parties, the Council is mandated to ensure that these arrangements are in conformity with the relevant provisions of the conventions regarding their financial mechanisms and include procedures for jointly determining the aggregate GEF funding requirements for the purpose of the conventions. The relationship between the GEF governing structure and that of the relevant MEAs is outlined in paragraph 6 of the GEF Instrument as follows: "the GEF shall function under the guidance of, and be accountable to, the Conferences of the Parties which shall decide on policies, program priorities and eligibility criteria for the purposes of the conventions."

GEF major funding policies include paying for agreed incremental costs (see below) to generate GEB; funding projects and programs which are country driven and based on national priorities designed to support sustainable development; and being guided by and accountable to the COPs of the conventions.

The application of the concept of incremental cost has been recognized as complex and not always transparent. The GEF Council meeting of June 2007 approved the "Operational Guidelines for the Application of the Incremental Cost Principle." The guidelines enhance the transparency of the determination of incremental costs of a project during the preparation period, as well as its implementation through:

- Determination of the environmental problem, threat, or barrier, and the "business as-usual" scenario (or: What would happen without the GEF?).
- Identification of the GEB and fit with GEF strategic programs and priorities linked to the GEF focal area.
- Development of the result framework of the intervention.
- Provision of the incremental reasoning and GEF's role.
- Negotiation of the role of cofinancing.

GEF funds the "incremental" or additional costs associated with transforming a project with national benefits into one with GEB; for example, choosing solar energy technology over coal or diesel fuel meets the same national development goal (power generation), but is more costly. GEF grants cover the difference or "increment" between a less costly, more polluting option and a costlier, more environmentally friendly option.

Since 2005, the GEF has introduced the Resource Allocation Framework (RAF) to provide each recipient country at the outset of each replenishment period an indicative level of resources available during that period. The RAF is based on the potential of countries to generate GEB and the capacity, policies, and practices to successfully implement GEF-(co)funded projects. The RAF is currently only applicable to projects in the focal areas of biodiversity and climate change but will be extended to land degradation focal area during GEF-5.

Since 1992, the GEF has provided US$ 8.6 billion in grants and leveraged another US$ 36 billion for 2400 projects in 165 developing countries and CEIT.

The GEF is replenished every four years. The preparation of the fifth replenishment (for the period from July 1, 2010 to June 30, 2014-GEF 5) started in November 2008 and was successfully concluded May 2010. Any GEF replenishment is preceded by an independent review of the performance of the GEF in the current period and the results of the review are used as reference in the negotiations of the new replenishment. At the same time, strategies for each focal area are developed to assess the funding needs of each focal area and reviewed by the conventions concerned, the GEF agencies and other stakeholders.

GEF Funding on POPS/Chemicals Management

GEF's involvement in POPS and chemicals dates back to 1995, with the introduction of the International Waters Operational Strategy and its component to reduce pollution from chemicals and pesticides. In the late 1990s, GEF began to develop a portfolio of strategically designed projects, including regional assessments and pilot demonstrations that addressed a number of pressing POP issues. These initial activities allowed the GEF to quickly respond to requests for support from the negotiators of the Stockholm Convention for its implementation. This in turn led to the adoption of guidelines by the GEF Council for POPS-enabling activities in May 2001, the same month the Convention was adopted.

In the ensuing years, the GEF has committed US$ 360 million to projects in the POPS focal area. This cumulative GEF/POPS allocation had leveraged some US$ 440 million in cofinancing to bring the total value of the GEF POPS portfolio to US$ 800 million.

The GEF Cofinancing comprises the total of cash and in-kind resources committed by governments, other multilateral or bilateral sources, the private sector, NGOs, the project beneficiaries, and the concerned GEF agency, all of which are essential for meeting the GEF project objectives.

Adequate cofinancing is expected for GEF-financed Medium Size Projects (MSPs) (<US$ 1,000,000) and Full Size Projects (FSP) (>US$ 1,000,000); 1:4 cofinancing ratio is the average (1:1 is minimum benchmark targeted (1:1 means 50% GEF; 1:4 means 20% GEF). Finding adequate resources for cofinancing is, however, often a major challenge for less developed countries.

Ensuring an adequate cofinancing constitutes a key principle underlying GEF's success in its efforts to have significant positive impact on the global environment and is a key indicator of the strength of the commitment of the counterparts, beneficiaries, and donors and GEF Agencies.

Finally, it helps ensure the success and local acceptance of those projects by linking them to sustainable development, and thereby maximizes and sustains their impacts.

Under GEF-3 (July 2002–June 2006), efforts focused on supporting National Implementation Plans (NIP) development in eligible countries. Under GEF-4, activities are characterized by a shift from preparation to implementation with the GEF supporting projects submitted by partner countries to implement their NIPs. The shift from NIP preparation to NIP implementation has been materialized through implementation and elaboration of a wide range of projects, based on priority activities identified in the countries' NIPs. These projects include innovative projects on integrated POPS management and introduction of Best Available Technologies and Best Environmental Practices (BAT/BEP) in selected industrial sectors and for the reduction of unintentional POPS releases from open burning of municipal wastes. Management and disposal of obsolete pesticides and PCB projects remain the largest part of the POPS portfolio. Projects also include capacity strengthening, monitoring, and reporting to help countries comply with their obligations under the Stockholm Convention.

GEF-5 Perspectives on Chemicals Management

Since the time of the GEF-4 replenishment, the international chemicals agenda has expanded considerably in quantity and scope, requiring an enhanced response from the GEF: SAICM was adopted in 2006 with the International Conference on Chemicals Management at its second session in May 2009 (ICCM-2) "urging the GEF [...] to consider expanding its activities related to the sound management of chemicals to facilitate SAICM implementation [...];" negotiations for a legally binding agreement on mercury were launched in 2009; the linkages between the ozone-depleting substances (ODS) and climate-forcing greenhouse gases (GHGs) have been emphasized; and the synergy process currently taking place within the Stockholm, Rotterdam, and Basel COPs creates demand and opportunity for a more comprehensive approach that extends support beyond POPS and ODS.

In the field of chemicals, the GEF's mandate as the financial mechanism of the Stockholm Convention will require addressing the newly listed chemicals under the Convention. There are complex and challenging issues related to these chemicals throughout their life cycle and eligible countries will require assistance to address these. This extends to environmentally sound disposal of POPS-containing waste. The GEF will also continue to support cost effective efforts to phase out ODS in CEIT to meet their Montreal Protocol compliance obligations.

The GEF Instrument provides that "the agreed incremental costs of activities to achieve GEB concerning chemicals management," as they relate to the GEF focal areas, are eligible for funding. Many substances apart from POPS are of global concern, even if they are not yet covered by global treaties. Mercury releases are relevant to the biodiversity and international waters focal areas, and there are potentials for synergies in relation to GHG emissions. The positive experiences from GEF's early work before the Stockholm Convention was finalized indicate that early action to build capacity for reducing releases of mercury will also achieve good results.

Many of the challenges concerning the management and phase-out of POPS are similar to the steps that countries need to take to comply with the Basel, Bamako, and Rotterdam Conventions. Sound management of wastes will also be needed to address several of the newly listed Stockholm Convention chemicals and will be important in the context of a future Mercury Convention. Therefore, the existing GEF policy that support to the Stockholm Convention and Montreal Protocol implementation could build upon and contribute to strengthening a country's basic capacities for sound chemical management more generally will be actively pursued so that these activities in support of POPS and ozone depleting substances (ODS) are designed to also benefit implementation of SAICM at the country level, and attainment of the 2020 target of the Johannesburg World Summit.

Taking the above into consideration, and based on the POPS allocation (US$ 420 million) decided at the last replenishment meeting, the GEF will assist countries to address chemicals in an integrated manner in their national planning, and help mobilize other sources of finance for projects and programs for sound chemicals management to achieve global benefits. To achieve this, the three following objectives are proposed for Chemicals under GEF-5: Phase out POPS and reduce POPS releases; Phase out ODS and reduce ODS releases; and Pilot sound chemicals management and mercury reduction.

THE MLF FOR THE IMPLEMENTATION OF THE MONTREAL PROTOCOL

The MLF was established to provide financial and technical cooperation, to eligible Parties operating under paragraph 1 of Article 5 of the Montreal Protocol to enable their compliance with the control measures set out in the Protocol. The MLF is replenished on a three-year basis. The contributions to the Fund come from donors from both developed and developing countries and are assessed according to the UN scale of assessment.

The Meeting of the Parties of the Montreal Protocol which convenes once per year is responsible for deciding general policies such as the broad scope and categories of activities to be funded, the membership of the Executive Committee and the replenishment period of the MLF.

The Executive Committee meets three times per year and is in charge of developing and implementing operational policies of the Fund. It considers and approves projects and programs, and exercises oversight on funded activities to ensure cost-effectiveness and consistency with the overall policies set by the Meeting of the Parties. The MLF's major funding policies include the principle of covering the incremental costs of phasing out the consumption and production of ODS and performance-based fund disbursement where funds are paid out only upon independent verification of ODS reduction targets being achieved as planned.

From 1991 to 2008, about US$ 2.5 billion has been disbursed to fund about 6000 projects and programs in 144 countries. The activities include providing institutional support in each recipient country, ozone networks covering seven regions, the preparation and implementation of national ODS phase out strategies, funding technology transfers to industries to convert from ODS-based to non-ODS technologies, and compensating for closing down ODS production.

The MLF also provides each year from US$ 30,000 to US$ 450,000, depending on the size and consumption of the country, to support national ozone units (NOUs). This has significantly improved the rate of annual data reporting by countries to the Montreal Protocol Ozone Secretariat and facilitated the communication between countries and international organizations.

SAICM

The recently developed SAICM (see Chapter 17 of this book) sets out a comprehensive policy framework for the achievement of chemicals management objectives, including in relation to multilateral environment agreements, and the financing of their implementation. A full range of financial arrangements to support the broad chemicals management objectives of SAICM are set out in its Overarching Policy Strategy. These include supporting the initial capacity-building activities for the implementation of SAICM objectives under the SAICM "Quick Start Programme" (QSP) and its time-limited trust fund from voluntary contributions. The QSP is designed to support initial capacity-building activities in developing countries and CEIT for the implementation of SAICM objectives.

The QSP's governing structure comprise the Executive Board and the Trust Fund Implementation Committee. The membership of both reflects the multisectoral composition that corresponds to the multistakeholder nature of SAICM. The QSP Executive Board consists of two government representatives of each of the UN regions and the bilateral and multilateral donors and other contributors to the Programme. The QSP Trust Fund Implementation Committee consists of representatives of participating organizations of the Inter-Organization Programme for the Sound Management of Chemicals (IOMC), and the UNDP. The Executive Board reports to the International Conference on Chemicals Management (ICCM).

Other financial arrangements envisaged to support implementation of SAICM include:

- Actions at the national or subnational levels to support financing of Strategic Approach objectives.
- Enhancing industry partnerships and civil society participation to advancing SAICM objectives.
- Integration of SAICM objectives into multilateral and bilateral development assistance cooperation.
- Making more effective use of and building upon existing sources of relevant global funding, such as the GEF and the MLF for the Implementation of the Montreal Protocol.

As of May 14, 2010 the Trust Fund Committee approved 82 projects for a total funding of approximately US$ 16 million. In addition, 51 projects were recommended for further development and resubmission. The approved projects will be implemented by 74 Governments and 12 civil society organizations and will involve activities in 76 countries, including 35 least-developed countries and small island developing States.

BILATERAL OFFICIAL DEVELOPMENT ASSISTANCE

Generally, Official Development Assistance (ODA) funding for environment and especially for chemicals and wastes activities remains low in comparison to funding available to other sectors. Experience shows that development aid rarely focuses on chemicals and wastes management, which is not always included in countries' requests. Funds for chemicals and waste activities may be more indirectly provided through funding for broader areas, such as natural resources management.

Funding Agencies for chemicals or waste programs include: Australian Aid agency for POPS management and disposal in Pacific Islands Countries; French Development Agency in the Pacific region for the management of solid and hazardous wastes, Dutch supported project on the elimination of acute risks of obsolete pesticides in Moldova, Armenia, and Georgia.

Other bilateral agencies, including GTZ (German Technical Cooperation), JICA (Japan International Cooperation), USAID (US Agency for International Development), provide technical and financial assistance to a number of developing countries in the field of environmental protection. However, the share related to chemicals and wastes management remains very limited.

FLEXIBLE INSTRUMENTS

Several initiatives pertaining to sound chemical management and related to SAICM objectives have recently emerged from the private sector. These include: (1) The "Responsible Care Global Charter" of the International Council of Chemical Associations (ICCA). The charter focuses on new and important challenges facing the chemical industry and society including sustainable development, effective management of chemicals along the value chain, greater industry transparency, and increased global harmonization and consistency among Responsible Care programs around the world; and (2) The ICCA "Global Product Strategy," which is designed to enhance product stewardship within the chemical industry and with customers throughout the chain of commerce. Key components of this strategy include: Guidelines for product stewardship, to share best practices within the chemical industry and with customer industries; a tiered process for completing risk characterization and risk management actions for chemicals in commerce; greater transparency, including ways to make relevant product stewardship information available to the public.

New instruments practiced in some OECD countries and being demonstrated in developing countries include the concept of Chemical Leasing (ChL), (see Section V, Chapter 31 of this book, UNIDO). ChL aims at increasing the efficient use of chemicals while reducing the risks of chemicals and protecting human health. It improves the economic and environmental performance of participating companies and enhances their access to new markets.

Other initiatives include solving the E-waste Problem (StEP) Initiative, established in 2007 to start up and foster partnerships between companies, governmental and nongovernmental organizations and academic institutions on meeting the challenges that result from the production, usage, and disposal of electrical and electronic equipment. As a public–private partnership initiative founded by various UN organizations, StEP is uniquely positioned to contribute to the formulation of basic principles, policies and strategies, and the development of technologies and projects for action.

The Basel and the Stockholm Conventions recently initiated two innovative initiatives on mobile phones and PCB management. The Mobile Phone Partnership Initiative (MPPI) was established as a sustainable partnership on the environmentally sound management of used and end-of-life mobile telephones and the PCB Elimination Network which is a voluntary arrangement for information exchange and which aims at improving coordination and cooperation among stakeholders from different sectors with an interest in the environmentally sound management of PCBs.*

ECONOMIC INSTRUMENTS

The use of economic instruments in the field of chemicals and wastes management is still in the experimental stage for many countries. The process obviously needs improvement and the stakeholders, including relevant private sector entities and government institutions will need to have ownership of it. Economic instruments can be used to internalize the environmental externalities and provide finance for the implementation of obligations under the relevant agreements. Economic instruments for chemicals and wastes include: waste generation fees, essentially similar to a utility charge; waste disposal/tipping fees; environmental product levies, on items that are difficult to dispose of including bulky or hazardous items; deposit refund programs, involving a deposit/levy paid by the importer to the government, with a percentage of the deposit paid as a refund when the product is disposed of; and tax incentives and disincentives, including granting subsidies and concessions to environmentally sound products and alternatives.

UNEP Chemicals is currently working on producing guidance for national policymakers on cost recovery instruments for financing chemicals management that covers much of this discussion.

In researching the present application of economic instruments for chemicals and wastes management, it was found that cost internalization is not often a priority for the instruments being used–there is little indication that the fee or tax structures are designed specifically to internalize externalities from poor chemicals management. More often the concern is simply to charge fees that cover the cost of providing public chemicals management services, that is, inspections, extension services.

* See http://www.pops.int

OTHER EXISTING INSTRUMENTS

Generally speaking, these instruments are related to private Foundations and International NGOs focusing their work on chemicals and waste management. The Kapor Foundation*—a private foundation with the mission to ensure fairness and equity, especially in low-income communities of color—supports the development of the International POPS Elimination Network (IPEN). The Ford Foundation† is an independent, nonprofit nongovernmental organization, supporting among others, development of natural resource policies and programs that give poor communities more control over these resources and a stronger voice in decision making on land use and development. It has made a US$ 2.2 million grant to Vietnam in 2006 to bring critical health services to people living with dioxin-caused long-term disabilities. A number of newer partnerships are also emerging among donors and UN agencies that provide a model for resource mobilization for chemicals and wastes work. One such partnership is the Global Alliance for Vaccine and Immunization Fund (the GAVI Fund). The GAVI Fund, established in 2000 to give developing country children increased access to immunization, is a public–private partnership with participation from donor governments of both developing and developed countries, international organizations such as the World Bank, UNICEF, WHO, and also philanthropic partners, principally the Bill and Melinda Gates Foundation. The target countries eligible to receive funds from the GAVI Fund are those whose annual per capita income is less than US$ 1000. The total number of eligible countries currently stands at 72 and represent half the world's population. The GAVI Fund is an interesting model for resource mobilization for the chemicals and waste because it brings public and private partners together, and utilizes novel fund-raising mechanisms.

Another partnership example is the Earth Fund launched at the end of 2007 by the GEF and the International Finance Corporation to "support innovative solutions for the most pressing environmental challenges in developing countries."

The Earth Fund was conceived to engage the private sector in its activities and particularly to link donor funds into private sector creativity, investment, and participation. It is set up to operate as a venture capital entity to provide grants, soft loans and equity participation to fund promising innovations in areas such as second-generation biofuels, water treatment, and clean energy. The Earth Fund remains open to innovative programs and projects dealing with sound management of chemicals and wastes.

On a small scale, it is worth mentioning activities undertaken by the Blacksmith Institute;‡ an International nonprofit organization focusing its efforts on the world's most polluted areas. Blacksmith Institute provides financial and technical assistance to local partners in designing and implementing remediation strategies tailored to the specifics of the sites in question. Key programs include decontamination of lead poisoning caused by improper recycling of used car battery in Senegal, Dominican Republic and of mercury poisoning caused by artisanal gold mining in Senegal, Mozambique, Indonesia, and Cambodia.

---

* http://www.mkf.org/about/index.html

† http://www.fordfound.org/issues/sustainable-development/overview

‡ http://www.blacksmithinstitute.org/

CONCLUSION

The review of the above existing financial mechanisms and instruments clearly demonstrates that important efforts through dedicated and/or voluntary mechanisms are being deployed by the international community toward addressing sound management of chemicals and wastes. An effective and efficient use of these instruments and mechanisms requires the creation or enhancement of capacities of the recipient countries, in particular the least developed countries that are facing difficulties in designing and processing project proposals through the existing mechanisms.

It should be recognized that resources devoted to this area of international cooperation remain relatively modest compared to resources dedicated to other areas, such as biodiversity and climate change, taking also into account the magnitude of adverse impacts associated with the mismanagement of chemicals and wastes, in particular in the poorest countries that are also the most vulnerable. Furthermore, the scattering of limited financial resources, not to mention their unpredictability, constitutes a major constraint in addressing adequately chemicals and waste management in developing countries.

It goes without saying, as expressed by developing countries and CEIT on several occasions during international forums on chemicals and wastes management that without dedicated and predictable financial resources, it would be quite illusory to correctly address the challenges associated with chemicals and wastes management. In the meantime, a process of harmonization among the existing chemicals and wastes treaties and the financing sources is urgently needed. Furthermore, developing countries will need to demonstrate that chemicals and waste programs are among the priority sectors of their socioeconomic development agenda.

FURTHER READING

Blacksmith Institute–KEY PROGRAMS. 2010. Available at: http://www.blacksmithinstitute.org/our-programs-and-projects.html

GEF (Global Environmental Facility). 2009. *About the GEF. The Instrument and Outlook for GEF-5.*

GEF. 2009. *Chemicals Strategy, GEF-5 Programming Document.*

ICCA (International Council for Chemicals Associations). 2006. *ICCA's Position and Contribution to the SAICM Implementation*—Dubai, 2006.

SAICM (Strategic Approach to International Chemical management). 2006. *Report of the International Conference on Chemicals Management*, Dubai, 2006.

UNEP (United Nations Environment Programme). 2009. *Desk Study Resulting from the Consultative Process on Financing Options for Chemicals and Wastes.*

50 Information Resources Supporting Global Chemicals Policy and Management

Asish Mohapatra and Philip Wexler

CONTENTS

INTRODUCTION

Information is the foundation of knowledge. The conferences, conventions, and organizations discussed elsewhere in this book rely upon information to undertake their activities and, in turn, generate it as outcomes of their efforts. Paper resources, while by no means extinct, have largely given way to a robust spectrum of digital tools—documents, databases, portals, and compilations of information, virtually all online via the Web and, to a large extent, free. In the twenty-first century, this information is also increasingly graphical, interactive, collaborative. Sound science is the bedrock of effective chemicals management and policy. At the outset, this chapter will highlight some issues in the use of informatics tools for knowledge generation and sharing. It will take a brief look at how new technologies and social networking applications (Web 2.0) have fostered the creation, sharing, and distribution of this data, and speculate about what lies ahead in the realm of informatics applications. Further, it will offer a selective overview of global and select country-specific information resources in toxicology and risk assessment, which support decision making in chemicals management and policy.

As broadly defined by the U.S. Society of Toxicology, toxicology is the study of the adverse effects of chemical, physical, or biological agents on living organisms and the ecosystem, including the prevention and amelioration of such adverse effects. It is the *chemical* portion of this definition which is primarily emphasized in the subject matter of this book. The majority of toxicology and chemical databases tend to be in English, and most of them originate in North America and Europe, although the information can be accessed by users from virtually every any country. In the list below, resources in chemical toxicology, computational (predictive) toxicology, and toxicological analysis platforms facilitated by various emerging informatics tools have been highlighted.

Risk, too plays an important role in toxicology. With applications in worlds as disparate as chemical toxicology, transportation, and finance, risk is a function of hazard and exposure. In relation to toxicology, *hazard* typically refers to the inherent *toxicity* of substances. However, outside the laboratory, in human populations, involved in everyday activities, it is usually the exposure portion of the equation which is of more concern. That is, by what means and to what extent are people exposed to substances? Highly toxic substances do not pose a high risk to health if people do not come into contact with them.

CHALLENGES AND OPPORTUNITIES OF INFORMATICS TOOLS IN TOXICOLOGY AND RISK-ASSESSMENT COLLABORATIONS

English versus Non-English Language Access

Language is still a barrier which needs to be adequately addressed. For example, access to toxicology-related information may be limited to those who are English speakers, since most of the databases are in English. Purely from a language access perspective, scientists and chemical policy makers should work closely to ensure that consistent language be used in knowledge translation. Some translation tools, such as Google Translate (http://translate.google.com) can be used to translate Web sites of one language to another. However, users should be cautious in accepting the output without further checks, given the particular difficulties of translating technical language.

One example of a multilingual discussion list related to environmental health and toxicology issues is RETOXLAC (http://bvsde.per.paho.org/bvstox/i/retoxlac/intro.html), a discussion list sponsored by the Pan American Health Organization (PAHO/WHO) and the Pan American Center for Sanitary Engineering and Environmental Sciences (CEPIS/PAHO). It is a forum that disseminates toxicological information in Spanish, Portuguese, and English. The World Library of Toxicology, discussed later, takes another multilingual approach to information dissemination.

Chemical and Toxicological Language Processing

Natural Language Processing (NLP) is the process of converting human language into a machine readable format that can be understood by computers. Recently, the United States Environmental Protection Agency (US EPA), in its Health and Environmental Research Online (HERO) Database, introduced these technologies to make information access and extraction more efficient. They have used algorithms (sequences of machine understandable instructions) for training and deploying models to enable rapid extraction, retrieval, and organization of data. Sophisticated data management in tandem with full-text retrieval and semantic data processing can ensure that users have access to the highest quality of data. Furthermore, there have been recent developments in the automatic Information Extraction (IE) area. By using these emerging technologies, information from unstructured sources can be extracted that would allow new opportunities for querying, organizing, and analyzing scientific data, which can be integrated with information gathered from biomedical and environmental research.

Financing/Funding and Use of Emerging Informatics Tools

Computerized information requires extensive financial and human resources. The funding aspect of database management is always a challenge. If a particular database project is mandated by regulations, and countries at a regional level and multiple countries on a global scale get together to share and collaborate on information resources (e.g., REACH-related information sharing; World Library of Toxicology), the process of maintaining and sharing becomes more efficient. However, if countries have other priorities on a science and policy level, then some of the database management aspects become inefficient. The new informatics tools, automatic IE protocols and the Semantic Web may provide additional tools and resources to ensure that frequent updates are provided to users. For example, every month, the Comparative Toxicogenomics Database (http://www.ctd.mdbil.org) undergoes update and the users generally get notifications of these updates via email and other means.

WEB EVOLUTION AND SEMANTIC WEB

The World Wide Web has changed the way we access chemical, toxicology, and environmental health-related information. One model imagines the Web evolving from Web 1.0 (connecting information) to Web 2.0 (connecting people via social networking, for example) to Web 3.0 (connecting knowledge and thus creating knowledge bases) to Web 4.0 (connecting intelligence) (http://colab.cim3.net/file/work/SICoP/2007-04-25/InternetTo2020.pdf). The evolution of Web 1.0–4.0 and their respective applications to environmental health, toxicology, and chemistry would bolster efficient information dissemination.

The Semantic Web is making it possible for machines to understand in an intelligent manner what a particular user is looking for, and to subsequently satisfy users' requests (Berners-Lee et al., 2001, excerpted from The Semantic Web, Scientific American Magazine, W3C Semantic Web Frequently Asked Questions, available at: http://www.w3.org/2001/sw/SW-FAQ). It can be defined as a set of evolving formats and processing languages that specifically looks for user defined information and analyzes it (Feigenbaum et al., 2007, in Scientific American Article—The Semantic Web in Action).

New computing technologies have been gradually applied to various Web-based databases. For example, Hakia (http://hakia.com/) integrates searching of PubMed with other sources such as images, videos, Wikipedia, Twitter, Blogs, News, and so on. (As the Web evolves, more users are expected to take advantage of these new technologies to make their Web content more dynamic and efficient for users.)

Another emerging area in biomedical sciences is semantic publishing, which has been defined as "anything that enhances the meaning of a published journal article, facilitates its automated discovery, enables its linking to semantically related articles, provides access to data within the article in actionable form, or facilitates integration of data between papers. Among other things, it involves enriching the article with appropriate metadata that are amenable to automated processing and analysis, allowing enhanced verifiability of published information and providing the capacity for automated discovery and summarization." For example, ChemSpider (http://www.ChemSpider.com), a product of the Royal Society of Chemistry (RSC) is a free-to-access collection of compound data from across the Web. It aggregates chemical structures and their associated information into a single searchable repository. ChemSpider builds on the collected sources by adding additional properties, related information and links back to original data sources. It offers text and structure searching to find compounds of interest and provides services to improve this data by curation and annotation and to integrate it with users' applications.

SELECTED RESOURCES

Intergovernmental and other Global Agencies

SAICM Clearinghouse (*http://www.saicm.org/ich*)—One of the major Web sites directly addressing the information contained within, and driven by, a multilateral environmental policy framework, the Clearinghouse is an outgrowth of the Strategic Approach to International Chemicals Management (SAICM). The Clearinghouse's organization is built around tabs keyed to SAICM's Objectives: risk reduction, knowledge and information, governance, capacity-building, and illegal international traffic.

UNEP Chemicals (*http://www.unep.ch*)—Provides a network for the sound management of hazardous chemicals across the globe by providing access to information on toxic chemicals and by facilitating global actions to reduce or eliminate chemicals risks and by assisting countries in building their capacities for safe production, use and disposal of hazardous chemicals.

UNITAR National Profiles in Chemical Management (*http://www2.unitar.org/cwm/nphomepage/np2.html*)—A National Profile report is a comprehensive overview and assessment of the existing national legal, institutional, administrative, and technical infrastructure related to the sound management of chemicals in the context of Chapter 19 of *Agenda 21*. In developing countries and countries with economies in transition, National Profiles have served as a useful basis for identifying national chemicals management priorities and for initiating targeted and coordinated follow-up action in developing countries and emerging economies. By following the recommendations issued by the Intergovernmental Forum on Chemical Safety (IFCS) and the SAICM, and based on the IFCS-endorsed UNITAR/IOMC National Profile Guidance Document of 1996, countries have initiated preparation of National Chemicals Management Profiles with the involvement of those multi-stakeholders. These efforts have been recognized as a key element of SAICM implementation and are endorsed by three important SAICM reports—the Overarching Policy Strategy (OPS), the Global Plan of Action (GPA), and the ICCM Resolution that sets out the strategic priorities of the Quick Start Programme (QSP). The Web-based reports provide direct access to the National Profiles that have been developed through cooperation of UNITAR and the European Chemicals Bureau (ECB) of the European Commission in Ispra, Italy. The "National Profiles Homepage"—is maintained and updated by UNITAR and ECB with information provided by UNITAR, the IFCS Secretariat, and participating countries.

UNEP's GEO DATA Portal (*http://geodata.grid.unep.ch*)—The GEO data portal of the UNEP is the authoritative source for datasets in the Global Environmental Outlook report and integrated environmental assessments. The Web-based database holds more than 500 variables covering regional, national, and global statistics and geospatial maps of freshwater resources, population, forestry, emissions, climate, disasters, health, and so on.

UNFCC Data Portal (*http://unfccc.int/ghg_data/items/4133.php*)—The UNFCC has a vast array of datasets and databases pertaining to Green House Gases (GHGs). Some of the relevant datasets and databases include: Overview of GHG emission trends in Annex I Parties, UNFCCC data on GHG emissions in individual Annex I and non-Annex I Parties, UNFCCC data under the Kyoto Protocol in individual Annex I Parties that are Parties to the Kyoto Protocol, Data on population and gross domestic product (GDP), and References to non-UNFCCC data sources.

Online Access to Research on the Environment (*OARE*) (*http://www.oare-sciences.org/en/*)—OARE is an international public–private consortium coordinated by the United Nations Environment Programme (UNEP), Yale University, and leading science and technology publishers, and enables developing countries to gain access to one of the world's largest collections of environmental science research databases.

WHO IPCS INCHEM Web Portal—Information pertaining to the International Programme on Chemical Safety: Chemical Safety cards, Poison information monographs, Environmental Health Criteria, IARC Cancer Monographs, pesticide datasheets, and so on.

International Labor Organization (*ILO*) (*http://www.ilo.org/public/english/protection/safework/cis/index.htm*)—The ILO provides a database of Web links to exposure limits information from various countries. Additionally, information related to specific ILO publications such as the SafeWork Bookshelf, which is a collection of key Occupational Safety and Health documents produced by the ILO. It was compiled by the information arm (CIS) of the SafeWork Programme of the ILO.

OECD Portal and Databases

eChemPortal (*http://webnet3.oecd.org/echemportal*)—eChemPortal provides free public access to information on chemical properties and direct links to collections of information prepared for government chemical review programs at national, regional, and international levels. It is an effort of the Organization for Economic Cooperation and Development (OECD) in collaboration with the European Commission, the United States, Canada, Japan, the International Council of Chemical Associations, the Business and Industry Advisory Committee, the World Health Organization's International Programme on Chemical Safety, the United Nations Environment Programme on Chemicals and environmental nongovernmental organizations.

Pollutant Release and Transfer Registers (*PRTR*) *database* (*http://www.prtr.net/en/about/*)—The OECD PRTR is a tool for member countries governments to provide data to the public about potentially toxic releases to the environment. It is also a database or register of the quantities of potentially harmful chemicals released to air, water and soil and/or transferred off-site for treatment. The objective is to share PRTR data as widely and as effectively as possible. http://www.prtr.net/en/links/ provides links to various OECD member countries PRTR tool and Web sites.

Integrated High Production Volume (*HPV*) *Chemicals Database* (*http://www.oecd.org/dataoecd/55/38/33883530.pdf*)—In this database, the status of all HPV chemicals are recorded and it contains the list of all HPV chemicals together with any annotations on each chemical. Each chemical's stage in the assessment process is identified. The list of HPV Chemicals is a priority list from which chemicals are selected for data gathering and testing, and initial hazard assessment process. The list is compiled by the OECD Secretariat based on regular submissions by the member countries in which they are produced or imported in amounts greater than 1000 tonnes per annum.

International Directory for Emergency Response (ER) Centres (http://www.oecd.org/dataoecd/0/39/1933386.pdf)—This directory is a jointly published by three organizations: the OECD; the UNEP-DTIE (United Nations Environment Programme—Division of Technology, Industry and Economics); and the Joint UNEP/OCHA (Office for the Coordination of Humanitarian Affairs) Environment Unit. These ER centers are located in both OECD and non-OECD countries. The objective of such a directory is to increase and facilitate information access and assistance provided by ER centers located globally. The 2002 version of the directory can be accessed from the Web link provided above.

EUROPE

Several European countries have their National Country Profiles on Chemical Management, databases, regulatory and nonregulatory resources listed at the UNITAR Web site (http://www2.unitar.org/cwm/nphomepage/np2.html).

Selected European Union and European Commission Toxicology and Risk-Assessment Databases

European Chemical Substances Information System (ESIS): This is a Web portal maintained by the Consumer Products Safety and Quality unit of the EU. The ESIS includes information pertaining to the European inventory of Existing Chemicals (EINECS), the European List of Notified Chemical Substances (ELINCS), High Production Volume Chemicals (HPVCs) and Low Production Volume Chemicals (LPVCs), Classification and Labeling, International Uniform Chemical Information Database (IUCLID) and the EU's Chemical Risk-Assessment process. IUCLID (version 5.2.2 as of 09/08/2010 available at http://iuclid.eu/) is the primary software tool for data collection and evaluation within the EU Risk-Assessment Programme as well as OCED's Existing Chemical Programme.

European Pollutant Release and Transfer Register (http://www.eea.europa.eu/data-and-maps/data/eper-the-european-pollutant-emission-register-3)—This publicly available database was developed as a register of industrial emissions to air and water. EU member countries report emissions on over 50 contaminants from industrial facilities into the air and water. The information in the database can be accessed according to pollutant, industry sectors, air emissions or water discharges and country-specific information. The information can also be presented visually on a map.

REACH Substance Information Exchange Forum (SIEF) (http://echa.europa.eu/doc/reachit/sief_key_principles.pdf)—This forum was created to help REACH registrants of the same chemical substance share information so that duplication (e.g., toxicology testing, etc.) can be minimized. This forum has a critical role in terms of collaboration and data sharing so that the use of toxicology information becomes more efficient via a collaborative platform. The REACH Information Exchange System (SIES) provides a platform for joint registration dossiers.

REACH IT (*http://echa.europa.eu/reachit_en.asp*)—The REACH-IT portal is the main channel for companies to submit data to the European Chemicals Agency (ECHA). The portal provides an online platform to submit data and dossiers (late pre-registration, registration, etc.) on chemical substances. REACH-IT Industry User Manuals have step-by-step instructions on how to use REACH-IT. The manual can be accessed from the REACH-IT Web portal. The portal remains open on weekdays and can be accessed at: https://reach-it.echa.europa.eu/

European Commission (EC) Joint Research Centre (JRC) Institutes (*http://ec.europa.eu/dgs/jrc/index.cfm?id=5270*)—The EC's JRC is a technical source of information in toxicology, exposure assessments, and risk assessments. This is also a reference centre for European policy makers, European Commission. There are seven institutes under the JRC (http://ec.europa.eu/dgs/jrc/index.cfm?id=5270), which cover a wide area of technical areas in chemical toxicology, human health risk assessment, health and consumer protection, product chemical safety, research and development, energy, and materials science. The European Centre for Validation of Alternative Methods (ECVAM) is a vigorous proponent of reducing animal use in chemical testing by means of *in vitro* toxicity testing and computational tools use in toxicological data analysis.

SWEDEN

The Swedish Chemical Agency (*KemI*) is a central agency under the Swedish government which works in the areas of chemical control and management, and ensuring that the industries manufacturing and importing chemicals take the responsibility for the safety of the chemicals available to consumers. For additional information on Keml, refer to http://www.kemi.se/upload/Trycksaker/Pdf/Broschyrer/international_secretariat.pdf

Representative databases available from KemI:

N-class: Environmental Hazard Classification (*http://apps.kemi.se/nclass/default.asp*)—Contains chemical classification information for more than 7000 substances. The primary date refers to environmental effects as the basis of these classifications. Fire and health-related information is also included in the database.

PRIO (*http://kemi.se/templates/PRIOEngframes_4144.aspx*)—A Web-based tool primarily used to evaluate environmental and human health risk reduction and prevention from environmental chemical exposures.

RiskLine: A bibliographic toxicology database that contains more than 7000 bibliographic references and peer-reviewed information pertaining to 3000 chemicals. Information on both environmental effects, ecotoxicology and human health effects can be found in this database. Last updated in 2007.

FINLAND

KAMAT (*Information cards on exposure to chemicals in metal and car industry*) (*http://www.ttl.fi/Internet/partner/kamat*)—This database is maintained by the Finnish Institute of Occupational Health and contains information on exposure to chemicals in the metal and automobile industries in Finland. Resources listed are based on job profile, chemical groups, health risks, diseases related to specific occupations, and exposure and mitigation of risks.

KETU (*The Product Register of the Chemicals Register*) (*http://www.sttv.fi/kemo/TURE/In-English.htm*)—This database lists chemical products in commerce in Finland and is maintained by the National Product Control Agency for Welfare and Health (STTV).

NETHERLANDS

Historically, the Netherlands has shown exceptional leadership in promoting the science of toxicology and risk assessment. Chemicals Management has been a subject of critical discussion from the perspective of assessment and management of chemicals. In order to protect public health and the environment, both regulatory and nonregulatory agencies in Netherlands have taken various initiatives in the areas of chemical management.

Chemical Zone (*http://www.chemiezone.nl/*)—This Web link provides extensive information on chemical exposure, risk, and public health and contains several online databases in these areas.

Chemical Substances Bureau (*http://www.rivm.nl/bms/english*)—The Bureau performs regulatory work with regard to new and existing chemicals. Other projects from RIVM include publication and maintenance of various guidelines and toxicity profiles of chemicals. Various international toxicological databases have recognized the importance of RIVM databases and have included these databases as a major source of peer-reviewed toxicological values. F or a list of RIVM-related publications and other resources, see http://www.rivm.nl/en/.

RUSSIA

With an emerging economy, there has been rapid progress in various industrial sectors in Russia, especially in the chemical and petrochemical sectors.

Russian Register of Potentially Hazardous Chemical and Biological Substances (*The Federal State owned Establishment of Public Health, FSEH*) (*http://www.rpohv.ru/lang/en/*)—Since 2004, state registration certificates have been issued by the federal service for surveillance on Consumer Rights Protection and Human Wellbeing based on toxicological and hygienic expertise carried out by the Russian Register of Potentially Hazardous Chemical and Biological Substances (RRPHCBS). The information card provided by RRPHCBS contains physical, chemical, toxicological, and environmental properties of chemicals.

AFRICA

South Africa

National Chemical Monitoring Programme (*NCMP*) (*http://www.dwaf.gov.za/iwqs/water_quality/NCMP/default.htm*)—This programme provides reporting on the chemical quality of South Africa's surface waters.

National Toxicity Monitoring Programme (*http://www.dwaf.gov.za/iwqs/water_quality/ntmp/index.htm*)—This monitoring programme measures and evaluates reports on the status and trends of the nature and extent of potentially toxicants in South African waters.

Pesticide Registration Information for South Africa (*http://www.pesticideinfo.org/Detail_Country.jsp?Country=South%20Africa*)—This database lists pesticide chemicals registered for use in South Africa. Registration, toxicity, and other relevant information for those chemicals are provided. Additional information on the status of pesticides as legally registered for use, banned, cancelled or severely restricted are available.

Chemical and Allied Industries Association (*http://www.caia.co.za/index.htm*)—CAIA was established in 1993. Since 1994, CAIA launched the "Responsible Care Initiative" to address public concerns related to manufacture, storage, transport, use, and disposal of chemicals.

Toxicology Society of South Africa (*TOXSA*) (*http://www.toxsa.up.ac.za*)—TOXSA was established in 2001 to promote and advance the study and application of toxicology in South Africa.

AUSTRALIA

Australian Inventory of Chemical Substances (*AICS*) (*http://www.nicnas.gov.au/industry/AICs*)—The chemical inventory is maintained by National Industrial Chemicals Notification and Assessment Scheme (NICNAS) with a list of more than 38,000 chemicals that are currently in use in Australia. These chemicals can be imported and manufactured without any special notification to NICNAS, unless they are exempted.

National Industrial Chemicals Notification and Assessment Scheme (*NICNAS*) (*http://www.nicnas.gov.au*)—NICNAS assesses applications and information pertaining to new chemicals that are planned to be introduced into the Australian Environment. NICNAS also reviews existing priority chemicals that are listed in the AICS. The advisory group on Chemical Safety (AGCS) has been established under NICNAS.

Australian OHS Resources (*http://www.ohs.anu.edu.au/links/index.php*)—The National Research Centre for Occupational Health and Safety Regulation lists OHS-related Web links that provide information on Australian OHS legislation, authorities, OHS centers, and so on.

Australian Centre for Environmental Contaminants Research (*CECR*) (*http://www.clw.csiro.au/cecr*)—This center conducts contaminants research in various environmental matrixes (e.g., soil, water, air, sediment).

NORTH AND SOUTH AMERICA

United States

National Library of Medicine

National Library of Medicine (NLM), the world's largest biomedical library (http://www.nlm.nih.gov), is part of the National Institutes of Health (NIH), and a forerunner and innovator in the utilization of modern and emerging information technologies for the creation and distribution of biomedical information. NLM's Division of Specialized Information Services (SIS) houses the Toxicology and Environmental Health Information Programme (TEHIP) {http://sis.nlm.nih.gov/enviro.html}. This program offers a wide array of scientific databases of use to scientists, among others, for helping to assess chemical hazards. Just a few of the relevant chemical and toxicological databases are highlighted here.

ChemIDplus—Identifying chemicals, of course, is a critical first step to knowing how to deal with them. ChemIDplus provides access to structure and nomenclature authority files used for the identification of chemical substances cited in NLM databases, including the TOXNET (toxnet.nlm.nih.gov) system. ChemIDplus also provides structure searching and direct links to many biomedical resources at NLM and on the Internet for chemicals of interest. The database contains some 400,000 chemical records, of which about 300,000 include chemical structures, and is searchable by Name, Synonym, CAS Registry Number, Molecular Formula, Classification Code, Locator Code, Structure, Toxicity, and/or Physical Properties.

Hazardous Substances Data Bank (HSDB)—HSDB focuses on the toxicology of potentially hazardous chemicals and radionuclides. It is enhanced with information on human exposure, industrial hygiene, and emergency handling procedures, environmental fate, regulatory requirements, and related areas. All data are referenced and derived from a core set of books, government documents, technical reports, and selected primary journal literature. HSDB is peer-reviewed by the Scientific Review Panel (SRP), a committee of experts in the major subject areas within the data bank's scope. HSDB is organized into individual chemical records, and contains over 5000 such records. Recent efforts have been made to enhance HSDB with information on nanoparticles.

TOXLINE—A bibliographic file, such as TOXLINE, offers access to literature citations, typically with index terms and abstracts. Part of the broader TOXNET system, TOXLINE records provide bibliographic information covering the biochemical, pharmacological, physiological, and toxicological effects of drugs and other chemicals. It contains more than 3 million bibliographic citations, most with abstracts and/or indexing terms and CAS Registry Numbers. TOXLINE references are drawn from various sources organized into component subfiles which are searched together but which may be used to limit searches as well.

Household Products Database—The Second International Conference on Chemicals Management, held in Geneva in 2009, identified Chemicals in Products as an emerging issue. The Household Products Database links over 8000 consumer brands to health effects from Material Safety Data Sheets (MSDS) provided by manufacturers and allows users to research products based on chemical ingredients. The database includes information on chemical ingredients and their percentages in specific brands, manufacturer name and contact information, acute and chronic effects of chemical ingredients, and links to additional information on the ingredients from other NLM databases.

HazMap—An occupational health database designed for health and safety professionals and for consumers seeking information about the health effects of exposure to chemicals and biologicals at work. Haz-Map links jobs and hazardous tasks with occupational diseases and their symptoms.

International Toxicity Estimates for Risk (*ITER*)—ITER is a product of Toxicology Excellence for Risk Assessment (TERA) and contains over 650 chemical records with key data from the Agency for Toxic Substances & Disease Registry (ATSDR), Health Canada, National Institute of Public Health & the Environment (RIVM)—The Netherlands, U.S. Environmental Protection Agency (US EPA), the International Agency for Research on Cancer (IARC), NSF International, and independent parties whose risk values have undergone peer review.

US Environmental Protection Agency

Risk-Assessment Portal (*http://www.epa.gov/risk_assessment/*)—The Risk-Assessment portal provides basic information about environmental health risk-assessment information for public use. Additionally, the portal provides comprehensive guidance documents, reports, risk-related databases and Web-based tools.

Integrated Risk Information System (*IRIS*) (*http://www.epa.gov/IRIS/*)—US EPA IRIS contains over 500 chemical records. IRIS data, focusing on hazard identification and dose–response assessment, is reviewed by working groups of EPA toxicologists, risk assessors, and scientists. Among the key data provided in IRIS are EPA carcinogen classifications, unit risks, slope factors, oral reference doses, and inhalation reference concentrations. IRIS is also available via NLM's TOXNET system.

Human Exposure Database System (*HEDS*) (*http://epa.gov/heds/index.htm*)—This is an integrated Web-facilitated database which includes chemical measurements, questionnaire responses, reports and documents and other information resources related to EPA research studies of the exposure of populations to environment contaminants.

Human Exposure Factors Handbook (*http://cfpub.epa.gov/ncea/cfm/recordisplay.cfm?deid=20563*)—The EPA has been actively researching exposure pathways related to various environmental components (soil, air, water, food, etc.) for the purposes of human health risk evaluations.

National Centre for Environmental Assessment (*http://www.epa.gov/ncea*)—The National Centre for Environmental Assessment (NCEA) of US EPA's ORD office provides guidance on risk assessment to better protect human health and environment. To achieve these broad objectives, NCEA develops various tools, software, models, and databases to facilitate human health risk assessments. The Exposure Factors Handbook provides statistically valid data on various human exposure factors and assumptions used in the exposure analysis stages of the risk-assessment project. The 2008–2009 updated version of the handbook provides up-to-date information on various exposure factors.

Toxics Release Inventory (*TRI*) (*http://www.epa.gov/tri*)—A publicly available EPA database that contains information on toxic chemical releases and waste management activities reported annually by certain industries as well as federal facilities. The TRI program compiles the TRI data each year and makes it available through downloadable files and several data access tools. The goal of TRI is to provide communities with information about toxic chemical releases and waste management activities and to support informed decision making at all levels by industry, government, nongovernmental organizations, and the public. Also available via NLM.

Health and Environmental Research Online (*HERO*) (*http://www.epa.gov/hero/*)—The HERO database contains up-to-date and key studies that the EPA uses to develop environmental risk assessments to characterize the nature and magnitude of health risks to humans and the ecosystem from exposure to chemicals in the environment. One can browse through bibliographic references in HERO by technical topic or scientific assessment. Articles and reports can be searched by author, title, and so on.

Toxicity testing for the 21st Century related Projects and toxicological informatics resources (*http://www.epa.gov/spc/toxicitytesting/index.htm*)—In light of toxicity testing for the twenty-first century projects from the National Research Council (NRC), the National Academy of Sciences (NAS), and the emerging applications of REACH regulation and its impact on global chemical policy, there is a strong desire among the toxicological and risk-assessment community of practice to standardize medium and high throughput toxicological assays to understand the biological and toxicological pathway basis of chemical toxicology. Ultimately, this information would be applied to the next generation of chemical risk assessments to understand population level risks associated with exposure. These emerging approaches and information gathered from these approaches may eventually influence and refine global chemical policy. Currently, significant work is taking place in chemical and toxicological informatics and predictive (computational) toxicology.

AcTOR (http://www.epa.gov/actor/)—ACToR (Aggregated Computational Toxicology Resource) is a collection of databases developed by the US EPA National Center for Computational Toxicology (NCCT). More than 200 sources of environmental chemical information were compiled together that are searchable by chemical name, structure, and other unique identifiers such as physico-chemical values, *in vitro* assay data and *in vivo* toxicology data.

Agency for Toxic Substances and Disease Registry

In the United States, the Comprehensive Environmental Response, Compensation, and Liability Act (CERCLA) mandates that the Agency for Toxic Substances and Disease Registry (ATSDR) (http://www.atsdr.cdc.gov) determines whether adequate information on health effects is available for the priority hazardous substances. Under this public health mandate, toxicology and public health-related information on chemicals are provided to prevent harmful exposure and diseases. These databases and toxicology profiles are extensively peer-reviewed and used by various agencies around the world.

Toxicological Profile Database (*http://www.atsdr.cdc.gov/toxpro2.html*)—These profiles are based on the hazardous substances found at National Priorities List (NPL) contaminated sites and are ranked based on frequency of occurrence of those chemicals at NPL sites, toxicity, and potential for human exposure. The ATSDR also prepares toxicological profiles for the Department of Defense (DOD) and the Department of Energy (DOE) on substances related to federal sites. The profiles are also used to derive more succinct documents such as ToxFAQs and Public Health Statements.

Interaction Profile Database (*http://www.atsdr.cdc.gov/interactionprofiles/*)—One of the emerging issues in toxicology is the interaction between different chemicals. To address this from a human health effects perspective, ATSDR has published various toxicological profiles looking into the interaction between chemicals within a chemical mixture. This database evaluates information on the toxicology of the "whole" priority mixtures (if available) and on the joint toxic action of the chemicals in the mixture and recommends various approaches for the exposure assessment of the potential hazards to public health.

A Few More US Databases

American Conference of Governmental Industrial Hygienists (ACGIH) Database (http://www.acgih.org/TLV/)—ACGIH databases include guidance documents, manuals and chemical specific information related to Threshold Limit Value (TLVs) and Biological Exposure Indices (BEI). These occupational exposures-related documents and databases summarize information based on the chemical identity, physical, chemical, and toxicological properties, and occupational exposure scenarios.

National Institute for Occupational Safety and Health (NIOSH) Databases (http://www.cdc.gov/niosh/database.html)—The NIOSH Web Site lists various databases and collections of guidance manuals and information related to Chemicals, Injury, Illness and Hazards Data and Information, Respirators and other Personal Protective Equipment, Agriculture, Construction-related occupational exposure and toxicology information. Some relevant databases are the International Chemical Safety Cards, NIOSH Pocket Guide to Chemical Hazards, and NIOSHTIC-2. NIOSHTIC-2 is a bibliographic database of occupational safety and health publications, documents, and grant reports supported in whole or in part by the NIOSH.

Occupational Safety and Health Administration (OSHA) Database (http://www.osha.gov/web/dep/chemicaldata/)—The OSHA/EPA Occupational Chemical Database is a database portal jointly developed and maintained by these U.S. Federal agencies as a reference for the occupational safety and health communities. This database consists of information from several government agencies and organizations. Information can be searched according to the requirements of an occupational practitioner, and covers areas such as "Physical Properties," "Exposure Guidelines," "NIOSH Pocket Guide," and "Emergency Response Information."

RAIS (Risk-Assessment Information System) (http://rais.ornl.gov)—Sponsored by the U.S. Department of Energy, the Risk-Assessment Information System is a tool kit and database that lists toxicity profiles of chemicals, toxicological values, preliminary remediation goals for screening of chemicals from contaminated sites, various risk models, ecological benchmarks, background values of chemicals, gamma radiation instrument response tool, and radionuclide decay chain tool. In addition to chemical specific risk-assessment information, various guidance documents and US EPA tools are also listed to help risk assessors to conduct both screening level and site-specific risk-assessment projects.

Canada

Government of Canada—Chemical Substances Chemical Management Plan (http://www.chemicalsubstanceschimiques.gc.ca/en/index.html)—To better protect the health of Canadians, the Government of Canada has undertaken this ambitious plan to effectively manage chemical risk in the Canadian environment. Several collaborations are underway between various federal departments in Canada (e.g., Health Canada and Environment Canada) and industry partners to assess, manage and communicate risks.

CCOHS Web-based information services (*http://ccinfoweb.ccohs.ca/about.html*)—The Canadian Centre for Occupational Health and Safety maintains various OHS-related databases as part of their Web-based information services.

Health Canada—Consumer Product and Safety (*http://www-hc-sc.gc.ca/cps-spc/index_e.html*)—Using a risk-based framework, Health Canada protects the health of Canadians by researching, assessing and collaborating in the management of the health risks and safety hazards associated with the many consumer products, including pesticides and other chemicals.

Health Canada's Pesticides and Pest Management Programme (*http://www.hc-sc.gc.ca/cps-spc/pest/index-eng.php/index-eng.php*)—Manages pest control products and ensures that public health is protected based on scientific evaluations of those chemicals and products.

Also note: Health Canada's product safety program provides enforcement and compliance-related services and conducts assessment and collaborative work in the management of the health risk and safety hazards associated with children's products, household products (including chemical products), cosmetics and personal care products, new chemicals substances, biotechnology products, and workplace-related chemicals. Furthermore, Canada has also worked with other countries to harmonize existing hazard communication systems on chemicals in order to develop a single, Globally Harmonized System (GHS) to address classification of chemicals in terms of their hazards, and to communicate risk-based information through labels and safety data sheets (http://www.hc-sc.gc.ca/ahc-asc/intactiv/ghs-sgh/index-eng.php).

Environment Canada—National Pollutant Release Inventory (*NPRI*) (*http://www.ec.gc.ca/inrp-npri/*)—The NPRI is a legislated countrywide and publicly available chemical emission inventory in Canada. The database contains information related to annual release of chemical contaminants to air, water, and land. Additional information related to chemical waste disposal and recycling from all sectors such as industrial activities, government and commercial sectors is also included.

Federal Contaminated Sites Inventory (*FCSI*) (*http://www.tbs-sct.gc.ca/fcsi-rscf/home-accueil.aspx*)—In Canada, the Federal Contaminated Sites and Solid Waste Landfills Inventory Policy requires that government departments (otherwise known as custodian departments) and agencies that are responsible for those contaminated sites establish and maintain a database. As a minimum requirement, relevant information is submitted at least once a year to the FCSI database. The revised version of the FCSI released in March 2006, contains a more complete set of data on federal sites as well as enhanced reporting capabilities, including mapping technologies for better visual presentation of contaminated sites across Canada.

Arctic Contaminants Database (*http://www.hc-sc.gc.ca/ewh-semt/pubs/eval/inventory-repertoire/arcticdb-eng.php*)—This database was set up in 1996 to monitor several heavy metals and other contaminants in tissues (blood and breast milk) in indigenous arctic populations. Similar projects have been ongoing to monitor chemical contaminants under the Northern Contaminants Programme led by Indian and Northern Affairs Canada in collaboration with other federal departments and researches (http://www.ainc-inac.gc.ca/nth/ct/ncp/index-eng.asp).

Mexico

SOMTOX (Sociedad Mexicana de Toxicologia, A.C—Mexican Society of Toxicology) (http://www.somtox.com.mx)—SOMTOX was founded in 1994 by members of the Mexican chapter of the Latin American Society of Toxicology (ALATOX). Based on the objectives of SOMTOX, it shares and collaborates on toxicology and risk-assessment-related project information.

Brazil

SINITOX—National System of Toxi-Pharmacological Information (http://www.fiocruz.br/sinitox)—The National Poison Information (SINITOX) coordinates the collection, compilation, analysis, and dissemination of intoxication and poisoning cases reported in Brazil. The records are held by the National Network of Information Centers and Toxicological Assistance (Renaciat), which comprises of 35 units located in 19 states. The results of the study are published annually.

Pesticide Information System—SIA (http://www4.anvisa.gov.br/agrosia/asp/default.asp)—This Web site includes pesticide information in Portuguese.

Chemical Regulation, Toxicology, and Health Sciences Resources—For various chemicals, agrochemicals, cosmetics, and drugs legislation in Brazil, readers can refer to http://www.anvisa.gov.br. In the Virtual Library of Health—Toxicology (http://tox.anvisa.gov.br/html/pt/home.html) readers can access various toxicology- and health-related resources. Brazil is also a collaborating country for the Latin American and Caribbean Centre on Health Sciences Information (http://www.bireme.br).

Chile

The following lists of resources from Chile are representative of chemical toxicology, risk assessment, and management in the country.

Prevention and Management of Poisonings caused by chemical substances in major incidents: http://epi.minsal.cl/epi/html/public/bioter/aquim/aqpreven.doc

Epidemiological Situation of Acute Poisoning by Pesticides in Chile: http://epi.minsal.cl/epi/html/vigilan/revep/intox1998.pdf

Poison Information and Drug Information Centre, Chile: (http://www.cituc.cl/)

Costa Rica

Government-Related Resources—Resources from the Ministry of Environment, Energy, and Telecommunications can be obtained from http://www.minae.go.cr/

acerca/. Also, the Costa Rican government has several bilateral agreements with countries like Canada on chemical and environmental cooperation. Resources related to those agreements can be obtained from http://www.ec.gc.ca/caraib-carib/default.asp?lang =En&n=AFD03174-1.

Online Resources Listing Multilateral Environmental Agreements

European Commission's Treaties Office Database—http://ec.europa.eu/world/agreements/default.home.do

This Database contains all the bilateral and multilateral treaties or agreements concluded by the European Union, the European Atomic Energy Community (EAEC), and the former European Coal and Steel Community (ECSC), and those concluded under the Treaty on European Union. The database has a summary and the full text of each international treaty/agreement as well as analytical search facilities. Users can access the treaties database by several categories (bilateral, multilateral, counties, organization, and activity).

ENTRI—Environmental Treaties and Resource Indicators (*http://sedac.ciesin.org/entri/*)—ENTRI is an online service that provides easy access to multilateral environmental treaties information to users. It lists status updates, texts, and other information resources relevant for science and policy communities and public users. The "decision search tool" tab allows users to find decision level information from relevant treaties from conferences of the parties (CoP). Other relevant information is related to country profiles, country-level data and specific types of treaties related various environmental and chemical agreements.

International Environmental Agreement (*IEA*) *Database Project* (*http://iea.uoregon.edu/page.php?file=home.htm&query=static*)—The IEA database includes 1000 Multilateral Environment Agreements, over 1500 Bilateral Environment Agreements, and over 250 other environmental agreements. It is grouped by Date, Subject, and "Lineage" of legally related agreements (e.g., those related to the Montreal Protocol). "Other" includes environmental agreements between governments and international organizations or nonstate actors, rather than two or more governments.

Chemical Policy (*http://www.CHEMICALPOLICY.org*)—The Chemical Policy Web site lists state level regional (USA), continent and global chemical policy, treaties and regulations information.

Canada Treaty Information (*http://www.treaty-accord.gc.ca/*)—This Government of Canada portal for Department of Foreign Affairs and International Trade lists various treaties and bilateral and multilateral agreements where Canada is a party to the treaty and/or agreement.

Environment Canada (*International Affairs*) (*http://www.ec.gc.ca/international/multilat/mea_e.htm*)—Environment Canada is the federal government's environmental department. In its international affairs Web site, a list of MEAs related to air, oceans, biodiversity, chemicals, hazardous wastes, and environmental cooperation can be found. Under the chemicals category, the Rotterdam Convention and the Stockholm Convention on Persistent Organic Pollutants (POPS) are described.

HEALTH AND ENVIRONMENT COLLABORATION EFFORTS AND THE ROLE OF INFORMATICS TOOLS

Health and Environment Linkages Initiatives (HELI)—A Global effort led by the World Health Organization (WHO) and the United Nations Environment Programme (UNEP) to promote action in developing countries to reduce environmental health threats and to support sustainable development objectives. At ICCM 2009 in Geneva, discussions on collaborations between the health sector and implementation of SAICM took place. At the time of writing this chapter, the SAICM Secretariat had already developed a concept note on engagement of the global health sector. The draft strategy will be made available for consultations with all SAICM stakeholders and the strategy would then be considered for review by the third session of the ICCM in 2012.

From the informatics tools application perspective, effective information sharing and collaboration between and utilization of various toxicology and risk-assessment resources by global communities can further enhance and link policy and science communities.

An example is provided below to emphasize information sharing related to chemical exposure and environmental health.

Chemical Health Monitor (CHM) Project (http://www.chemicalshealthmonitor.org/)—The CHM project is from the Health and Environment Alliance in collaboration with other European agencies and organizations. It contributes to the tools and structures essential for understanding the REACH labyrinth and provides necessary information and views about key decisions voiced. Information related to diseases and chemical exposure, human biomonitoring, chemical policies (including REACH), and other information resources related to toxicology, environmental health risks, and chemical policies can be found on this Web site.

THE NEXT GENERATION OF CHEMICAL AND TOXICOLOGICAL INFORMATICS

With recent advances in toxicology and informatics tools to support computational and predictive toxicology, efforts are in progress in various countries (e.g., US EPA TOXCAST, EU OPENTOX, etc.). TOXCAST and related predictive toxicology initiative and programs have been previously described. Below, the EU OpenTox project is briefly described to further show the relevance of informatics tools to evaluating animal and human health effects from exposure to chemicals.

EU OpenTox Project (http://www.opentox.org/)—OpenTox is a predictive toxicology framework that is based on a toxicology service delivery-oriented architecture that provides user access to toxicological resources including data, computer models, and validation and reporting. The collaborative platform of the OpenToX involves a combination of academia and government research groups and some industries to design and build the initial OpenTox framework. Furthermore, numerous organizations associated with industries, regulatory or toxicoinformatics experts provide guidance and recommendation to further refine the framework. The goal of the OpenTox Community of Practice (CoP) is to expand and include various toxicology and risk-assessment projects that may enable experts and users of the framework to be involved in further developments and refinements and capacity building in predictive toxicology. Further collaboration on data standards, data integration, toxicological ontologies (i.e., consistent language and classification of various toxicological and related terminologies), and integration of toxicological effects data from different methodologies, and testing and validation are envisioned.

CEBS—Chemical Effects on Biological Systems Knowledgebase (*http://www.niehs.nih.gov/research/resources/databases/cebs/index.cfm*; *http://tools.niehs.nih.gov/cebs3/ui/*)—This public resource accepts data from academic, industrial, and governmental laboratories and is designed to provide data from the perspective of biology and study design, and also allows data integration across various studies for unique meta-analysis.

CTD—Comparative Toxicogenomics Database (http://ctd.mdibl.org/)—The CTD knowledgebase includes manually curated data that describes cross-species chemical–gene/protein interactions and chemical– and gene–disease relationships to emphasize and/or hypothesize various molecular mechanisms underlying susceptibility and environmentally influenced disease incidences.

EDKB—Endocrine Disruptor Knowledgebase (http://edkb.fda.gov/)—The EDKB consists of a biological activity database, peer-reviewed literature, computational models, and models for health risk assessments. This knowledge base is designed to help researchers and regulators, and other stakeholders to prioritize endocrine disrupting chemical testing by using the existing knowledgebase. These analyses reduce traditional animal toxicology experiments where the process of analysis may be slow and inefficient.

THE WORLD LIBRARY OF TOXICOLOGY

The World Library of Toxicology, Chemical Safety, and Environmental Health is a free global Web portal (http://www.wltox.org) that provides the scientific community and public with links to major government agencies, nongovernmental organizations, universities, professional societies, and other groups addressing issues related to toxicology, public health, and environmental health. It seeks to overcome barriers to the sharing of information between countries by creating a partnership of national and international organizations contributing high-quality scientific information resources in toxicology and environmental health to a borderless data depository. The overall goal is to improve global public health. The World Library is compiled by an international group of scientists (Country Correspondents) who are responsible for the selection and maintenance of reliable, quality resources for their respective countries.

REFERENCES

Berners-Lee, T., J. Hendler, and O. Lassila. 2001. "The Semantic Web". Scientific American, May 17, 2001, available at: http://www.scientificamerican.com/article.cfm?id=the-semantic-web

Feigenbaum, L., I. Herman, T. Hongsermeier, E. Neumann, and S.Stephens. 2007. The Semantic Web in Action. *Scientific American*, 297: 90–97. Available at: http://www.thefigtrees.net/lee/sw/sciam/semantic-web-in-action, reproduced with permission. Accessed August 30, 2010.

US EPA. 2010. Environmental assessment process. United States Environmental Protection Agency, Health and Environmental Research Online. Available at: http://cfpub.epa.gov/ncea/hero/index.cfm?action=content.assessment. Accessed August 30, 2010.

Section IX

Future Outlook

51 Future Outlook and Challenges

Jan van der Kolk and Ravi Agarwal

CONTENTS

SUMMARY OF ACHIEVEMENTS

Since the 1972 Stockholm Conference (see Section II, Chapter 2 of this book), major achievements have been realized in international cooperation on chemicals management, including the adoption of international legally binding treaties, as part of the broader environmental agenda.

First, the international community has established a variety of frameworks and convened fora for discussing and agreeing on a common agenda (see Section V, Chapter 21 of this book on IFCS and Section IV, Chapter 17 of this book on SAICM) and has developed and accepted a number of important agreements that cover part of the overall agenda (see Section III, Chapters 7 through 16 of this book on the different instruments). These instruments have all found their ways into national legislation or policies, with differing levels of implementation, and are supported by scientific bodies or advisory mechanisms as a basis for decision making.

Second, clear goals have been set, in particular in Rio de Janeiro in 1992 (Section II, Chapter 3 of this book) and in Johannesburg during the World Summit on Sustainable Development (WSSD) (Section II, Chapter 4 of this book), most significantly "by 2020, to use and produce chemicals in ways that do not lead to significant adverse effects on human health and the environment.[. . .]" Some of these goals have translated into treaties such as the Basel, Rotterdam, and Stockholm Conventions (see Section III, Chapter 8, Section III, Chapter 14, and Section III, Chapter 15 of this book).

A third major achievement is that most mechanisms have been developed with increasing levels of multistakeholder participation. This has given form to one of the major principles of the Rio Declaration, *Principle* 10, on access to information, participation in decision making, and access to justice (see Section III, Chapter 7 on Aarhus and Section V, Chapter 19 of this book on Governance). The increasing commitment of diverse stakeholder groups to the implementation of common instruments and the further development of a common agenda is, indeed, a major achievement (see Section V, Chapter 20 on ICCA; Section V, Chapter 24 on IPCP; Section V, Chapter 25 on IPEN, and Section V, Chapter 28 of this book on professional societies).

SAICM has developed ways of monitoring progress in the objectives it has outlined, and has developed mechanisms to include emerging issues in the discussion.

MEETING THE JOHANNESBURG 2002 WSSD GOAL FOR 2020

One of the major outcomes of the WSSD 2002 in Johannesburg, South Africa, is the following goal:

> By 2020, to use and produce chemicals in ways that do not lead to significant adverse effects on human health and the environment, using transparent science-based riskassessment procedures and science-based risk management procedures, taking into account the precautionary approach, as set out in *Principle* 15 of the Rio Declaration on Environment and Development, and support developing countries in strengthening their capacity for the sound management of chemicals and hazardous wastes by providing technical and financial assistance.

Meeting this goal is a major driving force for current developments, and is a cornerstone for SAICM, the overarching policy for sound chemicals management. SAICM has been the international community's answer to the WSSD 2020 goal.

As we are now half way between Johannesburg 2002 and 2020, we can see the achievements to date and the gaps remaining to fully realize the objective. It is very clear that the current instruments at the current speed of their development and implementation cannot achieve that goal. Their implementation is lagging behind considerably in many countries and regions. The pace of development of further instruments is very resource intensive, and the interested parties do not seem to have a clear and common understanding of how all the different steps that are being taken would, together, achieve the goal set by the international community.

If the 2020 goal is to be achieved, much more effort should be put into implementation of existing agreements, including SAICM, and maximizing their potential, rather than stretching thin, limited capacity in new agreements of limited scope, such as those which focus on a single chemical, like mercury, rather than agree on framework agreements which have the potential to include other similar chemicals in the future or open up a broader scope for replacement of a range of hazardous products.

It has to be recognized that most countries have very limited capacity for implementation of the broader chemicals agenda and that further prioritization is necessary. This holds for all stakeholders—governments, private sector, NGOs, and academia. The limited energy might best be used by exploiting to the fullest extent possible the available resources for strengthening the implementation of existing instruments. Further, the integration of chemical safety into everyday business practices, and throughout the product life cycle is needed, rather than dealing with it as an externality.

MAJOR CHALLENGES

In spite of the many achievements, major challenges remain if the WSSD 2020 goal is to be met.

The Link to the General Sustainable Development Agenda is Weak

Chemicals management has not found its way to the core of the sustainable development agenda. Although in many ways chemicals management can contribute to poverty alleviation and to the realization of several of the Millennium Development Goals (MDGs) (see Section V, Chapter 29 of this book on UNDP), it seems that the chemicals management issues still get limited attention on those agendas as well as on those related to health and labor. This, notwithstanding the fact that chemicals play an important role in all sectors of life, with both positive and negative potential (see among others Section V, Chapter 18 of this book on FAO, and Section V, Chapter 20 of this book on ICCA). Also, concrete links to other major environmental topics, such as Biodiversity, Biosafety, and Climate Change have hardly been explored, although the use of chemicals is often closely related to problems or solutions in these areas.

The Chemicals Agenda is only Weakly Linked to other Major Development Agendas and Financing Mechanisms

Chemicals and hazardous waste management has mainly been considered a niche area, whereas there are clearly mutual influences between it and other fields, such as climate change or biodiversity, and their respective instruments.

If chemicals and hazardous waste management were more closely linked to other items on the sustainable development agenda, such as industrial development, agriculture, health care and preventive health, education, or climate change, it might obtain a more prominent place on the donor's agenda, stronger donor support and a higher political profile. But chemicals management, although part of virtually all sectors of development, is often considered as a technical area that holds no appeal to those involved in discussions on development assistance. Financial mechanisms to support its implementation have hardly been developed. Only a limited GEF (Global Environment Facility) window is available for certain areas of chemicals management, but its effectiveness is greatly limited by burdensome requirements on co-financing (see Section VIII, Chapter 49 of this book on global financial instruments).

The Need for a Sustainable Financing Mechanism

The need for a sustainable financing mechanism has been debated on many occasions (see Section II, Chapter 6 of this book on ICCM-2). Significant resources are needed, by some estimates more than US$ 9 billion may be needed for the implementation of the Stockholm Convention alone.* Large sums and capacities are also needed to remove obsolete stocks, cleanup contaminated sites and remove old abandoned factories and waste dumps. However, no solution has been found to date. The request for a widening GEF window may not be adequate because of the hurdles that GEF sets up on cofinancing. Bureaucratic mechanisms, as well, can put a heavy burden on implementing bodies (see among others Section III, Chapter 15 of this book on the Stockholm Convention). However, if no sustainable financial mechanism is to be developed, the implementation of the common agenda in many countries with the aim of meeting the 2020 goal will be severely affected.

One should also realize that any long-term financing mechanism for chemical safety can only be achieved if it is internalized as part of the product life cycle, and such costs are included as part of product cost. The chemical industry has sales of more than US$ 1.5 trillion (including pharmaceuticals) per year, and accounted for an estimated 7% of global income and 9% of international trade, in 2006.† These figures are increasing, and a tiny fraction of these amounts can subsidize chemical safety globally.

The Gap between Industrial and Nonindustrial Countries and the Need for Capacity Building and Implementation

Many times a widening gap between industrialized countries, including recently industrializing countries, mostly producers of chemicals, and other, less or unindustrialized countries has been mentioned. In several agreements and conventions, capacity building has been addressed and taken up as an obligation for industrialized countries.

* UNEP-led Consultative Process on Financing Options for Chemicals and Wastes Second consultative meeting, Bangkok, October 25–26, 2009 Desk study on financing options for chemicals and wastes.

† Mainstreaming the Sound Management of Chemicals into Development Planning: Background and Rationale, pp. 13 at: http://www.chem.unep.ch/unepsaicm/mainstreaming

But the reality is that such support is lagging behind significantly and that there is an ever-widening implementation gap for sound chemicals management between these industrialized countries and the others. Intergovernmental Organizations (IGOs) have not been able to sufficiently address this gap, due to lack of funding and perhaps also because of not always putting the convincing arguments on the table to attract more donor support. IGOs have tried to create mechanisms to increase the synergies between them and to integrate their programs, but much more could be achieved if the UN is to deliver as one UN (see Section V, Chapter 23 of this book on IOMC). There is currently a risk that the existing instruments and the new ones being developed will be implemented in a limited number of countries, while others lag behind with too limited capacity to even implement the currently existing instruments.

The Integration between Different International Instruments is Still Weak

Recently, synergies between the three major conventions—Basel, Rotterdam, and Stockholm—have become a major issue and have led to steps toward integration. However, the potential for greater integration and synergy has been far from being fully exploited. Major steps in this area are necessary, both to intensify integration between these three conventions and between them and other instruments and upcoming new agreements (see Section VIII, Chapter 48 of this book on Emerging Issues). A potential challenge to such synergies is the example of a new convention currently being negotiated on mercury. It is far from clear, for example, that it would offer the opportunity to include other chemicals if needed, or whether such chemicals would be part of yet further, time- and resource-consuming, separate negotiations.

Another example of the need for integration is a call for concrete national policies to absorb the complex relationship between chemicals and hazardous waste management and other global environmental issues. The enhancement of cooperation among the Basel, Rotterdam, and Stockholm Conventions is a step in that direction provided it is translated into national policies and action. It means that countries would need to integrate waste and chemical management issues into their sustainable development plans and strategies. It would also require a shift in the way global and regional funding and development institutions organize their work, to include waste and chemicals, as part of both their thinking and operation. Improving environmental governance is dependent on improving global governance. Environmental governance will enrich social and economic development and will be strengthened through enhanced cooperation, synergies and partnerships among the key international public bodies responsible for promoting and helping to achieve the United Nations MDGs.

CoPs have become Politicized on Technical Issues

In recent years, there has been a clear tendency toward discussions and decisions by Conferences of Parties (CoP) to a Convention, that are not essentially based on sound science, including the scientific advice from their own advisory bodies, but are more driven instead by political or economic arguments. This has been the case for example with certain substances such as chrysotile asbestos and endosulfan which have not been included in the Rotterdam Convention, notwithstanding convincing scientific arguments for the risk they present under practical conditions and the need to inform countries and users of these risks, which is the very purpose of the Rotterdam Convention.

CoPs should Give More Attention to Implementation Issues

Practice in many countries has shown many obstacles for real implementation of agreed upon instruments (see also Section VI, Chapters 35 through 42 of this book). Ratifying a Convention is an important step, but if this is not followed by adequate implementation, there is little actual change. The Conference of Parties (CoP) to a Convention needs to discuss in depth implementation issues, opportunities, and obstacles. They need to investigate causes for major gaps in implementation. Taking the Stockholm Convention as an example, while many countries have completed National Profiles (NP) and National Implementation Plans (NIP), very few have taken the second step to real implementation, because of multiple constraints. Similarly, the Conventions have little teeth to penalize for nonimplementation, nor are able to effectively enforce liability regimes. The Vienna Convention and Montreal Protocol (see Section III, Chapter 16 of this book) are examples of good results in implementation, giving it important focus and adequate funding.

Increasing Illegal Trade

One of the problems that is increasing in its size and its negative impacts on chemicals management, and therefore, potentially on health and environment, is that of illegal trade. Ever-increasing quantities of very hazardous chemicals, including pesticides, find their way to the world market without adequate control mechanisms. Although the problems are well known and have been well documented, there has been no adequate answer to these practices so far. The problems may be most significant in countries with limited implementation capacities and insufficient customs control.

Increased Synergies between IGOs

In recent years, "delivering as one UN" has received wide attention. This approach is meant to overcome fragmentation and increase impact and efficiency. If this intention is translated into practical, down-to-earth, collaborative approaches of the various IGOs, using each others' strengths, and proven methods and methodologies, and is based on clearly expressed country demands, this might be a promising development. So far, though, there is still room for major improvement in exploiting synergies of the various organizations.

Eliminate the Burden of the Past

In many places of the world, more or less controlled stocks of obsolete chemicals, pesticides, and other materials are stored under sometimes unacceptable conditions, if one considers protection of health and environment. The World Bank and the Food and Agricultural Organization (FAO) have initiated a major program to eliminate obsolete pesticides from a number of countries, but it does not come close to addressing the many other stockpiles present. Some other parties, including private players, have also taken action in this regard. Some of the stockpiles also result from former, now closed or abandoned industries or war periods. A significant number of abandoned industrial sites is present in Eastern European countries. Many of them have been reported in the NP and NIP under the Stockholm Convention. The laudable efforts to deal with part of the stockpiles need to be supplemented by a more comprehensive approach, covering many more countries and sites and not only pesticides.

Pay Sufficient Attention to Governance Issues

Sound management of chemicals is not a technical challenge only. It needs to be part of wider governance considerations, in terms of priority setting awareness raising, decision making, stakeholder' involvement, access to information, and to justice (see Section V, Chapter 19 of this book). A common language between technicians and policy makers and other stakeholders' needs to be developed that translates technical issues for a nontechnical audience and for decision makers.

Investigate Different Approaches to Sound Chemicals Management

Traditionally, chemicals are being sold through a trade chain from producer to end user. This means, in the best of cases, that adequate information is passed on at every step in the supply chain. As the distance between producer and user increases, the risk becomes greater that the end user will only have the label (ideally in conformity with GHS, Section III, Chapter 11 of this book) as a means of communication, or, in many cases, even not a label but an unlabeled container.

It may be of interest to investigate other ways of use of chemicals. Chemical Leasing, for example, is a new and innovative instrument to promote sustainable management of chemicals and close the material cycles between suppliers and users of chemicals ("closing the loops"), but needs further testing in practice (see Section V, Chapter 31 of this book on UNIDO).

AREAS WHICH MAY REQUIRE INCREASED ATTENTION

The Exposure of Vulnerable Groups to Hazardous Chemicals

Chemicals impact everyone, but some types of chemicals affect some people more. Such vulnerable population groups include women, especially pregnant women or women of child-bearing age, children, and workers, especially those who work directly with chemicals and immunocompromised individuals and subpopulations. The reasons for vulnerability are both biological (e.g., children's bodies are still biologically immature and growing, which may result in incomplete mechanisms to excrete or transform chemicals, as well as higher impacts owing to the body being in a formative state), and also economic. While women of child-bearing age as well as children may be more affected, agricultural and production workers often have lower physical resistance due to a lower nutritional status* and higher exposures. Not only in developing, but also in developed countries, such impacts are "hidden" and data are often scanty. For example, rural women working in agricultural fields especially, but not only, in developing countries, and workers in small-scale textile and chemical production units often have high exposures to chemicals and pesticides, but there are few studies to show the impacts. The number of affected people in such groups can be very high, and in many developing countries more than 90% of the agricultural workforce could be migrant, from poor socioeconomic backgrounds, and often a large percentage of them are female. Reaching out to such populations to decrease their risks from health impacts owing to chemical exposures is a great challenge and will need special efforts from stakeholders across the divide, from both the developed and developing worlds.

Exposure to a Cocktail of Different Chemicals

Typically, in the research and policy arenas, only single chemicals are considered for their impacts on human health and the environment. However, chemicals do not exist in isolation, but as a cocktail of chemicals. This could significantly change their biological behavior and potentially magnify synergistic impacts. A study commissioned by the EU in 2007 reviewed current scientific knowledge and regulatory approaches to dealing with "chemical cocktails," and suggested that they could be one of the key future challenges on the global chemicals agenda.†

Substitution of More Hazardous Chemicals: Drivers, Mechanisms, Costs, and Benefits

Shifting the emphasis from end-of-pipe management to upstream substitution especially for the most hazardous chemicals is probably the most sustainable long-term solution for ensuring chemical safety. There could be technical and cost related implications toward this, and substitution may require investments in both products and processes, but it would also stimulate research into more sustainable solutions.

Drivers for such substitution can be global as well. Legislation such as the European REACH (see Section VII, Chapter 44 of this book) and RoHS‡ are already promoting the shift in this direction, and act as market-based policy and legal drivers. Such drivers are, however, not yet a part of most national chemical safety frameworks, though introduction of life-cycle approaches has been initiated in some countries.

---

* http://www.epa.gov/civilrights/womenandgirls/health.html, http://www.ourstolenfuture.org/newscience/oncompounds/phthalates/phthalates.htm and http://www.cancer.gov/cancertopics/factsheet/Risk/asbestos

† *State of the Art report on Mixtures Toxicity* commissioned by the Environment Directorate General (DG) of the European Commission February 9, 2010 and *Combination Effect of Chemicals*, Council Conclusions, 17820/09, December 23, 2009, Council of the European Union at: http://register.consilium.europa.eu/pdf/en/09/st17/st17820.en09.pdf. Also see *Risk Assessment for Chemical Mixtures*, U.S. Environment Protection Agency at: http://epa.gov/hhrp/quick_finder/chemical.html#04

‡ http://www.rohs.eu/english/index.html

Other drivers which promote substitution of hazardous chemicals relate to "greening" of supply chains of products (e.g., textiles) when consumers are concerned about the environmental or social damage (like using child labor) these products may have caused during their manufacture. Such products, especially when sourced globally and imported from other countries with weaker chemicals safety practices, can help put pressure in those countries to ensure "cleaner" production.

Dealing with global supply chains, and substitution, through national policies and initiatives, as well as market-based drivers, will be a significant challenge in the coming years.

Innovative Approaches in Risk Assessment

Current approaches to hazard and risk assessment have serious shortcomings, such as being extremely resource intensive and considering chemicals one by one, whereas man and environment are exposed to many chemicals simultaneously. Innovative approaches may be necessary to significantly speed up the number of chemicals assessed and to increase the relevance of these assessments for real life exposure of man and environment whilst decreasing the need for animal testing. Recent computational toxicology and predictive toxicology efforts and mode of action frameworks for chemicals are increasingly moving toward evaluating upstream effects (e.g., cellular, tissue, genetics, and organ level), to prevent downstream health effects.

CONCLUSION

Major achievements have been accomplished in promoting sound chemicals management. The international community has, over the last 50 years, and in particular since the Rio 1992 UNCED (United Nations Conference on the Environment and Development), taken major steps to develop and implement, at least partially, important instruments to give solid support to sound chemicals management. A major achievement has also been that this has been done in close collaboration with various stakeholders.

A critical review of achievements and gaps shows, however, that in order to make significant further progress, there needs to be enhanced mainstreaming of chemicals into the sustainable development agenda, strengthening of implementation, capacity reinforcement, and better options for sustainable financial instruments. A committed and focused attention to these areas, particularly within the SAICM context, may allow the world to come close to the Johannesburg goals when 2020 finally arrives, as it inevitably will.

Appendix A: Abbreviations

ABNT	Association of Technical Standards (Brazil)
ACC	American Chemistry Council
ACGIH	American Conference of Governmental Industrial Hygienists
ACS	American Chemical Society
AcTOR	Aggregated Computational Toxicology Resource
ADI	Acceptable Daily Intake
AGCS	Advisory Group on Chemical Safety (Australia)
AGEE	Advisory Group on Environmental Emergencies
AGRA	A Green Revolution in Africa
AICS	Australian Inventory of Chemical Substances
AITUC	All India Trade Union Congress
AMAP	Arctic Monitoring and Assessment Programme
APEC	Asia Pacific Economic Cooperation
ARfDs	Acute Reference Doses
ASP	Africa Stockpiles Programme
ATSDR	Agency for Toxic Substances & Disease Registry (USA)
BAN	Basel Action Network
BAT	Best Available Techniques
BCM	Bromochloromethane
BEI	Biological Exposure Indices
BEP	Best Environmental Practices
BFRs	Brominated Flame Retardants
BIAC	Business and Industry Advisory Committee
BIS	Bureau of Indian Standards
BLG	Biodiversity Liaison Group
BPA	Bisphenol A
BSR	Banned or Severely Restricted chemicals
BTWC	Biological and Toxin Weapons Convention
CAC	Codex Alimentarius Commission
CAS	Chemical Abstracts Service
CBD	Convention on Biological Diversity
CCECA	Canada–Chile Environmental Cooperation Agreement
CCOHS	Canadian Centre for Occupational Health and Safety
CCOL	Committee of the Ozone Layer
CCPR	Codex Committee on Pesticide Residues
CDM	Clean Development Mechanism
CEC	Commission for Environmental Cooperation (Australia)
CECR	Centre for Environmental Contaminants Research
CEE	Central and Eastern Europe
CEFIC	Chemical Industry Council
CEHA	Centre for Environmental Health Activities

CEHAPE	European Children's Environmental Health Action Plan
CEHI	Children's Environmental Health Indicators
CEIT	Countries with Economies In Transition
CEPA	Canadian Environmental Protection Act
CEPIS	Pan American Center for Sanitary Engineering and Environmental Sciences
CERCLA	Comprehensive Environmental Response, Compensation, and Liability Act (USA)
CERs	Certified Emission Reduction units
CETPs	Common Effluent Treatment Plants
CFCs	Chlorofluorocarbons
CG/HCCS	Cordinating Group for the Harmonization of Chemical Classification Systems
CGIAR	Consultative Group on International Agricultural Research
ChAMP	Chemical Assessment and Management Program
CHEMRAWN	CHEMical Research Applied to World Needs
ChL	Chemical Leasing
CHM	Chemical Health Monitor
CIAT	International Centre for Tropical Agriculture
CIB	Central Insecticide Board (India)
CICAD	Concise International Chemical Assessment Document
CIEN	Chemicals Information Exchange Network
CIPMA	Centro de Investigación y Planificación del Medio Ambiente (Chile)
CITU	Centre for Trade Unions (India)
CLAs	Coordinating Lead Authors
CLRTAP	Convention on Long-Range Transboundary Air Pollution
CMP	Chemicals Management Plan
CoP	Conferences of the Parties
CORS	Chemicals Office of the Republic of Slovenia
CP	Cleaner Production
CRC	Chemical Review Committee
CSA	Chemical Safety Assessment
CSD	Commission on Sustainable Development
CSDS	Chemical Safety Data Sheets
CSIR	Council for Scientific and Industrial Research (Ghana)
CSJ	Chemical Society of Japan
CSOs	Civil Society Organisations
CSR	Chemical Safety Report
CTD	Comparative Toxicogenomics Database
CWC	Chemical Weapons Convention
DDT	Dichloro-diphenyl-trichloroethane
DFID	Department for International Development (UK)
DGD	Decision Guidance Document
DGMS	Directorate General of Mines Safety (India)
DNA	Designated National Authority under a UN Convention

DSL	Domestic Substances List (Canada)
DWCPs	Decent Work Country Programmes (ILO)
EAEC	European Atomic Energy Community
EAM	Environmental Assessment Module
EAPCCT	European Association of Poison Control and Clinical Toxicologists
EC	Environment Canada
EC	European Commission
ECB	European Chemicals Bureau
ECEH	European Centre for Environment and Health
ECHA	European Chemicals Agency
ECOSOC	Economic and Social Council
ECSC	European Coal and Steel Community
ECVAM	European Centre for Validation of Alternative Methods
ED	Executive Director
EDKB	Endocrine Disruptor Knowledgebase
EDs	Endocrine Disruptors
EEA	European Environment Agency
EEAA	Egyptian Environmental Affairs Agency
EECCA	Eastern Europe, Caucasus and Central Asia
EFTA	European Free Trade Association
EHC	Environmental Health Criteria
EHS	Environment, health, and safety
EHSC	Environment, Health, and Safety Committee
EIA	Environmental Impact Assessment
EIF	Entry-Into-Force date
EINECS	European Inventory of Existing Commercial Chemical Substances
EIT	Economies in transition
EJ	Environmental justice
ELINCS	European List of Notified Chemical Substances
ELVs	Emission Limit Values
EMEP	European Monitoring and Evaluation Programme
EMERCHEM	Emerging Issues in Chemicals Management in Developing Countries
EMG	Environment Management Group
EMPRES	Emergency Prevention System for Transboundary Animal and Plant Pests and Diseases (FAO)
ENTRI	Environmental Treaties and Resource Indicators
EPA	Environmental Protection Agency
EPAP	Egyptian Pollution Abatement Program
EPR	Extended Producer Responsibility
ER	Emergency Response
ES	Exposure Scenario
ESD	Emission Scenario Documents
ESIS	European Chemical Substances Information System

ESM	Environmentally Sound Management
EU MIC	European Union Monitoring and Information Centre
EU	European Union
EUROTOX	European Society of Toxicology
ExCOPs	Extraordinary Meeting of the Conference of Parties
FAO	Food and Agriculture Organization of the United Nations
FCSI	Federal Contaminated Sites Inventory (USA)
FFDCA	Federal Food, Drug and Cosmetic Act (USA)
FIFRA	Federal Fungicide and Rodenticide Act (USA)
FSC	Forum Standing Committee (IFCS)
FSP	Full Size Project
FTA	Free Trade Agreement
GAEC	Ghana Atomic Energy Commission
GAIA	Global Alliance for Incinerator Alternatives
GAPs	Good Agricultural Practice
GAR	Global Alert and Response
GAVI Fund	Global Alliance for Vaccine and Immunization Fund
GC	Governing Council
GCDPP	Global Collaboration for Development of Pesticides for Public Health
GCLA	Government Chemist Laboratory Agency (Tanzania)
GDCh	Gesellschaft Deutscher Chemiker (Germany)
GDP	Gross Domestic Product
GEB	Global Environmental Benefits
GEF	Global Environment Facility
GHG	Greenhouse Gas
GHS	Globally Harmonized System for Classification and Labeling of Chemicals
GLP	Good Laboratory Practice
GMA	Global Mercury Assessment
GMEF	Global Ministerial Environment Forum
GPA	Global Plan of Action
GPS	Global Product Strategy
GSN	Green Suppliers Network
GTZ	German Society for Technical Cooperation
HBCD	Hexabromocyclododecane
HBFCs	Hydrobromofluorocarbons
HC	Health Canada
HCB	Hexachlorobenzène
HCBD	Hexachlorobutadiene
HCFC	Hydrochlorofluorocarbon
HCH	Hexachlorocyclohexane
HCWH	Health Care Without Harm
HEDS	Human Exposure Database System
HELI	Health and Environment Linkages Initiatives (WHO)
HEMA	Health and the Environment Ministers of the Americas

HERO	Health and Environmental Research Online
HFC-23	Hydroflurocarbon-23
HFCs	Hydrofluorocarbons
HHPs	Highly Hazardous Pesticides
HM	Heavy Metals
HPA	Hazardous Products Act (Canada)
HPMPs	HCFC Phase-out Management Plans
HPV	High Production Volume Chemicals
HSDB	Hazardous Substances Data Bank (USA)
HSE	Health, Safety, and Environment
HSGs	Health and Safety Guides
IACSD	Interagency Committee on Sustainable Development
IAEA	International Atomic Energy Agency
IANS	Indo-Asian News Service
IARC	International Agency for Research on Cancer
ICAWEB	Virtual Integrated Crisis Advise Web site (the Netherlands)
ICC	Indian Chemical Council
ICCA	International Council of Chemical Associations
ICCM	International Conference on Chemicals Management
ICCS	International Conference on Chemical Safety
ICRC	International Committee of the Red Cross
ICSC	International Chemical Safety Cards
ICT	International Congress of Toxicology
IEA	International Energy Agency
IEA	International Environmental Agreement
IEG	International Environmental Governance
IFCS	Intergovernmental Forum on Chemical Safety
IFDC	International Centre for Soil Fertility and Agricultural Development
IFG	International Fertiliser Group
IGOs	Intergovernmental Organizations
IHP	International Humanitarian Partnership
IHR	International Health Regulations
IIASA	International Institute for Applied Systems Analysis
IIED	International Institute for Environment and Development
IISD	International Institute for Sustainable Development
ILO	International Labor Organization
IMO	International Maritime Organization
INAC	Indian and Northern Affairs Canada
INC	Intergovernmental Negotiating Committee
INCD	Intergovernmental Negotiating Committee Desertification
INCRAM	Intergovernmental Mechanism for Chemical Risk Assessment and Management
IOCC	Inter-Organization Coordinating Committee
IOE	International Employers Organization

IOMC	Inter-Organization Programme for the Sound Management of Chemicals
IPCC	Intergovernmental Panel on Climate Change
IPCP	International Panel on Chemical Pollution
IPCS	International Programme on Chemical Safety
IPEN	International POPS Elimination Network
IPEP	International POPS Elimination Project
IPF/IFF	Intergovernmental Panel on Forests and by Intergovernmental Forum on Forests
IPM	Integrated Pest Management
IPNM	Integrated Plant Nutrient Management
IPPC	International Plant Protection Convention
IRIS	Integrated Risk Information System
IRPTC	International Registry of Potentially Toxic Chemicals
ISDE	International Society of Doctors for the Environment
ISEE	International Society for Environmental Epidemiology
ISG	Intersessional Group
iTE	Inherent Toxicity in the Environment
ITER	International Toxicity Estimates for Risk
iTH	Inherent Toxicity for Humans
ITUC	International Trade Union Confederation
IUCLID	International Uniform Chemical Information Database
IUPAC	International Union of Pure and Applied Chemistry
IUTOX	International Union of Toxicology
IVM	Integrated Vector Management
JECFA	Joint Expert Committee on Food Additives
JEU	Joint UNEP/OCHA Environment Unit
JICA	Japan International Cooperation
JMPM	Joint Meeting on Pesticides Management
JMPR	Joint Meeting on Pesticide Residues
JPOI	Johannesburg Plan of Implementation
JRC	Joint Research Centre
KemI	Swedish Chemicals Agency
LCA	Life Cycle Analysis
LMOs	Living Modified Organisms
LPVCs	Low Production Volume Chemicals
LRI	Long-Range Research Initiative
LRTAP	Long-Range Transboundary Air Pollution
LSAP	Lead Smelter Action Plan
MAD	Mutual Acceptance of Data program (OECD)
MAFF	Ministry of Agriculture, Forestry and Food
MDGs	Millennium Development Goals
MEA	Multilateral Environmental Agreement
MEA-REN	Multilateral Environmental Agreements Regional Enforcement Network
MESP	Ministry of the Environment and Spatial Planning (Slovenia)

METI	Ministry of Economics, Trade and Industries (Japan)
MIC	Methyl Isocyanate
MLF	Multilateral Fund
MN	Manufactured Nanomaterials
MOEF	Ministry of Environment and Forests (India)
MoFA	Ministry of Food and Agriculture (Ghana)
MoU	Memorandum of Understanding
MPPI	Mobile Phone Partnership Initiative
MPV	Medium Production Volume
MRL	Maximum Residue Limit
MSDS	Material Safety Data Sheets
MSPs	Medium Size Projects
NAAEC	North American Agreement on Environmental Cooperation
NABL	National Board for Laboratories (India)
NAFTA	North American Free Trade Agreement
NARAPs	North American Regional Action plans
NAS	National Academy of Sciences
NASA	National Aeronautic and Space Administration
NATO–EADRCC	North Atlantic Treaty Organisation–Euro Atlantic Disaster Response Coordination Centre
NCCT	National Center for Computational Toxicology (USA)
NCEA	National Centre for Environmental Assessment (USA)
NCMP	National Chemical Monitoring Programme
NCMP	National Chemicals Management Profile
NCP	Northern Contaminants Program (Canada)
NCPCs	National Cleaner Production Centres
NCSP	National Chemical Safety Programme
NEAC	Northeast Asian Conference on Environmental Cooperation
NECs	National Emission Ceilings
NELs	No-Effect-Levels
NGO	Nongovernmental Organization
NHANES	National Health and Nutrition Examination Survey (USA)
NICNAS	National Industrial Chemicals Notification and Assessment Scheme (Australia)
NIH	National Institutes of Health (USA)
NIOSH	National Institute for Occupational Safety and Health
NIP	National Implementation Plan
NLM	National Library of Medicine (USA)
NMVOCs	Non-Methane Volatile Organic Compounds
NP	National Profiles
NPEs	Nonylphenol Ethoxylates
NPL	National Priorities List
NPRI	National Pollutant Release Inventory
NPRM	Notice of Proposed Rule Making
NRC	National Research Council

Nrg4SD	Network of Regional Governments for Sustainable Development
OARE	Online Access to Research on the Environment
OCHA	Office for the Coordination of Humanitarian Affairs
OctaBDE	Octabromodiphenylether
ODA	Official Development Assistance
ODS	Ozone-Depleting Substances
OECD	Organisation for Economic Cooperation and Development
OELTWG	Open-Ended Legal and Technical Working Group
OEWG	Open-Ended Working Group of Governments
OPCW	Organisation for the Prohibition of Chemical Weapons
OPPTS	Office of Pollution Prevention and Toxics (USA)
OPS	Overarching Policy Strategy
OSH	Occupational Safety and Health
OSHA	Occupational Safety and Health Act (USA)
OSHA	Occupational Safety and Health Administration
OSHA	Occupational Safety and Health Agency
PAG	Program Advisory Group
PAHO	Pan American Health Organization
PAHs	Polycyclic Aromatic Hydrocarbons
PAN	Pesticide Action Network
PBBs	Polybrominated Biphenyls
PBDEs	Polybrominated Diphenyl Ethers
PBTs	Persistent, Bioaccumulative or Toxic chemicals/substances
PCB	Pentachlorobenzene
PCBs	Polychlorinated Biphenyls
PCDDs	Polychlorinated Dibenzodioxins
PCDFs	Polychlorinated Dibenzofurans
PCNs	Polychlorinated Naphthalenes
PCPA	Pest Control Products Act (Canada)
PCTs	Polychlorinated Terphenyls
PDSs	Pesticide Data Sheets
PeBDE	Pentabromodiphenylether
PEPAS	Promotion of Environmental Planning and Applied Studies (WHO)
PFAS	Perfluoroalkyl Sulfonates
PFCAs	Perfluoroalkyl Carboxylates
PFCs	Perfluorocarbons
PFCs	Perfluorinated Compounds
PFOA	Perfluorooctanoic Acid
PFOS	Perfluorooctane Sulfonate
PFOSA	Perfluorooctanesulfonyl Fluoride (POSF), Perfluorooctanesulfonamide
PFOS-F	Perfluorooctane Sulfonyl Fluoride

PHARE	Poland and Hungary Assistance for Reconstruction of the Economy
PIC	Prior Informed Consent
PID	Photo-Ionization Detector
PIL	Public Interest Litigations
PIMs	Poisons Information Monographs
PM	Particulate Matter
POPRC	POPS Review Committee
POPS	Persistent Organic Pollutants
POs	Participating Organizations
POSF	Perfluorooctanesulfonyl Fluoride
PPA	Pollution Prevention Act (USA)
PPP	Plant Protection Products
PPPs	Public–Private Partnerships
PPRSD	Protection and Regulatory Services Directorate (Ghana)
PRTR	Pollutant Release and Transfer Registers
PSL	Priority Substances List
PTC	Pesticides Technical Committee (Ghana)
PTS	Persistent Toxic Substances
QSARs	Quantitative Structure-Activity Relationships
QSP	Quick Start Programme (SAICM)
R&D	Research & Development
RAIPON	Russian Association of Indigenous Peoples of the North
RAIS	Risk-Assessment Information System
RASS	Risk Assessment Summer Schools
RAW	Risk Assessment Workshops
REACH	Registration, Evaluation, Authorisation and Restriction of Chemicals
RECIEL	Review of European & International Environmental Law
RECP	Resource-Efficient and Cleaner Production
REDD	Reducing Emissions from Deforestation and forest Degradation
REEEP	Renewable Energy and Energy Efficiency Partnership
REN21	Renewable Energy Network for the 21st century
RIPs	REACH Implementation Projects
RIVM	National Institute of Public Health & the Environment (the Netherlands)
RLPWHTM	Regulation on the Labor Protection in Workplaces Handling Toxic Materials (Canada)
RMM	Risk Management Measures
RoHS	Restriction of Hazardous Substances
RRPHCBS	Register of Potentially Hazardous Chemical and Biological Substances (Russia)
RSC	Royal Society of Chemistry (UK)
SAARC	South Asian Association for Regional Cooperation
SACEP	South Asian Cooperative Environment Programme

SAICM	Strategic Approach to International Chemicals Management
SC	Stockholm Convention
SCCP	Short-Chain Chlorinated Paraffins
SCEGHS	Subcommittee of Experts on the GHS
SCPI	Sustainable Crop Production Intensification
SDS	Safety Data Sheet
SEE	South-Eastern Europe
SETAC	Society of Environmental Toxicology and Chemistry
SHPF	Severely Hazardous Pesticide Formulations
SIDS	Screening Information Data Set
SIDS	Small Island Developing States
SIEF	Substance Information Exchange Forum (UNEP)
SINITOX	National System of Toxi-Pharmacological Information (Brazil)
SIS	Specialized Information Services
SLAPP	Strategic Lawsuit Against Public Participation
SMC	Sound Management of Chemicals
SME	Small and Mediumsize Enterprise
SMOC	Sound Management of Chemicals
SOMTOX	Sociedad Mexicana de Toxicologia, A.C—Mexican Society of Toxicology
SOT	Society of Toxicology
SPCBs	State Pollution Control Boards (India)
SPHFs	Severely Hazardous Pesticide Formulations
SPM	Summaries for Policy Makers
SRP	Scientific Review Panel
StEP	Solving the E-waste Problem
STTV	National Product Control Agency for Welfare and Health (Finland)
SVHCs	Substance of Very High Concern
TBTs	Tributylin compounds containing products
TCSA	Toxic Substances Control Act (USA)
TEFs	Toxic Equivalency Factors
TEHIP	Toxicology and Environmental Health Information Program (USA)
TEMM	Tripartite Environment Ministers Meeting (China, Japan, Rep. of Korea)
TERA	Toxicology Excellence for Risk Assessment
TLVs	Threshold Limit Value
TOC	Total Organic Carbon
TOR	Terms of Reference
TOXSA	Toxicology Society of South Africa
TPAWU	Tanzania Plantation Workers Union
TPRI	Tropical Pesticide Research Institute (Tanzania)
TRI	Toxics Release Inventory (USA)
UN JIU	UN Joint Inspection Unit

UN OCHA	United Nations Office for the Coordination of Humanitarian Affairs
UN SCETDG	United Nations Economic and Social Council's Sub Committee of Experts on the Transport of Dangerous Goods
UN	United Nations
UNCCD	United Nations Convention to Combat Desertification
UNCED	United Nations Conference on Environment and Development
UN-CETDG	UN Committee of Experts on Transport of Dangerous Goods
UNCETDGGHS	UN Committee of Experts on the Transport of Dangerous Goods and on the Globally Harmonized System of Classification and Labelling of Chemicals
UNCTAD	United Nations Conference on Trade and Development
UNCTs	UN Country Teams
UNDAC	United Nations Disaster Assessment and Coordination
UNDAF	UN Development Assistance Framework
UNDG	United Nations Development Group
UNDP	United Nations Development Programme
UNECE	United Nations Economic Commission for Europe
UNEG	United Nations Environment Organization
UNEP	United Nations Environment Programme
UNEP-DTIE	United Nations Environment Programme—Division of Technology, Industry and Economics
UNESCO	United Nations Educational, Scientific, and Cultural Organization
UNFCCC	United Nations Framework Convention on Climate Change
UNGA	United Nations General Assembly
UNIDO	United Nations Industrial Development Organisation
UNITAR	United Nations Institute for Training and Research
UNOPS	United Nations Office for Projects Services
UNRTDG	UN Recommendations on the Transport of Dangerous Goods, Model Regulations
UNSCEGHS	UN Sub-Committee of Experts on the Globally Harmonized System of Classification and Labelling of Chemicals
UNSCETDG	UN Sub-Committee of Experts on the Transport of Dangerous Goods
USAID	US Agency for International Development
US EPA	U.S. Environment Protection Agency
VCCEP	Voluntary Children's Chemical Evaluation Program (USA)
VOC	Volatile Organic Compounds
WBCSD	World Business Council on Sustainable Development
WCC	World Chlorine Council
WECF	Women in Europe for a Common Future
WEEE	Waste Electrical and Electronic Equipment
WFP	World Food Programme
WFPHA	World Federation of Public Health Associations

WGSR	Working Group on Strategies and Review (UNECE)
WHA	World Health Assembly
WHO	World Health Organization
WHOPES	WHO Pesticides Evaluation Scheme
WLT	World Library of Toxicology
WMC	Waste Minimization Circles
WMO	World Meteorological Organization
WPMN	Working Party on Manufactured Nanomaterials (OECD)
WPN	Working Party on Nanotechnology
WSSD	World Summit on Sustainable Development
WTO	World Trade Organization

Appendix B: Web Sites

Note: The list below displays the major organizations, conventions, and conferences mentioned in this book and their associated Web sites.

Aarhus Convention—http://www.unece.org/env/pp
Basel Convention—http://www.basel.int
Center for International Environmental Law—http://www.ciel.org
Chemical Weapons Convention—http://www.opcw.org/chemical-weapons-convention
Commission for Environmental Cooperation (of North America)—http://www.cec.org
Convention on Long-Range Transboundary Air Pollution—http://www.unece.org/env/lrtap
Earth Summit 2012—http://www.earthsummit2012.org
Food and Agricultural Organization of the UN—http://www.fao.org
Globally Harmonized System of Classification and Labelling of Chemicals—http://www.osha.gov/dsg/hazcom/ghs.html
Intergovernmental Forum on Chemical Safety—http://www.who.int/ifcs/en
Intergovernmental Panel on Climate Change—http://www.ipcc.ch
International Code of Conduct on the Distribution and Use of Pesticides—http://www.fao.org/docrep/005/y4544e/y4544e00.htm
International Conference on Chemicals Management 1, ICCM-1—http://www.saicm.org/index.php?menuid=8&pageid=7
International Conference on Chemicals Management 2, ICCM-2—http://www.saicm.org/index.php?content=meeting&mid=42&def=1&menuid=9
International Council of Chemical Associations—http://www.icca-chem.org
International Labour Organization—http://www.ilo.org
International Panel on Chemical Pollution—http://www.ipcp.ch
International POPS Elimination Network—http://www.ipen.org
International Programme on Chemical Safety—http://www.who.int/ipcs/en
Inter-Organization Programme for the Sound Management of Chemicals—http://www.who.int/iomc
Johannesburg 2001: The World Summit on Sustainable Development—http://www.un.org/jsummit/html/basic_info/basicinfo.html
Kiev Protocol—http://www.unece.org/env/pp/prtr.htm
Kyoto Protocol—http://unfccc.int/kyoto_protocol/items/2830.php
OECD—http://www.oecd.org
Ozone Secretariat (including Vienna Convention & Montreal Protocol)—http://ozone.unep.org
REACH—http://echa.europa.eu/reach_en.asp
Rio 1992: The UN Conference on Environment and Development (The "Earth Summit")—http://www.un.org/geninfo/bp/enviro.html
Rotterdam Convention—http://www.pic.int

SAICM—http://www.saicm.org

South Asian Association for Regional Cooperation (SAARC)—http://www.saarc-sec.org

Stockholm 1972: Conference on the Human Environment, Report—http://www.unep.org/Documents.Multilingual/Default.asp?documentid=97

Stockholm Convention on Persistent Organic Pollutants—http://chm.pops.int

Tripartite Environment Ministers Meeting (Japan, China, Korea) (TEMM)—http://www.env.go.jp/earth/coop/coop/english/dialogue/temm.html

United Nations Development Programme (UNDP)—http://www.undp.org

United Nations Environment Programme (UNEP)—http://www.unep.org

United Nations Environment Programme (UNEP Chemicals)—http://www.unep.org/themes/chemicals

United Nations Industrial Development Organization (UNIDO)—http://www.unido.org

United Nations Institute for Training and Research (UNITAR)—http://www.unitar.org

World Health Organization—http://www.who.int/en

Index

Note: n = Footnote

A

B

F

G

H

I

M

Q

R

T

X

Z

Milton Keynes UK
Ingram Content Group UK Ltd.
UKHW021937071024
449327UK00022B/1835